INTRODUCTION TO THE ART OF PROGRAMMING USING SCALA

CHAPMAN & HALL/CRC TEXTBOOKS IN COMPUTING

Aims and Scope

This series covers traditional areas of computing, as well as related technical areas, such as software engineering, artificial intelligence, computer engineering, information systems, and information technology. The series will accommodate textbooks for undergraduate and graduate students, generally adhering to worldwide curriculum standards from professional societies. The editors wish to encourage new and imaginative ideas and proposals, and are keen to help and encourage new authors. The editors welcome proposals that: provide groundbreaking and imaginative perspectives on aspects of computing; present topics in a new and exciting context; open up opportunities for emerging areas, such as multi-media, security, and mobile systems; capture new developments and applications in emerging fields of computing; and address topics that provide support for computing, such as mathematics, statistics, life and physical sciences, and business.

Published Titles

Pascal Hitzler, Markus Krötzsch, and Sebastian Rudolph, Foundations of Semantic Web Technologies

Uvais Qidwai and C.H. Chen, Digital Image Processing: An Algorithmic Approach with MATLAB®

Henrik Bærbak Christensen, Flexible, Reliable Software: Using Patterns and Agile Development

John S. Conery, Explorations in Computing: An Introduction to Computer Science

Lisa C. Kaczmarczyk, Computers and Society: Computing for Good

Mark J. Johnson, A Concise Introduction to Programming in Python

Paul Anderson, Web 2.0 and Beyond: Principles and Technologies

Henry M. Walker, The Tao of Computing, Second Edition

Mark C. Lewis, Introduction to the Art of Programming Using Scala

CHAPMAN & HALL/CRC
TEXTBOOKS IN COMPUTING

INTRODUCTION TO THE ART OF PROGRAMMING USING SCALA

Mark C. Lewis

CRC Press is an imprint of the
Taylor & Francis Group, an **informa** business
A CHAPMAN & HALL BOOK

CRC Press
Taylor & Francis Group
6000 Broken Sound Parkway NW, Suite 300
Boca Raton, FL 33487-2742

International Standard Book Number: 978-1-4398-9666-2 (Paperback)

Library of Congress Cataloging-in-Publication Data

Lewis, Mark C., author.
Introduction to the art of programming using Scala / Mark C. Lewis.
pages cm. -- (Chapman & Hall/CRC textbooks in computing)
Includes bibliographical references and index.
ISBN 978-1-4398-9666-2 (pbk.)
1. Scala (Computer program language)--Textbooks. 2. Computer programming--Textbooks. I. Title.

QA76.73.S28L49 2013
005.1'14--dc23 2012025084

Visit the Taylor & Francis Web site at
http://www.taylorandfrancis.com

and the CRC Press Web site at
http://www.crcpress.com

Contents

List of Figures

List of Tables

Acknowledgments

There are many people who have played a role in the creation of this book, and who deserve to have that role acknowledged. First and foremost is my wonderful wife, Cat. Writing a book of this length requires a fair bit of time. I could not have done it without her love and understanding. I also have to acknowledge my daughters, because if they were not such great children, there is no way I would have been able to do this.

Of course, one does not become a professor who writes his own teaching material in a vacuum. I have to thank my parents for raising me to appreciate learning and teaching. On the topic of teaching, I certainly had a great many teachers who influenced me as well. I am quite certain that my high school English teachers, who probably felt like they were pulling teeth to get me to write a five-page paper, never saw this coming.

Developing the Scala-based material for this book has taken about two years, and during that time I have taken full advantage of my students to help refine it. I truly appreciate these students who acted as guinea pigs along the way. In particular Kelsey Daniels who sent me over 140 typos from the first half of the book alone. I would also like to thank my mother-in-law for both advice and proofreading, as well as my wife's friend, Jen Hogan, for assistance with promotion.

This book would not have ever gone to press had it not been for Peter DePasquale mentioning the book to Randi Cohen of Taylor and Francis and her support for getting this book in print.

Lastly, I would like to thank Martin Odersky and the other creators of Scala for providing what I consider to be a beautiful language that makes writing real programs a joy.

Preface

Welcome to "An Introduction to the Art of Programming Using Scala." This book is intended to be used in first and second semester college classrooms to teach students beginning programming. To accomplish this task, the Scala[1] programming language is used. The book was constructed with a focus on the topics students need to know. These topics were then woven into a format that was deemed the best way to communicate those ideas. Because there are two very different audiences who might be reading this, the rest of the preface is split into two sections.

To the Student

Welcome to the world of programming. You are about to embark on a field of study that will hopefully open your mind to new ways of thinking and impact the way you view everything in the world around you. For students who intend to major in Computer Science and make careers working with computer technology, this is your first step in learning how to communicate with the computer and to instruct the computer in how it should solve problems for you.

For those who are not planning to make a career in a computing field, a course in computer programming can still be remarkably beneficial. Computer programming is, fundamentally, about problem solving. It is about figuring out how to solve problems and express them in ways that the computer can understand. Your entire future life is, in one way or another, going to deal with how to solve problems. You will need to have approaches to solving these problems well formed in your own head and be able to communicate them to other people in non-ambiguous ways so that they understand what you want and can act on it. Learning to program computers will help you develop and hone this ability.

There are more direct benefits as well. The world you live in is already extremely dependent on computing. Many of the things you own that you do not call a computer include microchips and digital processing. Those devices/objects run programs to complete their tasks. Many of the tasks you do during an average day are also enabled by computer power and the ability to process and store information quickly. Activities like browsing the web for work or leisure explicitly involve this. Other activities, such as financial transactions made with anything other than cash, implicitly involve it. This ubiquitous presence of digital processing is only going to grow over time. So even if you never have to directly write programs as part of your future life, you can only benefit from having an understanding of what is going on under the hood. Even if you do not write the software yourself, there is a good chance that you will interact with those who do, perhaps as you or your company strives to gain a competitive advantage by improving the information technology you use

[1]Code samples in this book were compiled and tested using Scala 2.9.x.

to do your work. Even a basic knowledge of programming will help you understand what these people are doing and communicate more effectively with them.

Even those who do not intend to make programming into a career might occasionally come across problems where they realize that they could save a lot of time if they could get a computer to do it for them. When that happens, even a basic knowledge of programming can make your life a whole lot better. Data is becoming a bigger and bigger part of the business world. The vast amounts of data created and collected by modern technology can not be processed by humans manually. To sift through this data to find meaning requires using machines and that means using software. You can either pay someone else to write it or you can write it yourself. That latter is only an option if you know how to program.

While many might be tempted to view programming as an advanced concept, the reality is this book is going to take you back to your early education quite a few times. The reason for this is you spent a lot of time in those early years learning how to solve certain basic problems. The approaches became second nature to you and you no longer think about them, you just do them. In the context of programming you have to go back and examine how you solve problems, even ones you have been doing for a long time. The reason for this is now you have to tell a computer how to do those same things and the computer will need detailed instructions. The reality is, the programming presented in this book is not an advanced topic, it is basic logic and problem solving, done at a level you probably have not worked at since you were much younger.

The title of this book refers to programming as an art. Programming shares many features with traditional arts like creative writing and painting or sculpture. The programmer first creates a vague image in his/her mind of what he/she wants to bring into existence. This is followed by a period of work bringing that mental image into a more real form. The end result is something new, born from the imagination of the creator, that can be experienced by others. The digital medium might, in some sense, seem less real than paint, clay, or stone, but it is also remarkably dynamic. You can create things in a computer that you have no chance of creating in the real world by hand. You can also create things that are remarkably useful. All the software you interact with every day, whether to get things done or just for entertainment, was written by programmers. The exercises and projects in this book have been created with the goal of giving you the ability to express your creativity.

Programming bears another characteristic in common with other creative arts, if you want to be good at it, you need to practice. There are many ways to program the solution to any problem. That introduces a lot of nuance. This book will strive to instruct you in regards to the strengths and weaknesses of different approaches, but to really understand how to write good code to solve a problem, you need to have the experience of solving similar ones. Imagine an art major who never draws or paints except for class projects, not even doodling in notes, or a creative writing major who never writes a story except the ones required for class. Hopefully those seem silly to you. A Computer Science major who never writes code beyond what is assigned for class is exactly the same. So explore the art of programming. Have fun with it and try to do something that interests you.

Using this Book

The goal of this book is to help you in your efforts to learn how to program. In addition to the practice that was just mentioned, there are some suggestions for how to get the most out of this book. Clearly working as many exercises and projects as you can counts as practice. In addition, you should really read descriptions in this book and make sure you understand the code samples shown in it. All too often, students treat the task of reading course material as sounding out all the words in your head. While that fits the denotative definition, the connotation is that you string the words together and understand what they mean together.

Read for understanding, not just so you can say you have gotten through a chapter. Gaining understanding of early material is especially important in this topic because the material in any given chapter is typically highly dependent on the chapters before it. If you fail to understand one chapter, you are not likely to understand the ones that follow it either. As such, effort spent on the early chapters will pay off later on. Fortunately, there are resources that accompany this book that can help with this.

Students coming into this course who have a background in programming should consider taking a "fast path" to chapters 6 and 7. You do not want to skip them completely as there are aspects of Scala that will likely differ from what you have done before. One way to approach this is the go to the end of any given chapter and try to do the exercises. Refer back to the chapter as needed to complete them. Once you can do the exercises and a project or two, you should feel comfortable moving on.

Book Website

The author has posted a number of different supplements to this book at `www.programmingusingscala.net` that can help you work through the material. Some of the material is code or prose that could not be put into the book due to length restrictions. The most useful things on the website are things that can not be represented in static text.

Many different sections have videos on the web that work through the material. Programming is a very non-linear process. This book typically shows completed pieces of code that do what we want. It is hard to demonstrate the process of building that code in the format of a book. The videos show construction of code from the ground up and include descriptions of the programmers thoughts as it is being done. This type of "live coding" is invaluable in learning to program as it lets you into the mind of a more experienced programmer where you can see the thought processes associated with the development of the code.

The website also includes solutions of some exercises for you to reference along with sample implementations of certain projects. Some of the projects, especially in the second half of the book, can be challenging to describe in text. To give you a better idea of what is expected, the author has implemented sample solutions that you can run to see what they do. The site also includes some additional exercises and links to data files and remote sites that make the exercises more relevant.

The second half of the book also includes large projects. For these, there are UML diagrams of the author's sample solution to help students see possible ways of breaking their problem down.

To the Instructor

If you are reading this, it likely means that you are already aware of many of the features of Scala that make it a great programming language.[2] The flexibility of Scala means that things can be covered in many different ways. The approach taken in this book might be summarized as semi-functional/semi-imperative with objects later. It is worth describing exactly what is meant by that.

This book takes advantage of the aspects of Scala that support programming in the

[2] If you are currently unconvinced of the benefits of Scala for teaching CS1 and CS2, there is a complete discussion at `www.programmingusingscala.net`.

small. It is expected that students will operate mainly in the Read, Evaluate, Print Loop (REPL) and in a scripting environment through at least the first 10 chapters and possibly through the first 15. The benefit of this is that you can focus on logic instead of extra keywords and scoping rules. Students are easily overwhelmed early on and this approach helps to flatten the learning curve a bit.

Scala is purely object-oriented, so students will be using objects and calling methods on objects very early on. However, the construction of classes to build their own objects is postponed until later. Fitting with this approach, programs are written as scripts, not as applications, in the early part of the book. So the `object` keyword and things like `def main(args: Array[String]) : Unit` are postponed until students are ready to write programs that are big enough that they need the organization provided by doing proper object-orientation (OO) across multiple files.

The approach is described as semi-functional because there is significant use of higher order functions and function literals and immutable style is often preferred. However, this is not enforced in a zealotous way and mutable state is introduced early on. Students are shown mutable and immutable approaches to a number of problems and the benefits and pitfalls of each approach are discussed.

While the ideal way to go through this book is in a linear fashion, particularly if you want to use the end-of-chapter projects as assignments, strict linearity is not required and you can choose to move some topics around so that concepts like class declarations are covered earlier. Though it would be a bit more challenging, it would be possible to use this book in a more purely functional manner by simply having students avoid var declarations (See chapter 3), assignments into mutable collections like arrays, and the use of while loops.

Using this Book

This book is intended to cover CS1 and CS2. It is divided into two parts with Part I covering introductory programming concepts and Part II covering object-orientation, abstraction, and data structures. While the break between Part I and Part II falls close to where the semester break between CS1 and CS2 will fall, the two don't have to be the same. Indeed, in my own use of this book I typically cover the first two chapters of Part II during the first semester for a variety of pedagogical reasons.

When looking at the chapters in Part I, you might feel that chapters 11 and 12 are outliers in certain ways. They definitely break from the straight algorithmic, problem-solving nature of the other chapters. This break was placed at this point in the book at the request of students who often find visual interfaces more compelling than text-based ones. Thanks to the syntax of Scala, it is possible to build GUIs without explicitly declaring classes or using keywords associated with it, such as `extends`. The same is true of graphics with components that draw themselves with the exception that the `override` keyword appears on the `paint` method. It is possible to skip these, but not advised as graphics play a role in many of the examples later in the book.

If you have a desire to cover full OO concepts earlier, chapters 16 and 17 can be moved to closely follow chapter 10. There are also a number of chapters through the book that contain material that can be skipped completely. For example, chapter 14 covers XML. Scala makes this fairly easy to do and it fits in well with the projects that are given to have students work on stored data in a formatted manner. However, if you don't want to spend time on that topic it can be skipped with little impact on later content other than projects and occasional usage in later chapters.

To help you decide what you can and can't skip, figure 1 shows rough dependencies for different chapters in the book. Arrows point from one chapter to later ones that have direct

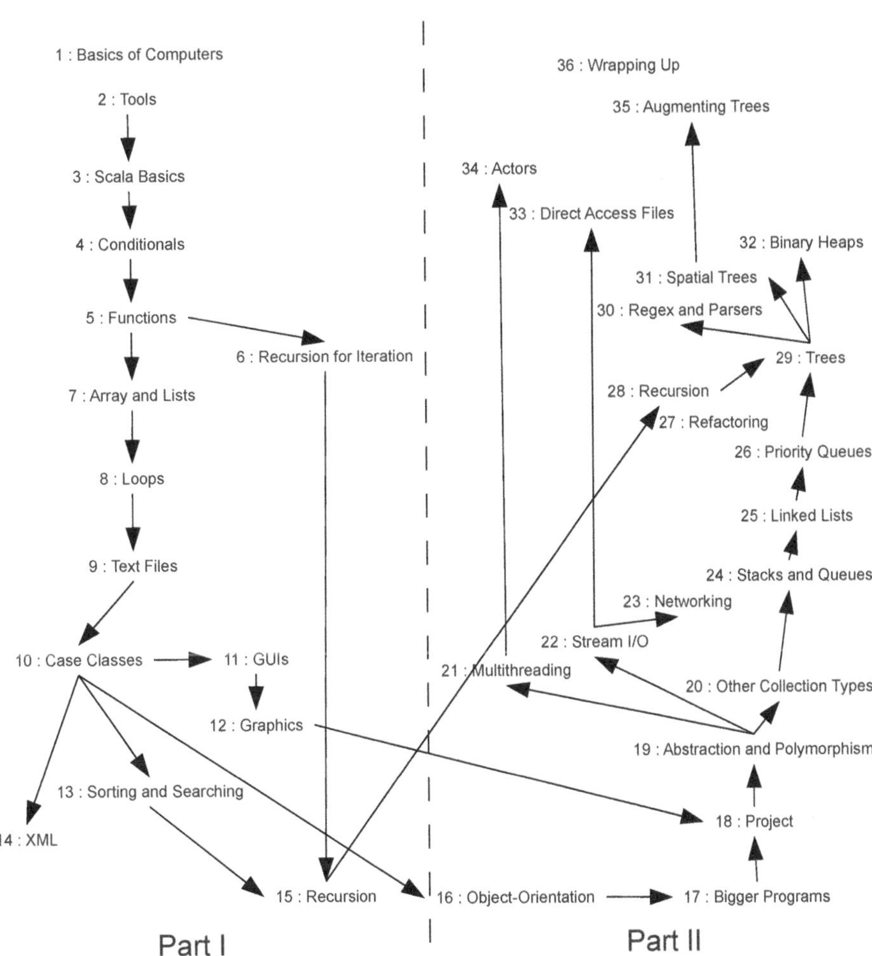

FIGURE 1: This figure shows the dependencies between different chapters in the book so that you can plan what you want to cover or to allow you to intelligently choose alternate paths through the book. Arrows point toward later chapters that use material from an earlier chapter. Chapters from Part I of the book go down the left and those from Part II go up the right.

dependencies. There will be occasional references to material that is not in the line of the dependency arrows, but the material should still be understandable.

The course website, `www.programmingusingscala.net`, includes a number of different types of material that can be helpful for instructors. There are solutions to certain exercises so you should check there before using exercises for grading purposes. Additional exercises with recent links and data sets are also available on the web.

There are executable JAR files for some of the projects to help students, and instructors, understand what is being asked for from the student. In the second half of the book, the projects get bigger and the website shows UML diagrams of the author's solutions. Code for projects is never given so they are safe for graded material.

In addition, there are videos posted for many sections of the text. These are generally "live coding" sessions. Instructors should feel free to use these as pre-lectures or to use with an inverted-lecture format for certain class meetings.

In the end, one of the major goals of this text is to be a book that CS students find worthy of keeping on their bookcases after they are done with the course. This is a challenging task for a CS1 and CS2 text. It is a significant part of the reason that this book, especially in the second half, includes a number of more advanced topics. It is quite possible that you will choose to not cover several of the later chapters in this book, however, they are important concepts from computing that can be difficult to find in a single source. The hope is that those topics, combined with clear and concise code solutions to many problems, will make this a book that is worth space on a bookshelf.

Part I

Introductory Concepts

Chapter 1

Basics of Computers, Computing, and Programming

In all things it is good to understand some foundational material before going into detail. This helps to ground you and give you some context. Inevitably, you already have some experience with computers and that does give you a bit of context to work in. However, it is quite possible that your experience is limited to rather recent technology. One of the factors that shapes the world of Computer Science is that it is fast moving and ever changing. Knowing how we got to where we are can perhaps help us see where we will be going.

1.1 History

One might think that Computer Science is a field with a rather short history. After all, computers have not existed all that long and have been a standard fixture for even less time. However, the history of Computer Science has deep roots in math that extend far back in time. One could make a strong argument that the majority of Computer Science is not even really about computers. This is perhaps best exemplified in this quote by Edsger Dijkstra, "Computer Science is no more about computers than astronomy is about telescopes."[8] Instead, Computer Science is about ALGORITHMs. An algorithm is a formal specification for stating a method to solve a problem. The term itself is a distortion of the name al-Khwārizmī. He was Persian mathematician who lived in the 11th century and wrote *Treatise on Demonstration of Problems of Algebra*, the most significant treatise on algebra written before modern times. He also wrote *On the Calculation with Hindu Numerals*, which presented systematic methods of applying arithmetic to algebra.

One can go even further back in time depending on how flexibly we use the term computation. Devices for facilitating arithmetic could be considered. That would push things back to around 2400 BCE. Mechanical automata for use in astronomy have also existed for many centuries. However, we will focus our attention on more complete computational devices, those that can be programmed to perform a broad range of different types of computation. In that case, the first real mechanical computer design would have been the Analytical Engine which was designed by Charles Babbage and first described in 1837. Ada Lovelace is often referred to as the first programmer because her notes on the Analytic Engine included what would have been a program for the machine. For various reasons, this device was never built and as such, the first complete computers did not come into existence for another 100 years.

It was in the 1940s that computers in a form that we would recognize them today came into existence. This began with the Zuse Z3 which was built in Germany in 1941. By the end of the 1940s there were quite a few digital computers in operation around the world including the ENIAC, built in the US in 1946. The construction of these machines was influenced in large part by more theoretical work that had been done a decade earlier.

One could argue that the foundations of the theoretical aspects of Computer Science began in 1931 when Kurt Gödel published his incompleteness theorem. This theorem, which proved that in any formal system of sufficient complexity, including standard set theory of mathematics, would have statements in it that could not be proved or disproved. The nature of the proof itself brought in elements of computation as logical expressions were represented as numbers and operations were transformations on those numbers. Five years later, Alan Turing and Alonzo Church created independent models of what we now consider to be computation. In many ways, the work they did in 1936 was the true birth of Computer Science as a field and it enabled that first round of digital computers.

Turing created a model of computation called a Turing machine. The Turing machine is remarkably simple. It has an infinite tape of symbols and a head that can read or write on the tape. The machine keeps track of a current state, which is nothing more than a number. The instructions for the machine are kept in a table. There is one row in the table for each allowed state. There is one column for each allowed symbol. The entries in the table give a symbol to write to the tape, a direction to move the tape, and a new state for the machine to be in. The tape can only be moved one symbol over to the left or right or stay where it is at each step. Cells in the table can also say stop in which case the machine is supposed to stop running and the computation is terminated.

The way the machine works is that you look up the entry in the table for the current state of the machine and symbol on the tape under the head. You then write the symbol from the table onto the tape, replacing what had been there before, move the tape in the specified direction, and change the state to the specified state. This repeats until the stop state is reached.

At roughly the same time that Turing was working on the idea of the Turing machine, Church developed the lambda calculus. This was a formal, math-based way of expressing the ideas of computation. While it looks very different from the Turing machine, it was quickly proved that the two are equivalent. That is to say that any problem you can solve with a Turing machine can be solved with the lambda calculus and the other way around. This led to the so-called Church-Turing thesis stating that anything computable can be computed by a Turing machine or the lambda calculus, or any other system that can be shown to be equivalent to these.

1.2 Hardware

When we talk about computers it is typical to break the topic into two parts, hardware and software. Indeed, the split goes back as far as the work of Babbage and Lovelace. Babbage designed the hardware and focused on the basic computation abilities that it could do. Lovelace worked on putting together groups of instructions for the machine to make it do something interesting. Her notes indicate that she saw the further potential of such a device and how it could be used for more than just doing calculations.

This split still exists and is significant because most of this book focuses exclusively on one half of the split, software. To understand the software though, it is helpful to have at least some grasp of the hardware. If you continue to study Computer Science, hopefully you will, at some point, have a full course that focuses on the nature of hardware and the details of how it works. For now, our goal is much simpler. We want you to have a basic mental image of how the tangible elements of a computer work to make the instructions that we type in execute to give us the answers that we want.

Modern computers work by regulating the flow of electricity through wires. Most of those wires are tiny elements that have been etched into silicon and are only tens of nanometers across. The voltage on the wires is used to indicate the state of a bit, a single element of storage with only two possible values, on or off. The wires connect up transistors that are laid out in a way that allows logical processing. While a modern computer processor will include literally hundreds of millions of transistors, into the billions, we can look at things at a much higher level and generally ignore that existence of those individual wires and transistors.

In general, modern computers are built on the von Neumann architecture with minor modifications. John von Neumann was another one of the fathers of computing. One of his ideas was that programs for a computer are nothing more than data and can be stored in the same place as the data. This can be described quite well with the help of the basic diagram in figure 1.1. There is a single memory that stores both the programs and the data used by the program. It is connected to a Central Processing Unit (CPU) by a bus. The CPU, which can be more generally called a processor, has the ability to execute simple instructions, read from memory, and write to memory. When the computer is running, the CPU loads an instruction from memory, executes that instruction, then loads another. This happens repeatedly until the computer stops. This simple combination of a load and execute is called a cycle.

One of the things the CPU does is to keep track of the location in memory of the next instruction to be executed. We call a location in memory an ADDRESS. By default this moves forward each time by the appropriate amount to get to the next instruction. Different types of computers can have different instructions. All computers will have instructions to read values from memory, store values to memory, do basic math operations, and change the value of the execution address to jump to a different part of the program.

The individual instructions that computers can do are typically very simple. Computers get their speed from the fact that they perform cycles very quickly. Most computers now operate at a few gigahertz. This means that they can run through a few billion instructions every second. There are a lot of complexities to real modern computers that are needed to make that possible which are not encompassed by this simple image of a computer. In many ways, these are details that we do not have to worry about too much at the beginning, but they are important in professional programming because they can have a profound impact on the performance of a program.

We will spend just a few paragraphs talking about two of the complexities in modern

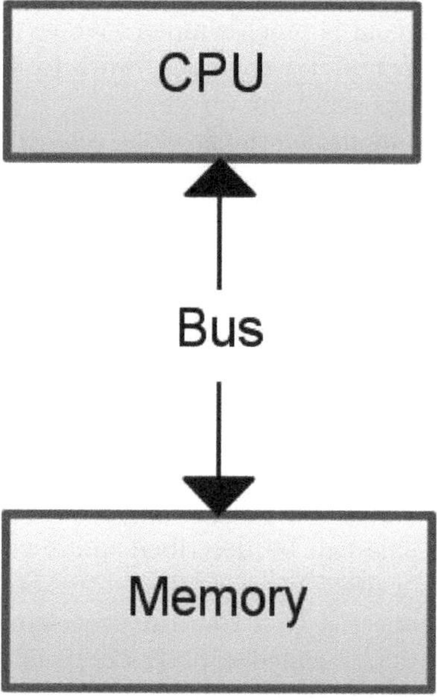

FIGURE 1.1: This is the most basic view of the von Neumann shared memory architecture. A CPU and the memory are connected by a bus. The memory stores both data and the programs themselves. The CPU can request information from memory or write to memory over the bus.

processors as they are probably the most significant. One of them will even impact material later in this book. The first is the arrangement of memory. While the simple model shows a single block of memory connected to the CPU, modern processors have an entire memory hierarchy. This hierarchy allows machines to effectively have access to very large amounts of storage while generally giving very fast access to what is really needed.

When you talk about memory on a computer, there are two things that make it into the normal machine specifications that are advertised most of the time. Those are the amount of RAM (Random Access Memory) and the amount of disk space. The latter is generally much larger than the former. Typically speed goes inverse of size for memory, and this is true of disks and RAM. Disk drives are significantly slower to access than RAM, though what is written to them stays there even when the machine is turned off. RAM is faster than the disk, but it still is not fast enough to respond at the rate that a modern processor can use it. For this reason, processors generally have smaller amounts of memory on them called cache. The cache is significantly faster than the RAM when it comes to how fast the processor can read or write values. Even that is not enough anymore though and modern processors will include multiple levels of cache referred to as L1, L2, L3, etc. Each level up is generally bigger, but also further from the processor and slower.

Some applications have to concern themselves with these details because the program runs faster if it will fit inside of a certain cache. If a program uses a small enough amount of memory that it will fit inside of the L2 cache, it will run significantly faster than if it does not. We won't generally worry about this type of issue, but there are many professional developers who do.

The other significant difference is that modern processors have multiple cores. What that means in our simple picture is that the CPU is not doing one instruction at a time, it is doing several. This is what we call parallel processing. When it happens inside of a single program it is referred to as multithreading. This is a huge issue for programmers because programs have not historically included multithreading and new computers require it in order to fully utilize their power. Unfortunately, making a program multithreaded can be difficult. Different languages offer different ways to approach this. This is one of the strengths of Scala and we will discuss it at multiple points in Part II of the book.

1.3 Software

The programs that are run on hardware are typically called software. The software is not a physical entity. However, without some type of software, the hardware is useless. It is the running of a program that makes hardware useful to people. As with the hardware, software can be seen as having multiple parts. It also has a layering or hierarchy to it. At the base is a layer that gives the computer the initial instructions for what to do when it is turned on. This is often called the BIOS (Basic Input/Output System). The BIOS is generally located on a chip in the machine instead of on the more normal forms of memory like the disk or RAM. Instructions stored in this way are often called firmware. This term implies that it is between the software and the hardware. The firmware really is instructions, just like any other software. In a sense it is less soft because it is stored in a way that might be impossible to write to or which is harder to write to.

The BIOS is responsible for getting all the basic functionality started up on the machine. Sitting on top of the BIOS is the Operating System (OS). The Operating System is responsible for controlling the operations of the machine and how it interacts with the user.

The OS is also responsible for writing files to disk and reading files from disk. In addition, it has the job of loading other programs into memory and getting them started. Over time, the amount of functionality in operating systems has grown so that they are also expected to present nice interfaces and have all types of other "basic" functionality that really is not so basic.

At the top level are the programs that the user runs. When the user instructs the operating system to run a program, the operating system loads that program into memory and sets the execution address so that the computer will start running the program. The breadth of what programs can do is nearly unlimited.[1] Everything that runs on every digital device in the world is a program. You use them to type your papers and do your e-mail. They likely also run the fuel injection system in your car and control the lights at intersections. Programs regulate the flow of electricity through the power grid and the flow of water to your indoor plumbing. Programs do not just give you the apps on your phone, they are running when you talk on the phone to compress your speech into a digital form and send it out to a local tower where another program examines it and sends it on toward the destination. On the way, it likely passes through multiple locations and gets handled by one or more programs at each stop. At some point on the other end, another program takes the digital compressed form and expands it back out to analog that can be sent to a speaker so the person you are talking to can hear it. Someone wrote each of those programs and over time more programs are being written that serve more and more different purposes in our lives.

In the last section we mentioned that newer processors have multiple cores on them. The availability of multiple cores (and perhaps multiple processors) is significant for software as well. First, they give the OS the ability to have multiple things happening at one time. All but the simplest of operating systems perform multitasking. This allows them to have multiple programs or processes running at one time. This can be done on a single core by giving each process a short bit of time and then switching between them. When there are multiple cores present, it allows the programs to truly run all at once.

Each process can also exploit its own parallelism by creating multiple threads. The OS is still generally responsible for scheduling what threads are active at any given time. This allows a single program to utilize more of the resources of a machine than what is present on a single core. While this does not matter for some specialized applications, the use of multiple cores has become more and more commonplace and the core count on large machines as well as smaller devices is currently climbing at an exponential rate. As a result, the need to multithread programs becomes ever more vital.

This increasing number of cores in machines has led to another interesting development in the area of servers. Servers are more powerful computers that are used by companies to store large amounts of data and do lots of processing on that data. Everything you do on the web is pulling data from servers. If you are at a University, odds are that they have at least one room full of servers that act as a control and data storage center for the campus.

The nature of the work that servers do is often quite different from a normal PC. A large fraction of their job is typically just passing information around and the workload for that can be very unevenly distributed. The combination of this and multiple cores has led to an increase in the use of virtualization. A virtual machine is a program that acts like a computer. It has a BIOS and loads an OS. The OS can schedule and run programs. This whole process happens inside of a program running potentially on a different OS on a real machine. Using virtualization, you can start multiple instances of one or more operating systems running on a single machine. As long as the machine has enough cores and memory,

[1]There are real limitations to computing that are part of theoretical Computer Science. There are certain problems that are probably not solvable by any program.

it can support the work of all of these virtual machines. Doing this cuts down on the number of real machines that have to be bought and run, reducing costs in both materials and power consumption.

1.4 Nature of Programming

Every piece of software, from the BIOS of each device to the OS and the multitude of applications they run is a program that was written by a programmer. So what is this thing we call programming and how do we do it? How do we give a computer instructions that will make it do things for us? In one sense, programming is just the act of giving the computer instructions in a format that it can work with. At a fundamental level, computers do nothing more than work with numbers. Remember the model in figure 1.1. Each cycle the computer loads an instruction and executes it. There was a time when programming was done by writing the instructions that the machine executes. We refer to the language of these instructions as machine language. While machine language is really the only language that the computer understands, it is not a very good language for humans to work in. The numbers of machine language do not hold inherent meaning for humans, and it is very easy to make mistakes. For this reason, people have developed better ways to program computers than to write out machine language instructions.

The first step up from machine language is assembly language. Assembly language is basically the same as machine language in that there is an assembly instruction for each machine language instruction. However, the assembly instructions are entered as words that describe what they do. The assembly language also helps to keep track of how things are laid out in memory so that programmers do not have to actively consider such issues the way they do with machine language. To get the computer to understand assembly language, we employ a program that does a translation from assembly language to machine language. This program is called an assembler.

Even assembly language is less than ideal for expressing the ideas that we want to put into programs. For this reason, other languages have been created. These higher level languages use more complete words and allow a more complex organization of ideas so that more powerful programs can be written more easily. The computer does not understand these languages either. As such, they either employ compilers that translate the higher level languages into assembly then down to machine language or interpreters that execute the instructions one at a time without ever turning them into machine language.

There are literally hundreds of different programming languages. Each one was created to address some deficiency that was seen in other languages or to address a specific need. This book uses the Scala programming language. It is hard to fully explain the benefits of Scala, or any other programming language, to someone who has not programmed before. We will just say that Scala is a very high level language that allows you to communicate ideas to the computer in a concise way and gives you access to a large number of existing libraries to help you write programs that are fun, interesting, or useful.

Early on in your process of learning how to program you will likely struggle with figuring out how to express your ideas in a programming language instead of the natural language that you are used to. Part of this is because programming languages are fundamentally different than natural languages in that they don't allow ambiguity. In addition, they typically require you to express ideas at a lower level than you are used to with natural language. Both of these are actually a big part of the reason why everyone should learn how to program. The true benefits of programming are not seen in the ability to tell a computer how to do something. The real benefits come from learning how to break problems down.

At a very fundamental level, the computer is a stupid machine. It does not understand or analyze things.[2] The computer just does what it is told. This is both a blessing and a curse for the programmer. Ambiguity is fundamentally bad when you are describing how to do things and is particularly problematic when the receiver of the instructions does not have the ability to evaluate the different possible meanings and pick the one that makes the most sense. That means that programs have to be rigorous and take into account small details. On the other hand, you can tell computers to do things that humans would find incredibly tedious, and the computer will do it repeatedly for as long as you ask it to. This is a big part of what makes computers so useful. They can sift through huge amounts of data and do many, many calculations quickly without fatigue-induced errors.

In the end, you will find that converting your thoughts into a language the computer can understand is the easy part. Yes, it will take time to learn your new language and to get used to the nuances of it, but the real challenge is in figuring out exactly what steps are required to solve a problem. As humans, we tend to overlook many of the details in the processes we go through when we are solving problems. We describe things at a very high level and ignore the lower levels, assuming that they will be implicitly understood. Programming forces us to clarify those implicit steps and, in doing so, forces us to think more clearly about how we solve problems. This skill becomes very important as problems get bigger and the things we might want to have implicitly assumed become sufficiently complex that they really need to be spelled out.

One of the main skills that you will develop when you learn how to program is the ability to break problems down into pieces. All problems can be broken into smaller pieces and it is often helpful to do so until you get down to a level where the solution is truly obvious. This approach to solving problems is called a top-down approach because you start at the top with a large problem and break it down into smaller and smaller pieces until, at the bottom, you have elements that are simple to address. The solutions you get are then put back together to produce the total solution.

Another thing that you will learn from programming is that while there are many ways to break down almost any problem, not all of them are equally good. Some ways of breaking the problem down simply "make more sense". Granted, that is something of a judgment call and might differ from one person to the next. A more quantifiable metric of the quality of how a problem is broken down is how much the pieces can be reused. If you have solved a particular problem once, you do not want to have to solve it again. You would rather use the solution you came up with before. Some ways of breaking up a problem will result in pieces that are very flexible and are likely to be useful in other contexts. Other ways will give you elements that are very specific to a given problem and will not be useful to you ever again.

There are many aspects of programming for which there are no hard and fast rules on how things should be done. In this respect, programming is much more an art than a science. Like any art, in order to really get good at it you have to practice. Programming has other similarities to the creative arts. Programming itself is a creative task. When you are programming you are taking an idea that exists in your head and giving it a manifestation that is visible to others. It is not actually a physical manifestation. Programs are not tangible. Indeed, that is one of the philosophically interesting aspects of software. It is a creation that other people can experience, but they can not touch. It is completely

[2]At least they don't yet. This is the real goal of the field of Artificial Intelligence (AI). If the dreams of AI researchers are realized, the computers of the future will analyze the world around them and understand what is going on.

virtual. Being virtual has benefits. Physical media have limitations on them imposed by the laws of physics. Whether you are painting, sculpting, or engineering a device, there are limitations to what can be created in physical space. Programming does not suffer from this. The ability of expression in programming is virtually boundless. If you go far enough in Computer Science you will learn where there are bounds, but even there the results are interesting because it is possible that the bounds on computation are bounds on human thinking as well.

This ability for near infinite expression is the root of the power, beauty, and joy of programming. It is also the root of the biggest challenge. Programs can become arbitrarily complex. They can become too complex for humans to understand what they are doing. For this reason, a major part of the field of Computer Science is trying to find ways to tame the complexity and make it so that large and complex ideas can be expressed in ways that are also easy for humans to follow and determine the correctness of.

1.5 Programming Paradigms

The fact that there are many ways to break up problems or work with problems has not only led to many different programming languages, it has led to whole families of different approaches that are called paradigms. There are four main paradigms of programming. It is possible others could come into existence in the future, but what appears to be happening now is that languages are merging the existing paradigms. Scala is one of the languages that is blurring the lines between paradigms. To help you understand this we will run through the different paradigms.

1.5.1 Imperative Programming

The original programming paradigm was the imperative paradigm. That is because this is the paradigm of machine language. So all the initial programs written in machine language were imperative. Imperative programming involves giving the computer a set of instructions that it is to perform. Those actions change the state of the machine. This is exactly what you get when you write machine language programs.

The two keys to imperative programming are that the programmer specifically states how to do things and that the values stored by the computer are readily altered during the computation. The converse of imperative programming would be declarative programming where the programmer states what is to be done, but generally is not specific about how. The Scala language allows imperative style programming and many of the elements of this book will talk about how to use the imperative programming style in solving problems.

1.5.2 Functional Programming

Functional programming was born out of the mathematical underpinnings of Computer Science. The first functional languages were actually based very heavily on Church's lambda calculus. In a way this is in contrast to imperative programming which bares a stronger resemblance to the ideas in the Turing machine. "Programming" a Turing machine is only loosely correlated writing machine language, but the general ideas of mutable state and having commands that are taken one after the other are present on the Turing machine. Like the lambda calculus, functional languages are fundamentally based on the idea of functions in mathematics. We will see a lot more of the significance of mathematical functions on programmatic thinking in chapter 5.

Functional languages are typically more declarative than imperative languages. That is to say that you typically put more effort into describing what is to be done and a lot less in describing how it is to be done. For a language to be considered purely functional the key is that it not have a mutable state, at least not at a level that the programmer notices. What does that mean? It means that you have functions that take values and return values, but do not change anything else along the way. The only thing they do is give you back the result. In an imperative language, little traces of what has been done can be dropped all over the place. Certainly, functional programs have to be able to change memory, but they always clean up after themselves so that there is nothing left behind to show what they did other than the final answer. Imperative programs can leave alterations wherever they want, and there is no stipulation that they set things back the way they found it. Supporters of functional programming often like to say that functional programs are cleaner. If you think of a program as being a person this statement makes for an interesting analogy. Were it to enter your room looking for something, you would never know, because the functional program would leave no trace of its passing. At the end, everything would be as it had started except that the function would have the result it was looking for. An imperative program entering your room might not change anything, but more than likely it would move the books around and leave your pillow in a different location. The imperative program would change the "state" of your room. The functional one would not.

Scala supports a functional style of programming, but does not completely enforce it. A Scala program can come into your room and it is easy to set it up so that it does not leave any trace behind, but if you want to leave traces behind you can. The purpose of this combination is to give you flexibility in how you do things. When a functional implementation is clean and easy you can feel free to use it. However, there are situations where the functional style has drawbacks and in those cases you can use an imperative style.

1.5.3 Object-Oriented Programming

Object-oriented programming is a relative newcomer to the list of programming paradigms. The basic idea of object-oriented programming first appeared in the SIMULA67 programming language. As the name implies, this dates it back to the 1960s. However, it did not gain much momentum until the 1980s when the Smalltalk language took the ideas further. In the 1990s, object-orientation really hit it big and now virtually any new language that is created and expects to see wide use will include object-oriented features.

The basic idea of object-orientation is quite simple. It is that data and the functions that operate on the data should be bundled together into things called objects. This idea is called encapsulation and we will discuss it a fair bit in Part II of this book. This might seem like a really simple idea, but it enables a lot of significant extras that do not become apparent until you have worked with it awhile.

Object-orientation is not really independent of the imperative and functional paradigms. Instead, object-oriented programs can be done in either a functional or an imperative style. The early object-oriented languages tended to be imperative and most of the ones in wide use today still are. However, there are a number of functional languages now that include object-oriented features.

Scala is a purely object-oriented language, a statement that is not true of many languages. What this means and the implications of it are discussed in detail in later chapters. For most of the first half of this book, the fact that Scala is object-oriented will not even be that significant to us, but it will always be there under the surface.

1.5.4 Logic Programming

The fourth programming paradigm is logic programming. The prime example language is Prolog. Logic programming is completely declarative. As the name implies, programs are written by writing logical statements. It is then up to the language/computer to figure out a solution. This is the least used of the paradigms. This is in large part because of significant performance problems. The main use is in artificial intelligence applications. There is some indication that logic programming could reappear in languages where it is combined with other paradigms, but these efforts are still in the early stages.

1.5.5 Nature of Scala

As was indicated in the sections above, Scala provides a mix of different paradigms. It is a truly hybrid language. This gives you, as the programmer, the ability to solve problems in the way that makes the most sense to you or that best suits the problem at hand. The Scala language directly includes imperative, functional, and object-oriented elements. The name Scala stands for Scalable Language, so it is not completely out of the question for them to add a library in the future that could support logic programming to some extent if a benefit to doing so was found.

1.6 End of Chapter Material

1.6.1 Summary of Concepts

- Computing has a much longer history than one might expect with roots in mathematics.
- Hardware is the term used for the actual machinery of a computer.
- Software is the term used for the instructions that are run on hardware. Individual pieces of software are often called programs.
- The act of writing instructions for a computer to solve a problem in a language that the computer can understand or that can be translated to a form the computer understands is called programming.
- Programming is very much an art form. There are many different ways to solve any given problem. and they can be arbitrarily complex. Knowing how to design good programs takes practice.
- There are four different broad types or styles of programming called paradigms.
  - The imperative programming paradigm is defined by explicit instructions telling the machine what to do and a mutable state where values are changed over time.
  - The functional paradigm is based on Church's lambda calculus and uses functions in a very mathematical sense. Mathematical functions do not involve a mutable state.
  - Object-orientation is highlighted by combining data and functionality together into objects.

- Logic programming is extremely declarative, meaning that you say what solution you want, but not how to do it.
- Scala is purely object-oriented with support for both functional and imperative programming styles.

1.6.2 Exercises

1. Find out some of the details of your computer hardware and software.
   (a) Processor
      i. Who makes it?
      ii. What is the clock speed?
      iii. How many cores does it have?
      iv. How much cache at different cache levels? (optional)
   (b) How much RAM does the machine have?
   (c) How much non-volatile storage (typically disk space) does it have?
   (d) What operating system is it running?
2. If you own a tablet computer, repeat exercise 1 for that device.
3. If you own a smartphone, repeat exercise 1 for that device.
4. Briefly describe the different programming paradigms and compare them.
5. List a few languages that use each of the different programming paradigms.
6. Do an Internet search for a genealogy of programming languages. Find and list the key languages that influences the development of Scala.
7. Go to a website where you can configure a computer (like `http://www.dell.com`). What is the maximum number of cores you put in a PC? What about a server?
8. Go to `http://www.top500.org`, a website that keeps track of the 500 fastest computers in the world. What are the specifications of the fastest computer in the world? What about the 500th fastest?

1.6.3 Projects

1. Search on the web to find a list of programming languages. Pick two or three and write descriptions of the basic properties of each including when they were created and possibly why they were created.
2. Compare and contrast the following activities: planning/building a bridge, growing a garden, painting a picture. Given what little you know of programming at this point, how do you see it comparing to these other activities?
3. Make a timeline of a specific type of computer technology. This could be processors, memory, or whatever. Go from as far back as you can find to the modern day. What were the most significant jumps in technology and when did they occur?

4. One of the more famous predictions in the field of computing hardware is Moore's law. This is the term used to describe a prediction that the number of components on an integrated circuit will double roughly every 18 months. The name comes from Intel® co-founder Gordon Moore who noted in 1965 that the number had doubled every year from 1958 to that point and predicted that would continue for at least a decade. This has now lead to exponential growth in computing power and capabilities for over five decades.[3] Write a short report on the impact this has had on the field and what it might mean moving forward.

5. Investigate and write a short report on some developments in future hardware technology. Make sure to include a things that could come into play when the physical limits of silicon are reached.

6. Investigate and write a short report on some developments in future software technology. What are some of the significant new application areas that are on the horizon? This should definitely include a look at artificial intelligence and robotics.

7. As computers run more and more of our lives, computer security and information assurance become more and more significant. Research some of the ways that computers are compromised and describe how this is significant for programmers writing applications and other pieces of software.

Additional exercises and projects can be found on the website.

[3]You can see the exponential growth of supercomputer power on plots at Top500.org.

Chapter 2

Getting to Know the Tools

Everything you do on a computer involves the running of software. At the most basic level, your computer has some operating system on it that controls everything that happens, organizes data, and lets you execute other programs inside of it. What operating system and tool set you choose to use has a significant impact on how you go about doing things. This chapter walks you through some tools that you might use and might not be familiar with. It then moves on to look at some of the tools that are part of a standard Scala distribution that we will use through the rest of the book.

Your choice of tools begins with your choice of operating system. It is very likely that you are familiar with common graphical environments on a Windows PC or Macintosh environment. For that reason, we are not going to take time to introduce them. Instead, this chapter talks about how you interact with various operating systems using a text interface. There are a number of reasons why it is good to be able to use this type of interface. For our purposes, the most significant is that is it the way we will interact with some of the Scala tools that we will be using.

2.1 Unix/Linux (includes Mac OS X)

This section covers the environment you will encounter in a Linux and Unix environment. Mac OS X is built on top of BSD Unix so everything here applies to the Mac. If you are on a Mac you can open a console to get access to the command-line tools.

2.1.1 Command-Line

Most people these days are used to the point and click approach of a Graphical User Interface, GUI. This is not how people have always interacted with computers. Indeed, a GUI takes a significant amount of processing power to run. For many years, computers had to have a simpler interface. Even today, there are times when you will need to interact with a computer without a GUI. This can happen when something goes wrong with the computer or if you need to use a computer that is located far away. There are also situations where people will choose to not run a GUI on a computer simply because they don't want to waste the processing power.

Even when a computer is running a GUI, you can choose to interact using a command-line interface. The GUI can give you the ability to keep multiple command prompts open at once and using the command-line itself provides you with abilities that are not easily accomplished with a GUI. The reason why many people choose not to use a command prompt is that it does have a bit of a learning curve. You have to know some basic commands to be able to do anything. If you are going to spend much time working on computers it is worth putting in the small amount of effort to learn the basic commands for the command prompt interface on your OS. Odds are that you will find that not only does it speed things up on a regular basis, but there will be times when it allows you to do things that you simply could not do otherwise.

Figure 2.1 shows a terminal window under Linux that gives a command prompt. Your command prompt might look different than this. In general, the prompt should display some useful information followed by a character that signifies the end of the prompt. In this figure the prompt shows the user name and computer name followed by the current directory and a $. It is the $ that signifies the end of the prompt. Commands that you enter appear after the $. Anything the command displays will be printed out below the prompt. After the command finishes, another prompt will be given.

2.1.1.1 Files and Directories

In a GUI you organize your files in folders. You have icons that represent different file types or folders and you can place files/folders inside of folders to organize them. These concepts came from the systems already in place on the command-line. The only change is terminology. The term folder worked well from a graphical standpoint. On the command-line they have long been called directories.

The first set of commands we will learn about allow us to work with files and directories so that we can navigate around and do the types of things you are used to doing by clicking, double-clicking, or drag and dropping. Here is a list of commands with a brief description of each. You will notice that in general they are abbreviations for what they do.

- `cat` – Display the contents of one or more files.
- `cd` – Change directory.
- `cp` – Copy one file to another name or location.

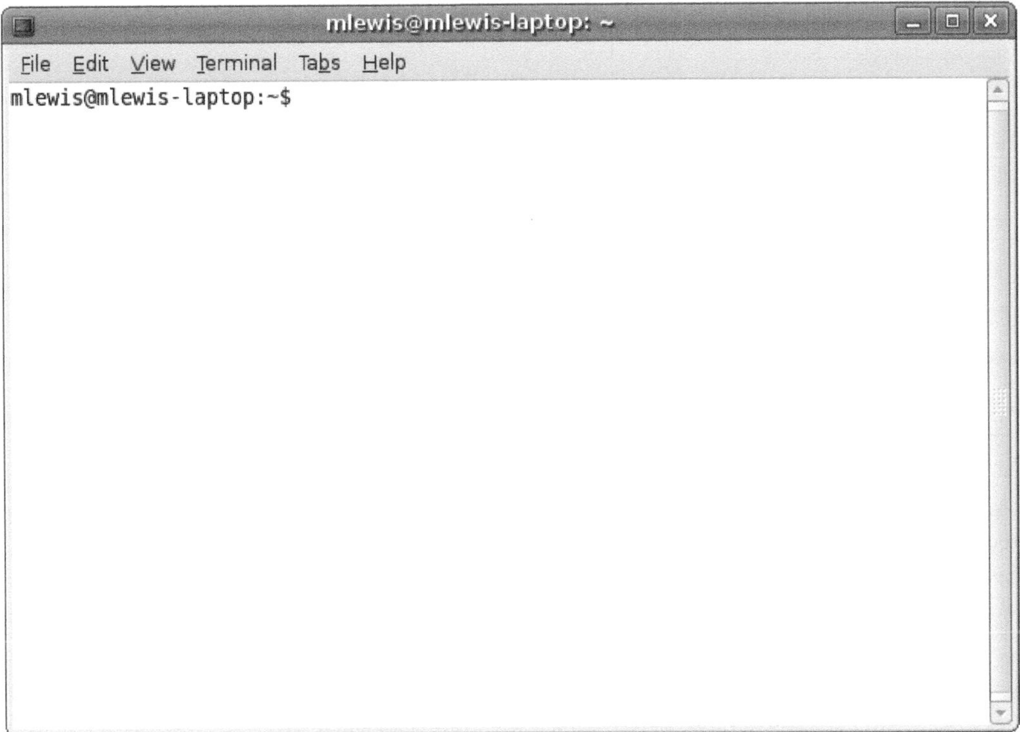

FIGURE 2.1: An example of a Linux terminal with a command prompt.

- `less` – Display out the contents of one or more files with the ability to move forward and backward, search, etc. (Less is more than more.)
- `ls` – List the contents of a directory.
- `mkdir` – Make a directory.
- `more` – Display the contents of one or more files with the ability to page through or search.
- `mv` – Move a file to a new name or location.
- `pwd` – Stands for "print working directory". This prints out the directory you are currently working in.
- `rm` – Remove a file.
- `rmdir` – Remove a directory if it is empty.

The big advantage of the command-line is that each command can be followed by arguments. Just entering a command is like double-clicking on a program icon. It is the arguments that let you specify additional information. Most of these commands do not do anything unless you give them one or more file/directory names to work on.

To see how you go about using these commands we will begin with `pwd`. It is a very simple command and can come in handy when you are starting off, or if you have lost track of where you are in the directory structure. At the command prompt simply type in `pwd` and hit Enter. Exactly what you get will depend on many things, but it should print out one line of output showing you the current directory. It could look something like this.

```
mlewis@mlewis-laptop:~$ pwd
/home/mlewis
mlewis@mlewis-laptop:~$
```

In this case, the directory is `/home/mlewis`. After printing the output we get a new prompt to enter the next command.

> **Aside**
>
> In Unix and Linux systems, the base of the entire file system is `/`. That is why it appears at the beginning of the `/home/mlewis` directory. A directory that begins with / is an absolute directory.

The next command we want to enter is the `ls` command. Type in `ls`, followed by an Enter and you will get a listing of all of the files in your current directory. This is the most basic usage of the command. You can also use it to see the contents of some other directory by following the `ls` command with a directory name. Here we see a listing of the contents of the root directory.

```
mlewis@mlewis-laptop:~$ ls /
bin dev initrd lib lost+found opt sbin tmp vmlinuz
boot etc initrd.img lib32 media proc srv usr vmlinuz.old
cdrom home initrd.img.old lib64 mnt root sys var
```

If you list a directory and it does not start with / then it will be a relative directory. So what you list will be appended to the end of the current directory. You might have a directory called Desktop in the current directory. Typing in `ls Desktop` will list the contents of the Desktop directory under the current directory.

```
mlewis@mlewis-laptop:~$ ls Desktop
AdobeReader.desktop QtCreator.desktop
```

You can also specify multiple files or directories and they will all be listed as seen here.

```
mlewis@mlewis-laptop:~$ ls Desktop /
/:
bin dev initrd lib lost+found opt sbin tmp vmlinuz
boot etc initrd.img lib32 media proc srv usr vmlinuz.old
cdrom home initrd.img.old lib64 mnt root sys var

Desktop:
AdobeReader.desktop QtCreator.desktop
```

You might not think that you would normally want to list multiple things at once, but it is a feature people use all the time with wild card characters. If you put a * into a name, the system will replace it with all the files or directories that match if the * is replaced with zero or more characters. Here is an example of that usage.

```
mlewis@mlewis-laptop:~$ ls /vmlinuz*
/vmlinuz /vmlinuz.old
```

For the first file, the * is replaced with nothing. For the second it is replaced with "`.old`". You can also use a `?` to represent any one character. These wild cards are not ever seen by the `ls` command of other commands. The command shell replaces them with all of the matching files and it is the multiple file names or directory names that get passed to the command.

The `ls` command has a lot of other possible options as well. The options are all preceded by a hyphen. The most commonly used option is `-l`, which tells ls to use a long display format. The following is an example of that.

```
mlewis@mlewis-laptop:~$ ls -l Desktop
total 8
-rwxrwxrwx 1 root root 1261 2008-11-28 12:21 AdobeReader.desktop
-rw-r--r-- 1 mlewis mlewis 250 2009-12-22 11:03 QtCreator.desktop
```

The long format puts each file or directory on a separate line. In addition to the file names it shows a number of other pieces of information. The first ten characters show permissions on the files. Permissions are discussed in section 2.1.1.4. After the permissions is a number showing how many links there are to this file. After that are the user name of the owner of the file and the group that the file belongs to. The last two pieces of information are the size of the file, measured in bytes, and the date and time of the last modification of that file.

The next most commonly used option is `-a`. This tells `ls` to list all the files, including the hidden files. Hidden files in Unix and Linux are files whose names begin with a period. We see here a usage of this combined with the long option. Only one hyphen is needed when options are combined.

```
mlewis@mlewis-laptop:~$ ls -al Desktop
total 16
drwxr-xr-x 2 mlewis mlewis 4096 2010-02-27 20:02 .
drwxr-xr-x 106 mlewis mlewis 4096 2010-08-09 16:57 ..
-rwxrwxrwx 1 root root 1261 2008-11-28 12:21 AdobeReader.desktop
-rw-r--r-- 1 mlewis mlewis 250 2009-12-22 11:03 QtCreator.desktop
```

This adds two additional lines. The letter `d` at the far left edge tells us that these are actually directories. The `.` directory is the name for the current directory. The `..` directory is the name of the directory in which the current directory sits. We might also refer to it as the parent directory.

By default, commands work on the things in the current directory. For this reason, one of the commands you will use the most is the command that changes the current directory, `cd`. The `cd` command is followed by a single directory name. The directory will be the new current directory. As we just saw, each directory has `.` and `..` as directory names inside of it. If you want to move up a level in the directory structure, simply do "`cd ..`". You can also enter `cd` without a directory at any time to change back to your home directory.

Other tasks that you do frequently when you are manipulating files in a GUI are to copy, move, or delete. The commands for these are `cp`, `mv`, and `rm`. The `cp` and `mv` commands are used in roughly the same way. One way to use them is to enter two file names where the first file will be copied to the second name or renamed to the second name. A second usage is to enter one or more file names and then end with a directory name. In that usage all of the files listed are either copied or moved to the directory listed at the end. If you include a name that has a wild card in it, it will be expanded to all the file names. We can see these along with `cd` in use here.

```
mlewis@mlewis-laptop:~$ cd Desktop
```

```
mlewis@mlewis-laptop:~/Desktop$ mv ../*.txt .
mlewis@mlewis-laptop:~/Desktop$ cp test.txt test2.txt
```

We start by changing to the `Desktop` directory. Note the prompt used in this example changes so that we can see we are in this new directory. We then move all the files from our home directory that end with `.txt` into the current directory. This line with `mv` uses both `..` to refer to the directory above the current one and `.` to refer to the current directory. In this case, one of those files was named `test.txt`. The last command makes a copy of that file and gives the copy the name `test2.txt`.

If you want to remove a file, use the `rm` command. The `rm` command can be followed by one or more files. Again, wild cards can be used to match multiple files. Here we are getting rid of all of the files that end with `.txt`.

```
mlewis@mlewis-laptop:~/Desktop$ rm *.txt
```

There is a word of warning about removing files on most Linux-based systems with the command-line, it is not forgiving. This is true whether you are using `rm` or if you happen to `mv` or `cp` to a file name when you already had a file of that name. While most of the GUIs that you have used probably move files you delete to some form of trash can, really just a different directory, and then allow you to clear that out later, doing `rm` on a file really deletes it then. You can not go digging in the trash to find it. In general when you delete a file with `rm` on these systems, it is gone and your only chance of getting it back is if you have a backup.[1]

This can be especially dangerous with the way that the command-line deals with wild cards. Take the `rm` command above. If you accidentally insert a space between the asterisk and the dot, you will delete every file in the current directory. This is because the `*` alone matches everything.[2] So be careful when you are using `rm` and look over the command before you hit Enter to make sure you did not mistype anything or insert any unwanted spaces.

Directories exist to keep files organized. For this reason, you should probably make directories for each different major grouping of work that you have. You might also nest these to further refine the organization scheme. You can make a new directory with the `mkdir` command. Simply follow the command with the names of one or more directories you want to create. For example, it is probably a good idea to make a directory to store all of the work for this book or the course you are taking. You might make subdirectories in there for different assignments/projects or the code you write in class.

By default, `rm` does not remove directories. To remove a directory use the `rmdir` command. Like `mkdir`, you follow it with the name of one or more directories you want to remove. `rmdir` will only remove a directory if it is empty. So you would need to remove the contents to the directory first before you can remove it.

2.1.1.2 Aside

If you really want to delete an entire directory and all of its contents, including potentially other directories, there is an option for `rm` that will do that. The `-r` option

[1] There is a `-i` option for `rm`, `mv`, and `cp` that will cause the program to prompt you before any file is removed. This is not practical if you want to remove a lot of files, but it could be considered a good default for most of what you do.

[2] There is one minor safety in place. The `*` alone does not match files that start with a `.`, so hidden files would not be deleted in that situation. That will not make you feel much better though if you have just erased five different multihour coding projects.

tells rm or cp to recursively[3] run down into a directory and either remove or copy all of its contents. This is very helpful if you want to copy a directory that has a lot of things in it. It can also be of great help if you want to remove such a directory. Remember to tread carefully though. Using rm with the -r flag has the possibility to wipe out a lot of files that you might have wanted to keep around. The command below uses -r with a copy to copy over an entire directory and all of its contents into the current directory.

```
mlewis@mlewis-laptop:~/Desktop$ cp -r ../Music/ .
```

The last thing you might want to do with your files is actually look at what is in them. There are many ways to do this. Indeed, a big part of most of the programs you have on your computer is the ability to let you view the contents of specific types of files. The commands we will consider here work best with plain text files. We will be dealing with a lot of those in this book. The most basic way to look at such a file is with the cat command, which simply prints the contents of one or more files to standard output. A step above that is the more command which will display one screen at a time and let you move down one line by hitting enter or a whole screen by hitting space. You can also search for things by typing what you want to search for after pressing /. Yet another step up is the less command, which allows you to move around freely using arrow keys or page-up and page-down as well as search as you could with more.

2.1.1.3 Helpful Tips

Many people who are not familiar with the command-line are initially turned off a bit by the amount of typing they have to do. People who use command-line all the time do not necessarily like typing more, they know more tricks to do things with the keyboard that do not require a lot of typing. Here are some handy tricks to help you use the command-line more efficiently.[4]

Typing in complete file names can be tedious and worse, is often error prone. If you mistype something it will not work and you have to enter the whole thing again. Worse, it might work and do something you did not really want to do. Because of this, tab completion is probably the most helpful feature you will find in your command-line environment. If you type in the first few letters of a file or directory then hit Tab, the shell will fill in as many characters as it can. If there is only one file that starts that way, it will give you the whole file. If there is more than one, it will fill in as much as it can until it gets to a point where you have to make a choice. If you double tab it will print out the different options that fit.

You should try this by going back to your home directory (remember you can do this at any time by typing in cd and hitting enter without giving it a directory) and then typing cd De and hitting Tab. Odds are good that the word Desktop will be completed for you with a / at the end because it is a directory. Use ls to see if there are two files/directories that start with the same first few letters. Type in cat followed by the first two letters and hit Tab. The shell will complete as much as it can then stop. Hit Tab twice quickly and it will show you the different options. It might look something like this.

```
mlewis@mlewis-laptop:~$ cd Do
Documents/ Downloads/
```

[3]Recursion is a topic that you probably are not familiar with yet, but which we will hit on several times during this book.

[4]Tricks like those discussed in this section vary depending on the exact command shell you are using. This describes features of the bash shell, which is standard for many Linux installs.

Not only does tab completion save you from typing a lot of extra characters, it never misspells a file or directory name. Use tab often and it will save you a lot of key strokes and a lot of little mistakes.

It is not uncommon to want to enter the same command or a very similar command more than once. The easiest way to do this, if it is a command you entered recently, is to use the up and down arrow keys to navigate backward and forward through the command history. When you get to a command, you can edit it if you want to make changes.

Using the arrow keys is not all that convenient if the command was something you entered a long time ago. The simplest way to repeat an old command exactly is to start the line with an exclamation point (often read "bang") and follow it by some of the command. This will go back through the history and find the most recent command that started with those characters. That command will be executed without the chance to edit it. The limitation that you can not edit it or see exactly what the command is you will execute before you execute it means that `!` is most useful for commands that start in a rather unique way.

You can also press `Ctrl-r` to get the ability to search through history. After pressing `Ctrl-r`, start typing in characters that occur consecutively in the command you want. It will bring up the most recent command that includes what you have typed in and you can edit it.

Another command that is very useful in general is the `man` command. This is short for manual and basically functions as help pages for commands or installed programs on the systems. The most basic usage is to type in `man` followed by the command that you want information on. You can also put in a `-k` option to do a search for something in the `man` pages.

2.1.1.4 Permissions

All modern operating systems have permissions that control who can get access to different types of files. What files you can access depends on who you are and what permissions you have. The act of logging into a machine determines who you are. You can use the command `whoami` to see who you are logged onto the machine as.

```
mlewis@mlewis-laptop:~$ whoami
mlewis
```

In this case the prompt also displays the user name, but that will not always be the case. Each user can also be a member of various groups. You can use the `groups` command to see what groups you are a member of.

```
mlewis@mlewis-laptop:~$ groups
mlewis adm dialout cdrom floppy audio dip video plugdev fuse lpadmin admin
```

The combination of who you are and the groups that you are in will determine what you have access to on a machine.

On Unix and Linux every file has read, write, and execute permissions for their owner, their group, and others. Those, along with the owner and the group of a file are displayed by `ls` when you use the `-l` option. Let us go back to the `Desktop` directory and look at the long listing of the files again.

```
mlewis@mlewis-laptop:~/Desktop$ ls -l
total 8
-rwxrwxrwx 1 root root 1261 2008-11-28 12:21 AdobeReader.desktop
-rw-r--r-- 1 mlewis mlewis 250 2009-12-22 11:03 QtCreator.desktop
```

There are two files here. The first ten characters tell us the permissions on the file. The first one is either 'd' or '-'. The 'd' would tell us it is a directory. Neither is a directory so both of these begin with '-'. After that are three groups of `rwx` where any of those can be replaced by '-'. The letters stand for read permission, write permission, and execute permission with '-' being used when that permission is not granted. The first set of `rwx` is the permissions of the user the file belongs to. The second set is for the group the file belongs to. The third is for others. The first file listed here gives full permissions to all three. The second gives read and write to the user and only read to the group and others. Shortly after the permissions appear two names. For the first file they are both `root`. For the second they are both `mlewis`. These are the user and group owners for the file. The name `root` is the superuser on Unix and Linux machines.

To make this more interesting, let us use the `mkdir` command to make a new directory and then list the contents of the directory again.

```
mlewis@mlewis-laptop:~/Desktop$ mkdir NewDir
mlewis@mlewis-laptop:~/Desktop$ ls -l
total 12
-rwxrwxrwx 1 root root 1261 2008-11-28 12:21 AdobeReader.desktop
drwxr-xr-x 2 mlewis mlewis 4096 2010-08-11 19:59 NewDir
-rw-r--r-- 1 mlewis mlewis 250 2009-12-22 11:03 QtCreator.desktop
```

This new directory is also owned by `mlewis` and is part of the group `mlewis`. Note the letter 'd' first thing on the line telling us that this is a directory. The owner, `mlewis`, has full read, write, and execute permissions. Anyone in group `mlewis` will have read and execute permissions as do other users on the machine. If you try to use a file/directory in a way you do not have permissions for, you will get an error message. To go into a directory and see what is inside of it you need both read and execute permissions.

So now that you know how to tell what permissions a file has, the next question is how do you change them. The `chmod` command is used to change permissions on a file. There are quite a few different ways to use `chmod`. We'll just introduce one of them here. After `chmod` you specify how you want to change or set permissions, then give a list of the files you want to make the changes to. The simplest way to specify changes is to use character codes for which set of users and the permissions involved and separate them with a '+', '-', or '=' to say if you want to add them, remove them, or set them. The different permission sets are specified by 'u' for the user, 'g' for the group, 'o' for others, and 'a' for all. The rights are specified by the letters 'r', 'w', and 'x' as we have already seen.

Let us say that we are going to put things in this new directory that we do not want anyone but the user to have the ability to see. In that case we would want to remove the read and execute permissions from the group and others. We can see how that is done and the result of doing it here.

```
mlewis@mlewis-laptop:~/Desktop$ chmod go-rx NewDir/
mlewis@mlewis-laptop:~/Desktop$ ls -l
total 12
-rwxrwxrwx 1 root root 1261 2008-11-28 12:21 AdobeReader.desktop
drwx------ 2 mlewis mlewis 4096 2010-08-11 19:59 NewDir
-rw-r--r-- 1 mlewis mlewis 250 2009-12-22 11:03 QtCreator.desktop
```

Had we wanted to give everyone full permissions we could have used `a+rwx` as the second argument to `chmod`.

Less common than changing permissions is changing ownership. If you own the file you can change the user and the group of the file with the `chown` command. The most likely scenario for this is if you want to change the group to a shared group so that a select set of users can have access to something. If the group were called `project` then you would type in "`chown :project files`".

Aside

As with the `cp` and `rm` commands, there are times when you will want to change the permissions of a large number of files in an entire directory. Both `chmod` and `chown` accept the `-R` option so that they will do this.

2.1.1.5 Compression/Archiving

If you spend much time on a system, you will likely need to interact with compressed or archived files at some point. Both of these play the role of taking many separate files and turning them into one large file. If it is compressed, that large file should be smaller than the sum of the files that went into it. Many programs that you might want to put on your computer come in the form of compressed files. If you have assignments or projects that span multiple files, your instructor might want you to combine them into a single file to turn them in. There are many different compression and archiving utilities. We'll just talk about the main ones here.

- `tar` – Archive files to a single tarball.
- `gzip/gunzip` – Compress one or more individual files to the gzip format.
- `zip/unzip` – compress multiple files to a single zip file. The zip format is used extensively on many operating systems.

The first two are used in the Unix/Linux world. They do archiving and compression separately, but are often used together to do both. The archiving program is called `tar`. It was originally created for archiving files from disk to tape. It simply collects multiple files together into a single file called a tarball. It can also extract files from a tarball. As with our other programs, `tar` is controlled by passing it arguments. You can use the `man` pages to see the many different options for `tar`. We will only discuss the basic ones.

The three main ways you will interact with a tarball are to create one, view the contents of one, or extract files from one. These different options are given with the characters `c`, `v`, and `x`. Most of the time you will want to interact with a file, and for that you will give the `f` option. The next thing after the `f` option should be the file name of the tarball you are creating, viewing, or extracting. Tarball files should typically end with the extension `.tar`. If you are creating a tarball, that will be followed by the file and directory names of the things you want to put into the tarball. Here are some sample invocations.

```
tar cf assign1.tar header.txt Assign1/
tar tf assign1.tar
tar xf assign1.tar
```

The first command is what you might execute to create a tarball called `assign1.tar`, which contains a header file and all the contents of the directory `Assign1`. The second command might be used to verify that everything you want is in the file or to check what is there before it is extracted. The last command would extract the contents.

The counterpart to `tar` is `gzip`. This command will compress one or more files. Simply follow the `gzip` command with a set of file names to compress. That will create a new set of compressed files that end with the extension `.gz`. The `gunzip` command will unzip any file to give you back the original file. You can use the `-r` option to recursively descend into a directory.

The combination of `tar` and `gzip` is so common that you can get `tar` to automatically zip or unzip files using `gzip`. Simply put the `z` flag in the options to `tar` and either `gzip` or `gunzip` will be invoked. Files created that way are typically given the `.tgz` extension.

Both `tar` and `gzip` are rather specific to the Unix/Linux world. If you have to deal with compressed files going to or from other systems, you might want to use the more broadly used `zip` format. The `zip` command can be followed by the name of a zip file and the set of file names to zip up into the specified zip file. If you want it to zip up a whole directory structure you will need to use the `-r` option. There is an `unzip` command that will extract the contents of a zip file.

2.1.1.6 Remote

One of the great strengths of Unix/Linux is the true multiuser capabilities that they have. In Windows you are probably used to only having one person at a time logged into a computer. That person will be sitting in front of the computer and if no one is at a computer, it is not being used. This is not true with Unix/Linux. These systems will allow multiple people to be logged in at once. A computer could be in full use even if no one is sitting at it. In fact, this is a big part of why you should be familiar with the command-line. Running graphical interfaces on machines across long distance Internet connections can be very unpleasant. However, the plain text form of a command-line will be quite responsive even across slow networks.

So how do you log into one computer from another one? If you are on a Unix/Linux box and you want to log into another there are several commands that you could use to make a connection. The most basic, and likely least useful, is `telnet`[5]. Simply type in `telnet` followed by the name or IP address of the machine that you want to connect to. This opens a very simple network connection between the machines. Because `telnet` connections are not secure and are, in a way, too flexible, you will probably find this approach blocked by the security setting on most machines.

Next up the scale is `rsh`. This stands for remote shell and can be invoked in the same way as `telnet`. The `rsh` connection is also not secure and as such, is blocked on many systems. However, for various reasons, it might be allowed on your system if you are logged into a trusted machine inside of the local firewall.

The connection type that you probably should use by default is `ssh`. This stands for secure shell and, as the name implies, all communication across a `ssh` connection is encrypted using public key cryptography. As such, if a system is going to let you log in remotely, this is the method that most system administrators are likely to leave open for you.

When you connect remotely to a machine you will be asked for your password and, assuming you get it right, you will be dropped at a command prompt on that machine. Everything that is discussed in this section will work on the remote machine just as well as the local one, no matter how far away the machine is. When you are done entering commands and you want to come back to the current machine, type `logout` or `exit` to terminate the remote session.

[5]You will have occasion to use `telnet` in the second half of the book when we start writing our own networked code. One of the project ideas from the second half of the book is a networked text-based game called a MUD that users would connect to with `telnet`.

Aside

If your username on the remote machine is different from that on the current machine you can use "`ssh` *username*`@`*machine*" to specify the remote username. Also, the `ssh` command has one argument that will be particularly helpful once you get past chapter 11. The `-Y` option allows the remote machine to send windows back to the current machine. Use "`ssh -Y` *machine*" to activate this.

What if you are on a Windows machine? How can you connect then? Your windows machine should likely have `telnet` on it if you bring up the command prompt. As was mentioned above, this type of connection will likely be blocked for security reasons. Thankfully, there are free `ssh` programs that you can get for Windows. One of these is called `putty` and it will allow you to `ssh` into a remote machine and use it from a Windows box. If you decide you like the Unix/Linux command prompt you can also install Cygwin under Windows and get `rsh`, `ssh`, and all the other commands we will talk about on your Windows machine.

Aside

You can get remote windows sent back to your machine through `putty` using X-forwarding. This is a bit more complex so you should look at the help for `putty` to see how to do it.

An even better approach if you have a Windows machine is to run Linux on the machine. There are versions of Linux such as Ubuntu® and Linux Mint® which are specifically targeted at novice users or people who want to just get things running without much difficulty. You can download these from their websites at `http://www.ubuntu.com`, and `http://linuxmint.com`, respectively. If you do not want to dual boot your machine, you can even run Linux in a window under Windows using free virtualization software such as VirtualBox®(`http://www.virtualbox.org`) or VMWare Player®(`http://www.vmware.com/products/player/`).

`telnet`, `rsh`, and `ssh` all give you ways to log into a remote machine and execute commands on that machine. Sometimes what you need is to move files from one machine to another. For this you should probably use `scp` or `sftp`.[6] The `scp` command stands for secure copy. You use it much like you would the normal `cp` command. The only difference is that your file names can be prepended with a machine name and a colon to say that they come from or are going to a different machine. A sample invocation is shown here.

```
scp cs.trinity.edu:Desktop/fileINeed.txt .
```

This copies the file `fileINeed.txt` from the `Desktop` directory on the machine `cs.trinity.edu` to the current directory on the current machine. If the second argument has a machine name with a colon then the file will go to that remote location. As with `cp`, the `-r` option can be used to copy an entire directory structure.

[6]There are also `rcp` and `ftp` programs, but just like `rsh` and `telnet`, secure systems likely will not let these through.

The `sftp` command is a secure version of the classic `ftp` program. `ftp` stands for File Transfer Protocol. It allows you to navigate local and remote directory structures, see what files are on the remote machine, and then move them across. The full usage of `sftp` is beyond the scope of what we want to cover here, but you should be aware it exists in case you need to move files around and are not certain what you want to move or where it should be moved to or from.

2.1.1.7 Other Commands

There are many other commands that are available to you on the Unix/Linux command-line. You do not have to be familiar with all of them to get things done with the command-line. In fact, after this chapter you should have enough to get through the tasks that you will need for this book. There are just a few more commands you might find helpful that we will list here.

- `clear` – Clears the terminal so you get a fresh screen. This is only cosmetic, but it can be helpful at times.
- `df` – Stands for disk free. This will list all of the different volumes on the current disk system and show you information on their usage.
- `du` – Stands for disk usage. This will give you a rundown of how much disk space is being used in the current directory and its subdirectories.
- `echo` – Prints whatever follows the command back to the terminal.
- `find` – Find files that have different properties specified on the command-line.
- `grep` – Searches for text inside of the specified files.
- `head` – Print the first lines of a file.
- `ps` - Show information about processes running on the machine.
- `tail` – Print the last lines of a file.
- `touch` – Updates the edit time on the specified file to the current time.
- `top` – Lists the top resource consuming programs currently running on the machine.
- `w` – Tells you who is logged onto your machine, where they are coming from, and what they are running.
- `wget` - You give this command a URL and it will download the file at that URL to the current directory.
- `where` - This command should be followed by the name of an executable. It will tell you the full path to that executable if it is in the current path.

As a final tip, you can follow any command with `&` to make it run in the background.

2.1.2 I/O Redirection

So far we have been talking about the different commands we can run from the command-line. We have typed input into the console when it was needed and the commands have output to the terminal so we could see it. The Unix/Linux command-line gains a significant amount of power from the fact that you can redirect input and output. The simple forms of redirection have the output of a program go to a file or have the input come from a file. These will be of use to you in many of the projects in later chapters even if you do not see the benefit right now.

To send the output of a program to a file you put a greater than, >, followed by a file name after the command. The output will go into that file. Here is an example.

```
mlewis@mlewis-laptop:~/Desktop$ ls -l > list.txt
mlewis@mlewis-laptop:~/Desktop$ ls -l
total 16
-rwxrwxrwx 1 root root 1261 2008-11-28 12:21 AdobeReader.desktop
-rw-r--r-- 1 mlewis mlewis 259 2010-08-12 22:32 list.txt
drwx------ 2 mlewis mlewis 4096 2010-08-11 19:59 NewDir
-rw-r--r-- 1 mlewis mlewis 250 2009-12-22 11:03 QtCreator.desktop
mlewis@mlewis-laptop:~/Desktop$ cat list.txt
total 12
-rwxrwxrwx 1 root root 1261 2008-11-28 12:21 AdobeReader.desktop
-rw-r--r-- 1 mlewis mlewis 0 2010-08-12 22:32 list.txt
drwx------ 2 mlewis mlewis 4096 2010-08-11 19:59 NewDir
-rw-r--r-- 1 mlewis mlewis 250 2009-12-22 11:03 QtCreator.desktop
```

The first call to `ls` does not print to screen. Instead, it sends it to the file `list.txt` because of the greater than followed by that file name. We can then do another `ls` and let it print to screen to verify the new file is there. Indeed, we have a new file that is 259 bytes in length. We can also use `cat` to view the contents of the file and you see the output you expect from `ls`. There is one interesting aspect to it. The file we are outputting to is there, but it has a size of zero. This is because the file is created as soon as we execute the command, but it does not get contents written to it immediately. That happens over time as parts of the output get big enough that they have to be moved from a buffer in memory to the file.

Using > for redirecting output will create a new file if none is there and will wipe out any existing file if one is there. If you want a command to add onto an existing file, use >> instead. This will redirect the output, but append to any existing file.

We can also tell a command to take its input from a file instead of from standard input using the less than, <, symbol. None of the commands we have looked at require additional input from standard input unless it requires a password, which you do not want to put in an unencrypted file. Many of the programs that you write will need to have you type text into them. You will often have to run these programs multiple times and it can get very tedious to input the same values over and over while you try to get the program to work. Having it take the input from the file can save you time as you only have to enter it into the file once and then use the file when you are testing the program.

The real power of redirection power comes from the the ability to send the output of one command into the next command as the input. This is called "piping" and it is done with the vertical bar, |, a symbol that is read as pipe. Perhaps the most common usage of the pipe is to send an output that is long into `grep` to search for something in it. This example here will find any files in the current directory that were last edited in December of 2009. For the directory listings we have done so far this type of thing would not be of much use, but it is more helpful in large directories like the one this example was executed on.

```
mlewis@mlewis-laptop:~$ ls -l | grep 2009-12
drwxr-xr-x 6 mlewis mlewis 4096 2009-12-22 11:03 qtsdk-2009.05
drwxr-xr-x 4 mlewis mlewis 4096 2009-12-23 17:33 QtWorkspace
-rwx------ 1 mlewis mlewis 2291366 2009-12-28 12:21 SwiftVis-0.3.0.jar
```

Another example is when some programs that are long running produce very long output files. Even if you use `grep` to cut the output down to just lines that have what you are interested in, that might still be many pages of material. You could pipe that output to programs like `head`, `tail`, or `less` to allow you to look at them in a manner that is easier to handle. For example, if you have a log file for a program that includes certain lines that have the word "Step" that you care about and you only want to see the last few you could do this.

```
grep Step log | tail
```

2.1.3 Text Editors (vi/vim)

Most programming languages have programs that are written in plain text. As such, you can use any simple text editor that you want to edit your programs. Word processors, such as Microsoft Word® or Open Office® are not plain text editors. You should not use them to edit your programs. Word processors store all types of additional information such as the font and format of text in addition to the text itself. Programs need only the straight text. On a Windows machine, Notepad is an example of a text editor.

You could use Notepad to edit Scala programs, but not all text editors are equally useful for programming. Due to the fact that programming is a fairly common usage for plain text editors, many text editors have features in them that are specifically aimed at programming. On Unix/Linux a primary example of this is the `vi` editor. While there are lots of different editors that one can choose from on Unix/Linux, we will work with `vi` because it is lightweight and is installed on virtually all such machines.[7] Many other editors will bring up separate windows or are not part of a default installation. If you decide you do not like `vi` or have another preference, you are more than welcome to use whatever editor you want in going through this book.

To start running `vi`, simply type in `vi` followed by the name of the file you want to edit. If the file exists, you will see its contents on your screen. If not, you will get a mostly blank screen with tildes down the left side and some extra information at the bottom.

The first thing to know about `vi` is that is has two main modes. When you first start it up, you are in command mode. To type things you need to be in an editing mode. When you are in an editing mode you can type just like you would in Notepad. What gives `vi` the extra power that is helpful for programmers is the ability to use the command mode. To get from command mode into a edit mode, type one of the following.

- `i` – Insert before the current character.
- `I` – Insert at the beginning of the current line.
- `a` – Append after the current character.
- `A` – Append at the end of the current line.

[7]Many Linux installs technically come with `vim` instead of `vi`. `vim` stands for "`vi` improved" and everything covered in regards to `vi` here will work for `vim`. On those systems executing `vi` will typically run `vim` by default so it is likely you will never notice a difference. In fact, a few features described in this section, like multiple undo and redo, only work in `vim`, not in standard `vi`.

- `R` – Start replacing characters from the current position.

Most of the time you will probably just use `i`, but there are occasions when the others can be helpful. The line at the bottom of your terminal should change to show you that you are now in an edit mode. After you get into edit mode you start typing. In `vim` you can also use the arrow keys along with **`Home`**, **`End`**, **`Page Up`**, and **`Page Down`** to move around. If you are in a true `vi` install that will not work, instead you will have to return to command mode to move the cursor around.

To get back into command mode, simply hit escape (**`Esc`**). The bottom line will let you know that you are back in command mode. In command mode you can move the cursor around. In `vim` the special keys will still work for this. If you are in a true `vi` install, you will need to use the following keys to move around.

- `h` – Move left.
- `j` – Move down.
- `k` – Move up.
- `l` – Move right.
- `Ctrl-d` – Page down.
- `Ctrl-u` – Page up.

You can do other edits while you are in command mode. Here are some of the other keys that do things for you in command mode.

- `x` – Delete a character.
- `dd` – Delete a line and place on the clipboard. (Precede with a number for multiple lines.)
- `yy` or `Y` – Yank a line. This copies to the clipboard. (Precede with a number for multiple lines.)
- `p` – Paste the clipboard after the current line.
- `P` – Paste the clipboard before the current line.
- `r` – Replace a single character.
- `J` – Join lines.
- `/` – Search for something. Enter the search string after the /.
- `n` – Repeat the last search.
- `cw` – Change the current word. This removes up to the next white space and goes into insert mode.
- `.` – Repeat the last command.
- `u` – Undo the last command.
- `Ctrl-r` – Redo the last undone command.

If you type in a number before any command in `vi`, that command will be repeated that number of times. This includes going into an edit mode. So if you type 100 and 'A', any text you add in will be added 100 times when you hit `Esc`. If that is not what you wanted, you will probably find that the 'u' command comes in handy.

Some commands work well in patterns. For example, if you want to do a find and replace where you look at each instance before doing the replace, you might use '/' to search for the first instance, then use 'cw' to replace the word in question and hit `Esc`. After that, you use 'n' to move to the next instance of the string and '.' to repeat the 'cw' command if you want to replace that instance. The '.' command is something that would only have minimal value in writing a term paper, but comes in handy a lot more with programming tasks.

There are also a whole set of commands that you invoke by typing a colon. For example, you might have wondered how you save files or how you get out of `vi` all together. Here are just a few of the colon commands that you might find helpful.

- :n – Jump to the nth line of the file where n is a number.
- `:w` – Save the file as it is.
- `:q` – Safe quit of `vi`. This won't work if you have made changes since the last save.
- `:q!` – Unsafe quit of `vi`. This will quit and throw away any unsaved changes.
- `:wq` – Save the file and quit `vi`.

You should spend some time getting used to these commands. If you really want to see what `vi` can do, read the `man` page. Many students who are new to `vi` fall into the pit of hitting 'i' as soon as they start up and then using it like Notepad until they are ready to save. This approach ignores all the powerful features of `vi` and, in the long run, slows you down. Try to get used to hitting `Esc` whenever you pause in your typing. If nothing else, it will allow you to also hit ':w' so you save your changes frequently. On systems with `vim`, you can run `vimtutor` to get a guided tour.

`vi` Settings

You can change settings in `vi` by editing a file in your home directory with the name "`.exrc`". In particular, you should consider adding this line.

```
set tabstop=2
```

This makes it so that tabs display as four characters. You can use 4 if you want, but the default is 8 which is a bit too large to work well with Scala.

2.2 Windows

Based on installation statistics, odds are good that the computer you use most of the time is running some form of Microsoft Windows®. That will not prevent you from programming or doing any of the activities described in this book. One of the advantages of the Scala

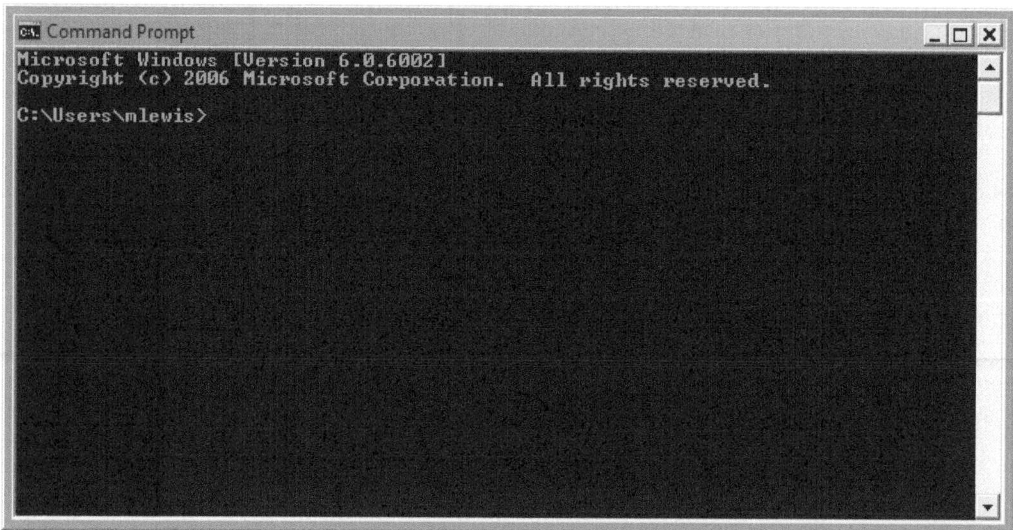

FIGURE 2.2: This figure shows the Windows command prompt. This is the command-line interface for Windows.

language is that it is platform independent and will work equally well on many operating systems, including Windows.[8]

2.2.1 Command-Line

Most people interact with Windows using the GUI. This is so much the case that you might not even realize that there is a command-line interface. If you look under the `Accessories` option on the `Programs` menu you will see an option for "`Command Prompt`". If you select and run this, you should see a window that looks like figure 2.2. Like the terminal for Unix/Linux, this is a text interface that gives you some type of prompt with basic information. In the figure it shows you the current directory followed by a '>'. In this case we are in the directory "`C:\Users\mlewis`".

While the Windows command prompt looks and acts much like a Unix/Linux terminal, it has a different set of commands. The history of Windows and its command prompt are rooted in DOS (Disk Operating System) and the commands that are used in the command prompt for Windows today are largely the same as for that purely command-line based operating system. One of the differences between Unix/Linux and Windows that should be noted immediately is that unlike Unix/Linx, the Windows system is not case sensitive. Commands will be presented here in lowercase, but they will work fine in uppercase or mixed case.

2.2.1.1 Files and Directories

As with the Unix/Linux environment, the folders that you are used to in a GUI are called directories on the command-line and they are used to organize files on the computer. There are a number of different commands that can be used for working with directories and files with the Windows command prompt.

[8] Even though you can do your coding under Windows, you might want to consider installing Cygwin or running Linux through virtualization or as a dual-boot configuration just to get experience with something new.

- `cd/chdir` – Changes the current directory.
- `copy` – Copy files from one location/name to another.
- `del`/erase – Deletes files. This doesn't move them to Trash like doing a deleter from the GUI. Once this has been executed the files can not be recovered.
- `dir` – Lists the contents of the specified directory or the current directory if none is listed.
- `md/mkdir` – Make the specified directory.
- `more` – Print the contents of a file one screen at a time.
- `move` – Moves a file to a different directory.
- `path` – Let's you see the current path or change the path.
- `ren/rename` – Rename an existing file.
- `rd/rmdir` – Remove a specified directory.
- `tree` – Display the contents of a directory or drive as an ASCII-based tree.

There are some significant differences between directories under Windows and those under Unix/Linux. The first is that directories are separated by a backslash, '\', instead of a forward slash, '/'. This is something you have to keep in mind, but it does not significantly impact how any commands are used. The other difference is that Windows machines have different drives, each with its own directory tree.

The prompt above showed the current directory as being "`C:\Users\mlewis`". The `C:` at the beginning of this specifies the drive. The drives on a Windows machine are specified as single capital letters. By default, `C:` is the master hard-disk of the machine. If you want to deal with locations on other drives you need to be certain to specify the drive name in the path you are using. In addition, by default the `cd` command will either change drives or move you to a different directory. To do both you have to give the '`/d`' option.

Aside

The `C:` standard for the hard drive arose early on. The first DOS machines did not have hard-drives. Instead, they typically had one or two floppy disk drives. These were named `A:` and `B:`. When hard drives were added on, they took the next letter in the alphabet. Even though it is very rare to find a computer with a floppy disk drive these days, the main disk still gets the `C:` distinction and other drives use higher letters in the alphabet.

By contrast, different drives in Unix/Linux systems are represented as directories under the main directory tree that starts with `/`. Removable drives are typically mounted in subdirectories of `/media`.

Figure 2.3 shows what the command prompt looks like after the `dir` command has been run. This shows a lot of the same information that you get from running "`ls -l`" on a Unix/Linux machine.

```
Command Prompt

C:\Users\mlewis>dir
 Volume in drive C has no label.
 Volume Serial Number is F86B-59CE

 Directory of C:\Users\mlewis

05/29/2011  10:52 AM    <DIR>          .
05/29/2011  10:52 AM    <DIR>          ..
05/29/2011  10:53 AM                15 .scala_history
09/18/2008  09:47 AM    <DIR>          .VirtualBox
09/17/2008  02:32 PM    <DIR>          Contacts
10/03/2008  02:26 PM    <DIR>          cygwin
06/10/2010  11:25 AM    <DIR>          Desktop
10/29/2009  10:50 AM    <DIR>          Documents
05/29/2011  09:33 AM    <DIR>          Downloads
09/17/2008  02:32 PM    <DIR>          Favorites
09/17/2008  02:32 PM    <DIR>          Links
09/17/2008  02:32 PM    <DIR>          Music
09/17/2008  02:32 PM    <DIR>          Pictures
09/17/2008  02:32 PM    <DIR>          Saved Games
09/17/2008  02:32 PM    <DIR>          Searches
09/29/2008  05:36 PM    <DIR>          Videos
               1 File(s)             15 bytes
              15 Dir(s)   9,867,812,864 bytes free

C:\Users\mlewis>_
```

FIGURE 2.3: This figure shows the command prompt from Windows after the `dir` command has been run.

2.2.2 Text Editors

No matter what the operating system, programs are still generally written in plain text files so you need a plain text editor for doing that. In this section we'll look at some of the options you have for this in Windows.

2.2.2.1 Edit

Working on the command-line in 32-bit Windows the command "`edit`" will bring up a text-based program that works much like a GUI. You can see what it looks like in figure 2.4. It has a set of menus that you can access by hitting `Alt`. These let you do the normal things you would expect from an editor. The `edit` program is much more of a general purpose editor and is much more user friendly than `vi`. As a result, you can use it without too much introduction. The downside is that it lacks features that are specifically beneficial for programming. If you have a newer Windows install it is probably 64-bit and there will not be an install of `edit`. You can use a GUI-based text editor instead.

2.2.2.2 Notepad

Of course, you do not have to edit your text files in the command prompt. You can feel free to bring up Notepad on a Windows machine and edit files that way. Then you can use the command prompt just for running the programs that need to use that interface. The simple GUI of Notepad has the same benefits and drawbacks of `edit`. It will work fine for small things, but it was not built for programming and will slow you down at a certain point.

2.2.2.3 Others

There are many other text editors that you could use on Windows which are more directly aimed at programming. Attempting to list all such editors and provide any type

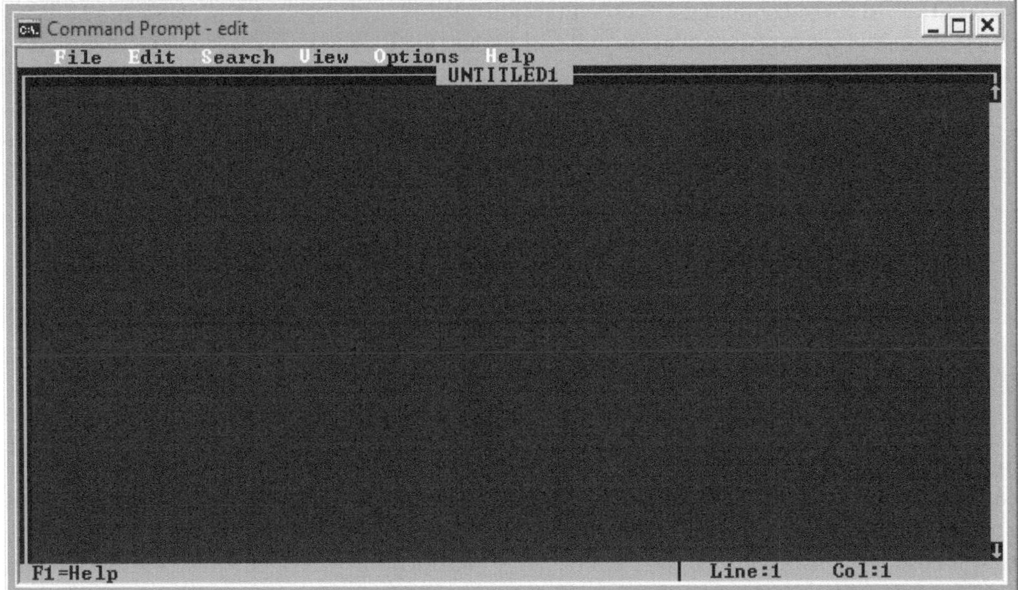

FIGURE 2.4: This figure shows a command prompt with the `edit` program running.

of comparison or judge their relative merits in this space would be pointless. Not only are there too many of them, they can change over time. Doing a web search for "text editor" will likely provide you with many different options. In addition, other people have done the work of compiling a list with comparisons on Wikipedia®. Simply go to `http://en.wikipedia.org/wiki/Comparison_of_text_editors` and you can see tables comparing different editors and check out other information on the ones that seem interesting.

2.2.3 Other Commands

There are a number of other commands that you might find it helpful to know when working with the Windows command prompt.

- `cls` - Clears the screen and puts you at a blank prompt. This command only impacts appearance, but can be useful at times.
- `echo` - Like the Unix/Linux command, this will print out whatever text is passed in to it.
- `exit` - This will close the command prompt.
- `find` - This is the equivalent of `grep` in Unix/Linux. It will find strings in a file for you.
- `help` - This gives you help on commands. If you enter it without specifying a command it will list the possible commands for you.
- `where` - Takes one argument which is the name of a command. It will tell you the full path of that command.
- `whoami` - Tells you the name of the user account you are logged in as.

In addition to these commands, I/O redirection with '<', '>', '>>', and '|' work in the Windows command prompt just like they did in Unix/Linux. You can also use tab completion in Windows to auto-complete file names and save yourself some typing or the possibility of typos. The use of Tab in Windows is particularly helpful because many files and directories have spaces in them that have to be handled properly. If you use tab completion, the system will automatically add double quotes around things so that spaces are dealt with appropriately.

2.3 Scala Tools

When you install Scala on your machine there are several different programs that get installed in the `bin` directory under the Scala installation. To begin with, we will only concern ourselves with two of these: `scala` and `scalac`.[9] The `scala` command actually runs `scala` programs. The `scalac` command is used to compile `scala` text files into bytecode that is compatible with either the Java or .NET platform. It will be a while before we begin writing our own classes and using `scalac` to compile larger programs. However, we will begin using the `scala` command immediately.

Scala on Your Machine

If you only use Scala on a machine in a computer lab, hopefully everything will have been set up for you so that you can simply type the name of a command and it will run. To run Scala on your own machine may take a little more setup.

Installation

The first thing you will have to do is install Scala on your computer. Scala requires Java® to run so if you don't have Java installed you should go to `http://java.oracle.com` and download then install the most recent version of the Java SE JDK. When you install you can go with the default install locations.

Once you have Java installed you can install Scala. To download Scala go to `http://www.scala-lang.org`. On that site download the latest version of Scala. There are typically several installers to choose from. The IzPack installer for all platforms is typically the easiest to work with even if it is a bit larger to download. It is suggested that when you run that you use the default install location.

Dealing with the PATH

If you are using Scala on your own machine it is possible that entering the command "`scala`" or "`scala.bat`" could produce a message telling you that the command or program `scala` could not be found. This happens because the location of the installed programs are not in your default `PATH`.

The `PATH` is a set of directories that are checked whenever you run a command. The first match that is found for any executable file in a directory in the `PATH` will be run. If none of the programs in the `PATH` match what you entered, you get an error.

[9] On a Windows system these commands should be followed by "`.bat`".

When you installed Scala, a lot of different stuff was put into the install directory. That included a subdirectory called "`bin`" with different files in it for the different executables. If you are on a Windows machine, odds are that you installed the program in `C:\Program Files\scala` so the `scala.bat` file that you want to run is in `C:\Program Files\scala\bin\scala.bat`. You can type in that full command or you can add the `bin` directory to your `PATH`. To do this right click on the Computer and go to the Advanced Settings and change the Environment Variables.

Under Unix/Linux you can do this from the command-line. Odds are that Scala was installed in a directory called `scala` in your user space. To add the `bin` directory to your path you can do the following.[10]

```
export PATH=$PATH:/home/username/scala/bin
```

Replace "`username`" with your username. To make it so that you do not have to do this every time you open a terminal, add that line to the `.bashrc` file in your home directory.

There are three ways in which the `scala` command can be used. If you just type in `scala` and press Enter you will be dropped into the Scala REPL (Read-Execute-Print Loop). This is an environment where you can type in single Scala commands and immediately see their values. This is how we will start off interacting with Scala and it is something that we will come back to throughout the book because it allows us to easily experiment and play around with the language. The fact that it gives us immediate feedback is also quite helpful.

To see how this works, execute `scala`. It should print out some information for you, including telling you that you can get help by typing in `:help`. It will then give you a prompt of the form `scala>`. If you type in `:help` you will see a number of other commands you could give that begin with a colon. At this time the only one that is significant to us is `:quit`, which we will use when we are done with the REPL and want to go back to the normal command prompt.

It is customary for the first program in a language to be Hello World. So as not to break with tradition, we can start by doing this now. Type the following after the `scala>` prompt.

```
println("Hello World!")
```

If you do this you will see that the next line prints out "Hello World!". This exercise is less than exciting in the REPL because it always prints out the values of things, but it is a reasonable thing to start with. It is worth asking what this really did. `println` is a function in Scala that tells it to print something to standard output and follow that something with a newline character. In this case, the thing that was printed was the string "Hello World!". You can make it print other things if you wish. One of the advantages of the REPL is that it is easy to play around in. Go ahead and test printing some other things to see what happens. We will come back and do a lot more with this in the next chapter.

The second usage of the `scala` command is to run small Scala programs as scripts. The term script is generally used in regards to short programs that perform specific tasks. There are languages that are designed to work well in this type of usage and they are often called scripting languages. The design of Scala makes it quite usable as a scripting language. Unlike most scripting languages, however, Scala also has many features that make it ideal for developing large software projects as well. To use Scala for scripting, simply type in a

[10]This syntax assumes you are using the bash shell. If it does not work for you, you can do a little searching on the web for how to add directories to your path.

little Scala program into a text file that ends with "`.scala`" and run it by putting the file name after the `scala` command on the command-line.

2.4 End of Chapter Material

2.4.1 Summary of Concepts

- The software you use to interact with your computer comprises a tool set. The nature of these tools is extremely important to software developers.
- The first half of this book is written with the expectation the reader is using command-line tools. These vary between operating systems.
- Linux and Unix based operating systems, including Mac OS X, include a powerful command-line that is often utilized.
  - There are a number of different commands that you use to give instructions to the computer using the command-line. You will likely commit these to memory over time as you use them. The real power of these commands is that they can be given arguments to make them behave in specific ways or do specific things.
  - Files are organized in directories. Some of the most important commands you will need to learn are for dealing with directories and navigating the directory structure. The name `.` always represents the current directory and `..` represents the parent directory.
  - Input and output for commands can be redirected in different ways.
    * Use `>` to send the output of a program to a file, deleting anything that is there.
    * Use `>>` to send the output of a program to a file, appending to the end of current contents.
    * Use `<` to have the contents of a file act as the standard input for a program.
    * Use `|` to have the output of one program sent to the input of another. You can chain multiple programs together with this technique called "piping".
  - Programs are typically written in plane text with a text editor. The text editor described in this chapter is `vi`. It is very common on Linux/Unix installs and the capabilities of the command mode make it a good editor for programming.
- Windows also incorporates a command prompt that resembles the older DOS interface.
  - There are a different set of commands that you can use in the Windows command prompt. I/O redirection works like it does in Linux/Unix.
  - The command-line text editor is called `edit`. It is not included in 64-bit Windows installs. You can use Notepad or some other GUI text editor. Microsoft Word and other word processing programs are not text editors.
- When you install Scala on your computer you get a number of different executable commands.
  - The `scala` command can run scripts or applications. If no argument is given it opens up the REPL for you to type in individual statements.

   – The `scalac` command is used to compile Scala source code to bytecode. This usage is more significant in the second half of the book.

2.4.2 Exercises

1. Make a directory for the work associated with this book in your user directory.

2. Enter the "Hello World" program in the directory you created. Make sure the file name that you use ends with `.scala`. Run it using the `scala` command.

3. (Linux/Unix/Mac) Use `wget` to download the file for the primary page of Google. This should give you a file called `index.html`. Move this to the name `google.html`. Count how many times the word "google" appears in that file. Describe how you did this and how many times it occurred.

4. Ask your instructor if you have a web space on the machines you work on. If so, make a directory in that space called `CS1`. Put a file called `persinfo.txt` in that directory that tells your instructor a little about you and why you are taking the course. Set permissions on the directory and the file so that they are visible for the web (`a+rX` for Linux). You can test if you did this correctly by pointing a browser at your web space to make certain the files are there.

5. (Linux/Unix/Mac) The `ps` command can tell you what is running on a machine. Run "`ps -ef`" and send the output to a file called `allprocs.txt`. Look at the file to see what all is running.

6. (Linux/Unix/Mac) Do a long listing (`ls -l`) of the `/etc` directory. Pipe the output to `grep` to see how many files contain your first and last initials (e.g., `jd` for John Doe).

Additional exercises can be found on the website.

Chapter 3

Scala Basics

Now that we have explored the tools that we will be using, it is time to put them to use and begin our journey learning how to program with the Scala language. We will spend most of this chapter working in the REPL environment and learning some of the basics of Scala and programming in general. Toward the end we will write our first little script to solve a problem and run that as well to see how basic programs function.

3.1 Expressions, Types, and Basic Math

All programming languages are built from certain fundamental parts. In English you put together words into phrases and then combine phrases into sentences. These sentences can be put together to make paragraphs. To help you understand programming, we will make analogies between standard English and programming languages. These analogies are not perfect. You cannot push them too far. However, they should help you to organize your thinking early in the process. Later on, when your understanding of programming is more mature, you can dispense with these analogies as you will be able to think about programming languages in their own terms.

The most smallest piece of a programming language that has meaning is called a TOKEN.

A token is like a word or punctuation mark in English. They are the smallest pieces with any meaning. If you break up a token, you change the meaning of that piece just like breaking up a word is likely to result in something that is no longer a word and does not have any meaning at all. Indeed, many of the tokens in Scala are words. Other tokens are symbols like punctuation. At the end of the last chapter we looked at a line of code like the following.

```
println("Hello World!");
```

This line contains a number of tokens: `println`, `(`, `"Hello World!"`, `)`, and `;`.

When you think of putting words together, you probably think of building sentences with them. A sentence is a grouping of words that stands on its own in written English. The equivalent of a sentence in Scala, and most programming languages, is the STATEMENT. A statement is a complete and coherent instruction that we can give the computer. When you are entering "commands" into the REPL, they are processed as full statements. If you enter something that is not a complete statement in the REPL, instead of the normal prompt, you will get a vertical bar on the next line telling you that you need to continue the statement. The command listed above is a complete statement, which is why it worked the way it did.

Note that this statement ends with a semicolon. In English you are used to ending sentences with a period, question mark, or exclamation point. Scala follows many other programming languages in that semicolons denote the end of a statement. Scala also does something called semicolon inference. Put simply, if a line ends in such a way that a semicolon makes sense, Scala will put one there for you. As a result of this, our print statement will work just as well without the semicolon.

```
println("Hello World!")
```

You should try entering this into the REPL to verify that it works. Thanks to the semicolon inference in Scala, we will very rarely have to put semicolons in our code. One of the few times they will really be needed is when we want to put two statements on a single line for formatting reasons.

While you probably think of building sentences from words in English, the reality is that you put words together into phrases and then join phrases into sentences. The equivalent of a phrase in Scala is the EXPRESSIONS. Expressions have a far more significant impact on programming languages than phrases have in English, or at the least programmers need to be more cognizant of expressions than English writers have to be of phrases. An expression is a group of tokens in the language that has a value and a TYPE. Just like some phrases are made from a single word, some tokens represent things that have values on their own and as such, they are expressions themselves. The most basic of these are what are called LITERALS. Our sample line was not only a statement, without the semicolon it was also an expression. In Scala, any valid expression can be used as a statement, but some statements are not expressions. The "Hello World!" part of our full statement was also an expression. It is something called a string literal, which we will learn more about later in this chapter.

Let us take a bit of time to explore these concepts in the REPL. Run the Scala command without any arguments. This will put you in the REPL with a prompt of `scala>`. In the last chapter we typed in a line that told Scala to print something. This was made from more than one token so we want to start simpler here. Type in a whole number, like 5, followed by a semicolon and hit Enter. You should see something like this:

```
scala> 5;
res0: Int = 5
```

The first line is what you typed in at the prompt. The second line is what the Scala REPL printed out as a response. Recall that REPL stands for Read-Evaluate-Print Loop. When you type something in, the REPL reads what you typed, then evaluates it and prints the result. The term loop implies that this happens over and over. After printing the result, you should have been given a new prompt.

So what does this second line mean? The REPL evaluated the statement that you input. In this case, the statement is just an expression followed by a semicolon and the REPL was printing out the value of the expression you entered. As was mentioned above, the REPL needs you to type in full statements so that it can evaluate it. In this case, we typed in a very simple statement that has an expression called a NUMERIC LITERAL followed by a semicolon. This semicolon will be inferred if you do not add it in. We will take advantage of that and leave them out of statements below.

The end of the output line gives us the value of the expression which is, unsurprisingly, 5. What about the stuff before that? What does `res0: Int` mean? The `res0` part is a name. It is short for "result0". When you type in an expression as a statement in the Scala REPL as we did here, it does not just evaluate it, it gives it a name so that you can refer back to it later. The name `res0` is now associated with the value 5 in this run of the REPL. We will come back to this later. For now we want to focus on the other part of the line `:Int`. Colons are used in Scala to separate things from their types. We will see a lot more of this through the book, but what matters most to us now is the type, `Int`. This is the type name that Scala uses for basic numeric integers. You can try typing in a few other integer values to see what happens with them. Most of the time the results will not be all that interesting, but if you push things far enough you might get a surprise or two.

What happens if you type in a number that is not an integer? For example, what if you type in 5.6? Try it and you should get something like this:

```
scala> 5.6
res1: Double = 5.6
```

We have a different name now because this is a new result. We also get a different type. Instead of `Int`, Scala now tells us that the type is `Double`. In short, `Double` is the type that Scala uses by default for any non-integer numeric values. Even if a value technically is an integer, if it includes a decimal point, Scala will interpret it to be a `Double`. You can type in 5.0 to see this in action. Try typing in some other numeric values that should have a `Double` as the type. See what happens. Once again, the results should be fairly mundane unless you push far enough.

So far all of the expressions we have typed in have been single tokens. Now we will build some more complex expressions. We will begin by doing basic mathematical operations. Try typing in "`5+6`".

```
scala> 5+6
res2: Int = 11
```

This line involves three topics. Each character in this case is a separate token. If you space things out, it will not change the result. However, if you use a number with multiple digits, all the digits together are a single token and inserting spaces does change the meaning.

There should not be anything too surprising about the result of `5+6`. We get back a value of `11` and it has a type of `Int`. Try the other basic arithmetic operations of `-`, `*`, and `/`. You'll notice that you keep getting back values of type `Int`. This makes sense for addition, subtraction, and multiplication. However, the result of `5/2` might surprise you a little bit. You normally think of this expression has having the value of `2.5` which would be a `Double`. However, if you ask Scala for the result of `5/2` it will tell you the value is the `Int` 2. Why is this and what happened to the `0.5`? When both operands are of type

Int, Scala keeps everything as Ints. In the case of division, the decimal answer you might expect is truncated and the fractional part is thrown away. Note that it is not rounded, but truncated. Why is this? It is because in integer arithmetic, the value of 5/2 is not 2.5. It is 2r1. That is to say that when you divide five by two, you get two groups of two with one remainder. At some point in your elementary education, when you first learned about division, this is how you were actually told to think about it. At that time you only had integers to work with so this is what made sense.

Scala is just doing what you did when you first learned division. It is giving you the whole number part of the quotient with the fractional part removed. This fractional part is normally expressed as a remainder. There is another operation called modulo that is represented by the percent sign that gives us the remainder after division. Here we can see it in action.

```
scala> 5%2
res3: Int = 1
```

The modulo operator is used quite a bit in computing because it is rather handy for expressing certain ideas. You should take some time to re-familiarize yourself with it. You might be tempted to say that this would be your first time dealing with it, but in reality, this is exactly how you did division yourself before you learned about decimal notation for fractions.

What if you really wanted 2.5 for the division? Well, 2.5 in Scala is a Double. We can get this by doing division on Doubles.

```
scala> 5.0/2.0
res4: Double = 2.5
```

All of our basic numeric operations work for Doubles as well. Play around with them some and see how they work. You can also build larger expressions. Put in multiple operators, and use some parentheses.

One last thing before we move on to other types is to see what happens when you combine a Double and an Int in an expression. Consider this example:

```
scala> 5.0/2
res5: Double = 2.5
```

Here we have a Double divided by an Int. The result is a Double. When you combine numeric values in expressions, Scala will change one to match the other. The choice of which one to change is fairly simple. It changes the one that is more restrictive to the one that is less restrictive. In this case, anything that is an Int is also a Double, but not all values that are Doubles are Ints. So the logical path is to make the Int into a Double and do the operation that way.

3.2 Objects and Methods

One of the features of the Scala language is that all the values in Scala are OBJECTS. The term object in reference to programming means something that combines data and the functionality on that data. In Scala we refer to the things that an object knows how to do as METHODS. The normal syntax for calling a method on an object is to follow the object by a period (which we normally read as "dot") and the name of the method. Some methods need extra information, which we called arguments. If a method needs ARGUMENTS then those are put after the method name in parentheses.

In many programming languages, numeric literals like the ones we have used so far would not be objects. They would be simple values, called primitives, that we could not call methods on. In Scala though, even the most basic literals are treated as objects in our program and we can therefore call methods on them. An example of when we might do this is when we need to convert one type to another. In the sample below we convert the `Double` value `5.6` into an `Int` by calling the `toInt` method. In this simple context we would generally just use an `Int` literal, but there will be situations we encounter later on where we are given values that are `Doubles` and we need to convert them to `Ints`. We will be able to do that with the `toInt` method.

```
scala> 5.6.toInt
res6: Int = 5
```

One thing you should note about this example is that converting a `Double` to an `Int` does not round. Instead, this operation performs a truncation. Any fractional part of the number is cut off and only the whole integer is left.

We saw at the beginning of this chapter that Scala is flexible when it comes to the requirement of putting semicolons at the end of statements. Scala will infer a semicolon at the end of a line if one makes sense. This type of behavior makes code easier to write and allows it to flow better. In Scala, the dot between an object and a method is also optional. So we can write the following instead:

```
scala> 5.6 toInt
res7: Int = 5
```

Even the parentheses on the arguments to a method are optional assuming that there is only one argument. This type of flexibility makes certain parts of Scala more coherent and provides the programmer with significant flexibility. Though you did not realize it, you were using this fact in the last section. To see this, go into Scala and type "`5.`" then press Tab. The Scala REPL has tab completion just like the command-line, so what you see is a list of all the methods that could be called on the `Int`. It should look something like the following.

```
scala> 5.
!=           #            %            &            *
+            -            /            <            <<
<=           ==           >            >=           >>
>>>          ^            asInstanceOf equals       hashCode
isInstanceOf toByte       toChar       toDouble     toFloat
toInt        toLong       toShort      toString     unary_+
unary_-      unary_~      |
```

We will not go into any detail on most of these here, but some you have already seen and used. We just finished using `toInt` on a `Double`. We can call `toDouble` on an `Int` as well. The things that might stand out though are the basic math operations that were used in the previous section. The `+`, `-`, `*`, `/`, and `%` we used above are nothing more than methods on the `Int` type. The expression `5+6` is really `5 .+ (6)` to Scala. In fact, you can type this into Scala and see that you get the same result.

```
scala> 5 .+ (6)
res8: Int = 11
```

The space between the 5 and the . is required here because without it Scala thinks you want a `Double`. Instead of spacing this out, you can put the 5 in parentheses.

So when you type in 5+6, Scala sees a call to the method + on the object 5 with one argument of 6. We get to use the short form simply because Scala allows both the dot and the parentheses to be optional in cases like this.

3.3 Other Basic Types

Not everything in Scala is a number. There are other non-numeric types in Scala which also have literals. We will start simple and move up in complexity. Perhaps the simplest type in Scala is the `Boolean` type. Objects of the `Boolean` type are either `true` or `false` and those are also valid literals for `Booleans`.

```
scala> true
res9: Boolean = true

scala> false
res10: Boolean = false
```

We will see a lot more on `Booleans` and what we can do with them in chapter 4 when we introduce Boolean logic.

Another type that is not explicitly numeric is the `Char` type. This type is used to represent single characters. We can make character literals by placing the character inside of single quotes like we see here.

```
scala> 'a'
res11: Char = a
```

The way that computers work, all character data is really numbers and different numbers correspond to different characters. We can find out what numeric value is associated with a given character by using the `toInt` method. As you can see from the line below, the lowercase "a" has a numeric value of 97.

```
scala> 'a'.toInt
res11: Int = 97
```

Because characters have numeric values associated with them, we can also do math with them. When we do this, Scala will convert the character to its numeric value as an `Int` and then do the math with the `Int`. The result will be an `Int`, as seen in this example.

```
scala> 'a'+1
res12: Int = 98
```

In the last section you might have noticed that the `Int` type has a method called `toChar`. We can use that to get back from an integer value to a character. You can see from the following example that when you add 1 to `'a'` you get the logical result of `'b'`.

```
scala> ('a'+1).toChar
res13: Char = b
```

An object of the `Char` type can only be a single character. If you try to put more than one character inside of single quotes you will get an error. It is also an error to try to make a `Char` with empty single quotes. However, there are lots of situations when you want to be able to represent many characters, or even zero characters. This includes words, sentences, and many other things. For this there is a different type called a `String`. String literals are formed by putting zero or more characters inside of double quotes like we see in this example.

```
scala> "Scala is a programming language"
res14: java.lang.String = Scala is a programming language
```

Notice that the type is listed as `java.lang.String`. Scala integrates closely with Java and uses some of the Java library elements in standard code. It also allows you to freely call code from the Java libraries. The type `String` has a full name of `java.lang.String`, but Scala will automatically provide the `java.lang` part so we will generally leave it out.

Certain operations that look like mathematical operations are supported for `Strings`. For example, when you use `+` with `Strings`, it does string concatenation. That is to say it gives back a new string that is the combined characters of the two that are being put together as shown here:

```
scala> "abc"+"def"
res15: java.lang.String = abcdef
```

This type of operation works with other types as well. This example shows what happens when we concatenate a `String` with an `Int`. The `Int` is converted to a `String`, using the `toString` method, and normal string concatenation is performed.

```
scala> "abc"+123
res16: java.lang.String = abc123
```

This works whether the `String` is the first or second argument of the `+`.

```
scala> 123+"abc"
res17: java.lang.String = 123abc
```

In addition to concatenation, you can multiply a string by an integer and you will get back a new string that has the original string repeated the specified number of times.

```
scala> "abc"*6
res18: String = abcabcabcabcabcabc
```

This can be helpful for things such as padding values with the proper number of spaces to make a string a specific length. You can do this by "multiplying" the string `" "` by the number of spaces you need.

There are other types that are worth noting before we move on. One is the type `Unit`. The `Unit` type in Scala basically represents a non-value or a meaningless value. The equivalent in many other languages is called `void`. We have actually seen an example of code that uses `Unit`. The first program we saw in chapter 2 used a function called `println`. When we called `println` Scala did something, but did not give us back a value. This is what happens when we type in an expression that gives us back a value of `Unit` in the REPL.

The last type in Scala that we will deal with here is a grouping of different types called a TUPLE. A tuple is a sequence of a specified number of specific types. It is basically, a collection of values that is strict about how many and what type of values it has. We can make tuples in Scala by simply putting values in parentheses and separating them with commas as seen in the following examples.

```
scala> (5,6,7)
res19: (Int, Int, Int) = (5,6,7)

scala> ("book",200)
res20: (java.lang.String, Int) = (book,200)

scala> (5.7,8,'f',"a string")
res21: (Double, Int, Char, java.lang.String) = (5.7,8,f,a string)
```

The tuples in Scala provide a simple way of dealing with multiple values in a single package and they will come up occasionally through the book. Note that the way we express a tuple type in Scala is to put the types of the values of the tuple in parentheses with commas between them, just like we do with the values to make a tuple object.

Tuples with only two elements can have special meanings in some parts of Scala. For that reason, there is an alternate syntax you can use to define these. If you put the token `->` between two values, it will produce a 2-tuple with those values. Consider the following example.

```
scala> 3 -> "three"
res22: (Int, java.lang.String) = (3,three)
```

The `->` will only produce tuples with two elements though. If you try using it with more than two elements you can get interesting results.

```
scala> 4 -> 5 -> 6
res23: ((Int, Int), Int) = ((4,5),6)
```

So if you want tuples with more than two elements, stick with the parentheses and comma notation.

Once you have a tuple, there are two ways to get things out of them. The first is to use methods named _1, _2, _3, etc. So using `res21` from above we can do the following.

```
scala> res21._1
res24: Double = 5.7

scala> res21._3
res25: Char = f
```

The challenge with this method is that method names like _1 are not very informative and can make code difficult to read. We will see an alternative approach in section 3.7 that requires a bit more typing, but can produce more readable code.

3.4 Back to the Numbers

Depending on how much you played around with the topics in section 3.1 you might or might not have found some interesting surprises where things behaved in ways that you were not expecting. Consider the following:

```
scala> 1500000000+1500000000
res9: Int = -1294967296
```

Mathematicians would consider this to be an error. It is actually a reflection of the way that numbers are implemented on computers. The details of this implementation can impact how your programs work, so it is worth taking a bit of time to discuss it.

At a fundamental level, all information on computers is represented with numbers. We saw this with the characters being numbers. On modern computers all these numbers are represented in BINARY, or base two. The electronics in the computer alternate between two states that represent 1 and 0 or on and off. Collections of these represent numbers. A single value of either a 0 or a 1 is called a BIT. It is a single digit in a binary number. The term BYTE refers to a grouping of 8 bits, which can represent 256 different numbers. In Scala these will be between -128 and 127. To understand this, we need to do a little review of how binary numbers work.

You have likely spent your life working with decimal numbers, or base ten. In this system, there are ten possible values for each digit: 0, 1, 2, 3, 4, 5, 6, 7, 8, and 9. Digits in different positions represent different power of ten. So the number 365 is really $3*10^2+6*10^1+5*10^0$. There is nothing particularly unique about the base ten other than perhaps it relates well to the number of digits on human hands. You can just as well use other bases in which case you need an appropriate number of symbols for each digit and each position represents a power of that base.

Binary uses a base of two. For this we only need two different digits: 0 and 1. This is convenient on computers where the electronics can efficiently represent two states. The different positions represent powers of two: 1, 2, 4, 8, 16, 32, ... So the number 110101=32+16+4+1=53. This example shows how you convert from binary to decimal. Simply add together the powers of two for which the bits are one. A byte stores eight bits that would represent powers of two from 128 down to 1. The word "would" is used here because there is a significant nuance to this dealing with negative numbers that we will discuss shortly.

There are two basic approaches to converting from decimal to binary. One involves repeated subtraction of powers of two while the other involves repeated division by two. We will start with the first one and use the value 296 in decimal for the conversion. We start by finding the largest power of 2 that is smaller than our value. In this case it is $256 = 2^8$. So we will have a one in the 2^8 position or the 9th digit.[1] Now we subtract and get $296 - 256 = 40$ and repeat. The largest power of 2 smaller than 40 is $32 = 2^5$. So the digits for 2^7 and 2^6 are 0. Subtract again to get $40 - 32 = 8$. We now have $8 = 2^3$ so the final number in binary is 100101000. This procedure is written out the way you might actually do it in figure 3.1.

The other approach is a bit more algorithmic in nature and is probably less prone to error. It works based on the fact that in binary, multiplying and dividing by 2 moves the "binary point" the same way that multiplying or dividing by 10 moves the decimal point in the decimal number system. The way it works is you look at the number and if it is odd you write a 1. If it is even you write a 0. Then you divide the number by 2, throwing away any remainder or fractional part, and repeat with each new digit written to the left of those before it. Do this until you get to 0. You can also think of this as just dividing by two repeatedly and writing the remainder as a bit in the number with the quotient being what you keep working with.

The number 296 is even so we start off by writing a 0 and divide by 2 to get 148. That is also even so write another 0. Divide to get 74. This is also even so write another 0. Divide to get 37. This is odd so write a 1. Divide to get 18, which is even so you write a

[1] Remember that the first digit is $2^0 = 1$.

```
Value      Power of 2    Digit
 296          256          1
  40          128          0
  40           64          0
  40           32          1
   8           16          0
   8            8          1
   0            4          0
   0            2          0
   0            1          0
```

FIGURE 3.1: Illustration of the conversion from decimal to binary using the subtraction method. This method works from the top down. To get the number in binary just read down the list of digits.

```
Value      Digit
   1          1
   2          0
   4          0
   9          1
  18          0
  37          1
  74          0
 148          0
 296          0
```

FIGURE 3.2: Illustration of the conversion from decimal to binary using the repeated division method. This method works from the bottom up so you get the bits in the result starting with the smallest.

0. Divide to get 9 and write a 1. Divide to get 4 and write a 0. Divide to get 2 and write a 0. Divide to get 1 and write that one. The next division gives you zero so you stop. This procedure is illustrated in figure 3.2.

3.4.1 Binary Arithmetic

Now that you know how to go from binary to decimal and decimal to binary, let's take a minute to do a little arithmetic with binary numbers. It is certainly possible to do this by converting the binary to decimal, doing the arithmetic in decimal, then convert back to binary. However, this is quite inefficient and not worth it because it really is not hard to work in binary. If anything, it is easier to work in binary than in decimal. Let us begin with the operation of addition. Say we want to add the numbers 110101 and 101110. To do this you do exactly what you would do with long addition in decimal. The difference is that two in binary is 10 so there is a lot more carrying.

```
  110101
+ 101110
 -------
 1100011
```

Multiplication in binary can also be done just like in decimal and you have a lot fewer multiplication facts to memorize. Zero times anything is zero and one times anything is that number. That is all we have to know. Let us do multiplication with the same numbers we just worked with. First we will get all the numbers that need to be added up.

```
      110101
    * 101110
  ----------
     1101010
    11010100
   110101000
  11010100000
```

Adding these numbers is best done in pairs. The reason is that as soon as you add together 3 or more numbers in binary you have the capability to have to do something you are not accustomed to doing in decimal: carry a value up two digits. In decimal you would have to have a column sum up to one hundred or more for this to happen. However, in binary you only have to get to four (which is written as 100 in binary). That happens in this particular instance in the 6th digit. To reduce the odds of an error, it is better to add the values two at a time as we have shown here.

```
     1101010
  + 11010100
  ----------
   100111110
 + 110101000
  ----------
  1011100110
+11010100000
  ----------
100110000110
```

You can do division in the same way that you do long division with integers, but we will not cover that here.

3.4.2 Negative Numbers in Binary

We still have not addressed the question of how we represent negative numbers on a computer. The description that we have given so far only deals with positive values. Numbers that are interpreted this way are called UNSIGNED. All the numeric types in Scala are SIGNED so we should figure out how that works.[2] To do this, there are two things that should be kept in mind. The first is that our values have limited precision. That is to say that they only store a certain number of bits. Anything beyond that is lost. The second is that negative numbers are defined as the additive inverses of their positive counterparts. In other words, $x + (-x) = 0$ for any x.

To demonstrate how we can get negative numbers, let's work with the number 110101 (53 in decimal). Unlike before though, now we will limit ourselves to a single byte. So we have 8 digits to work with and the top digits are zeros. Our number stored in a byte is really `00110101`. So the question of what should be the negative is answered by figuring out what value we would add to this in order to get zero.

[2]The `Char` is actually a 16-bit unsigned numeric value, but the normal numeric types are all signed.

```
  00110101
+ ????????
  --------
  00000000
```

Of course, there is nothing that we can put into the question marks to make this work. However, if we go back to our first fact we can see what we must do. We do not need zero, we need eight digits of zero. So in reality, what we are looking for is the following.

```
  00110101
+ ????????
  --------
 100000000
```

This problem is solvable and the top `1` will be thrown away because we can only store 8 bits in a byte. So the answer is given here.

```
  00110101
+ 11001011
  --------
 100000000
```

Note that top bit is on in the negative value. The top bit is not exactly a sign bit, but if a number is signed, the top bit will tell us quickly whether the number is positive or negative. This style of making negatives is called TWO'S COMPLEMENT. In the early days of digital computing other options were tried, such as adding a sign-bit or a method called ones' complement where the bits are simply flipped. However, two's complement is used in machines today because it allows numeric operations to be done with negative numbers using the same circuitry as is used for positive numbers.

This process gives us the correct answer and is based on the proper definition of what a negative number is. Finding negatives using the definition of what a negative value is works and can be a fallback, but there is a simpler method. To get the two's complement negative of a binary number of any size, simply flip all the bits and add one. You can verify that this approach works for our example above.

3.4.3 Other Integer Types

There are larger groups of bits beyond the 8-bit bytes that have meaning in Scala. In fact, if you go back to section 3.2 and you look at the different methods on an `Int`, you will see that `toDouble` and `toChar` are not the only conversions we can do. Scala has other integer types of `Byte`, `Short`, and `Long`. A `Byte` in Scala is an 8-bit number. A `Short` is a 16-bit number. The `Int` that we have been using is a 32-bit number. The `Long` type is a 64-bit number. The reason for the odd behavior that was demonstrated at the beginning of this section is that we added two numbers together whose sum is bigger than what can be stored in the lower 31 bits of an `Int` and the OVERFLOW, as it is called, wrapped it around to a negative value. Table 3.1 shows the minimum and maximum values for each of the different integer types.

Occasionally you will need to use literals that are bigger than what an `Int` can store. You can do this with a `Long`. Making a numeric literal into a `Long` is done by simply adding an `L` to the end. You can see this here.

```
scala> 5000000000L
res8: Long = 5000000000
```

Type	Bits	Min	Max
Byte	8	-128	127
Short	16	-32768	32767
Int	32	-2147483648	2147483647
Long	64	-9223372036854775808	9223372036854775807

TABLE 3.1: Integer types with their sizes and ranges.

The value five billion is not a valid `Int`. If you leave off the `L` here you get an error. The `L` can be lower case, but then it looks a lot like the number one so it is better to use the upper case.

We talked about binary above and Scala has a method that will let you see the binary form of a number. This method works on the four normal numeric types and `Char`. Here we use it to see the binary representation for 53 and -53 for the values as both `Int` and `Long` types.

```
scala> 53.toBinaryString
res15: String = 110101

scala> (-53).toBinaryString
res16: String = 11111111111111111111111111001011

scala> 53L.toBinaryString
res17: String = 110101

scala> (-53L).toBinaryString
res18: String =
  1111111111111111111111111111111111111111111111111111111111001011
```

The `toBinaryString` method does not display leading zeros, so the positive values only show six digits in both formats. However, the negative form has many leading ones and all of these are printed.

3.4.4 Octal and Hexadecimal

Binary is what the machine uses, but it really is not that useful to humans. This is in large part due to the fact that the number of digits in a binary number is often large, even if the number itself is not what we consider large. There are two other bases that are commonly seen in programming and dealing with computers. They are base 8, OCTAL, and base 16, HEXADECIMAL or HEX. Like decimal, these bases allow you to represent fairly large numbers with relatively few digits. Unlike decimal, converting from octal or hex to binary and back is trivial. The reason for this is that eight and 16 are powers of two.

To see this, let us start with octal. When working in base 8, the digits can be between `0` and `7` with each subsequent digit being a higher power of 8. The ease of converting to and from binary comes from the fact that eight is 2^3. In binary the values `0` to `7` are represented with three bits between `000` and `111`. The fourth and subsequent bits represent values that are multiples of eight. Because of this, we can convert a binary number to an octal number by grouping the bits into groups of three and converting those groups. The only catch is that this grouping has to start with the ones digit. So our favorite binary number, `110101` is `65` in octal. The lowest three bits, `101`, convert to `5` and the next three, `110`, convert to `6`. We can use the `toOctalString` method to confirm this.

```
scala> 53.toOctalString
res21: String = 65
```

To go the other way, from octal to binary, we simply convert the octal digits to three digit binary numbers. So the octal value, 3726 converts to 011111010110. We can emphasize the groupings of bits by spacing them out: 011 111 010 110.

Moving between hexadecimal and binary is similarly simple. The only catch is that now a single digit needs to have 16 possible values. So the 0-9 that we are used to will not suffice. It is typical to augment the normal digits with the letters A-F where A is ten and F is fifteen. Because $16 = 2^4$, we use groups of 4 bits when converting between hexadecimal and binary. Once again, you start the process with the lower bits and work up. So 110101 becomes 35. The large number we got from octal above, 0111 1101 0110 becomes 7D6 in hex. Again, there is a method called toHexString that can be used on the numeric types to quickly get the hexadecimal representation of a number.

While toOctalString and toHexString give us octal and hexadecimal representations of numeric values that we have in decimal, it is sometimes helpful to be able to enter values into programs using octal or hexadecimal. It is possible to enter literals in Scala using these bases. To get an octal literal, prefix the number with a leading zero. To get a hexadecimal literal, prefix it with 0x. The following uses of this confirm the conversions that we did for the larger number above.

```
scala> 03726.toBinaryString
res23: String = 11111010110

scala> 0x7D6.toBinaryString
res24: String = 11111010110
```

Values in hex or octal are particularly significant when you want to make sure that you know how many bits will be in the binary representation of the number.

3.4.5 Non-Integer Numbers

You now have a rather complete description of the integer number types and how they are represented in the machine. What about non-integer numeric values? We saw previously that if we type in a numeric value that includes a decimal point Scala tells us that it has type Double. The Double literal format is more powerful than just including decimal points. It also allows you to use scientific notation to enter very large or very small numbers. Simply follow a number by an e and the power of ten it should be multiplied by. So 15000.0 can also be written as 1.5e4.

The name Double is short for double precision floating point number. The full name includes information about the way that these numbers are stored in the memory of a computer. Like all values in a computer, the Double is stored as a collection of bits. To be specific, a Double uses 64-bits. This size is related to the double precision part. There is another type called Float that is a single precision floating point number and only uses 32-bits. In both cases, the internal representation uses floating point format. In many ways, this is similar to scientific notation, but in binary instead of decimal. The bits in a floating point number are grouped into three different parts. We will call them s, e, and m and the value of the number is given by $(-1)^s * (1 + m) * 2^{(e-bias)}$. The first bit in the number is the sign bit, s. When that bit is on, the number is negative and when it is off it is positive. After the sign bit is a group of bits for the EXPONENT, e. Instead of using two's compliment for determining if the exponent is negative, the exponent is biased by a value that is picked to match with the number of bits in the exponent. Using a bias instead of two's compliment

Type	e Bits	m Bits	bias	Min	Max
Float	8	23	127	-3.4028235E38	3.4028235E38
Double	11	52	1023	-1.7976931348623157E308	1.7976931348623157E308

TABLE 3.2: Floating point types with sizes and ranges.

means that comparisons between floating point values can be done with the same logic used for integer values with the same number of bits. All remaining bits are used to store a MANTISSA, m. The stored mantissa is the fractional part of the number in normalized binary. So the highest value bit is $\frac{1}{2}$, the next is $\frac{1}{4}$, and so on. Table 3.2 below gives the number of bits used for e and m, the *bias*, and the range of numbers they can represent in the `Double` and `Float` types. The E notation is short for multiplication by 10 to that power.

As we have seen, floating point literals are by default considered to be of type `Double`. If you specifically need a `Float` you can append an `f` to the end of the literal. There are many other details associated with floating point values. There is only one main point that will be stressed here though. That is the fact that floating point values, whether `Double` or `Float`, are not Real numbers in the sense you are used to in math with arbitrary precision. Floating point numbers have limited precision. Like the integers, they can be overflowed. Unlike the integers, they are fundamentally imprecise because they represent fractional values with a finite number of bits.

To understand this, let us look at a situation where you should have seen this before. Consider the simple fraction, $\frac{1}{3}$. Now consider the representation of this value as a decimal. The decimal representation is 0.33333... In order to write this fraction accurately in decimal, you need an infinite number of digits. In math we can denote things like this by putting in three dots or putting a line over the digits that are repeated. For floating point values though, the digits simply cut off when you get to the end of the mantissa. As such, they are not exact and the circuitry in the computer employs a rounding scheme to deal with this. This imprecision is not visible most of the time, but one immediate implication of it is that you can not trust two floating point numbers to be equal if they were calculated using arithmetic. It also means that you should not use floating point numbers for programs that involve money. The decimal value `0.1` is a repeating fraction in binary and as such, is not perfectly represented. That is generally considered a bad thing when dealing with people's money. Instead you should use an integer type and store cents instead of dollars.

3.5 The `math` Object

While on the topic of numbers, there are quite a few standard functions that you might want to do with numbers beyond addition, subtraction, multiplication, and division. There are a few other things you can get from operators that we will discuss later. Things like square root, logarithms, and trigonometric functions are not operators. They are found as methods in the `math` object. You can use tab completion in the REPL to see all the different methods that you can call on `math`.

```
scala> math.

BigDecimal                   BigInt
```

```
E                      Equiv
Fractional             IEEEremainder
Integral               LowPriorityOrderingImplicits
Numeric                Ordered
Ordering               PartialOrdering
PartiallyOrdered       Pi
ScalaNumber            ScalaNumericConversions
abs                    acos
asin                   atan
atan2                  cbrt
ceil                   cos
cosh                   exp
expm1                  floor
hypot                  log
log10                  log1p
max                    min
package                pow
random                 rint
round                  signum
sin                    sinh
sqrt                   tan
tanh                   toDegrees
toRadians              ulp
```

Many of these probably do not make sense right now and you should not worry about them. However, many of them should be identifiable by the name. So if we wanted to take a square root of a number, we could do the following.

```
scala> math.sqrt(9)
res15: Double = 3.0
```

You would use a similar syntax for taking cosines and sines. The functions provided in the math object should be sufficient for the needs of most people. Only two of the contents of math that start with capital letters are worth noting at this point. `Pi` and `E` are numeric constants for π and e.

```
scala> math.Pi
res16: Double = 3.141592653589793
```

3.6 Details of Char and String

There is a lot more to `Char` and `String` than we covered in section 3.3. Some of it you really should know before we go further. We saw how we can make character literals or string literals that contain keys that appear on the keyboard and that go nicely into a text file. What about things that we can not type as nicely or that have other meanings? For example, how do you put double quotes in a `String`? Typing the double quote closes off the string literal instead of putting one in. How would you make a character of a single quote? How do you get a new line? You are not allowed to have a normal string break across lines.

We can do all of these things and more with ESCAPE CHARACTERS. These are denoted by a backslash in front of one or more characters. For example, if you want to put a double

Literal	Meaning	Unicode Hex Encoding
\b	backspace	\u0008
\f	form feed	\u000C
\n	line feed	\u000A
\r	carriage return	\u000D
\t	tab	\u0009
\"	double quote	\u0022
\'	single quote	\u0027
\\	backslash	\u005C

TABLE 3.3: Table of special character escape sequences in Scala.

quote in a string, simply put a backslash in front of the double quote. The same thing is true for a single quote in a character. You can insert a newline with a \n. What if you want to insert a backslash itself? Simply put in two backslashes. Table 3.3 shows other commonly used escape characters.

In addition to escape characters, the backslash can be used to put any type of special character into a string. If you know the Unicode value for a special character, you can put \u followed by four hexadecimal digital in a string to specify that character. For characters from the lower 256 characters that are represented in ASCII you can follow the backslash by an octal number between 0 and 377.

There are some times when using the escape characters becomes a pain. For example, there are times when you need to build strings that have a number of backslashes. Each one you want requires you to put in two. This can get unwieldy. In addition, if you have a long, multiline string, it can be difficult to format the string the way you want. For these types of situations, Scala includes a special form of string that begins and ends with three double quotes. Anything you type between the set of three double quotes is taken to be literally part of the string without alteration. The following shows an example of using this to enter a long string in the REPL.

```
scala> """This is a long string.
| It spans multiple lines.
| If I put in \n and \\ or \" they are taken literally."""
res8: java.lang.String =
This is a long string.
It spans multiple lines.
If I put in \n and \\ or \" they are taken literally.
```

3.7 Naming Values and Variables

We have seen enough that you can solve some simple problems. For example, if you were given a number of different grades and asked to find the average, you could type in an expression to add them all up and divide by the number of them to get the average. We basically have the ability to use Scala now to solve anything we could solve with a calculator. We will develop a lot more over time, but we have to start somewhere. As it stands we are not just limited to solving problems we could do with a calculator, we are solving them the way we would with a calculator. We type in mathematical expressions the way we would write them on paper and get back an answer. Real programming involves tying together multiple lines of instructions to solve larger problems. In order to do this, we need to have a way to give names to the work we have done and refer back to them.

There are three keywords in Scala that give names to values. We will see the first two here: `val` and `var`. To begin with, let us look at the full syntax of `val` and `var` in two samples. Then we can pull them apart, talk about what they do, see how they are different, and discuss what parts of them are optional.

```
scala> val age:Int=5
age: Int = 5

scala> var average:Int=(2+3+4+5)/4
average: Int = 3
```

Syntactically the only difference between these two is that one says `val` and the other says `var`. That is followed by a name with a colon and a type after it. The rules for names in Scala are that they need to start with a letter or an underscore followed by zero or more letters, numbers, and underscores. Scala is also case sensitive. So the names `AGE`, `age`, `Age`, and `agE` are all different. In general it is considered very poor style to use names that differ only in capitalization. Scala borrows a standard naming convention from Java called camel case. The names of values begin with a lower case letter and the first letter of subsequent words are capitalized. For example, `theClassAverage` is a name that follows this convention. Type names use the same convention except that the first letter is capitalized. This is called camel case because the capital letters look like humps.

The types in both of these examples are followed by an equals sign and an expression. Unlike many other programming languages, this is not optional in Scala. In Scala, when you declare a `val` or `var`, you must give it an initial value at that point.[3]

While the initial value is not optional, the type generally is. This too is unlike most programming languages. As we have already seen, Scala is able to figure out the types of things for us in many situations. If we leave off the colon and the type, Scala will simply use whatever type it determines is appropriate for the expression in the initial value. Most of the time, the type that it gives us will be exactly what we want. Using this we could instead have the following shorter forms of these declarations.

```
scala> val age=5
age: Int = 5

scala> var average=(2+3+4+5)/4
average: Int = 3
```

The reason for using a `val` or `var` declaration is that they give a name to the value that we can refer back to later. For example, we could now type in `age+10` and Scala would give us `15`. The names serve two purposes. They prevent us from typing in expressions over and over again. They also help give meaning to what we are doing. You should try to pick names that help you or other readers figure out what is going on with a piece of code.

So far we have discussed all the similarities between `val` and `var` and you might be wondering in what way they are different. The declarations themselves are basically identical. The difference is not in the syntax, but in the meaning, or semantics. A `val` declaration gives a name to a reference to a value. That reference can not be changed. It will refer to

[3]There are very good reasons for requiring initialization of variables. Even in languages that do not require it, a programmer can make his/her life a lot easier by initializing all variables at creation. The declaration and initialization should ideally happen at the point where you have a real value to put into the variable. This prevents many errors and as a result, can save you a lot of time in your programming.

the thing it was originally set to forever. In the REPL, you can declare another `val` with the same name, but it does not do anything to change the original. A `var` declaration, on the other hand, allows the reference to change. In both cases we are not naming the value, we are naming a box that stores a reference to the value. The significance of this will be seen in section 7.8.

The act of changing the reference stored in one of these boxes we call variables is referred to as an assignment. Assignment in Scala is done with the same equal sign that was used to set the initial value. In an assignment though there is no `val` or `var` keyword. If you accidentally include either `var` or `val` you will be making a new variable, not changing the old one.

```
scala> average=8
average: Int = 8

scala> average=2*average
average: Int = 16
```

The first assignment causes the box named average to change from referring to the object 3 to the object 8. The second one uses the previously referred to value and multiplies it by two, then stores a reference to that new value back into the variable.

There is a bit more to the initialization of `val` and `var` declarations than was mentioned above. Technically, the assignment is able to do something called PATTERN MATCHING that we will get to in detail in chapter 6. For now, the only aspect we will care about is that we can put tuples on the left-hand side of the equals sign where we would normally put just a variable name. First, let us see what happens if we do a `val` declaration with a tuple on the right-hand side.

```
scala> val t=(100,5.7)
t: (Int, Double) = (100,5.7)
```

Note that `t` refers to the tuple and has a type `(Int,Double)`. This is exactly what we would expect. The new power that pattern matching provides is that if you put multiple names inside of a parentheses on the left of the equals, much like a tuple, all the names will be bound. That type of behavior is shown here.

```
scala> val (price,weight)=(100,5.7)
price: Int = 100
weight: Double = 5.7
```

The same can be done with a `var` and then all the names will have the ability to change what they refer to. This is the second way of getting values out of tuples. It is more readable because we can pick meaningful names for the variables.

Let us use the ability to name values to do a little problem solving. We are given a total time in seconds and we want to know what that is in hours, minutes, and seconds. We then want to print that out in a reasonable format of "hh:mm:ss". The first step in solving this problem is to figure out how to go from just seconds to hours, minutes, and seconds. Once we have that, we can worry about formatting it to get the right string value.

How do we get from seconds to hours, minutes, and seconds? First, how do you get from seconds to minutes? That is fairly easy, you simply divide by 60. Thanks to the fact that integer division truncates, you will get the proper number of whole minutes. Here are two lines that define a number of total seconds as well as a number of total minutes.

```
scala> val totalSeconds=123456
totalSeconds: Int = 123456
```

```
scala> val totalMinutes=totalSeconds/60
totalMinutes: Int = 2057
```

That number of minutes is not exactly the amount of time we want though. There are seconds left over. How do we figure out how many seconds we should display? We could do `totalSeconds-(60*totalMinutes)`, but a simpler expression is used here.

```
scala> val displaySeconds=totalSeconds%60
displaySeconds: Int = 36
```

The modulo gives us the remainder after we have gotten all the full groups of 60. That is exactly what we want. Now how do we get the number of hours and the number of minutes to display? The math is the same because there are 60 minutes in each hour.

```
scala> val displayMinutes=totalMinutes%60
displayMinutes: Int = 17

scala> val displayHours=totalMinutes/60
displayHours: Int = 34
```

What we see from this is that 123456 seconds is 34 hours, 17 minutes, and 36 seconds. We could repeat this same process for a different number of seconds if we used a different value for `totalSeconds`.

Now that we have these values, we want to figure out how to get them into a string with the format "hh:mm:ss". A first attempt at that might look like the following.

```
scala> val finalString=displayHours+":"+displayMinutes+":"+
| displaySeconds
finalString: java.lang.String = 34:17:36
```

For this particular number of seconds, this works just fine. However, if you play around with this at all, you will find that it has a significant shortcoming. If the number of minutes or seconds is less than 10, only one digit is displayed when we want two. So we need to come up with a way to get a leading zero on numbers that only have one digit. To do this, we will break the problem into two steps.

The first step will be to get the number of minutes and seconds as `Strings`.

```
scala> val min=displayMinutes.toString
min: java.lang.String = 17

scala> val sec=displaySeconds.toString
sec: java.lang.String = 36
```

This might seem odd, but the string version has something that the number itself does not, an easy way to tell how many digits/characters are in it. When there is only one digit, we want to add an extra zero. When there is not, we leave it as is. We can get this effect by using the `*` method on the `String` and a little math. The short names were selected to keep our expression shorter for formatting, but that is not required.

```
scala> val finalString=displayHours+":"+("0"*(2-min.length))+min+":"+(
| "0"*(2-sec.length))+sec
finalString: java.lang.String = 34:17:36
```

The result for these values is the same, but we could force some different value into `min` and `sec` to see that this does what we want.

```
scala> val min="5"
min: java.lang.String = 5

scala> val sec="2"
sec: java.lang.String = 2

scala> val finalString=displayHours+":"+("0"*(2-min.length))+min+":"+(
| "0"*(2-sec.length))+sec
finalString: java.lang.String = 34:05:02
```

3.8 Sequential Execution

Working in the REPL is great for certain tasks, but what if you have a sequence of things you want to do, and you want to do it multiple times. Having to type in the same set of instructions repeatedly is not a very good option. What we just did is a perfect example of that. If we want to do this for a different number of seconds, we have to repeat all the commands we just performed. Indeed, you can not really say that you have programmed until you put in a fixed set of instructions for solving a problem that you can easily run multiple times. That is what a program really is. So now it is time to write our first program of any significance.

We have used the REPL to enter commands one at a time. This is a great way to test things out in Scala and see what a few commands do. A second way of giving commands to Scala is to write little programs as SCRIPTs. The term script is used to describe small programs that perform specific tasks. There are languages, called scripting languages, that have been created specifically to make the task of writing such small programs easier. Scala is not technically a scripting language, but it can be used in that way. The syntax was created to mirror a lot of the things that are commonly put into scripting languages and if you run the `scala` command and give it the name of a file that ends in `.scala`, that file will be run as a script where the statements in it are executed in order.[4] We will use Scala as a scripting language until chapter 17 when we move up to a different style that helps us to organize larger pieces of code. The script for our time conversion looks like this.

```
val totalSeconds=123456
val displaySeconds=totalSeconds%60
val totalMinutes=totalSeconds/60
val displayMinutes=totalMinutes%60
val displayHours=totalMinutes/60
val sec=displaySeconds.toString
val min=displayMinutes.toString
val finalString=displayHours+":"+("0"*(2-min.length))+min+
":"+("0"*(2-sec.length))+sec
println(finalString)
```

If you put this into a file called `TimeScript.scala` and then run `scala TimeScript.scala`, you will get the output `34:17:36`. The `println` statement is required for the script because

[4]We will see later that the statements are not always executed in order because there are statements that alter the flow of control through the program. Since we have not gotten to those yet though, execution is completely sequential at this point.

unlike the REPL, the script does not print out values of all statements. You can run through this code in the REPL using the `:load` command. If you do "`:load TimeScript.scala`" you will see it print out all of the intermediate values as well as the result of the `println`.

This script allows us to run the commands repeatedly without retyping. By editing the value of `totalSeconds`, we can test other total times fairly quickly. However, a better solution would be to allow a user to tell us how many seconds to use every time we run the script. We can easily get this behavior by replacing the top line of the script we had with these two lines.

```
print("Enter the number of seconds. ")
val totalSeconds=readInt()
```

The first line does nothing more than print a prompt to let the user know that we are waiting for something to be input. After that we have altered the initialization of `totalSeconds` so that instead of giving it a fixed number, it is initialized to the value returned by `readInt`. This calls a function that reads in a single integer from the user. If you make this change and run the script, you will be able to enter any number of seconds, assuming it is a valid `Int`, and see it converted to hours, minutes, and seconds.

3.8.1 Comments

When writing programs in files, not in the REPL, it is often useful to include plane English descriptions of parts of the code. This is done by writing comments. If you are writing code for a course you likely need to have your name in the code. Your name is most likely not valid Scala so it should go in a comment. Different instructors and companies will have different commenting standards that you should follow. In a professional setting, comments are used primarily for two reasons. The first is to indicate what is going on in the code, particularly in parts of the code that might be difficult for readers to understand. The second is for documentation purposes as we will see in section 17.3.

There are two basic comment types in Scala, single-line comments and multiline comments. Single-line comments are made by putting // in the code. Everything after that in the line will be a comment and will not be ignored when the program is compiled and run. Multiline comments begin with /* and end with */. You can put anything you want between those and they can be spaced out across many lines. Code shown in this book will have limited commenting as descriptions of the code appear in the text of the book and there is little point in duplicating that content.

You have now been introduced to Scala and programming. This idea of giving specific instructions in the order that we need them to happen to solve a problem underpins everything we will do in this book. However, this is only a start. There is a lot more to explore and we begin to open the door to these possibilities in the next chapter.

3.9 A Tip for Learning to Program

In many ways, learning to program, whether in Scala or any other programming language, is very much like learning a new natural language. The best way to learn a natural language is through immersion. You need to practice it and be surrounded by it. The key is to not simply memorize the rules and vocabulary, but to put them into use and learn them through regular usage. You should strongly consider approaching programming in the same way.

So what does it mean to immerse yourself in a programming language? Clearly you are not going to have conversations in it or enjoy television or radio broadcasts in it. The way to immerse yourself in a programming language like Scala is to take a few minutes every day to write in it. You should consider trying to spend 15-30 minutes each day writing code. The REPL in Scala is an excellent tool for you to enter in statements to see what they do. Try to play around with the language. Instead of approaching it as memorizing keywords and rules, try to put things into use. The things that you use frequently will stick and become natural. Those things that you do not use regularly you will have to look up, but that is normal. Programmers, even professional programmers with many years of experience in a language, still keep references handy.

Over time, the number of lines of code that you write in these short time intervals each day will grow as the basics become second nature and you begin to practice more advanced concepts. By the end of this book you might find yourself writing a hundred lines or so of functional code on certain days during that time span. By that time you will hopefully also pick up a "pet project", something that you are truly interested in programming and that you will think about the structure and logic of much more frequently.

Especially early on, you might find it hard to think of anything that you can write. To help you with this, many of the chapters in this book contain a "Self-Directed Study" section, like the one below. Use these as a jumping-off point for the material in each chapter. After that will come a set of exercises and often a set of larger problems called projects. Remember that one of the significant goals of learning to program is improving your problem-solving skills. While the Self-Directed Study section will help you to familiarize yourself with the details presented in a chapter, the exercises and projects are actual problems that you are supposed to solve in a formal way using Scala. You should use these to help provide you with the immersion you need to learn the language.

3.10 End of Chapter Material

3.10.1 Problem-Solving Approach

Many students who are new to programming struggle with putting the English descriptions for solving a problem that they have in their head into whatever programming language they happen to be learning. The reality is that for any given line of code, there are a fairly small number of "productive" things that you could write. In the REPL you can test out any statement that you want, but in a script, an expression like `4+5` does not do much when used alone as a statement. These sections will appear at the end of a number of chapters as we introduce new concepts that might stand alone as statements, or which alter statements we have talked about previously in a significant way. The goal of these sections is to help focus your thinking so you can narrow down the list of possibilities any time that you are trying to decide what to put into the next line of code.

Given what we have just learned, there are only three types of statements that you would put in a script that stand alone:

1. A call to `print` or `println` to display information to the user. The function name should be followed with parentheses that contain the *expression* you want to print.

2. A variable declaration using `val` or `var`. A `val` declaration would look like `val` *name*

= *expression*. The *name* could be followed with a colon and a type though most of the time it won't be.

3. An assignment into a previously declared `var` of the form *name* = *expression*. The *expression* must have a type that agrees with the type the variable was created with.

If you want to read information using a function like `readLine()`, `readInt()`, or `readDouble`, that should appear as part of an *expression* in one of the above statements.

3.10.2 Summary of Concepts

- Programming languages have relatively simple rules that they always follow with no ambiguity.
  - Tokens are the smallest piece with meaning. They are like words in English.
  - Expressions are combinations of tokens that have a value and a type.
  - Statements are complete instructions to the language. In Scala, any expression is a valid statement.
  - The simplest expressions are literals.
    - `Int` literals are just numbers with no decimal points like 42 or 365.
    - Adding an `L` to the end of an integer number makes a `Long` literal.
    - Numbers that include decimal points or scientific notation using `e` syntax are of the type `Double`.
    - Adding an `f` to the end of a number makes it a `Float`
    - `Char` literals are single characters between single quotes.
    - `String` literals can have multiple characters between double quotes. Raw strings start and end with three double quotes and allow newlines.
- An object is a combination of information and functionality that operates on that information.
  - The information is called data members, fields, or properties.
  - The functionality is called methods.
  - All values in Scala are objects.
  - Methods are normally invoked using the "dot" notation. Arguments go in parentheses after the method name.
    - Scala allows the `.` to be left out.
    - Parentheses are also optional for argument lists of length zero or one.
    - Operators are really method calls. So `4+5` is really `(4).+(5)`.
- Numbers in computers are not exactly like numbers in math and you need to know some of the differences so you will understand when they lead to unexpected behavior.
  - All values stored in a computer are stored in binary, base 2, numbers. Each digit is called a bit. Different types use different numbers of bits for storage. The finite number of bits means that there a minimum and maximum values that can be stored in each type.
  - Negative integer values are stored using two's compliment numbers.

  - Binary numbers require a large number of digits, though they are all either 0 or 1, and converting to and from decimal is non-trivial. For this reason, computer applications frequently use base 8, octal, and base 16, hex. You can make an octal literal by putting a leading `0` on an integer. You can make a hexadecimal literal by putting a leading `0x` on an integer.
  - Non-integer numeric values are stored in floating point notation. This is like scientific notation in binary. These use the types `Float` and `Double`. Due to finite precision, not all decimal numbers can be represented perfectly in these numbers and there are small rounding errors for arithmetic.

- Additional mathematical functions, like trigonometric functions and square root are methods of the `math` object.
- You can declare variables using the keywords `val` and `var`. The name of a variable should start with a letter or underscore and can be followed by letters, underscores, or numbers. A `var` declaration can be reassigned to reference a different value.
- Instructions can be written together in scripts. The default behavior of a script is for lines of code to execute sequentially. Script files should have names that end with `.scala`. You run a script by passing the filename as a command-line argument to the `scala` command.
- Learning a programming language is much like learning a natural language. Do not try to memorize everything. Instead, immerse yourself in it and the things you use frequently will become second nature. Immersion in a programming language means taking a few minutes each day to write code.

3.10.3 Self-Directed Study

Enter the following statements into the REPL and see what they do. Try some variations to make sure you understand what is going on. Not all of these will be valid. You should try to figure out why.

```
scala> val a=5
scala> val b=7+8
scala> var c=b-a
scala> a=b+c
scala> c=c*c
scala> b=a/c
scala> b%2
scala> b%4
scala> b%a
scala> 0.3*b
scala> val name = "Your name here."
scala> name.length
scala> name+a
scala> println("Hi there "+name)
scala> println("\n\n\n")
scala> println("""\n\n\n""")
scala> 'a'+5
scala> ('a'+5).toChar
scala> math.Pi/2
scala> math.sqrt(64)-4.0
```

```
scala> math.sqrt(1e100)
scala> math.cos(math.Pi)
scala> 3000000000
scala> 3000000000L
scala> 3000000000.0
scala> 3e9
scala> 1/0
scala> 1.0/0.0
scala> 0.0/0.0
```

3.10.4 Exercises

1. What are the types of the following expressions?
   (a) 1
   (b) 1.7
   (c) 1.0
   (d) 'h'
   (e) "hi"
   (f) 5/8
   (g) 1+0.5
   (h) 7*0.5
   (i) "hi".length
2. Do the following 8-bit binary arithmetic by hand.
   (a) $10101101_2 + 11010100_2$
   (b) $00111110_2 + 00111011_2$
   (c) $01001010_2 * 00110010_2$
3. Convert the following decimal values to binary (8-bit), hex (2-digit), and octal (3-digit) by hand.
   (a) 7
   (b) 18
   (c) 57
   (d) 93
   (e) 196
4. Convert the following hex values to binary and decimal by hand.
   (a) 0x35
   (b) 0x96
   (c) 0xA8
   (d) 0x7F
5. Convert the following decimal values to binary (8-bit) and hex (2-digit) by hand.
   (a) -87

(b) -32

(c) -105

(d) -1

6. Write a script that will calculate how far a projectile will go given a launch speed and an angle ignoring friction. Assume that the projectile is launched from ground level with a certain speed in m/s and at a certain angle in radians. Use the fact that acceleration due to gravity is $9.8m/s^2$. The steps in doing this would be to calculate the speed parallel and perpendicular to the ground with math.sin and math.cos, then figure out how long it takes for the projectile to slow to a vertical speed of zero ($v = v_0 - at$), and use double that time as how long it stays in the air.

7. Using Scala as a calculator, figure out how much you have to make each year to bring home $100,000 assuming a 27% tax rate.

8. In the REPL, declare a variable with the type `String` that has the name `str`. Give it whatever value of string you want. On the next line, type `str.` then hit Tab to see the methods for `String`. By playing around with the REPL, try to figure out what the following methods do.
   - `toUpperCase`
   - `trim`
   - `substring` – This method takes two `Int` arguments.
   - `replace` – This method can be called with two `Char` arguments or two `String` arguments.

9. Kepler's third law of planetary motion says that $P^2 \propto a^3$, where P is the orbital period and a is the semi-major axis of the orbit. For our Sun, if you measure P in years and a in Astronomical Units (AU), the proportionality becomes equality. Look up the semi-major axis values for three bodies in our solar system other than the Earth and use Scala as a calculator to find the period according to Kepler's third law.

10. In this option you will write a little script that does part of a 1040-EZ. We have not covered enough for you to do the whole thing, but you can write enough to ask questions and do most of the calculations for lines 1-13. Do what you can and remember this is a learning exercise.

11. Your goal for this exercise is to write a script to calculate the cost of a simple, 3-ingredient, recipe. You start by asking the user for the names and amounts of the three ingredients. Then prompt them for the cost, per unit, of each ingredient. Output the total cost of each ingredient and for the whole recipe. To make things simple, feel free to use the `Double` type.

Chapter 4

Conditionals

We solved some basic problems in the last chapter, but the techniques that we have access to are fundamentally limited at this point. The real problem is that all the lines in a script execute in the same order from top to bottom every time the script is run. In most real problems we need to be able to do what is called conditional execution, where something happens only in certain situations. In this chapter we will learn the most fundamental method for doing conditional execution in Scala and see how we can use it to solve more complex problems.

In order to do this properly, we need to develop a formalized way to express logic and put it into our programs. This system, called Boolean logic, will allow us to state the conditions under which various parts of our program will or will not happen.

4.1 Motivating Example

You have been asked to write a program that will calculate charges for people visiting a local amusement park. There are different charges for adult vs. child, whether they are bringing in a cooler, and whether they want to also get into the water park. We need to write code that will tell us how much the person pays. We will have the user input the needed information such as an `Int` for the persons age, a `Boolean` for whether they have a cooler, and another Boolean for whether they want to get into the water park.

This is something that we could not do last chapter because we did not have a way of performing logic. We could not say that we wanted to do something only in a particular situation. This ability to do different things in different situations is called conditional execution and it is a very important concept in programming. It is also critical for problem solving in general. Conditional execution gives you the ability to express logic and to solve much more complex problems than you could do without it.

4.2 The if Expression

Virtually all programming languages have a construct in them called `if`. The idea of this construct is you have a condition where one thing should happen if the condition is true. If the condition is false, then either nothing or something different will happen. In non-functional languages, the `if` construct is a statement. It has no value and simply determines what code will be executed. In Scala and other functional languages, the `if` is an expression which gives back a value. Scala allows you to use it in either style.

Let us start with an example, then we can broaden it to the more general syntax. Take the ticket price example and consider just the person's age. Say that we want to consider whether a person should pay the $20 children's rate or the $35 adult rate. For our purposes, a child is anyone under the age of 13. We could make a variable with the correct value with the following declaration using an `if` expression.

```
val cost = if(age<13) 20 else 35
```

This assumes that age is an `Int` variable that has been defined prior to this point. The first part of this line is a basic variable declaration as discussed in chapter 3. The new part of this is the use of the if expression which checks the age and gives back one of two values depending on where `age<13` is true of false.

This example uses the `if` as an expression. So we care about the value it gives us. In this type of usage, we always need to include an `else` because there has to be a value that is used if the expression, is false. This same type of behavior can be also accomplished with a `var` using the `if` as a statement.

```
var cost=20
if(age>=13) cost=35
```

This code creates a `var` and gives it the initial value of 20. It then checks if the age is greater than or equal to 13 and if so, it changes the value to 35. Note with this usage, the `else` clause is not required. Here we only want to do something if the condition is true. In Scala you should generally prefer the first approach. It is shorter, cleaner, and leaves you with a `val`, which you can be certain will not be changed after it is created. The first approach is a functional approach while the second is an imperative approach.

In general, the format of the `if` is as follows:

```
if(condition) expr1 [else expr2]
```

The square brackets here denote that the `else` and the second expression are optional. They are not part of the syntax. The condition can be replaced by any expression of type `Boolean`, while `expr1` and `expr2` can be any expression you want.

In the situation where the expression is complex, you can use curly braces to create a block of code. You can put any code inside of curly braces. This block is itself an expression and the value of the expression is the value of the last statement in the curly braces. You can play with this in the REPL with expressions like this example.

```
scala> {
     | println("First line")
     | 4+5
     | }
First line
res0: Int = 9
```

For any `if` expression that does not fit on a single line, you should consider using a format like this.

```
if(condition) {
  ...
} else {
  ...
}
```

Being consistent with this format can help make your code more readable and prevent errors.

Common Bug

One common error that novice programmers encounter with the `if` statement occurs when they put more than one expression after the `if` or the `else` without using curly braces. Here is an example of how that can happen. We start with the following code that calculates an area.

```
val area = if(shape=="circle")
  math.Pi*radius*radius
else
  length*width
```

Here the `if` is broken across multiple lines so there is not one long line of code. This can make it easier to read. There are no curly braces, but they are not needed because each part of the `if` has only a single expression.

Now consider the possibility that the program this is part of does not work. To help figure out what is going on, you put print statements in so that you know when it uses the circle case and when it uses the rectangle case. The error comes when the code is changed to something like this.

```
val area = if(shape=="circle")
  println("Circle")
  math.Pi*radius*radius
else
  println(Rectangle)
  length*width
```

Now both branches have two expressions and we have a problem. This code needs curly braces, but they are easy to forget because the indentation makes the code look fine. Scala does not care about indentation. That is only for the benefit of humans. In reality, this code should look something like the following.

```
val area = if(shape=="circle") {
  println("Circle")
  math.Pi*radius*radius
} else {
  println("Rectangle")
  length*width
}
```

What about the other parts of our admission park entrance cost? We also wanted to check if the person had a cooler or if they wanted to get into the water park. These should both be variables of type `Boolean`. We might call them `cooler` and `waterPark`. Let us say it costs an additional \$5 to bring in a cooler and \$10 to go to the water park. If we used the `if` as an expression we can type in the following:

```
val cost = (if(age<13) 20 else 35)+(if(cooler) 5 else 0)+
   (if(waterPark) 10 else 0)
```

Here we are adding together the results of three different `if` expressions. This format is somewhat specific to functional languages. It would be more common in most languages to see this instead:

```
var cost=20
if(age>=13) cost=35
if(cooler) cost=cost+5
if(waterPark) cost=cost+10
```

In this second form, we use `if` as a statement instead of an expression and have the body of the `if` change, or mutate, the value of the variable cost.

Note on Style

While there are differences between the functional and imperative versions of code in different applications in regard to things like performance and the likelihood of errors, at this stage those should not be your top concerns. You should pick the style that makes the most sense to you. Later you can evaluate the differences and pick the approach that best fits the task.

In the second and third `if` statements, the name `cost` is repeated. This type of repetition is often avoided in programming. Many languages, including Scala, include operations that allow us to avoid it. When the duplication is caused like this with an assignment and a variable appearing on both sides of the equals sign, it is possible to use an abbreviated syntax where the operator is placed in front of the equals sign like this:

```
var cost=20
if(age>=13) cost=35
if(cooler) cost+=5
if(waterPark) cost+=10
```

4.3 Comparisons

The first `if` statement shown in the previous section uses `>=` to do a comparison between two values. You likely had no problem figuring out that this can be read as greater than or equal to. Your keyboard does not have a $\geq$ key so instead we use two characters in a row. All of the normal comparisons that you are used to exist in Scala, but some, like the greater than or equal to, differ from how you are used to writing them on paper.

The simplest comparison you can do is to check if two things are the same or different. You read this as saying that two things are equal or not equal to one another. In Scala we represent these with `==` and `!=` respectively. Note that the check for equality uses two equal signs. A single equal sign in Scala stands for assignment, which we have already seen stores a value into a variable. The double equal sign checks if two expressions have the same value and produces a `Boolean` value with the result of the comparison. Here are a few example of the use of this in the REPL.

```
scala> 2==3
res1: Boolean = false

scala> 7==7
res2: Boolean = true

scala> 'a'=='a'
res3: Boolean = true

scala> "hi"=="there"
res4: Boolean = false

scala> "mom".length=="dad".length
res5: Boolean = true
```

The `!=` operator is basically the opposite of `==`. It tells us if two things are different and should be read as not equal. As we will see, it is customary in Scala to read the exclamation point, pronounced "bang" in many computing contexts, as not. Any of the examples above could use `!=` instead of `==` and the result would have been the opposite of what is shown.

In addition to equality and inequality, there are also comparisons of magnitude like the `age>=13` that we used above. The comparisons of magnitude in Scala are done with `<`, `>`, `<=`, and `>=`. These also give us back a value of the `Boolean` type so it will be either `true` or `false`. The order of the characters in `<=` and `>=` is significant. They are in the same order that you say them, "less than or equal to" and "greater than or equal to", so it will not be hard to remember. If you reverse the order, Scala will not be able to figure out what you mean and will return an error.

```
scala> 5=<9
<console>:6: error: value =< is not a member of Int
       5=<9
        ^
```

You can use `==` or `!=` on any of the different types in Scala, both those we have talked about and everything that we have not yet talked about. This is not the case for the magnitude comparisons. While you can use `<`, `>`, `<=`, and `>=` for many of the types that we have seen so far, not every type has an order where these comparisons makes sense. We will have types for things like colors, shapes, and fruits. Magnitude comparisons will not make sense with these and they will not work if you try to use them.

Equality vs. Identity (Advanced)

If you have experience with Java, you might find the behavior of == confusing. This is because in Scala, == does what most people expect, it checks for equality between values. For anyone who has not programmed in Java he/she might wonder what other options there are. We will see later that objects with the same value are not necessarily the same objects. If you go to a store and pick up two boxes of the same type of item, they are basically equal as far as you can tell, but they are not the same object. Each one has its own identity.

There are times in programming when you don't want to just check if two things are equal, you want to actually know if they are the same thing. This requires doing a check for identity. In Scala we use `eq` to check if two things are the same thing. We can show an example of where `eq` and == return different result.

```
scala> "sssss" eq "s"*5
res8: Boolean = false

scala> "sssss" == "s"*5
res9: Boolean = true
```

4.4 Boolean Logic

Imagine if the theme park had a policy where seniors are charged the same amount as children. So now anyone over 65 or anyone under 13 should pay the reduced rate. We could accomplish this by having separate `if` statements in either the functional or the imperative manner.

```
val cost = if(age<13) 20 else if(age>65) 20 else 35
```

or

```
var cost=20
if(age>=13) cost=35
if(age>65) cost=20
```

Both of these are verbose, potentially inefficient, and prone to errors. The errors occur because in both we had to enter the number 20 in two places. What we would really like to say is exactly what we said in English. We want to say that we use the lower rate if the person is under 13 or over 65 and use the higher rate otherwise. Boolean logic gives us the ability to say this.

There are four different BOOLEAN OPERATORS we can use to build complex Boolean expressions from simple ones. These are shown in table 4.1. Table 4.2 shows the value that results from applying these different operators on different arguments.

Description	Usage	Meaning
and	`a && b`	True if both a and b are true.
or	`a || b`	True if a or b is true. Allows both being true.
exclusive or (xor)	`a ^ b`	True if either a or b is true, but not both.
not	`!a`	True if a is false and false if a is true.

TABLE 4.1: This table shows the Boolean operators for Scala and what each one does.

a && b	a=true	a=false
b=true	true	false
b=false	false	false

a || b	a=true	a=false
b=true	true	true
b=false	true	false

a ^ b	a=true	a=false
b=true	false	true
b=false	true	false

	!a
a=true	false
a=false	true

TABLE 4.2: Truth tables for the different Boolean operators.

We can use the || operator just like we used or in our English description of what we wanted above. If we do this, our functional approach would simplify to this.

```
val cost = if(age<13 || age>65) 20 else 35
```

We can use && to say and.

```
var cost=20
if(age>=13 && age<=65) cost=35
```

The first one reproduces the English description and uses an or to give a single Boolean expression for when the lower rate is charged. The second one is the converse and uses an and to determine when the higher rate should be charged. The second expression could be written instead with an || and a ! to make it more explicit so that it is the converse of the first one.

```
var cost=20
if(!(age<13 || age>65)) cost=35
```

The extra set of parentheses is required here so that the not is done for the whole expression.

It is worth noting that || is not the "or" or normal English. In normal usage, when you say "or" you mean the logical exclusive or, ^. For example, when a parent offers their child the option of "cake or cookies", the parent is not intending for the child to take both. The inclusive or, ||, allows both. The exclusive or, ^, does not.

Another example would be code that tells us if two squares intersect. Each square is defined by its top-left corner as x and y coordinates along with the edge lengths of the two squares. We want a result that is a Boolean telling us whether or not the squares intersect.

Before we can write a program to do this, we need to figure out how we would do this independent of a program.

Your first inclination might be to say that given two squares it is obvious whether or not they intersect. Indeed, novice programmers are tempted to give that type of description to many problems. This is not because the solution really is obvious, it is because novice programmers have not developed their skills at analyzing what they do when they solve a problem. This is something that you will develop over time and it is a requirement for any real programmer.

So what are you missing when you say that a problem like this is obvious? Given a set of six numbers as is the case here, most people would not find the solution obvious. To make it obvious they would draw the squares and look at them. This use of your visual processing is basically cheating and implicitly brings into play a large amount of processing that your brain does automatically. For this reason, we will avoid any type of solution where you would be tempted to say you would "look at" something.

Even if we did take the approach of drawing the squares, that is not as straightforward as you might picture. When you picture drawing squares, you likely picture squares of nearly the same size that are in easy-to-draw coordinates. Plotting gets a lot harder if one of the squares is millions of times larger than the other. If your program is trying to see if an ant has entered a building, that is not at all out of the question. So we can not settle for "just look at it" or "it is obvious". That means we have to figure out what it really means for two squares to be intersecting just using the numbers.

While looking at it is not allowed as a solution, it can be helpful to figure out what we really mean. Draw a number of different squares on paper. Label them with the numbers 1 and 2. What are different possible relative positions for the squares? What are cases when they do not intersect? What has to happen for them to intersect? These are not rhetorical questions. Go through this exercise and come up with something before you read on. The ability to break down a problem that is "obvious" into the real steps that humans go through on a subconscious level is a cornerstone of programming. It is also useful for those who do not intend to be programmers as it can help you to understand your thinking on all manner of topics.

When you went through the processing of drawing the squares, one of the things you might have found was that squares can overlap in either an x or y direction, but they only intersect if they overlap in both x and y. That gives us something to work with. Our answer could say something like, `overlapX && overlapY`. All we have to do now is figure out what it means to overlap in a given direction. Even this has lots of different possibilities, but if you play with them you will see that any situation where there is an overlap satisfies the following: the minimum of the first range has to be less than the maximum of the second range and the maximum of the first range has to be greater than the minimum of the second range. Go back to your drawings and verify that this is true.

At this point we have the ability to say what we want. There are many ways that we could do so. We are going to pick an approach which breaks the problem down into smaller pieces. This will be easier for people to read. We said above that the squares overlap if their ranges overlap in both x and y. So we can write code that checks to see if two ranges overlap in one dimension, then the other. We can then combine these to see if the squares overlap.

```
val overlapX = x1<x2+size2 && x1+size1>x2
val overlapY = y1<y2+size2 && y1+size1>y2
val squareOverlap = overlapX && overlapY
```

The `Boolean` value, `squareOverlap`, tells us the answer to the question.

Short-Circuit Operators

One other significant factor about the `Boolean` `&&` and `||` operators is that they are short-circuit operators. This means that if the value they will give is known after the first argument is evaluated, the second argument will not be evaluated. For `&&` this happens if the first argument is `false` because no matter what the second argument is, the final value will be `false`. Similarly, if the first argument of `||` is `true`, the final value will be `true` so there is no point spending time to evaluate the second argument. This will be significant to us later on when we get to expressions that can cause errors. The only thing we could do now that would cause an error is to divide by zero. We can use that to demonstrate how short circuiting can prevent an error and that the `^` operator is not short circuited.

```
scala> val n=0
n: Int = 0

scala> 4/n
java.lang.ArithmeticException: / by zero
      at .<init>(<console>:7)
      at .<clinit>(<console>)
      at RequestResult$.<init>(<console>:9)
      at RequestResult$.<clinit>(<console>)
      at RequestResult$scala_repl_result(<console>)
      at sun.reflect.NativeMethodAccessorImpl.invoke0(Native Method)
      at sun...

scala> n!=0 && 4/n==6
res3: Boolean = false

scala> n==0 || 4/n==6
res4: Boolean = true

scala> n==0 ^ 4/n==6
java.lang.ArithmeticException: / by zero
      at .<init>(<console>:7)
      at .<clinit>(<console>)
      at RequestResult$.<init>(<console>:9)
      at RequestResult$.<clinit>(<console>)
      at RequestResult$scala_repl_result(<console>)
      at sun.reflect.NativeMethodAccessorImpl.invoke0(Native Method)
      at sun...
```

4.5 Bigger Expressions

So far all of our expressions have been fairly small. Other factors can come into play when they get large. One factor that becomes significant is precedence. This is the order in

First Character
(other special characters)
* / %
+ -
:
= !
< >
&
^
|
(all letters)
(all assignment operators)

TABLE 4.3: Table of operator precedence in Scala. The precedence of an operator is determined by the first character.

which operations are done. You know from math that multiplication happens before addition and Scala follows that rule. What happens when we start including these other operations for comparing values and Boolean operators? These are not things you have been taught in math class, but for the most part they are understandable. All of the operators introduced in this section happen after all the mathematical operators we worked with before. This makes sense as otherwise we could not type something like `a+5<b*2` without using parentheses. Similarly, comparisons have higher precedence than the Boolean operators. The expressions for `overlapX` and `overlapY` used both of these facts as they combined addition, comparison, and `&&`.

As was mentioned earlier, operators are really just methods in Scala. The ones we have talked about so far are simply methods that are defined in the Scala libraries on the appropriate types. So `&&` and `||` are defined on the `Boolean` type. The comparison operators are defined on numeric types, etc. When you use operator notation with a method, the precedence is determined by the first character in the operator. Table 4.3 shows the order.

There is another interesting point that we have used implicitly already, but is worth noting explicitly. This is that `if` expressions can be nested inside of one another. We saw this when we first tried to add senior citizens at a lower cost.

```
val cost = if(age<13) 20 else if(age>65) 20 else 35
```

The contents of the `else` on the first `if` is itself an `if`. This is a general property of most programming languages. The `if` in Scala needs some form of expression inside of it for the `true` and `false` possibilities if it is an expression in Scala. As such, the `if` itself makes a perfectly valid expression to nest. So you can nest `if`s inside of one another as much as it makes sense for the problem you are solving.

To make this more explicit, let us go back to our theme park and this time consider concessions. The menu is not broad and is standard for fast food. They have drinks, fries, and various main course items like hamburgers and hot dogs. You can also get a combo, which has one of each. For any of these you can specify a size. The cost is specified by the simple matrix shown in table 4.4.

We need to convert this table into code. We will start by reading in two strings. The first is the item type and the second is the size, both as `String`s. In the end we want the variable `cost` as a `Double`.

```
println("What item are you ordering?")
```

Item	S	M	L
Drink	\$0.99	\$1.29	\$1.39
Side	\$1.29	\$1.49	\$1.59
Main	\$1.99	\$2.59	\$2.99
Combo	\$4.09	\$4.99	\$5.69

TABLE 4.4: Theme park food item costs.

```
val item = readLine()
println("What size do you want?")
val size = readLine()
val cost = if(item=="Drink") {
    if(size=="S") 0.99
    else if(size=="M") 1.29
    else 1.39
  } else if(item=="Side") {
    if(size=="S") 1.29
    else if(size=="M") 1.49
    else 1.59
  } else if(item=="Main") {
    if(size=="S") 1.99
    else if(size=="M") 2.59
    else 2.99
  } else {
    if(size=="S") 4.09
    else if(size=="M") 4.99
    else 5.69
  }
```

This code has a top level set of `if`s that pick the item type. Inside of each is an `if` statement that picks from the different sizes. The way this was written, it will default to a combo if the item is not recognized and to a large if the size is not recognized. There are better ways to deal with this, but this will work for now.

This method demonstrates a standard formatting style used with this type of structure where the only thing in the `else` is another `if`. Instead of putting another set of curly braces after the `else` and indenting everything, leave off the curly braces and just put the `if` there. This prevents the indentation from getting too deep.

Bit-wise Arithmetic

The bits in a binary number are just like `Boolean` values. We can perform Boolean logic on the bits in integer values the way we would on standard `true` and `false` values. To do this we use slightly different operators. We use `&` and `|` instead of `&&` and `||`. The versions with a single character are not short circuit. The concept of short circuit does not make sense for this type of operation because a bit can not be a complex operation to evaluate. The other difference is that we use `~` instead of `!` for the bit-wise negation.

If you think back to chapter 3 you will remember that every number on a computer is represented in binary. We store values of one or zero to represent different powers of two.

When we do bit-wise operations we simply take the bits in the same position and do the specified operations on them to get the bits in the result. To see how this works let us run through a set of examples on the binary operators where we use the four bit numbers 1100 and 1010.

```
  1100
& 1010
------
  1000

  1100
| 1010
------
  1110

  1100
^ 1010
------
  0110
```

Negation is pretty straightforward: `~1001=0110`. Of course, these bit combinations all have numeric decimal values. We can put commands into Scala to do these same things. Our first value is 12 and our second is 10. The value we took the negation of is 9. Here are the operations performed in Scala. Check that they match the answers we got.

```
scala> 12 & 10
res3: Int = 8

scala> 12 | 10
res4: Int = 14

scala> 12 ^ 10
res5: Int = 6

scala> ~9
res6: Int = -10
```

The last one is interesting. When you check if it makes sense, remember that negative values are represented using two's complement and an `Int` has 32 bits.

While on the topic of bit-wise operations there are two others that are worth mentioning. They are left-shift and right-shift and are written in Scala as `<<` and `>>`. These operators shift the bits in an integer value the specified number of positions. If you think about the meaning of bits, this is like multiplying or dividing by powers of two in the same way that adding zeros to the end of a decimal number is the same as multiplying by ten. Some simple examples show you how we can use this to get powers of two.

```
scala> 1 << 0
res7: Int = 1

scala> 1 << 1
res8: Int = 2

scala> 1 << 2
res9: Int = 4
```

```
scala> 1 << 3
res10: Int = 8
```

There is a second version of right-shift written as >>>. The normal version does not move the sign bit so that signs are preserved. The second version does not do this and will shift the sign bit down along with everything else.

You might wonder why you would want to do these things. Using a single integer value as a collection of `Boolean` values is common in libraries based on the C language and frequently appear in operating system code. There is another usage that could come into play in projects later in this book.

If you have adjusted display settings on a computer you have probably seen that colors can be represented as 8-bit, 16-bit, 24-bit, or 32-bit values. Given the abundance of memory on modern computers, most of the time people will use 32-bit color. Have you ever wondered what that means? If you have written a web page or looked at a web page you have seen colors represented as six hexadecimal digits. The first two digits specify how much red, the next two specify how much green, and the last two specify how much blue. This is called RGB for obvious reasons and is exactly what 24-bit color gives you. 32-bit color uses an extra 8-bits because computers can operate more quickly with memory in 32-bit chunks because they are typically wired to deal with a 32-bit integer. The additional 8-bits stores the alpha channel which can be used for transparency. It does not matter much for your screen, but it is something that can be used to nice effect in 2-D graphics which we will discuss in chapter 12.

32-bit color is often called ARGB because it has alpha, red, green, and blue values all packed into 32 bits. Each gets 8 bits or a byte. This is where bit-wise operations come into play. You might be given four different values for alpha, red, green, and blue for a color and need to pack them together into a single 32-bit integer. Alternately, you might be given a single ARGB value as a 32-bit integer and have to figure out the individual components. Indeed, both of these appear as exercises below.

4.6 End of Chapter Material

4.6.1 Problem-Solving Approach

The `if` can be used as a statement. In the situation where an `if` includes side effects, you can use it as a top level command. This is a 4th option for statements that you can put into a script. So when you are considering what your next line of code is you should be thinking of one of these four possibilities.

1. Call `print` or `println`.
2. Declare a variable with `val` or `var`.
3. Assign a value to a variable.
4. Write an `if` statement. Note that to make sense as a statement, the expressions for the true and false possibilities should include one or more statements from this list. You should use an `if` when the code needs to do different things in different situations. When the word "if" fits in your English description of the solution to the problem, odds are good that it fits into the code as well.

4.6.2 Summary of Concepts

- Constructs that allow different pieces of code to be executed depending on different conditions in the code are called conditionals.
- The most basic conditional is `if`.
  - The syntax is `if(`*condition*`)` *trueExpression* `else` *falseExpression*.
  - The *condition* needs to be an expression with the `Boolean` type.
  - In Scala it can be used as an expression with a value or as a statement. As an expression, you need to have an `else` clause.
  - Curly braces can define blocks of code that function as a large expression. The value of the expression is the value of the last statement in the block.
- The `Boolean` expressions often involve comparisons of values.
  - Any values can be checked for equality or inequality using `==` or `!=`, respectively.
  - Values that have a natural ordering can also be compared using `<`, `<=`, `>`, or `>=`.
- More complex Boolean expressions can be built by combining simple expressions using Boolean logic.
  - The `||` operator is an inclusive or. This means that it is `true` if either of the two arguments are `true` as well as when both are `true`.
  - The `&&` operator represents a logical and. It is only true when both arguments are true.
  - The `^` operator represents exclusive or. This is "or" as used in normal English where the result is `true` if one argument or the other is `true`, but not both.
  - The `!` operator is logical negation.
  - The `||` and `&&` operators are short-circuit operators.
- When building large expressions, the order in which operators are applied is significant. This is called precedence. The precedence of operators in Scala depends on the first character. Table 4.3 gives a full list or precedence in Scala.
- Bits can be viewed as `Boolean` values. Bit-wise arithmetic is operations that work on numeric numbers as collections of bits instead of normal numbers.
  - `|` is bit-wise or.
  - `&` is bit-wise and.
  - `^` is bit-wise xor.
  - `~` is bit-wise negation.

4.6.3 Self-Directed Study

Enter the following statements into the REPL and see what they do. Try some variations to make sure you understand what is going on. Note that some lines read values so the REPL will pause until you enter those values. The outcome of other lines will depend on what you enter.

```
scala> val a = readInt()
scala> val b = readInt()
scala> val minimum = if(a<b) a else b
scala> if(minimum != (a min b)) {
  println("Oops, something went wrong.")
} else {
  println("That's good.")
}
scala> true && true
scala> true && false
scala> false && true
scala> false && false
scala> true || true
scala> true || false
scala> false || true
scala> false || false
scala> !true
scala> !false
scala> true ^ true
scala> true ^ false
scala> false ^ true
scala> false ^ false
scala> a < b || { println("a>=b"); a >= b }
scala> a < b && { println("a>=b"); a >= b }
scala> a match {
  case 7 => "That is a lucky number."
  case 13 => "That is an unlucky number."
  case _ => "I'm not certain about that number."
}
scala> 13 & 5
scala> a.toBinaryString
scala> a & 0xff
scala> (a & 0xff).toBinaryString
scala> a ^ 0xff
scala> (a ^ 0xff).toBinaryString
scala> (a << 3).toBinaryString
scala> ((a >> 8) && 0xff).toBinaryString
```

4.6.4 Exercises

1. Write Boolean expressions for the following:

    (a) Assume you have a variable called `age`. Tell if the person is old enough to legally drink.

    (b) Given a `height` in inches, tell if a person can ride an amusement park ride that requires riders to be between 48" and 74".

2. Determine if the following expressions are `true` or `false`. Assume the following, `a=1`, `b=2`, `c=3`, `d=true`, `e=false`.

    (a) `a==1`

    (b) `c<b || b>c`

    (c) `a<=c && d==e`

(d) `1+2==c`
(e) `d`
(f) `!e`
(g) `d || e`
(h) `6-(c-a)==b && (e || d)`
(i) `c>b && b>a`
(j) `a+b!=c || (c*a-b==a && c-a>b)`

3. Determine if the following expressions are true or false. Assume the following, `a=1`, `b=10`, `c=100`, `x=true`, `y=false`.

   (a) `x`
   (b) `x && y`
   (c) `a==b-9`
   (d) `a<b || b>a`
   (e) `!y && !x`
   (f) `(c/b)/b==b/b`
   (g) `(c+b+a==b*b+c/c) || y`
   (h) `a <= b && b <= c && c >= a`
   (i) `c/(b/b)==b/b`
   (j) `!(x || y)`

4. The `reverse` method can be called on a `String`. Use this to write a script where the user inputs a word (use `readLine`) and you tell them whether or not it is a palindrome.

5. It is customary to express colors on a computer as a combination of red, green, and blue along with another value called alpha that indicates transparency. A single `Int` has 32 bits or 4 bytes. The four different color values are often packed into a single `Int` as an ARGB value. The highest byte is the alpha and below that you have red, green, and blue in order. Each byte can store values between 0 and 255. For alpha, 0 is completely transparent and 255 is completely opaque.

   Write code that reads four `Int` values for alpha, red, green, and blue and calculates an `Int` with the combined ARGB value. If one of the numbers passed in is outside the 0 to 255 range, use 0 or 255, whichever it is closer to. Note that bitwise operations are appropriate for this exercise. You might find hexadecimal representation of numbers to be useful as well.

6. Repeat the previous exercise, but this time the input should be `Doubles` between 0.0 and 1.0 that you convert to `Ints` in the proper range.

7. Write code that does the opposite of what you did for exercise 5. It should take an `Int` with an ARGB value and calculate the four `Int` values with the component values between 0 and 255.

8. Write code that does the opposite of what you did for exercise 6. It should take an `Int` with an ARGB value and calculate four `Double` values with the component values between 0.0 and 1.0.

9. Write a script that has the user input a location as x and y coordinate values which tells whether the point specified is in the unit circle.[1]

[1]The unit circle is a circle centered on the origin with radius 1.

4.6.5 Projects

1. Write a script that asks the user for the coefficients of a quadratic equation (a, b, and c in $ax^2 + bx + c$). It should print the roots for those coefficients. It needs to handle different options for the roots and print one or two solutions with real or complex values as appropriate.

2. Write a script that tells you whether or not a rectangle overlaps with a circle. It needs to prompt the user for the required information for the two shapes that includes positions and sizes and it prints an appropriate message based on the Boolean result.

3. You have built a simple robot that is supposed to move along a straight line from wherever it is placed to a flag. The robot can move in one of two ways. It can roll any distance at a speed of one inch per second. It can jump and each jump takes one second and moves a predetermined amount. When it jumps it always jumps that amount, not a fraction of it. Write a script that reads the distance you start from the flag and the distance covered in each jump and prints how many seconds it will take to get to the flag using an optimal approach. (The trick with this problem is to make sure you consider all the possibilities.)

4. This problem starts a track of options that you can work on that build up to having a functioning ray tracer at the end of the semester that can render images in 3-D. For this first step you are to write a script that determines if and where a ray intersects either a plane or a sphere.

   The script should start by asking for information on the ray. That includes the start point of the ray, $\vec{r}_0$, and the direction vector for the ray, $\vec{r}$. It should then ask if you want to consider a plane or a sphere. The code will have an `if` that splits into two possibilities at that point. One will ask for information on the sphere, a center point and radius, and do the proper calculation for a sphere. The second will ask for information on a plane, a normal direction and distance, and do the proper calculation for the plane. The script finishes by printing out information about the intersection.

   A ray can be defined as a start point and a direction vector using the parametric equation $\vec{r}(t) = \vec{r}_0 + t * \vec{r}$, for $t \geq 0$. You find intersections by solving for the t that satisfies certain conditions like coming a certain distance from a point (for a sphere) or satisfying the equation $\vec{r}(t) \cdot \vec{n} = d$ (for a plane with normal $\vec{n}$ that is d units from the origin). You can view coming within a certain distance from a point as solving the equation $(\vec{r}(t) - \vec{n}) \cdot (\vec{r}(t) - \vec{n}) = radius^2$. With a little algebra you can reduce this to a quadratic equation. You only care about the solution with the smaller value of t. This option is a bit more conceptually challenging, but if you do all the ray tracing options the results can be impressive. The website at `www.programmingusingscala.net` includes a number of images made using a ray tracer the author wrote in Scala.

5. You have been going through your book of *Simple 2-Ingredient Recipes* looking for something to cook. The problem is, you are really running low on ingredients. You only have four items and they are in short supply. For this project you will write a script that will take the four items you have and how much of each you have. After that it will ask for two ingredients and how much of each is needed. It should output whether or not you have enough stuff to make that recipe.

   For the input, items will be strings and you do not care about units on the amount. Item names have to match exactly to be the same and assume the same units are used for any given item.

6. Imagine that you were considering taking a number of different majors or minors for college and you want to write a program to do a little math and help you see what different approaches would require. You have four major/minor interests. For each one the user should input a name and the number of hours it requires. You also have a number of hours for the core curriculum, a minimum number of hours for graduation, and how many hours you bring in from outside (AP/IB/Transfer). Prompt the user to enter those 11 values. Then run through the four major/minor interests and let them enter "yes" or "no" for whether they will do it to test an option. After that, print out the average number of hours that need to be taken each of the 8-semesters of college to complete that combination.

Chapter 5

Functions

Earlier in the book we made the statement that programming was all about problem solving. It is basically the art of giving a computer a set of instructions to solve a problem in terms that the computer can understand. One of the key foundational elements of good problem solving is problem decomposition. The idea of problem decomposition is that good solutions of large, complex problems are built out of solutions to smaller, more manageable problems. This is especially important in programming because the problems can be arbitrarily complex and, at the base, we have to break them down to the level where we can communicate them to the computer, which is fundamentally quite simplistic.

One of the key elements of problem decomposition in programming is the FUNCTION. A function is a collection of statements to solve a specific problem that we can call as needed. The scripts we have built to this point share some characteristics with functions. They were collections of statements we put together to do something repeatedly without retyping it. Functions give us that and much more. We'll even put functions in our scripts in this chapter.

5.1 Motivating Example

Imagine a script where you you had to calculate admission or the cost of food items in multiple places. You could accomplish this by repeating the code to do the calculation in multiple places, but there are many problems with this approach. The ability to cut and paste means that it will not take all that much extra effort, but it will make your code a lot longer and harder to manage. The primary problem arises if you realize that you have to change something in the calculation. If you have multiple copies of the code that do the calculation you have to find all of them and change them. This is an activity that is not only tedious, it is error prone because it is very easy to miss one or more of the copies.

The way to get around this is to use functions. A function lets you group together all the commands that are used in a particular calculation so that they can be called on from other parts of the code. You can specify what information needs to be provided for the function to do its work. In this way, functions provide flexibility. Instead of doing the copy and paste, you simply call that one function from different parts of the code, passing in the information that is needed at that point. If you realize something needs to change in the calculation, you simply change the function, and all the calls to it will use the modified version.

5.2 Function Refresher

Functions are things that you will be familiar with from many different math classes, going back to algebra. The simplest example of a function from math might be something like $f(x) = x^2$. This says that f is a function that takes a number, given the name x, and it has the value of that number squared. Note that when the function is written, the FORMAL PARAMETER, x, has no value. For a function like this in math, it was usually understood that x was a real number, but that does not have to be the case. Complex numbers work as well. For other functions you might want to limit yourself to integers.

The value of x is specified when we use the function. For example, if we say $f(3)$ then 3 is the ARGUMENT to the function and x takes on the value of 3 when we figure out the function's value. When we do this we find that $f(3) = 9$.

That was an example of a function of one variable. We can also have functions of two or more variables. For example, you might have $g(x, y) = x + y^2$. When this function is called, we provide two values as arguments and they are used as the value of x and y. So $g(3, 5) = 3 + 25 = 28$. The value of x is 3 and the value of y is 5. You know this because the order of the arguments matches the order of the formal parameters.

In a general sense, a function is a mapping that associates input values with output values. In your math classes you have likely focused on functions that map from numbers into numbers. We have seen that our programs can work with numbers, but there are other types that we can work with as well. Some simple examples of this include the following:

- A function that takes a String and returns how many vowels are in it.
- A function that takes a String and returns the first uppercase letter in it.
- A function that takes two points as (x, y) pairs and returns the distance between them.

All of these have parameters that are not simple numbers. In terms you used in algebra, we would say that their domain is not just the real numbers. They also give us back values from different types. In algebra this would be the range. The first example can only produce a non-negative integer. The second produces a character. The third is a real number. The generality of this idea is often shown with a diagram like figure 5.1.

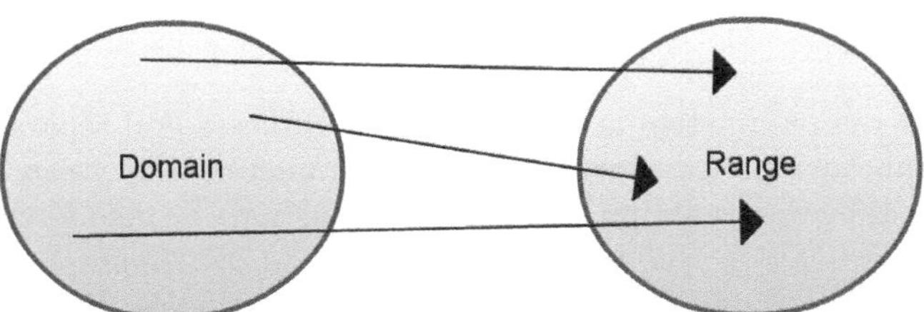

FIGURE 5.1: This figure shows a standard way of viewing what a function does. It maps from values in some domain to values in some range.

5.3 Making and Using Functions

There are a number of ways to make functions in Scala. We will begin with the standard definition format. In chapter 3 we saw that we can use the keywords `val` and `var` to declare variables. In a similar way, we use the keyword `def` to define a function. We will start with the simple mathematical examples that were used in the previous section and create Scala functions that match them in the REPL.

```
scala> def square(x:Double):Double = x*x
square: (x: Double)Double
```

In math, functions are given names like f and g. This works fine for a limited context, but it does not scale up well if you are using a lot of different functions. We use the name `square` here because it describes what the function does. The name is written after the `def` and after that we have a set of parentheses that contain a name and a type separated by a colon. This is like a `val` declaration except that it does not include an initial value. The value is specified when this function is called, just like for the math function. The contents of the parentheses are the formal parameters of the function.

After the parentheses is another colon and the type that this function gives us back, or returns. In Scala we refer to this as the result type. In this case, because we are mirroring the mathematical function, the type is `Double`. After the result type is an equal sign and the expression `x*x`. Once this has been entered, the REPL tells us that we have created something called `square` with the type `(x: Double)Double`. This is Scala's way of telling us that we have a function that takes a `Double` and results in a `Double`.

Now that we have defined square we can use it. In the last section we used our function f by doing $f(3)$. The syntax in Scala is the same.

```
scala> square(3)
res0: Double = 9.0
```

We give the name of the function followed by parentheses with the value we want to use for `x`. When we do this with the value 3, Scala gives us back a value of 9.0 as a `Double`. This particular value could also be represented as an `Int`, but the function says it works with type `Double` and so the `Int` that we passed in was converted to a `Double` and the result is also a `Double`.

Let us now look at the second example math function. In this case we will use the short name `g` and make it work with the `Int` type instead for demonstration purposes.

```
scala> def g(x:Int,y:Int):Int = x+y*y
g: (x: Int,y: Int)Int
```

Note that both of the parameters are put in the parentheses and separated by a comma. In general, you can have an arbitrary number of parameters. They can all have different types, and they will be separated by commas. The function is called in the way one would expect by giving the function name with values separated by commas.

```
scala> g(3,5)
res1: Int = 28
```

In this example we used simple literals for the arguments to the function, but the arguments can be any expression with the proper type. For example, we could do the following:

```
scala> val a=5
a: Int = 5

scala> g(a+2,square(3).toInt)
res2: Int = 88
```

Here we show that the arguments to `g` can include a reference to a `val` or a call to the `square` function. The call to `square` needs to be followed by a conversion to an `Int` because `g` is declared to take type `Int` and the result of `square` is `Double`. If you leave that out, you will get an error like this:

```
scala> g(a+2,square(3))
<console>:9: error: type mismatch;
 found  : Double
 required: Int
       g(a+2,square(3))
             ^
```

With the `val` and `var` declarations, we saw that we could typically leave off the types and Scala would figure them out for us. With functions, the types on the parameters are required. The result type is not required as it can often be inferred, but it is considered better style to specify it.[1] The reason this is preferred is that if you tell Scala a type and make a mistake in the function that causes it to result in a different type, Scala can tell you earlier that something has gone wrong.

These functions were limited in scope. We can write longer functions. To do this, we need to employ code blocks using curly braces. Going back to the English analogy, a block of code probably resembles a paragraph. This is a weak analogy because a block is not only composed of statements, it is a statement itself and can be an expression. When used as an expression, the value you get from it is the value of the last expression in the block. To see how this works, we could take the statements that we had in our script at the end of chapter 3 and put them in a function. The function might look something like this if we type it into the REPL.

```
scala> def secondsToTimeString(totalSeconds:Int):String = {
     | val displaySeconds=totalSeconds%60
     | val totalMinutes=totalSeconds/60
     | val displayMinutes=totalMinutes%60
     | val displayHours=totalMinutes/60
```

[1]The exception to this is recursive functions where the result type is required. More on that in chapter 6.

```
     | val sec=displaySeconds.toString
     | val min=displayMinutes.toString
     | displayHours+":"+("0"*(2-min.length))+min+":"+
     | ("0"*(2-sec.length))+sec
     | }
secondsToTimeString: (totalSeconds: Int)String
```

This function takes a single `Int` of the number of seconds. It results in a `String`. After the equal sign is an open curly brace that opens a block of code. This block is closed at the end of the function after an expression that gives us the String we want to return.

The advantage of this approach is we can easily call the function using different numbers of seconds, even in the REPL.

```
scala> secondsToTimeString(123456)
res0: String = 34:17:36

scala> secondsToTimeString(654321)
res1: String = 181:45:21
```

In this way, functions allow us to organize code and give useful names to things that we might want to do frequently.

This code also does something else that we have never seen before, it declares variables inside of a function. When we do this the variables are called LOCAL VARIABLES and they can only be accessed inside of the function. If, after typing in this function, you try to use `sec`, or any of the other local variables, outside the function, you will get an error message.

```
scala> sec
<console>:6: error: not found: value sec
       sec
       ^
```

The part of the code over which a name can be accessed is called the SCOPE of that name. When you declare a variable or a function in a script outside any block of code, it has a scope that goes through the rest of the script. If you declare it in a block of code, it has a scope that goes from the point of the declaration to the end of the block it is declared in. A block is denoted with curly braces, so you can use the name until you get to the right curly brace that closes off the pair the name is declared in. As a general rule, you want to minimize the scope of variables so they can only be accessed where they are needed. This reduces the amount of code you have to look through if something goes wrong with that variable.

These examples utilize the mathematical concept of a function as something that takes input values and maps them to outputs values. In programming it is possible for functions to do things other than give back a value. We have already seen an example of such a function. Our first program called the function `println`. You might notice that if you use `println` in the REPL, a value is printed, but there is not a result value shown. This is because `println` gives us back a value of type `Unit`, which was discussed in chapter 3. When you write a function that results in `Unit`, you can use the normal syntax and put `:Unit` after the parameter list. An alternate, shorter syntax is to leave out the equals sign, and just have a block of code. The following is a simple example of this.

```
scala> def printMultiple(s:String,howMany:Int) {
     | println(s*howMany)
     | }
printMultiple: (s: String,howMany: Int)Unit
```

```
scala> printMultiple("hi ",3)
hi hi hi
```

We'll see more useful examples later.

Before moving on, we will do another example of a function that is more interesting, like calculating an average in a course. The course average is calculated by combining the average of a number of different grades. For example, you might have tests, assignments, and quizzes. Each of these might contribute a different fraction of the total grade. The tests might be worth 40%, assignments worth 40%, and quizzes worth 20%. You will also have a different number of each of these grades. For example, you might have 2 tests, 3 assignments, and 4 quizzes. In many courses, part of the course average is computed by dropping a lowest grade and taking the average of what remains. In this case, let us assume that the lowest quiz grade is dropped.

What we want to do is write a function that takes all the grades, and returns the average for the course. To start with, it might look something like this:

```
def courseAverage(test1:Double,test2:Double,assn1:Double,
    assn2:Double,assn3:Double,quiz1:Double,quiz2:Double,
    quiz3:Double,quiz4:Double):Double = {
  ...
}
```

This function takes in nine `Doubles` and gives back an average for the full class as a `Double`. We want to calculate the averages for each of the different parts of the grade separately, then combine them with the proper percentages. Filling in a bit more gives the following.

```
def courseAverage(test1:Double,test2:Double,assn1:Double,
    assn2:Double,assn3:Double,quiz1:Double,quiz2:Double,
    quiz3:Double,quiz4:Double):Double = {
  val testAve=(test1+test2)/2
  val assnAve=(assn1+assn2+assn3)/3
  val minQuiz=...
  val quizAve=(quiz1+quiz2+quiz3+quiz4-minQuiz)/3
  testAve*0.4+assnAve*0.4+quizAve*0.2
}
```

All that is left to do is to figure out the minimum quiz grade. To accomplish this, we will use a method called `min` that is defined on the `Int` type. If you have two `Int` values, `a` and `b`, then the expression `a min b` will give you the smaller of the two values. This is a call to the `min` method on `Int` and could be written as `a.min(b)`. However, the operator syntax where the dot and parentheses are left off is superior when we have multiple values because we can put the values in a row to get the smallest of all of them. This gives us a complete version of the function which looks like the following.

```
def courseAverage(test1:Double,test2:Double,assn1:Double,
    assn2:Double,assn3:Double,quiz1:Double,quiz2:Double,
    quiz3:Double,quiz4:Double):Double = {
  val testAve=(test1+test2)/2
  val assnAve=(assn1+assn2+assn3)/3
  val minQuiz=quiz1 min quiz2 min quiz3 min quiz4
  val quizAve=(quiz1+quiz2+quiz3+quiz4-minQuiz)/3
  testAve*0.4+assnAve*0.4+quizAve*0.2
}
```

If you put this into the REPL, you can call it and get output as shown here.

```
scala> courseAverage(90,80,95,76,84,50,70,36,89)
res4: Double = 81.93333333333334
```

5.4 Problem Decomposition

As we have said, programming is about problem solving. One of the major aspects of problem solving is breaking hard problems into smaller pieces. There are a number of reasons to do this. The most obvious reason is that big problems are hard to solve. If you can break a big problem into smaller pieces, it is often easier to solve the pieces and then build an answer to the full problem from the answers to the pieces. This can be repeated on each subproblem until you get to a level where the problems are simple to solve.

A second advantage to breaking problems up is that the solutions to the pieces might be useful in themselves. This allows you to potentially reuse pieces of code. The ability to reuse existing code is critical in programming. There are many different ways that you can break up a problem. A good way to tell if one is better than another is if one gives you functions that you are more likely to be able to reuse.

A third advantage to decomposing a problem is that the resulting code can be easier to understand and modify. If you give the functions good names, then people can easily read and understand top-level functions and have a better idea of what is going on in the low-level functions. In addition, if you decide at some point that you need to modify how something is done, it is easier to determine where the change needs to be made and how to make it if the problem has been broken down in a logical way.

We looked at two different problems and saw how they could be programmed as a set of instructions that we can put inside of a function. The question now is how could we break these things up? If you go back to the original discussions, the descriptions of how to solve the problems was done piece by piece in a way that worked well for breaking them apart.

Let us begin with the grading program. There are four basic parts to this problem. We find the average of the tests, the assignments, and the quizzes. Once we have those, we combine them using the proper percentages. Each could be broken out into a separate function. If we write this into a file it might look like the following.

```
def testAve(test1:Double,test2:Double):Double=(test1+test2)/2

def assnAve(assn1:Double,assn2:Double,assn3:Double):Double =
  (assn1+assn2+assn3)/3

def quizAve(quiz1:Double,quiz2:Double,quiz3:Double,quiz4:Double):Double = {
  val minQuiz=quiz1 min quiz2 min quiz3 min quiz4
  (quiz1+quiz2+quiz3+quiz4-minQuiz)/3
}

def fullAve(test:Double,assn:Double,quiz:Double):Double =
  test*0.4+assn*0.4+quiz*0.2

def courseAverage(test1:Double,test2:Double,assn1:Double,
    assn2:Double,assn3:Double,quiz1:Double,quiz2:Double,
    quiz3:Double,quiz4:Double):Double = {
```

```
  val test=testAve(test1,test2)
  val assn=assnAve(assn1,assn2,assn3)
  val quiz=quizAve(quiz1,quiz2,quiz3,quiz4)
  fullAve(test,assn,quiz)
}
```

Once you have this in a file you can either use this as part of a script or use the `:load` command in the REPL to load it in and call the functions directly. Note that this version requires a lot more typing than what we had before. That will not always be the case, but it is for this example and many small examples where we can not reuse functions. Even though this version requires more typing, it has the advantages of being easier to understand and alter because the functionality is broken out into pieces.

This example is one where we could potentially have some reuse if we just knew a bit more Scala. Both `testAve` and `assnAve` do nothing more than take the average of numbers. Because one averages two numbers while the other averages three, we do not yet know how to write a single function to handle both cases. We will fix that in chapter 7.

Our other example was for converting a number of seconds into a properly formatted time. When we originally discussed this, we broke it into two problems. First, we had to figure out how many seconds, minutes, and hours a given number of seconds was equal to. After that we had to format the string properly. We will maintain that separation of work with functions. If we write this in a file, it might look like the following.

```
def calcHMS(totalSeconds:Int):(Int,Int,Int) = {
  val displaySeconds=totalSeconds%60
  val totalMinutes=totalSeconds/60
  val displayMinutes=totalMinutes%60
  val displayHours=totalMinutes/60
  (displayHours,displayMinutes,displaySeconds)
}

def formatHMS(numHours:Int,numMinutes:Int,numSeconds:Int):String = {
  val sec=numSeconds.toString
  val min=numMinutes.toString
  numHours+":"+("0"*(2-min.length))+min+":"+ ("0"*(2-sec.length))+sec
}

def secondsToTimeString(totalSeconds:Int):String = {
  val (h,m,s)=calcHMS(totalSeconds)
  formatHMS(h,m,s)
}
```

This code does something that we have not seen a function do yet. The `calcHMS` function returns a tuple. This function needs to give us back three values for the hours, minutes, and seconds. This type of occurrence is common in programming. How do you get functions to return multiple values? Different languages have different solutions to this problem. In Scala, the most direct solution is to have the function return a tuple as we see here.

This approach of taking a bigger problem and breaking it down into pieces is often called a top-down approach. The mental image is that you have the full problem at the top and in each level below it you have smaller pieces that are combined to make a solution to the level above them. This structure stops at the bottom when you get to something that can be solved relatively easily. In contrast to top-down, it is also possible to approach problems from the bottom-up. This approach works when you have a certain familiarity with solving a particular type of problem and you know what types of pieces you will need

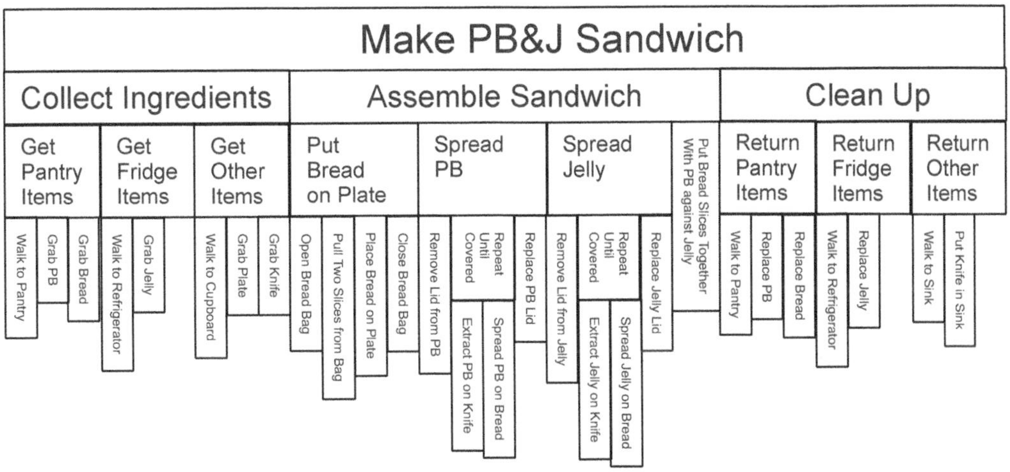

FIGURE 5.2: This figure shows a graphical breakdown of making a peanut butter and jelly sandwich.

or when your top-level problem is not completely well defined and you have to try some different approaches. This can happen when you are building software for a customer, as the customer may have only a vague idea of what they want their program to look like. In that situation, you can build pieces you know will be useful and try putting them together in different ways until you find something that the customer is happy with.

To help you understand this, consider the common example of making a peanut butter and jelly sandwich shown in figure 5.2. Often this example is used to help illustrate the explicitness required for programming. Here the main focus is on how the problem can be broken down and solved piece-by-piece. At the top is the full problem. That is broken into three subproblems, which are each broken down further. In order to be truly explicit, the steps would have to be broken down much further. Imagine the instructions were going to a robot instead of a human. To have it "Grab Jelly" you would have to instruct it on moving its arm to the proper location, grasping, etc.

This example also shows that the problem probably contains some reusable parts. There are several boxes at the bottom that start with "Grab" or "Walk to". Ideally we would find a way to write a function of walking that could be passed information on where to go.

Instead of viewing this in the graphical way shown in the figure, it can also be viewed as an outline like the one shown below. Nesting shows how one action is broken down into others.

- Make PB&J Sandwich

1. Collect Ingredients
    (a) Get Pantry Items
        i. Walk to Pantry
        ii. Grab Peanut Butter
        iii. Grab Bread
    (b) Get Fridge Items
        i. Walk to Fridge
        ii. Grab Jelly

   (c) Get Other Items
      i. Walk to Cupboard
      ii. Grab Plate
      iii. Grab Knife

2. Assemble Sandwich
   (a) Put Bread on Plate
      i. Open Bread Bag
      ii. Pull Two Slices from Bag
      iii. Place Bread on Plate
      iv. Close Bread Bag
   (b) Spread Peanut Butter
      i. Remove Lid from Peanut Butter
      ii. Repeat Until Bread is Covered
         A. Extract Peanut Butter on Knife
         B. Spread Peanut Butter on Bread
      iii. Replace Peanut Butter Lid
   (c) Spread Jelly
      i. Remove Lid from Jelly
      ii. Repeat Until Bread is Covered
         A. Extract Jelly on Knife
         B. Spread Jelly on Bread
      iii. Replace Jelly Lid
   (d) Put Bread Slices Together with Peanut Butter Against Jelly

3. Clean Up
   (a) Return Pantry Items
      i. Walk to Pantry
      ii. Replace Peanut Butter
      iii. Replace Bread
   (b) Return Fridge Items
      i. Walk to Refrigerator
      ii. Replace Jelly
   (c) Return Other Items
      i. Walk to Sink
      ii. Put Knife in Sink

This outline format and the indentation that comes with it looks more like the code that we will write in Scala.

The problems that we wrote in Scala above had solutions such that the decomposed solution required more text than the original, monolithic solution. We have expressed that this is not such a bad thing as the decomposed solutions have certain advantages. There is another factor that you should keep in mind when decomposing problems. The discussion of the top-down approach requires that you keep breaking the problem down until you get to something that is fairly simple to solve. This is not just practical for the problem solving, it turns out that there are good reasons to keep functions short. Long functions are much more likely to include errors than small functions are. For this reason, it is generally advised that programmers keep their functions relatively short. How short you make your functions can be a matter of style and personal preference, but there is a good rule you can follow that is backed up by research. The rule is that functions should be kept short enough that they fit completely on your screen at one time. When a function gets long enough that it does not all fit on the screen, the programmer begins having to rely upon his/her memory of what is in the parts that are off the screen. Human memory is not a great thing to rely upon, and as a result, the rate of errors goes up significantly when there are parts of the function that you can not see on the screen.

We can also employ the `if` construct that we learned in chapter 4 inside of functions. Consider the code we wrote to calculate the cost of an item purchased at the theme park. This can be converted to the following function.[2]

```
def itemCost(item:String, size:String):Double = {
  if(item=="Drink") {
    if(size=="S") 0.99
    else if(size=="M") 1.29
    else 1.39
  } else if(item=="Side") {
    if(size=="S") 1.29
    else if(size=="M") 1.49
    else 1.59
  } else if(item=="Main") {
    if(size=="S") 1.99
    else if(size=="M") 2.59
    else 2.99
  } else {
    if(size=="S") 4.09
    else if(size=="M") 4.99
    else 5.69
  }
}
```

This function does not include reading the input from the user. Instead, it is assumed that the code that calls it will read the values and pass them in. In general, passing values into functions and returning values back is more flexible than reading them and printing results to output in the function. A function that does reading and/or printing internally can not be reused if the values come from something other than standard input or if you want to format the output in a different way. Alternately, you might want to add up all the costs for a day. You can not add things that are printed. That requires having the values returned.

To understand, consider the simple example of wanting to know how much should be paid for two items.

```
println("What is the first item?")
val item1 = readLine()
println("What size?")
val size1 = readLine()
```

[2]The `Double` type really should not be used for money because it rounds. Remember that numbers in the computer are represented in binary. The value 0.1 in decimal is an infinite repeating binary number. That number gets truncated after a certain number of bits so the value 0.1 can not be represented perfectly in a `Double`. That is fine for most applications, including scientific applications. It is not desirable for an application that deals with real money.

```
println("What is the second item?")
val item2 = readLine()
println("What size?")
val size2 = readLine()
val totalCost = itemCost(item1,size1)+itemCost(item2,size2)
println("The total cost is "+totalCost)
```

The line where we calculate `totalCost` shows the real benefit of putting our code in a function that returns a value.

5.5 Function Literals

In chapter 3 we saw that the simplest form of expressions for `Int`, `Double`, `String`, and some other types was the literal form. For example, just the number 5 is a literal of type `Int`. Literals allow you to have an expression of a type that you can write in a short format without declaring a name for it. Scala, because it includes elements of functional programming, allows you to express functions as literals too.

The syntax for a function literal starts with a set of parentheses that have a comma separated list of arguments in them followed by an arrow made from an equals sign and a greater than sign with the body of the function after that. This type of arrow is often read as "rocket". So if we go back to our first two functions we might write them in this way.

```
scala> (x:Double)=>x*x
res9: (Double) => Double = <function1>

scala> (x:Int,y:Int)=>x+y*y
res10: (Int, Int) => Int = <function2>
```

In this type of usage, the type of the arguments is required. As we will see, there are many usages where it is not required. If the function were longer, one could use curly braces to hold multiple statements. In theory one can use function literals for functions of any length, in practice this is a style issue and you will not want to use function literals that are too long. The function literals will be embedded in other functions and you do not want those functions to exceed a screen size in length so the literals should not be more than a few lines at most.

Scala has an even shorter form for some special case function literals. This form uses underscores to represent parameters and skips the parameter list and rocket completely. It can only be used when each parameter is used only once and in the same order they are passed in. In this form, the function literal `(x:Int,y:Int)=>x+y` could be represented as `((_:Int)+(_:Int))`. When Scala can figure out the types, we can use the even shorter form of `(_+_)`. Note that this can not be used for x^2 written as `x*x` because we use the `x` twice. It has significant limitations, but there are many situations where it is useful and convenient to use. For that reason, we will make use of it in later chapters.

5.6 Non-Functional Functions/Procedures

The title of this section might seem odd. How could a function not be functional? In purely functional programming, a function takes inputs and calculates a value. The value it returns depends only on the arguments and the function does not do anything but give back that value. Often in programming, functions will do other things as well, or might return different values for the same inputs. These types of functions are not technically "functional". When they do other things we refer to that as SIDE EFFECTs. This terminology is like that for medications. You take a medication because it has a certain desired effect. However, it might have certain other effects as well called the side effects. While the side effects of a medication typically are not desired, there are times when side effects of functions are desired. The simplest example of this is printing. The print statement does not return anything. Instead it sends information to output. That is a side effect.

It is possible to have functions that return a value and have side effects. However, quite frequently you will find that functions with side effects do not return any data. To Scala that means their return type is `Unit`. Here is a simple example of such a function.

```
scala> def introduce(name:String):Unit = {
     | println("Hi, my name is "+name+".")
     | }
introduce: (name: String)Unit

scala> introduce("Mark")
Hi, my name is Mark.
```

This is common in non-functional programming. Indeed, some languages include a special construct, the PROCEDURE, for this. Scala does not have a separate type of function, but it does allow the shorter syntax we mentioned earlier. That makes our earlier function look like this instead:

```
scala> def introduce(name:String) {
     | println("Hi, my name is "+name+".")
     | }
introduce: (name: String)Unit
```

This shortcut syntax is handy and you will likely find that you use it quite often. Indeed, it looks more like the function syntax in many other languages. However, it does introduce a type of syntax error[3] that you should be aware of because it can lead to odd error messages and be hard to locate. The syntax error is to leave off an equals sign from a function declaration that should have one. This omission can lead to one of two problems. If you follow the recommendation and put a return type on functions that do return values, Scala will be confused by the lack of an equals sign and give you an error message. In some cases, the error message will say that you might be missing an equals sign. If you do not follow the style recommendation and allow Scala to figure out return types, then leaving out the equals sign will give your function a return type of `Unit` because no return type and no equals sign is the shorthand syntax for a `Unit` result. This problem will not be identified at the function declaration at all. It will only pop up when you try to call the function and do something with the returned value.

[3]A syntax error is an error caused by you typing something that is not valid in the language.

Higher-Order Functions

Function literals would be nothing more than a novelty in Scala or any other programming language if it were not for higher-order functions. A higher-order function is a function that operates on other functions. This means that either we pass other functions into it, or it results in a function. Higher-order functions are a functional programming concept and they can be challenging to get ones head around, but we will in the coming chapters how they can be useful.

As an example of a higher-order function here, we will use the idea of composition. If we have two functions, $f(x)$ and $g(x)$, then the composition of f with g is $f(g(x))$. We can write a Scala function that does this like follows.

```
scala> def compose(f:Double => Double,g:Double => Double):Double => Double =
     | (x)=>f(g(x))
compose: (f: (Double) => Double,g: (Double) => Double)(Double) => Double
```

Note that we write the function types themselves using arrows. We could call this with either functions we build using `def` or function literals. We will do it here with the following two functions that have been defined using `def`.

```
scala> def plus5(x:Double):Double = x+5
plus5: (x: Double)Double

scala> def square(x:Double):Double = x*x
square: (x: Double)Double
```

Now we want to use compose to build new functions from these two using the compose function and then see our new functions working.

```
scala> val h=compose(plus5,square)
h: (Double) => Double = <function1>

scala> val j=compose(square,plus5)
j: (Double) => Double = <function1>

scala> h(3)
res0: Double = 14.0

scala> j(3)
res1: Double = 64.0
```

The function $h(x) = x^2 + 5$ while $j(x) = (x + 5)^2$. You can see that when we call these with an argument of 3, we get values of 14 and 64 respectively, just as would be expected.

One limitation of our compose function, as written here, is that it only works with normal math-style functions that take `Doubles` and return `Doubles`. We will learn in chapter 19 how we can make this more general so these functions can return other types. The only real restriction on the types for composite should be that the output type of `g` matches the input type of `f`.

5.7 type Declarations

Scala provides a mechanism that allows you to provide alternate names for types. This can provide significant power when combined with concepts like abstraction that we will cover later in the book. However, for now, it can be used to help give you short meaningful names for tuple types that you use commonly.

The syntax of a `type` declaration begins with the keyword `type`. This is followed by the name you want to give to use. As with other names, you should pick something that is meaningful to you so that when you are reading through code, having this name helps it to make more sense. Unlike `var`, `val`, and `def`, it is the general style for type names to being with capital letters. You have probably noticed that the names of all the types we have encountered so far, like `Int` and `Double`, start with capital letters. This is a style choice, not something that Scala enforces, but it is highly recommended that you follow it. After the name is a equals sign and then the type you are giving a new name to.[4]

As an example of this, you might put something like the following at the top of a script if you were going to write a lot of functions dealing with vectors in 3-D.

```
type Vect = (Double, Double, Double)
```

In the rest of your code, you could refer to the type `Vect` instead of having to type out `(Double, Double, Double)`. This does not change how you interact with the tuple. You would still use the methods _1, _2, and _3 or a `val` with a tuple pattern to get the values out of the `Vect`.

There can even be a value to using a `type` declaration to give a different name to a standard type like `Int` like this.

```
type Counter = Int
```

If you carefully use the `Counter` type through your program in places that call for it, you can easily change the size of the integer type that is being used. If you find that you need your counters to be able to go over `Int.MaxValue`, you can switch to using a `Long`. On the other hand, if your counters are always small and you have a need to save memory, you could consider going down to a `Short`. This type of usage is not common is Scala, but it is used significantly in C when writing libraries so that it is easy to modify types to fit the platform you are compiling for.

5.8 Putting it Together

In the last chapter we also wrote conditional code for calculating the admission cost to the theme park. This makes a good example of a function as well. Here is that converted to a function.

```
def entryCost(age:Int, cooler:Boolean, waterPark:Boolean): Double = {
  (if(age<13 || age>65) 20 else 35) +
  (if(cooler) 5 else 0) +
  (if(waterPark) 10 else 0)
```

[4]The old name continues to work. You just get to use this one as well to refer to the same thing.

```
}
```

The information needed for the calculation is passed in and this function only does a calculation with the data, giving back a number.

Now we can put these pieces of code together and write a script that will tell you the cost of a group of people with up to four members including the purchase of meals. We will assume that the whole group either brings a cooler or does not, that all would either go to the water park or not, and that they all order some size of combo meal. The following code could be put in a script with the `entryCost` and `itemCost` functions.

```
def individualCost(cooler:Boolean, waterPark:Boolean): Double = {
  println("What is the person's age?")
  val age = readInt()
  println("What size combo are they ordering?")
  val size = readLine()
  entryCost(age,cooler,waterPark)+itemCost("Combo",size)
}

println("How many people are in your group? (1-4)")
val numPeople = readInt()
if(numPeople<1 || numPeople>4) {
  println("This script can not handle "+numPeople+" people.")
} else {
  println("Is the group bringing a cooler? (Y/N)")
  val cooler = readLine()=="Y"
  println("Will they be going to the water park? (Y/N)")
  val waterPark = readLine()=="Y"
  val totalCost = individualCost(cooler,waterPark) +
    (if(numPeople>1) individualCost(cooler,waterPark) else 0) +
    (if(numPeople>2) individualCost(cooler,waterPark) else 0) +
    (if(numPeople>3) individualCost(cooler,waterPark) else 0)
  println("The group cost is $"+totalCost+".")
}
```

Note that this code is not in a function. When you write a script, function declarations do exactly that, they declare a function. You need to have code outside of the functions that makes something happen, typically by calling one or more of the declared functions.

This script does have one significant limitation. It only works for 1-4 people. We do not yet have the ability to make a piece of code happen an arbitrary number of times. That is something we will learn how to do in the next chapter.

Ways to Run a Scala Program (Aside)

One thing that students occasionally struggle with is figuring out how they should run their Scala programs. There are three different ways that you can run Scala. Two of these you have seen: the REPL and scripts. A third, compiling to applications, will be dealt with in the second half of the book. To use the REPL you enter `scala` on the command-line with no file to execute. To run a script you enter `scala` on the command-line and give it a file name that ends with `.scala`. At first it might seem that these two approaches are completely distinct. However, the REPL has a `:load` command that you can use to load in code from a Scala file. Using this, it is possible to load a script file into the REPL. There are times when this can be handy, but you need to understand the difference between these two approaches.

When you run a file as a script, the only thing that the user sees is the values that are printed out. If your file doesn't contain a print statement, nothing will be shown on screen and it will appear to the user that it didn't work. If all the file has in it are declarations, statements beginning with `val`, `var`, or `def`,[5] the script doesn't really do anything. You have to have statements that use the declarations for them to matter in the script.

This isn't the case when you load a file into the REPL. In that usage, feedback is shown for every declaration in the form of a line that is printed with the type and value of the declaration. If you declared functions in the file, you can call them in the REPL after the file has been loaded. For this reason, files you load into the REPL don't always need statements at the end that call functions and they often don't include print statements to display output.

Basically, the REPL gives you the ability to "play around" with the code. Using a script simply runs it and does what it says. You should ask your instructor about this, but odds are good that any code you submit for grading should be in the form of a script that includes calls that demonstrate the functionality of the script and prompt users for input. In this case, you can use the `:load` option in the REPL when you are working on the program, but before you turn it in, make sure it runs properly and does what you want when used as a script.

5.9 End of Chapter Material

5.9.1 Problem-Solving Approach

The material learned in this chapter adds significant flexibility to our approach to problem solving. It also adds a number of choices you can consider when you think about what you might write for any particular line of code. You can now write functions. Technically, function declarations can go anywhere, including inside of other functions. It is best that they have some logical organization to them. In the text we will tend to group top-level functions together at the top of scripts.

We also learned in this chapter about functions that are called only for their side effects. Technically this was part of our list previously in the form of calls to `print` and `println`. We now know that these are special cases of a general class of options. Given what we have learned, here is a revised list of what any productive line of code might be doing.

1. Call a function just for the side effects. Previously `print` or `println` were the only examples we had of this, but we can now be more general.

2. Declare something:

    - A variable with `val` or `var`.
    - A function with `def`. Inside of the function will be statements that can pull from any of these rules.

3. Assign a value to a variable.

[5]There are other types of declarations that we will learn about later, but these three cover what we know at this point.

4. Write an `if` statement.

5.9.2 Summary of Concepts

- Functions are used to break problems up into smaller pieces.
  - Help solve problems.
  - Informative names make code more understandable.
  - Smaller functions are easier to work with.
- The functions in programming are similar to those from math. Information is passed in through formal parameters. The value of the parameters is determined at the time the function is called when arguments are passed in.
- Functions are declared in Scala using `def`. This is followed by the function name, an argument list, a result type, and then an equals sign with the expression for the function. Result type can be left off and inferred, but it is recommended you include it anyway.
- Functions can also be written as literals in Scala.
  - The rocket notation has argument list and body separated by `=>`.
  - Shorter notation uses underscores for arguments. Only works for certain functions.
- Functions that are called only for their side effects and do not need to return a value return `Unit`. This is done enough that there is a simplified syntax where the colon, result type, and equals sign are left off and curly braces go after the argument list.
- If you have functions that often take or return tuples, it can be useful to use a `type` declaration to give shorter, meaningful names to the tuples.

5.9.3 Self-Directed Study

Enter the following statements into the REPL and see what they do. Some will produce errors. You should figure out why. Try some variations to make sure you understand what is going on.

```
scala> def succ(n:Int):Int = n+1
scala> succ(1)
scala> succ(succ(1))
scala> var cnt = 0
scala> def inc { cnt=cnt+1 }
scala> cnt
scala> inc
scala> cnt
scala> inc
scala> inc
scala> cnt
scala> val f = (x:Double,y:Double)=>math.sqrt(x*x+y*y)
scala> f(3,4)
scala> def doThreeTimes(g:(Double)=>Double,x:Double) = g(g(g(x)))
```

```
scala> doThreeTimes(y=>y+1,1)
scala> doThreeTimes(y=>y*2,1)
scala> doThreeTimes(a=>a-5,100)
scala> doThreeTimes(_+1,1)
scala> doThreeTimes(_*2,1)
scala> def incChar(c:Char,offset:Int) = (c+offset).toChar
scala> incChar('a',1)
scala> incChar('a',2)
scala> incChar('z',-3)
scala> def tnp1(n:Int):(Int,Int) = {
    val odd = n%2
    (n/2*(1-odd),(3*n-1)*odd)
}
scala> odd
scala> var name = ""
scala> def introduction {
    println("What is your name?")
    name = readLine()
    println("Hello "+name)
}
scala> name
```

5.9.4 Exercises

1. What would be the types of the parameters and result of functions that do the following tasks. (Note that you are not supposed to write the functions, just say what their types are.)
   - Tell you how many words are in a person's name.
   - Take a person's name as normally written and give it back in "last, first" format.
   - Take three words and return the TLA (Three Letter Acronym) for them.
   - Take ten points in 2-space (x-y coordinates) and return how many are in the unit circle.
   - Take a number and tell how many distinct prime factors it has.
   - Take a number and tell has many positive factors it has.
   - Take a value in Fahrenheit and give back the equivalent in Celsius.
   - Take a value in Fahrenheit and give back the equivalent in Celsius and Kelvin.
2. Write functions that return a `Boolean` for the different parts of exercise 1 (p.85).
3. Write a function that takes a `String` parameter and returns a `Boolean` telling if the argument is a palindrome.
4. Write a function for the solution to exercise 5 (p.86).
5. Write a function for the solution to exercise 6 (p.86).
6. Write a function for the solution to exercise 7 (p.86).
7. Write a function for the solution to exercise 8 (p.86).

8. Convert the code you wrote to for exercise 9 (p.86) into a function that takes the x and y values are arguments and returns a `Boolean` for whether a point is in the unit circle.

9. Write a function to convert a temperature from Fahrenheit to Celsius.

10. Write a function to convert from miles to kilometers.

11. Write a function to convert from seconds to years.

12. Write a function to convert from AU (Astronomical Units) to miles.

13. Write two functions that takes four numbers and returns the smallest. The first can use `min`, but the second can not.

14. Write a function that takes three numbers and returns the median.

5.9.5 Projects

1. Write a set of functions to do the following operations on a 2-tuples of `Int` as if they were the numerator and denominator of rational numbers.
   (a) Addition
   (b) Subtraction
   (c) Multiplication
   (d) Division

2. Write a set of functions to do the following operations on 3-tuples of `Doubles` as if they were vectors.
   (a) Addition
   (b) Subtraction
   (c) Dot product
   (d) Cross product

3. Write a set of functions to do the following operations on a 2-tuples of `Doubles` as if they were complex numbers.
   (a) Addition
   (b) Subtraction
   (c) Multiplication
   (d) Division
   (e) Magnitude

4. Use your solution to exercise 3 to make a function that takes two complex numbers, z and c and returns the value $z^2 + c$.

5. Write a function to solve the quadratic equation for real values. Your solution should return a `(Double, Double)` of the two roots.

6. Enhance the solution to exercise 5 so that it returns a `((Double, Double),(Double, Double))`. This is a 2-tuple of 2-tuples. This represents the two roots and each root can be a complex number expressed as a `(Double, Double)`.

7. The hyperbolic trigonometric function, sinh, cosh, and tanh, are all defined in terms of power of e, the base of the natural logarithms. Here are basic definitions of each:

$$\begin{aligned} \sinh(x) &= \tfrac{1}{2}(e^x - e^{-x}) \\ \cosh(x) &= \tfrac{1}{2}(e^x + e^{-x}) \\ \tanh(x) &= \tfrac{\sinh(x)}{\cosh(x)} \end{aligned}$$

   Write functions for each of these. You can use math.exp(x) to represent e^x.

8. Write two functions for converting between Cartesian and polar coordinate systems. The first takes `x:Double` and `y:Double` and returns a `(Double, Double)` with r and θ. The second takes `r:Double` and `theta:Double` and returns `(Double, Double)` with x and y. You can use the `math.tan2(y:Double, x:Double)` function to get an angle. This avoids the problem of using division when x is zero.

9. This option has you doing a scientific calculation. We can not do that much yet, but we will work our way up. We are going to play with calculating the non-greenhouse temperatures for planets with moderate to fast spin rates. This might seem like a complex thing to do, but it is not difficult. You need two pieces of information and a little algebra. The first piece of information is the Stefan-Boltzmann Law ($j^* = \sigma T^4$, $\sigma = 5.670400 * 10^{-8} \left[\frac{J}{s * m^2 * K^4}\right]$) for the amount of energy given off in thermal radiation by any body. The second is the fact that intensity of radiation drops off as $1/r^2$.

   To calculate the non-greenhouse temperature of a planet you need the following pieces of information. You should prompt the user for values to these. To keep things simple use mks units and keep temperatures in Kelvin.

   - Radius of the star.
   - Surface temperature of the star.
   - Orbital semimajor axis of the planet.
   - Albedo of the planet.

   Use the Stefan-Boltzmann law to determine the energy output per square meter on the stars surface. Make that into a function that takes the needed values. Using the inverse square relationship you can calculate the intensity at the location of the planet (use the ratio of the planet's orbit distance to the stellar radius for this). Make this another function. The star light will cover the planet with an area of πr^2 where r is the planets radius. A fraction of that, determined by the albedo, is reflected. What is not reflected warms the planet. The planet cools through its own thermal radiation from a surface of $4\pi r^2$. Setting the absorbed and emitted values equal allows you to solve the temperature. (Note that the planetary radius should cancel out.) Make a function that takes the incident power and albedo and gives you back a temperature.

10. If you wrote code for project 4 you can convert that to use functions. Write one function for a sphere and another for a plane. Each should take information for a ray as well as for the geometry that is being intersected. They should return the value of t at the point of intersection. An appropriate negative value can be used if they do not intersect.

11. In project 5, you wrote conditional statements to determine if you had enough ingredients for some simple recipes. The numbers were kept small there in large part because of the length of code and problems with duplication. Redo that problem for the situation where you have five items in your pantry and the recipe involves four ingredients. Use functions to make this manageable.

12. Convert your code from project 6 to use functions.

13. Neglecting air resistance, objects that are thrown or fired into the air travel on a parabolic path of the form $x(t) = v_x t$, $y(t) = -\frac{1}{2}gt^2 + v_y t + h$, where v_x and v_y are the components of the velocity, g is the acceleration due to gravity, and h is the initial height. Write a function that is passed the speed, the angle relative to the ground, and the initial height of a projectile and results in the distance the projectile will go before it hits the ground with $y(t) = 0$.

Chapter 6

Recursion for Iteration

Gaining conditionals provided us with a lot of power. We are now able to express logic in our code. Adding functions gave us the ability to break problems into pieces and repeat functionality without retyping code. There is still something very significant that we are missing. Now when we write a piece of logic, it happens once. We can put that logic into a function and then call the function over and over, but it will only happen as many times as we call it. We can not easily vary the number of times that something happens. There is more than one way to make something happen multiple times in Scala. One of these ways, RECURSION, we can do with just functions and conditionals, constructs that we have already learned.

6.1 Basics of Recursion

Recursion is a concept that comes from mathematics. A mathematical function is recursive if it is defined in terms of itself. To see how this works, we'll begin with factorial. You might recall from math classes that $n!$ is the product of all the integers from 1 up to n. We might write this as $n! = 1 * 2 * ... * n$. More formally we could write it like this.

$$n! = \prod_{i=1}^{n} i$$

Both of the formal and informal approaches define factorial in terms of just multiplication, and assume that we know how to make that multiplication happen repeatedly. We can be more explicit about the repetition if we write the definition using recursion like this.

$$n! = \begin{cases} 1 & n < 2 \\ n * (n-1)! & otherwise \end{cases}$$

In this definition, the factorial function is defined in terms of itself. To describe what factorial is, we use factorial.

To see how this works, let us run through an example and take the factorial of 5. By our definition we get that $5! = 5 * 4!$. This is because 5 is not less than 2. Subsequently, we can see that $4! = 4 * 3!$ so $5! = 5 * 4 * 3!$. This leads to $5! = 5 * 4 * 3 * 2!$ and finally to $5! = 5 * 4 * 3 * 2 * 1$.

This definition and its application illustrate two of the key aspects of recursion. There are two possibilities for the value of this function. Which one we use depends on the value of n. In the case where n is less than 2, the value of the factorial is 1. This is called a BASE CASE. All recursive functions need some kind of base case. The critical thing about a base case is that it is not recursive. When you get to a base case, it should have a value that can be calculated directly without reference back to the function. Without this you get what is called INFINITE RECURSION. There can be multiple different base cases as well. There is no restriction that there be only one, but there must be at least one.

To see why the base case is required, consider what would happen without it. We would still get $5! = 5*4*3*2!$, but what would happen after that? Without a base case, $2! = 2*1!$. Technically there is not a problem there, but what about 1!? $1! = 1 * 0!$ in the absence of a base case and $0! = 0 * (-1)!$. This process continues on forever. That is why it is called infinite recursion.

The second case of the recursive definition demonstrates the recursive case. Not only does this case refer back to the function itself, it does so with a different value and that value should be moving us toward a base case. In this case, we define $n!$ In terms of $(n-1)!$. If the recursive case were to use the same value of n we would have infinite recursion again. Similarly, if it used a value greater than n we would also have infinite recursion because our base case is for small numbers, not large numbers.

What else could we define recursively? We could define multiplication, which is used by factorial, recursively. After all, at least for the positive integers, multiplication is nothing more than repeated addition? As with the factorial, we could write a definition of multiplication between positive integers that uses a math symbol that assumes some type of repetition like this.

$$m * n = \sum_{i=1}^{n} m$$

This says that $m*n$ is m added to itself n times. This can be written as a recursive function in the following way.

$$m * n = \begin{cases} 0 & n = 0 \\ m + m * (n-1) & otherwise \end{cases}$$

This function has two cases again, with a base case for a small value and a second, recursive case, that is defined in terms of the value we are recursing using a smaller value of that argument.

We could do the same type of things to define exponentiation in terms of multiplication. We could also use an increment (adding 1) to define addition by higher numbers. It is worth taking a look at what that would look like.

$$m + n = \begin{cases} m & n = 0 \\ 1 + m + (n-1) & otherwise \end{cases}$$

While this seems a bit absurd, it would be less so if we named our functions. Consider the following alternate way of writing this.

$$add(m,n) = \begin{cases} m & n = 0 \\ 1 + add(m, n-1) & otherwise \end{cases}$$

Now it is clear that as long as we can do increment (+1) and decrement (-1) we could write full addition. With full addition we could write multiplication. With multiplication we can write exponentiation or factorial. It turns out that this is not all you can do. It might be hard to believe, but if you have variables, recursion (which simply requires functions and an `if` construct), increment, and decrement, you have a full model of computation. It can calculate anything that you want. Of course, we do not do it that way because it would be extremely slow. Still, from a theoretical standpoint it is very interesting to know that so much can be done with so little.

6.2 Writing Recursion

We have seen the mathematical side of recursion and have written some basic mathematical functions as recursive functions. Now we need to see how we write these things and more in Scala. The translation from math functions to programming functions is not hard. In fact, little will change from the math notation to the Scala notation.

As before, we will begin with the factorial function. Here is the factorial function written in Scala in the REPL.

```
scala> def fact(n:Int):Int = if(n<2) 1 else n*fact(n-1)
fact: (n: Int)Int
```

We have called the function `fact`, short for factorial. The body of the function is a single `if` expression. First it checks the value of `n` to see if it is less than 2. If it is, the expression has a value of 1. Otherwise it is `n*fact(n-1)`. We can see the results of using this function here:

```
scala> fact(5)
res1: Int = 120
```

We see here that it correctly calculates the factorial of 5.

One significant difference between recursive and non-recursive functions in Scala is that we have to specify the return type of recursive functions. If you do not, Scala will quickly let you know it is needed.

```
scala> def fact(n:Int) = if(n<2) 1 else n*fact(n-1)
<console>:6: error: recursive method fact needs result type
       def fact(n:Int) = if(n<2) 1 else n*fact(n-1)
                                          ^
```

Factorial is an interesting function that is significant in Computer Science when we talk about how much work certain programs have to do. Some programs have to do an amount of work that scales with the factorial of the number of things they are working on. We can use our factorial function to see what that would mean. Let us take the factorial of a few different values.

```
scala> fact(10)
res2: Int = 3628800

scala> fact(15)
res3: Int = 2004310016

scala> fact(20)
res4: Int = -2102132736
```

The first two show you that the factorial function grows very quickly. Indeed, programs that do factorial work are referred to as intractable because you can not use them for even modest-size problems. The third example though shows something else interesting. The value of 20! should be quite large. It certainly should not be negative. What is going on here?

If you remember back to chapter 3, we talked about the way that numbers are represented on computers. Integers on computers are represented by a finite number of bits. As a result, they can only get so large. The built in number representations in Scala use 32 bits for `Int`. If we changed our function just a bit to use `Long` we could get 64 bits. Let's see that in action.

```
scala> def fact(n:Long):Long = if(n<2) 1L else n*fact(n-1)
fact: (n: Long)Long

scala> fact(20)
res5: Long = 2432902008176640000

scala> fact(30)
res6: Long = -8764578968847253504
```

The 64 bits in a `Long` are enough to store 20!, but they still fall short of 30!. If we give up the speed of using the number types hard wired into the computer, we can represent much larger numbers. We can then be limited by the amount of memory in the computer. That is a value not measured in bits, but in billions of bytes. To do this, we use the type `BigInt`.

The `BigInt` type provides us with arbitrary precision arithmetic. It does this at the cost of speed and memory. You do not want to use `BigInt` unless you really need it. However, it can be fun to play with using a function like factorial which has the possibility of getting quite large. Let us redefine our function using this type and see how it works.

```
scala> def fact(n:BigInt):BigInt = if(n<2) 1L else n*fact(n-1)
fact: (n: BigInt)BigInt

scala> fact(30)
res7: BigInt = 265252859812191058636308480000000

scala> fact(150)
res8: BigInt = 5713383956445854590478932865261054003189553578601126418
25483758331798291248453983931265744886753111453771078787468542041626
66250198684504466355949195922066574942592095735778929325357290444962
47240541679072211844543712226967552000000000000000000000000000000000
00
```

Not only can this version take the factorial of 30, it can go to much larger values as you see here. There are not that many applications that need these types of numbers, but they can certainly be fun to play with.

Now that we have beaten factorial to death, it is probably time to move on to a different example. The last section was all about examples pulled from mathematics. The title of this chapter though is using recursion for iteration. This is a far broader programming concept than the mathematical functions we have talked about. So let us use an example that is very specific to programming.

A simple example to start with is to write a function that will "count" down from a certain number to zero. By count here we it will print out the values. Like the factorial we will pass in a single number, the number we want to count down from. We also have to have a base case, the point where we stop counting. Since we are counting down to zero, if the value is ever below zero then we are done and should not print anything.

In the recursive case we will have a value, `n`, that is greater than or equal to zero. We definitely want to print `n`. The question is, what do we do after that? Well, if we are counting down, then following `n` we want to have the count down that begins with `n-1`. Indeed, this is how you should imagine the recursion working. Counting down from `n` is done by counting the `n`, then counting down from `n-1`. Converting this into code looks like the following:

```
def countDown(n:Int) {
  if(n>=0) {
    println(n)
    countDown(n-1)
  }
}
```

The way this code is written, the base case does nothing so we have an `if` statement that will cause the function to simply return if the value of `n` is less than 0. You can call this function passing in different values for `n` to verify that it works.

Now let us try something slightly different. What if I want to count from one value up to another? In some ways, this function looks very much like what we already wrote. There are some differences though. In the last function we were always counting down to zero so the only information we needed to know was what we were counting from. In this case though, we need to be told both what we are counting from and what we are counting to. That means that our function needs two parameters passed in. A first cut at this might look like the following.

```
def countFromTo(from:Int,to:Int) {
  println(from)
  if(from!=to) {
    countFromTo(from+1,to)
  }
}
```

This function will work fine under the assumption that we are counting up. However, if you call this with the intention of counting down so that `from` is bigger than `to`, you have a problem. To see why, let us trace through this function. First, let us see what happens if we call it with 2 and 5, so we are trying to count up from 2 to 5. What we are doing is referred to as tracing the code. It is the act of running through code to see what it does. This is an essential ability for any programmer. After all, how can you write code to complete a given task if you are not able to understand what the code you write will do? There are lots of different approaches to tracing. Many involve tables where you write down the values of different variables. For recursive functions you can often just write down each call and what it does, then show the calls it makes. That is what we will do here. We will leave out the method name and just put the values of the arguments as that is what changes.

```
(2,5) => prints 2
  ↓
(3,5) => prints 3
  ↓
(4,5) => prints 4
  ↓
(5,5) => prints 5
```

The last call does not call itself because the condition `from!=to` is `false`.

Now consider what happens if we called this function with the arguments reversed. It seems reasonable to ask the function to count from 5 to 2. It just has to count down. To see what it will do though we can trace it.

```
(5,2) => prints 5
  ↓
(6,2) => prints 6
  ↓
(7,2) => prints 7
  ↓
(8,2) => prints 8
  ↓
...
```

This function will count for a very long time. It is not technically infinite recursion because the `Int` type only has a finite number of values. Once it counts above $2^{31} - 1$ it wraps back around to -2^{31} and counts up from there to 2 where it will stop. You have to be patient to see this behavior though. Even if it is not infinite, this is not the behavior we want. We would rather the function count down from 5 to 2. The question is, how can we do this? To answer this we should go back to the trace and figure out why it was not doing that in the first place.

Looking at the code and the trace you should quickly see that the problem is due to the fact that the recursive call is passed a value of `from+1`. So the next call is always using a value one larger than the previous one. What we need is to use `+1` when we are counting up and `-1` when we are counting down. This behavior can be easily added by replacing the 1 with an `if` expression. Our modified function looks like this.

```
def countFromTo(from:Int,to:Int) {
  println(from)
  if(from!=to) {
    countFromTo(from+(if(from<to) 1 else -1),to)
  }
}
```

Now when the `from` value is less than the `to` value we add 1. Otherwise we will add -1. Since we do not get to that point if the two are equal, we do not have to worry about that situation. You should enter this function in and test it to make sure that it does what we want.

6.3 User Input

We are still limited in what we can realistically do because we are stuck using values that we can easily calculate. For example, we can go from one number to the one before it or the one after. We could use those values as inputs to more complex functions, but we still need them to be calculated from numbers we can nicely iterate through. These are shortcomings that will be fixed in the next chapter. For now, we can look at how to make the user (or more generally standard input) into a source of more interesting values.

We saw back in chapter 3 that we can call the function `readInt` to read an integer from standard input. Now we want to read multiple values and do something with them. We will start by taking the sum of a specific number of values. We can write a function called `sumInputInts`. We will pass this function an integer that represents how many integers we want the user to input and it will return the sum of those values. How can we define such a function recursively? Well, if we want to sum up 10 numbers, we could say that sum is the first number, plus the sum of 9 others. The base case here is that if the number of numbers we are supposed to sum gets below 1, then the sum is zero. Let us see what this would look like in code.

```
def sumInputInts(num:Int):Int = {
  if(num>0) {
    readInt()+sumInputInts(num-1)
  } else {
    0
  }
}
```

The `if` is being used as an expression here. It is the only expression in the function so it is the last one, and it will be the return value of the function. If `num`, the argument to the function, is not greater than zero, then the function returns zero. If it is, the function will read in a new value and return the sum of that value and what we get from summing one fewer values.

What if we don't know in advance how many values we are going to sum? What if we want to keep going until the end is reached? We could do this. One problem is determining what represents the end. We need to have the user type in something distinctly different that tells us they have entered all the values they want to sum. An easy way to do this would be to only allow the user to sum positive values and stop as soon as a non-positive value is entered. This gives us a function that does not take any arguments. We do not have to tell it anything. It will return to us an integer for the sum of the numbers entered before the non-positive value. Such a function could be written as follows.

```
def sumInputPositive():Int = {
  val n=readInt()
  if(n>0) {
    n+sumInputPositive()
  } else {
    0
  }
}
```

This time we read a new value before we determine if we will continue or stop. The decision is based on that value, which we store in a variable, `n`. Empty parentheses have been added after the function name for both the declaration and the call. This is a style issue because they are not required. It is considered proper style to use parentheses if the function has side effects and to leave them off if it does not. You will recall from chapter 5 that side effects are the way in which a function changes things that go beyond just returning a value. What does this function do that causes us to say it has side effects? The side effects here are in the form of reading input. Reading input is a side effect because it can have an impact beyond that one function. Consider having two different functions that both read input. The order in which you call them is likely to change their behavior. That is because the second one will read input in the state that is left after the first one.

This function does a good job of letting us add together an arbitrary number of user inputs, but it has a significant limitation, it only works with positive values. That is because we reserve negative values as the stop condition. There could certainly be circumstances where this limitation was a problem. How could we get around it? What other methods could we use to tell the function to stop the recursion? We could pick some particular special value like -999 to be the end condition. While -999 might not seem like a particularly common number, this is really no better than what we had before because our function still can not operate on any valid integer value. We'd like to have it where the termination input is something like the word "quit". Something special that is not a number.

We can do this if we do not use `readInt`. We could instead use the `readLine` function, which will read a full line of input and returns a `String`. You might be tempted to create a method like this:

```
def sumUntilQuit():Int = {
  val n=readLine()
  if(n!="quit") {
    n+sumUntilQuit()
  } else {
    0
  }
}
```

If you enter this into a file and then load it into the console, though, you will get the following error.

```
<console>:8: error: type mismatch;
 found : java.lang.String
 required: Int
        n+sumUntilQuit()
        ^
```

This is because the function is supposed to return an `Int`, but `n` is a `String` and when we use `+` with a `String` what we get is a `String`. Why is `n` a `String`? Because that is the type returned by `readLine` and Scala's type inference decided on the line `val n=readLine()` that `n` must be a `String`.

This problem can be easily fixed. We know that if the user is giving us valid input, the only things which can be entered are integer values until the word "quit" is typed in.[1] So we should be able to convert the `String` to an `Int`. That can be done as shown here.

```
def sumUntilQuit():Int = {
  val n=readLine()
  if(n!="quit") {
```

[1]In the first half of this book, we generally assume that user input will be valid. Dealing with invalid input requires exception handling, which is discussed in chapter 22.

```
    n.toInt+sumUntilQuit()
  } else {
    0
  }
}
```

Now we have a version of the function which will read one integer at a time until it gets the word "quit". If you do some testing you might figure out there was another slight change to the behavior. The earlier functions worked whether the numbers were placed on one line with spaces between them or each on separate lines. This version only works in the second case. We will be able to fix that problem in the next chapter.

Summing up a bunch of numbers can be helpful, but it is a bit basic. Let us try to do something more complex. A tiny step up in the complexity would be to take an average. The average is nothing more than the sum divided by the number of elements. In the first version of the function when we enter how many number would be read, this would be trivial to write. We do not even need to write it. The user knows how many numbers there were, just divide by that. Things are not so straightforward for the other versions though. It is not clear we know how many values were input and we do not want to force the user to count them. Since we need both a sum and a count of the number of values to calculate an average, we need a function that can give us both.

This is another example of a function that needs to return two values and as before, we will use a tuple to do the job. So we will write a new function called `sumAndCount`, which returns a tuple that has the sum of all the numbers entered as well as the count of how many there were. We will base this off the last version of `sumUntilQuit` so there are no restrictions on the numbers the user can input. Such a function might look like the following:

```
def sumAndCount():(Int,Int) = {
  val n=readLine()
  if(n!="quit") {
    val (s,c)=sumAndCount()
    (s+n.toInt,c+1)
  } else {
    (0,0)
  }
}
```

If you load this function into the REPL and call it, you can enter a set of numbers and see the return value. If, for example, you enter 3, 4, 5, and 6 on separate lines followed by "quit", you will get this:

```
res0: (Int, Int) = (18,4)
```

This looks a lot like what we had before, only every line related to the return of the function now has a tuple for the sum and the count. We see it on the first line for the return type. We also see it on the last line of both branches of the `if` expression for the actual return values. The last place we see it is in the recursive branch where the return value from the recursive call is stored in a tuple. This syntax of an assignment into a tuple is actually doing pattern matching which will be discussed later in this chapter.

Now we have both the sum and the count. It is a simple matter to use this in a different function that will calculate the average. The function shown below calls `sumAndCount` and uses the two values that are returned to get a final answer.

```
def averageInput():Double = {
```

```
  val (sum,count)=sumAndCount()
  sum.toDouble/count
}
```

The one thing that you might at first find odd about this function is that it has two places where `Double` appears. That is given as the return type and in the last expression the `toDouble` method is called on `sum`. This is done because averages are not generally whole numbers. We have to call `toDouble` on `sum` or `count` because otherwise Scala will perform integer division which truncates the value. We could convert both to `Doubles`, but that is not required because numerical operations between a `Double` and an `Int` automatically convert the `Int` to a `Double` and result in a `Double`.

6.4 Abstraction

What if, instead of taking the sum of a bunch of user inputs, we want to take a product? What would we change in `sumAndCount` to make it `productAndCount`? The obvious change is that we change addition to multiplication in the recursive branch of the `if`. A less obvious change is that we also need the base case to return 1 instead of 0. So our modified function might look like this.

```
def productAndCount():(Int,Int) = {
  val n=readLine()
  if(n!="quit") {
    val (s,c)=productAndCount()
    (s*n.toInt,c+1)
  } else {
    (1,0)
  }
}
```

This is almost exactly the same as what we had before. We just called it a different name and changed two characters in it. This copying of code where we make minor changes is something that is generally frowned upon. You might say that it does not smell right.[2] There are a number of reasons why you would want to avoid doing this type of thing. First, what happens if your first version had a bug? Well, you have now duplicated it and when you figure out something is wrong you have to fix it in multiple places. A second problem is closely related to this, that is the situation where you realize you want a bit more functionality so you need to add something. Again you now have multiple versions to add that into. In addition, it just makes the code base harder to work with. Longer code means more places things can be messed up and more code to go through when there is a problem. For this reason, we strive to reduce code duplication. One way we do this is to include abstraction. We look for ways to make the original code more flexible so it can do everything we want. We will see abstraction comes up a lot in this book. It is one of the most important tools in Computer Science and a remarkably powerful concept that you will want to understand. Here we are starting with a fairly simple example.

In order to abstract these functions to make them into one, we focus on the things that were different between them and ask if there is a way to pass that information in as

[2]Indeed, the term "smell" is the actual terminology used in the field of refactoring for things in code that are not quite right and should probably be fixed.

arguments to a version of the function that will do both. For this, the changing of the name is not important. What is important is that we changed the operation we were doing and the base value that was returned. The base value is easy to deal with. We simply pass in an argument to the method that is the value returned at the base. That might look like this.

```
def inputAndCount(base:Int):(Int,Int) = {
  val n=readLine()
  if(n!="quit") {
    val (s,c)=inputAndCount(base)
    (s*n.toInt,c+1)
  } else {
    (base,0)
  }
}
```

The argument base is passed down through the recursion and is also returned in the base case. However, this version is stuck with multiplication so we have not gained all that much.

Dealing with the multiplication is a bit harder. For that we need to think about what multiplication and addition really are and in particular how they are used here. Both multiplication and addition are operators. They take in two operands and give us back a value. When described that way, we can see they are like functions. Indeed, what we need is a function that takes two `Int`s and returns an `Int`. That function could be multiplication or addition and then the `inputAndCount` function would be flexible enough to handle either a sum or a product. It might look like this.

```
def inputAndCount(base:Int,func:(Int,Int)=>Int):(Int,Int) = {
  val n=readLine()
  if(n!="quit") {
    val (s,c)=inputAndCount(base,func)
    (func(s,n.toInt),c+1)
  } else {
    (base,0)
  }
}
```

The second argument to `inputAndCount`, which is called `func`, has a more complex type. It is a function type. It is a function that takes two `Int`s as arguments and returns an `Int`. As with `base`, we pass `func` through on the recursive call. We also used `func` in place of the `*` or the `+` in the first element of the return tuple in the recursive case. Now instead of doing `s+n.toInt` or `s*n.toInt`, we are doing `func(s,n.toInt)`. What that does depends on the function that is passed in.

To make sure we understand this process we need to see it in action. Let us start with doing a sum and use the longest, easiest-to-understand syntax. We define a function that does addition and pass that in. For the input we type in the numbers 3, 4, and 5 followed by "quit". Those values are not shown by the REPL.

```
scala> def add(x:Int,y:Int):Int = x+y
add: (x: Int,y: Int)Int

scala> inputAndCount(0,add)
res3: (Int, Int) = (12,3)
```

In the call to `inputAndCount` we used the function `add`, which was defined above it as the second argument. Using a function defined in this way forces us to do a lot of typing. This is exactly the reason Scala includes function literals. You will recall from chapter 5 that a function literal allows us to define a function on the fly in Scala. The normal syntax for this looks a lot like the function type in the definition of `inputAndCount`. It uses a `=>` between the parameters and the body of the function. Using a function literal we could call `inputAndCount` without defining the `add` function. That approach looks like this.

```
scala> inputAndCount(0,(x,y)=>x+y)
res4: (Int, Int) = (12,3)
```

One thing to notice about this is that we did not have to specify the types on `x` and `y`. That is because Scala knows that the second argument to `inputAndCount` is a function that takes to `Int` values. As such, it assumes that `x` and `y` must be of type `Int`. Not all uses of function literals will give Scala this information, but many will, and when they do we can do less typing by letting Scala figure out the types for us.

If you remember back to the section on function literals, you will recall there is an even shorter syntax for declaring them that only works in certain situations. That was the syntax that uses _ as a placeholder for the arguments. This syntax can only be used if each argument occurs only once and in order. In this case that happens to be true, so we are allowed to use the shorthand. That simplifies our call all the way down to this.

```
scala> inputAndCount(0,_+_)
res5: (Int, Int) = (12,3)
```

Of course, the reason for doing this was so that we could also do products without having to write a second function. The product function differed from the sum function in that the base case was 1 and it used `*` instead of `+`. If we make those two changes to what we did above, we will see that we have indeed created a single function that can do either sum or product.

```
scala> inputAndCount(1,_*_)
res6: (Int, Int) = (60,3)
```

Not only did we succeed here, we did so in a way that feels satisfying to us. Why? Because our abstraction can be used in a minimal way and only the essential variations have to be expressed. We do not have to do a lot of extra typing to use the abstraction. It is not much longer to call `inputAndCount` than it was to call `sumAndCount` or `productAndCount`. In addition, the only things we changed between the two calls were the changing of the 0 to 1 and the `+` to `*`. Those were the exact same things we had to change if we had done the full copy and paste of the functions. This means that the concept we want to express is coming through clearly and is not obfuscated with a lot of overhead.

The `inputAndCount` function is what we call a higher-order function. That is because it is a function that uses other functions to operate. We provide it with a function to help it do its job. This type of construct is seen mostly in functional languages. It is possible to create this type of effect in other languages, but the syntax is typically much longer.

You might say we only had to duplicate the code once to have the sum and the product. Is it really worth the effort of our abstraction to prevent that? Is this really all that smelly? The answer with only a sum and a product is probably "no". A single code duplication is not the end of the world. However, if I next ask you to complete versions that return the minimum or the maximum, what do you do then? Without the abstraction, you get to copy and paste two more versions of the code and make similarly minor changes to them. With the abstraction, you just call `inputAndCount` with different arguments. The question of whether it is worth it to abstract really depends on how much expansion you expect in the future. If the abstraction does not take much effort it is probably worth doing to start with because it is often hard to predict if you will need to extend something in the future. You might not feel this in courses, but it becomes very significant in professional development when you often are not told up front exactly what is wanted of you and even when you are, the customer is prone to change their mind later in the process.

6.5 Matching

The `if` expression is not the only conditional construct in Scala. There is a second, far more expressive, conditional construct, `match`. While the `if` construct picks between two different possibilities, based on whether an expression is `true` or `false`, the `match` construct allows you to pick from a large number of options to see if a particular expression matches any of them. The term "matches" here is vague. Indeed, the power of the `match` construct comes from something called PATTERN MATCHING.

The syntax of the `match` expression in its simplest form is as follows.

```
expr match {
  case pattern1 => expr1
  case pattern2 => expr2
  case pattern3 => expr3
  ...
}
```

The value of the expression before the `match` keyword is checked against each of the different patterns that follow the `case` keywords in order. The first pattern that matches will have its expression evaluated and that will be the value of the whole match expression.

We can use this to repeat our example from chapter 4 related to the cost of food at a theme park.

```
def foodPriceMatch(item:String,size:String):Double = item match {
  case "Drink" => size match {
    case "S" => 0.99
    case "M" => 1.29
    case _ => 1.39
  }
  case "Side" => size match {
    case "S" => 1.29
    case "M" => 1.49
    case _ => 1.59
  }
  case "Main" => size match {
    case "S" => 1.99
    case "M" => 2.59
    case _ => 2.99
  }
  case _ => size match {
    case "S" => 4.09
    case "M" => 4.29
    case _ => 5.69
  }
}
```

When an entire function is a single `match`, it is customary to put the start of the `match` after the equals sign as done here. Inside we have the four different `cases`, each with its own `match` on the `size`. The one thing that might seem odd here is the use of the underscore. An underscore as a pattern matches anything. This is done so that the behavior would agree with what we had in the `if` version where it defaulted to "Combo" as the item and "large" as the size.

This example shows that the `match` expressions can be nested and the _ can be used to match anything. There is a lot more to the `match` expression though. The following example shows how to use match to give responses to whether you might buy different food items.

```
def buy(food:(String,Double)):Boolean = food match {
  case ("Steak",cost) if(cost<10) => true
  case ("Steak",_) => false
  case (_,cost) => cost<1
}
```

This is a very limited example, but it demonstrates several aspects of `match` that are worth noting. First, `food` is a tuple and the cases pull the two items out of that tuple. That is part of the pattern matching aspect. We will use this later on when we have other things that work as patterns. The second thing we see is that if we put a variable name in the pattern, it will match with anything and it will be bound to the value of that thing. In the example, the second element of the tuple in the first and third `cases` is given the name `cost`. That variable could appear in the expression for the `case` as in the last `case` where we will buy anything for under a dollar. It can also be part of an `if` that "guards" the case. The pattern in the first `case` will match anything that has the word "Steak" as the food. However, the `if` means we will only use that `case` if the cost is less than 10. Otherwise it falls down and checks later `cases`.

`match` versus `switch`

While this is a simple little example, it hopefully demonstrates the significant power we can get from a `match` expression. If you are familiar with other programming languages you might have heard of a `switch` statement before. On the surface, `match` might seem like `switch`, but `match` is far more powerful and flexible in ways allowing you to use it more than a `switch`.

For this chapter, an appropriate example of `match` would be to demonstrate recursion using it instead of an `if`. We can start with a simple example of something like `countDown`.

```
def countDown(n:Int) = n match {
  case 0 =>
  case i =>
    println(i)
    countDown(i-1)
}
```

This function is not quite the same as what we had with the `if` because it only stops on the value 0. This makes it a little less robust, but it does a good job of illustrating the syntax and style of recursion with a `match`. The recursive argument, the one that changes each time the function is called, is the argument to `match`. There is at least one case that does not involve a recursive call and one case that does.

A more significant example would be to rewrite the `inputAndCount` function using a `match`.

```
def inputAndCount(base:Int,func:(Int,Int)=>Int):(Int,Int) = readLine() match {
  case "quit" =>
    (base,0)
  case n =>
    val (s,c)=inputAndCount(base,func)
    (func(s,n.toInt),c+1)
}
```

Here the call to `readLine` is the argument to match. This is because there is not a standard recursive argument for this function. The decision of whether or not to recurse is based on user input instead. So the user input needs to be what we match on.

6.6 Putting it Together

Back to the theme park. Let's take some of the functions we wrote previously, and put them to use in a recursive function that is part of a script that we could use at the front of the park where they sell admission tickets. The recursion will let us handle multiple groups paying for admission with any number of people per group. It will then add up all the admissions costs so we can get a daily total. In addition, it will keep track of the number of groups and people that come in. The code for this makes the same assumptions we made before about groups in regards to coolers and the water park.

```
def entryCost(age:Int, cooler:Boolean, waterPark:Boolean): Double = {
  (if(age<13 || age>65) 20 else 35) +
  (if(cooler) 5 else 0) +
  (if(waterPark) 10 else 0)
}

def individualAdding(num:Int, cooler:Boolean, waterPark:Boolean): Double = {
  if(num>0) {
    println("What is the persons age?")
    val age = readInt()
    entryCost(age, cooler, waterPark)+individualAdding(num-1, false, waterPark)
  } else 0.0
}

def groupSizeCost(): (Int,Double) = {
  println("How many people are in your group?")
  val numPeople = readInt()
  println("Is the group bringing a cooler? (Y/N)")
  val cooler = readLine()=="Y"
  println("Will they be going to the water park? (Y/N)")
  val waterPark = readLine()=="Y"
  (numPeople,individualAdding(numPeople,cooler,waterPark))
}

def doAdmission(): (Int,Int,Double) = {
  println("Is there another group for the day? (Y/N)")
```

```
  val another = readLine()
  if(another=="Y") {
    val (people, cost) = groupSizeCost()
    val (morePeople, moreGroups, moreCost) = doAdmission()
    (people+morePeople, 1+moreGroups, cost+moreCost)
  } else (0,0,0.0)
}

val (totalPeople, totalGroups, totalCost) = doAdmission()
println("There were "+totalPeople+" people in "+totalGroups+
  " groups who paid a total of "+totalCost)
```

This code has two different recursive functions, one for groups and one for people. The `doAdmission` function recurses over full groups. It asks each time if there is another group and if there is it uses `groupSizeCost` to get the number of people and total cost for the next group, then adds those values to the return of the recursive call. The `groupSizeCost` function uses the recursive `individualAdding` function to run through the proper number of people, ask their age, and add them all up.

One interesting point to notice about `individualAdding` is that the function takes a `Boolean` for whether or not there is a cooler, but when it recursively calls itself, it always passes `false` for that argument. This is a simple way to enforce our rule that each group only brings in one cooler. If the initial call uses `true` for `cooler`, that will be used for the first person. All following calls will use a value of `false`, regardless of the value for the initial call.

Bad Input, Exceptions, and the `try/catch` Expression

At this point, you have probably noticed that if you call `readInt` and enter something that is not a valid `Int` it causes the program to crash. If you looked closely at the output when that happens you have noticed that it starts with `java.lang.NumberFormatException`. This is the error type reported when the program is expecting a number and gets something that does not fit that requirement. The `NumberFormatException` is just one type of `Exception`. There are many others and we will learn in section 22.3 that you can make your own.

For the first half of this book we generally assume that users will input appropriate values. As such, we do not include code for handling situations when they do not. If you want to be able to make more flexible code or simply deal nicely with users giving invalid inputs you need to use the `try/catch` expression. This begins with a `try` block that does what the name implies, it is going to try to execute a piece of code with the knowledge that it might fail. This is followed by a `catch` block with different cases for things that could go wrong. Here is a very basic example where we try to read an integer and give back the value that is read if it works, otherwise it gives us zero.

```
val num = try {
  readInt()
} catch {
  case _ => 0
}
```

As this shows, the `try/catch` is an expression with a value. One point to keep in mind about this is that the values given back by different cases generally need to match the type of the last expression of the try block.

You could also see the normal printout you get from an exception by giving a name to the exception in the case associated with it. All exception objects have a method called printStackTrace.

```
val num = try {
  readInt()
} catch {
  case e =>
    e.printStackTrace
    0
}
```

This can be especially useful during debugging and that stack trace includes a significant amount of useful information.

Both of these examples have pitfalls that they give back a value of zero when anything goes wrong. If you really need to have an integer read, and only want to handle NumberFormatExceptions, you might consider code like this.

```
def readIntRobust():Int = try {
  readInt()
} catch {
  case e:NumberFormatException =>
    println("That was not an integer. Please try again.")
    readIntRobust()
}
```

This recursive function will call itself repeatedly until the user enters a valid integer. Here the case only matches an exception of the NumberFormatException type. So if some other type of exception were to occur, that would still cause a crash. Such behavior is typically want you want. You should only handle errors that you know how to deal with at that point in the code.

6.7 Looking Ahead

This will not be our last look at recursion in this book. Indeed, we have just scratched the surface. We have only used recursion in this chapter to repeat tasks, as a model for repetition. The real power of recursion comes from the fact that it can do a lot more than just repetition. Recursive calls have memory. They do not just know where they are, they remember what they have done. This really comes into play when a recursive function calls itself more than once. That is a topic for later, but before we leave this first encounter with recursion here is a little brainteaser for you to think about.

Below is a little bit of code. You will notice that it is nearly identical to the countDown function that we wrote near the beginning of this chapter. Other than changing the name of the method the only difference is that two lines have been swapped. Put this function into Scala. What does it do? More importantly, why does it do that?

```
def count(n:Int) {
  if(n>=0) {
    count(n-1)
    println(n)
  }
}
```

6.8 End of Chapter Material

6.8.1 Problem-Solving Approach

While recursion gives us much greater flexibility with the ability to repeat code an arbitrary number of times, it used the same constructs we had learned before and, for that reason, it does not increase the list of options we have for any given line. It only adds new ways to think about and use options we had before. This chapter did include one new type of statement/expression, the `match` conditional expression. That is now included along with `if` as a conditional option.

1. Call a function just for the side effects. Previously `print` or `println` were the only examples we had of this, but we can now be more general.
2. Declare something:
   - A variable with `val` or `var`.
   - A function with `def`. Inside of the function will be statements that can pull from any of these rules.
3. Assign a value to a variable.
4. Write a conditional statement:
   - An `if` statement.
   - A `match` statement.

6.8.2 Summary of Concepts

- The concept of recursion comes from mathematics where it is used to refer to a function that is defined in terms of itself.
  - All recursive functions must have at least one base case that does not call itself in addition to at least one recursive case that does call itself.
  - Lack of a base case leads to infinite recursion.
- In a programming context, a recursive function is one that calls itself.
  - Recursion is used in this chapter to provide repetition.
  - Typically an argument is passed in to the function, and on each subsequent call the argument is moved closer to the base case.

- If the argument is not moved toward the base case it can result in an infinite recursion.

- Recursion can also be done on user input when we do not know in advance how many times something should happen.
  - Reading from input is a mutation of the state of the input.
  - Functions that use this might not take an argument if they read from standard input.
  - The base case occurs when a certain value is input.
- It is inefficient to make copies of code that only differ in slight ways. This can often be dealt with by introducing an abstraction on the things that are different in the different copies.
  - When used with recursion, this is done by passing in other arguments that tell the code what to do in the parts that were different in the different copies.
  - Values can easily be passed through with parameters of the proper value type.
  - Variations in functionality can be dealt with using parameters of function types. This makes the abstract versions into higher-order functions.
- There is another type of conditional construct in Scala called `match`.
  - A `match` can include one or more different `case`s.
  - The first `case` that matches the initial argument will be executed. If the `match` is used as an expression, the value of the code in the `case` will be the value of the `match`.
  - The `case`s are actually patterns. This gives them the ability to match structures in the data and pull out values.
    * Tuples can be used as a pattern.
    * Lowercase names are treated as `val` variable declarations and bound to that part of the pattern.
    * You can use _ as a wildcard to match any value that you do not need to give a name to in a pattern.
  - After the pattern in a `case` you can put an `if` guard to further restrict that `case`.

6.8.3 Self-Directed Study

Enter the following statements into the REPL and see what they do. Try some variations to make sure you understand what is going on. Note that some lines read values so the REPL will pause until you enter those values. The outcome of other lines will depend on what you enter.

```
scala> def recur(n:Int):String = if(n<1) "" else readLine()+recur(n-1)
scala> recur(3)
scala> def recur2(n:Int,s:String):String = if(n<1) s else recur2(n-1,s+readLine())
scala> recur2(3,"")
scala> def log2ish(n:Int):Int = if(n<2) 0 else 1+log2ish(n/2)
scala> log2ish(8)
```

```
scala> log2ish(32)
scala> log2ish(35)
scala> log2ish(1100000)
scala> def tnp1(n:Int):Int = if(n<2) 1 else
  1+(if(n%2==0) tnp1(n/2) else tnp1(3*n+1))
scala> tnp1(4)
scala> tnp1(3)
scala> def alpha(c:Char):String = if(c>'z') "" else c+alpha((c+1).toChar)
scala> alpha('a')
```

6.8.4 Exercises

1. Write functions that will find either the minimum or the maximum value from numbers input by the user until the user types in "quit".
2. Use `inputAndCount` to find the minimum and maximum of numbers that the user enters.
3. Write exponentiation using multiplication. Your function only has to work for positive integer values.
4. If you did 3, most likely you have a function where if you raise a number to the Nth power, it will do N (or maybe $N-1$) multiplications. Consider how you could make this smarter. It turns out that you can make one that does a lot fewer multiplications, $\log_2 N$ to be exact. Think about how you would do this and write code for it.
5. Write a recursive function that will print powers of two up to some power.
6. Write a recursive function that will print powers of two up to some value.
7. Write recursive functions that will print a multiplication table up to 10s. Try to get it running first, then consider how you could make everything line up.
8. Describe the behavior of the last count function in the chapter.
9. Write a recursive function called `isPrime` that returns a `Boolean` and lets you know whether or not a number is prime.
10. Write a function that prints the prime factors of a number.
11. An efficient method of finding the greatest common divisor, *gcd*, of two integers is Euclid's algorithm. This is a recursive algorithm that can be expressed in mathematical notation in the following way.

$$gcd(a,b) = \{ \begin{array}{ll} a & b=0 \\ gcd(b, a \mod b) & otherwise \end{array}$$

    Convert this to Scala code.

12. Certain problems can use a bit more information than just the *gcd* provided by Euclid's algorithm shown in exercise 11. In particular, it is often helpful to have the smallest magnitude values of x and y that satisfy the equation $gcd = xa + by$. This information can be found efficiently using the extended Euclid's algorithm. This is the math notation for that function.

$$eEuclid(a,b) = \{ \begin{array}{ll} (a,1,0) & b=0 \\ (d,x,y) = eEuclid(b, a \mod b), (d, y, x - \lfloor a/b \rfloor * y) & otherwise \end{array}$$

Convert this to Scala. Note that the $\lfloor a/b \rfloor$ operation is naturally achieved by the truncation of integer division for positive a and b.

6.8.5 Projects

1. For this option I want you to write functions that do the basic math operations of addition, multiplication, and exponentiation on non-negative `Int`s. The catch is that you can not use `+`, `*`, or any functions from `math`. You only get to call the successor and predecessor functions shown here.

   ```
   def succ(i:Int):Int = i+1
   def pred(i:Int):Int = i-1
   ```

   These functions basically do counting for you. So you will define addition in terms of those two, then define multiplication in terms of addition and exponents in terms of multiplication.

   Put the functions that you write in a script and have the script prompt for two numbers. Using your functions, print the sum and the product of those two numbers.

2. Write a function that will return a `String` with the prime factorization of a positive `Int`. The format for the print should be `p^e+p^e+...` Here each `p` is a prime number and `e` is the how many times it appears in the prime factorization. If you call this function with 120 it would return `2^3+3^1+5^1` because $120 = 2*2*2*3*5$. Remember that a number `n` is divisible by `i` if `n%i==0`. For the prime numbers start counting up from 2. If you pull out all the factors of lower numbers, you will not find any factors that are not prime. For example, when you pull `2^3=8` out of 120 you get 15, which only has factors of 3 and 5 so you can not get 4.

3. This project builds on top of project 5 (p.108). You are using the functions from that project and putting them into a format where they can be used for multiple geometric objects. The user inputs a ray first and after that is a series of spheres and planes. You want a function that returns the first object hit by the ray, smallest t value, and the parameter (t) of the hit.

4. An interesting twist in biology over the past few decades is the ability to look at the populations of different species and how they interact with one another. Often, the way in which different populations vary over time can be approximated by simple mathematical expressions. In this project you will use your basic knowledge of conditionals and functions with recursion to examine a simple case where you have two different populations that interact in a predator-prey manner.

   The simplest form of this problem is the rabbit and fox scenario. The idea is that each summer you count the population of rabbits and foxes in a certain region. This region is fairly well isolated so you do not have animals coming in or leaving. In addition, the climate is extremely temperate and there is always enough grass so environmental factors do not seem to impact the populations. All that happens is each year the rabbits try to eat and have babies while not getting eaten and the foxes try to catch rabbits. We will make up some formulas for what happens to the population from one year to the next and you will write a program to produce this sequence.

   Over the course of each year, the rabbit population will be impacted in the following ways. Some rabbits will be born, some rabbits will die of natural causes, and some rabbits will be eaten. Similarly some foxes will be born and some will die. The number of rabbits eaten depends upon the population of foxes (more foxes eat more rabbits) and the number of foxes who are born and die depends on the number of rabbits because foxes can not live long or have young without finding rabbits to eat. We can combine these things to come up with some equations that predict the numbers of foxes and rabbits in a given year based on the number in the previous year.

$$R_{n+1} = R_n + A * R_n - B * R_n * F_n$$
$$F_{n+1} = F_n - C * F_n + D * R_n * F_n$$

Here we assume that the natural tendency of rabbit populations is to increase without foxes around and the natural tendency of fox populations is to decrease without rabbits around. The four constants should have positive values. A represents the normal increase in rabbit population without predation. B is the predation rate and is multiplied by both the rabbit population and the fox population because if either one is small, the predation rate is small. C is the rate at which foxes would normally die out without being able to bear young (if they did not have enough food). D is the rate at which fox will bear young when they do have rabbits to feed on. In reality, foxes and rabbits only come in whole numbers, but for numeric reasons, you should use `Doubles` in your program.

The input for your program is the initial rabbit population, R_0, the initial fox population F_0, and the four constants. To start you off, you might try values of 100, 10, 0.01, 0.001, 0.05, and 0.001. The last four numbers are A, B, C, and D respectively. You can play with these values to try to find some that produce interesting results. Print out the first 1000 iterations. To make it so that you can see your results easily, output only numbers. Never prompt for anything. The advantage of this is that you can create a file that is easy to plot. For plotting, you can input the values into a spreadsheet like Excel. Under Linux you could also use gnuplot. When you run the program you redirect the output to a file then you can run `gnuplot` and plot it to see what it looks like. If you print 3 numbers per line, "n R F", and put it in a file called "pop.txt" then you can plot that in gnuplot with a command like "plot 'pop.txt' using ($1):($2), 'pop.txt' using ($1):($3)". There are many other options in `gnuplot` and you can use the Help command to see them.

Write this as a script that has the user enter R_0, F_0, A, B, C, and D without any prompts. It should then output only the numbers for plotting so that they can be sent to a file that can be plotted without editing.

5. Write a script that will calculate a student's GPA with each course properly weighted by the number of hours of credit it counts for. It should prompt for a grade and a number of hours for each class. The grades are letter grades that are converted to a 4-point GPA scale according to the following table. When "quit" is entered as the grade you should stop reading values. The script should then print the cumulative GPA.

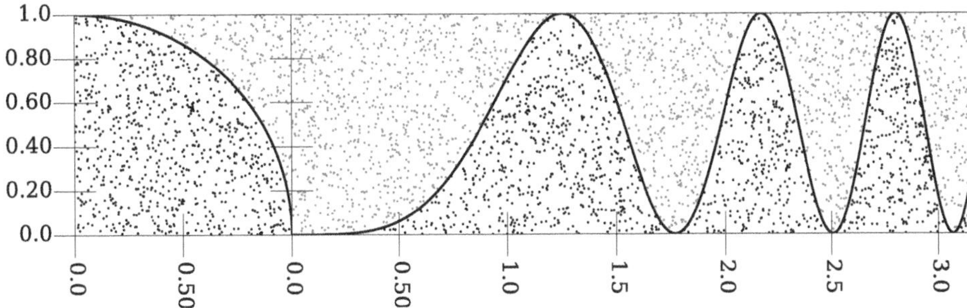

FIGURE 6.1: This figure is associated with project 7. This shows how you can use picking random numbers to estimate the area under curves. This figure shows two curves and random points in square regions that bound those curves. The points that are below the curves are darker than those above. The fraction of points below is proportional to the area under the curve.

Grade	GPA
A	4.000
A-	3.666
B+	3.333
B	3.000
B-	2.666
C+	2.333
C	2.000
C-	1.666
D+	1.333
D	1.000
F	0.000

6. The Mandelbrot set is a fractal in the complex plane. It is described by the simple equation $z_{n+1} = z_n^2 + c$ where $z_0 = 0$ and c is the point in the plane. If the sequence is bounded, the point is in the set. If it heads off to infinity, it is not. Use the solution to exercise 5.3 to write a function that will take a value of c, and a value of z and return how many iterations it takes for the magnitude of z_n to become greater than 4.0 or 1000 if it does not get that big in 1000 iterations.

7. Using a computer, you can estimate the value of π by picking random numbers. In fact, you can find numeric solutions for most integration problems by picking random numbers. This approach to solving such problems is called the Monte-Carlo method. For this option, you will write code to do that using calls to `math.random()` to generate the random numbers and use that to integrate the area under a curve. You will have a script that uses this method to solve two problems, one of which is getting an estimate for the value of π.

   The approach is illustrated in figure 6.1. If you can draw a bounding box around a region of a function, you can randomly select points inside of that box and the area under the curve is estimated by

   $$A\frac{N_{below}}{N_{total}},$$

   where A is the area of the box. The more points you draw randomly, the more accurate the estimate.

   Write a recursive function with the following signature.

```
def countUnder(nPnts:Int,xmin:Double, xmax:Double, ymin:Double,
    ymax:Double,f: Double=>Double):Int
```

This function should generate the specified number of random points in the rectangle with the given bounds and return how many of them were below the curve defined by the function `f`. Note that you can get a random number in the range $[min, max)$ with code like this.

```
val r = min+math.random()*(max-min)
```

You want to do this for the x and y axes.

To estimate π use a function of $\sqrt{1-x^2}$ with x and y in the range of $[0, 1)$. The full rectangle has an area of 1 while the area under the curve is $\pi/4$. If you multiply the fraction of points under the curve by 4, you get an estimate of *pi*.[3]

In addition to printing out your estimate of π, you should print out an estimate of

$$\int_0^\pi (\sin x^2)^2$$

for the same number of random points.

8. If you did project 1 (p.108), you can now add another option to it. Write a function that takes a rational number as an `(Int,Int)` and returns an `(Int,Int)` where all common factors have been removed so that it is in the lowest form.

   With that added in, write a script that uses a recursive function to do calculations on rational numbers. The script presents the user with a menu with options for add, subtract, multiply, divide, and quit. If one of the first four options is selected, it should ask for a fraction and perform that operation between the "current" fraction and the one entered. When the program starts, the "current" value will be $\frac{1}{1}$ represented by `(1,1)`. After each operation you print the new current value, making sure it is always in the lowest form.

The website provides additional exercises and projects with data files that you can process using input redirection.

[3]Be sure to pay attention to whether you are doing integer or floating point arithmetic.

Chapter 7

Arrays and Lists in Scala

Adding conditionals and logic expanded our capabilities dramatically. Combining it with recursion gives us the ability to do almost anything we want. Technically, we have full computing power. However, there are some ideas that are challenging to express with what we currently know and there are better ways to say them in Scala and programming languages in general.

One of these ideas is the ability to refer to many different objects. In chapter 3 we gained the ability to refer to objects by names with `val` and `var` declarations. Using those, we could make as many names as we want and hard code in references to them. We could even have big match statements that allow us to select particular variables based on some value. These approaches are verbose and not particularly useful. In this chapter we will begin learning about COLLECTIONS in Scala. Collections are types that allow us to store and look up many different values using a single name. Most of this chapter will focus on the most basic types, ARRAYs and LISTs. They are two different types that on the surface seem to provide similar functionality, but we will see later on that they are actually very different.

7.1 Making Arrays

The most basic collections of data in Scala or other languages are arrays and lists. Virtually every language will include one as something of a fundamental aspect of the language. Scala happens to include both as easy to access parts of the library. The array and the list are what Scala refers to as SEQUENCEs. That means that they store a number of different values in a specific order and you can get to the elements in them by an integer index. This first section will slowly build up why such things are significant and the most basic usage of them in arrays.

In the last chapter we used recursion to do things such as sum or take the product of a bunch of numbers entered by the user. What if we wanted to do both? We could have done both a sum and a product if the user entered the numbers twice, but that would be inefficient. We had one significant problem with what we were doing. We took the numbers from the user and performed operations on them, but we did not really store them so that we could use them again. It was all a one-shot deal. The reason for this was that the types we have so far store a single value or, in the case of a tuple, a fixed number of values. We do not really have a good way to store a variable number of values so that we can work with them and do multiple operations on them. That is exactly what collections will allow us to do. To really understand this, it helps to look at some examples.

We will start by making some arrays that have values in them.

```
scala> Array(1,2,3,4)
res0: Array[Int] = Array(1, 2, 3, 4)

scala> Array(1.0,2.0,3.0,4.0)
res1: Array[Double] = Array(1.0, 2.0, 3.0, 4.0)

scala> Array('c','a','t')
res2: Array[Char] = Array(c, a, t)

scala> Array("This","is","a","test")
res3: Array[java.lang.String] = Array(This, is, a, test)
```

Here we have created four different arrays of different types. The syntax for making an array with values in it is to follow the word `Array` with a list of parameters for the values we want in the array. As always, after each expression the REPL tells us the name used for this, the types of the expression, and the value of the expression. The type of the expression here shows something that we have never seen before. The type is not just `Array`. The type name `Array` is followed by square brackets that have another type in them. This is called a PARAMETERIZED TYPE. All the collection types in Scala are parameterized. We will talk about this in more detail later in this chapter, but the meaning should be clear. The type `Array[Int]` is an array that can hold integer values in it. As usual, we did not have to tell Scala that the first array was an array of type `Int`, it figured that out. You can override Scala's type inference and specifically tell it the type you want. The example below makes an `Array[Double]` even though all the values in it are integers. This is perfectly valid because all values in `Int` are also valid values of `Double`.

```
scala> Array[Double](1,2,3,4)
res4: Array[Double] = Array(1.0, 2.0, 3.0, 4.0)
```

Such forcing has to match what you pass in. As you can see below, Scala will not allow you to do this for an invalid conversion.

```
scala> Array[Int]("Does","this","work")
<console>:6: error: type mismatch;
 found : java.lang.String("Does")
 required: Int Array[Int]("Does","this","work")
                         ^

<console>:6: error: type mismatch;
 found : java.lang.String("this")
 required: Int Array[Int]("Does","this","work")
                                ^

<console>:6: error: type mismatch;
 found : java.lang.String("work")
 required: Int Array[Int]("Does","this","work")
                                       ^
```

We can also make arrays that have the default initial value. This is helpful for making larger arrays. It is not unusual for arrays to have thousands or even millions of elements. We certainly do not want to have to type in values for those by hand. Here is the syntax for that.

```
scala> new Array[Int](101) res7: Array[Int] = Array(0, 0, 0,
0, 0, 0, 0, 0, 0, 0, 0, 0, 0, 0, 0, 0, 0, 0, 0, 0, 0, 0, 0,
0, 0, 0, 0, 0, 0, 0, 0, 0, 0, 0, 0, 0, 0, 0, 0, 0, 0, 0, 0,
0, 0, 0, 0, 0, 0, 0, 0, 0, 0, 0, 0, 0, 0, 0, 0, 0, 0, 0, 0,
0, 0, 0, 0, 0, 0, 0, 0, 0, 0, 0, 0, 0, 0, 0, 0, 0, 0, 0, 0,
0, 0, 0, 0, 0, 0, 0, 0, 0, 0, 0, 0, 0, 0, 0, 0, 0, 0)
```

The default value for numeric types is zero. The default for the `Boolean` type is `false`. For most other types, many of which we have not learned about, the default is `null`.[1]

7.2 Using Arrays

So now you know how to make arrays, but what can you do with them? The first things we need be able to do are get the values stored in the array and change the values stored in the array. This is done by following the name of the array with parentheses and a number that is the index of the thing we want. Arrays, like with most things in most modern programming languages, are zero indexed. So the first element in the array is at index zero and if there are N things in the array, the last one is at index N-1. Here is an example.

```
scala> val arr=Array(7,4,6,3,9,1)
arr: Array[Int] = Array(7, 4, 6, 3, 9, 1)

scala> arr(0)
res8: Int = 7

scala> arr(1)
```

[1]The value `null` represents an invalid reference or a reference to nothing. Trying to do something with a `null` value will cause an error. For this reason, we will typically avoid this approach to making arrays. We will generally prefer to use `Array.fill` or `Array.tabulate`, which will be introduced later in this chapter.

```
res9: Int = 4

scala> arr(5)
res10: Int = 1
```

The first statement creates a new variable named **arr** that is an array of integers and gives that array six values to store. The next three commands basically pull out values from that array. The expression `arr(0)` gets the first elements, `arr(1)` gets the second, and `arr(5)` gets the last.

What goes in the parentheses does not have to be a simple integer either. It can be any expression of type `Int`. You could do something much more complex to pull out a value.

```
scala> val i=2
i: Int = 2

scala> arr(2*i-2)
res12: Int = 6
```

The same type of expression can also be used in an assignment expression. So we can alter the values that are stored in an array as we see in the following example.

```
scala> arr(4)=99

scala> arr
res13: Array[Int] = Array(7, 4, 6, 3, 99, 1)
```

Here we see how we can use assignment to change the value of one of the elements of the array. This might surprise you because we originally declared `arr` to be a `val`. Previously we said that you can not change what is referred to by a variable that is created with `val`. This does not actually break that rule. You see, the name `arr` still refers to the same array. What has changed is the value in the array. An analogy might be that an array is a house and the values are the people in it. The variable `arr` refers to a particular house. People might come and go from the house, but the house itself is the same. We can demonstrate this by trying to change what array the variable `arr` references.

```
scala> arr=Array(2,7,5)
<console>:6: error: reassignment to val
       arr=Array(2,7,5)
          ^
```

We call the Array type a MUTABLE type because the values in it can be changed. Types whose internal values can't be changed are called IMMUTABLE types. This distinction is very significant to us in Scala and in programming in general, as it alters how we deal with data. We will talk more about it because the `List` type, which we will also look at in this chapter, happens to be immutable. Indeed, all of the types that we have seen so far are immutable. Once they are created, the values they have never change. The `Array` is our first example of something that is mutable.

Not everything about an array can be changed. As we have seen, we can change the values stored in an array. However, we can not change how many things are stored in an array. The number of things an array holds can be called the length or the size of the array. In fact, the `Array` type has methods called `length` and `size` which give us this information.

```
scala> arr.length
res5: Int = 6
```

```
scala> arr.size
res6: Int = 6
```

When you create an array you have to specify a length and that length will never change. If you use a `var` style variable, you can make a new array with a different length and have the name refer to it, but to do that you create a completely new array, you do not alter the size of the old one.

If you try to use an index value that is either negative or too large, Scala will give you an error message saying that you have gone outside the bounds of the array. We can see what that looks like here with attempts to access and change indexes out of bounds.

```
scala> arr(100)
java.lang.ArrayIndexOutOfBoundsException: 100
      at .<init>(<console>:7)
      at .<clinit>(<console>)
      at RequestResult$.<init>(<console>:9)
      at RequestResult$.<clinit>(<console>)
      at RequestResult$scala_repl_result(<console>)
      at sun.reflect.NativeMethodAccessorImpl.invoke0(Native Method)
      at sun...

scala> arr(100)=0
java.lang.ArrayIndexOutOfBoundsException: 100
      at .<init>(<console>:7)
      at .<clinit>(<console>)
      at RequestResult$.<init>(<console>:9)
      at RequestResult$.<clinit>(<console>)
      at RequestResult$scala_repl_result(<console>)
      at sun.reflect.NativeMethodAccessorImpl.invoke0(Native Method)
      at sun...
```

Now that we know the ground rules for using arrays, we can write some functions that take advantage of them in a useful way. In the last chapter we used recursion to read in a set of numbers that we could then do operations on. One motivation we used for collections was that we could not store the values to do two or more operations on them. So two useful functions might be to have one that fills an array from user input and another that does the types of operations we did before, but this time on the array.

We will start with the function to fill the array from user input. This function needs to be passed the array that it will fill. When we are using an array, it will be easiest to use the style where we specify at the beginning how many numbers are to be read. This is because arrays have a fixed size. The array knows its size so that information does not need to be passed. What does need to be passed is the index that the next number will be read into. The termination condition would be that we are trying to read into a location beyond the end of the array. The function to do this might look like this.

```
def fillArray(arr:Array[Int],index:Int) {
  if(index<arr.length) {
    arr(index)=readInt()
    fillArray(arr,index+1)
  }
}
```

This function is straightforward, but it is worth noting that it does not return anything. The fact that arrays are mutable means that we can pass in an array and have the function mutate the contents. As a result, nothing needs to be returned. The results of this function are held in the modified contents of the array. The following shows this function in action.

```
scala> val numbers=new Array[Int](4)
numbers: Array[Int] = Array(0, 0, 0, 0)

scala> fillArray(numbers,0)

scala> numbers
res3: Array[Int] = Array(2, 3, 4, 5)
```

A new array of integers is created that can hold four values. We call `fillArray` and enter the values 2, 3, 4, and 5. After doing that we can inspect numbers and see that it now holds those values.

Now we need to perform an operation on the contents of the array. We will skip the step of making a special function for just doing addition or multiplication and jump straight to the more abstract and flexible method where we pass in a function to operate on the contents. In addition to the function doing the operation, we also need an array and an integer for the current index. Unlike what we did in the previous chapter, we do not need to pass in a base case value because we know when we are at the beginning or end of the array. The function could look like this.

```
def operateOnArray(arr:Array[Int],index:Int,func:(Int,Int)=>Int):Int = {
  if(index<arr.length-1) {
    func(arr(index),operateOnArray(arr,index+1,func))
  } else {
    arr(arr.length-1)
  }
}
```

If an index at or beyond the last element of the array is passed in, this function returns the last element of the array. Otherwise, it applies the function to the current element and the result of the recursive function on subsequent elements. We can see this in action on the previously defined array, `arr`, in these commands.

```
scala> operateOnArray(arr,0,_+_)
res0: Int = 30

scala> operateOnArray(arr,0,_*_)
res1: Int = 4536
```

Already this allows us to see the added power we get from using an array. Having the values stored gives us the ability to operate on them multiple times without having to input them multiple times.

7.3 Lists

Arrays are the built-in collection of choice in most non-functional languages. An array is typically stored as a single block of memory. This makes them fast and efficient for a lot of operations. As we saw though, arrays are mutable. Functional languages tend to lean away from mutation and other side effects. If you do not allow mutation, arrays become much less efficient. If you want to change a single value in the array, you have to make a complete copy of the array. For this reason, functional languages tend to prefer lists. Technically these are linked lists, a data structure that will be discussed later in the book. You do not have to understand the structure in detail to see how to use it.

We can build a list using the same syntax we used to build an array with initial values.

```
scala> List(7,4,6,3,9,1)
res2: List[Int] = List(7, 4, 6, 3, 9, 1)
```

Like the array type, the list type is parametric and Scala will figure out the best type if you use this syntax. There is no syntax for making an uninitialized list. That is because lists are immutable. Once you have created a list, the values in it can not be changed. As such, there is not much point in creating a long list with some default value. Changing it would require making a new list.

However, there is another way to put lists together when we do not know initially all of the values that will be stored in them. We can efficiently build lists one element at a time if we add elements to the front of the list. To add elements to a list we use the "cons" operator, `::`. Here is an example of adding a single element to an existing list.

```
scala> val lst=List(2,3,4)
lst: List[Int] = List(2, 3, 4)

scala> 1::lst
res3: List[Int] = List(1, 2, 3, 4)
```

We begin by creating a list that contains 2, 3, and 4, then cons a 1 to the beginning of the list. This operation did not alter the original list. Instead it gave us a new list with an extra element at the beginning. Because of the way that lists work, this is efficient. It did not actually have to copy the whole list to make it work. We can look at `lst` again to see that it has not changed.

```
scala> lst
res4: List[Int] = List(2, 3, 4)
```

If you build a list with the cons operator, it is common to start with an empty list. There are two ways to represent an empty list in Scala. One is to use what we did before, but not put any arguments in the parentheses. The other is to use `Nil`. So we can build the same list we had before this way.

```
scala> 1::2::3::4::Nil
res5: List[Int] = List(1, 2, 3, 4)
```

You have to place the `Nil` at the end because the `::` operator needs to have a list on the right-hand side. Notice that for this to work, the `::` operator is right associative. So `1::2::3::Nil` is the same as `1::(2::(3::Nil))`. This is the opposite of the normal mathematical operators, which are left associative. In Scala any operator that ends with a colon will be right associative.

We can use the cons operator to write a function that builds a list of numbers from user input. This is a recursive method that will read in numbers until it gets to "quit". Going until "quit" works well for lists because we can easily and efficiently add new elements to a list. That was not the case for arrays and we needed to have the size of the array set when we began. The method for doing this is quite simple.

```
def inputList():List[Int] = {
```

```
  val in=readLine()
  if(in=="quit") Nil else in.toInt::inputList()
}
```

We can see this at work as well if we run it and type in 3, 4, 5, and “quit” on separate lines.

```
scala> inputList()
res1: List[Int] = List(3, 4, 5)
```

It is possible to access the elements of a list the same way you do an array by putting an index inside of parentheses. However, for a list this is very inefficient. The preferred method, especially in a recursive function, is to use the methods `head` and `tail`. The `head` method will give you back the first element of the list. The `tail` method gives you a list of all the elements after the first element. Using these methods we can write an `operateOnList` function that mirrors the `operateOnArray` function like this.

```
def operateOnList(lst:List[Int],func:(Int,Int)=>Int):Int = {
  if(lst.tail==Nil) lst.head else
    func(lst.head,operateOnList(lst.tail,func))
}
```

Note that we do not require an index to be passed in. We do not have any `+1` or `-1` in this function either. That type of behavior comes from the fact that when we recurse, we pass in `lst.tail`. We can see this function in action here.

```
scala> val lst=List(7,4,6,3,9,1)
lst: List[Int] = List(7, 4, 6, 3, 9, 1)

scala> operateOnList(lst,_+_)
res0: Int = 30

scala> operateOnList(lst,_*_)
res1: Int = 4536
```

This function was written using an `if` statement. When working with lists, it is also common to use pattern matching. The `::` can be used in a pattern to indicate a list with different parts. This particular function can be rewritten as shown here.

```
def operateOnList2(lst:List[Int],func:(Int,Int)=>Int):Int = lst match {
  case h::Nil => h
  case h::t => func(h,operateOnList2(t,func))
  case _ => 0
}
```

You might wonder about the last case. This is not required, but if we leave it out we will get a warning telling us that the match is not exhaustive. This is not just the compiler being overly picky either. It turns out that the original method that uses an `if` expression is not completely safe. Try calling it with an empty list and you will see why.

7.4 Standard Methods

One of the strengths of Scala is that it has rich interfaces. These are interfaces with a lot of different methods in them. We looked at `length` and `size` on the `Array` and `head` and `tail` on the `List`, but this was only scratching the surface. You can actually call either of those on either `Lists` or `Arrays`. However, `length` and `size` are not that efficient for `Lists` while `tail` is inefficient on the `array`. In this section we will run through a sampling of the other methods that are available to us when working with `Lists` and `Arrays`. We will start with the simple ones.

7.4.1 Basic Methods

The methods are broken into a few groups based on what they do. Inside of each group the methods are in alphabetical order. The methods that say they give you a new collection return a collection of the same type that it is called on. So if you call them on an `Array` you will get back an `Array`. If you call them on a `List` you will get back a `List`. Short examples are shown for each using the `lst` variable defined above. The type `Seq` appears occasionally. You can think of this as an `Array` or a `List`.

- Methods that give you part of a collection
  - `drop(n:Int)` – Takes an `Int` and gives you back a new collection with all the elements after skipping that many.

    ```
    lst.drop(2)
    res0: List[Int] = List(6, 3, 9, 1)
    ```

  - `init` – Takes no arguments and returns a new collection with all the elements except the last.

    ```
    scala> lst.init
    res0: List[Int] = List(7, 4, 6, 3, 9)
    ```

  - `last` – Takes no arguments and returns the last element in the collection.

    ```
    scala> lst.last
    res0: Int = 1
    ```

  - `slice(from:Int, until:Int)` – Takes two arguments which are both integer indexes. It returns a new collection with all the elements beginning with the index of the first argument and ending with the one before the index of the second value.

    ```
    scala> lst.slice(2,4)
    res0: List[Int] = List(6, 3)
    ```

  - `splitAt(n:Int)` – Takes an `Int` for the index of the split location. It returns two new collections where the first has the first `n` elements and the second has the rest.

    ```
    scala> lst.splitAt(3)
    res0: (List[Int], List[Int]) = (List(7, 4, 6),List(3, 9, 1))
    ```

  - `take(n:Int)` – Takes an `Int` and gives back a new collection with that many elements from the beginning of this collection.

    ```
    scala> lst.take(3)
    res0: List[Int] = List(7, 4, 6)
    ```

  - `takeRight(n:Int)` – Like `take`, but pulls the last `n` elements.

    ```
    scala> lst.takeRight(3)
    res0: List[Int] = List(3, 9, 1)
    ```

- Boolean tests

  - `contains(elem:Any)` – Takes an element and returns whether or not the collection contains an element equal to it.

    ```
    scala> lst.contains(8)
    res0: Boolean = false
    ```

  - `endsWith(that:Seq[B])` – Takes a collection of elements and tells whether the current collection ends with elements equal to those in the collection passed in.

    ```
    scala> lst.endsWith(List(3,9,1))
    res0: Boolean = true
    ```

  - `isEmpty` – Returns whether or not the collection is empty.

    ```
    scala> lst.isEmpty
    res0: Boolean = false
    ```

  - `nonEmpty` – The opposite of `isEmpty`.

    ```
    scala> lst.nonEmpty
    res0: Boolean = true
    ```

  - `startsWith` – Takes a collection of elements and tells whether the current collection starts with elements equal to those in the collection passed in.

    ```
    scala> lst.startsWith(List(7,5,6))
    res0: Boolean = false
    ```

- Search for something

  - `indexOf(elem:A)` – Takes an element and returns the index of the first element in the collection equal to the value passed in. Returns `-1` if no element is found.

    ```
    scala> lst.indexOf(3)
    res0: Int = 3
    ```

  - `lastIndexOf(elem:A)` – Takes an element and returns the index of the last element in the collection equal to the value passed in. Returns `-1` if no element is found.

    ```
    scala> lst.lastIndexOf(4)
    res0: Int = 1
    ```

- Other simple methods of note
  - `diff(that:Seq[A])` – Takes an argument that is a collection of the same type as what this is called on and returns the multiset difference between the two. This means that it will give you back all the elements that were in the original collection that do not have a match in the argument collection.

    ```
    scala> lst.diff(List(1,2,3,4))
    res0: List[Int] = List(7, 6, 9)
    ```

  - `mkString` – Can be called with zero, one, or three arguments. It builds a single long string from the string representations of the elements. If no argument is provided then nothing is put between the strings for the elements. If one argument is specified, it should be a string that is used to separate the element strings. If three arguments are specified the middle is a separator and the first and last are strings to put before and after the elements.

    ```
    scala> lst.mkString("; ")
    res0: String = 7; 4; 6; 3; 9; 1
    ```

  - `reverse` – Takes no arguments and returns a new collection with the elements in the reverse order.

    ```
    scala> lst.reverse
    res0: List[Int] = List(1, 9, 3, 6, 4, 7)
    ```

  - `toArray`, `toList` – Take no arguments and makes a new collection of the type specified with the elements in the current collection.

    ```
    scala> lst.toArray
    res0: Array[Int] = Array(7, 4, 6, 3, 9, 1)
    ```

  - `zip(that:Iterable[B])` – Takes another collection as an argument and returns a collection of tuples where the first element comes from the collection this is called on and the second comes from the collection passed in. The length of the result is the shorter of the two.

    ```
    scala> lst.zip(lst.reverse)
    res0: List[(Int, Int)] = List((7,1), (4,9), (6,3), (3,6), (9,4), (1,7))
    ```

  - `zipWithIndex` – Returns a new collection of tuples where the first is an element from the collection and the second is its index.

    ```
    scala> lst.zipWithIndex
    res0: List[(Int, Int)] = List((7,0), (4,1), (6,2), (3,3), (9,4), (1,5))
    ```

The methods listed above will work on any type of sequence. So they will work on a `List[Int]`, a `List[String]`, an `Array[Double]`, or a `List[Array[Double]]`. There are a few methods provided that have some special requirements for the type of things in the list. They require that certain operations be defined. These methods, which are self-explanatory, are `min`, `max`, `sum`, and `product`. The `min` and `max` methods will work for types that can be ordered. That includes not just things like `Int` and `Double`, but also `Strings` and many other types where an ordering makes sense. The `sum` and `product` methods require that the type of the collection be numeric.

So while we wrote operateOnList and operateOnArray to do sums and products of those collections, Scala would have allowed us to simply call the sum or product methods as is seen here.

```
scala> lst.sum
res2: Int = 30

scala> lst.product
res3: Int = 4536
```

The requirement that the values be numeric means that while you can concatenate Strings with +, you can not put them together with sum. For a List[String] or Array[String], you should use mkString to concatenate the values.

7.4.2 Higher-Order Methods

While you might feel like the list of methods shown here is quite a lot and gives us many capabilities, we have not yet hit on the real power of the Scala collections. All of these methods have taken normal values for arguments. Just like our first recursive methods, they can be made more powerful by adding some abstraction and making them higher-order methods. Below is a list of many of the higher-order methods that are part of the sequences in Scala. The type A is the type that is contained in the List or Array. The type B could be any other type.

- count(p:(A)=>Boolean) – Takes a function that will operate on an element and return a Boolean. Returns the number of elements in the collection for which this returns true.

  ```
  scala> lst.count(_>5)
  res0: Int = 3
  ```

- dropWhile(p:(A)=>Boolean) – Takes a function that will operate on an element and return a Boolean. Returns a new collection that contains all elements after the first group for which that function is true.

  ```
  scala> lst.dropWhile(_>3)
  res0: List[Int] = List(3, 9, 1)
  ```

- exists(p:(A)=>Boolean – Takes a function that will operate on an element and return a Boolean. Returns true if there is some element in the collection for which the function is true.

  ```
  scala> lst.exists(x => x>4 && x<7)
  res0: Boolean = true
  ```

- filter(p:(A)=>Boolean) – Takes a function that will operate on an element and return a Boolean. Returns a new collection that contains only the elements for which the function is true.

  ```
  scala> lst.filter(_<5)
  res0: List[Int] = List(4, 3, 1)
  ```

- `filterNot(p:(A)=>Boolean)` – Takes a function that will operate on an element and return a `Boolean`. Returns a new collection that contains only the elements for which the function is `false`.

```
scala> lst.filterNot(_<5)
res0: List[Int] = List(7, 6, 9)
```

- `flatMap(f:(A)=>Seq[B])`[2] – Takes a function the will operate on an element and return a collection. Returns a new collection built from all the result collections appended together.

```
scala> lst.flatMap(n => if(n<6) lst.take(n) else Nil)
res0: List[Int] = List(7, 4, 6, 3, 7, 4, 6, 7)
```

- `forall(p:(A)=>Boolean)` – Takes a function that will operate on an element and return a `Boolean`. Returns `true` if the function is `true` for all elements of the collection.

```
scala> lst.forall(_>2)
res0: Boolean = false
```

- `foreach(f:(A)=>Unit)` – Takes a function that operates on an elements and applies it to all elements in the collection. Returns nothing. This method is called only for the side effects.

```
scala> lst.foreach(n=>println(2*n))
14
8
12
6
18
2
```

- `indexWhere(p:(A)=>Boolean)` – Takes a function that will operate on an element and return a `Boolean`. Returns the index of the first element for which the function is `true`.

```
scala> lst.indexWhere(_%2==0)
res0: Int = 1
```

- `lastIndexWhere(p:(A)=>Boolean)` – Takes a function that will operate on an element and return a `Boolean`. Returns the index of the last element for which the function is `true`.

```
scala> lst.lastIndexWhere(_%2==0)
res0: Int = 2
```

- `map(f:(A)=>B)` – Takes a function that operates on an element and returns something. It returns a new collection that contains the results of applying that function to all the contents of the original collection.

[2]The return type of the function `f` that is passed into `flatMap` is technically a `GenTraversableOnce[B]`. This is more general than a `Seq[B]`, but for now the difference is not important.

```
scala> lst.map(_*2)
res0: List[Int] = List(14, 8, 12, 6, 18, 2)
```

- partition(p:(A)=>Boolean) - Takes a function that will operate on an element and return a Boolean. Returns a tuple with two new collections. The first contains only the elements for which the function is true and the second is the rest.

```
scala> lst.partition(_<5)
res0: (List[Int], List[Int]) = (List(4, 3, 1),List(7, 6, 9))
```

- takeWhile(p:(A)=>Boolean) – Takes a function that will operate on an element and return a Boolean. Returns a new collection that contains all elements all the elements at the beginning for which that function is true.

```
scala> lst.takeWhile(_>3)
res0: List[Int] = List(7, 4, 6)
```

While the first set of methods was straightforward, this group could use a bit more explanation. We will focus on map and filter because they are very standard in the functional programming world. Imagine we have a list or an array of people's names. This list might be very long, but for our sample code we will use just a few so you can see the illustration. The names are in the normal format that people write them: "first middle last" with the middle name being optional or potentially having multiple middle names. For programming purposes we would like them to be in the format "last, first middle" because we often care more about the last name. What we have described here is the application of a function to every element of a sequence to get back a new sequence with the modified elements. That is exactly what the map method does for us. All we have to do is figure out how we would write a function to do that transformation and then we can use it with map. The transformation we want to do is to take the last word in the name and move it to the front with a comma and a space between it and the first name. Basically, we want to split up the string on the last space. The methods listed above could do this nicely, if only they worked on a String. As it turns out, they do. We can use a String as a sequence of characters. So we can use lastIndexOf to find the last space and splitAt to make two Strings. We can see this on a particular string here.

```
scala> val name="Mark C. Lewis"
name: java.lang.String = Mark C. Lewis

scala> name.splitAt(name.lastIndexOf(' ')+1)
res45: (String, String) = (Mark C. ,Lewis)
```

Now all we need to do is put that back together in the opposite order with a comma and a space.

If we were going to be reorganizing names like this frequently, we could put this code in a stand-alone function. If not, we can use a function literal to define it on the fly. For this example, we will use the latter approach.

```
scala> val names=List("Mark C. Lewis","Jason C. Hughes",
   "Glen R. Stewart","Jen Hogan")
names: List[java.lang.String{}] = List(Mark C. Lewis, Jason C. Hughes,
   Glen R. Stewart, Jen Hogan)

scala> val lnames=names.map(n => {
```

```
   val (rest,last)=n.splitAt(n.lastIndexOf(' ')+1) last+", "+rest } )
lnames: List[java.lang.String] = List(Lewis, Mark C. , Hughes, Jason C. ,
   Stewart, Glen R. , Hogan, Jen )
```

So the use of `map` allowed us to complete a task in a quick and short way that could have taken us a fairly large amount of code using other approaches.

If you had a long list of these names, you might want to do something like find all the people who have last names beginning with a particular letter. For example, we might want to find everyone whose last names starts with an 'H'. For this we would use the `filter` function. The `filter` function will select out values that satisfy some condition and give us a sequence with only those values.

```
scala> lnames.filter(_.startsWith("H"))
res48: List[java.lang.String] = List(Hughes, Jason C. , Hogan, Jen )
```

Thanks to `filter` and the `startsWith` method, this is a very simple one-liner.

There are some other methods that were not listed above because they take a bit more explanation. They should not be too hard to understand if you have followed up to this point because they do exactly what we did ourselves earlier in this chapter with the `operateOnArray` and `operateOnList` functions. We will start with the `reduce` methods, `reduceLeft` and `reduceRight`. These methods take a function that operates on two elements of the sequence and returns a value of that type. The `reduce` methods repeatedly apply the function to successive elements the same way the `operateOnArray` and `operateOnList` methods did. The difference between the two is whether they apply the operations moving from left to right or from right to left. If the operation is commutative, like addition or multiplication, then the order does not impact the result, only potentially the efficiency. For non-commutative operations it can matter. Here are a few examples using the `lst` variable that we defined earlier.

```
scala> lst.reduceLeft(_+_)
res49: Int = 30

scala> lst.reduceLeft(_*_)
res50: Int = 4536

scala> lst.reduceLeft(_-_)
res51: Int = -16

scala> lst.reduceRight(_-_)
res52: Int = 14
```

The first two calls do what we have seen before taking the sum and product of the list. The last two use a difference, which is not commutative, and show how the results of `reduceLeft` and `reduceRight` are different. The `reduceLeft` method gives us $((((7-4)-6)-3)-9)-1$. The `reduceRight` method gives us $7-(4-(6-(3-(9-1))))$.

In the last chapter we made a recursive method that could do these types of operations with user input. That differed from what we did in this chapter in that we had to pass in an extra value to use when the user terminated the recursion. We did not do anything special with that, but it could have opened other possibilities. The `reduce` operations have to operate on the element type of the collection for both arguments because that is what happens in the most deeply nested application. If we specify the base value, the function can take one argument of a different type than the elements in the sequence as long as it returns that type as well. If we say that the sequence has type `A` and we want a function that will produce type B then the function has the form (B,A)=>B and we provide a first value of type B. We can run through the sequence applying this function to each of the A values until we get a final B at the end.

This functionality is provided by the foldLeft and foldRight methods. These methods use something we have not seen yet, CURRYING. A curried function is a function that takes one set of arguments and returns a function that takes another set of arguments.[3] Here is a simple example of making a curried function using the normal Scala function type and function literal syntax.

```
scala> def add(x:Int):(Int)=>Int = (y:Int)=>x+y
add: (x: Int)(Int) => Int

scala> val plus5=add(5)
plus5: (Int) => Int = <function1>

scala> plus5(6)
res54: Int = 11
```

The first input defines a function named add that takes an integer and returns a function which takes an integer and returns an integer. The add function is a higher-order function because it returns a function. In the second line we call add and pass it the argument 5. We store the result of this in a variable named plus5. Note that plus5 is actually a function that takes an integer and returns an integer. At the end we call plus5 and pass it the value 6, which gives us back 11.

This example shows what it means to curry a function. There can be times when this approach is quite useful. For that reason, Scala, and a number of other languages, provide a shortcut for doing it. If you define a function and follow it with more than one argument list, the function is curried. So the above could be entered like this instead.

```
scala> def add(x:Int)(y:Int) = x+y
add: (x: Int)(y: Int)Int

scala> val plus5=add(5)_
plus5: (Int) => Int = <function1>

scala> plus5(6)
res55: Int = 11

scala> add(5)(6)
res56: Int = 11
```

One big difference here is the underscore where the second argument list would go. This tells Scala that you really meant to only specify the first argument.

The topic of currying is introduced here because the foldLeft and foldRight methods are curried. The first argument list takes the base value to use on the first application. The second list is the function to apply. The types A and B do not have to be different so we can use foldLeft to sum up a list of integers like this.

```
scala> lst.foldLeft(0)(_+_)
res57: Int = 30
```

[3]The term currying comes from the name of Haskell Curry, a mathematician and logician.

However, unlike the reduce methods, we can do other things like count up the total number of characters in a sequence of Strings. A reduce method on a sequence of Strings could only give us back a String, but the fold methods can give us back another type. Here we do exactly that. We show both the longer function literal syntax as well as the shorter version.

```
scala> val wordList=List("How","many","characters","do","we","have")
wordList: List[java.lang.String] = List(How, many, characters, do, we, have)

scala> wordList.foldLeft(0)((count,word)=>count+word.length)
res58: Int = 25

scala> wordList.foldLeft(0)(_+_.length)
res59: Int = 25
```

7.4.3 Combinatorial/Iterator Methods

There are some other methods on sequences that, for reasons of efficiency and memory limitations, work a bit differently. These methods give us an object that provides access to multiple different collections that are created from the original collection. The object that they give us is called an Iterator, and specifically they give us an Iterator[List[A]] or an Iterator[Array[A]]. The Iterator type is more primitive than the List or Array. You can only go through it once because it is consumed as you go through it. The reason for using an Iterator on these methods is generally for performance and memory benefits. The fact that the Iterator consumes things as it goes means that it does not have to store all the contents at once. In fact, only one needs to exist at any given time. When it moves from one to the next, it can forget the last and make the next one.

The methods in this category are listed here. To show the values of the outputs a call to foreach(println) is used. Without this, all that would be shown is Iterator[List[Int]] = non-empty iterator.

- combinations(n:int) – Generates all combinations of the elements of this sequence of length n.

  ```
  scala> lst.combinations(3).foreach(println)
  List(7, 4, 6)
  List(7, 4, 3)
  List(7, 4, 9)
  List(7, 4, 1)
  List(7, 6, 3)
  List(7, 6, 9)
  List(7, 6, 1)
  List(7, 3, 9)
  List(7, 3, 1)
  List(7, 9, 1)
  List(4, 6, 3)
  List(4, 6, 9)
  List(4, 6, 1)
  List(4, 3, 9)
  List(4, 3, 1)
  List(4, 9, 1)
  List(6, 3, 9)
  List(6, 3, 1)
  ```

```
List(6, 9, 1)
List(3, 9, 1)
```

- `grouped(size:Int)` – Runs through the sequence, grouping items into groups of the specified `size`.

```
scala> lst.grouped(2).foreach(println)
List(7, 4)
List(6, 3)
List(9, 1)

scala> lst.grouped(3).foreach(println)
List(7, 4, 6)
List(3, 9, 1)
```

- `inits` – Provides an iterator going from the full sequence to an empty one, removing elements from the end.

```
scala> lst.inits.foreach(println)
List(7, 4, 6, 3, 9, 1)
List(7, 4, 6, 3, 9)
List(7, 4, 6, 3)
List(7, 4, 6)
List(7, 4)
List(7)
List()
```

- `permutations` – Lets you run through all the permutations of the sequence. Note that the example here only does permutations of the first three elements as there are $6! = 720$ different ones for the full list.

```
scala> lst.take(3).permutations.foreach(println)
List(7, 4, 6)
List(7, 6, 4)
List(4, 7, 6)
List(4, 6, 7)
List(6, 7, 4)
List(6, 4, 7)
```

- `sliding(size:Int)` – Provides an iterator that gives the effect of sliding a window of a certain `size` across the sequence.

```
scala> lst.sliding(2).foreach(println)
List(7, 4)
List(4, 6)
List(6, 3)
List(3, 9)
List(9, 1)

scala> lst.sliding(3).foreach(println)
List(7, 4, 6)
List(4, 6, 3)
List(6, 3, 9)
List(3, 9, 1)
```

- `tails` – Gives an iterator that runs through sublists starting with the full list and ending with an empty one removing one element from the left end each step.

```
scala> lst.tails.foreach(println)
List(7, 4, 6, 3, 9, 1)
List(4, 6, 3, 9, 1)
List(6, 3, 9, 1)
List(3, 9, 1)
List(9, 1)
List(1)
List()
```

If you want to do something more involved with the values given by the `Iterator`, you can convert it to a `List` or an `Array` with `toList` or `toArray`. However, if there are many elements, as is the case for `lst.permutations` you might want to filter it or do something else to narrow things down before doing that conversion.

These methods provide an easy way to do some more interesting work with data. For example, using `sliding`, you can quickly smooth noisy data by averaging a data series over a window. Using `combinations` or `permutations`, you could run through all possible options of some type and find the one that was the best in some way.

At some point you might have wondered where you would find out about all of these methods. Even with everything discussed in this section, our list is incomplete. The complete list of all the types available and methods for them can be found in the Applications Programmer Interface (API). You can find the Scala API at the main Scala website. This has a lot of elements that go beyond your current knowledge, but one of the main objectives of this book is that you will be able to look up information in the API when you have questions about how to do things in Scala.

7.5 A World of Types

We have only begun to scratch the surface when it comes to types in Scala. By the end of this book we will see that not only are there many, many types provided by the Scala language/libraries, we can define our own types and give them names. Still, it is worth being more explicit about what a type is and some of the details of types in Scala.

Let us begin with the question of what is a type. A type is a set of values and the associated operations that can be performed on those values. Consider the type `Int`. There are a set of different values that we can represent with an instance of `Int`. They are all the integer values between -2147483648 and 2147483647. There is also a set of operations that you can perform on an `Int` or on pairs of them such as addition and subtraction.

In this chapter we have begun to see that there is more to types in Scala than the basics like `Int`. The types `List` and `Array` are not complete on their own. They are parametric types. To be complete they have to have type parameters. So `List[Int]` is a complete type as is `Array[String]`. `List` and `Array` are far from the only parametric types. Many of the types in the Scala library are parametric. In the second half of this book we will make our own parametric types. For now it will be enough to talk about them in the existing library code and how they can be used with functions.

7.5.1 The Option Type

If you start digging through the API much you will find that the type `Option` comes up a fair bit. The idea of an `Option` type is that they should be used when you are not certain if you will have a value or not. An `Option[A]` can either be `Some[A]` or `None`. If it is `Some` it will have a value in it that can be retrieved with `get`. If there was no value then you get `None` and know that it does not have a value. An example of a method that does this is the `find` method on a sequence. You can call `find` on either a `List` or an `Array` and pass it a function that takes an element and returns a `Boolean`. The `find` method is supposed to return the first element for which the function returns `true`. That description sounds simple enough, but what if there is not one? What happens if `find` runs through the whole sequence and nothing makes the function return `true`? For that reason, the `find` method returns `Option[A]` where `A` is the type contained in the sequence. So if nothing works, it can return `None`.

Let us run through two examples to see how this works using our list of integers we have been using throughout this chapter.

```
scala> lst.find(_<4)
res20: Option[Int] = Some(3)

scala> lst.find(_<1)
res21: Option[Int] = None
```

The list does contain elements that are less than 4. The first of these is 3 so we get back `Some(3)`. However, it does not contain anything less than 1 so that call returns `None`. You might wonder how you can use this type of return value. The answer is to use `match`. This little example takes the result of a find and tells us what it found or else tells us that nothing was found.

```
scala> lst.find(_<4) match {
     | case Some(i) => "Found "+i
     | case None => "Nothing found"
     | }
res22: java.lang.String = Found 3
```

7.5.2 Parametric Functions

Types like `List`, `Array`, and `Option` are not the only things that can have parameters. Functions can have parameters as well. This is something that we will revisit later on, but it can help you to read the API if you are introduced to the syntax. We would make a function parametric if we want it to work with multiple different types. The simplest example of a PARAMETRIC FUNCTION is the identity method shown here.

```
scala> def ident[A](a:A)=a
ident: [A](a: A)A
```

This method takes an argument of any type and returns the value passed in. While there are not many situations where you would want to do this, it demonstrates the syntax of parametric functions. You simply place a type parameter name in square brackets between the function name and the normal argument list. It also demonstrates the power of parametric functions, especially if we put it to use as we do here.

```
scala> ident(3)
```

```
res1: Int = 3

scala> ident(3.3)
res2: Double = 3.3

scala> ident("3.3")
res3: java.lang.String = 3.3
```

First, this one function works just fine with an Int, a Double, and a String. That is pretty good. Even better, it worked with all those types without us telling it the types. Parametric functions can almost always infer the types of the parameters.

Here is a slightly more complex example though it really does not do much more, a function that takes two arguments and returns a tuple with those two values in it.

```
scala> def makeTuple[A,B](a:A,b:B) = (a,b)
makeTuple: [A,B](a: A,b: B)(A, B)
```

This demonstrates how a function can have multiple parameters. They appear as a comma separated list in the square brackets. A last simple example would be a function that makes Lists with three elements.

```
scala> def threeList[A](a1:A,a2:A,a3:A) = List(a1,a2,a3)
threeList: [A](a1: A,a2: A,a3: A)List[A]
```

The main reason for introducing these is that they help us understand something we saw earlier in this chapter, the fold methods. We said that fold was very much like one of the recursive functions we wrote in the last chapter where we pass in both a base value and a function. However, the function that we wrote only works with Ints. The fold methods work on sequences of any type and what is more, they can return a different type. With the use of parameters we can write a function with this same capability.

```
def ourFold[A,B](lst:List[A],base:B)(f:(A,B)=>B):B = {
  if(lst.isEmpty) base
  else f(lst.head,ourFold(lst.tail,base,f))
}
```

Like the one in the API, this method is curried. It turns out that doing so helps with type inference and allows us to not have to specify types on the function. We can see this working with lst using two different processing functions.

```
scala> ourFold(lst,0)(_+_)
res0: Int = 30

scala> ourFold(lst,"")(_+" "+_)
res2: java.lang.String = 7 4 6 3 9 1
```

The first one takes the sum as we have seen several times already. This does not really exercise the ability to have different types because everything involved is an Int. The second example though puts those Ints together in a String, effectively making use of that second type.

There was another example we saw previously that could benefit from parametric types. Back in chapter 5 we looked at doing function composition. At the time we only worked with mathematical functions and limited ourselves to the Double type. By this point you should see there is a lot more to functions than numbers. We really should be able to compose two functions, f and g, as long as the output of function g is a type that we can pass into function f. So there are three types involved here, the type passed into g, the type returned by g then passed into f, and the type returned by f. Thanks to parametric types, we can write such a function in one line of Scala.

```
def compose[A,B,C](f:(B)=>A,g:(C)=>B):(C)=>A = (x)=>f(g(x))
```

7.5.3 Subtyping

So far we have talked about types as if they are completely independent and unrelated. We have written functions that might works with an Int or a String or a Tuple. By adding parameters we were able to make functions that could work with any type, but this still did not imply any relationship between the types. In reality, Scala, like most object-oriented languages, supports SUBTYPING. A type B is a SUBTYPE of type A if any place where we would want to use an object of type A, we can use an object of type B instead.

We will go into a lot more detail about subtyping when we talk about inheritance in chapter 19, but for now you should be aware of the term because there will be times when it will come up. Some of the general subtype relationships in Scala are shown in figure 7.1. At the top of the figure there is an ultimate supertype called Any. Every object in Scala is an instance of Any. Since all values in Scala are objects, everything is an instance of Any, either directly or indirectly. The types like Int and Double that we learned about back in chapter 3 are on the left-hand side of figure 7.1, a type called AnyVal which is a subtype of Any. On the right-hand side there is another type called AnyRef that has a bunch of unspecified types below it. All the other types we have or will talk about fall somewhere under AnyRef. At the bottom of the diagram is a type called Nothing which is a subtype of all other types in Scala. There is no value of type Nothing, The type exists to make the type system complete and to handle situations when functions do not return.

We will not be making significant explicit use of parametric functions and subtyping until significantly later in the book in chapter 19. However, these concepts will come up in implicit ways before then so you should be aware of their existence.

7.6 Variable Length Argument Lists

Have you wondered about constructing Lists and Arrays with calls like this?

```
val lst=List(7,4,6,3,9,1)
```

There are some details that we will not go into, but one aspect of this type of call is something that we now know enough to talk about. We are able to pass a variable number of arguments in this call. In this case we have passed six arguments, but it would work just as well with any other number. The functions we have written can not do this. Every function we have written so far has had an argument list with a specific number of values in it. So how can we make it so our function will let us pass in a variable number of arguments? In fact, it is not hard to do. Simply add an asterisk after the last type of the last argument in the parameter list. This will allow the caller to pass zero or more of that type.

A convenient place to do this would be an average function like we might use for calculating a grade. Such a method might look something like this.

```
def average(nums:Double*):Double =
```

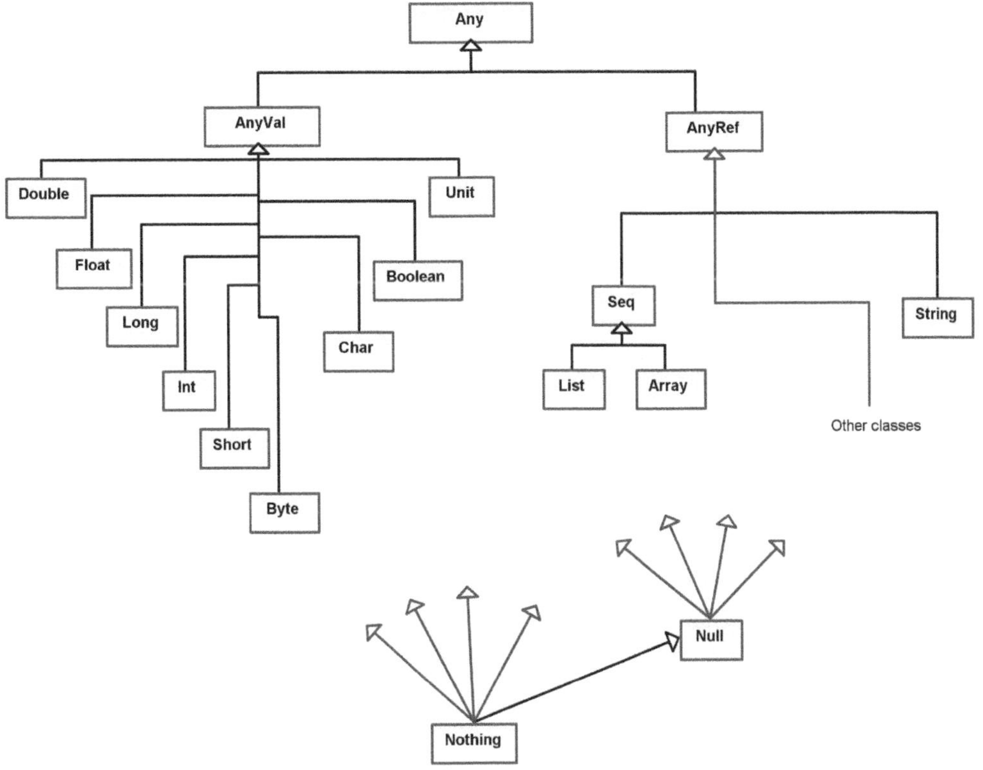

FIGURE 7.1: Diagram of general subtype relationships in Scala. This figure has been adapted from a similar figure in *Programming in Scala* by Odersky, Spoon, and Venners [9]. While `Array` is not actually a subtype of `Seq`, it can be used as such because of implicit conversion, a topic for covered in appendix B.

```
  ...
```

The question is, how do we use `nums` inside of the function? As you might guess, this topic goes into this chapter because we treat `nums` just like a `List` or an `Array`. Technically, it is a `Seq`, a supertype of both, and all of the methods discussed in this chapter are available on it. These methods will make it very easy to write the average function.

```
def average(nums:Double*):Double =
  nums.sum/nums.length
```

That is it. Nothing more is needed because we can use `sum` on a sequence of `Double` values and we can get the `length` of any sequence.

What if we want the average with the minimum value dropped? That does not add much complexity because there is a `min` method that we can call as well.

```
def averageDropMin(nums:Double*):Double =
  (nums.sum-nums.min)/(nums.length-1)
```

We could use these along with the `fullAve` function from chapter 5 to make a revised version of `courseAverage`.

```
def courseAverage(test1:Double,test2:Double,assn1:Double,
    assn2:Double,assn3:Double,quiz1:Double,quiz2:Double,
    quiz3:Double,quiz4:Double):Double = {
  val aveTest=average(test1,test2)
  val aveAssn=average(assn1,assn2,assn3)
  val aveQuiz=averageDropMin(quiz1,quiz2,quiz3,quiz4)
  fullAve(aveTest,aveAssn,aveQuiz)
}
```

This gives us less total code than before because we were able to reuse the average function. However, now that we know about `Lists` and `Arrays`, the way we call this method is a little bit unpleasing. This takes a fixed number of test grades, assignment grades, and quiz grades. This would be a lot more flexible if we passed in `Lists` of these different values instead. Then the function would look more like this.

```
def courseAverage(tests:List[Double],assns:List[Double],
    quizzes:List[Double]):Double = {
  ...
}
```

Unfortunately, if you try to write this in the first way that comes to mind you will find that Scala does not like it.

```
def courseAverage(tests:List[Double],assns:List[Double],
    quizzes:List[Double]):Double = {
  val aveTest=average(tests)
  val aveAssn=average(assns)
  val aveQuiz=averageDropMin(quizzes)
  fullAve(aveTest,aveAssn,aveQuiz)
}
```

You will get an error that looks like this.

```
<console>:10: error: type mismatch;
 found   : List[Double]
```

```
 required: Double
       val aveTest=average(tests)
                          ^

<console>:11: error: type mismatch;
 found   : List[Double]
 required: Double
       val aveAssn=average(assns)
                          ^

<console>:12: error: type mismatch;
 found   : List[Double]
 required: Double
       val aveQuiz=averageDropMin(quizzes)
                                 ^
```

This is because `List[Double]` is not the same type as `Double*`. However, the two are very similar in practice and it seems that you should be able to quickly and easily use the `Lists` in a place that calls for a variable length argument. Indeed, you can. You have to tell Scala that is what you are doing. You do this with a syntax much like specifying the type of something with a colon after the name followed by the type. You do not use `Double*` as the type though, instead you use `_*` because Scala really does not care about the specific type.

```
def courseAverage(tests:List[Double],assns:List[Double],
    quizzes:List[Double]):Double = {
  val aveTest=average(tests:_*)
  val aveAssn=average(assns:_*)
  val aveQuiz=averageDropMin(quizzes:_*)
  fullAve(aveTest,aveAssn,aveQuiz)
}
```

Now we have a function that computes an average with `Lists` so it is flexible in the number of grades passed in and rather simple to call and use in a larger program. If you do enough testing on this code you will find there there is a bug. You can make it produce incorrect output. We leave it as an exercise for the reader to find this bug.

7.7 Mutability and Aliasing

In this chapter we have seen that `List` is immutable while `Array` is mutable. It has been stated that the functional style will tend to use immutable and that while mutable data has significant benefits for some operations, it is also potentially less safe. You might wonder why it is less safe though. After all, it should not be too hard to make sure that you do not make changes to an `Array`. That is true in small programs, but it gets harder and harder as the programs get larger and for reasons that you probably do not fully grasp at this point.

A big part of the problem comes from something called ALIASING. This is when two different names refer to the same object. To understand this, we need to refine our view of what is happening in the memory of the computer. Consider the following variable declarations.

```
var i1=5
var d1=5.9
var s1="Scala is fun!"
var l1=List(1,2,3,4)
var a1=Array(1,2,3,4)
```

FIGURE 7.2: Simple image of memory layout for a few variables. The objects on the right side are overly simplified, especially the list, but this portrays sufficient detail for understanding what is happening at this point.

All of these have been made as `var`s just to give us the option of altering them below. Normally we would want to use a `val` unless we explicitly found a reason why we had to change them. In this case the reason is just for illustration.

The memory layout looks something like what is shown in figure 7.2. This is an idealized representation. In reality memory is all linear and has structure to it that will be discussed in chapter 13. For now, this view is sufficient. On the left are boxes that represent the variables. In Scala the best mental better picture is to see the variables as boxes that store references to values. In some situations Scala will put the actual values in the boxes for the variables, but that is an optimization that you do not really have control over and it will not alter the behavior, so it is good to picture the memory like this.

What happens to this picture if we change the value of two of the variables? For example, say that now we execute these two lines of code.

```
i1=8
s1="New string."
```

The first one changes `i1` so that it references the value 8. The second one changes the `s1` so that it references this alternate `String`. What would this look like in memory? The result is shown in figure 7.3. The 5 and the `String` "Scala is fun!" have not been changed. They remain in memory just like before. Because nothing references them, these objects will be collected and disposed of automatically. What has changed is the references in `i1` and `s1`. They now point to new objects in new chunks of memory that hold these new values. Both the `Int` and `String` types are immutable. As a result, once the objects are created, nothing can change them. They do not allow any operations that would alter their values. If we had made these variables using `val` we would not be able to change where the arrows point either.

Indeed, all of the types used here are immutable with the exception of the `Array`. So `Array` is the only one where we can change the value in the box on the right. To see how

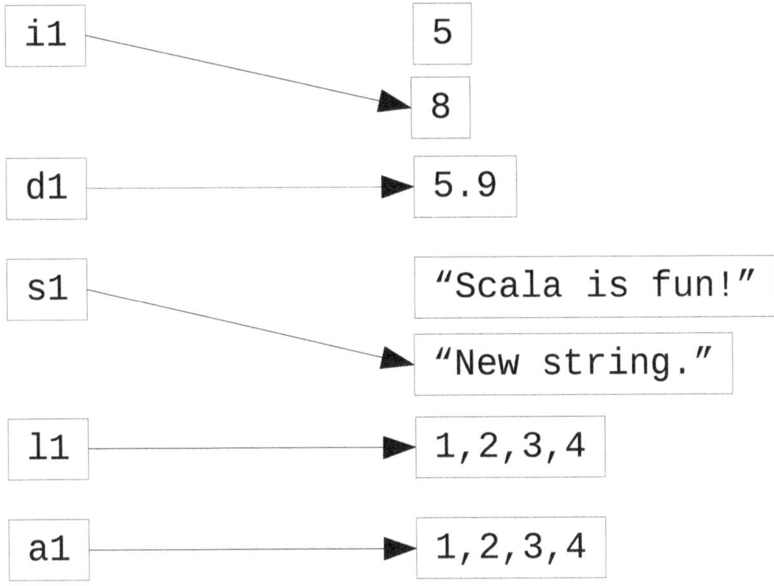

FIGURE 7.3: Modified view of the memory after two variables have been assigned different values.

this can matter we will consider only the `List` and the `Array` variables and add two more lines of code to the picture.

```
val l2=l1
val a2=a1
```

This gives us a slightly different picture that is shown in figure 7.4. Now we have two different variables that point to each of the `List` and `Array` objects. These second references to the same object are often called aliases. Aliasing is where mutability starts to cause problems. If you have not figured out why yet, perhaps this code will make it clear.

```
scala> a2(3)=99

scala> a1
res1: Array[Int] = Array(1, 2, 3, 99)
```

In the first line we change the value in the last slot in `a2`. On the next line we look at the value of `a1`. Note that it is changed. We did not explicitly do anything with `a1`, but the value is changed because `a2` is an alias of `a1`. You can not do this with `l2` because the `List` type is immutable.

There are times when the ability to make an alias and change it is great. There are some tasks for which this can make things much more efficient. However, if you lose track of what is going on and what values are aliases of what values, you can run into serious problems.

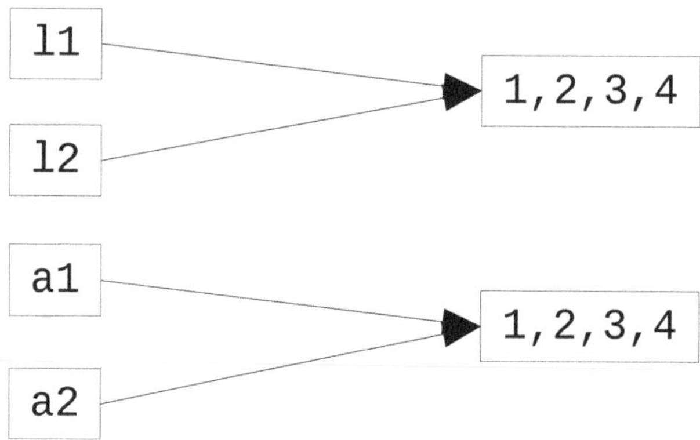

FIGURE 7.4: What the memory looks like after the introduction of l2 and a2.

7.8 Basic Argument Passing

Now that we have looked at the way memory is organized for variable declarations and the aliasing issue, we should take another look at what happens when you call a function. The two are actually very similar. When you pass values into a function, the function has local vals with the local argument names, but they reference the same objects that the outside variables referenced. Consider the following code from a script.

```
def getAndClear(arr:Array[Int],index:Int):Int = {
  val ret=arr(index)
  arr(index)=0
  ret
}

val numbers=Array(7,4,9,8,5,3,2,6,1)
val place=5
val value=getAndClear(numbers,place)
// Other stuff
```

The function is passed an Array of Ints and an index for a location in that array. It is supposed to return the value at that location and also "clear" that location. The meaning of "clear" here is to store a zero at that location. To see how this works look at figure 7.5 which shows the arrangement of memory. The function is defined and the variables numbers and place are both declared and initialized. We get new objects for them to reference.

When the getAndClear function is called, numbers and place are passed in to be the values of the arguments arr and index. While the code is in getAndClear, arr is an alias for the object referred to by numbers and index is an alias for the object referred to by place. This is significant because the getAndClear method has a side effect. It modifies one of the values stored in the Array that is passed to it, when the code gets down to "Other stuff", the memory looks a bit different as shown in figure 7.6. At that point, the function has finished so its local variables are no longer needed. In addition, the 6th element in the array numbers has been changed. The function managed to change the value for the Array

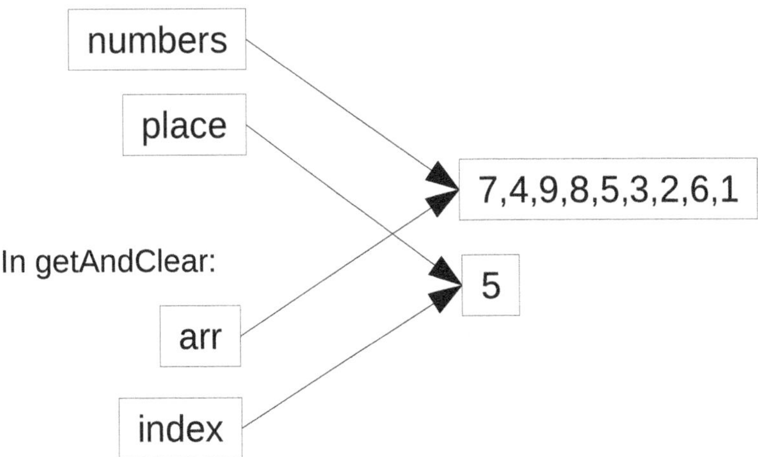

FIGURE 7.5: The memory layout from the `getAndClear` script. The arguments passed into the function become aliases for the objects created at the top level of the script.

as seen outside that function. With a name like `getAndClear` you could probably figure out that this type of side effect might happen. However, when you pass any mutable value into a function, you need to be aware that the function could change your value, even if you do not really want it to.

How would this be different with a `List`? Well, the `List` type is immutable so it would be impossible for the function to change the list. If you wanted it changed, the function would need to return a tuple including both the value gotten and a new `List` with that value changed. The code below shows how that could be done with some of the methods we have introduced in this chapter.

```
def getAndClear(lst:List[Int],index:Int):(Int,List[Int]) = {
  (lst(index),lst.zipWithIndex.map( tup => {
    val (n,i)=tup
    if(i==index) 0 else n} ) )
}
```

There is only one expression in this function, the tuple to return. The first element in the tuple is `lst(index)`. It looks up the proper value and uses the value there. Remember that looking things up in this way on a `List` is not really efficient, but we have to do so if given a `List` and an index. The second part of the tuple is a bit more complex. It called the method `zipWithIndex` on `lst`. This returns a new `List` where each element is a tuple that contains an elements from the `List` along with the index of that element. We have to do this because we need the index in order to decide if it is the one to be cleared or not. If we were clearing based on value instead of position this would not be needed.

The method `map` is immediately called on the `List` of tuples and it is given a function literal that takes one argument called `tup`, short for tuple. This function literal uses curly

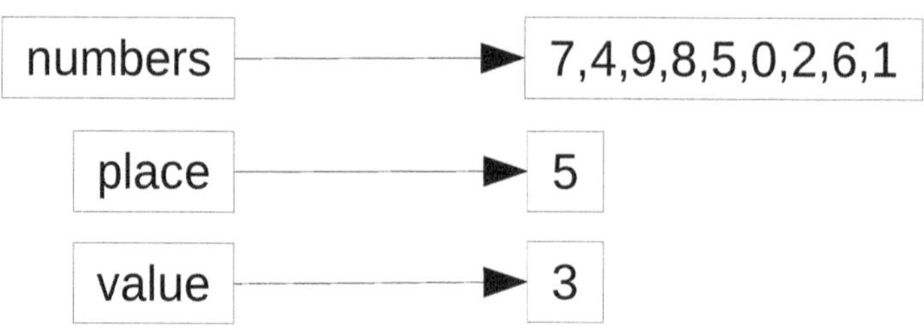

FIGURE 7.6: The configuration of memory after `getAndClear` has executed.

braces and has two lines. The first line is a `val` that pulls the two parts out of the tuple. The number from the `List` is called `n` and the index is called `i`. The function literal finished with an `if` expression that checks the index and gives the value of 0 or `n` based on if `i` is the same as `index`. There are other ways to write this. One that could be more efficient when `index` is small compared to the length of the `List` is shown here.

```
def getAndClear2(lst:List[Int],index:Int):(Int,List[Int]) = {
  (lst(index),lst.take(index):::(0::lst.drop(index+1)))
}
```

This version uses the cons operator, `::`, which we saw before. It adds one new element to the front of a `List` and gives that back. It also uses an operator that we have not seen previously, the `:::` operator. This does roughly the same thing as `::` except that the argument on the left is a full `List` that should be appended to the front of the second `List`.

Of course, because the `List` version is having to rebuild the `List`, it could go a step further and truly clear the value. It could remove it from the `List` instead of overwriting it with zero. This would be done with `filter` instead of `map` on the first version. The second version is even easier to change. We leave that as an exercise for the reader to create and try.

An interesting question now becomes which of these is better? Is it better to use an `Array` which can be mutated or a `List` which can not? The answer is that it depends on whether you want a function to be able to mutate the data. If the function is supposed to mutate the data then there can be a speed benefit to passing something mutable. However, if you do not want the data mutated then there is a problem with passing mutable data. When you pass mutable data into a function, you have no control over whether it is mutated. In situations where you do not want your data mutated, this requires that you make a DEFENSIVE COPY. To protect your version of the data, you make a copy of it and pass the copy into the function.

For this reason, the question of whether you want to use a mutable or immutable structure depends a lot on the way in which you use that data. If it is going to be mutated frequently then it might make sense for it to be mutable. However, if it is not going to be mutated frequently then immutable likely makes sense so you do not have to make defensive copies. Even in situations where there might be a fair bit of mutation, you need to consider how much defensive copying will be needed if you choose mutable.

7.9 Pass-By-Name

Earlier we looked at the way in which arguments are passed in Scala. This is a style called PASS-BY-VALUE. The function gets a copy of what the variable was in the calling code. The function can not change that variable, it can only change what it refers to if what it refers to is mutable. Some languages provide an alternate way of passing things called PASS-BY-REFERENCE. When something is passed by reference, the function has the ability to change the variable in the calling function itself. The fact that all variables in Scala basically are references blurs the line between the two, but fundamentally Scala only allows pass-by-value and the function can only modify things seen to the outside if the variable refers to a mutable object.

While Scala does not provide a true pass-by-reference, it does provide a passing style that few other languages provide, PASS-BY-NAME. The idea of pass-by-name is that the argument is passed not as a value, but as a THUNK that is basically a set of code that will be executed and give a value when the parameter is used in the function. You can imagine pass-by-name as automatically creating a function that takes no argument and returns a value that will be executed each time the argument is used. To help you understand this and see the syntax, we will give a simple example. We will start making a basic increment function in the way we are used to doing.

```
scala> def incr(n:Int):Int = {
     | println("About to increment.")
     | n+1
     | }
incr: (n: Int)Int
```

The print statement is just to help us keep track of what is happening when. Now we will write the same function again, but this time pass the argument by name instead.

```
scala> def incrByName(n : =>Int):Int = {
     | println("About to increment.")
     | n+1
     | }
incrByName: (n: => Int)Int
```

The syntax for passing an argument by name is to put a rocket before the type of the argument. This is the same arrow used in function literals. If we were to put empty parentheses in front of it to get `()=>` we would have the type of a function that takes no arguments and returns an `Int`. The by-name argument is much the same only when you call the function you do not have to explicitly make a function. To start off, we will call this function in the simplest way we possibly could.

```
scala> incr(5)
About to increment.
res0: Int = 6

scala> incrByName(5)
About to increment.
res1: Int = 6
```

No surprises here. They appear to both do the same thing. Both print the statement and give us back 6. However, the two are not really doing the same thing. To make that clear, we will call them again and have the argument be a block that includes a print.

```
scala> incr({println("Eval"); 5})
Eval
About to increment.
res2: Int = 6

scala> incrByName({println("Eval"); 5})
About to increment.
Eval
res3: Int = 6
```

Now it is clear that they are not doing the same thing. When you pass an argument by value, it is evaluated before the function starts so that you can have the value to pass through. When you pass by name, the evaluation happens when the parameter is used. That is why the line "Eval" printed out after "About to increment." in the second case. The `println` in the function happens before the value of `n` is ever accessed.

Using pass-by-name gives you the ability to do some rather interesting things when mutations of values comes into play. To see that, we can write a slightly different function.

```
scala> def thriceMultiplied(n : => Int):Int = n*n*n
thriceMultiplied: (n: => Int)Int
```

It might seem that this method should be called cubed, but as we will see, this might not always apply for it because it uses pass-by-name. To start with, let us call it with an expression that gives a simple value, but prints something first.

```
scala> thriceMultiplied({println("Get value."); 5})
Get value.
Get value.
Get value.
res4: Int = 125
```

Note that the `println` statement happened three times. This is because the value `n` was used three times in the function. Had the parameter not been passed by-name, that would not have happened. It would have printed once, when the function was called. Try this for yourself.

This particular call still gave us a valid cube though. So why was the function not simply called cube? The reason is that if the code in the argument does the right thing, the result won't be a cube. Here is an example.

```
scala> var i=5
i: Int = 5

scala> thriceMultiplied({i+=1; i})
res5: Int = 336
```

336 is not a perfect cube. So how did we get this value? In this example we introduced a `var`. The code in the pass-by-name argument alters this `var`. As a result, every time that `n` is used, the value given by `n` is different. The first time we use `n` the value is 6 because the original 5 was incremented and then returned. The next time it is 7 because the 6 that it was set to the previous time is incremented again. The last time it is 8. So the answer is $6 * 7 * 8 = 336$.

These calls with both parentheses and curly braces might seem odd. The creators of Scala thought so. As a result, they made it so that if a function or method takes an argument list with only one value in it, that value can be in curly braces instead of parentheses. This allows a shorter syntax for the call that looks like this.

```
scala> thriceMultiplied {i+=1; i}
res6: Int = 990
```

You might be tempted to think that leaving off the parentheses changed what happened because the result is different. It was not the parentheses though. Remember that after the last call `i` had been incremented up to 8. So this time the result was $9 * 10 * 11 = 990$.

7.10 Fill and Tabulate

One of the reasons for introducing pass-by-name in this chapter is that there are some other ways to make `Array`s and `List`s. One of them uses pass-by-name. Previously we saw how we could follow `Array` or `List` with a list in parentheses to get back either an `Array` or a `List` with those values in it. We could also use `new` to get a large array of default values or `::` to put together `List`s. There are also methods we can call on `Array` and `List` called `fill` and `tabulate`. Both can be used to make a new `Array` or `List`. They both work well for making large `Array`s and `List`s, unlike the version where you have to specify the individual values. Let us look at how these work.

The `fill` method is a curried method that takes a first argument specifying how many elements to make and a second argument, which is a pass-by-name argument giving the value. The simplest invocation of this might look like the following.

```
scala> Array.fill(10)(4)
res3: Array[Int] = Array(4, 4, 4, 4, 4, 4, 4, 4, 4, 4)

scala> List.fill(6)(23)
res4: List[Int] = List(23, 23, 23, 23, 23, 23)

scala> List.fill(6)("hi")
res5: List[java.lang.String] = List(hi, hi, hi, hi, hi, hi)
```

If the second argument is a constant, then the call simply fills the whole `Array` or `List` with that value. However, that argument is passed by-name so it is actually re-evaluated for each element of the sequence and if it has side effects, they will happen multiple times.

The easiest way to see this is to have the second argument print something.

```
scala> Array.fill(4){ println("Evaluating argument"); 5 }
Evaluating argument
Evaluating argument
Evaluating argument
Evaluating argument
res6: Array[Int] = Array(5, 5, 5, 5)
```

Note that here the second argument has been put in curly braces only. Remember that Scala allows you to do that for an argument list that only has one argument. You can add an extra set of parentheses, but you will still need the curly braces because this is a block of code with two expressions in it. This example is illustrative, but not really all that useful. A more interesting example involves using a var.

```
scala> var i=1
i: Int = 1

scala> List.fill(5){ i*=2; i }
res7: List[Int] = List(2, 4, 8, 16, 32)
```

Here the var i is initialized to be 1. The value argument multiplies i by 2 and stores that value back in i. Then it gives back i as the value to use. You can see in the output that this gives us powers of 2. While there are situations where this can be helpful, this method of filling an array with powers of two is likely to confuse most people reading your code. An even more useful example is to have the value argument read input.

```
scala> Array.fill(5){ readInt() }
res8: Array[Int] = Array(6, 4, 8, 2, 3)
```

This is what we started off the chapter doing, writing functions to read values into an Array or List. It turns out that as long as we know how many elements we want, we can do this with a single line of code thanks to fill and pass-by-name.

It is often helpful, when calculating values for an Array or List, to know the index of the value you are calculating for. The fill method does not do this for us by default. We could use a var to keep track of that, in a manner similar to what we did above, but this is a pain and quite error prone. That is why the tabulate method exists. You call tabulate much like you do fill, only the second argument is not a pass-by-name variable, it is a function that takes an Int and returns the type the sequence should be filled with. So if you want to fill an array with the result of evaluating a polynomial on the index you could do the following.

```
scala> Array.tabulate(6)(x => 3*x*x+5*x-7)
res9: Array[Int] = Array(-7, 1, 15, 35, 61, 93)
```

This fills the array with values of $3x^2 + 5x - 7$ where x is the index in the array.

7.11 Complete Grades Script/Software Development

In chapter 5 we spent some time writing functions to calculate averages in a class. We now have the ability to put together everything that we have learned and write a script that will keep track of the grades for a single student and be able to calculate that student's average. This could be extended to multiple students, but we will not do that here.

When we solve a large problem there are certain steps that much be taken. The first is to get a clear description of what problem you are solving. This step is called ANALYSIS. The analysis step does not involve any coding or even thinking about coding. We just want a description that has enough detail that we know what it is we are supposed to create. Once we have that we can move on to the next step which is called DESIGN. In the design step we figure out how we are going to go about solving the problem described in the analysis. The design step does not involve coding either, but it specifies what we will do when we get to the coding. The design phase is where we do our problem decomposition. We think about how to break the problem down and what pieces we will need to put together to get a single, whole solution to the problem. After we have a design we get to the IMPLEMENTATION phase. This is where we will actually write the program. Ideally, by the time that we get to implementation we have worked out most of the problems we will run into. The ability to see the problems we will encounter requires a lot of experience. As novice programmers, you should try to spend time on design, thinking about how to break the problem down, but do not become so intent on having a complete design that you stall out and never get to implementation.

After you have an implementation you have to make sure it works and fix any problems. This is called the TESTING and DEBUGGING phase because it is where you test the code to find all the bugs and fix them. For real software, after all the bugs are fixed and the software has been rolled out, you worry about MAINTENANCE. This is where you continue to make alterations to the software to improve it or address customer requests.

At the level you are at now, you typically don't do much analysis or maintenance. You do not do much analysis because generally your instructor or this book will give you a fairly complete description of what you are doing. You do not do maintenance because you do not have a customer base that is using the software. A little of each of these can still creep into your work when you make sure you fully understand the description of a problem and then perhaps if you are given a later problem that incorporates code you had done previously.

You should spend a significant amount of time working on design and debugging. In fact, the implementation phase is often the shortest phase of the software creation process.[4] If done properly, most of the thinking goes into the design phase so implementation is simply typing in what you had designed. Testing and debugging is extremely variable in time. It is possible to write your program and have everything work perfectly the first time. That gets less likely as programs get longer and is especially rare for novice programmers, though the static type checking in Scala will help to make it so that if everything in your program checks out it has a better chance of working. When things do not work, debugging can take arbitrarily long periods of time and is often the longest phase of software development.

In real software development these different phases are not always so distinct, and they are rarely done in sequential order. It is common that in one phase of development you realize something that was missed in an earlier phase. Some approaches to software development mix the phases together so that the developer is involved in more than one at any given time. If you continue on in Computer Science, these topics should be covered in full detail in a course on Software Engineering.

At this point we have a very vague description of the problem that we are going to solve. For that reason, we need to do some analysis up front so that we have a clear description of exactly what we are going to solve. This script is going to use a text menu to give the user options. The options will include the ability to add grades of different types as well as an option to print the current grades and average. Types of grades are tests, quizzes, and assignment. For the final average, the tests and assignments count 40% each and the quizzes make up the remaining 20%. The lowest quiz grade should be dropped. The menu will have numbers by each option and users enter a number to select that option. That is probably a sufficient analysis for our needs here. We can start working on the design and if we hit anything we have questions on, we can go back and add some detail to the analysis before we proceed further with the design.

So how are we going to break this problem down? At the top level we have code that repeats over and over and each time a menu is printed, the user selects an option, and we execute code to perform the selected option. The only way we currently know how to do

[4]At least if you do things properly the implementation will be short. I like to tell my students that hours of coding can save minutes of thinking. The point of this is that if you do not sit and think about design before you go to the keyboard and write code you are likely to spend hours hashing out the design that you could have nailed down in a few minutes by thinking about it to begin with.

something repeatedly an unspecified number of times is with a recursive function. So we will need a recursive function that contains all this other functionality. That function will call itself until we hit the condition where it exits. There is a natural decomposition at this point. Printing the menu is an obvious function. We have already seen that we can write functions for taking averages or averages dropping the minimum grade as well as one to calculate the final average.

That breaks down the functionality. If you were to run to the keyboard now and start typing you would quickly realize that something significant was left out of the design, how we are storing the grades. There are many ways that we could do this, but given that our only functionality is to add grades and calculate averages, the solution is fairly clear. Our only real options at this point are `Arrays` and `Lists`. `Arrays` have a fixed size and are inefficient to add to. That is less than ideal for this application because we will be adding grades and have not specified how many there can be. So `Lists` make more sense. We can use three `Lists` that store the three different types of grades. If we do this in an imperative style, the `Lists` will be declared with `var` before the main recursive function. If we use a functional style, they will be passed into the next call for the iteration. We will look at both to illustrate the differences. Most of the functions will not be altered by this distinction.

That gives us a decent design for this little application. Now we turn to the implementation. We will reuse the course average function from section 7.6 and the functions that it calls for now. We need new functions for our main recursive function and printing the menu. We will start with an imperative version of the main function and look at a functional adaptation afterward.

```
def printMenu {
    println("""Select one of the following options:
 1. Add a test grade.
 2. Add a quiz grade.
 3. Add an assignment grade.
 4. Calculate average.
 5. Quit.""")
}

var tests = List[Double]()
var quizzes = List[Double]()
var assignments = List[Double]()

def mainGrades {
  printMenu
  readInt() match {
    case 1 =>
      println("Enter a test grade.")
      tests = readDouble() :: tests
    case 2 =>
      println("Enter a quiz grade.")
      quizzes = readDouble() :: quizzes
    case 3 =>
      println("Enter an assignment grade.")
      assignments = readDouble() :: assignments
    case 4 =>
      println("The average is "+courseAverage(tests,assignments,quizzes))
    case 5 =>
      return
    case _ =>
      println("Unknown option. Try again.")
```

```
  }
  mainGrades
}
```

This begins with the `printMenu` function which simply prints a menu. The only thing of interest here is the use of the triple quote string so that the string can be made to span multiple lines. If you do not do this, you will have to use `\n` to denote line breaks.

In this imperative version, the `printMenu` function is followed by the declaration of three variables. They are all `vars` and they are of the type `List[Double]`. The have to be `vars` because the `List` itself is immutable. So whenever something is added we have to change the variable reference and point to a new `List`. Remember, this can be done efficiently for the `List` type so that is fine.

If you type this code in and play with it a bit, you will probably find that you can make it behave poorly. In particular, you can get it so that when you ask for the average you get an output like this.

```
The average is NaN
```

This should lead you to two questions. First, what is this thing called `NaN`? Second, what is causing this? Technically the second matters more because we need to know that to fix it. However, knowing the answer to the first can help you figure it out if you were not able to do so from your testing. The `Double` type, as we have discussed previously, is used to represent numeric values that have fractional parts. It has a rather large range and good precision, but it is technically a fixed precision type. You can not represent any real number with it. In addition, some mathematical operations produce results that are not well defined. Consider the following.

```
scala> 1.0/0.0
res2: Double = Infinity

scala> -1.0/0.0
res3: Double = -Infinity
```

You should have learned in a math class that you can not divide by zero. If you do this with `Ints` you get an error. However, if you do it with `Doubles` you get what you see here. Any operation that goes outside the range of the `Double` will also get you either `Infinity` or `-Infinity`, depending on which side of the range it goes out on as is shown here.

```
scala> 1e150*1e200
res6: Double = Infinity
```

What does all of this have to do with `NaN`? Well, `NaN` stand for "Not-a-Number" and it is the result of an operation when it is not clear what the result should be. One way to get this is to add `Infinity` and `-Infinity` or do other operations with `Infinity` or `-Infinity` that the computer can not know the result of. Consider this example.

```
scala> 1e150*1e200-1.0/0.0
res10: Double = NaN
```

In general the computer can not know if this would be out of range on the positive or negative so it goes to `NaN` because it gives up. There is one other way to get a `NaN`: divide zero by zero.

```
scala> 0.0/0.0
res11: Double = NaN
```

This is what is happening to cause our error, and it happens when zero test or assignment grades or only one quiz grade have been entered.

If we have no test or assignment grade, the normal average will return 0.0/0, as the sum is the `Double` 0.0 and the length is the `Int` 0. If the list of quiz grades only has one element, then that one is the minimum. So the function returns `(x-x)/(1-1)` where `x` is the one grade as a `Double`. Clearly that is going to be a `NaN`. So how do we fix this? We need to make the `average` and the `averageDropMin` functions a bit smarter so that they have special cases in the instances when they should not do division. We can do this by putting `if` expressions in both.

```
def average(nums:Double*):Double =
  if(nums.isEmpty) 0 else nums.sum/nums.length

def averageDropMin(nums:Double*):Double =
  if(nums.isEmpty || nums.tail.isEmpty) 0 else
  (nums.sum-nums.min)/(nums.length-1)
```

If you use these versions of the functions things should work well.

Now to the functional alternative. The imperative version uses three `var` declarations outside of the function. These declarations can be seen by anything in the script that follows them. We typically refer to this as the GLOBAL SCOPE. For small scripts this might not be a problem, but you do not want to get into the habit of making things global. Anything that is global in a large program can be messed up by any line of code, making it very hard to track down certain types of errors. It is better to have our variables declared in a LOCAL SCOPE, inside of a function.

If we simply declare the variables inside of `mainGrades`, it will not work because when you make a recursive call, you get new versions of all of the local variables. Instead, we want to pass the three lists into the function as arguments. The arguments to a function act just like local `val` declarations except that the value is provided by the calling code. This new version will not include any `var` declarations, nor will it have any mutable values as everything is stored in immutable `Lists`. So other than the printing, it is completely functional. Here is what such a function might look like.

```
def mainGrades(tests:List[Double],assignments:List[Double],
    quizzes:List[Double]) {
  printMenu
  readInt() match {
    case 1 =>
      println("Enter a test grade.")
      mainGrades(readDouble() :: tests,assignments,quizzes)
    case 2 =>
      println("Enter a quiz grade.")
      mainGrades(tests,assignments,readDouble() :: quizzes)
    case 3 =>
      println("Enter an assignment grade.")
      mainGrades(tests,readDouble() :: assignments,quizzes)
    case 4 =>
      println("The average is "+courseAverage(tests,assignments,quizzes))
      mainGrades(tests,assignments,quizzes)
    case 5 =>
    case _ =>
      println("Unknown option. Try again.")
      mainGrades(tests,assignments,quizzes)
  }
}
```

Instead of mutating values, this version passes the newly built `List` into the recursive call where it is a completely new argument value for the next level down in the recursion. The options that do not change the `Lists` simply pass through the same values.

7.12 End of Chapter Material

7.12.1 Problem-Solving Approach

Despite the length of this chapter, no new valid statement types were introduced. Instead, a significant amount of flexibility was added in what you might do with the existing statement types. In particular, you can now use expressions with collection types and call methods on those collection types. You also learned that types themselves can be more complex with type parameters. You can make functions that take variable numbers of arguments and you can pass arguments into functions using pass-by-name.

7.12.2 Summary of Concepts

- `Arrays` and `Lists` are collections that allow us to store multiple values under a single name. They are sequences, meaning that the values have an order and each one is associated with an integer index. Indexes start at zero.
- Both can be created in a number of ways.
  - `Array(e1,e2,e3,...)` or `List(e1,e2,e3,...)` to make short collections where you provide all the values.
  - `new Array[Type](num)` will make an `Array` of the specified type with the specified number of elements. All elements will have a default value. This is the least common usage in Scala.
  - Use the `::` operator with `Lists` to append elements to the front and make a new `List`. Use `Nil` to represent an empty `List`.
  - `Array.fill(num)(byNameExpression)` will make an `Array` of the specified size with values coming from repeated execution of the specified expression. This will also work for `Lists`.
  - `Array.tabulate(num)(functionOnInt)` will make an `Array` with `num` elements with values that come from evaluating the function and passing it the index. Often function literals are used here. This will also work with `List`.
- `Arrays` have a fixed size and are mutable. Access and mutate using an index in parentheses: `arr(i)`.
- `Lists` are completely immutable, but you can efficiently add to the front to make new `Lists`. Can be accessed with an index, but using `head` and `tail` is typically more efficient.
- The Scala collections, including `Array` and `List`, have lots of methods you can call on to do things.

  - There are quite a few basic methods that take normal values for arguments and give you back values, parts of collections, of locations where things occur.
  - The real power of collections is found in the higher-order methods. The primary ones you will use are `map`, `filter`, and `foreach`.
  - There are a handful of methods that give you back various pieces of collections, including combinatorial possibilities such as `permutations` and `combinations`.

- You can make the last parameter of a function accept a variable number of arguments, var args, by putting an asterisk after the type. The name given to that parameter functions as a sequence. To pass a real sequence into a function in place of a var args parameter, follow it with `:_*` to tell Scala you want that seen as a `*` argument of some type.
- When two variables refer to the same object in memory we call them aliases.
- Mutable objects can cause problems with aliases because changes to the object can be made using any of the aliases and they will be reflected in all of them.
- By default, arguments in Scala are passed by-value. The value of the reference is passed into the function. This automatically makes an alias of the object. If you want to protect a mutable object from unintended changes when it is passed to a function you need to make a defensive copy.
- Scala also allows pass-by-name parameters. Instead of passing a value, these pass a chunk of code called a thunk that is evaluated every time the parameter is used.

7.12.3 Self-Directed Study

Enter the following statements into the REPL and see what they do. Try some variations to make sure you understand what is going on. Some of these are intended to fail. You should understand why.

```
scala> val arr = Array(1, 2, 3)
scala> arr(0)
scala> arr(1)
scala> arr(0) = 5
scala> arr.mkString(" ")
scala> val lst = List(2, 3, 5, 7, 11, 13, 17)
scala> lst.head
scala> lst.tail
scala> val bigArr = new Array[String](100)
scala> bigArr(0) = "A string"
scala> bigArr(1) = "Another string"
scala> bigArr(0).length
scala> bigArr(2).length
scala> val lst2 = 1 :: 2 :: 3 :: 4 :: 5 :: Nil
scala> lst.zip(lst2)
scala> lst.zipWithIndex
scala> val arr3 = Array.fill(5)(readLine())
scala> val lst3 = List.tabulate(20)(i => i*i)
scala> def printList(lst:List[Int]) = lst match {
  case h::t =>
    println(h)
```

```
    printList(t)
  case Nil =>
}
scala> printList(lst3)
scala> lst3.mkString(" ")
scala> lst3.map(i => math.sqrt(i))
scala> arr3.map(s => s.length)
scala> bigArr.filter(s => s!=null)
scala> arr.permutations.foreach(println)
scala> lst.sliding(4)
scala> def printWords(copies:Int,words:String*) {
  words.foreach(w => println(copies*w))
}
scala> printWords(3,"Hi")
scala> printWords(2,"This","is","a","test")
scala> val numWords = lst.map(_.toString)
scala> printWords(4,numWords)
scala> printWords(4,numWords:_*)
scala> val arrAlias = arr
scala> arrAlias.mkString(" ")
scala> arr.mkString(" ")
scala> arrAlias(0) = 42
scala> arr.mkString(" ")
scala> def alterArray(a:Array[Int]) {
  a(0) = 99
  a(1) = 98
  a(2) = 97
}
scala> alterArray(arr)
scala> arr.mkString(" ")
scala> def getIndex[A](index:Int,a:Array[A]):Option[A] = {
  if(index<a.length) Some(a(index)) else None
}
scala> getIndex(1,arr)
scala> getIndex(4,arr)
scala> lst.find(_ == 5)
scala> lst.find(_ == 9)
scala> def recursiveWhile(cond: =>Boolean)(body: =>Unit) {
  if(cond) {
    body
    recursiveWhile(cond)(body)
  }
}
scala> var i=0
scala> recursiveWhile(i<10) {
  println(i)
  i += 1
}
```

7.12.4 Exercises

1. Think of as many ways as you can to make an **Array[Int]** that has the values 1-10 in it. Write them all to test that they work. The first two should be easy. If you work you can get at least three others knowing what has been covered so far.

2. Think of as many ways as you can to make a `List[Int]` that has the values 1-10 in it. Write them all to test that they work. The first two should be easy. If you work you can get at least three others knowing what has been covered so far.

3. Make a `List[Char]` that contains `'a'-'z'` without typing in all the characters. (Use `toChar` to make this work.)

4. Given two `Array[Double]` values of the same length, write a function that returns the element-wise sum. This is a new `Array` where each element is the sum of the values from the two input arrays at that location. So if you have `Array(1,2,3)` and `Array(4,5,6)` you will get back `Array(5,7,9)`.

5. Write a function that produces a sequence of prime numbers. You can use the `isPrime` function that you wrote for exercise 9 (p.130). The function should take an argument for how many prime numbers need to be in the list.

6. Write a function that takes a number of values and returns the average excluding the largest and smallest values.

7. Write a function that takes a number of grades and returns the average with the lowest two grades dropped. Make sure this function behaves reasonably even for smaller input sets.

8. Write several versions of code that will take an `Array[Int]` and return the number of even values in the `Array`. Each one will use a different technique. To test this on a larger array you can make one using `Array.fill(100)(util.Random.nextInt(100))`

   (a) Use a recursive function.

   (b) Use the count higher-order method.

   (c) Use the filter higher-order method in conjunction with a regular method.

   (d) Use the map higher-order method in conjunction with a regular method.

9. Write a version of `getAndClear` that uses a `List` and completely removes the value at the specified index.

10. You can treat a `String` as a sequence of `Char` and use all the methods that we introduced here for `List` and `Array` on it. Using that, what code would you use to do the following? For all of these assume that `str` is a `String` provided by the user through `readLine()`.

    (a) Get a version of the `String` with all the spaces removed.

    (b) Get a version of the `String` with all vowels removed.

    (c) Split off the first word of the `String`. Write code that gives you a `(String, String)` where the first element is everything up to the first space in `str` and the second one is everything else.

    (d) Convert to a sequence that has the integer values of each of the characters in `str`.

    (e) Write code that will take a sequence of `Int`, like the one you just made, and give you back a String. (Note that if you get a sequence of `Char` you can use `mkString` to get a simple `String` from it.)

7.12.5 Projects

1. We can express a general polynomial as

$$A_n x^n + A_{n-1}x^{n-1} + A_{n-2}x^{n-2} + ... + A_2x^2 + A_1x + A_0$$

   where the A values are the coefficients on different terms. You can view a sequence of `Doubles` as a polynomial by saying that the value at index n is the value of A_n. Using this representation, `Array(1.0,0.0,2.0)` would represent $2x^2+1$. Note that the order when you look at the `Array` is the opposite of the order seen in the polynomial.

   Using this representation of polynomial, write functions that do addition and subtraction of two polynomials. You can assume that the `Arrays` passed in are the same length. Also write functions that gives you the derivative of a polynomial and the anti-derivative.[5] Note that the anti-derivative function will need and extra argument for the value of the constant term.

   To help you learn, try to write these functions using both recursion with an index and using standard methods on sequences. For the former option try doing it with `Lists` as well using `::` to build the `Lists` in the recursion. For the latter a hint when you write `add` and `subtract` is to consider using the `zip` method.

   If you want an extra challenge, try writing multiplication as well. This is a significantly harder problem, but if you can do it with both recursion and the built-in methods you should feel good about your understanding.

2. This project has you implement a simple form of encryption and decryption. You should write at least two functions called `encrypt` and `decrypt`.

   The form of encryption we are using is offset encryption where you simply offset letters in the alphabet. If you pick an offset of 1, then if there is an `'a'` in the original message, the encrypted message will have a `'b'`. For a first cut, write these functions so they just add or subtract a specified offset to all characters. (Note that when you do arithmetic on a `Char` you get an `Int`. Use the `toChar` method to get back that type. So `('a'+1).toChar` will give you `'b'`.) Your methods will both take a `String` and an `Int` and return a `String`. The `Int` passed in is the offset to apply to the characters.

   This approach is easy, but it produces non-letters in certain situations like when you do an offset on `'z'`. Refine your functions so that they skip any characters that are not letters and wraps around so that if the offset goes above `'z'` it comes back to `'a'`. You have to handle capital letters as well. (Note that the `Char` type has methods called `isLetter`, `isUpper`, and `isLower` that can help you with this. You can also do comparisons and math with the `Char` type.)

   If you want an extra challenge and you want to make it harder for someone to try to break your code, revise your methods so that they take an extra `Int` and insert random characters at intervals of that spacing. So if the value is 5, you will insert an extra random character between every fifth character from the message. You can generate a random letter character with the expression `('a'+util.Random.nextInt(26)).toChar`.

   For this whole problem remember that you can treat `Strings` as sequences. So you can use them like an `Array[Char]` except that they are not mutable.

3. For this project you will be doing some simple data analysis. If you go to the book website and look at the page for this chapter you will find a text file for this exercise that has values in it. It is weather data for 2/17/2010 in San Antonio, TX

[5] You might recognize this better as an indefinite integral.

from the Weather Underground (`http://www.wunderground.com`). It has one line for each hour and each line has a single number that it the temperature for that hour. Write a function that will read this into an `Array[Double]` using redirection and `readDouble`. Then write functions to calculate the following: standard deviation and change per hour. The standard deviation will return a `Double`. The change will return an `Array[Double]` that is one shorter than the original data set.

4. This project is to be built on top of project 6.3. For this one you make functions that will take a `List` or `Array` of the values used to represent spheres and planes and return the impact parameter of the first one the ray hits and the information on that geometry.

5. For this project, you will write a function that could help people playing Scrabble®. You will want to break this down so you have multiple functions in the end, but the last one is what really matters. If you go to the book's website, the page for this chapter will have a data file for this exercise called 2of12.txt. This is a file with a list of 41238 English words. There is one word per line. Using input redirection and `readLine` you will read the contents of this file into an `Array[String]`.

   Once you have this file read in, your goal is to give back words that the person could likely make using letters he/she has to play. Your end goal is a function that takes the list of the dictionary words and a `String` of the letters the person has to play, which returns all the words that can be made from their letters plus one other letter. The one other letter is something you do not know that should have been played somewhere on the board. For our purposes here we do not care what it is. You should give back any word that can be formed with some subset of the player's letters plus one extra.

   To do this you really want to use functions to break the problem down. Consider what higher order methods could be helpful and what types of functions you need to use with them. To make this all work with input redirection download and edit the 2of12.txt file and insert a top line where you put the letters the player has in Scrabble. The script will read that line to get the letters and then read the proper number of lines to get all the words.

   Lastly, to help you solve this problem consider using the `diff` method. This method tells you the difference between two sequences. Play around with calling `diff` using different strings to see what it does. Start with `"aabbcc".diff("abcde")` and `"abcde".diff("aabbcc")`.

6. Now that you can keep store multiple values in `List`s and `Array`s, you can push the recipe program to use an arbitrary number of items. Make a script that asks the user for items in their pantry. It should ask for item names and how much they have until the user enters a name of "quit". After that, the program will let the user run through recipes, entering one recipe name, followed by a number of ingredients, followed by the name and amount of each ingredient. It should then tell the user if they have enough to make that recipe and ask if they want to check another recipe.

7. Every semester you get to build a schedule of the courses that you will take in the following semester. This is the start of some projects at the end of which you write a program that will help you do so. For this first part, you want to have the user input courses he/she is interested in taking as well as how much they want to take those courses. The user should then tell the program how many courses should be taken the next semester and the minimum score for the "want to take" sum that will be accepted. The program should print out all of the schedule combinations with that

many courses that make the cut. (Hint: Use `combinations` to run through all the possibilities. For each possibility, you can use `map` and `sum` to get the total "score" for that option.)

8. Write a function that takes a `Double*` and returns the median value as a `Double`. Your solution is not expected to be efficient. This is more of a test of your ability to do logic using the methods that were presented in this chapter.[6]

9. Write a function that takes an `Int*` and returns the mode of the values. As with the previous project, the goal here is more for you to think about how to make a solution than to make an efficient solution.

There are additional exercises and projects with data files to work on posted on the book's website.

[6]The API includes a number of methods for sorting. Those could be used to make a more efficient and shorter solution to this problem, but the goal here is for you to do this without sorting.

Chapter 8

Loops

We saw in chapter 6 how we could use recursion to produce an iterative behavior where something was done multiple times. We also saw in chapter 7 that we can use collections to make certain operations happen multiple times. These approaches are the primary methods used in functional languages and they often work well to provide the functionality we need in Scala. Most languages, including Scala, provide other constructs called loops that are designed specifically for creating this iterative behavior. In this chapter we will explore the loop structures present in Scala and see how we can use these in our programs. We will start by repeating some of the things that we did previously using recursion.

8.1 `while` Loop

The most basic looping construct in Scala and many other languages is the `while` loop. The name tells you roughly what it does. You should keep repeating something while some condition is true. The syntax of the `while` loop is as follows.

```
while(condition) statement
```

The `condition` can be any expression that evaluates to a `Boolean` type. The statement is very commonly a block of code, so you will typically see the `while` followed by curly braces.

To see an example of this, let us use a `while` loop in a function that builds a `List` of numbers input by the user. The `List` will end when the user enters "quit". This is exactly like the example that we did with recursion. We could not do it with the collections because we did not know in advance how many numbers there would be.

```
def readInts:List[Int] = {
  var lst=List[Int]()
  var input=readLine()
  while(input!="quit") {
    lst=input.toInt :: lst
    input=readLine()
  }
  lst
}
```

This code is distinctly imperative. We have to declare two variables with `var` to make it work. In fact, the `while` loop and it's partner, the `do-while` loop that we will discuss in the next section, are only usable as statements. They are not expressions and can not be used in places that need to have a value. The fact that the `while` loop has to be used in an imperative way is somewhat implicit in the way it works. For the code inside of the `while` loop to execute, the condition must be `true` originally. In order for the loop to finish, the value of the condition has to change to `false`. This change requires the mutation of data at some point so it is impossible to use the `while` loop in a completely functional way.

Another thing to note about the `while` loop is that it is a PRE-CHECK loop. This means that the condition is checked before the body of the loop is executed. As a result, it is possible that the contents of the loop will never execute. If the condition is `false` when the loop is first reached, the body of the loop will never execute.

Let us look at another example of the `while` loop. One of our first examples of using recursion to get iteration was the factorial. We can re-write factorial using a `while` loop in the following way.

```
def factorial(n:Int):Int = {
  var product=1
  var i=1
  while(i<=n) {
    product*=i
    i+=1
  }
  product
}
```

We declare two variables named `product` and `i` at the beginning and initialize both to the value of 1. The condition on the `while` loop causes it to iterate as long as `i` is less than or equal to `n`. Inside the loop, the value product is multiplied by `i` and `i` is incremented by one.

The `*=` and `+=` operators are examples of assignment operators. They provide a handy shorthand for when you want to apply a mathematical operation to a value and store the result back in the original variable. You can follow any operator by an equal sign and Scala will see it as a compound operation that performs the specified operator and stores the value back. The storing of the value is a mutation operation. As such, these operators have to be used with mutable data. That either requires `var` declarations, or mutable constructs such as `Arrays`.

This function also shows another element that is common to most while loops and which can lead to a common form of bugs. The line i+=1 is what we often call the iterator of the loop. It is what moves us from one case to the next in the loop. The common bug is to accidentally leave out the iterator. Consider what happens in this code if you do that. Without the line i+=1, the value of i will never change. As a result, the condition will never change either and if it is true to begin with, it will be true forever. This leads to what we call an infinite loop, a loop that never exits. This type of error is easy to put into the code with a while loop, because the loop does not include anything in its structure to remind you to include the iterator.

8.2 do-while Loop

Scala provides a second construct that is very closely related to the while loop. It is the do-while loop. The syntax for the do-while loop is as follows.

```
do {
  statements
} while(condition)
```

The curly braces are not technically required here either, but it is rare to see a do-while loop without them. Again, the statement does very much what it says that it does. It will do the statements while the condition is true.

Given how very similar this sounds to the normal while loop, you might wonder what the difference is. The difference is implied by the layout. The normal while loop checks the condition before it executes the statements in the body of the loop. The do-while loop checks the condition after it executes the body of the loop. As a result, the body of a do-while loop will always execute at least once.

The do-while loop is not used that often in programming. The only times it is used are in situations where the POST-CHECK nature of the loop is helpful and you want the contents of the loop to always happen once. A common example of this is in menu based applications where you need to read what the user wants to do and then act on it. The decision of whether or not the loop should be executed again is based on what option the user picks.

The mainGrades function from the end of the last chapter was an example of this. In that chapter we wrote the program using recursion because that was the only method we knew for making the program execute the same code repeatedly. This function can be converted over to use a do-while loop and the result might look like the following.

```
def mainGrades {
  var tests=List[Double]()
  var assignments=List[Double]()
  var quizzes=List[Double]()
  var option=0

  do {
    printMenu
    option=readInt()
    option match {
      case 1 =>
```

```
        println("Enter a test grade.")
        tests ::= readDouble()
      case 2 =>
        println("Enter a quiz grade.")
        quizzes ::= readDouble()
      case 3 =>
        println("Enter an assignment grade.")
        assignments ::= readDouble()
      case 4 =>
        println("The average is "+
          courseAverage(tests,assignments,quizzes))
      case 5 =>
      case _ =>
        println("Unknown option. Try again.")
    }
  } while(option!=5)
}
```

Whether you use this or the code in section 7.11 is primarily a question of style. Most developers would probably write this version by default, but that is mainly because most developers have a background in imperative programming and will tend to favor this approach for reasons of familiarity.

8.3 for Loop

The `while` loop is the simplest loop, but it is not the most commonly used loop in most languages. In languages that provide a `for` loop, it is typically the most commonly used loop. The `for` loop in Scala is a bit different from that provided in many other languages, but you will probably find that it is the one that you turn to the most when you are putting iteration into your code.

In most languages, the natural usage of the `for` loop is to count so we will start with that. A `for` loop that counts from 1 to 10 and prints out the values would be written as follows in Scala.

```
for(i <- 1 to 10) {
  println(i)
}
```

The name `i` is a variable name just like you would get with a `val` declaration for Scala. As such, you can call it whatever you want. For counting loops it is very common to use names such as `i`, `j`, and `k`. However, anything will work and as with other variable names, it is better if your choice communicates something to those reading the code to make it easier for them to understand. After the variable name is an arrow pointing to the left made from a less than and a hyphen or minus sign. You will see this in all of your for loops in Scala. You can read the `<-` as "in". After that is a nice expression that you should be able to read. We will talk more about exactly what that means shortly. You can read this `for` loop as something like "for i in 1 to 10".

As we saw in chapter 7, the indexes in collections in Scala do not start counting at one. Instead, they start counting at zero. As such, you often would not count from 1 to 10. Instead, we would normally count from 0 up to 9. This could be expressed in Scala by replacing 1 to 10 with 0 to 9. However, it is very common that you want to start at zero and express the number of elements you want to go through. For this reason, the following also works in Scala.

```
for(i <- 0 until 10) {
  println(i)
}
```

Using until causes the counting to finish one before the last value listed.

The `for` loop in Scala is not just about counting though. Indeed, this usage is something of a special case in Scala. In general, the expression to the right of the `<-` in a `for` loop in Scala can evaluate to any type of collection. In other languages, this type of loop that runs through the elements of a collection is often called a FOR-EACH loop because it does something for each element of the collection. That might remind you of the `foreach` method from section 7.4. This is more than a passing resemblance, but that is something you do not need to understand at this point.

To illustrate this usage of the `for` loop, consider the following code in the REPL.

```
scala> List.tabulate(10)(i => i*i)
res0: List[Int] = List(0, 1, 4, 9, 16, 25, 36, 49, 64, 81)

scala> for(elem <- res0) {
     | println(elem)
     | }
0
1
4
9
16
25
36
49
64
81
```

In this case, the `for` loop actually does exactly the same thing that `foreach` does and runs the code inside the loop on each of the different elements in the list. What code you put in the `for` loop can be arbitrarily complex. The `println` statements shown here just make simple examples.

A more general description of the syntax of the `for` loop would be the following.

```
for(name <- collection) statement
```

The `name` can be any valid Scala name. The collection can be any expression that results in a collection. The statement can be anything that you want to have happen. Frequently a code block is used and you will see multiple statements in curly braces.

Let us use a `for` loop to evaluate a polynomial. We will treat an `Array[Double]` as the polynomial where each element in the array is a coefficient of a term in the polynomial. For example, the polynomial $3x^3 + 6x^2 - 4x + 7$ is represented as `Array(3.0,6.0,-4.0,7.0)`. We want the function to evaluate the polynomial for a particular value of x that will also be passed into the function. We will write this in several ways. The first one will use a counting loop.

```
def evalPolyCount(coefs:Array[Double],x:Double):Double = {
  var ret=0.0
```

```
  for(i <- 0 until coefs.length) {
    ret+=coefs(i)*math.pow(x,coefs.length-1-i)
  }
  ret
}
```

This will work, but it is particularly inefficient. The use of `math.pow` for small integer exponents is very inefficient. Walking through the `Array` with the index is not bad, but if we decided to use a `List` for the coefficients that would change.

Recall that the `for` loop is intended to go through the elements of a collection. As such, we could just run through the elements of `coefs` and perform the math. The only challenge in this is that we were using the index, `i`, to calculate the power of `x` as well. We could get rid of that and remove the use of `pow` if we simply went through the `Array` in the reverse order. Putting that logic into code produces this.

```
def evalPolyReverse(coefs:Array[Double],x:Double):Double = {
  var ret=0.0
  var power=1.0
  for(c <- coefs.reverse) {
    ret+=c*power
    power*=x
  }
  ret
}
```

This version does not count with an index. Instead it runs through the array elements. Each value in the `Array` is put into `c` and then the return value is incremented. A separate variable called `power` is created with a value of 1.0 and each time through it is multiplied by `x`. This provides us with a running power of `x` and removes the need to call `math.pow`.

This function is also perfectly correct. It's main drawback is that in order to do the powers of `x` properly, the `Array` had to be reversed. Given this usage, that will create a completely new `Array` and copy the elements across in the reverse order. While that is also inefficient, this does allow us to nicely illustrate the usage of the `for` loop to run through any collection, even one that is created through operations on other collections.

8.3.1 Range Type

Now that you have seen that the `for` loop really just runs through a collection, you might wonder about the counting usage with something like this.

```
for(i <- 0 until 10) { ...
```

To understand this it might help to type some expressions into the REPL and see what is really going on. Here we have done that.

```
scala> 1 to 10
res8: scala.collection.immutable.Range.Inclusive with
scala.collection.immutable.Range.ByOne = Range(1, 2, 3, 4, 5, 6, 7, 8, 9, 10)

scala> 0 until 10
res9: scala.collection.immutable.Range with
scala.collection.immutable.Range.ByOne = Range(0, 1, 2, 3, 4, 5, 6, 7, 8, 9)
```

We will not try to understand what this gives us in any great detail, but there is one significant thing to realize from this. The expressions `1 to 10` and `0 until 10` give us back values that actually are collections. These are instances of the `Range` type.

The `Range` type in Scala gives us a simple way to use the for-each style loop that Scala provides to do the counting operations that most programmers are used to using the `for` loop to accomplish. You can use `to` and `until` with other integer types including the `Char` type. So the expression `'a' to 'z'` will give you a collection of all of the lowercase letter.

What if you want to count down? The `Range` type has an operation `by` defined on it that will allow you to specify the step between elements in the `Range`. So if you want to go from 10 down to 1 you can use `10 to 1 by -1`. We can use this to get a version of the polynomial evaluation function the uses the index counting, but does not require `math.pow` and instead keeps track of the exponent of `x`.

```
def evalPolyCountDown(coefs:Array[Double],x:Double):Double = {
  var ret=0.0
  var power=1.0
  for(i <- coefs.length-1 to 0 by -1) {
    ret+=coefs(i)*power
    power*=x
  }
  ret
}
```

This version has most of the advantages of both of the previous versions. It does not have to reverse the `Array`, nor does it require the use of `math.pow`. The one downside is that it still would not translate well to a `List`.

If you use `by`, you can also use `Range`s of the `Double` type. Other, less standard, numeric types like `BigInt` will also work nicely with the `Range` type. The fact that the `Range` type is really a collection means that all of the methods that were discussed in the last chapter are available for them. This leads to a concise way of expressing factorial.

```
scala> (1 to 5).product
res21: Int = 120
```

In addition to `product` and `sum`, you can also apply `map`, `filter`, or other operations to instances of the `Range` type.

There are sometimes when you want to count through the indices of a collection like an `Array`. You could do this with code like the following assuming that you have an `Array` called `a`.

```
for(i <- 0 until a.length) ...
```

You can also use the `indices` method on the collection. Calling `a.indices` will give you a `Range` that goes from the first index to the last one. So this loop could also be expressed in this way.

```
for(i <- a.indices) ...
```

Not only is this shorter, it is slightly less error prone in that you can not accidentally start at 1 instead of 0, nor can you accidentally use `to` instead of `until`.

8.3.2 yield

The while loop is a statement only and can not be used as an expression. This is not true of the for loop. You can cause the for loop to produce a value so that is can be used as an expression. This is done by putting the yield keyword right after the close parentheses of the for loop. When you use yield, instead of a statement you need to have an expression. The result of the for expression is a new collection with all of the yielded values in it. The following shows a simple example of this.

```
scala> for(i <- 1 to 10) yield i*i
res22: scala.collection.immutable.IndexedSeq[Int] =
Vector(1, 4, 9, 16, 25, 36, 49, 64, 81, 100)
```

Another slightly different example shows how you could use a for loop to make a collection that is filled with values read in from input.

```
val nums=for(i <- 1 to 10) yield readInt()
```

This could work as an alternative to fill if you find it more readable.

You should note that the example gives a result with a type we have not seen before. The general type is listed as an IndexedSeq[Int] and the specific object is a Vector[Int]. Do not let these different types throw you off. For our purposes, we will use them just like we would the Array type. The difference between the Array and the Vector is that the Vector is immutable. You can index into it efficiently like an Array, but like a List, you are not allowed to change the values of the elements. All the standard functions that we saw earlier for Array and List will work on these types as well.

8.3.3 if Guards

The for loop in Scala also allows conditionals. After the first generator, you can put an if that has a condition on it. The for loop will only happen for those instances where the condition is true. This can lead to a more compact syntax than putting a if inside of the for loop. It can also be more efficient. Most importantly, it can be useful when you have a yield so that you do not add unwanted elements to the value of the expression.

As an example of this, and other aspects of the for loop, let us consider having a sequence of points in 2-D that are stored as (Double,Double). I want an expression that will give me back a sequence that has the distances to those points. The catch is that I only want the distances that are less than one. Without the if guard, this would require two steps. One would calculate the distances and a second would filter out the large values. The if guard lets us do this in a single loop.

```
for((x,y) <- points; if magnitude(x,y)<1.0) yield magnitude(x,y)
```

This example was written assuming a function called magnitude that might look like the following.

```
def magnitude(x:Double,y:Double):Double = math.sqrt(x*x+y*y)
```

The beginning of this loop illustrates how you can use a pattern on a tuple to pull out the two elements in the point. This is actually one of the great strengths of the for loop in Scala that helps simplify your programming.

Syntax Note

Note that you do not have to include parentheses after the `if` in an `if` guard. You can, but unlike a normal `if` it is not required.

The one significant drawback of this approach is that the magnitude function is called twice. The `sqrt` function can be expensive so this is less than ideal. We will see how to get around that shortly.

8.3.4 Multiple Generators

The `for` loop in Scala also supports the ability to iterate across multiple collections in a single for loop. This can be done by putting more than one `variableName <- collection` in the parentheses. Each of these that you put into the `for` loop will generate values from the collection to run through the logic. The first generator will pull the first value from its collection. A second generator will then run through all of its values before the first one goes on to the second option. So the number of times the body of the loop happens goes as the product of the number of elements in the collections for the generators, not the sum. To help you understand this, consider the following example.

```
scala> for(i <- 1 to 5; c <- 'a' to 'c') println(i+" "+c)
1 a
1 b
1 c
2 a
2 b
2 c
3 a
3 b
3 c
4 a
4 b
4 c
5 a
5 b
5 c
```

You can see that the character variable, `c`, goes through all of its values for each value that `i` takes on. As a result, there are 15 lines printed.

You might wonder why you would want to do this. Consider again the example of using a 2-tuple to represent points. You might want to make a collection of all the points on a grid in some range of x and y values with a particular spacing in the grid. You could do that with the following code.

```
val xmin = -1.5
val xmax = 0.5
val xstep = 0.01
val ymin = -1.0
val ymax = 1.0
val ystep = 0.01
val pnts=for(x <- xmin to xmax by xstep;
  y <- ymin to ymax by ystep) yield (x,y)
```

The output from the last line will appear in the REPL as something like the following.

```
pnts: scala.collection.immutable.IndexedSeq[(Double, Double)] =
Vector((-1.5,-1.0), (-1.5,-0.99), (-1.5,-0.98), (-1.5,-0.97),
(-1.5,-0.96), (-1.5,-0.95), (-1.5,-0.94), (-1.5,-0.93), (-1.5,-0.92),
(-1.5,-0.91), (-1.5,-0.9), (-1.5,-0.89), (-1.5,-0.88), (-1.5,-0.87),
(-1.5,-0.86), (-1.5,-0.85), (-1.5,-0.84), (-1.5,-0.83), (-1.5,-0.82),
(-1.5,-0.81), (-1.5,-0.8), (-1.5,-0.79), (-1.5,-0.78), (-1.5,-0.77),
(-1.5,-0.76), (-1.5,-0.75), (-1.5,-0.74), (-1.5,-0.73), (-1.5,-0.72),
(-1.5,-0.71), (-1.5,-0.7), (-1.5,-0.69), (-1.5,-0.68), (-1.5,-0.67),
(-1.5,-0.66), (-1.5,-0.65), (-1.5,-0.64), (-1.5,-0.63), (-1.5,-0.62),
(-1.5,-0.61), (-1.5,-0.6), (-1.5,-0.59), (-1.5,-0.58), (-1.5,-0.57),
(-1.5,-0.56), (-1.5,-0.55), (-1.5,-0.54), (-1.5,-0.53), (-1.5,-0.52),
(-1.5,-0.51), (-1.5,-0.5), (-1.5,-0.49), (-1....
```

In this case, the output is truncated before it even gets to the second value of `x`.

8.3.5 Patterns in `for` Loops

One example above used the pattern `(x,y)` to the left of the `<-` in a for loop. You can use any pattern that you want in that position of a `for` loop. Over time we will learn about more patterns to demonstrate just how powerful this is. What makes it truly useful in `for` loops is that any value that does not match the pattern is skipped.

The true usefulness of this will be exposed in the second half of the book. In the first half, the main usage will be to quickly and easily pull values out of tuples. However, we can present one other interesting usage here that uses the fact that collections can be used as patterns. Consider the following code that makes an `Array[Array[Double]]` where each of the contained `Arrays` has a variable length between 3 and 9.

```
scala> val twoD = Array.fill(100){
     | Array.fill(util.Random.nextInt(7)+3)(math.random)
     | }
twoD: Array[Array[Double]] = Array(Array(0.9714402463829903, 0.14015197447391436,
    0.8524582916143384, 0.6162004743306447, 0.620366190244299, 0.36698269639501,
    0.46318519546397396), Array(0.6436214632596926, 0.48145976017298175,
    0.5205354884596076, 0.20188086494076174, 0.9186534118857578,
    0.206412655336915), Array(0.41326520865491023, 0.5388572013936772,
    0.3835287127371739, 0.840667735649998, 0.5776048750341035, 0.8564378792435797,
    0.33358311231736193), Array(0.8386133676386185, 0.19634635871412187,
    0.85047321636848, 0.8920110191832437, 0.22432093122102714, 0.9053781210756321,
    0.7642421256500077), Array(0.7958975255688977, 0.30398364976466374,
    0.8810424486159291, 0.1328719423800543, 0.7129174104031204),
    Array(0.6067234631262645, 0.5276942206810142, 0.06504059155788122,
    0.4145379572950526...
```

You can imagine a situation where this is data that you got from some data set and you only care about the entries with only three values in them and for those you only want the average value. The following `for` loop would provide exactly that data.

```
scala> for(Array(d1,d2,d3) <- twoD) yield (d1+d2+d3)/3
res1: Array[Double] = Array(0.7844266684446944, 0.4057923197637461,
    0.44310232980470454, 0.5634809009372609, 0.576642991638965,
    0.3789396949661376, 0.5706536514773105, 0.5844720273258665,
    0.3445436835569556, 0.3547819380526076, 0.5996534540605474,
```

```
0.38416980809208406, 0.8018516553113365, 0.2244482193993954,
0.5098449834833878, 0.6578966352311121)
```

Note that the array `twoD` has a length of 100, but `res1` has only 16 elements. That is because the other 84 had more than three elements in them. The pattern `Array(d1,d2,d3)` matches only `Arrays` that have exactly three elements in them. Those three elements are bound to the names `d1`, `d2`, and `d2`.

8.3.6 Variable Declarations

`if` guards, multiple generators, matching patterns, the `for` loop seems like a dream, but wait! There's more! You can define variables inside of the `for` loop. This is helpful for situations like we had earlier where we do not want to have to calculate the magnitude twice for each iteration.

```
for((x,y) <- points; val dist=magnitude(x,y); if dist<1.0) yield dist
```

In this sample, the magnitude is calculated and stored in dist. That value is then used in the two different locations where it had been calculated before.

The generators, `if` guards, and value declarations can be combined in any way given that a generator comes first. This provides a significant amount of power in a single construct. Just do not abuse it to make code that no one can understand.

For Comprehensions

In reality, the `for` loop in Scala is just a form of "syntactic sugar". When you write a `for` loop, it is converted to appropriate calls to `foreach`, `map`, `flatMap`, and `filter`. In this way, the implementation can be optimal for the collection type in question.

Also, because it is common to have multiple generators, `if` guards, and variable declarations in `for` loops, Scala allows you to leave out semicolons and use newlines instead if you enclose them all in curly braces instead of parentheses.

8.4 Multidimensional Arrays

When we first introduced `Arrays` and `Lists` back in chapter 7, we saw that these types are parametric. That means that the type requires a type argument to be fully defined. So you can not have just an `Array` or just a `List`. Instead you have `Array[Int]` or `List[String]`. Each of these is a type in Scala. The parameter for these parametric types can be any type.[1] If you put these together you can build things like `Array[Array[Double]]`, `List[List[String]]`, or `List[Array[Int]]`. You do not have to stop there though. Scala will be perfectly happy with a `List[Array[List[List[Array[Int]]]]]`. It is not clear what you would want such a type for, but if you find a use, Scala will support it.

In the case of `Arrays` of `Array` types, we have a special term for them. They are called

[1]There can be restrictions on parametric types, but the collections generally allow any type to be used.

multidimensional arrays. This is because of how you might picture them in your head. You can picture a normal Array as a row with multiple bins that each store a value. An Array[Array[Int]] could be pictured as a table of integers that has rows and columns. Such a table could be said to be two-dimensional. If you had an Array[Array[Array[Int]]] you could picture it as a cube of values in three dimensions. In general all these things can be applied to Lists just as well as Arrays, but the term multidimensional list is not nearly as common.

So how can you create and use multidimensional arrays? The most basic syntax mirrors what we used to originally create normal Arrays.

```
scala> val tda1=Array(Array(1,2,3),Array(4,5,6))
tda1: Array[Array[Int]] = Array(Array(1, 2, 3), Array(4, 5, 6))
```

In this usage the number of elements in each subarray does not have to be the same. When the lengths are different, they are called ragged arrays. When they are all the same, they are called rectangular arrays.

If you are interested in making large rectangular arrays, you should use either the fill method or the tabulate method. This usage is shown below.

```
scala> val tda2=Array.fill(10,10)(0)
tda2: Array[Array[Int]] = Array(Array(0, 0, 0, 0, 0, 0, 0, 0, 0, 0),
Array(0, 0, 0, 0, 0, 0, 0, 0, 0, 0), Array(0, 0, 0, 0, 0, 0, 0, 0, 0, 0),
Array(0, 0, 0, 0, 0, 0, 0, 0, 0, 0), Array(0, 0, 0, 0, 0, 0, 0, 0, 0, 0),
Array(0, 0, 0, 0, 0, 0, 0, 0, 0, 0), Array(0, 0, 0, 0, 0, 0, 0, 0, 0, 0),
Array(0, 0, 0, 0, 0, 0, 0, 0, 0, 0), Array(0, 0, 0, 0, 0, 0, 0, 0, 0, 0),
Array(0, 0, 0, 0, 0, 0, 0, 0, 0, 0))

scala> val tda2=Array.tabulate(10,10)((i,j) => i*j)
tda2: Array[Array[Int]] = Array(Array(0, 0, 0, 0, 0, 0, 0, 0, 0, 0),
Array(0, 1, 2, 3, 4, 5, 6, 7, 8, 9), Array(0, 2, 4, 6, 8, 10, 12, 14, 16, 18),
Array(0, 3, 6, 9, 12, 15, 18, 21, 24, 27), Array(0, 4, 8, 12, 16, 20, 24, 28,
32, 36), Array(0, 5, 10, 15, 20, 25, 30, 35, 40, 45), Array(0, 6, 12, 18, 24,
30, 36, 42, 48, 54), Array(0, 7, 14, 21, 28, 35, 42, 49, 56, 63), Array(0, 8,
16, 24, 32, 40, 48, 56, 64, 72), Array(0, 9, 18, 27, 36, 45, 54, 63, 72, 81))
```

Note that the number of arguments you pass into the first argument list of tabulate or fill determines the dimensionality of the resulting structure. In the case of tabulate, the function that is passed in second needs to take as many arguments as there are dimensions.

You can also use tabulate to make non-rectangular arrays by building a 1-D array whose contents are arrays and using the index to determine the length. That technique is used here to create a triangular 2-D array.

```
scala> val tda3=Array.tabulate(10)(i => Array.fill(i+1)(0))
tda3: Array[Array[Int]] = Array(Array(0), Array(0, 0), Array(0, 0, 0),
Array(0, 0, 0, 0), Array(0, 0, 0, 0, 0), Array(0, 0, 0, 0, 0, 0),
Array(0, 0, 0, 0, 0, 0, 0), Array(0, 0, 0, 0, 0, 0, 0, 0),
Array(0, 0, 0, 0, 0, 0, 0, 0, 0), Array(0, 0, 0, 0, 0, 0, 0, 0, 0, 0))
```

Note that the output here has a different number of elements in each of the subarrays.

To access the elements of a 2-D Array, simply put two sets of parentheses after the Array name with two different indices in them. For example, we can pull values out of tda2 in this way.

```
scala> tda2(3)(4)
res0: Int = 12
```

The 2-D `Array` `tda2` was created to be something like a multiplication table. This particular expression pulled off the element at position 3,4 which is 12.

The advantages and restrictions of the `List` and `Array` types that were discussed previously apply to higher-dimensional cases as well. Once you have made an `Array`, the values can be changed, but the length can not. Similarly, you can make new `Lists` by efficiently adding to the head of old ones, but you can not mutate values in a `List` even if it has a higher dimension.

To bring these things back to the original topic of the chapter, how can you use `for` loops to produce higher-dimensional data structures? One might think that multiple generators would do this. However, that is not the case as you can see if you go back to our examples in that section. If you want to have a construct with higher dimensions, you need to have multiple nested for loops. As an example of this, we will use `for` loops to build the multiplication table that we built earlier with tabulate.

```
val multTable = for(i <- 0 until 10) yield {
  for(j <- 0 until 10) yield i*j
}
```

If you execute this code in the REPL you get the following result.

```
multTable: scala.collection.immutable.IndexedSeq[scala.collection.immutable.
IndexedSeq[Int]] = Vector(Vector(0, 0, 0, 0, 0, 0, 0, 0, 0, 0),
Vector(0, 1, 2, 3, 4, 5, 6, 7, 8, 9), Vector(0, 2, 4, 6, 8, 10, 12,
14, 16, 18), Vector(0, 3, 6, 9, 12, 15, 18, 21, 24, 27),
Vector(0, 4, 8, 12, 16, 20, 24, 28, 32, 36), Vector(0, 5, 10, 15,
20, 25, 30, 35, 40, 45), Vector(0, 6, 12, 18, 24, 30, 36, 42, 48, 54),
Vector(0, 7, 14, 21, 28, 35, 42, 49, 56, 63), Vector(0, 8, 16, 24,
32, 40, 48, 56, 64, 72), Vector(0, 9, 18, 27, 36, 45, 54, 63, 72, 81))
```

This has the same values as the `Array[Array[Int]]` that we made earlier and we can use it the same way despite the fact that it is technically a `Vector[Vector[Int]]`.

Parallel `for` Loops

Modern processors have the ability to do multiple things at the same time through a process called multithreading. We will cover this in detail in chapter 21. At this point we can provide a brief introduction to a simple way to include multithreading in your programs through parallel collections.

The primary motivation for multithreaded parallelism is to more fully utilize the hardware of modern processors and speed things up. So if you have a loop with a body that does a lot of work, you can make it parallel so that the processor does work on different parts at the same time. You can make an `Array` or a `Range` parallel by calling the `par` method.[2] Many calls to these parallel collections will get split up across multiple threads, as will for loops. To see the impact of this, consider the following.

```
scala> def fact(n:BigInt):BigInt = if(n<2) 1 else n*fact(n-1)
fact: (n: BigInt)BigInt

scala> for(i <- 30 to 15 by -1) println(fact(i))
265252859812191058636308480000000
```

[2]You can call the `par` method on a `List`, but it is not efficient.

```
8841761993739701954543616000000
304888344611713860501504000000
10888869450418352160768000000
403291461126605635584000000
15511210043330985984000000
620448401733239439360000
25852016738884976640000
1124000727777607680000
51090942171709440000
2432902008176640000
121645100408832000
6402373705728000
355687428096000
20922789888000
1307674368000

scala> for(i <- 30 to 15 by -1 par) println(fact(i))
1124000727777607680000
51090942171709440000
2432902008176640000
265252859812191058636308480000000
121645100408832000
6402373705728000
403291461126605635584000000
355687428096000
15511210043330985984000000
20922789888000
620448401733239439360000
304888344611713860501504000000
25852016738884976640000
10888869450418352160768000000
1307674368000
8841761993739701954543616000000
```

The factorial on `BigInt` is used in part because it is fairly slow. In the first version, everything runs in the way that you expect and the values of 30! down to 15! are printed in order. With the addition of `par`, the values are no longer printed in order. Instead, the work is broken up into different threads and each each value prints after it has been calculated. The biggest values take longer to calculate so they are not the first ones to print.

While you can use this to speed things up, it has to be done with care. To get some idea of why this is, consider the following few lines typed into the REPL.

```
scala> var count = 0
count: Int = 0

scala> for(i <- 1 to 1000000 par) count +=1

scala> count
res0: Int = 930420
```

At the end of this, `count` should be 1000000, but it is not. It is about 70000 shy of that. Were you to do this on your own machine, you would certainly get a different value. Doing it on the same machine a second time will even produce a different value. This code has what is called a race condition, a problem that is discussed in detail in chapter 21. For the time being, you can consider using `par` when you are not mutating values. If it is part of a `for` loop, you should probably be using `yield`.

8.5 Testing

We have now gotten to the point where you can write programs of reasonable complexity. You know most of the constructs that exist in the Scala programming language. As soon as you start writing larger programs, there are some new elements of the programming process that becomes more significant such as testing and debugging.

Just because you have written the code and it compiles and runs does not mean that it is correct. To determine if it actually does what you want it to you need to test it. This means that you need to run the code with a variety of different inputs to make sure that they all work and then fix the problems when they do not.

The first part of this, running the program with different inputs is called testing. The challenge in testing is trying to figure out what inputs make good tests. When you are testing code you are actually looking for inputs that will break what you have written. You want to give it some things that you know it will work on, but you also want to give it some things that you think might break it. In addition, a good set of tests will go through all of the functionality of the code.

Thinking of things that will break the code often involves looking for boundary conditions. Things that are right at the edge of valid input. For example, if you have a function that takes an `Array` or a `List`, what happens if the `List` or `Array` you pass in is empty? You want to check different small number input sets as well as large ones. If the code takes numeric values, does it work for both positive and negative? What about zero? Giving you back answers that are wrong or do not make sense can be worse than crashing. If the input is a `String`, what happens when the `String` is empty?

There are some situations where you will be certain that the input has a particular structure. In other situations the input will be coming from a user who might give you something that is not quite what you expected. Even when the input is coming from another part of the program and something that you think will have the right format, there can be cases that you have not considered. Good testing will help you find these situations.

If parts of your code require that certain things be true, you can use the `require` function in Scala to force the program to terminate if a condition is violated. You can call `require` with just a `Boolean` argument. If the `Boolean` is `false`, the program will terminate. The termination can be made more informative by providing a second argument that is a message to give the user if the requirement fails. The following shows how this might be used.

```
def weightedAverage(values:Array[Double],weights:Array[Double]):Double = {
  require(values.length==weights.length,
    "Must have same number of values and weights.")
  require(weights.length>0,"Average of zero elements not defined.")
  require(weights.sum!=0.0,"Sum of weights can't be zero.")
  (for((v,w) <- values.zip(weights)) yield v*w).sum/weights.sum
}
```

This function is intended to take the weighted sum of a set of values. There are a number of requirements on the values passed into it. There are three `require` calls that make sure that each of these is `true` before it calculates the value. This might seem like a lot of extra typing, but if you put calls to `require` in your code whenever there really is a requirement, you will find that it makes the debugging process a lot easier.

The other part of testing is coverage. Showing that the code works for one test input is not sufficient to show that it is really correct. How many tests do you have to write to feel confident that your code works? One of the challenges of Computer Science is that you can not, in general, prove that programs are correct. This was one of the earliest results of Computer Science and is still a fundamental aspect of the theory of the field.[3] Certainly, some programs can be proved to be correct, but generally the best we achieve is to show that our programs work across a broad range of inputs.

There are some criteria, beyond looking for boundary cases, you can use to determine if you have enough tests. The metric to determine this is called code coverage. You want to know what fraction of your code has been executed by the tests. There are a number of different code coverage metrics that can be used.

- Function coverage - Has every function been called?
- Statement coverage - Has every statement been executed?
- Decision coverage - Has every option in branching structures (`if` and `match`) been executed?
- Condition coverage - Has every Boolean subexpression evaluated as both `true` and `false`?
- Condition/decision coverage - Combination of the two above.
- Loop coverage - Has every loop been executed zero, one, and more than one times?
- Path coverage - Has every path through part of the code been executed?

The more complete the coverage your test set has, the more confident that you are that your code is correct. The levels of coverage higher in this list are basically minimal standards. If your tests have not gone to every function or every statement, then there are parts of the program that you simply have not tested. Going beyond those you start looking at different ways for things to happen. There are often several different places from which a function can be called. Covering decisions will make sure that you have called them from different locations. Covering conditions makes sure that all the possibilities for why different parts of the code might be reached have been exercised.

If you stop and think about it, you will probably realize that getting condition/decision coverage requires quite a few test cases. Even these options potentially leave a lot of possibilities unchecked as they do not force loops to happen different numbers of times.

[3]The proof itself was due to Alan Turing showing that you can not write a program that will take any program and an input and determine if the program terminates when run on that input. This is called the "Halting Problem". The implication is that you can not, in a completely general way, even show that your program will terminate, much less give the right answer assuming it does stop. There are ways of writing things that avoid errors, but no systematic way of demonstrating correctness. It is worth noting that one nice thing about `for` loops is that they do always terminate as long as they are run on finite collections.

It should also be mentioned that while there is no completely systematic way to prove programs correct, there is a significant amount of work that has gone into proofs of correctness. Unfortunately, proving a program or algorithm correct is often challenging so it is only done for small algorithms or when it is absolutely required. Making this more applicable to general programming could be a significant boost to a world that is increasingly dependent on the proper functioning of programs.

The ultimate form of coverage, path coverage, is generally unattainable for any program of even modest size. Having path coverage implies that you have tested every possible path that the execution could take through the code. Consider a simple function with three `if` statements one after the other. One path through the code would have all three evaluate to `true`. Another path might have the first two `true` and the last `false`. There are actually eight different paths through such a function. If you add another `if`, the number of paths doubles to 16. Path coverage requires exponentially many different cases be tested as conditional statements are added. If that was not bad enough, a single `while` loop creates an infinite number of different paths as the loop could execute zero, one, two, or more times. Each one is a different path through the code. As such, path coverage is generally viewed as an unattainable ideal for anything beyond fairly simple functions.

Due to the challenges of getting good coverage on large collections of code, it is common to do testing on small blocks at a time. This process is called UNIT TESTING. Each different unit of the code has a test suite written for it that checks its functionality independent of other parts of the code. These test suites are run over and over again during development to make sure that no new changes break code that was written earlier.

The real advantage of Unit testing is that in a small unit, one can hope to get fairly complete path coverage. However, it is not sufficient to only do Unit tests. As the different units are assembled, they have to be put through integration tests that test how the pieces work together. It is very possible for two units of code to work perfectly in isolation and fail miserably when they are put together.

Views (Advanced Topic)

The collection methods that we learned about in chapter 7 provide you with the ability to write concise expressions that have remarkable power. Unfortunately, if you string together many of these methods, the result can be inefficient code. Consider the following for some `List` of `Int`s.

```
numList.filter(_>70).map(_/10-6).sum
```

This expression makes two `List`s and runs through `List`s a total of three times. It first runs through `numList` with the `filter` and produces a new `List` of the elements that pass the `filter`. It then runs through that `List` and maps the elements to create another `List`. Finally it runs through that `List` and takes the sum of the elements. The multiple intermediate `List`s and the iteration through them is inefficient.

All of this extra work happens because the `List` type is a STRICT TYPE. That means that whatever it contains is truly kept in memory as a real structure. For expressions like this we would like the ability to have a non-strict representation of the `List`. In Scala such things are called `View`s. Most operations on a `View` accrue the operation without actually performing it. Later on, the operations can be forced which will cause them to actually happen and produce a strict representation.

To get a `View` call the `view` method of a collection. Operations like `map` and `filter` that are done on the `View` will give you a new `View` type that has a memory of what operations are to be performed, but the work will not have been done yet. You can force the operations to be applied to give you back a strict representation of the data with the `force` method. Some other methods, such as `sum`, which produce a final value, will also force the accrued operations to be performed. So the above could be done using `View`s in the following way.

```
numList.view.filter(_>70).map(_/10-6).sum
```

This code only runs through the collection once at the call to `sum` and does not create any intermediate `List`s. If `numList` were particularly long, this could provide a significant benefit.

8.6 Putting it Together

Going back to the theme park, imagine that you have the job of scheduling workers to operate rides. Your scheduling needs to take into account a number of different factors. Each ride needs a minimum number of operators and, on days when there are lots of people riding, that number needs to be increased. Also, the people who are working have to be trained to operate rides. Not everyone has been trained for every ride, so you have to make sure you have enough people scheduled who can operate each ride.

You have data from multiple weeks telling you how many people ride each ride on different days of the week. That is fairly consistent so you will use averages of those values to plan for each day. It is possible to write a program that will generate optimal schedules for an entire week. We are not yet at the point where we are ready to write such a program. Instead, we will write a program that outputs potential schedules for each day of the week. This will help you to build schedules, but will not complete the entire task for you.

The script needs to start by reading in all the data on rides and employees. There will need to be a fair bit of this so this is a script that should probably be run using input redirection and having the contents of a file put in as the standard input. The input will start by telling you how many rides there are followed by information for each ride. That information will include a name, the number of operators needed on a slow day, and the number of riders that qualifies as a heavy day. We will assume that on heavy days, one extra operator is needed. That will be followed by the number of employees. For each employee there will be a name, a number of rides they are trained on, and the names of those rides.

The last type of data in the input will be information on sample days. This will start by telling you how many samples there are. Each sample will have the name of a day, the name of the ride, and the total number of people who rode it that day. No assumptions will be made about the days or how many times each day appears.

Once the data has all been read in, the script should run through every day that there is data for, average the number of riders for each ride on that day, and list possible combinations of workers who can cover the rides that day. Any ride that does not have data for a given day can be assumed to be closed and does not need an operator.

The approach to finding possible groups of ride operators requires looping through the rides that are active on a given day and determining how many operators each one needs based on the average number of riders in the data. Our code will store this by having a single sequence with one entry for each operator needed on each ride. The length of that sequence tells us how many total operators are needed.

The `combinations` method is then used to pick all groupings of that many workers as our goal is to not bring in more people than we have to. For each combination, the code will run through permutations of the ride-operator list using `permutations`. It will check whether that permutation has operators who match up with rides they know how to run. If any permutation matches, that combination of operators is a possibility and it is printed. Code for doing this is shown here.

```
def readRide():(String,Int,Int) = {
  val name = readLine()
  val numOps = readInt()
  val heavyRiders = readInt()
  (name,numOps,heavyRiders)
}

def readEmploy():(String,List[String]) = {
  val name = readLine()
  val num = readInt()
  val rides = List.fill(num)(readLine())
  (name,rides)
}

def readDay():(String,String,Int) = {
  val day = readLine()
  val ride = readLine()
  val numRiders = readInt()
  (day,ride,numRiders)
}

val numRides = readInt()
val rideInfo = Array.fill(numRides)(readRide())
val numEmploys = readInt()
val employInfo = Array.fill(numEmploys)(readEmploy())
val numDays = readInt()
val daysInfo = Array.fill(numDays)(readDay())

val days = daysInfo.map(_._1).distinct
for(day <- days) {
  val thisDay = daysInfo.filter(_._1==day)
  val rides = thisDay.map(_._2).distinct
  val operatorRides = rides.flatMap(ride => {
    val nums = thisDay.filter(_._2==ride).map(_._3)
    val avg = nums.sum/nums.length
    val rideData = rideInfo.find(_._1==ride).get
    Array.fill(rideData._2+(if(avg>=rideData._3) 1 else 0))(ride)
  })
  val totalOps = operatorRides.length
  for(choice <- employInfo.combinations(totalOps)) {
    val perms = operatorRides.permutations
    var works = false
    while(!works && perms.hasNext) {
      val perm = perms.next
      if((perm,choice).zipped.forall((r,op) => op._2.contains(r)))
        works = true
    }
    if(works) {
      println(day+" : "+choice.map(_._1).mkString(", "))
    }
  }
}
```

The top of the code defines some functions for reading information, then reads in all the data. Once the data has been read in, the days we have data for is put into `days` using the `distinct` call to remove duplicates.

Inside of the loop running through the days the variable `thisDay` gets all the ride data for the day being considered. That is used to build `rides`, which contains the unique rides that we have data for on that day. The next step is to expand that so we have a sequence, called `operatorRides` with each ride duplicated a number of times equal to how many operators are needed for it. This is done using `flatMap` with a function that returns an `Array` of the proper size that is built using `fill`.

Another loop then goes through all combinations of employees with a length matching the number of operators needed. The selection of operators goes into `choice`. Permutations of `operatorRides` are then taken and a check is done to see if operators match with rides in that permutation. This is done with a `while` loop so that it can exist early if any match is found. If there is a match, the `choice` sequence with operator names is printed along with the day in question.

A sample input can be found at the website. The output from running that program on the sample input is shown here. This sample input had only four rides and ten employees, but it does show the basic functionality.

```
Fri : Mark, Amy, Madison, Kelsey, John, Jason
Fri : Mark, Amy, Madison, Kelsey, John, Kevin
Fri : Mark, Amy, Kelsey, John, Jason, Kevin
Fri : Mark, Madison, Kelsey, John, Jason, Jane
Fri : Mark, Madison, Kelsey, John, Kevin, Jane
Fri : Mark, Kelsey, John, Jason, Kevin, Jane
Sat : Mark, Amy, Madison, Amber, Kelsey, John, Jason, Kevin, Jane
Sat : Mark, Amy, Madison, Amber, Kelsey, John, Jason, Jim, Jane
Sat : Mark, Amy, Madison, Amber, Kelsey, John, Kevin, Jim, Jane
Sat : Mark, Amy, Madison, Kelsey, John, Jason, Kevin, Jim, Jane
Sat : Mark, Amy, Amber, Kelsey, John, Jason, Kevin, Jim, Jane
Sun : Mark, Madison, Amber, Kelsey, John, Jason, Kevin
Sun : Mark, Madison, Amber, Kelsey, John, Jason, Jim
Sun : Mark, Madison, Amber, Kelsey, John, Kevin, Jim
Sun : Mark, Madison, Kelsey, John, Jason, Kevin, Jim
Sun : Mark, Amber, Kelsey, John, Jason, Kevin, Jim
```

One of the significant aspects of this example is the use of `combinations` and `permutations` to run through various possibilities. We will explore alternate ways of solving problems like this that can be more efficient in chapter 15. For now, these methods give us the ability to solve complex problems that would otherwise be out of our reach.

8.7 End of Chapter Material

8.7.1 Problem-Solving Approach

This chapter added quite a few new constructs for you to pick from for any given line of code in the form of three different types of loops. These have been added below to what was given in chapter 6.

1. Call a function just for the side effects.

2. Declare something:
   - A variable with `val` or `var`.
   - A function with `def`. Inside of the function will be statements that can pull from any of these rules.
3. Assign a value to a variable.
4. Write a conditional statement:
   - An `if` statement.
   - A `match` statement.
5. Write a loop statement:
   - Use a `while` loop when you do not have a collection or know how many times something will happen, nor do you need to use it as an expression.
   - Use a `do-while` loop in a situation where you could consider a `while` loop, but you know that it should always happen at least once.
   - Use a `for` loop to run through the elements of a collection or to do simple counting.

8.7.2 Summary of Concepts

- The `while` loop is a pre-test conditional loop. It will repeat the body of the loop until a condition check returns `false`. The condition is checked before the first execution of the body and then before any subsequent executions. The `while` loop is used as a statement only. It has no value so it can not be used as an expression.
- The `do-while` loop is just like the `while` loop except that the condition is checked after each execution of the body. This means that the body of a `do-while` loop will always execute at least once.
- The most commonly used loop is the `for` loop. Scala's `for` loop is a for-each loop that iterates through each member of a collection. It has many options that give it a lot of flexibility and power.
  - A generator in a `for` loop has a pattern followed by a `<-` followed by a collection that is iterated through. The `<-` symbol should be read as "in".
  - To make counting easy, there is a `Range` type that can specify ranges of numeric values. The methods `to` and `until` can produce `Range`s on numeric types. The method `by` can adjust stepping. Floating point `Range`s require a stepping.
  - The `yield` keyword can be put before the body of a `for` loop to cause it to produce a value so that it is an expression. When you have a `for` loop yield a value, it produces a collection similar to the one the generator is iterating over with the values that are produced by the expression in the body of the loop.
  - The left side of a generator in a `for` loop is a pattern. This can allow you to pull values out of the elements of the collection, such as parts of a tuple. In addition, any elements of the collection that do not match the pattern are skipped over.
  - `if` guards can be placed in `for` loops. This is particularly helpful when using `yield` and the values that fail the conditional check will not produce an output in the result.

  - You can also place variable declarations in the specification of a `for` loop. This can help make the code shorter, easier to read, and faster.

- The type parameters on collections can themselves be other collections. This allows for the creation of multidimensional `Array`s and `List`s in Scala. The `fill` and `tabulate` methods can produce these by passing the proper number of arguments into the first argument list.
- Testing is an essential part of software development. This is where you run the program using various inputs to make certain that does not fail and produces the correct output. Proper testing should exercise all parts of the code. It is generally impossible to test all paths through the code, though good coverage for that is also ideal. Challenging test cases often include boundary values.

8.7.3 Self-Directed Study

Enter the following statements into the REPL and see what they do. Some will produce errors. You should try to figure out why. Try some variations to make sure you understand what is going on.

```
scala> var i = 0
scala> while(i<20) {
  println(i)
  i += 2
}
scala> while(i<30) {
  println(i)
}
scala> do {
  println(i)
  i -= 1
} while(i>0)
scala> var resp = ""
scala> do {
  println("Go again? (Y/N)")
  resp = readLine()
} while(resp=="Y")
scala> 1 to 10
scala> 1 to 10 by 2
scala> 0.0 to 1.0 by 0.1
scala> for(i <- 1 to 10) println(i)
scala> for(i <- 1 to 10) yield i
scala> for(i <- 1 to 5; j <- 2 to 4) println(i+" "+j)
scala> val tups = for(i <- 1 to 5; j <- 2 to 4) yield (i,j)
scala> for((n1,n2) <- tups) yield n1*n2
scala> val twoD = List.fill(6,4)(99)
scala> val mult = Array.tabulate(10,10)((i,j) => i*j)
scala> mult(3)(4)
scala> twoD(1)
```

8.7.4 Exercises

Many of these exercises are identical to ones that were given in chapter 6. The only difference is that those problems were to be solved with recursion and these are to be solved with loops.

1. Write the `isPrime` that returns a `Boolean` telling if a number is prime using a loop.
2. Write a function using a loop that will print powers of two up to some value.
3. Write a function using a loop that will print powers of two up to some power.
4. Write a function using loops that will print a multiplication table up to 10s. Try to get it running first, then consider how you could make everything line up.
5. Write a function that returns a `List[Int]` of the prime factors of a number using a loop.
6. Repeat exercise 6 (p.130) using a loop instead of recursion.
7. Write code that can take a `List[Int]` and give you back a new one where all the values have been doubled. Do this with a `while` loop, a `for` loop without a `yield`, and a `for` loop with a `yield`.
8. This problem is like exercise 8 (p.170) in that you are supposed to count the number of even values in an `Array[Int]`. The difference is that now you will do it once with a `while` loop and once with a `for` loop.
9. Another problem that is significant when doing real cryptography is solving linear equations under modulo arithmetic. That sounds complex, but it is really just solutions to the following:

   $$ax \equiv b \mod n,$$

   where we know a, b, and n and want to find x. To find the solutions to this, there can be more than one, you need to use the extended Euclid's algorithm for exercise 12 (p.130).

   You start off by calling the extended Euclid's algorithm on a and n, putting the returned values into d, x, and y. If b is not divisible by d then there is no solution. Otherwise make $x_0 = x(b/d) \mod n$. The solutions are given by $(x_0 + i(n/d)) \mod n$ for $i \in [0, d-1]$.
10. Try to write functions to do these different things with `String`s in the following ways: with a `while` loop and an index, with a `for` loop and an index, with a `for` loop and no index, with a `Range` and higher-order methods but no loops, and with only higher-order methods.

    - Determine if a `String` is a palindrome.
    - Count the number of times a letter occurs.
    - Remove all occurrences of a letter.
    - Replace all occurrences of a letter (without using any `replace` methods).
    - Count the number of occurrences of a substring.
    - Remove all occurrence of a substring.
    - Replace all occurrences of a substring (without using any `replace` methods).

- Count the number of vowels.
- Remove all vowels.
- Convert all characters to uppercase (without using `toUpper`).
- Convert all characters to lowercase (without using `toLower`).

8.7.5 Projects

1. This project builds on top of project 4 (p.178). For this you will fill in an entire grid of values with intersection parameters for a set of geometry. Most images on computers are made as grids of values where the values give the colors. We do not quite have the ability to introduce colors and images yet, but we are getting close.

   For now you will fill an `Array[Array[Double]]` with the t parameters for a collection of geometry. You should write a function that takes the following arguments: location of the viewer as a 3-D point, forward direction for the viewer as a 3-D vector, up direction for the viewer as a 3-D vector,[4] a sequence of geometry (spheres and planes), and the number of cells across the square grid should be. You will cast one ray for each cell in the grid and see if it hits anything and if so, how far out it hits. Fill the grid with the values for a minimum intersection parameter.

   The grid represents a plane in space that is one unit in front of the viewer position with a top-left corner that is one unit up and one unit to the left. (You can find a left vector by doing the cross product of the up and forward vectors.) The grid extends to one unit to the right and one unit down. This is the basic approach for building images using ray tracing.

2. One of the useful things that you learn in calculus is that functions can be approximated. Your calculus text will mention both the MacLaurin series approximation and the Taylor series approximation. They are basically the same other than MacLaurin series are always taken about $x = 0$ and this is what we will be working with here. The definition of the MacLaurin series is

$$f(x) \sim \sum_i \frac{f^{(i)}(0)}{i!} x^i$$

   So this is the sum from $i = 0$ up to some n (or infinity if you want to be really accurate). In the sum we have x raised to the i power times the ith derivative of $f(x)$ evaluated at 0 divided by i factorial. Obviously, this is a real pain to use on functions where taking the derivative is not easy. However, for some functions where the derivatives are straightforward, performing this approximation is very easy. Examples of that would be e^x, $\sin(x)$, and $\cos(x)$.

   Write a program that does a Maclaurin approximation of $\cos(x)$. That is not that hard because the derivative is $-\sin(x)$, which has a derivative of $-\cos(x)$ which goes to $\sin(x)$ then back to $\cos(x)$. Also note that you are always evaluating at $x = 0$ so all the terms for sin go to zero.

   The first few terms in this series are:

$$1 - 0 - \frac{x^2}{2!} + 0 + \frac{x^4}{4!} - 0 - \frac{x^6}{6!} + 0 + \frac{x^8}{8!} + ...$$

[4] For a standard projection the up and forward directions should be perpendicular. However, the math works as long as they are not parallel. You simply get a distorted view in that case.

For this project, you should ask the user for the x to use, as well as an accuracy. Use the `math.cos` function to get the "real" value of cosine at that value of x. Iterate until the difference between the series value and what that function gives you is less than the input accuracy. After the loop, print out the real value, the value you got from the series, and how many terms you had to sum to get that. (For an extra challenge, make your program use a Taylor series instead. This means inputing another value x_0 which the series is expanded around.)

3. Computers are used extensively for simulating physical systems, especially when the system is hard or impossible to build in a lab. For this project you will write a simple simulation of the gravitational Kepler problem. You will also explore the accuracy of what you are doing a little bit. Imagine you have a body in orbit around a star. We will assume that the star is much larger than the other body so it stays at the origin, $(0,0)$, of our coordinate system. The other body starts at some position (x, y) with a velocity (v_x, v_y). A simple "integrator" for a system like this can be constructed by a discretization of Newton's laws (a fancy way of saying that we avoid calculus and do things in a way that is more computer friendly). Newton's second law tells us $F_1 = m_1 * a_1$ and for gravity $F = -G\frac{m_1 * m_2}{d^2}$. We are going to simplify this for our toy system and just say that $a = -\frac{1}{d^2}$. We can break this into components and get $a_x = -\frac{x}{d^3}$ and $a_y = -\frac{y}{d^3}$. Now, the trick on the computer is to say that instead of moving smoothly, the particle jumps over certain time steps, Δt. So after one time step the new position is $x = x + \Delta t * v_x$ and $y = y + \Delta t * v_y$. Similarly, $v_x = v_x + \Delta t * a_x$ and $v_y = v_y + \Delta t * a_y$. Doing this in a loop "integrates" the motion of the body. (Use the `math.sqrt` function to calculate d.)

   This integrator is very simple, but far from perfect. If you start with your body at $(1,0)$ with a velocity of $(0,1)$ it should be in a nice circular orbit staying that distance forever. By measuring how that distance changes, you can get a measure of how accurate, or inaccurate, the integrator is. You can play with other positions and velocities to see what happens.

   You will write a program that takes the initial x, y, v_x, and v_y as well as a time step, Δt, as inputs. It should advance the system for a total time of 10.0 (so if $\Delta t = 0.1$ that requires 100 iterations). At the end of it you should measure the distance from the origin and print a message giving that and the original distance. Then check to see if the change is less than 1%. If it is, say that the integration was accurate enough, otherwise say it is not. In a comment in your code you should tell how small you had to make your time step for this to be reached given the coordinate `1 0 0 1`. (Note that this measure of accuracy is only good for circular orbits. We are not going to do enough physics to go beyond that, but if you happen to want to, the real objective is to conserve total energy. For an extra challenge, compare initial and final total energies of the system.)

   For fun, you can change it so it prints the x and y values during the simulation and see what is happening with a spreadsheet of using `gnuplot` in a manner similar to what is described in project 4 (p.131). This can also be helpful for debugging. Such plots are shown on the website.

4. An alternate physics problem that can be solved in the same way as that for the previous project is calculating the trajectory of a projectile. If you consider air resistance, the path of a body is not a simple parabola. Using a numerical integrator that was described in the previous project, you can figure out how far a projectile will go assuming there is air resistance.

The force of gravity near the ground can be approximated a $\vec{F}_g = -gm\hat{\jmath}$.[5] The friction force from the air can be approximated by $F_d = \frac{1}{2}\rho v^2 C_d A$, where ρ is the density of the fluid, C_d is the drag coefficient of the shape, and A is the cross-sectional surface area of the particle. The value of C_d for a smooth sphere is 0.1. The density of air is about $1.2kg/m^3$. This force is directed in the opposite direction of the motion.

Using a `while` loop write a script that will tell you how far a ball will go before it hits the ground with the user specifying the height from which it is thrown/launched, its initial speed, its initial angle, its radius, and its density. If you want a bit of extra challenge, allow the user to input a wind speed.

5. For this problem you will do some string parsing that has relationships to chemical formulas in chemistry. We are going to keep things fairly simple for this. The basic idea is that the user types in a string that is a chemical formula, and your program should parse that string and tell how many of each type of element are on each side of the equation. This is the first step in balancing a chemical equation. A later project will have you go through the process of doing the actual balancing.

   The format of the chemical equation will look something like this: `CH4+O2=H2O+CO2`. This is a reaction for burning/oxidizing methane. Note that it is not well balanced as there need to be coefficients in front of each term. Your program will assume a coefficient on each term in the equation as a lowercase letter starting with 'a' in alphabetical order from left to right and output how many of each element there are. So for this input the output would be:

```
C: a*1=d*1
H: a*4=c*2
O: b*2=c*1+d*2
```

   or if you want to clean it up a bit,

```
C: a=d
H: a*4=c*2
O: b*2=c+d*2
```

   This gives us three linear equations that we could try to solve (actually we have 3 equations for 4 unknowns so the system is under-determined, but that is often the case, so we will find the solution where `a`, `b`, `c`, and `d` have the smallest values possible and are all integers but you don't have to worry about that now). We will not be solving it in this project.

   To be more specific about the input, it has a sequence of terms that are separated by `+` or `=`. The reagents are in terms on the left-hand side of the `=` and the products are on the right-hand side of the `=`. Each term can have one or more element names, each followed by the number of that element in the given molecule. The element names will all be one character long and capitalized. Also, the number of elements will be just one digit. If no number is present you assume there is only one. Allowing elements with more than one letter (uppercase followed by lowercase) or numbers with more than one digit makes a nice project for anyone looking for an extra challenge.

   The output should have a separate line for each element that was present in the equation. It should list the symbol for the element followed by a colon and then the

[5] If you are not familiar with the notation, $\hat{\imath}$ and $\hat{\jmath}$ represent unit vectors in the x and y directions respectively.

equation that tells what the coefficients have to satisfy for that element to be balanced on both sides of the equation. You can choose either format above.

6. For this project you can keep working with recipes. You can think of this as an extension of project 6 (p.178), but you do not have to have completed that project to do this one. For this project you will write a text menu with the following options.

   (a) Add a pantry item.
   (b) Print pantry contents.
   (c) Check a recipe.
   (d) Cook recipe.
   (e) Quit

   If the user selects to add a pantry item you ask for the name of the item and how much they are adding then return to the menus. The option to check a recipe has them enter names and amounts until they give a name of "quit". It then tells them whether or not they can make the recipe based on what they currently have. The last option will subtract the appropriate amounts for the items in the last recipe that was successfully checked. If no recipe has been successfully checked, it should print an appropriate message.

7. For this project you can upgrade what you did for project 7 (p.178) so that there is a text menu with options.

   (a) Add course of interest.
   (b) Print current courses.
   (c) Remove course of interest.
   (d) Build a schedule.
   (e) Quit

   Adding a course should have them type in a unique `String` for the course number or description along with a numeric value for how much they want that course and an integer for the time slot.[6] When they select remove they should type in the unique ID and that course will be removed from consideration. The schedule building option should ask for how many hours they want to take that semester. It will then print out the three "best" schedules that match that number of hours and do not contain courses at conflicting times.

Additional exercises and projects, along with data files, are available on the book's website.

[6] Real time slots involve days and times. That would make this problem a lot harder. You can do that if you want the challenge, but to keep things simple you could use a number for each standard time slot in the schedule. So use 1 for the first MWF (Monday, Wednesday, Friday) slot, 2 for the next one and so on.

Chapter 9

Text Files

Most of the programs that you use on a regular basis would be far less useful if they did not have the ability to read from and write to files. Consider a word processor that could not save your files and let you load them back in later: One where you had to print the paper before you closed the program to have any record of what you had done. Such a program would not be very useful and as such, it would not be used much. That is true even if it provided you with a very full set of other capabilities.

The reason for this is that all work you do without a file sits in the temporary memory of the computer that is given to the application when it runs. When the application is stopped, for any reason, all of that memory is lost. It might be given over to other applications or used by the operating system. Whatever happens to it, you no longer have the ability to get to it in any useful way.

Files give you the ability to store values from one run of a program to the next. Files can also be used to store information that you do not want to have to write into the source code or just to store amounts of information larger than what will fit in the memory of the machine. Disk drives are much slower than memory so this last part has to be done with extreme caution.

9.1 I/O Redirection

In a certain way, you have already had the ability to make your program deal with files through input/output (I/O) redirection. When you run the program from the command-line as a script, you can have standard input come from a file using "<". You can also make the standard output go to a file using ">". This works very nicely if you are going to be entering the same values multiple times or want to preserve the output so that you can look at it.

The downside of using I/O redirection is that it is rather limited. You only get to read from one file or write to one file. What is more, if you decide to read from a file, you can not also have the user provide some of the input from the keyboard, or if you decide to have the output go to a file, your user will not see anything that is printed come up on the screen. These limitations make this approach impractical for most applications. As such, it is important for us to learn other ways.

9.2 Packages and `import` Statements

Before we can learn about reading from files and writing to files, we need to learn a little about how code is organized in large projects, and specifically in the standard libraries of Scala. Everything that we have used so far was available in Scala without us having to specify where to look for it. The basic parts of the language are simply available by default. File handling is not available by default. We will have to tell Scala where to go looking in the libraries for these things.

The Scala language uses packages to organize code, a feature inherited from Java and present in many other languages. To understand the reason for packages, consider the `List` type that we learned to use in chapter 7. This type is defined by a class written in the Scala libraries that is available to you by default because it is so commonly used. However, the word `List` is not an uncommon one. Without packages there could only be one type called `List`. If anyone tried to create another one, it would conflict with the first one. Packages allow you to have multiple types that all have the same base name like `List`, but which are differentiated by being in different package. In the case of `List`, the one you have been using is technically a `scala.collection.immutable.List`. However, the name `List` could also refer to a `java.util.List` or a `java.awt.List`.

These longer names are the fully specified names of the types. They specify both a package name and the specific type name. Packages are typically named in all lowercase and the dots separate subpackages in a hierarchy going from broadest to most specific. So the `List` that you have been using sits in the top-level package `scala`, which has a subpackage called `collection`, which inside of it a package called `immutable`, and the `List` type is inside of that subpackage. We will not worry about creating packages at this point. That is a topic that is better left for later on. Right now all you need to understand is that they exist and how to deal with code that is in packages.

To illustrate what you need to know, we will consider one of the types we will use for the topic of this chapter. To help us read from files we will use instances of the `scala.io.Source` type. While it is useful to know the fully specified name of this type as well as that of others, these full names are a real pain to type in all the time. Imagine if you had to type in `scala.collection.immutable.List` any time you wanted a `List` in your code. This is the reason that the `import` statement exists. An `import` statement gives Scala directions on how to find things. If it sees the name of a type or value that is not declared locally, it will look in things that have been imported to see if it can find it there. The basic syntax of an `import` statement is to have the keyword `import` followed by what you want to import. To make it easy to use `scala.io.Source` you can do the following.

```
import scala.io.Source
```

After putting in this line, you can just use the name `Source` instead of the fully specified name and Scala will know what you mean.

The `import` statement in Scala is very flexible and powerful. You can use `import` anywhere you want and it will be in effect from that point down to the end of the current code block. So `imports` have the same scope as other declarations. If it appear inside of curly braces, it will only be in effect inside those curly braces. In many instances, you will want to import everything in a package (or other structure). You can tell Scala that you want to do this by using an underscore as a wild card. For example, this line would bring everything in the `scala.io` package into scope.

```
import scala.io._
```

You can also import several things with a single line by grouping what you want to import in curly braces. If you wanted to have the `Source` class and the `BufferedSource` class from the `scala.io` package, but no other classes, you could use a line like this.

```
import scala.io.{Source,BufferedSource}
```

There are a number of other possibilities for `import` in Scala that we will consider later when they are significant to us.

By default, Scala always includes three `imports` for every file.

```
import java.lang._
import scala._
import Predef._
```

The first one being all the standard classes from Java into scope. This is the only default `import` in the Java language. It also imports the basic `scala` package and then all the contents of the object `scala.Predef`. This last one is the reason you can use functions like `readInt()` as easily as you can.

The last `import` demonstrates something else about Scala packages, they are truly nested. That last import could also be written in this way.

```
import scala.Predef._
```

However, because the line above it imports the contents of the `scala` package, that can be left off. The implicit `import` of `scala._` also means that we can refer to `scala.io.Source` as just `io.Source` in either an `import` or in normal usage. Because `io.Source` really is not that much longer than `Source`, you might not even feel a need to include the `import`.

You can see the full structure of packages for Scala by looking at the Application Programming Interface (API) on the Scala website (`http://www.scala-lang.org/`). The API is a very helpful resource in general. It lists all of the methods on everything that is in the standard libraries. By the time we are done with this book, you should have the ability to go to the API and look up any functionality that you might need. For now there are quite a few things that won't make sense because there is more for us to learn.

9.3 Reading from Files

Now that you have seen packages and **import** statements it is time to actually use `scala.io.Source` to help us read information from sources other than standard input. There are other ways that we could read from files that use the Java libraries, some of which we will discuss in chapters 22 and 33, but for now we will use `Source` because it provides us with sufficient capabilities. Any code in this section will assume that you have done an **import** of `scala.io.Source` so that we can refer to it by the short name.

There is an object called `Source` that has methods we can call to get instances of type `Source` that we will use to read from things. The simplest way to read from a file is to use the `fromFile` method of the `Source` object.

```
scala> val fileSource=Source.fromFile("ch8.txt")
fileSource: scala.io.BufferedSource = non-empty iterator
```

As you can see from this output, the object we got back was specifically of the type `scala.io.BufferedSource`. The `BufferedSource` type will provide better performance for files which is why we were given that. It happens that reading from a hard disk is one of the slowest things you can do on a computer. A big part of the slowdown is the result of having to move to the right part of the disk. As a result, it is much faster to read a bunch of data at once, even if you do not need it all yet, than it is to read each little piece of data one byte at a time. The `BufferedSource` does exactly this. It reads in a relatively large amount of data into a buffer and then gives you that data as you request it. When the buffer is emptied, it will read again.

When you are done with the `Source` object, you should call close on it. In this case that would look like the following.

```
fileSource.close
```

There are many reasons you should remember to do this, but two really stand out. The first is that a program is only allowed to have a certain number of files open at a given time. If you fail to close files for an extended period of time, you can run out of available file descriptor and then you will try to open a file and have it fail. It also signals to Scala that you are done with the file so that it can clean up things like the memory that was used for the buffer.

`FileNotFoundException` and `IOException`

File handling activities include a lot of operations where things can go wrong. In the case of `fromFile` you could have mistyped the file name or the file that is specified might not be one that you can read from. Either of these situations would result in a `java.io.FileNotFoundException`. When reading from files, you could also attempt to do something like read beyond the end of a file. This and other mishaps would result in a more general `java.io.IOException`. If we want to make your code deal with these situations gracefully, you will need to use the `try/catch` expression that was mentioned in chapter 6.

The full details of this construct are left for section 22.3. However, if this situation you might have code like the following.

```
def readStuffFromFile(fileName:String):SomeType = {
  try {
    val fileSource = Source.fromFile(fileName)
    // ...
    fileSource.close()
  } catch {
    case e:java.io.FileNotFoundException =>
      // An expression to build an empty SomeType
  }
}
```

9.3.1 Iterators

When the value of the `BufferedSource` is printed all it shows is "`non-empty iterator`". This is because the `BufferedSource`, and the normal `Source`, are both of a more general type called an `Iterator`. An `Iterator` is much like the `Array` and `List` types that we saw previously. In fact, virtually all the methods that we could call on the `Array` and the `List` are available on an `Iterator`. The difference is that a basic `Iterator` can only be used once. The `Array` and the `List` are said to be `Iterable`, which means that they can give us multiple `Iterators` to use to call those methods over and over again. In the case of an `Iterator`, once you call a method that runs through the whole thing, it is spent and you can not use it again.

```
scala> fileSource.mkString
res0: String = This is a sample text file that I have written to use
for chapter 9. There really is not all that much to this file. It is
simply supposed to be used to illustrate how we can read from files.

scala> fileSource.mkString
res1: String =

scala> fileSource
res2: scala.io.BufferedSource = empty iterator
```

The first call to `mkString` gives us back a string that has the contents of the file. The second call gives us back nothing. The reason for this is clear if we look at the value of `fileSource`. After reading through the file the first time, it has gone from a `non-empty iterator` to an `empty iterator`. There is nothing left for us to read. The `Source` type provides a method called `reset` that you do not get in most `Iterators`. The `reset` method gives you a new `Source` that is set back to the beginning for you to read again.

At a fundamental level, the `Iterator` type is based on two methods: `hasNext` and `next`. The `hasNext` method returns a `Boolean` and tells you whether or not there is something more in the `Iterator`. The `next` method will give you that next thing. You should be able to picture how these two methods can be used in either a `while` loop or with recursion to let you run through the entire contents of the `Iterator`.

The reason the `BufferedSource` is just an iterator is not hard to understand. Remember, reading files is extremely slow. You should do it once and get the information you want that time. Reading it over and over again would be inefficient. To help force you toward efficiency, you would have to explicitly reset the file or open it back up to iterate through it again.

With the List and Array types, you saw that they were parametric. We could make a List[Int] or a List[String]. This is also true of Iterators. The Source and BufferedSource types are specifically Iterator[Char]. This means that they operate naturally on individual characters. If you call functions like map or filter, or if you convert the Iterator to a List or an Array, the result will be a bunch of characters.

While technically all the data that you want is in the form of characters, it can be a bit difficult to operate with it in that format. For this reason, there is a method called getLines that will give you back a different Iterator. This new Iterator is of the type Iterator[String] and each element is a full line in the file without the newline character at the end.

9.3.2 String split Method

Even lines are not always all that useful because there might be multiple pieces of data on each line. There are many ways that you can split up a String into different pieces. For most purposes, the simplest of these is the split method. The split method takes a single argument that is a String, which should be the delimiter between the pieces you want split up. It will then return an Array[String] with everything that was separated by that delimiter. Here is a simple example.

```
scala> "This is a test.".split(" ")
res4: Array[java.lang.String] = Array(This, is, a, test.)
```

The String, "This is a test.", is split up using a single space as the delimiter. The result has each word from the String as a separate element.

Technically, the argument to split is a REGULAR EXPRESSION. We will not go into the details of regular expressions until chapter 30. There are just a few things that are worth mentioning at this point. The regular expression can have more than one character. Also, the characters '+' and '*' have special meanings. The '+' says that the character before it can occur one or more times while the '*' says that it can occur zero or more times. This is worth mentioning at this point because it is not uncommon for inputs to potentially have multiple spaces between words. To handle this, you will often call split with " +" instead of just a space.

Now we want to put the split method into action with the ability to read the file line-by-line. What we want to do is create code that will read in numeric data from a file into a 2-D data structure. In this case, we will create a 2-D Array because the direct access capability is useful for most of the applications we would want to use this in. We will do this in two different ways. We will break this up into two functions. The first function will work with any Iterator[String] and give us back the Array[Array[Double]] that we want. The Iterator[String] is helpful because that is what getLines will give us and that is more flexible than forcing it to be a file. The second function will take a file name and return the Array[Array[Double]]. It will not do much itself other than use a Source to read the file and pass that Iterator to the first function. Both functions will also take a delimiter for split so that the person calling them can choose what we split on.

```
def dataGrid(lines:Iterator[String],delim:String):Array[Array[Double]] = {
  (lines.map(s => s.split(delim).map(_.toDouble))).toArray
}

def fileToDataGrid(fileName:String,delim:String):Array[Array[Double]] = {
  dataGrid(Source.fromFile(fileName).getLines,delim)
}
```

Now we can demonstrate how this works by calling it on a file called `numbers.txt` that we have created. The file has the values for a 3x3 identity matrix with the values separated by commas as shown here.

```
1.0, 0.0, 0.0
0.0, 1.0, 0.0
0.0, 0.0, 1.0
```

The choice of format is particular because this is what you would get if you had Excel® write out a CSV format with only numbers. CSV stands for Comma Separated Values. It is more complex if we include non-numeric data. Anything that is not numeric is put inside of double quotes in a CSV file. We will ignore that for now and only deal with numbers.

We can read in this file and see the result in the REPL.

```
scala> fileToDataGrid("numbers.txt"," *, *")
res0: Array[Array[Double]] = Array(Array(1.0, 0.0, 0.0),
Array(0.0, 1.0, 0.0), Array(0.0, 0.0, 1.0))
```

Everything here should be pretty clear with the possible exception of the delimiter. For this file, it would have worked to use the delimiter `", "`. However, that delimiter is not all that robust. If there were any extra spaces either before or after the comma it would fail. It would also fail if a space was left out after the comma.

Even with this delimiter, the `dataGrid` method leaves some things to be desired. We are not trying to make it deal with `String`s so we do not consider it a problem if this function crashes when there is a `string` in the file that is not a number. More problematic is the fact that it also does not deal well with blanks. In a CSV file there are two types of blanks that can occur. One is when there are two commas with nothing between them and one is when there is a completely blank line. If either of these occurs in the file right now, our code will not respond well. We could fix this in our current version by adding some extra logic, but as long as we are good with just skipping the blanks, it is easier to do this with `for` loops and `if` guards.

```
def dataGrid(lines:Iterator[String],delim:String):Array[Array[Double]] = {
  (for(l <- lines; if(!l.trim.isEmpty)) yield {
    for(n <- l.split(delim); if(!n.trim.isEmpty)) yield n.trim.toDouble
  }).toArray
}
```

The `if` guards make it easy to skip things that do not fit what we want. In this case, what we want is a `String` that is not empty. The `trim` method on a `String` gives back a new `String` with all leading and trailing white space removed.

9.3.3 Reading from Other Things

One of the nice things about the `scala.io.Source` object is that it has more than just the `fromFile` method for us to use in reading files. Indeed, there are quite a few different methods that start with `from` in `scala.io.Source`. We will only consider one of them here, though you can certainly go to the API to look at others. The one you might find most interesting is the `fromURL(s:String)` method. URL stands for Uniform Resource Locator and they are the things that you type into the address bar of a web browser. You call this method passing it a URL as a `String`, very much like what you would put in the web browser.

This makes it remarkably easy to have your program read information off of the web.

For example, executing the following line will give you back the contents of my web page as a `String`.

```
Source.fromURL("http://www.cs.trinity.edu/~mlewis/").mkString
```

You can do everything with this `Source` that you would do with a `Source` from a file. Like with the file, the `Source` is an `Iterator` so you can only read through it once without calling `reset`. This also makes sense because if there is anything slower on a modern computer than reading from a hard disk, it is reading information from the web.

9.3.4 Other Options (Java Based)

The advantage of using a `scala.io.Source` with `getLines` is that you get a Scala based `Iterator` that has all the methods you have gotten used to in working with the `List` and `Array` types. However, that does not mean that `Source` is always your best option. For many situations, you might find that the `java.util.Scanner` class is better.

The `java.util.Scanner` class has methods for reading specific types of data as well as for checking if there are specific types of data available. Here is a sampling of the methods that are available.

- `hasNext():Boolean` - Check to see if there is another "word".
- `hasNextDouble():Boolean` - Check to see if there is a `Double` ready to read.
- `hasNextInt():Boolean` - Check to see if there is an `Int` ready to read.
- `hasNextLine():Boolean` - Check to see if there is a line of text ready to read.
- `next():String` - Read the next "word".
- `nextDouble():Double` - Read the next `Double`.
- `nextInt():Int` - Read the next `Int`.
- `nextLine():String` - Read the next line of text.
- `useDelimiter(pattern:String)` - Change the pattern used for the delimiter.

This list should be fairly self-explanatory. The only thing one might wonder about is why "word" is in quotes. The reason for this is reflected in the last method listed. When you want to read a "word" with a `Scanner`, it will read up to the next delimiter, whatever that happens to be. By default it is any grouping of white space.

In order to use a `Scanner`, you have to make one. To do this you probably want to `import java.util.Scanner` and use the following expression.

```
new Scanner(file)
```

The `file` needs to be an expression that has the type `java.io.File`, which we will also make with a new expression and probably also want to `import`. Putting this together, you could put something like this into a program to read and print a bunch of `Ints`.

```
import java.util.Scanner
import java.io.File

val sc = new Scanner(new File("numbers.txt"))
while(sc.hasNextInt) {
```

```
  println(sc.nextInt())
}
sc.close()
```

The choice of printing here comes from the fact that the Scanner does not nicely produce a Scala collection. You can get the values into a collection, but it will typically take a bit more effort. You should also remember to close your Scanners when you are done with them. If they link to a file, they are holding onto valuable resources.

9.4 Writing to File

Scala does not provide any functionality for writing to a file in its own libraries. That is something that is already well supported by Java and adding Scala libraries has not yet been seen as providing a significant benefit. There are many ways to set up a file to write to in Java. The easiest, and the one that we will use, is the java.io.PrintWriter. We can make a new PrintWriter with new and tell it the name of the file we want to write to as the only argument. So we can get a PrintWriter that we can use for doing output with this code.

```
import java.io.PrintWriter
val pw = new PrintWriter("output.txt")
```

Now we could call methods on pw that will cause things to be written to that file. You can use print and println methods to print to the file in much the same way that you have been using those functions in Scala to print to standard output.

Using this we could write the following code to print 100 random points with x and y values between 0 and 1.

```
for(i <- 1 to 100) {
  pw.println(math.random+" "+math.random)
}
pw.close()
```

The last line is critical because the PrintWriter is also using a file resource and buffering, just like the BufferedSource. However, it holds things in memory until there is enough to make it worth writing to the disk. If you do not close the file, the text in the buffer will not have been written out. If you are not done with the file, but you want to make sure that what you have written goes into the file, you can use the flush method.

```
pw.flush
```

When you call this, anything in the buffer will be written out. This way if the computer crashes, loses power, or something else goes wrong, that information will be out of the volatile memory.

9.4.1 Appending to File

Creating a PrintWriter in the way just described has the side effect that it will either create a new file or delete an existing one and write over it. If there is an existing file and you want to append to the end of it instead of overwriting it, there is one more step that needs to be added. Instead of simply telling the `PrintWriter` what file name to use, we tell it a `java.io.FileWriter` to use. When we create a `FileWriter` we can tell it the file name and also give it a `Boolean` value that tells it whether or not to append. If the `Boolean` value is `true`, the contents of the existing file will be left in place and anything else written will go to the end of the file. If it is `false` it will behave like what we had before. So if you wanted to append to the `output.txt` file you could do the following.

```
import java.io.{FileWriter,PrintWriter}
val pw = new PrintWriter(new FileWriter("output.txt",true))
```

`printf` and `format`

If you have an application where you have to be picky about the formatting of output there are some additional functions/methods that you should be aware of. These are the `printf` function for printing to a screen or file and the `format` method on `String`. These two methods provide you with a way of encoding how you either want something printed or how you want a string to look. They use a style of formatting that is the default way of doing things in the C programming language.

A full description of formatting is beyond the scope of this book and can be found in the Java API under `java.util.Formatter`. You can find the Java API through a web search or going to the `java.oracle.com` web site. As of this writing, Java 7 is the current version. You would want to look at the Java SE 7 version of the API or whatever newer version might be available. A brief introduction is presented here.

Let us start by creating a few variables and printing them using `println` the way we are accustomed to doing.

```
scala> val g = 6.67e-11
g: Double = 6.67E-11

scala> val name = "Mark"
name: java.lang.String = Mark

scala> val classSize = 20
classSize: Int = 20

scala> println(g+" "+name+" "+classSize)
6.67E-11 Mark 20
```

This approach requires using `+` a lot for `String` concatenation. It also gives us very little control over how the different values are printed. Both complaints can be addressed using `printf`.

The `printf` function uses variable arguments length parameters. The first argument is a format `String`. This is followed by arguments for that format. The format `String` can include format specifiers. These start with a percent sign and end with a character that specifies a type of conversion. Between these there can be some additional information that describe how the conversion should be applied. Some of the more common conversion types are:

- `d` – decimal formatting for an integer,
- `e` – scientific notation for a floating point value,

- f – standard decimal notation for a floating point value,
- g – uses either e or f formatting depending on the number, the precision, and the width,
- s – a string,
- x – hexadecimal formatting for an integer.

Here is an example using the variables above with spaces between them.

```
scala> printf("%e %s %d",g,name,classSize)
6.670000e-11 Mark 20
```

By default, the scientific notation displays a significant number of trailing zeros. It is possible to specify the width, in characters, of values by putting a number between the percent sign and the conversion. For fractional numbers, precision can also be specified by putting a dot and the number of digits of precision that are desired. Here are some examples showing that.

```
scala> printf("%.2e %s %d",g,name,classSize)
6.67e-11 Mark 20
scala> printf("%19.2e %s %d",g,name,classSize)
         6.67e-11 Mark 20
scala> printf("%19f %s %d",math.Pi,name,classSize)
         3.141593 Mark 20
```

By default, values are right-aligned when the width is larger than the number of characters displayed. This can be changed by putting a flag after the percent sign. The - flag says you want the value left-aligned. The comma flag says that you want proper regional separators put in long numbers. Examples of both of these are shown here.

```
scala> printf("%-19f %s %d",math.Pi,name,classSize)
3.141593            Mark 20
scala> printf("%,19d %s %d",Int.MaxValue,name,classSize)
      2,147,483,647 Mark 20
```

9.5 Use Case: Simple Encryption

To demonstrate the use of files, we will write some very basic encryption and decryption scripts. At this point we will not be using any serious cryptography methods. Over time we can build our way up to that. For now we will work with some simpler methods. They will probably stop others from reading your stuff, but you would not want to use them to access your bank accounts.

9.5.1 Command-Line Arguments

The approach that we are going to take is to write scripts that allow us to specify all the information needed, including the names of the input and output files, on the command-line. In order to do this we need to discuss how we can get the command-line arguments into our programs in Scala.

When you run a program as a script in Scala, anything that appears on the command-line after the name of the script file is available in the program as an element of the Array args, which has the type Array[String]. We can get the values out of this Array the same way we would any other Array. You can refer to args(0) to get the first argument after the file name, args(1) to get the second, and so on. A simple example to illustrate the use of arguments would be a script that converts two arguments to Doubles and adds them.

```
println(args(0).toDouble+args(1).toDouble)
```

If we put this into a file called add.scala we could call it with the following command on the command-line.

```
scala add.scala 88 12
```

This would print 100.0 as the sum. The decimal point appears because the Strings were converted to a Double and so the sum is a Double and prints as a Double.

9.5.2 Mapping a File

To start off with, we will write a script that uses two command-line arguments for the input file name and the output file name. The heart of the script is a function that will read a file and write a modified file using a function that is passed in as a parameter to map from the character read to the one printed.

To make it so that the formatting is reasonable, the transform function will only be called for characters that are in the alphabet, whether they are uppercase or lowercase. This will leave white space and punctuation intact. Such a function could be written in the following way.

```
def mapFile(inFile:String,outFile:String,trans:Char=>Char) {
  val pw=new PrintWriter(outFile)
  val in=Source.fromFile(inFile)
  for(c <- in) {
    pw.print(if(c>='a' && c<='z' || c>='A' && c<='Z') trans(c) else c)
  }
  in.close
  pw.close
}
```

Make sure you do the two calls to close at the end to clean up things before the function exits. Both pw and in are local variables so if you get out of this function you will not have any way to manually close them.

This function could be invoked with a simple identity function to make a copy of a file.

```
mapFile(args(0),args(1),c=>c)
```

The function c=>c simply returns whatever is passed into it. If you put this in a file that has imports for Source and PrintWriter, you can call it from the command-line specifying two file names, one that exists and one that does not, and the contents of the file that exists will be copied to the new file name.

9.5.3 Character Offset

A simple way to make it much harder for anyone to read what is in your file is to take the characters and offset them by a specified amount in the alphabet. The simplest code for doing this might look like the following.

```
val offset=args(2).toInt
mapFile(args(0),args(1),c=>(c+offset).toChar)
```

Here we convert the third argument to an `Int` and then add it to whatever character we are encrypting. This has to be turned back into a `Char` with `toChar` because doing arithmetic with the `Char` type implicitly results in an `Int`.

This is simple enough and works well for both encoding and decoding. To decode a message, simply use an offset that is the additive inverse of what you used originally. The only thing lacking in this is that letters can be shifted out of the alphabet making them look a bit odd when the text is printed. This problem can be resolved by having the characters wrap around the end of the alphabet. So if the offset would move the letter beyond `'z'`, it is wrapped back to `'a'`. Similarly, we need to handle the case where the offset is wrapped before `'a'`. The easy way to do this is with a modulo operator. The following code will work for any offset larger than -26.

```
val offset=args(2).toInt
mapFile(args(0),args(1),c=> {
  if(c.isLower) ('a'+(c-'a'+offset+26)%26).toChar
  else ('A'+(c-'A'+offset+26)%26).toChar
})
```

This code is a bit more complex because it has to differentiate between lowercase and uppercase letters so that it knows what to subtract from the character to get a value that we can take the modulo of.

This code also does something that might look a bit odd to you at first. The function literal now includes curly braces. Remember that the curly braces just make a code block, which is an expression whose value is the value of the last expression in the code block. This block only has one statement, an `if` expression, but putting it in a block helps to set it off and make it easier to read.

9.5.4 Alphabet Flip

A slightly different way to alter the letters is to flip the alphabet around. So a `'z'` will become an `'a'`, a `'y'` will become a `'b'`, and so on. This encoding does not need an extra argument for something like the offset. It also has an interesting side effect that the transformation is its own inverse. The code to do this might look like the following.

```
mapFile(args(0),args(1),c=> {
  if(c.isLower) ('a'+(25-(c-'a'))).toChar
  else ('A'+(25-(c-'A'))).toChar
})
```

This has the same basic structure of the normal offset version. It just does not need to get an offset and it uses an adjustment of 25 minus the characters position in the alphabet.

9.5.5 Key Word

Both of the methods described above are reasonably easy to crack assuming someone has some idea about what you are doing. For the offset method they just need to know what offset you used and they can probably figure that out if they get to look at a decent sample of encoded text. For the alphabet-flipping model all they need to know is what you are doing and they can decrypt any message you send.

To make things a little harder, you can use a key string to provide a variable offset. The encoding starts with the first letter in the key string and offsets the message character by the number of positions that character is above `'a'`. It then uses the second letter in the key string to offset the second character in the message. This repeats until the end of the key string is reached as which point it wraps back around and starts using characters at the beginning. Code that does this is shown here.

```
val key=args(2)
val factor=args(3).toInt
var keyIndex=0
mapFile(args(0),args(1),c=> {
  val offset=factor*(key(keyIndex)-'a')+26
  keyIndex=(keyIndex+1)%key.length
  if(c.isLower) ('a'+(c-'a'+offset+26)%26).toChar
  else ('A'+(c-'A'+offset+26)%26).toChar
})
```

This method needs both a `key` and a `factor`. The `factor` is required, because decoding is done by applying the same `key` with the negative of the `factor`. Most of the time the `factor` should be either 1 or -1. The way this code is written, the `key` should only include lowercase letters. After it uses two arguments to create the `key` and the `factor`, it then makes a mutable variable for the index in the `key` which begins at zero. The transforming function calculates an offset using the value of the `key` at the index and then increments the index using modulo so that the value wraps back around to zero. The code for calculating the new character is the same as that used for the first offset method.

9.5.6 Putting it Together

To make a more useful script, all three of these approaches can be put into a single script that takes an extra argument to specify the style to be used. The full contents of such a script are shown here including the `imports` and a repeat of the `mapFile` function.

```
import scala.io.Source
import java.io.PrintWriter

def mapFile(inFile:String,outFile:String,trans:Char=>Char) {
  val pw=new PrintWriter(outFile)
  val in=Source.fromFile(inFile)
  for(c <- in) {
    pw.print(if(c>='a' && c<='z' || c>='A' && c<='Z') trans(c) else c)
  }
  in.close
  pw.close
}

args(2) match {
  case "copy" =>
```

```
      mapFile(args(0),args(1),c=>c)
    case "offset" =>
      val offset=args(3).toInt
      mapFile(args(0),args(1),c=> {
        if(c.isLower) ('a'+(c-'a'+offset+26)%26).toChar
        else ('A'+(c-'A'+offset+26)%26).toChar
      })
    case "flip" =>
      mapFile(args(0),args(1),c=> {
        if(c.isLower) ('a'+(25-(c-'a'))).toChar
        else ('A'+(25-(c-'A'))).toChar
      })
    case "key" =>
      val key=args(3)
      val factor=args(4).toInt
      var keyIndex=0
      mapFile(args(0),args(1),c=> {
        val offset=factor*(key(keyIndex)-'a')+26
        keyIndex=(keyIndex+1)%key.length
        if(c.isLower) ('a'+(c-'a'+offset+26)%26).toChar
        else ('A'+(c-'A'+offset+26)%26).toChar
      })
}
```

This script uses the third command-line argument to tell it what type of transform function to use. Any information, like an offset or a key and factor, should be specified in the arguments after that.

9.5.7 Primes and Real Cryptography

The type of cryptography that you would want for your financial transactions involves significantly more math than what we have just covered. In particular, real cryptography makes extensive use of concepts from number theory. To understand and write cryptography algorithms does not require that you have a full and complete knowledge of number theory. There are some key concepts that you will need to understand though. We will start developing those a little here and add onto it in future chapters with the objective that you will be able to write code for the RSA public-key encryption and decryption system later in the book.

The first concept we need to cover is that of prime numbers. A prime number is a number that is only divisible by one and itself. So the sequence of primes begins as 2, 3, 5, 7, 11, 13, 17, ... All the values that were skipped in this sequence have other divisors. For example, all even numbers greater than two are divisible by two and are therefore not prime. There are an infinite number of primes. To show this, we can prove that there is no largest prime. This is done using a proof by contradiction where we start by assuming that there is a largest prime number, call it p_n. If that is the largest prime then we could list all of the primes as p_1, p_2, p_3, ..., p_n. Now consider this number.

$$P = 1 + \prod_{i=1}^{n} p_i = 1 + p_1 * p_2 * p_3 * ... * p_n$$

This number is not divisible by any of the p_i values. To see this, think of the value $P - 1$ and where other multiples of the primes are relative to it. The nearest multiples of 2 will be $P - 2$ and $P + 2$. The nearest multiples of 3 will be 3 away and so forth. As such, P

will not be divisible by any of the p_i values and must therefore be prime. Since $P > p_n$ the original assumption that p_n is the largest prime must be false.

Now that we have had a little refresher on primes, we can write some code that deals with primes. For this section it will be sufficient to just write a function called `isPrime` that tells us whether or not a number is prime. Thanks to the `Range` type and the higher-order methods present in Scala, this can be done in one line.

```
def isPrime(n:Long):Boolean = (2L until n).forall(n%_!=0)
```

This code takes all the numbers between 2 and `n-1` and checks whether `n` is divisible by any of them. If `a` divides evenly into `b` then `b%a==0`. As long as that is not true for any of the values between 2 and `n-1` then the number is prime. This code uses the type `Long` because we really want to have the ability to use it on very large numbers.

While this code is easy to write, it is also inefficient, particularly in the cases where `n` is prime. The reason is that we really do not need to go up to `n-1`. If there is not a divisor less than the square root of `n`, there can not be one above that. If `n` were really large, the difference between `n` and the square root of `n` would be significant. A more efficient function can be written in the following way.

```
def isPrime2(n:Long):Boolean = {
  var i=2
  while(i*i<n+1 && n\%i!=0) i+=1
  n%i!=0
}
```

A `Range` type could have been used with a square root function, but the `math.sqrt` function works with `Doubles` that introduce difficulties we did not want to deal with here.

9.6 End of Chapter Material

9.6.1 Summary of Concepts

- Files are important to real applications as values stored in memory are lost when a program terminates. Files allow data to persist between runs. They can also be used as a source of large data sets.
- Minimal file interactions can be accomplished with I/O redirection. This approach comes with significant limitations for interactive programs.
- Large groups of code, like the libraries for Scala and Java, have to be broken into pieces. Packages are groups of code that have common functionality.
- Names that include full package specifications can be very long. `import` statements can be used to allow the programmer to use shorter names.
- One way to read from files is using `scala.io.Source`.
  - A call to `Source.fromFile` will return an instance of `BufferedSource` that pulls data from the file.
  - `Source` is an `Iterator[Char]`. The `Iterator` part implies that it is a collection that is consumed as values are pulled off it. It gives individual characters.

    - The getLines method returns an Iterator[String] with elements that are full lines.

- The split method on String is useful for breaking up lines into their constituent parts. It takes a delimiter as a String and returns an Array[String] of all the parts of the String that are separated by that delimiter.
- You can use the Source.fromURL(url:String) method to get a Source that can read data from a source located on the web in the form of a URL.
- For some applications it is easier to read data with a java.util.Scanner. This does not provide a Scala-style collection, but it has methods for checking if certain types of values are present and reading them in a way that is more independent of line breaks.
- You can use the java.io.PrintWriter to write text data out to file. This type has print methods like those that you have already become familiar with.

9.6.2 Self-Directed Study

Enter the following statements into the REPL and see what they do. Some will produce errors. You should try to figure out why. Try some variations to make sure you understand what is going on.

```
scala> import java.io._
scala> import scala.io._
scala> val pw = new PrintWriter(new File("RandomMatrix.txt"))
scala> for(i <- 1 to 20) {
     | pw.println(Array.fill(20)(math.random).mkString(" "))
     | }
scala> pw.close()
scala> val nums = {
     | val src = Source.fromFile("RandomMatrix.txt")
     | val lines = src.getLines
     | val ret = lines.map(_.split(" ").map(_.toDouble)).toArray
     | src.close()
     | ret
     | }
scala> import java.util.Scanner
scala> val sc = new Scanner(new File("RandomMatrix.txt"))
scala> var nums = List[Double]()
scala> while(sc.hasNextDouble) {
     | nums ::= sc.nextDouble()
     | }
scala> nums
scala> val webPageSource = Source.fromURL("http://www.google.com")
scala> val webLines = webPageSource.getLines
scala> webLines.count(_.contains("google"))
scala> webPageSource.close()
```

9.6.3 Exercises

1. If you did project 5 (p.178), you can now modify it so that it does not have to use input redirection. Make a function that reads in the dictionary of words and then

write a script that will allow the user to input Scrabble letter sets until they enter a special value to stop the program.

2. Create a file with the letters of the alphabet on one line separated by spaces. The challenge is that you can not manually type in a `String` with that.

3. Create a file with the letters of the alphabet with one letter on each line.

4. Write scripts that copy a file using each of the following constructs. They should take two command-line arguments for the input file name and the output file name.

    - `while` loop
    - `for` loop
    - Higher-order methods
    - Recursion
    - You can repeat each of the above using `getLines` if you did not the first time or without it if you used it the first time.

5. Write scripts using each of the different methods from exercise 4 that capitalizes every letter.

6. Write scripts using each of the different methods from exercise 4 that shift each vowel up one. So "a" becomes "e" and so forth.

7. Write a script that takes a number of rows and columns as command-line arguments and outputs a file with a matrix of random numbers of that size.

8. Write a script that takes a file name as a command-line argument and reads in a matrix of numbers from that file. It should print out the row and column sums and averages.

9.6.4 Projects

1. There is a utility in Linux called `wc` that stands for "word count." You can run it on one or more files and it will tell you how many lines, words, and characters there are in the files. For this project you will do the same thing, only you should also count how many times each letter occurs in each file (regardless of case). The files you should work on will be specified on the command-line. Your program should read through each file and count up how many characters, words, and lines there are as well has how many times each letter occurs. You will print this information for each file followed by a grand total for all the files. Your program might be invoked as `scala wc.scala *.txt`, which would go through all the ".txt" files in the current directory.

    You will consider any string that is between white spaces to be a word. To make the counting of letters easier try doing this: `"abcde".map(c=> c-'a')` to see what it does and think about how that can help you.

2. Back in section 9.5 we went through a few simple forms of encryption. These might keep most people from reading your files, but you would not want to make changes to your bank account using them. The encryption techniques that are used for sending truly sensitive messages rely heavily on number theory. The most common systems today are what are called public-key cryptosystems. In this type of system, each user has both some information that they make public, the public-key, and some

information they keep private, the private-key. For this project, you will implement a simple version of the RSA cryptosystem.

The way the RSA cryptosystem works is the following:

(a) Select two prime numbers p and q such that $p \neq q$. The level of security of the system is dependent on how big p and q are. For good security you want them to be 512-1024 bit numbers, far larger than what you can store in an `Int` or even a `Long`. Finding large primes is beyond the scope of this book so to keep this simple you should use two primes that multiply to give you a number between 128 and 256.

(b) Compute $n = pq$.

(c) Select an small odd integer, e, that is relatively prime to $(p-1)(q-1)$.

(d) Compute d, the multiplicative inverse of e, modulo $(p-1)(q-1)$. The technique for finding this is described in exercise 9 (p.203).

(e) The pair (e, n) is the RSA public-key. Anyone who wants to send you a message know these values.

(f) The pair (d, n) is the private-key. You need to know this to decode a message and you don't want anyone else to have it.

(g) To encode a message, M, the sender uses $P(M) = M^e \bmod n$. For this to work you need to break up the message into chunks that can be encoded into numbers, M, that are less then n.

(h) To decode the message you use $S(C) = C^d \bmod n$, where C is the encrypted message you got from the sender.

3. The proper way to represent a recipe is using a file. The same thing is true for the contents of your pantry. For this project you will extend project 6 (p.207)to include options for building a cookbook along with file access for the pantry contents. The script should take two command-line arguments for a pantry file and a recipe file. Those files should be loaded in when the program starts. The menu should have the following options.

   (a) Add a pantry item.

   (b) Print pantry contents.

   (c) Add a recipe.

   (d) Check a recipe.

   (e) Cook recipe.

   (f) Quit

   When the user selects the "Quit" option. The files should be written to reflect new changes. You can decide what file format to use. You simply have to make the code that writes it work with the code that reads it.

   In this version, the "Add recipe" option should have the user type in a unique name of a recipe followed by item names and amounts. That should be remembered so that the "Check a recipe" option allows the user to type in just a name.

4. For this project you will extend what you did as part of project 7 (p.207) so that the courses you are interested in will be stored in a file for reuse. The script should take a single command-line argument for the name of the storage file. The file stores the course information along with the level of interest. Menu options should be as follows.

(a) Add course of interest.

(b) Print current courses.

(c) Remove course of interest.

(d) Modify course interest.

(e) Build a schedule.

(f) Quit

There is a new option to modify interest because if the program is run over a period of time, the user might become more or less interested in the course. Also, for the build option, you should add the option to have the user say they are taking one of the printed schedules and remove all those courses from the list of interests. Save the modified list of courses of interest to the file when the user quits.

Additional exercises and projects, along with data files, are available on the book's website.

Chapter 10

Case Classes

One of the things that you should have noticed by this point is that there are times when it can be helpful to group pieces of data together. We have seen two ways that we can do this. If all the data is the same type, you could use something like an `Array` or a `List`. The downside of this is that the compiler can not check to make sure that you have the right number of elements. For example, if you want a point in 3-D with x, y, and z coordinates then you would need an `Array[Double]` that has three elements in it. Having more or less could cause the code to break and Scala would not be able to check for that until the program was running. The alternative, which also works if the types of the values are different, is to use a tuple. That same 3-D point could be represented as a `(Double, Double, Double)`. While this works reasonably well for a 3-D point, it too has some significant limitations. The main limitation is that being a `(Double, Double, Double)` does not tell you anything about what those three `Double`s mean or how it should be used.

To illustrate this, consider some different things that could be represented as three `Double`s. One is a point in 3-D. Another closely related one is a vector in 3-D. Either the vector or the point could be represented in Cartesian coordinates, cylindrical coordinates, or polar coordinates and the tuple would not tell you which it is. The three `Double`s though could also represent three subaverages for a student's grade in a class. For example, they might be the test, quiz, and assignment averages. The tuple does not tell you which it is and does not help you at all with keeping things straight.

What is more, the tuple lacks some flexibility and the syntax for getting things out of them is less than ideal. Calling the `_1` and `_2` methods all through your code can make it difficult to read. Imagine if instead you wanted to represent a full student in a class. Then the tuple might have a `String` and a large number of grades, all of the same type. Keeping track of what numbers are associated with what grades would very quickly become problematic. To get around these limitations, we will consider the use of `case class`es for grouping data.

10.1 User-Defined Types

What we really need to break away from the limitations of using tuples to deal with groupings of data is the ability to define our own types that have meaningful names. Tuples definitely have their place, and they are very useful in those places. However, there are many times when it would be handy to create a type specifically for a particular purpose. Then we could give the type a name that indicated its purpose and have the compiler check for us that we were using that type in the proper way.

User defined types are a common feature in modern programming languages and have been for decades. Scala provides three constructs for creating user-defined types: `classes`, `traits`, and singleton `objects`. For this chapter, in order to keep things simple, we will only consider a specific type of one of these, the `case class`. The others will be covered in detail when we consider object-orientation in full.

Back in section 7.5 we defined a type as a collection of values and the operations that can be performed on them. The user-defined types typically take the form of being collections of other types. This makes them fundamentally similar to just using a tuple. Where they prove to be more than just a tuple is the control they give you in determining what can be done with the types. We will remain somewhat limited in this regard for now, but even with those limitations you should find that our user-defined types provide a significant boost to our programming capabilities.

10.2 Case Classes

The simplest way to start with user-defined types in Scala is with `case classes`. Perhaps the best way to introduce them is to show some examples. We'll start with two that were mentioned above: a 3-D point and a student with some grades.

```
case class Point3D(x:Double,y:Double,z:Double)

case class Student(name:String,assignments:List[Double],tests:List[Double],
  quizzes:List[Double])
```

The first one declares a type called `Point3D` that stores inside of it three different `Doubles` that are named `x`, `y`, and `z`. The second declares a type called `Student` that has a `name` as a `String` and three different `Lists` of `Double` to use as grades.

There can be more to the declaration of a `case class`, but for now we will limit ourselves to this syntax that begins with the keywords "`case class`". After that is the name you want to give the type. This could be any valid Scala name, but it is customary to begin type names with uppercase letters and use camel naming so all subsequent words also begin with uppercase letters. After that is a list of name/type pairs in parentheses. The format of these is just like the arguments to a function. The elements of this list give the names and types of the values stored in this new type.

10.2.1 Making Objects

After you have declared a type, you need to be able to create objects of that type. With a case class, you can do this with an expression that has the name of the class followed by an argument list, just like a function call. The two classes listed above could be created and stored in variables with the following lines of code.

```
val p = Point3D(1,2,3)
val s = Student("Mark",Nil,Nil,List(89))
```

The first line makes a point that has x=1, y=2, and z=3 and stores a reference to it in a variable names p. The next line makes a student with the name "Mark" who has no assignment or test grades, but who has an 89 on one test and stores a reference to it in a variable named s.

You could insert the word new and a space after the equals signs so that these lines look like the following.

```
val p = new Point3D(1,2,3)
val s = new Student("Mark",Nil,Nil,List(89))
```

The result of this would be exactly the same. The first syntax is shorter and works for all case classes so we will stick with that in our sample code.[1]

10.2.2 Accessing Elements

In order to be able to use these objects, we must be able to access the different elements in them. This is very simple to do, just use the dot notation to access the elements. So if you want the x value in the Point3D, p that we made above, you would just do this.

```
scala> p.x
res1: Double = 1.0
```

To get the name of the Student you would do this.

```
scala> s.name
res2: String = Mark
```

The dot notation in Scala simply means that you are using something from inside of an object. It could be a method or a value that is stored in the object. For now we will only be concerning ourselves with the values that we store in our case classes.

We could put this to use by writing a function to find the distance between two Point3Ds. It might look something like this.

```
def distance(p1:Point3D,p2:Point3D):Double = {
  val dx=p1.x-p2.x
  val dy=p1.y-p2.y
  val dz=p1.z-p2.z
  math.sqrt(dx*dx+dy*dy+dz*dz)
}
```

We could also use it to calculate and average for a Student with code like this.

```
def classAverage(s:Student):Double = {
```

[1]In chapter 16 you will learn that normal classes require the use of new by default. To get around this requires writing some code in a companion object, a technique also covered in that chapter.

```
  val assnAve=if(s.assignments.isEmpty) 0.0
    else s.assignments.sum/s.assignments.length
  val quizAve=if(s.quizzes.length<2) 0.0
    else (s.quizzes.sum-s.quizzes.min)/(s.quizzes.length-1)
  val testAve=if(s.tests.isEmpty) 0.0
    else s.tests.sum/s.tests.length
  0.5*assnAve+0.3*testAve+0.2*quizAve
}
```

The `if` expressions here prevent us from doing division by zero.

One of the things to note about `case classes` is that the elements in them are `vals`. As such, you can not change what they refer to. If you try to make such a change you get something like the following.

```
scala> p.x=99
<console>:8: error: reassignment to val
       p.x=99
          ^
```

Whether you can change anything in an object created from a `case class` depends on whether the things in it are mutable or not. In our two examples, all of the contents are immutable. As a result, the `case class` as a whole is immutable. Once you create a `Point3D` or a `Student`, the object you create can not change its value in any way. However, if one or more of the fields in the `case class` were an `Array`, then the values in the `Array` would be mutable. You would not be able to change the size of the `Array` without making a new object, but you could change the values stored in it.

10.2.3 Named and Default Arguments (Advanced)

A few options were added to Scala in version 2.8 in regards to function arguments and calling functions. Normally, Scala figures out which of the arguments passed into a function is associated with which formal parameter by their order. Consider this function.

```
def evalQuadratic(a:Double,b:Double,c:Double,x:Double):Double = {
  val x2=x*x
  a*x2+b*x+c
}
```

If you load this into the REPL you can execute it as follows.

```
scala> evalQuadratic(2,3,4,5)
res0: Double = 69.0
```

In this call, `a`=2, `b`=3, `c`=4, and `x`=5. This is because that is they order the arguments to appear in both the definition of the function and the call to it. For functions where there are a significant number of arguments that are all of the same type, this can lead to confusion. To get around this, Scala has NAMED ARGUMENTS. When you call the function you can specify the names you want the values associated with it. So the call above would be like the following:

```
scala> evalQuadratic(a=2,b=3,c=4,x=5)
```

```
res2: Double = 69.0
```

In this call it is now explicit what values are going to what parameters. One advantage of this is when you enter the arguments in a different order, the meaning will be what you want. For example, you might think that x was the first argument instead of the last. Without names arguments, this would lead to an error with no error message. You would simply get the wrong answer. However, if you use named arguments everything is fine because the names supersede the order.

```
scala> evalQuadratic(x=5,a=2,b=3,c=4)
res3: Double = 69.0
```

Here we see that even though x is first, the value we get is correct.

You can use named parameters without naming all the parameters. You can start the list with arguments that are based on position and then use names for the later ones. All the arguments after the first named one have to be named and they can not duplicate any that you gave using the position.

For some functions there are some arguments that will have a particular value a lot of the time. In that situation, it is nice to make it so that people calling the function do not have to provide them and they will use a DEFAULT VALUE. When you declare the function simply follow the type with an equals sign and the value you want to have for the default. If the caller is happy with the default value, then that argument can be left out. Default arguments at the end of the list can be simply omitted. If they are in the middle then you will have to use named arguments to specify any arguments after them in the list. Consider a function to add a grade to a Student.

```
def addGrade(name:String,grade:Int = 0):Student = ...
```

Here the default grade is a zero. So this function can be called in two ways.

```
addGrade("Jane",95)
addGrade("Joe")
```

The first call is like everything we have seen to this point. The second one leaves off the grade. As a result, Joe gets a 0 for whatever grade this was.

10.2.4 The copy Method

The fact that you can not mutate the values in a case class means that it would be helpful to have a way to make new case class objects that are only slightly changed from others. To see this, consider what happens when you want to add a new grade to a Student. The grades are in Lists, and it is easy to add to a List. The problem is, that does not mutate what is in the original List, it just gives us a new List that includes the new values as well as what was already there.

To help get around this problem, case classes come with a copy method. The copy method is intended to be used with the named arguments that were discussed in section 10.2.3. The arguments to copy have the same names as the fields in the class. Using named arguments, you only provide the ones you want to change. Anything you leave out will be copied straight over to the new object. So using the Student object we gave the name s above, we could use copy to do the following.

```
val ns = s.copy(tests = 99::s.tests)
```

This gives us back a new Student who is the same as the one we had in s, only it has a test grade of 99 in addition to the quiz grade of 89 it had originally.

You can specify as many or as few of the fields in the case class as you want. Whatever fields you give the names of will be changed to the value that you specify. If you leave the parentheses empty, you will simply get a copy of the object you have originally.

10.2.5 Case Class Patterns

Another capability that comes with case classes is that you can use them in patterns. This can be used as a simple way to pull values out of an instance or to select between objects in a match. As an example of pulling out values, consider the following code using Point3D.

```
for(Point3D(x,y,z) <- points) {
  // Do stuff with x, y, and z.
}
```

This is a for loop that runs through a collection of points. Instead of calling each point with a name like point, this pulls out the values in the point and gives them the names x, y, and z. That can make things shorter and more clear in the body of the loop.

As an example of limiting what is considered, we can use another for loop that goes through a course full of students.

```
for(Student(name,_,List(t1,t2,t3),_) <- courseStudents) {
  // Processing on the students with three test grades.
}
```

This does something with patterns that we have not seen before, it nests them. You can nest patterns in any way that you want. This is part of what makes them extremely powerful. In this case, the assignment and quiz grades have been ignored and the loop is limited to only considering students with exactly three test grades. Those grades are given the names t1 , t2, and t3. That could also have been specified with the pattern t1::t2::t3::Nil. Students who have more or fewer test grades will be skipped over by this loop.

10.3 Mutable Classes

case classes are not the only option you have when creating classes. We will go into the full details of classes in chapter 16. There are some applications for which you do not need the full power classes can provide, but you would like to have fields in the class that are mutable. Remember that every field in a case class is like a val so the case class is generally immutable. The only way the value of something in a case class can change is if one of the fields references a mutable value like an Array.

When you want to make objects with mutable fields, you need to leave off the case keyword and just define a class. To be able to use the fields, given what you currently know, each one will need to be preceded by either val or var. If you put val before a field its value will not be allowed to change. If you use var, the field will be allowed to change. So if you have objects that truly need to be altered because they change very rapidly and modified copies are not efficient, you can make a class using `var` for each field that needs to change.

Note that by leaving off the `case` keyword, you do more than just allow for `var` fields. You also lose the `copy` method, the ability to leave off `new` when making a new instance of the class, and the ability to do pattern matching. In general, there are many benefits to immutability so you should not take this approach without good reason.

10.4 Putting it Together

Now we want to use a `case class` along with other things that we have learned to create a small, text-based application. The application that we will write will use the `Student` that we defined earlier along with the `classAverage` function and other functions that we will write to make a grade book. This program will be run from a text menu and give us various options similar to what was done in section 8.2.

The program will also use the file handling capabilities that we have learned so that the grades of the students in the course can be saved off and then be loaded back in when we restart the program. The menu for the program will have the following options:

1. Add Test Grade
2. Add Assignment Grade
3. Add Quiz Grade
4. Print Averages
5. Save and Quit

The program will take a command-line argument for the file to load in. If none is given, the user will be asked how many students are in the section and their names along with the file name to save it under. When one of the first three menu options is selected, the program will list each student's name and ask for their grade. The "Print Averages" option will print out the names of each student along with their grades in each area, their average in that area, and their total average.

There is quite a bit to this program so it is worth breaking it up into different functions and then writing each of those. To do this we can outline what will happen when we run the program and use the outline to break things down then assign function names to things.

- Startup
  - load a file (`loadSection`)
  - or create a section (`createSection`)
- Main menu (`mainMenu`)
  - print the menu (`printMenu`)
  - act on the selection
    - add a test grade (`addTest`)
    - add an assignment grade (`addAssignment`)

  * add a quiz grade (addQuiz)
  * print the averages (printAverages)

- Save when done (saveSection)

Now that we have figured out roughly what we need to do, we can write these functions in any order that we want. In general the process of writing functions like this can be very non-linear. You should not feel any reason why you would have to go through the functions in any particular order. Often in a real project you would do things in a certain order as you figure out how to do them.

The more experience you gain, the more comfortable you will be in writing code and then you might decide to pick certain functions because they will give you functionality that you can test. One of the advantages of having the REPL to fall back on is that we can load in our file and test functions one by one, seeing the results along the way. Without that, the printAverages function would prove to be extremely important to us as it would be the only way that we could see what was going on.

For our purposes we will start with createSection and saveSection. These two functions pair well together and are closely related because we have to decide how we are going to represent a section both in the memory of the computer and in the file. We will start with createSection and the way in which things are represented in the memory of the computer.

We have already created a case class called Student that can be used to represent one student in the section. We just need a bunch of them. We also need to realize that they will change over time as grades are added. It would probably be sufficient to just keep an Array[Student]. However, there are benefits to actually wrapping the array inside of a different case class like this.

```
case class Section(students:Array[Student])
```

One advantage for our purposes here is that this is a chapter on case classes and this provides yet another example of one. In general, this can also provide greater flexibility. We might decide at some point that we want to attach data for a course name, semester, instructor, etc., to each Section. Those things can not be added to a simple Array. However, they could easily be added to the case class. It also has the advantage of providing extra meaning. This is not just a random collection of Students, it represents a section of a class.

Now that we know this, we can write the createSection function. This function will prompt the user for the information that is needed to create the Section. For now that is a file name to save it to, the number of students, and the names of the students. The function will return the file name and the Section.

```
def createSection:(String,Section) = {
  println("What file would you like to save this as?")
  val fileName=readLine()
  println("How many students are in the class?")
  val numStudents=readInt()
  println("Enter the student names, one per line.")
  (fileName,Section(Array.
    fill(numStudents)(Student(readLine(),Nil,Nil,Nil))))
}
```

The first five lines of this function are fairly self-explanatory with prompts being printed and values being read. After that is the return tuple, which includes a call to Array.fill that has a readLine in the pass-by-name parameter. This means that it not only makes the return value, it also includes the input of the names.

Now that we have created a new Section, we can consider what it will look like in a file. There are many different ways that the file could be formatted. The manner that we will pick here starts with the number of students in the class on a line. After that there are four lines for each student. They are the student's name followed by a line each with assignment, test, and quiz grades. This function can be written as follows.

```
def saveSection(fileName:String,section:Section) {
  val pw=new PrintWriter(fileName)
  pw.println(section.students.length)
  for(s <- section.students) {
    pw.println(s.name)
    pw.println(s.assignments.mkString(" "))
    pw.println(s.tests.mkString(" "))
    pw.println(s.quizzes.mkString(" "))
  }
  pw.close()
}
```

The function takes the file name and the Section. It then makes a PrintWriter with the fileName, which is closed at the end of the function, and prints the needed information. The use of mkString on the different Lists makes the code for doing this much shorter.

As you are writing these functions, you need to test them. One way to do that is to load them into the REPL and call them. Another way is to end the script with calls to them. At this point, the end of the script might look something like the following.

```
val (fileName,section)=createSection
saveSection(fileName,section)
```

This comes after the definition of both the case classes and the different functions. If you run the script with this in it, you should be prompted for the information on the Section and after you enter the the script should stop. You can then look at the file that you told it to save as and make sure it looks like what you would expect.

We will hold the loadSection function until the end and go into the main functionality with mainMenu and printMenu. You can write them in the following way.

```
def printMenu {
  println("""Select an option:
1. Add Test Grade
2. Add Assignment Grade
3. Add Quiz Grade
4. PrintAverages
5. Save and Quit""")
}

def mainMenu(section:Section) {
  var option=0
  do {
    printMenu
    option=readInt()
    option match {
      case 1 => addTest(section)
      case 2 => addAssignment(section)
      case 3 => addQuiz(section)
      case 4 => printAverages(section)
      case 5 => println("Goodbye!")
```

```
      case _ => println("Invalid option. Try again.")
    }
  } while(option!=5)
}
```

You can not test this code yet because mainMenu calls four other functions that have not been written yet. Once we have those written, we can put a call to mainMenu at the end of the script right before the call to saveSection.

The three different add functions will all look pretty much the same. We will only show the addTest function and let you figure out the others. It is worth thinking a bit about how that function will work. The Student type is immutable. All the fields in the case class are vals so they can not be changed. The String and the three different List[Int] values are all immutable so once a Student is created, it is set forever. Fortunately, the Section type stores the Students in an Array. This means we can change what Student objects are being referred to. We can use the copy capabilities of the case class to make new instances that are almost the same except for small variations. Using this, the addTest function could be written in the following way.

```
def addTest(section:Section) {
  for(i <- 0 until section.students.length) {
    println("Enter the grade for "+section.students(i).name+".")
    section.students(i)=section.students(i).
      copy(quizzes=readInt()::section.students(i).quizzes)
  }
}
```

This code works just fine, but it is a bit verbose because we have to type in section.students(i) so many times. We have to have the index because we need to be able to do the assignment to an element of the Array. The section.students(i) before the equals sign in the assignment is hard to get rid of because we have to mutate that value in the design of this code. The code could be shortened with appropriate use of imports, but there is another, more interesting solution.

```
def addTest(section:Section) {
  for((s,i) <- section.students.zipWithIndex) {
    println("Enter the grade for "+s.name+".")
    section.students(i)=s.copy(tests=readInt()::s.tests)
  }
}
```

This version uses zipWithIndex and a pattern on the tuple to give us both a short name for the student, s, and an index into the array, i. Both of these are equally correct so use the one that makes more sense to you and duplicate it for assignments and quizzes.

The next function in the menu is printAverages. A very basic implementation of this would just print student names and the course average. However, it could be helpful to see all the grades and the partial averages as well. That is what is done in this version.

```
def printAverages(section:Section) {
  for(s <- section.students) {
    println(s.name)
    val assnAve=if(s.assignments.isEmpty) 0.0
      else s.assignments.sum/s.assignments.length
    println(s.assignments.mkString("Assignments:",", "," = "+assnAve))
    val quizAve=if(s.quizzes.length<2) 0.0
```

```
      else (s.quizzes.sum-s.quizzes.min)/(s.quizzes.length-1)
    println(s.quizzes.mkString("Quizzes:",", ","," = "+quizAve))
    val testAve=if(s.tests.isEmpty) 0.0
      else s.tests.sum/s.tests.length
    println(s.tests.mkString("Tests:",", ","," = "+testAve))
    println("Average = "+(0.5*assnAve+0.3*testAve+0.2*quizAve))
  }
}
```

This function uses the code from the earlier class average function and inserts some print statements. The only thing in here that might seem odd is the use of a `mkString` method that takes three arguments instead of just one. With this longer version, the first string goes before all the elements and the third one goes after all the elements. The argument in the middle is the delimiter as it has been in previous usage.

```
def loadSection(fileName:String):(String,Section) = {
  val src=Source.fromFile(fileName)
  val lines=src.getLines
  val section=Section(Array.fill(lines.next().toInt)(Student(
    lines.next(),
    lines.next().split(" ").filter(_.length>0).map(_.toDouble).toList,
    lines.next().split(" ").filter(_.length>0).map(_.toDouble).toList,
    lines.next().split(" ").filter(_.length>0).map(_.toDouble).toList
  )))
  src.close
  (fileName,section)
}
```

This function includes three lines for handling the file. The meat of the function is in the declaration of the `section` variable, which calls `lines.next()` anytime that it needs a new line from the input file. The first time is to read how many students are in the section for building the `Array`. Each student pulls in four lines for the name and three different grade types. The lines of grades are `split`, `filter`ed, and them `map`ped to `Double`s before they are converted to a `List`. The `filter` is required for the situation where you have not entered any grades of a particular type.

You might wonder why the return type of this function includes the `fileName` that was passed in. Technically this is not required, but it makes this function integrate much more nicely at the bottom of the script.

```
val (fileName,section)=if(args.length<1) createSection
  else loadSection(args(0))
mainMenu(section)
saveSection(fileName,section)
```

Having `createSection` and `loadSection` return the same information greatly simplifies this part of the code as they can be called together in a simple `if` expression.

That is everything. You now have a full little application that could be used to store a grade book for some course. Try putting this code in and playing with it a while.

Tuple Zipped Type (Advanced)

Something that you need to do fairly frequently is to run through two collections at the same time, pulling items from the same location of each. One way to do this is to use the `zip` method to zip the collections together into a new collection of tuples. While this works well if you have two collections, especially if you use a `for` loop, it does not work as well for three or more collections, and it is fundamentally inefficient because the `zip` method will go through the effort of creating a real collection with a bunch of tuples in it.

To get around these limitations, the types for tuples of length 2 and 3 have a type associated with them called `Zipped`. The sole purpose of the `Zipped` type is to let you get the benefits of running through a zipped collection without actually doing the zipping. To get an instance of the `Zipped` type, simply make a tuple that has all the collections you want in it and call the `zipped` method. The `Zipped` type has some of the main higher-order methods that you are used to using on collections: `exists`, `filter`, `flatMap`, `forall`, `foreach`, and `map`. The difference is that in the `Zipped` type they take multiple arguments. Specifically, they take as many arguments as there are elements in the tuple. This is significant because if you call a function like `map` that is mapping a collection of tuples the function has to take one argument and go through some effort to pull the elements out of the tuple. With the `Zipped` type you do not have to do that as the function literals are supposed to take multiple arguments instead of a single tuple with the multiple values.

A comparison of the two approaches is shown here.

```
val l1=List(1,2,3)
val l2=List(4,5,6)
l1.zip(l2).map(t => t._1*t._2)
(l1,l2).zipped.map((v1,v2) => v1*v2)
```

For this example the first one is a bit shorter, but that typically will not be the case. More importantly, the first one relies on the _1 and _2 methods which will make the code hard to read and understand for anything with more logic. To get the benefit of easy to read names using `zip` you would have to do the following.

```
l1.zip(l2).map(t => {
  val (v1,v2)=t
  v1*v2
})
```

It remains an exercise for the reader to see what happens if you want to iterate over three collections using `zip`. Consider the `Zipped` type when you need to iterate over two or three collections at the same time.

10.5 End of Chapter Material

10.5.1 Summary of Concepts

- The act of grouping together data is very useful in programming. We have been doing this with tuples. The problem with tuples is that they do not provide meaning and their syntax can make code difficult to read and understand.
- User defined types let you create your own types that have meaning related to the problem you are solving.
- One way of making user-defined types is with `case classes`. We will use these to group values together giving them useful, easy-to-read names.
  - To create a `case class` follow those keywords with an argument list like that for a function with names and types separated by commas. Names for types typically start with a capital letter.
  - You create an instance of a `case class` give the name of the type followed by an argument list of the values it should store.
  - When you want to access the members of a case class, use the dot notation we have been using for other objects.
  - The members of a `case class` are all `val`s. As a result, instances of `case classes` tend to be immutable. The only way that will not be true is of a member is itself mutable.
  - To make new instances of `case classes` that are slightly different from old ones, use the `copy` method. This method is called with named arguments for any members that you want to have changed in the copy.
  - Another useful capability of `case classes` is that they can be used as patterns.
- If you need the ability to mutate values, you can make a normal `class` leaving off the `case` keyword and specifying whether the member is a `val` or a `var`.
  - Instances of these types will need to be created with `new`.
  - There is no `copy` method automatically defined.
  - These types will not implicitly work as patterns.

10.5.2 Self-Directed Study

Enter the following statements into the REPL and see what they do. Some will produce errors. You should try to figure out why. Try some variations to make sure you understand what is going on.

```
scala> case class Accident(dlNumber1:String,dlNumber2:String)
scala> case class
    Driver(name:String,dlNumber:String,dob:String,history:List[Accident])
scala> def wreck(d1:Driver,d2:Driver):(Driver,Driver,Accident) = {
     | val accident = Accident(d1.dlNumber,d2.dlNumber)
     | (d1.copy(history = accident::d1.history),
     |  d2.copy(history = accident::d2.history),
```

```
     | accident)
     | }
scala> var me = Driver("Mark","12345","long ago",Nil)
scala> var otherPerson = Driver("John Doe","87654","01/01/1990",Nil)
scala> val (newMe,newOther,acc) = wreck(me,otherPerson)
scala> me = newMe
scala> otherPerson = newOther
scala> println(me.name)
scala> println(otherPerson.dlNumber)
scala> println(me.history.length)
scala> otherPerson.name = "Jane Doe"
scala> case class Vect2D(x:Double,y:Double)
scala> def magnitude(v:Vect2D):Double = {
     | math.sqrt(v.x*v.x+v.y*v.y)
     | }
scala> def dot(v1:Vect2D,v2:Vect2D):Double = v1.x*v2.x+v1.y*v2.y
scala> def makeUnit(angle:Double):Vect2D = {
     | Vect2D(math.cos(angle),math.sin(angle))
     | }
scala> def scale(v:Vect2D,s:Double):Vect2D = Vect2D(v.x*s,v.y*s)
scala> val a = makeUnit(math.Pi/4)
scala> val b = makeUnit(3*math.Pi/4)
scala> dot(a,b)
scala> magnitude(a)
scala> magnitude(b)
scala> magnitude(scale(a,3))
```

10.5.3 Exercises

1. Write a `case class` to represent a student transcript.
2. Using your answer to the previous exercise, define a function that adds one semester of grades to the transcript.
3. Using your answer to 1, write a function that will return the student's GPA.
4. Write a `case class` to represent a recipe.
5. Using your answer to 4, write a function that takes a recipe and the name of an ingredient and returns how much of that ingredient is needed.
6. Write a `case class` to represent the information needed for a house in a Realtor posting.
7. Re-write the grade book program in a completely functional way so that it has neither `Array`s, `var`s or other mutable objects.
8. Play with `zip` using 3 collections. Compared it to using `zipped`.
9. Pick a favorite sport and make a `case class` that can be used to store player information.
10. Extends what you did on the previous exercise so you have a `case class` that stores the information for a team.

10.5.4 Projects

1. This is an extension on project 3 (p.205). You will use `Arrays` and `classes` to take the Keplerian simulator a bit further. In that program all you had was one body in motion about a "central mass" that was not moving at all. Now you can store and work with the positions and velocities of many bodies because you can store their component data in `Arrays` or `case classes`. That is to say you can have an `Array` for x positions as well as y, v_x and v_y, or an `Array` of some `case class` that stores those values. This allows you to simulate the motions of many particles at the same time which is much more fun. Earlier you only had to calculate the acceleration due to the central particle. Now you want to calculate accelerations due to all interactions between all the particles. You can also make the central particle one of the particles in the array or try not even having a central particle.

   With multiple particles, you need to have a nested loop (or a `for` loop with two generators) that calculates the accelerations on each particle from all the others and adds them all up. Keep in mind that if particle i pulls on particle j then particle j pulls back on i just as hard, but in the opposite direction. That does not mean the accelerations are the same though. The acceleration is proportional to the mass of the puller because the mass of the pullee is canceled out by its inertia. Earlier we had $a_x = -\frac{x}{d^3}$ and $a_y = -\frac{y}{d^3}$. When the particle doing the pulling is not at the origin, d is the distance between the particles and x and y are the distances between them in x and y directions. We also need a factor of m for the mass of the puller. You want to add up the accelerations from all other particles on each one and store that into arrays so

$$a_x(i) = -\sum_j \frac{(x_j - x_i) * m_j}{d_{ij}^3}$$

   There is a similar formula for y. The d_{ij} value is the distance between particle i and particle j. Also note that given a value, `c`, the best way to cube it is `c*c*c`, not `math.pow(c,3)`.

   When you write your formula in the code, think a bit about it. This formula causes a problem if we really use d because particles can get too close to one another. It is suggested that you make $d = \sqrt{d_x^2 + d_y^2} + \epsilon$, where *epsilon* is a small value. You can play with how small you want it to be because that depends on the scale of your problem. It is also recommended that you have your integrator not use the normal Euler method which calculates accelerations, then updates positions, then velocities. Make sure that it does the accelerations on velocities before applying the velocities to the positions. Keep in mind that you want to break this up into functions that fit together nicely and are helpful. It will hurt your brain more if you do not.

   The input for the program will have a first line that is the number of bodies, the timestep, the stopping time, and the number of steps between outputs. This will be followed by lines with the x, y, v_x, v_y, and *mass* values for each particle. A sample input file can be found on the book website on the page for this chapter. Note that the central mass in that file has a mass of 1 and all the others are much smaller.

   As output, you should write out the positions of all the particles in your simulation to a file once for every n steps, where n is a value given on the first line of the input. If you do this then you can run a spreadsheet of `gnuplot` to plot it. If you use `gnuplot` and give it the command `plot 'output'` it will make a little scatter plot showing you the paths of the particles in the simulation.

2. This project builds on top of project 8.1 (p.204). You have likely been using tuples or separate values to represent the geometry elements in your ray tracer. This is information that is much more naturally represented as a case class. For this project you should go through and edit your existing code so that it includes three different `case classes`, one for spheres, one for planes, and one for a scene which has a `List[Sphere]` and a `List[Plane]`.

3. This is the first installment for you building your own text adventure. Your program will read in from a map file that you will write by hand and let the user run around in that map by using commands like "north" to move from one room to another. The map file will have a fairly simple format right now and you will create your own map file using vi. Make sure when you turn in this program you turn in both the script and the map file so it can be tested with your map.

   The format of the map file should start with a line telling the number of rooms then have something like the following. You can change this if you want to use a slightly different format:

   ```
   room_number
   room_name
   long line of room description
   number_of_links
   direction1
   destination1
   direction2
   destination2
   ...
   ```

   This is repeated over an over. (The number of rooms at the top is helpful for storing things in an `Array` so that you know how big to make it.) Each room should have a unique room number and they should start at 0. The reason is that you will be putting all the room information into `Arrays`. There is a link on the book website to a sample map file, but you do not have to stick exactly to that format if you do not want to. You might deviate if you are thinking about other options you will add in later. Odds are good you will be refactoring your code for later projects.

   The interface for your program is quite simple. When you run the program it should read in the map file and keep all the map information stored in an `Array[Room]` where `Room` is a `case class` you have made to keep the significant information for each room.[2] You will start the user in room 0 and print the description of that room and the different directions they can go as well as where those exits lead to, then follow that with a prompt. You could just use > as the prompt to start with. It might get more complex later on when you have real game functionality. So when the user starts the game it might look something like this if you read in the sample map file.

   ```
   Halsell 228
   You are standing in a room full of computers and comfy chairs with
   a giant air conditioning unit hanging from the ceiling. While the
   surroundings are serene enough, you can't help feeling a certain amount
   of dread. This isn't just a fear that the air conditioning unit is
   going to fall either. Something in you tells you that this room is
   ```

[2]Many implementations also make a `case class` to represent an `Exit` from a room.

```
regularly used for strange rituals involving torture. You can only
wonder what happens here and why there isn't blood all over the place.
Your uneasiness makes you want to leave quickly.
The hallway is east.
>
```

The user must type in either a direction to move or "quit". If anything else is entered you should print an appropriate error message. The only goal of this project is to allow the user to move around the map. Collection methods such as `find`, `indexWhere`, `filter`, or `partition` can be extremely helpful for this project.

4. Convert the work you did for project 3 (p.227) to use `case classes` to bind data. So you should have a `case class` for an item with an amount, one for a recipe, one for pantry contents, etc.

5. Convert the work you did for project 4 (p.227) to use `case classes`. With this addition you can put more information into each course because you now have a better way to store it all together.

6. Sports have a tendency to produce a lot of information in the form of statistics. A `case class` is a good way to represent this information. For this project you need to find a text format of box scores for the sport that interests you. Then make a `case class` to represent that data. Put multiple box scores into a text file and read them in. Have menu options for calculating averages for players and team in the different relevant statistics.

Additional exercises and projects, along with data files, are available on the book's website.

Chapter 11

GUIs

So far, all of the programming that we have done has been part of a text interface. The programs print information to a terminal window and to give them input we enter text from the keyboard. These types of programs have their place in computing, but these days very few people actually use such programs. They typically run in the background instead, doing things that people do not see. The programs that you are likely used to working with are more graphical. They open up windows and you can interact with them through mouse clicks as well as typing.

This type of program is called a GUI (Graphical User Interface). In this chapter we will look at how we can write GUIs in Scala and get them to interact with the user.

11.1 GUI Libraries and History

There are reasons why we began doing our programming with the console. There are certain complications involved in programming GUIs. Some of that is inherent to the medium. In Scala, some of that is due to history. Scala relies on the underlying virtual machine and the libraries for it to do a lot of the work in a GUI. Scala simply adds things on top when it is possible to improve them in a significant way by doing so. In the implementation of Scala that we are using, this means that there is a dependence on the Java GUI libraries.

When Java was originally released, there was one GUI library called the Abstract Windowing Toolkit, AWT. The AWT library makes direct calls to the operating system or windowing environment for building elements of the GUI. This gave the AWT a significant benefit in speed. However, it also provided a significant limitation. Due to the cross-platform nature of Java, they only included elements in the AWT that were present on virtually all platforms. They are the most fundamental parts of GUI construction like windows, buttons, text boxes, etc. These elements were drawn by the underlying platform and always looked like the underlying platform.

Both the restriction in appearance and the limited number of component types led to the creation of a second GUI library for the Java standard. This second library was called Swing and it was a "pure" Java library. That means that all the code in Swing was written in Java. It uses calls to AWT to make things happen, but does not directly call anything that is not in Java itself. The Swing library includes a lot of different GUI elements that are more powerful and flexible than what was provided in AWT. It also includes the ability to have elements of the GUI rendered in different ways so they do not have to look just like the underlying system. Originally this flexibility came at a significant cost in speed. Over time various optimizations have been made and now Swing runs perfectly fine assuming the programmer does not do anything silly to slow it down.

Scala has its own library that is a wrapper around Swing. It is making calls to Swing and we will call it Swing, but we get to interact with it in a way that is unique to Scala. This makes things simpler and is a big part of the reason that GUIs are included this early in the text.

The way that the various GUI libraries are built on top of one another makes it so that we will be using a little bit from each one. The Scala Swing library is where we will focus most of the time. The types we use for it are in the `scala.swing` and `scala.swing.event` packages. Every so often we might have to go back down to the Java Swing libraries which are found in `javax.swing`[1] and various packages under it. There will even be a need to go down to AWT occasionally. We will be working with a fair number of AWT types in the next chapter. Those will come from the `java.awt` package and its subpackages.

11.2 GUI Components

The construction of GUIs is very modular. They are constructed by putting together different pieces called `Components`. The specific types that we use, such as `Button`, are subtypes of the more general `Component` type. Subtyping was discussed briefly in section 7.5.3 and will be discussed in more detail in chapter 19.

11.2.1 Frames and Windows

To display a GUI we need to start by bringing up a window. There is more than one type of window. The type that we will use as a main window for an application is called

[1]You might wonder about the `javax` in the package name. There are several libraries in Java that begin this way. These are libraries that started off as extension packages that were not part of the standard libraries. When they were brought into the standard libraries the original name was kept so as not to break existing code that used them.

FIGURE 11.1: This is a sample window brought up by a simple script. The window is 500x500 pixels and has the title "First GUI". When you close the window, the script will exit.

a `MainFrame`. Other options include a regular `Frame` and a `Dialog`. The advantage of the `MainFrame` is that it terminates the program when it is closed. The following is a little script that you can run to bring up a window.

```
import scala.swing._

val frame=new MainFrame {
  title = "First GUI"
  size = new Dimension(500,500)
}

frame.visible=true
```

When you run this it should pop up a window that looks like figure 11.1. There are quite a few things in this script that are worth describing. We start with an `import` statement. The underscore is a wild card that tells Scala to import everything in the `scala.swing` package. We do not have to do that here, but it will be more useful as we go on because we will use a lot of different types from this package.

Below the `import` we make the `Frame` and store it in a `val` named `frame`. The syntax here can be read in the following way. We are making a new `MainFrame` object and the curly braces go around stuff that we want to put into this object. Inside the curly braces we are changing some things so they will not be the defaults used for `MainFrame`. The first is to set the `title`. This appears on the top of the window. After that we set the `size`. The `size` has type `Dimension` (technically it is a `java.awt.Dimension`). In this script we make a new `Dimension` object with width and height both set to 500.

After the creation of `frame`, there is a line that sets the `visible` value in `frame` to be `true`. This causes the window to pop up and be visible.[2]

11.2.2 Components

In order to make a GUI do something, we need to add other components into it. There are many different components in the `scala.swing` package. A partial list is shown here with brief descriptions.

- `Button` - A standard button that you click to make things happen. Create by calling Button(text:String)(action : => Unit).
- `CheckBox` - A labeled box that you can select or de-select. Create with new CheckBox(label:String). Contains selected:Boolean that will tell you if the box is currently selected.
- `ComboBox` - This is a drop-box that allows the user to select from one of several options. Create with `new ComboBox(items:Seq[A])`. Has an object in it called `selection` that deals with the current selection. Use `selection.index` to get the index of the current selection.
- `EditorPane` - A complex text element that displays multiple lines and formatting. Create with `new EditorPane(contentType:String,text:String)`.
- `Label` - A simple, inactive, label. Create with `new Label(text:String)`.
- `ListView` - A display of a list of information that is selectable. Create with `new ListView(items:Seq[A])`. Has an object called `selection` in it to deal with making selections on the list. You can use the collection `selection.indices` to interact with the index values that are selected.
- `PasswordField` - A text field that hides the data input into it. Create with `new PasswordField` if there is no default or `new PasswordField(text:String)` if you want to start with a default. Has a member called `text:String` that will tell you the text that is currently entered.
- `ProgressBar` - A graphical display for progress. Has fields called `min:Int`, `max:Int`, and `value:Int` that are used to control/read the progress display.
- `RadioButton` - A selectable button that can be put in a `ButtonGroup` so only one can be selected at a time. Has a field called `selected:Boolean` that will tell you if this currently selected.
- `ScrollBar` - A horizontal or vertical bar intended for scrolling. The `ScrollBar` has a `minimum`, `maximum`, and `value`, which are all `Ints`, that let you set or inspect where it is. Few applications will use a `ScrollBar` directly. The functionality you generally want is part of the `ScrollPane` described in the next section.

[2]Prior to Scala 2.9, reaching the end of a script would cause it to terminate even if there were still windows open. To stop that you could put `while(true) { }` or `readLine()` at the end of the script.

- `Slider` - A horizontal of vertical component that can slide to select a value. Has fields for `min`, `max`, and `value`.
- `Table` - A table for displaying data. Create a new one with `new Table(rowData: Array[Array[Any]], columnNames: Seq[Any])`.
- `TextArea` - A multiline text display and input. Has a `text:String` member that will let you know what has been entered.
- `TextField` - A single line text display and input. Has a `text:String` member that will let you know what has been entered.

As you can see, there are quite a few different options for building GUIs in Scala. It is not the purpose of this book to document the use of all of the different GUI elements. The API can help you with that. We will just consider a few in this chapter so that you will get a general understanding of the concepts.

The easiest one to demonstrate is the `Button`. The line below can be added into the code block for the new `MainFrame` will cause it to have a single button in it that says "Press me".

```
contents=Button("Press me"){ println("Button pressed") }
```

When you press the button the program prints out "Button pressed". How the button looks depends a bit on where you put this line. If you put it before the line that sets the size, the window is still 500x500 and the button takes up the whole center area. On the other hand, if you put it after the size line, changing the contents causes the `Frame` to resize itself to an optimal size for the contents. In that case, the button is just big enough to look nice with the text in it and the window is just large enough to hold the button.

So what does this line do? The `Frame` only gets to hold one thing and that one thing is determined by whatever we set `contents` to be equal to. In this case, it is being set to a new `Button`. There are several ways to make a new `Button`. This one is the most direct for our purposes. We are making a call to the `Button` object and giving it arguments telling it the details we want for the new `Button`. The arguments are Curried and the second one is passed by-name. This is very similar to the way you call `Array.fill`. The first argument, which is in its own argument list, is a `String` that is the text on the button. The second one is code that you want to have happen any time the button is clicked. The second argument is put inside of curly braces here. This is just a shortcut that Scala allows. Curly braces are required if there are multiple statements but not in this case. You can put parentheses around the curly braces if you want.

This code works fine, but it has a significant limitation: we can only put one component in the frame. Real GUIs are not made from a single component. In order to include multiple components, we need to learn just a bit more.

11.2.3 Panels and Panes

There are other types of components that were not considered above that act as containers for holding multiple other components. These are called panels and panes. If we want the `Frame` that we created to have multiple components displayed in it, we put either a panel or a pane in the `Frame` and then add things into that panel or pane.

While it is possible to specifically place components in a GUI, such a practice is generally discouraged. The main reason is that doing so leads to GUIs that are not flexible. You can not easily change the size of windows, a factor that could be very limiting if you want to move to different platforms or just use computers with different-sized screens. Instead, there are components that can hold multiple other components and which control how things are laid out. These can be nested inside of one another to give the overall effect that the programmer wants.

The simple container components, which just position and size their contents, are called panels. There are also more complex containers that provide some user interaction as well. These are called panes. The list below shows the different panels and panes that are part of `scala.swing`.

- `BorderPanel` - Can hold up to five different components in the `North`, `South`, `East`, `West`, and `Center` positions. The components are added to the `layout` of the `BorderPanel` as a tuple of `(Component,Position)`. The `Positions` are part of `BorderPanel.Position` (e.g., `BorderPanel.Position.North`).

- `BoxPanel` - Can hold a variable number of components oriented either vertically or horizontally, each taking the space it needs. When you make a new `BoxPanel` you pass it an orientation (e.g., `new BoxPanel(Orientation.Vertical)`). The components can be added one at a time to `contents` (e.g. `contents+=Button("text")(action)`).

- `FlowPanel` - Can hold multiple components that are laid out from left to right wrapping when it gets to long like text in a word processor. You can pass a variable length list of `Components` as an argument at construction or add the components to `contents`.

- `GridBagPanel` - A flexible and powerful container that holds contents in a variable-sized grid. The `Components` are added added to the `layout` as a tuple of `(Component,GridBagPanel.Constraints)`. This panel is complex and will not be used in this chapter.

- `GridPanel` - Holds a regular grid of components. You specify how many rows and columns the grid has at creation (e.g., `new GridPanel(3,4)` for 3 rows and 4 columns). The `Components` inside are added to the `contents` of the panel.

- `ScrollPane` - A pane that holds a single `Component`, which can be passed in as an argument at construction. If the `content Component` is larger than what this can display, scroll bars are automatically added so the user can scroll around.

- `SplitPane` - A container that holds two `Components`. They are separated by a bar that the user can move as desired. At construction you should provide and orientation and the two children (e.g., `new SplitPane(Orientation.Horizontal,leftComp, rightComp)`).

- `TabbedPane` - This can hold multiple different `Components`. Only one `Component` is shown at a time, but tabs are always shown for all of them. The user can click a tab to bring one to the front. You can add `Components` to the `TabbedPane` by adding them to the `page` object inside of the `TabbedPane`. What you add to it are `Page` objects. Each `Page` object is constructed by giving it a `String` for the tab label and a `Component` that should go under that tab (e.g., `pages += new Page("A Tab",tabComponent)`).

None of these alone is sufficient to create a complex GUI. Instead, it is customary to nest panels and panes one inside of another to create the desired effect.

To illustrate this, we will lay out the GUI for a little program that we will construct in this chapter. The program is going to be something like a digital recipe book. The program should be able to store multiple recipes. For each recipe, it should allow the user to specify ingredients with amounts as well as have a written description of the cooking directions.

FIGURE 11.2: This shows a possible layout for the GUI that we want to build for editing recipes.

The GUI should give the user the ability to add and remove recipes. They should be able to select recipes and edit the various information that we have on them.

At this point in the analysis you would probably want to sketch out a little picture of what you want the GUI to look like. That is done for this GUI in figure 11.2. The left side is taken up by a `ListView` that shows all the recipes. When the user clicks on a recipe, the information for it should be shown on the right. That includes a list of all the ingredients and the directions. The ingredients can also be edited and there need to be some buttons to help with this. The options for adding and removing recipes will be in the menus, which will be discussed in the next section.

The question is how we should lay this out. Any GUI can be built in multiple different ways using the different panel and pane options that were listed above. For example, the separation between the recipe list and the information about the selected recipe could be created by a `BorderPanel`, a `BoxPanel` with a horizontal orientation, or a `SplitPane`. Which of these you pick will normally be determined by the way in which the panels would size things or whether you want to give the user the control of a `SplitPane`. We will use a `BorderPanel` for the top-level split. We will then use a `SplitPane` to separate the directions from the ingredient information. We will use two `BorderPanel`s for the ingredient information and the `Buttons` will be built using a `GridPanel`. The settings area will also need to be refined a bit more. It probably just needs the name of the ingredient along with the amount needed. That can be laid out using a combination of `BorderPanel` and `GridPanel`.

The code to do this is shown below. It is the full script and it is fairly long.

```
import scala.swing._
```

```
val recipeList=new ListView(List("Pop Tarts"))
val ingredientList=new ListView(List("Pop Tarts"))
val directionsArea=new TextArea("Toast the poptarts ...\nor don't.")
val ingredientNameField=new TextField("Pop Tart")
val amountField=new TextField("1 packet")

val frame=new MainFrame {
  title="Recipe Book"
  contents=new BorderPanel {
    layout += (new ScrollPane(recipeList)
      -> BorderPanel.Position.West)
    layout += (new SplitPane(Orientation.Horizontal,
      new BorderPanel {
        layout += (new BorderPanel {
          layout += (new GridPanel(1,2) {
            contents+=Button("Add") (println("Add ingredient"))
            contents+=Button("Remove") (println("Remove ingredient"))
          } -> BorderPanel.Position.North)
          layout += (new ScrollPane(ingredientList)
            -> BorderPanel.Position.Center)
        } -> BorderPanel.Position.West)
        layout += (new BorderPanel {
          layout += (new GridPanel(2,1) {
            contents += new Label("Name")
            contents += new Label("Amount")
          } -> BorderPanel.Position.West)
          layout += (new GridPanel(2,1) {
            contents += ingredientNameField
            contents += amountField
          } -> BorderPanel.Position.Center)
        } -> BorderPanel.Position.Center)
      }, new ScrollPane(directionsArea)
    ) -> BorderPanel.Position.Center)
  }
  size=new Dimension(800,600)
}

frame.visible=true
```

The interactive elements were stored as values with names at the beginning. The purpose of that will become clear in section 11.3. Indentation has been used to indicate the nesting of panels and panes. Alternately, each different part could have been given a name as a value and then the parts could have been put together. The choice between these options is an issue of style. This example is just big enough, and the panels nested just deep enough that it could be worth considering breaking it up some with named constructs for some of the more deeply nested sections. Here is a different version that uses that approach.

```
import scala.swing._

val recipeList=new ListView(List("Pop Tarts"))
val ingredientList=new ListView(List("Pop Tarts"))
val directionsArea=new TextArea("Toast the poptarts ...\nor don't.")
val ingredientNameField=new TextField("Pop Tart")
val amountField=new TextField("1 packet")
```

```
val ingredientListPanel = new BorderPanel {
  layout += (new GridPanel(1,2) {
    contents+=Button("Add")(println("Add ingredient"))
    contents+=Button("Remove")(println("Remove ingredient"))
  } -> BorderPanel.Position.North)
  layout += (new ScrollPane(ingredientList)
    -> BorderPanel.Position.Center)
}

val ingredientDataPanel = new BorderPanel {
  layout += (new GridPanel(2,1) {
    contents += new Label("Name")
    contents += new Label("Amount")
  } -> BorderPanel.Position.West)
  layout += (new GridPanel(2,1) {
    contents += ingredientNameField
    contents += amountField
  } -> BorderPanel.Position.Center)
}

val frame=new MainFrame {
  title="Recipe Book"
  contents=new BorderPanel {
    layout += (new ScrollPane(recipeList)
      -> BorderPanel.Position.West)
    layout += (new SplitPane(Orientation.Horizontal,
      new BorderPanel {
        layout += (ingredientListPanel
          -> BorderPanel.Position.West)
        layout += (ingredientDataPanel
          -> BorderPanel.Position.Center)
      },
      new ScrollPane(directionsArea))
      -> BorderPanel.Position.Center)
  }
  size=new Dimension(800,600)
}

frame.visible=true
```

This second version would likely be preferred because the deepest nesting has been cut down two levels and the named values can help readers figure out what different parts of the code are doing. This can be taken further if you desire.

When you run either of these scripts you get a window like what is shown in figure 11.3. In the code you might notice that each of the `ListViews` and the large `TextArea` are put inside of `ScrollPanes`. This is to make it so that if the lists get large or the directions get long, scroll bars will be added to allow the user to scroll around and see everything they need to.

11.2.4 Menus

All we lack in the appearance of the GUI now is the menus. The `MainFrame` has a field called `menuBar` that we can set to be an object of type `MenuBar`. The `MenuBar` type has

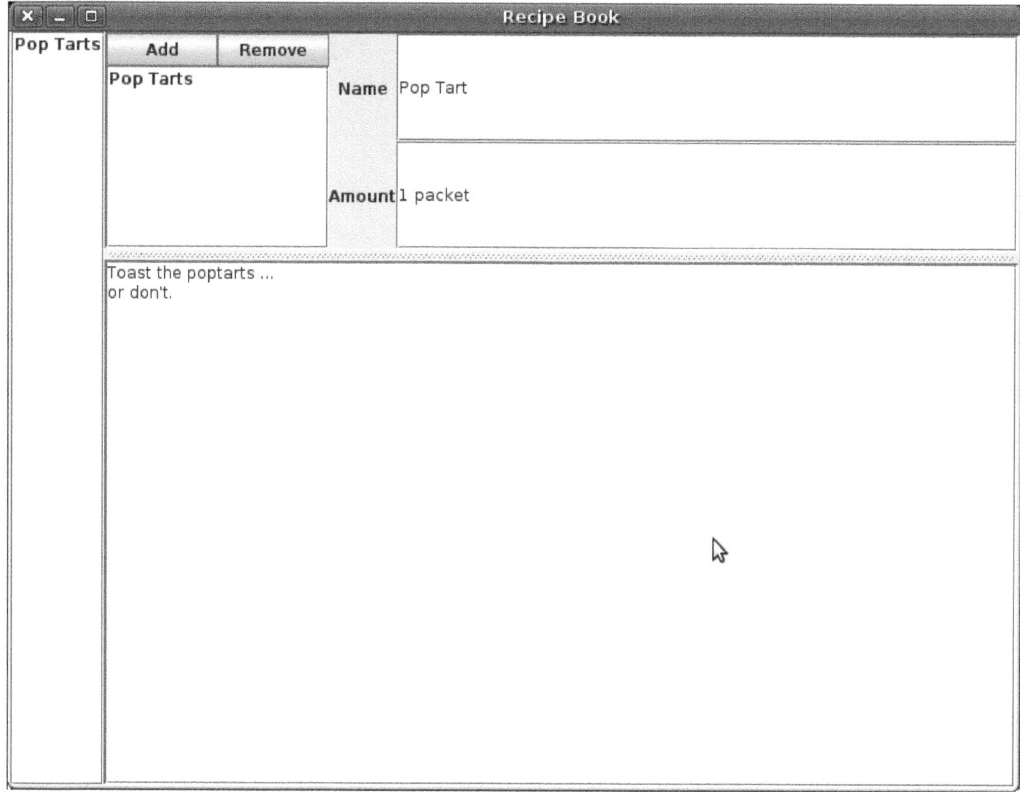

FIGURE 11.3: This is the window produced by the code to do the initial layout of the recipe book.

`contents`, like the panels and panes, but you should only add objects of the `Menu` type to a `MenuBar`.

The `Menu` objects are created with `new` and should be passed a `String` that you want as the title of the `Menu`. The `contents` of a `Menu` can be of one of five types.

- `MenuItem` - This is the most common thing you will put in a `Menu`. You create them with `new` and the easiest approach is to give them an `Action` object. The `Action` object can be made like a `Button` with a `String` and an action that is passed by-name. For example `new MenuItem(Action("Exit")(exit(0)))`.

- `Menu` - You can add a `Menu` to a `Menu`. This will appear as a submenu when you run the program.

- `CheckMenuItem` - `Menus` can also have options on them that are selected. To do this use the type `CheckMenuItem`. You make these with `new` and pass in a `String` for the text to display.

- `RadioMenuItem` - If you have multiple values and you only want to allow one to be selected use a `RadioMenuItem`. These are made like `CheckMenuItems`. All the `RadioMenuItems` and you want to have grouped together should be added to a single object of the `ButtonGroup` type.

- `Separator` - This is a simple component used to space items out to improve appearance and usability.

We can set up a simple menu system for our recipe script by adding the following code inside the `MainFrame` right before we set the `size`.

```
menuBar = new MenuBar {
  contents += new Menu("File") {
    contents += new MenuItem(Action("Open")(openFile))
    contents += new MenuItem(Action("Save")(saveFile))
    contents += new Separator
    contents += new MenuItem(Action("Exit")(exit(0)))
  }
  contents += new Menu("Recipe") {
    contents += new MenuItem(Action("Add")(addRecipe))
    contents += new MenuItem(Action("Remove")(removeRecipe))
  }
}
```

In order to get this to run, we have to put declarations of the four functions that are called in the script before we make `frame`. For now these can be left completely empty.

```
def openFile {
}

def saveFile {
}

def addRecipe {
}

def removeRecipe {
}
```

With these additions we now have a script that will bring up a fairly complete GUI. Unfortunately, it still does not do anything yet. We will get to that soon enough.

11.2.5 Tooltips

You have probably seen GUIs where, if you leave the mouse hovering above something for a while, a message will pop up telling you something about what it does. These are called TOOLTIPS. All the components have a field in them called `tooltip` that is a `String`. You can set the `tooltip` to be a message that you want to pop up. This will happen automatically if the user leaves their mouse pointer over that component for a period of time.

11.3 Basic Interactivity

The GUI that we have created for our recipe book looks fine and has all pieces that we need. Unfortunately, it does not do anything at this point. In order to make it interactive, we need to code that in. Interactivity in GUIs is significantly different from that in console programs. For a console program, the only real input is to have the user type in information. We were able to read that information using functions like `readInt()`, `readLine()`, `readDouble()`, etc. Similarly, the only output was to print things back to the console. In addition, while the user was typing stuff in, the program was supposed to be paused. We describe this by saying that user input is BLOCKING. So nothing else was happening when we called `readLine()`. That call would block the execution and it would not continue until the user had entered something.

When you move to a GUI environment, all the rules of interactivity change. There are now multiple components that the user can interact with. Each one might respond in different ways. For example, a `Button` can just be clicked. On the other hand, a `TextField` can have the user type into it and you need to know when they are done. Each different component has different ways the user can interact with them, and each of them needs to be waiting for the user to interact. No single component can block execution while it waits for the user. They all have to be active at the same time.

This section looks at the way we do user input in a GUI built with the `scala.swing` library.

11.3.1 Partial Functions

There is one feature of the Scala language that we have not talked about before that is required for doing interactions in GUIs: the PARTIAL FUNCTION. The idea of a partial function comes from mathematics. It is a function that can only operate on certain values of the arguments and is undefined for other arguments. The syntax for partial functions in Scala is basically like having a `match` expression without the argument and the `match`. You simply have curly braces with different `case`s inside of them.

```
{
  case 1 => // code goes here
  case 2 => // code goes here
}
```

This example partial function would operate on an `Int`, but would only be defined if the argument was 1 or 2.

This is more useful when we use more of the options for `case`s and the pattern matching they do. For GUIs in particular, it is most important to be able to `match` on different types. If the `case` includes a name that begins with a lowercase letter, it will use that as a variable name to store the value of the match in.

```
{
  case v => println(v) // this will happen for everything
}
```

Using the colon syntax that you have been using all along to specify the types of arguments to functions and the fields in `case classes`, you can also specify that a named value in a `case` has to have a specific type.

```
{
  case i:Int => println("Partial functions called on an Int "+i)
  case d:Double => println("Partial functions called on a Double "+d)
  case s:String => println("String = "+s)
}
```

This example partial function is only defined for `Int`, `Double`, and `String`. If you pass in any other type, it will fail. If you call it with one of those types, it will give that value to the name `i`, `d`, or `s`, then print the appropriate output.

11.3.2 Publishers and Reactors

The model for having users interact with GUIs in the `scala.swing` library is based on types called `Publisher` and `Reactor`. The idea is that the components in a GUI are `Publishers`. When something happens to them, let us call it an EVENT. The `Publisher` will publish that event. Code can be set up in a `Reactor` to pay attention to a particular `Publisher` and then it can react to these events in the proper way. There is a type called `Event` that represents something that can happen in the GUI. There are lots of different subtypes of `Event` that are part of the `scala.swing.event` package.

As we saw above, you do not have to do this for `Buttons` because when you create a `Button`, there is a simple syntax for providing the code that will be executed. Not everything is so simple though. For many components there are several different possible events that you might react to or there is other information that you need to know when the interaction occurs.

To see this, consider the example of a `ListView`. Two of these were placed in the GUI for the recipe book. When the user clicks on something in the `ListView`, the program needs to be able to display the appropriate information in other fields. To illustrate how this works, we will start with a simpler example that contains only a `ListView` and a `TextField`. When the user selects an item from the `ListView`, the text field should be set to the value of what was clicked. This example can be set up with the following code.

```
import scala.swing._
import scala.swing.event._

val options = "A list of words".split(" ")
val list = new ListView(options)
val field = new TextField
```

```
val frame = new MainFrame {
  title = "First GUI"
  contents=new BorderPanel {
    layout += (list -> BorderPanel.Position.Center)
    layout += (field -> BorderPanel.Position.South)
  }
  size=new Dimension(500,500)
}

frame.visible=true
```

To make this interactive we need to tell the `frame` to listen to the `selection` `Publisher` in the `ListView` and then give it a partial function to handle the events that we are interested in.

This can be accomplished by adding the following text inside the block of code for building the `frame`.

```
  listenTo(list.selection)
  reactions += {
    case e:SelectionChanged =>
      field.text=list.selection.items.mkString("; ")
  }
```

The first line of code has the `frame` listen to the `selection` `Publisher`. If you later decided that you did not want to listen to that anymore, you could call `deafTo` on the `frame`. After that, a partial function is added to the `reactions` of the `frame` that has a single `case` for handling the `SelectionChanged` event. The `selection` field of `ListView` has a collection in it called `items` that contains all the items that are selected from the `ListView`. This is a collection because the `ListView` allows multiple items to be selected at a time. The `text` for the field is set to be all of the `items` separated by semicolons.

The type `SelectionChanged` is part of the `scala.swing.event` package. There are a large number of different event types in this package for all the different things that can happen in a GUI. The list is long enough that they will not be put in here. Readers can go to the Scala API to see the list of different events. We will introduce a few more in this chapter for the recipe book example.

In general, your partial function for the reactions will have more than one case. This example simply did not need that.

11.3.3 Mnemonics

It is often helpful to put keyboard shortcuts into GUIs so that the user can execute certain commands without having to use the mouse to do so. These shortcuts are referred to as MNEMONICS and the different components in `scala.swing` that have simple modes of interaction, like `Buttons`, `MenuItems`, and `CheckBoxes` have a member called `mnemonic` that can be set to the keystroke that you want people to use. The following example demonstrates adding mnemonics to a short menu.

```
  menuBar = new MenuBar {
    contents += new Menu("File") {
      mnemonic = Key.F
      contents += new MenuItem(Action("Exit")(exit(0))) {
        mnemonic=Key.X
      }
```

```
    }
  }
```

This can be added into the simple GUI created above to add a "File" menu that you can bring down with Alt-F and an "Exit" item that can be executed by hitting "X" when the menu is down.

11.4 FileChooser

There is at least one other option in the GUI library that is worth pointing out because it is used so frequently, the file chooser dialog. Many programs have "Open" and "Save" options in their menu. When these options are selected, the program will typically pop up a dialog box that allows the user to navigate through the directory structure and pick files. You could write your own dialog to do this, but it would take a significant amount of effort and it is such a common task that it has been added to the library.

The `FileChooser` type is used to display a file selection dialog. You can make a new `FileChooser` with or without a single argument for a starting directory. If you omit the starting directory it uses a default directory to start in. There are three different methods in `FileChooser` that all start with "`show`" which can be used to pop up a dialog box: `showDialog`, `showOpenDialog`, `showSaveDialog`. All three take a first argument of the `Component` you want the dialog to pop up over. If you pass `null` for that argument, the dialog will appear in the middle of the screen. The `showDialog` method also takes a second argument that is a `String` that will be displayed on the approval button. These methods will return one of the following: `FileChooser.Result.Approve`, `FileChooser.Result.Cancel`, or `FileChooser.Result.Error`. If it gives back `Approve`, then the `FileChooser` will have valid values in the `selectedFile` and `selectedFiles` members.

To help understand this, consider the following example of an "Open" option added to the menu that was created in the last section. This option could be added to the menu by adding the following code in the block for the menu bar.

```
contents += new MenuItem(Action("Open")(openFile)) {
  mnemonic = Key.O
}
```

To make this work, we need an `openFile` function. This can be defined in the following way and placed after the declarations of `list` and `field` and before that of `frame`.

```
def openFile {
  val chooser = new FileChooser
  if(chooser.showOpenDialog(list)==FileChooser.Result.Approve) {
    list.listData = Source.fromFile(chooser.selectedFile).getLines.toList
  }
}
```

This code creates a `FileChooser` and displays an open dialog. If the user selects a file, that file is used to open a `Source` and the lines from that `Source` are converted to a `List`, which is set to be the data for the `ListView`.

11.5 Tables

One of the more complex GUI elements in Swing is the `Table`. The name does a good job of describing what this element does. It presents a 2-D grid of data cells the user can interact with. While the `scala.swing.Table` type can give you basic functionality easily, there is a lot more the `javax.swing.Table` type can do that requires you to go into the original Java libraries. For that reason, this section is devoted to this component.

When you make a `scala.swing.Table`, there are three options for arguments you pass to it. You can tell it how many rows and columns to put in the table as two `Int` arguments. This will make a simple table of the specified size where all the elements are blank. You can play with this by entering code like the following into the REPL.

```
val frame = new Frame {
  title = "Table Test"
  contents = new Table(3,4)
}
frame.visible = true
```

This code will pop up a window with a `Table` with 3 rows and 4 columns with all the cells blank. This example uses a `Frame` instead of a `MainFrame` because it is intended to be done in the REPL. If you use a `MainFrame`, when you close the window, the entire REPL will be terminated.

You can also pass in the arguments `rowData: Array[Array[Any]], columnNames: Seq[_]`. This will make a table that has the size of the `rowData` and starts with those values. Remember from figure 7.1 that `Any` is a type in Scala which can represent any other type. Your `Table` does not have to have `Strings` of data in it. They can be far more complex, but we will not be going into all of those details here. The second argument is a sequence that has the values you want to appear at the tops of columns labeling them. These labels will only appear if the table is nested in a `ScrollPane`.

To test how this works, try entering the following into the REPL.

```
val data = Array(Array(1,2,"hi"),Array(3,4,"bye"))
val colHeads = Seq("Num","Num","Word")
val frame = new Frame {
  title = "Table Test"
  contents = new Table(data,colHeads)
}
frame.visible = true
```

This will bring up a small window with a `Table` that has two rows and three columns. Note that the column headers do not appear in this version. To get the column headers, change the line of code setting the contents of the `Frame` to the following.

```
contents = new ScrollPane(new Table(data,colHeads))
```

Now when you make the `Frame` visible you get a much larger window with the same table except that now the column headers are visible.

The third way to build a `Table` does not pass any arguments in at all. If you pick this option, you will need to do more work setting things up so for now we will focus on the other options.

If you played with the `Tables` that you brought up in the REPL with the above examples, you probably found that you can select cells, move around with the arrow keys, and even edit the values in cells. To work with the Table in code, we need to have a name to refer to the Table by. We can get this by modifying the code slightly so we have a variable called table.

```
val table = new Table(data,colHeads)
...
  contents = new ScrollPane(table)
...
```

To get values out of the table, simply index it with a row and a column. So if you were to do the following line in the REPL you would see the value of the top-left cell.

```
println(table(0,0))
```

If you did this immediately after creating the table you would get back the value 1. However, if you edit that cell in the Table manually then print it, you will see the modified value.

You can modify values in the Table in the same manner. Simply do an assignment into the Table indexed by row and column. Try this by doing the following in the REPL.

```
table(1,1) = "New text"
```

You will see that the value in cell (1,1) of the Table is updated graphically to reflect the change. This approach can only be used to modify cells within the normal bounds of the Table. If you go beyond the size set when you first created the Table you will get an error.

If you try entering table(0,0) into the REPL you will get back a result with the type Any. This can be a bit of a problem because there are not many things that you can do with the Any type. For example, you might try doing the following.

```
table(0,0)+table(1,0)
```

Addition like this would work just fine for any numeric type or the String type as all of those types have the method + defined for them. However, if you try to do this you would get the following error message.

```
<console>:12: error: type mismatch;
 found   : Any
 required: String
       table(0,0)+table(0,0)
                 ^
```

That is because the type Any does not have a + method. This makes sense as there are some types that qualify as Any that you should not be able to use + with.

To do anything with the values we get from the table we have to call methods that exist on Any. The challenge is that there are not very many of these. You can check for equality or inequality, doing operations associated with multithreading that will be discussed in chapter 21, check the actual type of the object, or convert the object to a String. The last two can provide the functionality that we want. We will focus on the last one because it uses only methods that we have seen before.

To convert any object to a String call the toString method. You can try the following to see how this works.

```
table(0,0).toString
```

Once you have done this you can use + to do String concatenation. If you want to do numeric operations on the values in a `Table` you can use `toInt` or `toDouble` on the `String`. So if you want to multiply two table values you could do this.

```
table(0,0).toString.toDouble{*}table(1,0).toString.toDouble
```

This is a bit verbose, but it is the price we pay for being able to put all types of different things in a `Table`.

If you play around with this enough you will probably stumble across the primary downfall of this approach, if the contents of the cell are not a number this will fail. To get around this, you would need to use the `try/catch` construct. We will discuss `try-catch` in detail in section 22.3. For now we will give a brief introduction. You might have seen that when something goes wrong with a statement the error message you get typically starts by giving you a type name that includes the word "Exception". An EXCEPTION is an object that indicates a problem and when something goes wrong the code will THROW an exception. If you know this might happen and you want to deal with it, you tell Scala you want to TRY some code and then CATCH the exception if one occurs. The exceptions you can `catch` are specified with `cases`, just like a `match` expression.

The `try-catch` in Scala can be used as an expression. If there is no exception you get the value of the code you tried to do. If there is an exception that has a matching `case`, the value will be what happens in that `case`. To help you understand this, consider the following function.

```
def asDouble(a:Any):Double = {
  try {
    if(a==null) 0.0 else a.toString.toDouble
  } catch {
    case e:NumberFormatException => 0.0
  }
}
```

This function takes an `Any` object and returns a number for it. If the value is `null` it returns zero. Otherwise, it uses the `.toString.toDouble` method calls shown above. In this function that code is put inside of a `try` that will `catch` a `NumberFormatException`. As the name implies, this is an exception that occurs when you try to make a number of something that does not have the right format. If that happens, the `case` gives back the value 0.0. The sum above could now be expressed in this way.

```
asDouble(table(0,0))+asDouble(table(1,0))
```

This code is more robust in that if a `Table` cell contains something that is not a number, instead of stopping with an exception, the code uses 0.0 for that cell.

11.5.1 Table Actions

You now know how to set up `Tables` as well as how to read and alter the values in them. For many applications this will be sufficient. The actions in a GUI can very often be initiated by having the user click a button or select a menu item and have the code initiated by that action utilize the values in the table. There are some applications, however, when you need to actually respond to events that occur with the `Table` itself. In this situation you need to listen to the `Table` or possibly the selection object in the `Table`.

There is a significant list of events that can happen with `Tables`. You can look in the API for complete information, but here is a list of the main ones.

- `TableChanged` – Indicates that the structure of the `Table` stayed the same, but anything else could have changed.
- `TableColumnsSelected` – Indicates a change in the column(s) currently selected.
- `TableRowsAdded` – This event occurs when additional rows are added to the `Table`.
- `TableRowsRemoved` – This event occurs when rows are removed from the `Table`.
- `TableRowsSelected` – Indicates a change in the row(s) currently selected.
- `TableStructureChanged` – This event indicates the most general type of change. Anything in the `Table` could have changed including row and column counts.
- `TableUpdated` – Indicates that values in a specified part of the `Table` have changed.

To get the events that indicate changes in the selection you listen to `table.selection`. To get the others simply listen to the `table`.

If all that you are interested in is knowing when the value in a cell has been changed, you can use code like the following.

```
import scala.swing._
import scala.swing.event._

val sumLabel = new Label("0.0")
val table = new Table(8,5)

def asDouble(a:Any):Double = {
  try {
    if(a==null) 0.0 else a.toString.toDouble
  } catch {
    case e:NumberFormatException => 0.0
  }
}

val frame = new MainFrame {
  contents = new BorderPanel {
    layout += sumLabel -> BorderPanel.Position.North
    layout += table -> BorderPanel.Position.Center
  }
  size = new Dimension(500,400)
  listenTo(table)
  reactions += {
    case e:TableUpdated =>
      val sum = (for(i <- 0 until table.rowCount;
        j <- 0 until table.model.getColumnCount) yield {
          asDouble(table(i,j))
      } ).sum
      sumLabel.text = sum.toString
  }
}

frame.visible = true
```

This code starts with an empty `Table` and a `Label` that shows 0.0. Every time that the user edits a cell, the code runs through all the cells and adds up the values in each, using zero if there is not a valid number. It changes the `Label` to reflect this sum. Enter this code into a file and run it as a script. Make changes to see what else you can do.

11.5.2 Table Models

Changing size is a bit more difficult than simply changing values. Part of the flexibility of Tables comes from the fact that you can change how the data behind them is modeled. The full details of this are beyond the scope of this book, but to have full control over a Table you need to have at least some knowledge of how to utilize this. The Table stores and interacts with the data using a javax.swing.table.TableModel. You can create your own implementations of TableModel to limit the types of values that can go into columns or change how values are edited. We do not need that much power here, but it would be nice to have the ability to add rows or columns. These can be done reasonably well with the javax.swing.table.DefaultTableModel.

A DefaultTableModel can be created in much the same way as a the Table passing in either a size as two Ints or with the arguments rowData:Array[Array[AnyRef]] and columnNames:Array[AnyRef].[3] The Table has a member called model that you can assign to any new model you want, including a modified DefaultTableModel.

To help illustrate how this can be used to make an application with a Table that changes size under the control of the user, consider the following example. This program has a Frame with a Table Originally the Table is completely empty because it has no columns. There is a menu option for adding columns that allows the user to name them. In addition, editing the last row in the Table causes it to grow by one row.

```
import scala.swing._
import event._
import javax.swing.table._

var data:Array[Array[AnyRef]] = Array.fill(1,0)("")
var headers:Array[AnyRef] = Array.tabulate(0)(i => i.toString)

val table = new Table {
  model = new DefaultTableModel(data,headers)
  listenTo(this)
  reactions += {
    case tu:TableUpdated =>
      if(tu.column>=0) {
        tu.range.foreach(i => {
          data(i)(tu.column) = this(i,tu.column) match {
            case ar:AnyRef => ar
            case other => ""
          }
        })
      }
      if(tu.range.head==data.length-1) {
        data :+= Array.fill(headers.length)("":AnyRef)
        model = new DefaultTableModel(data,headers)
      }
  }
}

def addColumn {
  val header = Dialog.showInput(table,"What header do you want for the
      column","Header",Dialog.Message.Question,null,Nil,"")
  header match {
```

[3]The exact types for the second option is slightly different from that for the Table because the DefaultTableModel is Java code and Java does not have an exact equivalent for Any.

```
    case Some(name) =>
      data = data.map(_ :+ "")
      headers :+= header.get
      table.model = new DefaultTableModel(data,headers)
    case None =>
  }
}

val frame = new MainFrame {
  title = "Resizable Table"
  contents = new ScrollPane(table)
  menuBar = new MenuBar {
    contents += new Menu("File") {
      contents += new MenuItem(Action("Add Column")(addColumn))
      contents += new Separator
      contents += new MenuItem(Action("Exit")(sys.exit(0)))
    }
  }
  size = new Dimension(500,500)
}

frame.visible = true
```

The information for the `Table` is stored in `data` and `headers`. To match with the types needed for building a `DefaultTableModel`, they are an `Array[Array[AnyRef]]` and a `Array[AnyRef]` respectively. Unfortunately, these types can not be used with the normal Scala creation approach so instead of passing in the data, the declaration of `table` includes a line that sets the `model`.

The `table` listens to itself and reacts to `TableUpdate` events. The updated data is always stored back into the `data` array. Note that this requires a bit of extra work because data pulled from the table is of the `Any` type. A `match` is used to make certain we have an `AnyRef`. There is also a check to see if the row being changed is the last one. In that situation, a new row is appended to `data` and the model is updated.

The `addColumn` method is called from a menu option. It asks the user for a name using `Dialog.showInput`, then appends a blank to the end of every row and uses the result as the updated `data`. It also appends the new header name to the end of `headers`. The last thing it does is to update the table model. The full signature of `showInput` is as follows.

```
def showInput [A] (parent: Component, message: Any, title: String, messageType:
    Value, icon: Icon, entries: Seq[A], initial: A): Option[A]
```

In our usage, we have the dialog box pop up over the `table`. We tell it that we are asking a question, with no icon, no entries, and an empty `String` for the initial value. This lets the user type in a `String` of their choosing. If `entries` were not empty, the user would get a drop-down box instead of a text field.

Note that `showInput` returns an `Option` type. That is because the user could hit "Cancel" or close the dialog box. This is the reason for the use of `match`. If the user does not provide a proper response, the `Option` is `None` and nothing is done.

This code does not demonstrate removing rows or columns, but either operation could be done with `patch` in a manner very similar to what is shown here.

11.6 Putting it Together

We can now put together all of the various aspects of GUIs into the recipe book script. Earlier we had built a GUI and added menus to it. However, the GUI was not interactive. Clicking on various elements did not do anything to other elements of the GUI. Now that we know how to make GUIs interactive we can add the proper functionality to this script. In order to be able to do this, we need to add some groundwork first. In particular, we need a good way to store the information associated with a recipe. This calls for some `case classes`.

We definitely want a `case class` to represent a recipe. The fact that a recipe can contain multiple ingredients makes it advantageous to have a `case class` for that as well. This leads to something like the following.

```
case class Ingredient(name:String,amount:String)
case class Recipe(name:String,ingredients:List[Ingredient],directions:String)
```

You can put this near the top of the script as many other parts of the script will be referring back to it.

We also need to have a way to store multiple recipes. We will do this with a variable declaration of an `Array` of `Recipes`. One could also choose to use a `var List`, but some of the interactions will require swapping out values and others might require doing direct access. Both of those will be easier and more efficient with an `Array`. The adding of new recipes would be easier with a `List`, but that operation will probably occur less frequently and can be done almost as easily with an `Array`.

```
var recipes = Array(Recipe("Pop Tarts",List(Ingredient("Pop Tart","1 packet")),
  "Toast the poptarts ...\nor don't."))

val recipeList = new ListView(recipes.map(_.name))
val ingredientList = new ListView(recipes(0).ingredients.map(_.name))
val directionsArea = new TextArea(recipes(0).directions)
val ingredientNameField = new TextField(recipes(0).ingredients(0).name)
val amountField = new TextField(recipes(0).ingredients(0).amount)
```

This code includes a declaration of the `Array` as well as revisions of the active component declarations that use the one `Recipe` in the `Array`.

Now that we have this set up, we can start with the easy part of completing the functions that are called from the menus. We will start with the `saveFile` function as that defines the format that we will use in the file and it is a bit easier.

```
def saveFile {
  val chooser = new FileChooser
  if(chooser.showSaveDialog(recipeList)==FileChooser.Result.Approve) {
    val pw = new PrintWriter(chooser.selectedFile)
    pw.println(recipes.length)
    for(r <- recipes) {
      pw.println(r.name)
      pw.println(r.ingredients.length)
      for(ing <- r.ingredients) {
        pw.println(ing.name)
        pw.println(ing.amount)
      }
```

```
      pw.println(r.directions)
      pw.println(".")
    }
    pw.close()
  }
}
```

This function allows the user to select a file, then writes the size of the Array to the file. After the size it writes each recipe. The format of the recipes is fairly simple with a name, then the number of ingredients and alternating lines of ingredient names and amounts. The only thing that might seem a bit odd is the line with a period after the directions. This is required because the directions can be multiple lines long and we have to have a way to know when to stop reading them.

After we have written save and defined the format of the file, we can write the openFile function. This is slightly more complex, primarily because we have to deal with the directions.

```
def openFile {
  val chooser = new FileChooser
  if(chooser.showOpenDialog(recipeList)==FileChooser.Result.Approve) {
    val src = Source.fromFile(chooser.selectedFile)
    val lines = src.getLines
    recipes = Array.fill(lines.next.toInt)(Recipe(
      lines.next,
      List.fill(lines.next.toInt)(Ingredient(lines.next,lines.next)),
      {
        var dir = ""
        var line=lines.next
        while(line!=".") {
          dir += (if(dir.isEmpty) "" else "\n")+line
          line = lines.next
        }
        dir
      }
    ))
    src.close()
    recipeList.listData = recipes.map(_.name)
    recipeList.selectIndices(0)
    setFields(recipes(0))
  }
}
```

The code is not all that different from save other than it uses a source and builds an Array to store in recipes with fill. A fair bit of the length of the method comes from the block of code that is used to read the directions. This block is used as an expression. It builds up the directions in a variable named dir and uses a while loop to continue reading until a line is reached that contains only a single period.

After the Source is closed, there are two more lines. The first one changes to contents of the recipeList so that it matches what has just been read in. The second calls a yet undefined function named setFields that is supposed to set the other elements of the GUI to values for a particular recipe. This function could be written in the following way.

```
def setFields(r:Recipe) {
  ingredientList.listData = r.ingredients.map(_.name)
```

```
  directionsArea.text = r.directions
  if(r.ingredients.isEmpty) {
    ingredientNameField.text = ""
    amountField.text = ""
  } else {
    ingredientNameField.text = r.ingredients.head.name
    amountField.text = r.ingredients.head.amount
  }
}
```

This function will be useful later on when we include the code to let the user select different elements from the `recipeList`.

Now we can turn to the functions to add recipes or remove recipes. The `addRecipe` function might look like this.

```
def addRecipe {
  val name = Dialog.showInput(recipeList, "What is the name of this recipe?",
      "Input", Dialog.Message.Question, null, Nil, "")
  name match {
    case Some(n) =>
      recipes = recipes :+ Recipe(name.get,
        List(Ingredient("ingredient","amount")),"Directions")
      recipeList.listData = recipes.map(_.name)
      recipeList.selectIndices(recipes.length-1)
      setFields(recipes.last)
    case None =>
  }
}
```

This creates a new recipe array that includes a new default recipe. This uses an operator that we have not seen before. The `:+` operator can be used to add elements to the end of an `Array` or a `List`. This is not something that you want to use in loops that occur a lot because it is inefficient for either of these types, but this only happens when the user selects an option from the menu so it can not happen all that often for this code. This also uses a part of the `javax.swing` library. The `Dialog` object has methods that can be used to bring up simple input dialog boxes. For most of these, there is similar functionality in the `scala.swing.Dialog` object. The input dialog that allows the user to type in a `String` is not mirrored in the Scala library so we use the one from Java.

The `removeRecipe` function needs to take out the item from the recipe list that the user has selected. As part of this it needs to check if anything is selected. The code could be written like this.

```
def removeRecipe {
  if(!recipeList.selection.items.isEmpty) {
    recipes = recipes.patch(recipeList.selection.leadIndex,List(),1)
    if(recipes.isEmpty) {
      recipes = Array(Recipe("New recipe",
          List(Ingredient("ingredient","amount")),"Directions"))
    }
    recipeList.listData = recipes.map(_.name)
    recipeList.selectIndices(0)
    setFields(recipes(0))
  }
}
```

Here too there is one element that might stand out as odd. Instead of having a simple `remove` method to remove the item at a particular index, the Scala collections have a `patch` method that is far more flexible. The `patch` method can replace part of a collection with some other collection beginning at a certain index. When called with an empty collection, it will remove the specified number of elements at the specified location. This code also has a check to see if the last element was removed and if so, it adds in a default element. So the recipe book can never be empty.

With this written, the GUI now has the ability to react to all of our menu options. What it lacks is the ability to display information about different recipes or edit that information. For the first part, we need to be listening to the `ListView` to find out when selections are changed. We also need to react to it when it is changed. Having already written the `setFields` function makes the second part of this fairly easy.

```
listenTo(recipeList.selection)
reactions += {
  case e:SelectionChanged =>
    if(!recipeList.selection.items.isEmpty) {
      setFields(recipes(recipeList.selection.leadIndex))
    }
}
```

This code could be added into the block of the `frame` to give us the desired behavior.

All that remains is having the ability to edit the recipes. This requires reacting to the different events that can happen with the panels associated with the ingredients. This code can go in a number of different locations. There are two styles that can make sense. One is that you put the code in a single location in the program so it is easy to find all the different reactions. A second approach is to put the reactions close to the code that adds those elements into the GUI. The advantage of this second approach is that everything associated with a given component tends to stay in a certain part of the code. We will adopt the second style here.

The `Buttons` for adding and removing ingredients had earlier just printed lines saying they were clicked. Those need to be changed so that they call functions that do what we want. All of the functions that edit the contents of a recipe can only work if there is a recipe selected in the `recipeList`. So the first thing that happens is a check to make sure that something is selected. If nothing is selected, nothing will be done because we do not know what recipe we would be editing. Given this, the `addIngredient` function and a helper it uses could be written in this way.

```
def setIngredients(ingr:Ingredient) {
  ingredientNameField.text = ingr.name
  amountField.text = ingr.amount
}
def addIngredient {
  if(!recipeList.selection.items.isEmpty) {
    val index = recipeList.selection.leadIndex
    val oldIngrs = recipes(index).ingredients
    val newIngr = Ingredient("Stuff","Some")
    recipes(index) = recipes(index).copy(ingredients=oldIngrs:+newIngr)
    ingredientList.listData = recipes(index).ingredients.map(_.name)
    ingredientList.selectIndices(recipes(index).ingredients.length-1)
    setIngredients(recipes(index).ingredients.last)
  }
}
```

This code checks that something is selected, finds the index of the selection, creates a new default ingredient, then makes an altered recipe and adjusts the GUI. The `removeIngredient` function is similar, but it also needs to check if the is an ingredient selected from that `ListView`.

```
def removeIngredient {
  if(!recipeList.selection.items.isEmpty &&
      !ingredientList.selection.items.isEmpty) {
    val rIndex = recipeList.selection.leadIndex
    val iIndex = ingredientList.selection.leadIndex
    val shortIngr = recipes(rIndex).ingredients.patch(iIndex,List(),1)
    recipes(rIndex) = recipes(rIndex).copy(ingredients=shortIngr)
    ingredientList.listData = recipes(rIndex).ingredients.map(_.name)
    ingredientList.selectIndices(0)
    setIngredients(recipes(rIndex).ingredients.head)
  }
}
```

As we have seen earlier, this code uses some local variables to store values that are used frequently. It also takes advantage of the `patch` method to remove a element from the `List`.

The only things that are left are having it so that editing the `TextFields` or the `TextArea` alters the data stored in the recipes `Array` and selecting ingredients changes what is displayed. The easiest field to change is the directions because it is a field of the recipe itself. Making this happen requires listening to something that happens with the `directionsArea`. When users enter text in a `TextArea` or `TextField`, it fires a `ValueChanged` event. However, we do not really want to change things for every single character. We only need to know when they are done editing. For a text element, the user is done editing it when he/she moves focus to a different part of the GUI. For that reason, we will listen for the `FocusLost` event. We will also need to start listening to the different text elements. Below is what we need to change our listening and reaction code to in the `frame`.

```
listenTo(recipeList.selection,ingredientList.selection,
  directionsArea,ingredientNameField,amountField)
reactions += {
  case e:SelectionChanged =>
    if(e.source==recipeList) {
      if(!recipeList.selection.items.isEmpty) {
        setFields(recipes(recipeList.selection.leadIndex))
      }
    } else if(e.source==ingredientList) {
      if(!recipeList.selection.items.isEmpty &&
          !ingredientList.selection.items.isEmpty) {
        val rIndex = recipeList.selection.leadIndex
        val iIndex = ingredientList.selection.leadIndex
        setIngredients(recipes(rIndex).ingredients(iIndex))
      }
    }
  case e:FocusLost =>
    val rIndex = recipeList.selection.leadIndex
    if(e.source==directionsArea) {
      recipes(rIndex) = recipes(rIndex).
        copy(directions = directionsArea.text)
    } else if(e.source==ingredientNameField) {
      if(!recipeList.selection.items.isEmpty &&
          !ingredientList.selection.items.isEmpty) {
```

```
        val rIndex = recipeList.selection.leadIndex
        val iIndex = ingredientList.selection.leadIndex
        changeIngredientName(rIndex,iIndex)
      }
    } else if(e.source==amountField) {
      if(!recipeList.selection.items.isEmpty &&
          !ingredientList.selection.items.isEmpty) {
        val rIndex = recipeList.selection.leadIndex
        val iIndex = ingredientList.selection.leadIndex
        changeAmount(rIndex,iIndex)
      }
    }
}
```

This code works for the `directionsArea` and selecting ingredients. It also sets up things for the two different `TextFields` so all we have left is writing two functions. In the case for `FocusLost`, the code checks the source of the event. The source of an event is the component that it came from. If it is the `directionsArea`, it alters the recipe in the way shown here. The altering of the ingredients is done with these functions.

```
def changeIngredientName(r:Int,i:Int) {
  val ing = recipes(r).ingredients(i)
  recipes(r) = recipes(r).copy(ingredients = {
    recipes(r).ingredients.patch(i,List(Ingredient(
      ingredientNameField.text,ing.amount)),1)
  })
  ingredientList.listData = recipes(r).ingredients.map(_.name)
  ingredientList.selectIndices(i)
}
def changeAmount(r:Int,i:Int) {
  val ing = recipes(r).ingredients(i)
  recipes(r) = recipes(r).copy(ingredients = {
    recipes(r).ingredients.patch(i,List(Ingredient(
      ing.name,amountField.text)),1)
  })
}
```

`patch` is used again here, but this time the patched list is one that contains a revised ingredient.

All of these different pieces go together into a single script. To help you see how they all fit together, this is the full script.

```
import scala.swing._
import scala.swing.event._
import java.io.PrintWriter
import scala.io.Source

case class Ingredient(name:String,amount:String)
case class Recipe(name:String,ingredients:List[Ingredient],directions:String)

var recipes = Array(Recipe("Pop Tarts",List(Ingredient("Pop Tart","1 packet")),
  "Toast the poptarts ...\nor don't."))
val recipeList = new ListView(recipes.map(_.name))
val ingredientList = new ListView(recipes(0).ingredients.map(_.name))
val directionsArea = new TextArea(recipes(0).directions)
```

```
val ingredientNameField = new TextField(recipes(0).ingredients(0).name)
val amountField = new TextField(recipes(0).ingredients(0).amount)

def setFields(r:Recipe) {
  ingredientList.listData = r.ingredients.map(_.name)
  directionsArea.text = r.directions
  if(r.ingredients.isEmpty) {
    ingredientNameField.text = ""
    amountField.text = ""
  } else {
    ingredientNameField.text = r.ingredients.head.name
    amountField.text = r.ingredients.head.amount
  }
}

def setIngredients(ingr:Ingredient) {
  ingredientNameField.text = ingr.name
  amountField.text = ingr.amount
}

def addIngredient {
  if(!recipeList.selection.items.isEmpty) {
    val index = recipeList.selection.leadIndex
    val oldIngrs = recipes(index).ingredients
    val newIngr = Ingredient("Stuff","Some")
    recipes(index) = recipes(index).copy(ingredients=oldIngrs:+newIngr)
    ingredientList.listData = recipes(index).ingredients.map(_.name)
    ingredientList.selectIndices(recipes(index).ingredients.length-1)
    setIngredients(recipes(index).ingredients.last)
  }
}

def removeIngredient {
  if(!recipeList.selection.items.isEmpty &&
      !ingredientList.selection.items.isEmpty) {
    val rIndex = recipeList.selection.leadIndex
    val iIndex = ingredientList.selection.leadIndex
    val shortIngr = recipes(rIndex).ingredients.patch(iIndex,List(),1)
    recipes(rIndex) = recipes(rIndex).copy(ingredients=shortIngr)
    ingredientList.listData = recipes(rIndex).ingredients.map(_.name)
    ingredientList.selectIndices(0)
    setIngredients(recipes(rIndex).ingredients.head)
  }
}

val ingredientListPanel = new BorderPanel {
  layout += (new GridPanel(1,2) {
    contents += Button("Add")(addIngredient)
    contents += Button("Remove")(removeIngredient)
  } -> BorderPanel.Position.North)
  layout += (new ScrollPane(ingredientList)
    -> BorderPanel.Position.Center)
}

val ingredientDataPanel = new BorderPanel {
```

```
  layout += (new GridPanel(2,1) {
    contents += new Label("Name")
    contents += new Label("Amount")
  } -> BorderPanel.Position.West)
  layout += (new GridPanel(2,1) {
    contents += ingredientNameField
    contents += amountField
  } -> BorderPanel.Position.Center)
}

def openFile {
  val chooser = new FileChooser
  if(chooser.showOpenDialog(recipeList)==FileChooser.Result.Approve) {
    val src = Source.fromFile(chooser.selectedFile)
    val lines = src.getLines
    recipes = Array.fill(lines.next.toInt)(Recipe(
        lines.next,
        List.fill(lines.next.toInt)(Ingredient(lines.next,lines.next)),
        {
          var dir = ""
          var line = lines.next
          while(line!=".") {
            dir += (if(dir.isEmpty) "" else "\n")+line
            line = lines.next
          }
          dir
        }
    ))
    src.close()
    recipeList.listData = recipes.map(_.name)
    recipeList.selectIndices(0)
    setFields(recipes(0))
  }
}

def saveFile {
  val chooser = new FileChooser
  if(chooser.showSaveDialog(recipeList)==FileChooser.Result.Approve) {
    val pw = new PrintWriter(chooser.selectedFile)
    pw.println(recipes.length)
    for(r <- recipes) {
      pw.println(r.name)
      pw.println(r.ingredients.length)
      for(ing <- r.ingredients) {
        pw.println(ing.name)
        pw.println(ing.amount)
      }
      pw.println(r.directions)
      pw.println(".")
    }
    pw.close()
  }
}

def addRecipe {
```

```
  val name = Dialog.showInput(recipeList, "What is the name of this recipe?",
      "Input", Dialog.Message.Question, null, Nil, "")
  name match {
    case Some(n) =>
      recipes = recipes :+ Recipe(name.get,
        List(Ingredient("ingredient","amount")),"Directions")
      recipeList.listData = recipes.map(_.name)
      recipeList.selectIndices(recipes.length-1)
      setFields(recipes.last)
    case None =>
  }
}

def removeRecipe {
  if(!recipeList.selection.items.isEmpty) {
    recipes = recipes.patch(recipeList.selection.leadIndex,List(),1)
    if(recipes.isEmpty) {
      recipes = Array(Recipe("New recipe",
          List(Ingredient("ingredient","amount")),"Directions"))
    }
    recipeList.listData = recipes.map(_.name)
    recipeList.selectIndices(0)
    setFields(recipes.head)
  }
}

def changeIngredientName(r:Int,i:Int) {
  val ing = recipes(r).ingredients(i)
  recipes(r) = recipes(r).copy(ingredients = {
    recipes(r).ingredients.patch(i,List(Ingredient(
      ingredientNameField.text,ing.amount)),1)
  })
  ingredientList.listData = recipes(r).ingredients.map(_.name)
  ingredientList.selectIndices(i)
}

def changeAmount(r:Int,i:Int) {
  val ing = recipes(r).ingredients(i)
  recipes(r) = recipes(r).copy(ingredients = {
    recipes(r).ingredients.patch(i,List(Ingredient(
      ing.name,amountField.text)),1)
  })
}

val frame=new MainFrame {
  title = "Recipe Book"
  contents = new BorderPanel {
    layout += (new ScrollPane(recipeList)
      -> BorderPanel.Position.West)
    layout += (new SplitPane(Orientation.Horizontal,
      new BorderPanel {
        layout += (ingredientListPanel
          -> BorderPanel.Position.West)
        layout += (ingredientDataPanel
          -> BorderPanel.Position.Center)
```

```
      },
      new ScrollPane(directionsArea))
      -> BorderPanel.Position.Center)
  }
  listenTo(recipeList.selection,ingredientList.selection,directionsArea,
     ingredientNameField,amountField)
  reactions += {
    case e:SelectionChanged =>
      if(e.source==recipeList) {
        if(!recipeList.selection.items.isEmpty) {
          setFields(recipes(recipeList.selection.leadIndex))
        }
      } else if(e.source==ingredientList) {
        if(!recipeList.selection.items.isEmpty &&
            !ingredientList.selection.items.isEmpty) {
          val rIndex = recipeList.selection.leadIndex
          val iIndex = ingredientList.selection.leadIndex
          setIngredients(recipes(rIndex).ingredients(iIndex))
        }
      }
    case e:FocusLost =>
      val rIndex = recipeList.selection.leadIndex
      if(e.source==directionsArea) {
        recipes(rIndex) = recipes(rIndex).copy(directions=directionsArea.text)
      } else if(e.source==ingredientNameField) {
        if(!recipeList.selection.items.isEmpty &&
            !ingredientList.selection.items.isEmpty) {
          val rIndex = recipeList.selection.leadIndex
          val iIndex = ingredientList.selection.leadIndex
          changeIngredientName(rIndex,iIndex)
        }
      } else if(e.source==amountField) {
        if(!recipeList.selection.items.isEmpty &&
            !ingredientList.selection.items.isEmpty) {
          val rIndex = recipeList.selection.leadIndex
          val iIndex = ingredientList.selection.leadIndex
          changeAmount(rIndex,iIndex)
        }
      }
  }
  menuBar = new MenuBar {
    contents += new Menu("File") {
      contents += new MenuItem(Action("Open")(openFile))
      contents += new MenuItem(Action("Save")(saveFile))
      contents += new Separator
      contents += new MenuItem(Action("Exit")(sys.exit(0)))
    }
    contents += new Menu("Recipe") {
      contents += new MenuItem(Action("Add")(addRecipe))
      contents += new MenuItem(Action("Remove")(removeRecipe))
    }
  }
  size = new Dimension(800,600)
}
```

```
frame.visible = true
```

Take some time to run through this code and make sure you understand all of the pieces of it. The way it was presented in this section reflects how you might write it. Writing code is a very non-linear process. You can not expect to write a program from top to bottom straight through. You will generally jump from one part of the code to another, hitting on the areas that you need to complete. This code makes extensive use of functions to break things down. Doing this in your own code will help you keep things organized and make it easier to understand.

There are a number of different aspects of this code that are less than ideal. There are ways to put the GUI into inconsistent states. It also is not exactly robust when it comes to the reading of files and you can make it crash if you play with it some. Finding these shortcomings and fixing them is left as an exercise for the student.

11.7 End of Chapter Material

11.7.1 Summary of Concepts

- Most of the programs that people interact with these days have Graphical User Interfaces, GUIs.
- We can write GUIs in Scala using types from the `scala.swing` package. This package wraps around code in the `javax.swing` package, which itself sits on top of the `java.awt` package. Occasionally we will have to go back to the Java libraries to include certain types of functionality.
- The elements of a GUI are called `Components`. There are a number of different active and passive `Components` in the `scala.swing` library. You should consult the API to see everything that is available.
  - To represent windows you can use `Frame`, `Dialog`, or `MainFrame`.
  - There many different active `Components` ranging from the simple `Button` to the complex `Table`.
  - Complex GUIs are built using panel and pane `Components`. These `Components` can hold other `Components`, including other panels and panes. Different panels and panes have different rules for how their contents are placed and sized.
  - The window types can be gives a `MenuBar` that holds `Menus`. These can hold several types including `MenuItems` and other `Menus`.
  - You can make additional information appear when the user hovers over a `Component` by setting the `tooltip` member to an appropriate `String` value.
- User interactivity for GUIs is different from terminal applications which have a single point of input and can block when waiting for information.
  - The model in `scala.swing` uses publishers and reactors. Any `Component` that might take input and which would need to notify some other code to occur will be a `Publisher`. `Components` can be told to `listenTo` or be `deafTo` various `Publishers`. They then need to have appropriate `reactions` set to deal with what they are listening to.

- Reactions are given as partial functions. These look like the case portion of a match.
- GUI actions can be made faster and easier for the user using mnemonics that attach keystrokes to them.

- The FileChooser type provides a standard file selection dialog box.
- The Table component is an easy to use method of displaying a grid of data.
  - There are a number of different actions that can occur with Tables that you might want to react to.
  - To have complete control over a Table you would need to write a new TableModel. You can have the ability to resize Tables by simply creating new instances of DefaultTableModel with the modified size.

11.7.2 Self-Directed Study

Enter the following statements into the REPL and see what they do. Some might produce errors. You should try to figure out why. Try some variations to make sure you understand what is going on.

```
scala> import scala.swing._
scala> import event._
val frame = new Frame {
     | title = "Small"
     | contents = new Label("A Label")
     | }
scala> frame.visible = true
scala> frame.size = new Dimension(500,500)
scala> frame.dispose
scala> val frame = new Frame {
     | title = "Large"
     | contents = new Label("Another Label")
     | size = new Dimension(600,600)
     | }
scala> frame.visible = true
scala> frame.dispose
val frame = new Frame {
     | title = "4 Buttons"
     | contents = new GridPanel(1,4) {
     |   for(i <- 1 to 4) contents += Button("b"+i)(println(i))
     | }
     | }
scala> frame.visible = true
scala> frame.dispose
scala> val frame = new Frame {
     | title = "Append Numbers"
     | val field = new TextField
     | contents = new BorderPanel {
     |   layout += field -> BorderPanel.Position.North
     |   layout += new GridPanel(1,10) {
     |     for(i <- 1 to 10) {
     |       val str = (i%10).toString
     |       contents += Button(str)(field.text += str)
```

```
     |     }
     |   } -> BorderPanel.Position.Center
     | }
     | }
scala> frame.visible = true
scala> frame.dispose
```

11.7.3 Exercises

1. Using three ComboBoxes and a TextField set up a little GUI where the user makes simple math problems. The ComboBoxes should be in a row with the first and third having numbers and the middle on having math operators. When the user picks a different value or operation, the TextField should update with the proper answer.

2. Write a GUI for a simple text editor. Use a TextArea for the main editing field and put in menu options to open and save files.

3. Write a GUI that has a Table that can display the contents of a CSV file. You can take a normal spreadsheet and save to CSV format. You want to be able to load it in and have the values appear in your Table in the appropriate locations.

4. Write tic-tac-toe with a 3x3 grid of buttons. Have something print when each is clicked.

5. Make a GUI with one of each of the Component types mentioned in this chapter, excluding panels and panes. You can decide what, if anything, it does.

6. Find bugs in the recipe script given in this chapter and fix them.

7. Add menu options to the last table example for removing a row or a column and implement the code to carry out those operations.

8. Make a GUI that has one Slider and one TextField. When the user moves the Slider, the value it is moved to should appear in the TextField.

9. Write a GUI that has two ListViews with two Buttons in between them. The Buttons should say "Add" and "Remove".

10. Write a GUI that has three CheckBoxes and a ListView. The CheckBoxes should be labeled "Uppercase", "Lowercase", and "Digits" for uppercase letters, lowercase letters, and the numeric digits. When the state of a CheckBox is altered, the ListView should be changed to show only the values for the things that are selected.

11. Repeat the above exercise except use RadioButtons and have them in a ButtonGroup so that only one option can be selected at a time.

11.7.4 Projects

1. Write a functional calculator in a GUI. You need to have at least the basic four math operations. Feel free to put in other functionality as well.

2. If you did project 5 (p.206) on parsing chemical equations, you might consider doing this problem as well. The chemical parser is not required for this, so you can start here if you want. In that problem, the chemical equation was converted to a system of

linear equations. Systems of linear equations can be represented as matrix equations. For this problem you will build a GUI that represents matrices as tables then lets you edit them and do basic operations of addition and multiplication on them.

To keep things easy, the matrices will be square. You should have a menu option where the user can set the size of the matrices. They should be at least 2x2. The GUI will display three matrices that we will call `A`, `B`, and `C`. The user gets to enter values into `A` and `B`. The value of C is set when the user selects menu options for add or multiply.

3. Write a GUI to play a basic minesweeper game. You can use a `GridPanel` of `Buttons` for the display and the user interaction. The challenge in this problem is deciding how to store the mines and associate them with `Buttons`. There are many ways to do this. Students should think about different approaches instead of just rushing into one.

4. This project continues the line of ray tracing options. You should build a GUI that lets you edit a scene. You should have `ListViews` of spheres, planes, and light sources. Each one should have the geometric settings as well as color information. You need to have menu options to save to a file or load from a file.

5. Editing rooms for the map traversal in project 3 (p.244) using a text editor in a basic text file can be challenging and is error prone. For this reason, it could be helpful to have a GUI that lets you edit the rooms. It can display a `ListView` of rooms that you can select from and then other options for setting values on the selected room. You need to have menu options for loading a file, saving a file, and adding a room. For a little extra challenge consider putting in the option to remove a room. Note that when you do this, the indices of all the rooms after that one change.

6. For this project you will write a GUI that lets users build schedules of courses. It should read information on the courses that are being offered from a file. It needs to have GUI elements that let you pick courses you want to take and put them into a different `ListView`. You should also be able to remove courses from that `ListView`. Give the user menu options to save and load the build schedules. You do not have to deal with conflicts in times or even duplicates of courses right now. Just let the user build a schedule from options. Consider how to prevent different types of conflicts and do that if you want an extra challenge.

7. Make a program that displays a music database. You want to store significant information for different songs including the name of the song, the artist, the album, the year released, and anything else you find significant. The program should be able to display the information for all the songs as well as allow the user to narrow it down to certain artists, albums, years, etc.

8. For this project, we want to turn the menu driven script of project 4 (p.245) into a GUI driven program. You can use a `ListView` or a `Table` to display information related to the pantry and recipe items. You need to have menu options for saving and loading. Like the menu-based version, users should be able to add items to their pantries, add recipes to the recipe books, check if they can make a recipe, and tell the program that they have made a recipe and have the proper amount reduced from their pantry contents. The details of how this looks and whether you use `Buttons` or `MenuItems` for most of the functionality is up to you.

9. This chapter did not user the theme park example so it appears in this project idea instead. You should write a GUI for the theme park functionality using different tabs in a `TabbedPane` for different areas of the park. The GUI should interact with files for

saving activity in the park. You should keep at least the following three tabs, "Ticket Sales", "Food Sales", and "Ride Operation".

For the first two, you should have `Buttons` that the user can click on to add items to a purchase. On the ticket tab those would include people of different ages, whether they were going to the water park, and various numbers of coolers. For the food purchase there are different types of food and their sizes. You can find food prices on page 81. All purchases should be added to a `ListView`. There should also be a `Button` to remove selected items.

On the ride operation tab you need to have a `ComboBox` for selecting the ride, a `ListView` with all the possible operator names for the user to select from, a `TextField` for entering how many people rode in the last period of time, and a `Button` for submitting the data.

10. For this project you will convert the functionality of the text menu based program in project 5 (p.245) to a GUI. You can decide the exact format of the GUI, but you should have a `ListView` of the courses of interest with the ability to add and remove entries as well as the ability to edit values for entries. The courses should have at least a course number, name, interest level, and time-slot information associated with them.

    When the user chooses to generate schedules, you should show the different schedules in some format and allow the user the ability to select one that they will take. All courses from that selected schedule should be removed from the list of courses of interest. The course of interest information should be saved in a file for use from one run to another.

11. For this project you will extend the work you did for project 6 (p.245) so that there is a GUI. The GUI should include a `Table` for displaying the statistics from a single game as well as a `ListView` that shows the different games that are known. A second part of the GUI, perhaps a second tab in a `TabbedPane`, should have a `ListView` with all the games and give the user the ability to select multiple options and display average stats for what is selected.

    Note that the exact details of how this works will depend significantly on the sport that you have chosen. If your sport does not match this description well, make modifications to get a better fit.

12. You can extend project 6 (p.245) by displaying the information from a box score in a GUI. Depending on the sport and what you want to display, this can be done with a `Table` or using a complex layout including `GridPanel` with `Labels` for the different information. The user should also get a `ListView` of the different box scores that are available and clicking on an option should display the proper box score. In a separate section of the GUI you should display the average values that you have been calculating in the earlier project.

Additional exercises and projects, along with data files, are available on the book's website.

Chapter 12

Graphics

Not everything that you might ever want to put into a Graphical User Interface (GUI) is part of the standard GUI library. They have only added elements that are used fairly frequently. There are also times when you want to display your own graphics, something completely unique to your program, in a GUI. In order to do this, you have to be able to write graphics code.

Scala does not have its own graphics library. This is because the one in Java is quite complete and works well in a Scala style. The graphics capabilities of Java we will be using are collectively referred to as Java2D because they are the 2-D graphics libraries in Java. The types that we will use are primarily located in three different packages: `java.awt`, `java.awt.geom`, and `java.awt.image`.

12.1 Custom-Drawn Panels

To add your own graphics to a GUI, you need to create a `Panel` that draws what you want when it is displayed on the screen. The method for doing this is to make a new `Panel` and then change the way it is painted to the screen. The `Panel` type has a method in it called `paint`, which is passed a single argument of type `java.awt.Graphics2D`. We want to replace this method in our `Panel`. The act of changing what a method does is called OVERRIDING and we will do it with the `override` keyword. The syntax for this looks like the following.

```
val panel=new Panel {
  override def paint(g:Graphics2D) {
    // Draw to g
  }
}
```

If you try this in your code make sure that you `import java.awt.Graphics2D` so that it will know where to find that type.

This panel can then be added to a GUI in the same way that you would add any other panel to a GUI. For this chapter, we will simply make our custom-drawn `Panel` the entire contents of our `Frame`.

12.2 `java.awt.Graphics2D`

The main capabilities of the Java2D library are accessed through the `Graphics2D` class in the `java.awt` package. This class has a number of different methods that we can break into different groups. We list a few of them here. To see all of them you would need to go to the Java API.

- Draw/Fill Methods
  - `draw(s:Shape)`
  - `drawLine(x1:Int, y1:Int, x2:Int, y2:Int)`
  - `drawRect(x:Int, y:Int, width:Int, height:Int)`
  - `drawImage(img:RenderedImage, x:Int, y:Int, obs:ImageObserver)`
  - `drawImage(img:RenderedImage, x:Int, y:Int, width:Int, height:Int, obs:ImageObserver)`
  - `drawRenderedImage(img:RenderedImage, trans:AffineTransform)`
  - `drawString(str:String, x:Int, y:Int)`
  - `fill(s:Shape)`
  - `fillRect(x:Int, y:Int, width:Int, height:Int)`
- Set Methods
  - `setPaint(paint:Paint)`

- `setStroke(stroke:Stroke)`
- `setTransform(trans:Transform)`
- `setFont(font:Font)`
- `setClip(clip:Shape)`

• Get Methods

- `getPaint:Paint`
- `getStroke:Stroke`
- `getTransform:Transform`
- `getFont:Font`
- `getClip:Shape`

• Other Methods

- `rotate(theta:Double)`
- `rotate(theta:Double, x:Double, y:Double)`
- `scale(sx:Double, sy:Double)`
- `shear(shx:Double, shy:Double)`
- `translate(tx:Double, ty:Double)`

The `draw` and `fill` methods actually do the drawing. The `set` methods change how things are drawn while the `get` methods let us ask about the current settings. The other methods also impact the ways in which things are drawn.

To illustrate how this works we can replace the comment in the paint method with a single line that does the simplest drawing possible.

```
g.fillRect(20,20,200,200)
```

This line draws a rectangle whose top left corner is 20 pixels right of the left edge and 20 pixels down from the top.[1] It also has a width and a height of 200 pixels. The result of adding this line is shown in figure 12.1.

You could make similar calls to `drawLine` and `drawRect`. These methods that you call by passing in coordinates are part of the Java graphics library that predates Java2D. While they are simple to use, they lack flexibility and power.

12.2.1 Shapes

The Java2D library added a significant level or abstraction with having a `fill` method that takes a `Shape`. This one method is remarkably powerful, but we will only see a little bit of that power in this chapter. In order to use this method or the closely related `draw` method, we have to be able to create objects of the `Shape` type. The `java.awt.Rectangle` type is a `Shape`. In addition, most of the `java.awt.geom` package is dedicated to providing you with different types of `Shapes`. Here are a few of the ones you can choose from and how you would go about making them. If you are going to be using these you probably want to have `import java.awt.geom._` in your code.

[1]The term PIXEL is short for "picture element". It refers to a single dot in an image. The dots are typically arranged in a grid called a raster. How big these are depends on the resolution of the device you are using.

FIGURE 12.1: The result of doing a simple `fillRect` inside of `paint`.

- `Arc2D` – This is an arc of an ellipse. It can be created as an open arc or it can be closed either as a pie slice or using a chord. You could create one with `new Arc2D.Double(x:Double, y:Double, w:Double, h:Double, start:Double, extent:Double, type:Int)`. The type can be `Arc2D.OPEN`, `Arc2D.CHORD`, or `Arc2D.PIE`.
- `Area` – This is a very general shape type that you can initially create from any other `Shape` with `new Area(s:Shape)`. The power of this type is that you have methods like `add`, `intersect`, and `subtract` that will combine areas to create arbitrarily complex shapes.
- `CubicCurve2D` – Makes a curve described by cubic polynomials in x, y, and z. You can create this by calling `new CubicCurve2D.Double(x1:Double, y1:Double, ctrlx1:Double, ctrly1:Double, ctrlx2:Double, ctrly2:Double, x2:Double, y2:Double)`.
- `Ellipse2D` – An ellipse that you can make using `new Ellipse2D.Double(x:Double, y:Double, w:Double, h:Double)`. The specified `x` and `y` position is the top-left corner of the bounding rectangle for the ellipse.
- `Line2D` – A line that you can make with `new Line2D.Double(x1:Double, y1:Double, x2:Double, y2:Double)` or `new Line2D.Double(p1:Point2D, p2:Point2D)`.
- `Path2D` – A general path in 2-D space that you can build as empty or using an

existing Shape. You can add to the path with methods like moveTo, lineTo, quadTo, or curveTo.

- Point2D – A point in 2-D space you make with new Point2D.Double(x:Double, y:Double).
- QuadCurve2D – Makes a curve described by quadratic polynomials in x, y, and z. You can create this by calling new QuadCurve2D.Double(x1:Double, y1:Double, ctrlx1:Double, ctrly1:Double, x2:Double, y2:Double).
- Rectangle2D – Represents a rectangle. You make one with new Rectangle2D.Double (x:Double, y:Double, w:Double, h:Double) where (x, y) specifies the top-left corner.
- RoundRectangle2D – This is a rectangle with rounded corners. You make one with new RoundRectangle2D.Double(x:Double, y:Double, w:Double, h:Double, arcw: Double, arch:Double).

Most of these are fairly straightforward. The exceptions are the Area and Path2D types. Interested readers are directed to the Java API for more details on these and how to use them. For the types that end with 2D, they can be created with either .Float or .Double. We have used .Double in the list above because it makes things simpler.

We can demonstrate quite a few of these in just a few lines of code. In the body of the paint method put the following lines instead of the call to g.fillRect.

```
g.fill(new Rectangle2D.Double(20,20,50,80))
g.draw(new Ellipse2D.Double(0,200,500,300))
g.fill(new Arc2D.Double(250,0,250,250,45,270,Arc2D.PIE))
g.draw(new QuadCurve2D.Double(0,500,250,0,500,500))
```

When you run this code you will get a window like that shown in figure 12.2.

12.2.2 Settings

Drawing shapes provides a significant amount of power and flexibility. However, there are some clear limitations that are visible in figure 12.2. The most significant of these is the fact that everything is appearing in the same color on the default light gray background. Another limitation is that the shapes are being drawn with thin, single pixels lines. In order to get around these limitations, we need to be able to tell the Graphics2D object to draw things in different ways. This is done by changing various settings.

12.2.2.1 Paint

The most significant setting in most application is the Paint setting. Prior to Java2D, this was a simple color setting. However, when you draw something there are a lot more options than just having a single, solid color. The Paint type is a way of representing these different options, just like the Shape type is a way of generally representing the different types of things that you might want to draw. The main options you can use for a Paint are listed here.

- Color – Solid colors are still helpful and the Color type represents these. A lot of standard colors are predefined for you. So you can use Color.white, Color.black, Color.red, Color.green, Color.blue, etc. You can also "mix" your own colors by combining different amounts of red, green, and blue with either

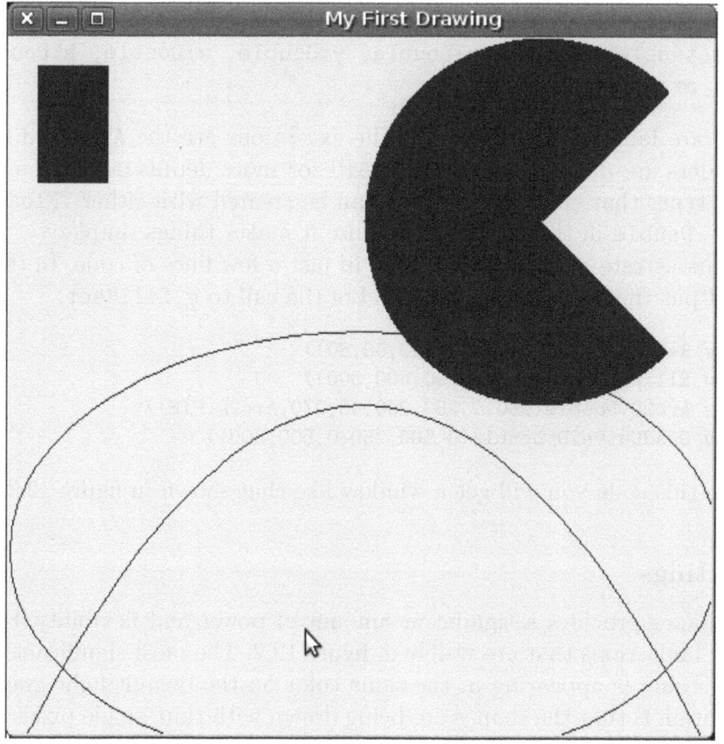

FIGURE 12.2: A sample drawing showing a rectangle, ellipse, arc, and quadratic curve made with calls to `fill` and `draw`.

`new Color(r:Float, g:Float, b:Float)`
or
`new Color(r:Int, g:Int, b:Int)`.
If you use the `Float` version, the values should be between `0.0f` and `1.0f`, inclusive. If you use the `Int` version they should be between `0` and `255`, inclusive. A fourth argument can be added to either of these that is the amount of ALPHA, which represents opacity. A zero alpha value makes the color completely transparent.

- Gradient Paints – There are a number of gradient paints you can use including these two.
  - LinearGradientPaint – This `Paint` allows the color to change gradually across a line. You can specify the endpoints of the line and multiple `Colors` that you want and where you want them on the line. There is no variation in color along the direction perpendicular to the line. You can make one of these with
    `new LinearGradientPaint(startX:Float, startY:Float, endX:Float, endY:Float, fractions:Array[Float], colors:Array[Color])`.
    The fractions should be between `0.0f` and `1.0f`, it should begin with `0.0f` and end with `1.0f`, and there should be as many elements in `fractions` as there are in `colors`.
  - RadialGradientPaint – This `Paint` style allows you to have colors that vary as you move outward radially from a given location. You can make one of these using
    `new RadialGradientPaint(cx:Float, cy:Float, radius:Float, fractions:Array[Float], colors:Array[Color])`
    or
    `new RadialGradientPaint(cx:Float, cy:Float, radius:Float, fx:Float, fy:Float, fractions:Array[Float], colors:Array[Color], cycleMethod:MultipleGradientPaint.CycleMethod)`.
    The latter form allows you to specify an off-center focal point. The `cycleMethod` is described next.
  - Cycle Options - For each of the gradient paints listed above you can provide an extra argument when you make them specifying what you want to have happen outside the normal bounds of the gradient. By default there is no cycling so a solid color is used in these areas. However, you can also specify that the colors repeat or reflect. To do this, use one of the following three constants.
    - `MultipleGradientPaint.CycleMethod.NO_CYCLE`
    - `MultipleGradientPaint.CycleMethod.REPEAT`
    - `MultipleGradientPaint.CycleMethod.REFLECT`
- `TexturePaint` – This will fill the painted area using a specified image that is stretched over an area and repeated outside that area. You can make one with
  `new TexturePaint(txtr:BufferedImage, anchor:Rectangle2D)`.
  The `BufferedImage` type will be described in section 12.3.1.

These different `Paint` styles give us a significant amount of power and control over the coloring of things that are drawn. We can use them to fill our panel with a blank, white background then fill in each of the different things we drew above with a different type of `Paint`.

The following code can be put in the `paint` method to illustrate the different types of `Paints` other than the `TexturePaint`.

FIGURE 12.3: This figure shows what we get when we add some `Paint` settings to the drawing of `Shapes` that we had before.

```
g.setPaint(Color.white)
g.fill(new Rectangle2D.Double(0,0,size.width,size.height))
g.setPaint(Color.red)
g.fill(new Rectangle2D.Double(20,20,50,80))
g.setPaint(Color.blue)
g.draw(new Ellipse2D.Double(0,200,500,300))
g.setPaint(new RadialGradientPaint(375,125,125,300,125,
  Array(0.0f,0.3f,0.8f,1.0f),
  Array(Color.yellow,Color.cyan,Color.magenta,Color.green),
  java.awt.MultipleGradientPaint.CycleMethod.REPEAT))
g.fill(new Arc2D.Double(250,0,250,250,45,270,Arc2D.PIE))
g.setPaint(new LinearGradientPaint(200,500,300,250,
  Array(0.0f,0.4f,0.7f,1.0f),
  Array(Color.white,Color.darkGray,Color.lightGray,Color.black)))
g.draw(new QuadCurve2D.Double(0,500,250,0,500,500))
```

This code will produce the output shown in figure 12.3. Each of the different `Shapes` is drawn with the paint that is specified before it.

12.2.2.2 Stroke

The main problem with what is shown in figure 12.3 is that you can not really tell what color the figures that were drawn, instead of filled, are. This is because they are drawn with lines that are only a single pixel in width. To fix this we need to introduce the next setting on `Graphics2D`, the `Stroke`. There is only one type of `Stroke` provided in the standard

Java libraries. It is called `BasicStroke`. The name is a bit misleading though because this one type of `Stroke` has all the capabilities you are likely to need.

There are several forms that you can use to create a new `BasicStroke` and they range from a simple adjustment of the width to enabling dotted lines with rounded ends. Here are the different options.

- `new BasicStroke(width:Float)`
- `new BasicStroke(width:Float, cap:Int, join:Int)`
- `new BasicStroke(width:Float, cap:Int, join:Int, miterlimit:Float)`
- `new BasicStroke(width:Float, cap:Int, join:Int, miterlimit:Float, dash:Array[Float], dashPhase:Float)`

While the first one is rather straightforward, some of the settings in the others require some explanation.

The `cap` value specifies the style in which you want the ends of lines to be handled. There are three options for this value.

- `BasicStroke.CAP_BUTT`
- `BasicStroke.CAP_ROUND`
- `BasicStroke.CAP_SQUARE`

The `CAP_BUTT` and `CAP_SQUARE` options both produce a square ending. The difference is that the square-cap style extends the end half the width of the line beyond the endpoint. The `CAP_ROUND` style puts a circle on the end with a radius that is half the width of the `Stroke`.

The `join` setting determines how the joints between lines are drawn. To see why this is needed, simply picture two segments as rectangles that have a common midpoint on one edge. If nothing is done to fill it in, there will be a wedge missing. There are three options for how to fill it in.

- `BasicStroke.JOIN_BEVEL`
- `BasicStroke.JOIN_MITER`
- `BasicStroke.JOIN_ROUND`

The `JOIN_BEVEL` setting simply connects the outer corners with a straight line. The `JOIN_ROUND` setting fills in the missing piece with a pie piece. The `JOIN_MITER` setting extends the outer edges until they meet to make a sharp point. One potential problem with the `JOIN_MITER` option is that if the lines come together at a very small angle, the miter can grow to be very long. This is why you can specify a miter limit to work as a maximum length for the miter.

The longest form of constructing a `BasicStroke` allows you the ability to make dashed or dotted lines. It takes an `Array` of `Floats` that specifies the dashing sequence. The numbers in the array tell how long the line will be drawn, then not-drawn, in a repeating fashion. The `dashPhase` allows you to have the dashing start somewhere in the middle of the sequence.

To illustrate some of these features, we will add two lines to our sample code. Before the `Ellipse2D` is drawn we add.this line.

```
g.setStroke(new BasicStroke(6f))
```

This will make the line that is used for the ellipse be six pixels across and far more visible. Before the `QuadCurve` is drawn we insert this line.

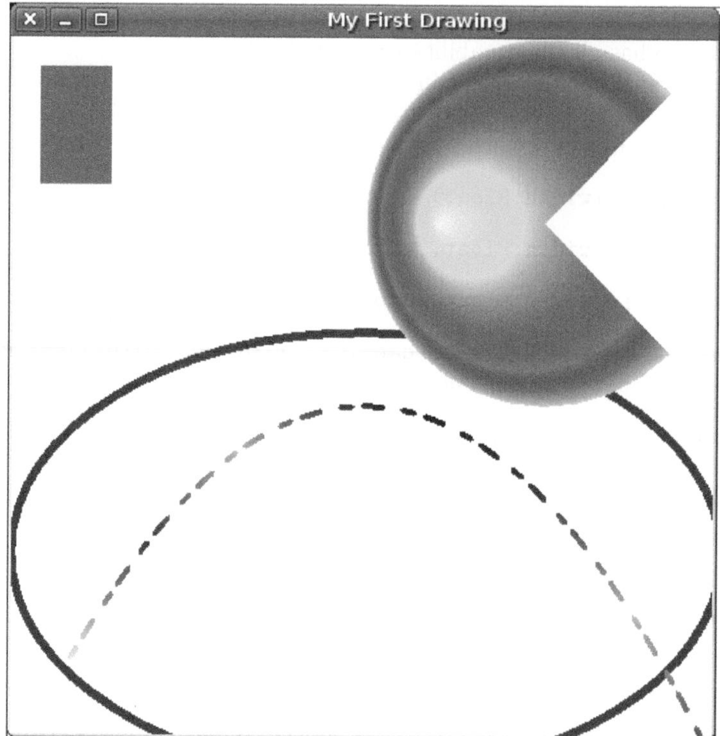

FIGURE 12.4: This is what the drawing looks like after we have added some `Stroke` settings for the two elements that are drawn instead of filled.

```
g.setStroke(new BasicStroke(4f,BasicStroke.CAP_ROUND,
  BasicStroke.JOIN_BEVEL,1f,
  Array(8f,8f,14f,14f),0f))
```

This makes a dashed line that is four pixels thick with round caps on the end of segments. The result we get when these are both added is shown in figure 12.4.

12.2.2.3 Font

When you call the `drawString` method, the text is written using the current `Font` at the specified location. It is worth noting what that location is. To help illustrate that, we will add the following code to the end of the `paint` method.

```
g.setPaint(Color.black)
g.drawString("A test string to print.",20,20)
g.setStroke(new BasicStroke(1.0f))
g.drawLine(20,20,200,20)
```

This sets the color to black, draws a `String`, then makes the `Stroke` narrow again and draws a line going from the same position as the `String` off to the left. The result of adding this code is shown in figure 12.5. What you can see from this is that the position for a `String` is not the top-left corner as it is for most of the shapes. For a `String`, the position would be the beginning of the bottom line if the text were to be written on lined paper. So letters like “g” and “p” dip below the y-coordinate, but most of the text is above it. Everything goes to the right of the x-coordinate.

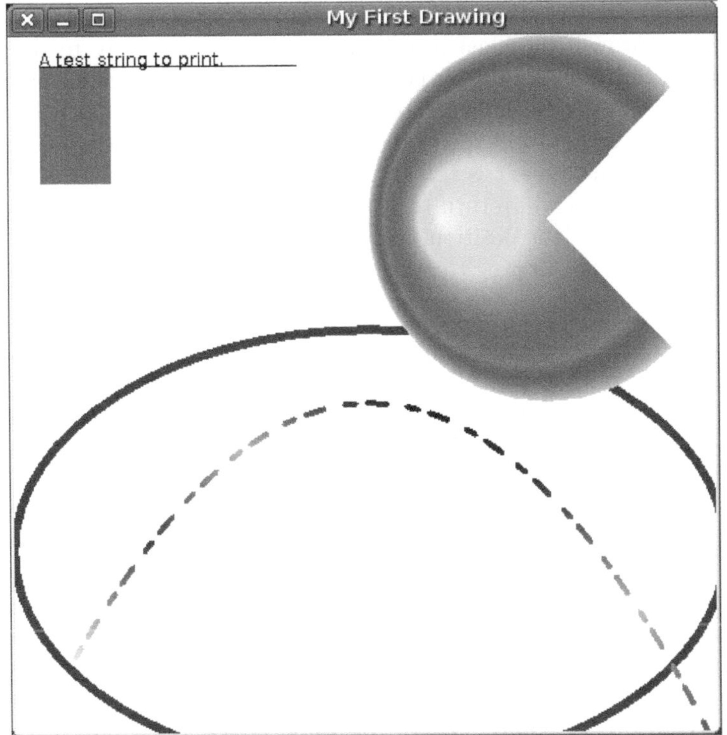

FIGURE 12.5: A simple example of writing a `String` to graphics.

You can change the `Font` that is used by calling the `setFont` method. This method takes a single argument of the type `java.awt.Font`. To build a `Font` we simple call `new Font(name:String, style:Int,size:Int)`. The `name` can be any valid `Font` installed on the machine. One potential problem is that if the program is run on a different computer you might not have the same fonts. To be safe there are a few defaults that are provided in the `Font` class.

- `Font.DIALOG`
- `Font.DIALOG_INPUT`
- `Font.MONOSPACED`
- `Font.SANS_SERIF`
- `Font.SERIF`

The `style` argument can be composed of the following values.

- `Font.BOLD`
- `Font.ITALIC`
- `Font.PLAIN`

You can combine these with a bitwise or, `|`, which was discussed in section 4.5. The `size` should be a point size.

One of the challenges in working with drawing `String`s to graphics is figuring out how large the `String` will be when it is drawn. The simplest way to do this is to use the `getStringBounds(str:String, frc:java.awt.font.FontRenderContext)` method of the `Font` type. If you are not making your own special `Font`, you can call the `getFont()` method on the `Graphics2D` object to get the one that is currently being used. The simplest way to get a `FontRenderContext` is to call `getFontRenderContext()` on the `Graphics2D` object. Getting the bounds will return to you a `Rectangle2D` that will bound the drawing of the `String`. You can get the position, `width`, and `height` of that rectangle to determine where you want to place the `String` to make it centered or whatever other effect you desire.

12.2.2.4 Clip

Normally the things that you draw to a `Graphics2D` will be cut off by the edge of whatever it is drawing to. From what we have done so far that edge would be the extent of the `Panel`. It could also be the bounds of an image as we will see in section 12.3. However, there are times when you want the bounds to be artificially clipped down to something smaller. An example of this would be when you are drawing `String`s or images that are bigger than the area you have allocated for them in your display. Instead of having them "leak" out into other areas, you want to clip them.

In Java2D, you can set the clipping region to any `Shape` object with the `setClip` method. This gives you significant flexibility. Most of the time you might want to clip to a rectangle, but you have the option of using an `Ellipse2D`, `RoundRectangle2D`, or even something far more complex like an `Area`.

Moving the clip around while drawing the same image can produce some interesting effects. If the clip is a circle it will look like you are moving a spotlight around. More complex clips can produce other, similar types of effects.

12.2.2.5 Transform

The last setting that we will consider is that of a transform. The next section is dedicated entirely to the topic of transforms so this will be the brief introduction. When you set the transform on a `Graphics2D` object, it will impact all subsequent elements that are drawn, including `String`s and images. The transformations can be as simple as moving things around or rotating elements or they can be more complex combinations. A simple example of when you would want to use a transform is to write text that is oriented in some direction other than left to right.

12.2.3 Affine Transforms

As was just mentioned, the things you draw to a `Graphics2D` object can be altered by transformations. More specifically, they can be drawn after being changed by an affine transform. Java2D defines the `java.awt.geom.AffineTansform` type to represent these.

An affine transform is one that preserves parallel lines. So if two lines are parallel before the transformation is applied, they must be parallel after it is applied as well. Implicit in this is that lines are preserved as well. There are four types of basic affine transformations.

- Translate
- Rotate
- Scale
- Shear

Any combination of these is also an affine transformation.

The `AffineTransform` is represented by a 3x3 matrix in the computer, the details of which we will generally ignore. Interested readers can look at the Application Programming Interface (API) page for some details or should consider a course in graphics, which should cover the details of matrix transforms for application in both 2-D and 3-D graphics. When you call `new AffineTransform()` you get an identity transform. This is a transform that does not change anything and is represented for the identity matrix. There are options for passing arguments that will let you build the transformation matrix in whatever manner you desire.

Generally, you do not want to go down to the level of the matrix elements. Normally you want to be able to specify how much of a rotation or translation you want. There are number of methods that you can call to do exactly this.

- `AffineTransform.getRotateInstance(theta:Double)`
- `AffineTransform.getRotateInstance(theta:Double, anchorx:Double, anchory:Double)`
- `AffineTransform.getScaleInstance(sx:Double, sy:Double)`
- `AffineTransform.getShearInstance(shx:Double, shy:Double)`
- `AffineTransform.getTranslateInstance(tx:Double, ty:Double)`

These methods return `AffineTransform` instances that produce the specified transformations. For the rotations, the angle, `theta`, should be given in radians.

`AffineTransforms` can be combined as well. This can be done in a general way with the `concatenate(tx:AffineTransform)` method. There are also methods to build up more complex transforms.

- `rotate(theta:Double)`
- `rotate(theta:Double, anchorx:Double, anchory:Double)`
- `scale(sx:Double, sy:Double)`
- `shear(shx:Double, shy:Double)`
- `translate(tx:Double, ty:Double)`

These methods will alter the transform object, which is mutable, to include this new transform. It should be noted that the `Graphics2D` object as a whole had nearly identical methods. Calling those methods will apply these changes to the `AffineTransform` that is being used for drawing.

To illustrate what a transform can do, we will add one line of code into our running example.

```
g.rotate(math.Pi/4)
```

This line should be inserted after the `QuadCurve` is drawn and before the `String` is drawn. The result of adding this is shown in figure 12.6. Notice how the orientation of the `String` has been changed. This version of rotate does the rotation about the origin which is at the top-left corner by default. We turned it through $\pi/4$ radians, which is the same as 45 degrees.

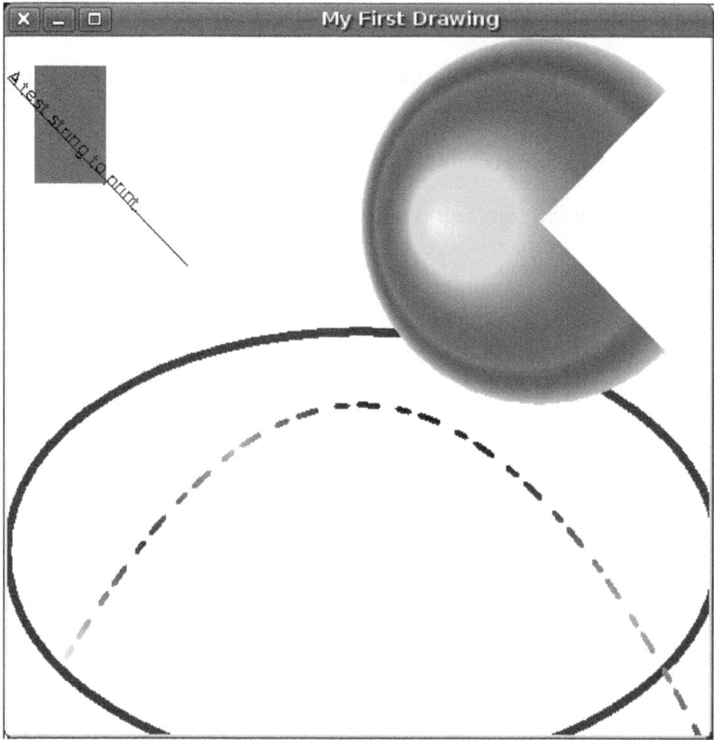

FIGURE 12.6: This figure shows what happens when we do a rotation before drawing our `String`.

12.3 Images

When we draw with a `Graphics2D` object, we have to be drawing to something. So far that something has always been a `Panel` that is displayed on the screen. That is not the only option, however, we can also draw to images. The `Image` type in Java is very general and flexible, but most of the time we will be using it to represent a raster-based graphic. A raster is a 2-D array of little points called pixels (short for picture element). The term pixel is used rather broadly these days as part of the term megapixel that can describe the resolution of digital cameras or graphical displays.

12.3.1 BufferedImage

The `java.awt.image.BufferedImage` is the primary raster-based image in the Java libraries. To create a blank image you can call `new BufferedImage(width:Int, height:Int, imageType:Int)`. The `width` and `height` are measured in pixels. The `imageType` can be any of a number of different values defined in `BufferedImage`. For our purposes we will always use `BufferedImage.TYPE_INT_ARGB`. This type gives you a broad color range and full transparency possibilities by packing alpha, red, green, and blue values into a single `Int`. You can look in the API for the other types that are available as well as other constructors for `BufferedImage`, but in this book we will limit ourselves to these possibilities.

Using this, we might create a 500x500 pixel image with the following code.

```
val img=new BufferedImage(500,500,BufferedImage.TYPE_INT_ARGB)
```

Once you have an image the next question is what can you do with it. There are quite a few different methods you can call on a `BufferedImage`. The ones that we find to be most significant are as follows.

- `createGraphics:Graphics2D` - Returns an object of type `Graphics2D` that can be used to draw to this image.
- `getHeight:Int` - Returns the height of the image in pixels.
- `getRGB(x:Int, y:Int):Int` - Returns a packed ARGB value for the pixel at location (x, y). This is helpful for determining what is at certain locations in the image. This value could be compared to an `Int` that you pack yourself or a hex literal. However, it can also be compared to the result of the `getRGB` method in `Color`.
- `getSubimage(x:Int y:Int, w:Int, h:Int):BufferedImage` - Returns a new image that is pulled from a rectangular region of this image.
- `getWidth:Int` - Returns the width of the image in pixels.
- `setRGB(x:Int, y:Int, rgb:Int)` - Will set the ARGB value at the specified (`x`, `y`) position. You can pack your own ARGB values, use hex, or use the `getRGB` method in the `Color` type. So to turn the top-left pixel black you might do this: `img.setRGB(0,0,Color.black.getRGB)`.

The `createGraphics` method is probably the most significant as it gives you back a `Graphics2D` object that you can use to draw to the image in the same way that we have been using them to draw to the panel on the screen. The `getRGB` and `setRGB` methods can also be very useful when you want to know exactly what is at a pixel or have full control over what color a pixel is drawn as.

Note that an ARGB value of 0 is a completely transparent black. So if you set all the pixels in an image to be 0 the image will be transparent. That way you can draw things on it and not have a background show up around them if you draw that image on top of something else.

12.3.2 Reading and Writing Images

In addition to creating blank images, you often want to be able to read images from files. In some situations it can also be nice to save back images to a file. Both of these activities can be accomplished easily with the `javax.imageio.ImageIO` type. For reading, there are two methods that are of interest.

- `ImageIO.read(input:java.io.File):BufferedImage`
- `ImageIO.read(input:java.net.URL):BufferedImage`

The first method should be passed a `File` object, much like what we have done previously with a `Scanner`. This will read in an image encoded in a standard format, such as a JPEG, GIF, or PNG, from the specified file.

The second method does basically the same thing, but not from a file. This method will read the image data from a URL so it can be located across the network on another machine. For obvious reasons, the `URL` class is part of the `java.net` package. You can make a new `URL` object with `new URL(spec:String)`. The `spec` should be a URL formatted just as it would be in your browser.

Once you have read in a file, you can draw it with a `Graphics2D` object to include it in what you are displaying on screen. You can also call `createGraphics` to get the `Graphics2D` object that will allow you to draw on this image to edit it in some way. Once you have edited it, you might want to write it back out to a disk. You can do this by calling `ImageIO.write(im:RenderedImage, formatName:String, output:File):Boolean`. A `BufferedImage` is a type of `RenderedImage`. The `formatName` is a `String` with the informal name of the format type you want to use for the file. To see all the formats that are supported by your computer and installed software you can run `ImageIO.getFormatNames:Array[String]`.

12.3.3 Double Buffering

Even in situations where you are drawing to a panel, it is often advisable to use an image. The technique is called DOUBLE BUFFERING and it improves performance and reduces flicker in animations. When you draw to the `Graphics2D` object that is passed into the `paint` method, the things that you here drawn are displayed on the screen, one element at a time. Normally it happens so fast that the user never notices. However, if there are a large number of elements in the drawing, or if you are repainting frequently for an animation, this can be visible and can cause flicker in the image. To speed things up and eliminate the flicker you can create an `Image` that is the same size as the `Panel`, do the drawing to the `Image`, then the draw that `Image` to the `Panel`. The following code shows what such a `Panel` might look like.

```
val panel=new Panel {
  var img:BufferedImage=null
```

```
  override def paint(g:Graphics2D) {
    if(img == null || img.getWidth < size.width ||
        img.getHeight < size.height) {
      img=new BufferedImage(size.width, size.height,
        BufferedImage.TYPE_INT_ARGB)
    }
    val g2=img.createGraphics
    // Draw to g2
    g.drawImage(img,0,0,null)
  }
}
```

The first line in the Panel creates a new variable to store images. The BufferedImage type specification is required here because otherwise Scala will infer a type of Null. In the paint method, the first thing that happens is a check to see if the image has been created and if it is large enough to cover the Panel. If not, a new image is created.

After we are certain that we have an image of sufficient size, the code gets the Graphics2D object for drawing to that image and anything that we want to draw is drawn to that image. The last thing to happen in the method is to draw the image to the Panel so it appears on screen.

12.3.4 Revisiting TexturePaint

Now that we know about images we can revisit one of the Paint types that was discussed earlier, TexturePaint. The idea of the TexturePaint is to fill in the drawn/filled shape with a specified image that it repeated on a certain grid. When you make a TexturePaint you provide it with an image and a rectangle that is referred to as the anchor. The image will appear inside of that rectangle and will be repeated in a tiling pattern around it. To illustrate this, we can enter the following code before setting the color to blue for the ellipse.

```
g.setPaint(new TexturePaint(ImageIO.read(new File("rtImg.png")),
  new Rectangle2D.Double(0,0,200,200)))
g.fill(new Ellipse2D.Double(0,200,500,300))
```

This fills in the same ellipse, but does so with a texture of a file that was loaded off disk. The result using this particular file is shown in figure 12.7.

12.4 Other Events

The standard components in the Swing library had code in them to publish events that were specific to the way you interacted with them or to make it very easy to provide the actions for those interactions. The Panel does not have any standard modes of interaction. The way in which the user interacts with it is completely up to you. As such, you have to handle events at a somewhat lower level. We need to listen to the mouse and keyboard actions directly and react to them appropriately.

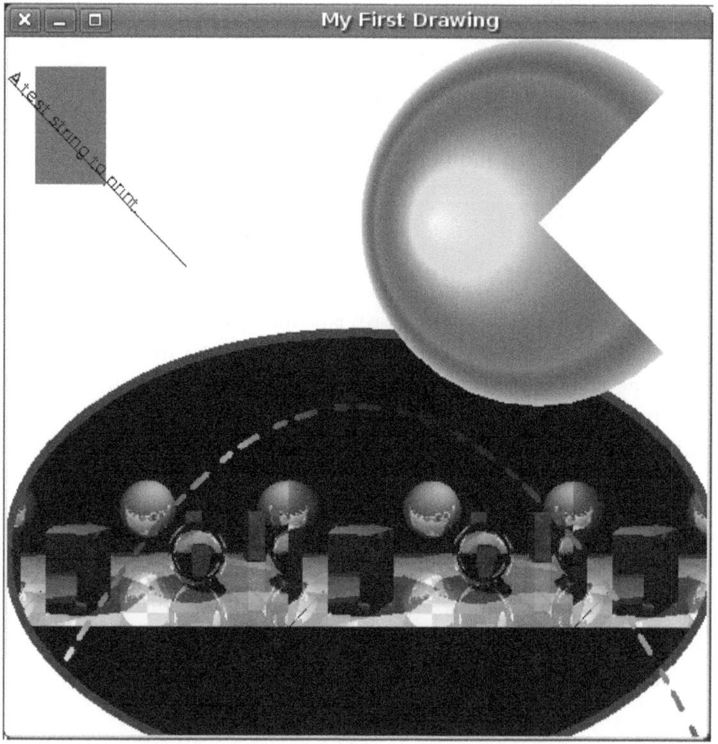

FIGURE 12.7: This image displays the use of a `TexturePaint`. The ellipse is filled with a tiling of an image that is loaded from a disk.

12.4.1 Mouse Events

The mouse is often the most natural way to interact with components of a GUI. At first you might think that the only thing you want to respond to with the mouse is the button clicks. However, if you take a minute to think about it you will realize that a lot of programs interact with the mouse in more complex ways than just when you click. There are eight different types of events that the mouse can fire. It is unlikely that you will care about all of them with any single panel, but you might need to handle several of them in order to produce the effect that you want.

These mouse events are available on all `Components`, even the standard ones that already handle things in their own ways. This allows you to include your own custom behavior. For this reason, all the `Components` contain an `object` called `mouse`. The `mouse object`, in turn, contains three `Publisher objects` called `clicks`, `moves`, and `wheel`. Each of these different `Publishers` will give you information about different things the mouse is doing. You need to listen to the one or ones that produce the events you need. So if you only need to know when the mouse is clicked, you would need to do `listenTo(mouse.clicks)`. If you wanted to know where the mouse was moving or being dragged to, you would add `mouse.moves` to the list of things you are listening to.

This is a list of the different events in `scala.swing.event` organized by which of the `Publishers` produces them.

- `MouseEvent` - All of the events listed below are all subtypes of `MouseEvent`.
- `mouse.clicks`
  - `MouseClicked`
  - `MouseEntered`
  - `MouseExited`
  - `MousePressed`
  - `MouseReleased`
- `mouse.moves`
  - `MouseMoved`
  - `MouseDragged`
- `mouse.wheel`
  - `MouseWheelMoved`

All of the mouse events contain `point`, `source`, and `modifiers` as members. The `point` member is a `java.awt.Point` and it has members in it for `x` and `y`. The `source` is a `Component` so that you can figure out what in your GUI the mouse is over when the event is produced. The modifiers are stored in an `Int` and can give you information about which mouse button or buttons are down.

Any time a mouse button goes down, a `MousePressed` event will fire. When the mouse button goes up, a `MouseReleased` event will fire. The `MouseClicked` event only fires if the mouse button goes down and back up without the mouse moving. If the mouse moves while it is down, instead of getting a `MouseClicked` event, you will get `MouseDragged` events through the movement.

Inside of the `reactions` to the event, you will likely change some mutable data that alters what is drawn. This change will not automatically be seen in the drawing, however.

To make it visible you need to tell the `Panel` that it should redraw itself. To do this, you call the `repaint()` method of the `Panel`. This method does not do the drawing, instead it schedules the drawing to be done in the future when the program is not too busy. The exact nature of how this type of thing works will be discussed in chapter 21. For now you only need to know that if you want a `Panel` to be drawn again you should call `repaint()`

12.4.2 Keyboard Events

You might also want to have your GUI respond to keys being pressed on the keyboard. An example of this might be to have something in your graphic that changes with letters or moves when arrow keys are pressed. The keyboard events are published by the `keys object` in components. So if you want to have `reactions` to the keyboard you add `listenTo(keys)` in the code.

Just like with the mouse, there are several types of different events that can come from the keyboard. All of them fall under the broad heading of `KeyEvent`.

- `KeyEvent`
- `KeyPressed`
- `KeyReleased`
- `KeyTyped`

This is simpler than the mouse. There are only four different types of events. The like the mouse, there are separate events for pressed and released. The `KeyTyped` event is similar to a `MouseClicked`, but there is not any equivalent to the mouse not moving. In this case, the `KeyTyped` event only fires when the key is a normal character key. It should also fire with the repeat rate you normally get when a key is held down. So if you want to react to arrow keys or other non-character keystrokes you need to use `KeyPressed` and `KeyReleased`.

When you are listening for `KeyTyped` you can use the `char:Char` field of the event to find out what character was typed. The `KeyPressed` and `KeyReleased` events can not have the key represented by a simple `Char` because they can handle things that are not normal characters. They have a field called `key` that has a value specified in the `scala.swing.event.Key object`. So if you want to check to see if the up arrow key was pressed you could put the following in your code.

```
if(e.key==Key.Up) ...
```

This assumes that the `KeyPressed` or `KeyReleased` variable is called `e`. The `Key object` has members for all the other keys that you can get on the keyboard as well. For a full list you should check in the API.

If you put in code that listens to the `keys Publisher` and then add a reaction to handle the events you are interested in, you will likely be quite disappointed when you run the program and nothing happens. Assuming that something is broken, you will probably start putting print statements into the code to try to figure out what is happening. What you will find is that you are not getting any key events at all, despite the fact that you are listening for them and have proper code to react to them. The problem relates to a concept called FOCUS.

The currently selected element in a GUI is said to have focus. Keystrokes only go to that component. To understand this, simply picture a GUI with several text fields in it. When you type, the characters should not go to all the text fields at once. Instead, your typing only appears in one, the one with focus. To get your `Panel` to receive the key strokes you have to give it the focus. You can do this by calling the `requestFocus` method on the `Panel`. It does not help to do this before the `Panel` is displayed so you can put it after you set the `frame` to be visible. It is probably also a good idea to have a mouse event that requests focus. Depending on the application, you might want to do this in a `MouseClicked` event or a `MouseEntered` event. The former will only get focus if the user clicks on the panel. The latter will give it focus any time the mouse enters the panel area.

12.5 Animation with Timer

It is also possible that you want something to happen without any keys being pressed or the mouse being used. An example of this would be animations. In order to accomplish this you should use a `javax.swing.Timer` object. `Timer` objects cause code to be executed at regular intervals. We build a `Timer` with `new Timer(delay:Int, listener:java.awt.event.ActionListener)`. The `delay` is a value given in milliseconds. The `ActionListener` type is used in the Java version of Swing as something that reacts to a simple action like clicking a button. These are used so much that the Scala version of Swing includes a simple method for making them, `Swing.ActionListener(action: (java.awt.event.ActionEvent) => Unit)`. Here again we see a type from the Java libraries. You will rarely have to use this type in your code though.

Putting this all together, you could make a `Timer` with the following code.

```
val timer=new javax.swing.Timer(100,Swing.ActionListener(e => {
  // code to mutate values used in drawing.
  panel.repaint()
}))
```

The call to `repaint` is needed for the same reason that was discussed in section 12.4.1. The user will only see the changes that have been made after the next time the screen is painted. The full name for the `Timer` was used here instead of an `import` statement because it is quite possible that this is the only place it will be used. If you find yourself typing it in more than once you would probably want to have an `import` statement.

The `Timer` has `start` and `stop` methods that allow you to do exactly what they say to it. The `Timer` will not be running initially. You have to call the `start` method to get it going. The `stop` method can pose an interesting problem if the reason for stopping is part of the action that is supposed to happen in the `Timer`. Using the style shown above, the action in the `Timer` is built while the `timer` variable is being declared. That effectively makes it a recursive declaration. As such, you can not call `timer.stop` inside of the action code using this approach. There are two ways to get around this. One is to specify a type on the `timer` declaration. This duplicates the type information and probably indicates an `import` should be used. This is shown here.

```
import javax.swing.Timer
val timer=new javax.swing.Timer(100,Swing.ActionListener(e => {
  // code that potentially includes timer.stop().
  panel.repaint()
}))
```

The other approach is to use a slightly different style of setting up the `timer` in which we first give it a null action, then add an ActionListener to it later. Code using this approach might look like the following.

```
val timer=new javax.swing.Timer(100,null)
timer.addActionListener(Swing.ActionListener(e => {
  // code that potentially includes timer.stop().
  panel.repaint()
})
```

12.6 Putting it Together

We finish this chapter with an example program that includes getting input from the mouse and keyboard and that uses a Timer in order to update things at regular intervals. To make the example interesting, we will use these features to make a little game. In the game the player controls a little green dot with the keyboard. Little red dots try to catch the green dot and if they succeed, the player loses. The mouse is used to "draw" puddles on the screen area that slow down both the player and the enemies. That is the basic idea.

There are a few more details in the analysis. Puddles fade over time and enemies only live so long. Enemies do regenerate though. As the game goes on, enemies start to regenerate faster and live longer. So the real objective is to see how long the player can stay alive. Here is code for this game.

```
import scala.swing._
import event._
import java.awt.{Graphics2D,Color}
import java.awt.geom._
import java.awt.image.BufferedImage
import javax.swing.Timer

case class Enemy(x:Int,y:Int,time:Int)
case class Player(x:Int,y:Int)

val PanelSize = 600

var enemies = List(Enemy(PanelSize-100,PanelSize-100,0))
var player = Player(300,300)
var leftPressed = false
var rightPressed = false
var upPressed = false
var downPressed = false
var currentTime = 0
var regenDelay = 100
var enemyLifespan = 100
val puddleImage = new
    BufferedImage(PanelSize,PanelSize,BufferedImage.TYPE_INT_ARGB)
for(i <- 0 until PanelSize; j <- 0 until PanelSize) {
  puddleImage.setRGB(i,j,0)
}

val panel = new Panel {
```

```
  override def paint(g:Graphics2D) {
    g.setPaint(Color.black)
    g.fillRect(0,0,size.width,size.height)
    g.drawImage(puddleImage,0,0,null)
    for(enemy <- enemies) {
      g.setPaint(new Color(1f, 0f, 0f,
          (currentTime-enemy.time)/enemyLifespan.toFloat))
      g.fill(new Ellipse2D.Double(enemy.x-5,enemy.y-5,10,10))
      val dx = enemy.x-player.x
      val dy = enemy.y-player.y
      if(dx*dx+dy*dy<25) {
        g.setPaint(Color.white)
        g.drawString("You Lose!",200,200)
        timer.stop
      }
    }
    g.setPaint(Color.green)
    g.fill(new Ellipse2D.Double(player.x-5,player.y-5,10,10))
  }

  val imageG = puddleImage.createGraphics()
  imageG.setColor(new Color(200,150,50))

  def drawPuddle(x:Int,y:Int) {
      val radX = util.Random.nextInt(5)+5
      val radY = util.Random.nextInt(5)+5
      imageG.fill(new Ellipse2D.Double(x-radX, y-radY, radX*2, radY*2))
      repaint
  }

  listenTo(mouse.clicks,mouse.moves,keys)
  reactions += {
    case mp:MousePressed => drawPuddle(mp.point.x, mp.point.y)
    case md:MouseDragged => drawPuddle(md.point.x, md.point.y)
    case me:MouseEntered => requestFocus
    case kp:KeyPressed =>
      if(kp.key==Key.Left) leftPressed = true
      if(kp.key==Key.Right) rightPressed = true
      if(kp.key==Key.Up) upPressed = true
      if(kp.key==Key.Down) downPressed = true
    case kp:KeyReleased =>
      if(kp.key==Key.Left) leftPressed = false
      if(kp.key==Key.Right) rightPressed = false
      if(kp.key==Key.Up) upPressed = false
      if(kp.key==Key.Down) downPressed = false
  }

  preferredSize = new Dimension(PanelSize,PanelSize)
}

def inMud(x:Int,y:Int):Boolean = {
  (puddleImage.getRGB(x,y) >>> 24) > 0
}

val timer:Timer = new Timer(100,Swing.ActionListener(e => {
```

```
  val speed = if(inMud(player.x,player.y)) 1 else 3
  if(leftPressed) player = player.copy(x = player.x-speed)
  if(rightPressed) player = player.copy(x = player.x+speed)
  if(upPressed) player = player.copy(y = player.y-speed)
  if(downPressed) player = player.copy(y = player.y+speed)

  enemies = for(e <- enemies; if e.time+enemyLifespan>currentTime) yield {
    val speed = if(inMud(e.x,e.y)) 1 else 4
    val dx = if(e.x<player.x) speed else if(e.x>player.x) -speed else 0
    val dy = if(e.y<player.y) speed else if(e.y>player.y) -speed else 0
    e.copy(x = e.x+dx+util.Random.nextInt(5)-2,
           y = e.y+dy+util.Random.nextInt(5)-2)
  }

  for(i <- 0 until PanelSize; j <- 0 until PanelSize) {
    val argb = puddleImage.getRGB(i,j)
    val alpha = argb >>> 24
    if(alpha>0) puddleImage.setRGB(i,j,(argb & 0xffffff) | ((alpha-1) << 24))
  }
  currentTime += 1
  if(currentTime%regenDelay==0) {
    val cx = util.Random.nextInt(2)
    val cy = util.Random.nextInt(PanelSize-10)
    enemies ::= Enemy(10+cx*(PanelSize-20), cy, currentTime)
    if(regenDelay>10) regenDelay -= 1
    if(math.random<0.1) enemyLifespan += 1
  }
  panel.repaint()
}))

val frame = new MainFrame {
  title = "Puddle Dash"
  contents = panel
  menuBar = new MenuBar {
   contents += new Menu("File") {
     contents += new MenuItem(Action("Start")(timer.start))
     contents += new Separator
     contents += new MenuItem(Action("Exit")(sys.exit(0)))
   }
  }
}

frame.visible = true
panel.requestFocus
```

There are effectively four broad sections to this code. At the top are various variable declarations. This is followed by the definition of a `Panel`, followed by a `Timer`, and finally a `MainFrame`. With the variables are declarations of two `case classes` called `Enemy` and `Player`. The enemies are stored in a `List` and there is a single `player` variable. There are also variables that keep track of what arrow keys are being held down. This allows the movement to be done by the `Timer` code, not the event handling. That allows us to control the relative speed of the player and the enemies. To make the game interesting, the player is roughly 25% slower than the enemies. There are also variables for keeping track of time and a BufferedImage that use used to draw and keep track of puddles.[2]

The Panel has a paint method that blanks the screen, draws the puddle image, draws the enemies, then draws the player. The puddleImage starts off completely transparent. The enemies also start off transparent and get more opaque until they die. This adds a bit of challenge as it is hard to see where they start, but makes the game more playable than it would be if enemies were transparent near their death. The Panel also has event handlers that call drawPuddle for MousePressed or MouseDragged as well as handlers for KeyPressed and KeyReleased that set the values of the Boolean vars which control player movement.

Most of the logic goes in the Timer. This moves the player based on keys and the presence of mud. It then moves the enemies toward the player with a speed based on the presence of mud. The enemy movement also has a random component to it to prevent things from being quite so predictable. After moving the enemies, every pixel of the puddleImage is run through and the alpha value, if greater than zero, is decremented. After that, the currentTime is incremented and checked to see if it is time to make a new enemy. If so, a random position along the edges is picked and the enemy is added there. At that time the delay between enemies is also decreased and there is a 10% chance the length of the enemy's life will go up.

The code ends with the creation of a frame, which hold the Panel and has a menu bar with a very basic menu. Enter this code and play the game a bit. For 125 lines of code it is a reasonably interesting game to play. Seeing it in action can also help you to understand what is going on.

12.7 End of Chapter Material

12.7.1 Summary of Concepts

- If you understand how to do graphics you can go beyond the standard GUI components and draw whatever you want to Panels.
- 2-D graphics are produced using the Java2D libraries. These are centered on the java.awt.Graphics2D type.
  - There are a number of different methods for drawing things in Graphics2D the most flexible are draw(s:Shape) and fill(s:Shape). The java.awt.Shape type is flexible and allows many types of shapes to be described. Examples can be found in java.awt.geom.
  - How things are drawn by a Graphics2D instance is determined by various settings.
    * The setPaint method can change how things are filled in. There are a number of different types of Paint including java.awt.Color, java.awt.LinearGradientPaint, and java.awt.TexturePaint.
    * The manner in which lines and curves are drawn is called the Stroke. You can change the stroke by calling setStroke and passing it a java.awt.BasicStroke.

[2]The option of making a Puddle type and keeping a list of those was considered. This path was taken as it is more efficient and it also demonstrates the use of BufferedImage nicely.

    * `String`s can be drawn with different fonts. This is set with a `java.awt.Font` object that is passed into `setFont`.
    * You can adjust the clipping area that is used for any part of the drawing by calling `setClip(s:Shape)`.
    * Everything drawing to a `Graphics2D` can be transformed using any affine transform.
  - An affine transform is any transformation that preserves parallel lines. They are represented by `java.awt.geom.AffineTransform`. There are four basic types of affine transforms and they can be combined in any way to create more complex affine transforms.
    * Translate moves what is drawn around in the 2-D space.
    * Rotate spins the drawing around a particular point in space.
    * Scale can change the size of the drawing in the x or y direction independently.
    * Shear can introduce part of one axis into the other.

- In addition to drawing to the screen, you can draw to images. The basic image type is `java.awt.Image`. It is extremely general and can represent many types of devices.
  - You will most commonly use a `java.awt.image.BufferedImage` which represents a raster of pixels.
  - You can read or write images from and to files with methods in the `javax.imageio.ImageIO` class.
  - One very significant use of images is double buffering, where drawing is done to an image off the screen, then the completed image is drawn to the screen.
- The `Panel` type does not have natural default forms of interaction like clicking a `Button` or typing into a `TextField`. Instead, you make them interactive by handling the mouse and keyboard events directly.
  - All `Components` have a `mouse object` that contains three `objects` in it for dealing with different types of mouse events.
    * `mouse.clicks` gives you `MouseClicked`, `MousePressed`, `MouseReleased`, `MouseEntered`, and `MouseExited` events.
    * `mouse.moves` gives you `MouseMoved` and `MouseDragged` events.
    * `mouse.wheel` gives you `MouseWheelMoved` events.
  - To get information on what keys are pressed you listen to the `keys` object in a `Component`.
    * Keyboard events only go to the `Component` with focus. Use the `requestFocus` method to give focus to the proper `Component`.
    * A `KeyTyped` event will be returned when a normal character key on the keyboard goes down and up. It has a `char` member that will tell you the `Char` for the key that was typed.
    * The `KeyPressed` and `KeyReleased` events let you know when a key goes down or up and works for any key, including non-character keys. To know what key way pressed or released, compare the `key` member to values in the `scala.swing.event.Key object`.
- To make things happen at regular intervals without requiring user interaction, for example animations, use a `javax.swing.Timer` object.

12.7.2 Exercises

1. Create a panel that draws several geometric shapes with different types of paints and strokes.
2. Use the `java.awt.geom.Area` type and create an interesting shape.
3. Use the `java.awt.geom.Path2D` type to make an interesting shape that includes at least one section that is not a line.
4. Use a gradient style paint and `Timer` to make a pattern where the colors move.
5. Use a dotted line on a wide stroke along with a `Timer` where the dotted line moves with the `Timer` ticks.
6. Write a script where you control the movement of a geometric figure with the arrows keys. Put another figure at the location of the mouse click draw them in a different color when the two intersect.
7. Polish up the the evade game from the end of the chapter.

12.7.3 Projects

1. If you did project 1 (p.243) you should have noticed that looking at the numbers to figure out what is going on is quite a pain. To see if particles are moving properly it really does help to have a plot of their motion. For this project you will add a GUI with a custom-drawn panel onto that program so that you can draw where particles are located.

   The only challenge in doing this is getting from the coordinates that your particles are at to coordinates that fit in a normal window. There are two ways to do this. One is to do the math yourself. Specify a minimum and maximum value for x and y and use linear equations to transform from the range (min, max) to $(0, width)$ or $(0, height)$. You can also do this with an `AffineTransform` using a combination of scale and translate. There is benefit to each approach and neither is significantly harder than the other.

   To make the GUI more functional you could include `TextFields` that allow the user to change the values of X_{min}, X_{max}, Y_{min}, and Y_{max}. You could also include a `CheckBox` that lets the user select whether the background is cleared each time particles are drawn. Without clearing, you will see the paths of the particles as they move. With clearing you will only see whatever shapes you use to draw the particles.

2. For this project you should write a script that reads in a text file of numbers and plots the data. You can pick what style of plotting you want to enable and what format the data should be in. As with the previous project, the main challenge in this is to convert from the coordinates that the data points are in to something that appears nicely on screen.

3. The movie Tron (`http://www.imdb.com/title/tt0084827/`), released in 1982, included a number of different early arcade-style games. One of these, light cycles, can be simplified down to a level where it is very easy to write. For this project, you will do just that.

   In the game, the players ride cycles that leave colored walls behind them. Running into a wall kills the player. The last player alive wins. For your implementation, the walls will be lines of pixels in a `BufferedImage` and the cycle will just be the end where the line is growing. You do not have to graphically represent the cycle. Two humans can play. One uses arrow keys and another uses letters. Only two keys are needed for each, one to turn left and one to turn right. Using an image to store the walls prevents you from having to keep that data in a separate array.

For an extra challenge, try putting in a computer-controlled player. This is not really all that hard. The easiest one to write is virtually impossible for a human to beat because it has perfect reflexes. Throw in occasional random turns to make things more even.

If you are having a hard time picturing what this is supposed to look like, the website includes an executable JAR that you can run to play a version written by the author.

4. A step up from the Tron Light Cycles game is the far more common Snake game. In this game a single person controls a snake that goes around eating apples or something else. Each time one is eaten, the snake grows longer. This continues until the snake runs into itself or an edge of the screen.

   The reason this is more challenging than Light Cycles is that the cycles leave walls that can stay until the program is done. The walls can also be a single pixel wide. With Snake, the body parts that follow the head have to "move" because the snake does not simply get longer all the time.

5. One of the early educational programming languages, called Logo, made graphics easy to use by implementing a turtle graphics system (`http://en.wikipedia.org/wiki/Turtle_graphics`). The idea of turtle graphics is that you have a cursor, typically called the turtle, that has a position, an orientation, and pen settings. The turtle can turn and move. When it moves, it can either draw or not. Simple systems only allow the turtle to move along straight lines and that is what you will do for this project.

   A simple way to encode instructions for a turtle is with a `String`. Different characters tell the turtle to do different things. A 'F' tells the turtle to move forward while drawing. A 'f' tells it to move forward without drawing. The '+' and '-' characters tell it to turn to the left and right, respectively. Other characters can be added to give the system more power and later projects will give you the opportunity to do so. The amount that the turtle moves or turns for each character is considered to be a fixed parameter. Using this system, one could draw a square by setting the angle to 90 degrees and using the string "F+F+F+F". Two squares that are separated could be made with "F+F+F+FfF+F+F+F".

   Make a GUI that has a `TextField` where the user can enter a `String`. You might also have fields for segment length and turn angle. There should be a `Panel` that, when painted, will draw the appropriate set of lines. Simply ignore any characters that are not 'F', 'f', '+', or '-'.

6. For this project you will model the process of crystal growth. This might sound like something of a challenge, but it is not really all that hard. Crystals grow when material dissolved in a solution meets a surface of the crystal and sticks. The dissolved material moves around due to Brownian motion and is basically a random walk. You start the process by putting a seed crystal in the solution for stuff to stick to.

   For our purposes, the crystal is simply represented as one color on top of a background that is another color. Use a `BufferedImage` to store this so that you can get and set pixels. This image should be drawn on a `Panel` that is in the GUI. If the user clicks on the panel it should add a new "seed" at the click location (simply set the color at that point in the image to the crystal color).

There should be either a `Button` or a menu option to release more particles. When a particle is released, it should start at one of the edges. You just need to keep track of the x, y location of it. Using a `while` loop, have the particle move randomly around until the move would put it on top of a spot that already has a crystal. At that point you change the pixel at the location the particle had been to the crystal color.

To move the particle randomly you could use `util.Random.nextInt(4)` to get a number in the 0-3 range and move either up, down, left, or right depending on that value. If a move would put a particle outside of the image, simply ignore that move. The menu option should probably run through a loop that drops 100 or so particles and lets each run until it sticks. You only need to `repaint` after all have found their place.

Note that especially early on when the crystal seed is small, it can take a long time for a particle to run around until it hits that seed.

7. If you have been doing ray tracing options in earlier projects, it is time for you to finally see something. You can use the code you wrote for project 2 (p.244) that makes a grid of parameter values and simply set colors in a `BufferedImage` based on the t intersect parameters. To do this you will probably want to build your own colors. You can make a new `Color` object by saying `new Color(r:Int, g:Int, b:Int)` where `r`, `g`, and `b` are in the 0-255 range. Values outside of that range will cause an error. Simply display the image in a `Panel` that you have put in the GUI to see depth information for your ray trace.

8. Write a simple 2-D game of your choosing with simple geometry or sprites (little images for characters). Your game should have a `Timer` to make actions occur as well and take input from the user in the form of key pressed or mouse movement. At this point you know enough to write many types of games, but it is suggested that you strive to keep things simple. Have movement and some collision checking.

9. This project has you write another simple "game" that you are likely familiar with that involves sliding tiles in a 4x4 grid where one piece is missing. The tiles might have numbers or parts of a picture. Your goal is to move then around by sliding pieces into the one open space until they are in the desired order.

   Write a GUI with a `Panel` that shows the 4x4 grid. You can choose if you want to use numbers of parts of an image. If you want to use parts of an image, you can get the small sections by loading the full image, then drawing it into smaller images with offsets so that only the part you want for a given piece appears in the smaller image.

   The user should interact with the game by clicking on the piece that they want to slide. In the simplest implementation, they have to click on a square immediately adjacent to the open square. For a bit of an extra challenge you can make it so that they can click on any square in the same row or column as the empty one and multiple pieces can be moved over at the same time.

10. If you did project 8 (p.281) then you already have a GUI set up for keeping track of what a person has in their pantry and checking on recipes. However, this GUI is fairly bland. Now that you know how to put in graphics you have the ability to make it more visually appealing. For this project you should add onto your previous GUI the ability to have images associated with ingredients and recipes. You should also add in directions for how to make the recipe in the GUI.

A basic implementation should specify a file name for images. Image files are to be copied into the directory that the script is run from. If you want a bit more of a challenge, use the ability to `ImageIO.read` to take a `URL` object. Then the paths to images can be given as a full URL and loaded off the web. To take this a step further, let the user type in a URL the first time, then save that locally as an image. You can save the image using the `write` method from `ImageIO`. Details are in the API.

11. An alternative upgrade to project 8 (p.281) is to make it so that it can help you organize your shopping trip. To make this happen, you need a second data file that tells you where items are located in the store. This can be as simple as listing rows that items are in. If you want to put some information for roughly where on an aisle each item is, you can.

    You need to give the user that ability to build a shopping list using the items that are in the store file. They should be able to do this while looking at the current pantry contents, or at least flip between the two views quickly. The grocery list should have the ability to specify an item and an amount. The user should be able to say that they are buying that list and have those items added to the pantry.

    The row information is used to build a simple graphical representation of the store. You can draw rectangles for rows and put labeled dots for the items in each row. The details are left up to you.

12. For this option, you can add a graphical representation to the schedule that is built in project 10 (p.282). The goal is to have what looks like a fairly standard "week view" of the schedule. You will have a column for each day, and the courses should be displayed as labeled rectangles that span the section of the day that course would occupy. This makes it easier to see when there are overly long blocks of class time or too many small breaks spread around.

    How you specify times is left up to you. In the previous project in this line, it was suggested that each time in the schedule be given a unique integer number. To stay with that, you need to hard code the times and days for those time slots. The alternative is to add functionality so that the user can enter specific times and days. Taking that route will require more complex code in the schedule building to make sure that no two courses are scheduled for the same time.

13. The Mandelbrot set is a famous fractal in the complex plane discovered by Benoit Mandelbrot. A web search will provide you with lots of information on this beautiful structure. Part of what is so remarkably about the Mandelbrot set is that it has such a simple definition, but contains infinite complexities. The set is defined as the points, c, in the complex plane for which the equation $z_{n+1} = z_n^2 + c$, where $z_0 = 0$, gives a bounded sequence.

    In practice, if the value of $|z|$ ever goes beyond 4, the sequence will diverge. So programs to explore the Mandelbrot set count the number of iterations to reach that point and assign a color based on how many iterations it takes. Because the points in the set itself will never diverge, there has to be some maximum number of iterations at which point you simply say it is in the set.

    For this project you will write a GUI program that displays the Mandelbrot set. You have a view on the set with minimum and maximum values in the real and complex axes. Allow the user to zoom in by clicking.

14. Most applications that allow you to keep track of your music, like that written for project 7 (p.281) will show cover art for albums. Put that into your solution so that

when albums (or possibly songs) are selected, the cover art for the album is displayed. Cover art can be stored in the same directory as the script and data file or you can use URLs.

15. Most interactive box scores do more than just present the statistics in the way you did for project 12 (p.282). They also use graphics to display things like team logos and photos of players. For this project you should extend what you did for project 12 so that it displays this information. You should also make it interactive so that the images of things like players can be changed by user actions. For example, you could listen to `MouseEntered` events on labels of their names. If the box scores involve different teams the logo displays should change with what teams are in the current box score.

Additional exercises and projects, along with data files, are available on the book's website.

Chapter 13

Sorting and Searching

Some of the most fundamental things that we do on computers are to order data according to certain values and to search through data for various things. These activities are known as SORTING and SEARCHING, respectively. Due to their importance in solving many different types of problems, a lot of work has gone into doing them as quickly and efficiently as possible. In this chapter we will start your exploration of the topics by working through some of the simpler approaches.

13.1 Basic Comparison Sorts

There are many different ways in which to sort data. The most general ones do not have many requirements on the nature of the data. All they require is the ability to compare elements to see if one should come before another. These comparison sorts are the ones that are used most frequently, mainly because they are so flexible. There are limits on how fast they can be because they are not customized to the data, but typically their generality trumps speed.

At this point we are not even going to consider the fastest comparison-based sorts. Instead, we will start with some of the conceptually simpler sorts. Before we do that though, you should take a second to think about how you sort things as a human. There are some different scenarios you could consider. When you were younger, odds are that you were given assignments in school where you had to sort a list of words into alphabetical order. What approach did you take to doing that? Could you write down a step-by-step procedure to explain to a small child who was doing it for the first time how he/she might go about it?

The procedure that you use might vary depending on what you have to work with. In the list of words, you might need to write it down as a sorted list and you do not want to have to erase and move things around if at all possible. You might be sorting folders in a filing cabinet. The folders can slide backward and forward. They do not have any set position, and moving a folder is not all that hard to do. On the other hand, imagine the silly situation of having to sort cars in a parking lot by their license plates. The cars do not move easily. Each move takes effort and they certainly do not slide around.

Take some time to consider how you would sort in these different scenarios. How does the type of object impact your approach? Space can also be an issue. If you are sorting folders in a big room you might pick a different approach than if you are in a small closet with a file cabinet. For the cars, you might pick one approach if you are moving the cars from one long row to a completely different row than you would if there was only one open space for you to park cars in as you move things around.

The sorts that we are going to work with here not only work with a basic comparison operation, they are also intended to work with arrays and can work IN-PLACE. This implies that they can be done using just a single array. You do not have to make a copy. Not all sorts have this property. So this is like the parking lot with only a few open spaces.

13.1.1 Bubble Sort

The first sort that we will describe is something of a classic in Computer Science. It is not efficient. In fact, you would be hard pressed to convince any person to use this sort of method when sorting things by hand. It works for computers because they will do whatever you tell them, even if it is extremely repetitive and pedantic. It is only taught broadly because it is so remarkably simple to describe and code.

The basic idea of the BUBBLE SORT is that you want to run through the array and look at items that are next to one another. If two items are out of order, you swap them. One pass through the array will not get it sorted unless it was very close to sorted to begin with. However, if you repeat this process over and over, the array will eventually get to a situation where it is sorted. Using this description, there are a few ways in which the bubble sort can be written. They all have an inner loop that looks something like the following.

```
for(i <- 0 until a.length-1) {
  if(a(i) > a(i+1)) {
    val tmp=a(i)
    a(i)=a(i+1)
    a(i+1)=tmp
  }
}
```

The end index of this loop can change over different iterations to make it more efficient, but the idea is the same. The value `i` runs through the array comparing each element to the one after it. If the first is larger than the second, then they do a swap.

The swap code is the three lines inside of the `if`. To picture what this is doing, imagine the analogy of cars in a parking lot. The `tmp` variable is an extra space where you can move one car. So you start by moving the first car to the empty space. After that you move the second car into the space the first one had been in. To finish it off, you pull the car from the extra space into the second spot. The car analogy is not perfect. When we move a car from one spot to another, the spot that is vacated is left empty. When we do an assignment, the variable or `Array` location we are getting the value from does not lose the value. Instead, we have two memory locations that both have references to the value. That would be like having a second version of the car. Much of the time this is only a technicality, but there are times when it can be significant as those extra references can hold onto memory that you really are not using anymore.

In order to do a full sort, this loop needs to be repeated over and over again. How many times must it be repeated? To answer that we can observe that each time through, the largest unsorted value is pushed all the way to where it belongs. If there are n elements in the `Array`, then after $n-1$ passes everything will certainly be in place.[1] Using this logic, we can write the following code.

```
def bubbleSort(a:Array[Double]) {
  for(j <- 0 until a.length-1) {
    for(i <- 0 until a.length-1-j) {
      if(a(i) > a(i+1)) {
        val tmp=a(i)
        a(i)=a(i+1)
        a(i+1)=tmp
      }
    }
  }
}
```

In this code, the outer loop executes $n-1$ times as the value `j` counts up from 0 to $n-2$. You will notice one other change in the inner loop. The end value for `i` has been changed from `a.length-1` to `a.length-1-j`. This is based on the observation that after one pass, the largest element is in the correct location at the end. After two passes, the second largest element is also in the correct place and so on. For this reason, the inner loop can stop one spot earlier each time through because there is no need to check those values that we already know are sorted. The sort would still work fine with subtracting off `1`, but it would do roughly twice as much work.

The other thing that this full sort includes that was not apparent from just the inner loop is that this function has been written to sort an `Array` of `Doubles`. As written, this will not work if you pass in anything other than an `Array[Double]`. If you need to sort something else, you have to write a different sort. There are many ways in which this is less than ideal, especially considering that everything except the type declaration would work equally well for `Array[Int]`, `Array[String]`, or an `Array` of anything else that works with greater than. We will talk about how to get around this in chapter 19.

Technically, this version of bubble sort might get the data sorted and then keep working on it. Smaller elements only move forward one slot on each pass. If the data is not too far out of order so that nothing has to move all that far forward, the sort might get the data in order before the $n-1$ iterations are complete. For this reason, you might want to use a FLAGGED BUBBLE SORT. This is a small variation on the normal bubble sort that uses a `while` loop for the outer loop. It exits after $n-1$ iterations or if it goes through a pass of the inner loop without doing any swaps as that implies that everything is already sorted.

```
def flaggedBubbleSort(a:Array[Double]) {
```

[1]We do not have to put n things in place because if there are only n places and $n-1$ are in the right place, the last one most also be in the right place as it is the only one left.

```
  var flip=true
  var j=0
  while(flip && j<a.length-1) {
    flip=false
    for(i <- 0 until a.length-1-j) {
      if(a(i) > a(i+1)) {
        val tmp=a(i)
        a(i)=a(i+1)
        a(i+1)=tmp
        flip=true
      }
    }
    j+=1
  }
}
```

This code adds a few more lines to deal with the flag and the fact that we have to do the counting ourselves with a `while` loop. The flag is called `flip` and it starts off as `true`. However, before the inner loop starts it is always set to `false` and only get set back to `true` when a swap is done.

13.1.2 Selection Sort (Min/Max)

The next sort we want to consider is one that humans are far more likely to employ, especially in a situation like sorting the cars where moving cars around takes a significant amount of effort. The sort is the SELECTION SORT and it is called this because it runs through the values and selects one and places it where it belongs. In order for this to work you have to know where the value should go. This is easy if the value is either the largest or the smallest in the collection. As such, the selection sort is typically implemented as either a minimum sort (MIN SORT) or a maximum sort (MAX SORT). In theory one can also write a min-max sort that picks both and moves them to where they belong.

We will write a min sort here. It is fairly trivial to convert this to a max sort. The inner loop of the min sort will run through all of the currently unsorted collection and find the minimum element in that. If that element is not already where it should be, it is swapped into place. For an `Array`, the fact that this is a swap is significant. To understand this, think about the cars. Shifting all the cars over one is hard. Swapping two is much easier. The `Array` is like the parking lot with fixed locations. Certain other data structures could act more similar to file folders which slide around easily and might allow shifting. If you use an `Array` and a shift instead of a swap, the code you produce will be both longer and significantly slower.

```
def minSort(a:Array[Double]) {
  for(j <- 0 until a.length-1) {
    var min=j
    for(i <- j+1 until a.length) {
      if(a(i) < a(min)) min=i
    }
    if(min!=j) {
      val tmp=a(j)
      a(j)=a(min)
      a(min)=tmp
    }
  }
```

```
}
```

As with the standard bubble sort, the outer loop happens $n - 1$ times as it has to swap $n - 1$ elements into the locations they belong. The contents of the inner loop here are quite short, just a check that changes the `min` variable if needed. The swap itself is longer than the loop.

The selection sort is not really more efficient than the bubble sort in general. However, because the swap is outside of the inner loop, this sort can be much more efficient in those situations where the swaps are expensive. This situation can not happen for normal `Arrays` in Scala, but it can be true for `Arrays` in some other languages or if the values being sorted are contained in a file instead of the memory of the computer.

13.1.3 Insertion Sort

The last sort we will discuss in this section is another one that humans might consider using. It is called the INSERTION SORT because the way it works is to build up a sorted group of elements at the beginning of the `Array` and each new element is inserted into the proper location in that group. The advantage of this sort is that it can do less checking. Once it finds the place to put the element it can stop and move to the next element. It is particularly efficient if the `Array` is fairly close to sorted as each element typically does not have to be moved very far.

The code for an insertion sort can be written in the following way.

```
def insertionSort(a:Array[Double]) {
  for(j <- 1 until a.length) {
    var i=j-1
    val tmp=a(j)
    while(i>=0 && a(i)>tmp) {
      a(i+1) = a(i)
      i -= 1
    }
    a(i+1) = tmp
  }
}
```

The outer loop here still executes $n - 1$ times, but in a different way and for a different reason. Instead of starting at 0 and stopping one short of the last element, this loop starts at 1 and goes to the last element of the `Array`. The reason for this is that an `Array` of one element can be viewed as a sorted `Array`. So we take the first element as being where it should be in a group of 1. Then we start with the second element and potentially move it forward if needed.

The code inside the outer loop can use some explaining as well, as this is a bit different from what we saw with the other two sorts. It begins with the declaration of the variable `i`, which begins at `j-1`. This is the index of the next element that we need to compare to the one we are moving. We also declare a temporary variable called `tmp` that stores the value in the `Array` that we are moving forward. Having the temporary variable here prevents us from doing full swaps and makes the code both shorter and faster. To understand this, consider again the example of moving cars in a parking lot. Consider you have ten cars, the first eight of which are already sorted. You need to move the ninth one forward to the place it belongs, potentially several spots down. You would not do this by swapping cars one after the other. Instead, you would pull that car out and put it in the one empty spot to the side. Then you would move the sorted cars down, one at a time, until you had cleared out the proper spot. At that point you would drive the car from the spot on the side into the correct position.

That mental image is exactly what this is doing. The `while` loop executes as long as we have not gotten down to the beginning of the `Array` and the element we are considering is greater than the one we are inserting. As long as that is true, we move the element up and shift our attention down to the next element. When we are done, we put the temporary value where it belongs.

13.1.4 Testing and Verifying Sorts

The last three subsections simply presented sorts as working products. In reality, sorts are like any other code and have to be tested. Fortunately, sorts are fairly easy to test. After the sort has completed, you should be able to run through the elements and check that each element is less than or equal to the one after it. As long as this is true for all of the elements, the array has been properly sorted. An imperative approach to doing this might look like the following.

```
def isSorted(a:Array[Double]):Boolean = {
  for(i <- 0 until a.length-1) {
    if(a(i)>a(i+1)) return false
  }
  true
}
```

This code does something that we have not seen before. It uses a `return` statement. We have seen that normally, the last statement in a function should be an expression and the value of that expression is what the function returns. It is also possible to force a function to return earlier using a `return` statement as is done here. If at any point we find consecutive elements that are out of order, there is a forced return of `false`. If it gets all the way through the array it returns `true`.

This style is frowned on by some because it complicates things for those reading the code. Normally when you see a `for` loop, you know that it will run through the entire collection and the code will exit when it gets to the end of the collection. The `return` statement forces the `for` loop to exit earlier and as such, forces the reader to look more closely to figure out what is going on here. Code that behaves in slightly unexpected ways like this is a common source of errors, especially if the code gets modified later on.

You could get around having the `return` statement by using a `while` loop. This would remove the requirement of a `return` statement, but it does not really make the code more functional or easier to read. If anything, it might make it a bit harder to read and deal with. An alternative to this is to use the `forall` method of collections. The only challenge in this is picking a way to compare sequential elements using `forall`, as that method only works on one argument at a time. One way to get around this is to use `forall` on a `Range` that is the index. This is logically very similar to the original `for` loop except that your part of the code is completely functional.

```
def isSorted(a:Array[Double]):Boolean = {
  (0 until a.length-1).forall(i => a(i) <= a(i+1))
}
```

The fact that random access is a fast operation on an `Array` allows this code to be efficient. A second approach, which is also functional, is to use `forall` on a `zipped` tuple as is shown here.

```
def isSorted(a:Array[Double]):Boolean = {
  (a,a.view.tail).zipped.forall(_ <= _)
}
```

This code is a bit shorter, but significantly more advanced. In fact, it uses advanced topics from sections 8.5 and 10.4. The approach here is to make a tuple that contains the original elements first and everything but the first element second. If we `zipped` these two collections with the `zip` method, the result would be a sequence of tuples with each element and the one after it in a tuple and the first and last element appear only once in the first and last tuples. That `zipped` collection could then be dealt with using `foreach`. Such a solution would be very inefficient however, as it would actually construct a lot of new objects and do a lot of copying. The advanced topics are used to make it more efficient.

The first advanced topic is the `view` from section 8.5. A `view` is a representation of a collection that does not make a full copy unless forced to do so. In this code, taking a `tail` of a normal `Array` would produce a whole new `Array` and copy a lot of data over into it. However, the `tail` of the `view` is just a small object that will interact with code the same way the `tail` would, but through logic instead of making a real copy.

The second advanced topic is the use of the tuple `zipped` method from section 10.4. This is a method of the 2-and 3-tuple types that allows you to more efficiently run common functions that need to go through two or three collections at the same time. Instead of making a whole new `zipped` collection of tuples, this gives you back an object that has methods like `map`, `filter`, and `forall` declared so they take as many arguments as are in the tuple. So we get a version of `forall`, which needs a function of two arguments we can easily provide using the shorthand syntax.

Whichever approach you choose, you can test a particular sort with code like the following.

```
val nums=Array.fill(args(0).toInt)(math.random)
flaggedBubbleSort(nums)
assert(isSorted(nums))
```

This code creates an `Array` of random `Doubles` with a length determined by a command-line argument. It then calls a sort on that `Array` and then uses the assert function to make sure it is true. Assert is a standard function in Scala that can be called with one or two arguments. The first argument is a `Boolean` that should be `true`. If it is not `true`, an error results terminating the program. A second argument can be provided that is passed by-name and should provide a useful message if the `assert` fails.

The `assert` function can be used generally in your code. There is also a `require` function that performs similarly, but should be used for argument values. So at the top of a function, if the function will only work on certain argument values, you can use a call to `require` to make it so the code fails in an informative manner when bad arguments are passed in.

13.1.5 Sort Visualization

It can help you understand sorts if you get to see them in action. There are many different websites that have animations that will show you how these and other sorts work. One advantage of having already studied graphics is that we can write our own code to visualize what the sorts are doing. This is part of why the sorts above were set up to work with the `Double` type. It makes it easy to generate with the call to `math.random` and to draw because the random numbers are uniformly distributed between 0.0 and 1.0.

To set things up for the visualization, we will use the `nums Array` from above and put in some other code before the sort is called. We will start with the basics.

```
val img=new BufferedImage(nums.length,500,BufferedImage.TYPE_INT_ARGB)

val panel=new Panel {
  override def paint(g:Graphics2D) {
    g.drawImage(img,0,0,null)
  }
  preferredSize=new Dimension(img.getWidth,img.getHeight)
}

val frame=new MainFrame {
  title="Sort Animation"
  contents=panel
  resizable=false
}
```

This creates an Image that has a width of the number of values we have in the Array with a height of 500. It then makes a Panel that draws the Image and has a preferredSize equal to the size of the Image. Lastly, we make a Frame and add the Panel into it. We also use a flag in the Frame that we have not used before to make it so the Frame can not be resized.

With this framework in place, we can define a method that will render an Array of Doubles to the Image. The function should also have the ability to draw two marker lines to help indicate what is happening in a selection sort where we really do not move things very often, but spend a lot more time searching for the smallest element.

```
def renderSort(a:Array[Double],m1:Int = -1,m2:Int = -1) {
  import java.awt._
  val g=img.createGraphics
  g.setPaint(Color.white)
  g.fillRect(0,0,a.length,500)
  g.setPaint(Color.red)
  for(i <- 0 until a.length) {
    g.drawLine(i,500,i,500-(a(i)*400).toInt)
  }
  g.setPaint(Color.black)
  g.drawLine(m1,0,m1,50)
  g.setPaint(Color.green)
  g.drawLine(m2,0,m2,50)
  panel.repaint()
  Thread.sleep(20)
}
```

Starting from the top, this function takes the Array and two Ints. The Ints, locations for the markers, have default values of -1. This mean that you can call this function and pass only the Array if you do not need the markers, and they will be drawn out of bounds. The first line in the function is a local import of the java.awt package. Local imports are something that Scala allows so that you do not have to pollute the namespace as much importing everything at the top of the file. Instead you can import what you need, when and where you need it.

The middle of the function is basic drawing. It gets the Graphics2D for the Image and clears the Image with white. Then it draws in red, drawing one line for each element of the Array. The lines go from the bottom up and are taller for larger values. After drawing the Array values the markers are drawn in two different colors and the panel is told to repaint.

The last line of this method is something that we have not seen before, but which is required if we want to actually be able to see what is going on. The `Thread.sleep(millis:Long)` method causes the current execution to pause for the specified number of milliseconds. The number 20 was chosen to give 50 updates every second. This is slow enough you can see what is going on, but fast enough that you can sort a decent number of elements in a reasonable amount of time. We will talk much more about threads in chapter 21.

Now all that remains is to add some calls to `renderSort` in the sort algorithms. Putting them into the inner loop works best. For `minSort` you need to pass `i` and `min` as the marker values. The others do not need to draw the markers.

13.1.6 Order Analysis

We said previously that the sorts presented above are not the most efficient sorts. These sorts are presented first because they are fairly simple and straightforward. However, it is a reasonable question to ask how "fast" they really are. This is a challenging question to answer because speed can depend on many different things including the computer you are running on and exactly what you are sorting. To address this, algorithms are generally described in terms of their "order" in a particular operation. The order is a rough functional description of how the number of operations scales with the size of the input. We will talk in more detail about order in section 24.3. For now we will use the term fairly loosely.

So the idea of order analysis is to pick one or more operations that are particularly important to the algorithm and try to find a function that models how the number of those operations changes with the size of the input. In the case of sorts, there are two operations that are particularly significant, comparisons and assignments. Most of the time we worry about comparisons and for the code we have written here, that is appropriate.

So how many comparisons does the full `bubbleSort` do on an `Array` with n elements? The first pass through the loop does $n-1$ comparisons. The second pass does $n-2$. The third does $n-3$ and so on. In math terms we can write the number of comparisons in this way:

$$C = \sum_{i=1}^{n-1} i = \frac{n(n-1)}{2} = \frac{1}{2}n^2 - \frac{1}{2}n.$$

This gives the exact number of comparisons between elements that will be done. It happens to be be exactly the same for the selection sort. We typically refer to this as $O(n^2)$. For order analysis we generally ignore coefficients and everything except the highest power. This does not provide a perfect picture of performance, but it works very well in general, especially when you talk about larger inputs.

The most important thing to note about an $O(n^2)$ algorithm is that if you double the number of elements, there will be four times as many comparisons done. As a result, you can generally expect the sort to take four times longer on twice as much stuff. The nature of the scaling is what we often care about the most. Different computers might be faster or slower, but just knowing this scaling and a runtime for a problem of a certain size gives you a good feel for what the runtime will be if you change the size of the problem.[2]

The number of comparisons in the selection sort and the bubble sort depends only on the size of the data set being sorted. It does not depend on the nature of the data. This is not the case for the flagged bubble sort or the insertion sort. These sorts can do very different numbers of comparisons depending on the data that is given to them. For example, if you

[2]Real hardware can break down these scaling arguments at certain critical points. For example, if the input set becomes larger than the cache of the machine, you will typically see a slowdown that does not scale quite as $O(n^2)$.

give a flagged bubble sort an array that is already sorted, it will only run through it once, doing $n-1$ comparisons. We would call that $O(n)$. Granted, that is a BEST-CASE scenario. The best-case is rarely of interest. Instead, we typically worry about AVERAGE-CASE and WORST-CASE situations.[3]

So what are the average and worst-case behaviors of insertion sort? The outer loop clearly happens $n-1$ times. So we can do a summation much like what we did above. However, the inner loop happens a variable number of times. In the worst-case, which would be getting data in reverse order, the inner loop will run i times and we get exactly the same behavior as with the bubble and selection sorts. In the best-case, which is data in the proper order, the inner loop does only one check. This gives linear performance, $O(n)$, because

$$C = \sum_{i=1}^{n-1} 1 = n - 1.$$

On average we expect something halfway between these extremes, we expect the inner loop to do $i/2$ comparisons. This gives

$$C = \sum_{i=1}^{n-1} \frac{i}{2} = \frac{n(n-1)}{4} = \frac{1}{4}n^2 - \frac{1}{4}n.$$

While this is better than the behavior of the other sorts by a factor of two, it is still $O(n^2)$. This is because doubling the size of the input still makes it take four times longer and that is what we generally care about the most.

So how good or bad is $O(n^2)$? Functions like this that grow as a polynomial function are referred to in a very general way as being tractable, because you can use them for fairly large values of n and still expect the calculation to finish in a reasonable time. We will see later on that not all algorithms have this behavior. However, $O(n^2)$ is not great if the value of n gets extremely large. We have no problem using these sorts for `Arrays` with 100 or 1000 values. However, as you continue to add zeros, these methods will prove to be too slow. Each factor of 10 in the size causes the program to do 100 times as many comparisons and generally this leads to it taking 100 times longer to complete. As such, these methods become unusable if n gets big enough. Fortunately, there are alternatives.

13.1.7 Shell Sort (Diminishing Gap Sort)

The first alternative sort we will look at is the SHELL SORT, also called the diminishing gap sort. This sort was first proposed in 1959 by Donald Shell and it was one of the first general sort algorithms developed that is faster than $O(n^2)$, though some minor changes had to be made in order to get that performance for the worst case. The basic idea of this sort is one that might seem a bit counterintuitive. It performs an insertion sort repeatedly on different subsets of the full array. To start with, the subsets are taken to be groups of elements that are widely spaced. The spacing between the elements in each subset is called the "gap". As the alternate name implies, the gap is then decreased in size for each run through the sort.

The counterintuitive aspect of this sort is that performing insertion sorts multiple times should sort the array faster than doing doing it outright. This can work because a sort with a smaller gap size maintains the ordering at the larger gap size so work is not undone, and the insertion sort is very efficient on partially ordered data. The sorts with large gaps do

[3]The exception to this is if we happen to know that we will very often have inputs that produce best-case behavior.

very little work compared to the smaller gaps because they do not contain many elements, but once you get to the smaller gaps the data is mostly sorted so you get close to the best-case performance of the insertion sort. The sort always ends with a gap size of 1, which is doing a normal insertion sort, so you know that the result is fully sorted.

The only trick to the Shell sort is figuring out what to do about the gap. The initial suggestion was to start at half the size of the `Array` and decrease it by factors of 2. This often works well, but spacings that are factors of 2 apart will keep even and odd subsets separate until you get down to a gap of 1 and in certain cases that can actually lead to $O(n^2)$ behavior. For this reason, we use a slightly different factor.

```
def shellSort(a:Array[Double]) {
  var gap=a.length/2
  while(gap>=1) {
    for(j <- gap until a.length) {
      var i=j-gap
      val tmp=a(j)
      while(i>=0 && a(i)>tmp) {
        a(i+gap) = a(i)
        i -= gap
        renderSort(a)
      }
      a(i+gap) = tmp
      renderSort(a)
    }
    gap=(gap/2.2).round.toInt
  }
}
```

This code was created by defining the `var gap` and writing the `while` loop, then cutting and pasting the insertion sort inside of the `while` loop. Once there, all the places where `i` was incremented or decremented by 1 were changed so the 1 was `gap`. The factor of 2.2 gives better spacings though you do need to `round` and convert back to an `Int`. Two calls to `renderSort` were left in this code so that you can easily visualize it. Of the sorts we have discussed so far, this one is probably the most important to visualize to get a grasp on what it is doing and why the process of reducing the gap is significant.

The exact order of the Shell sort is harder to pin down. This is largely because it varies with the selection of gap scaling and can range from $O(n^2)$ at the poor end all the way to $O(n \log^2 n)$ at the good end. Reasonable implementations like the one shown here will tend to give $O(n^{3/2})$ performance. This might not seem like a tremendous improvement, but if n gets large, the difference can be quite dramatic.

13.2 Searching

While sorting data is something that we do a fair bit with computers, searching for data is a far more common task. A lot of the things that you do with computers on a regular basis involve searching for information either just to retrieve it or to modify information related to what is found. This makes sense, as running through data is something that computers do particularly quickly and efficiently.

13.2.1 Sequential Search (Linear Search)

The most basic form of search is the SEQUENTIAL or LINEAR SEARCH. This involves running through data one element at a time and checking each one to see if it matches what you are looking for. If it finds what you are looking for, it either returns the data or an index where it can be found. If it gets to the end without finding it, it will return something to indicate that the data was not there. The following code is a linear search through an `Array` of `Int`s.

```
def linearSearch(a:Array[Int],value:Int):Int = {
  var i=0
  while(i < a.length && a(i)!=value) {
    i+=1
  }
  if(i > a.length) -1 else i
}
```

This search returns the index of the first element in the `Array` whose value is the same as what was passed in. It does this with a `while` loop so that it can stop early if it is found.

The `if` statement is needed because the common idiom when returning an index is to return -1 when the value is not found. We can get rid of this conditional at the end if we count backwards.

```
def linearSearchForLast(a:Array[Int],value:Int):Int = {
  var i=a.length-1
  while(i >= 0 && a(i)!=value) {
    i-=1
  }
  i
}
```

This modified version of the code starts at the end and goes backwards. It is a bit simpler, but it fundamentally alters the description of the function as we now find the last occurrence instead of the first.

There are quite a few different methods on the Scala collections that do searching. For the collections that we have learned about, these are all performing linear searches. They include the following, roughly as they are defined in `Seq[A]`:[4]

- `def find(p: (A) => Boolean): Option[A]`
- `def indexOf(elem: A, from: Int): Int`
- `def indexOf(elem: A): Int`
- `def indexOfSlice(that: Seq[A], from: Int): Int`
- `def indexOfSlice(that: Seq[A]): Int`
- `def indexWhere(p: (A) => Boolean, from: Int): Int`
- `def indexWhere(p: (A) => Boolean): Int`
- `def lastIndexOf(elem: A, end: Int): Int`

[4]The exact signatures of some of the methods have been simplified so that they make sense at this point in the book.

- `def lastIndexOf(elem: A): Int`
- `def lastIndexWhere(p: (A) => Boolean, end: Int): Int`
- `def lastIndexWhere(p: (A) => Boolean): Int`

Many of these methods come in pairs where one of the two takes an extra `Int` argument for an index in the collection that it should start working from. Only the first method returns an element from the collection, all the others return indices. The one that does return an element wraps it in an `Option` so that if no match is found it can return `None`. If no match is found for the methods that return an index, they will return -1.

13.2.2 Binary Search

Even though computers are very fast, linear search is far from ideal, mainly because searching is something that is done so very frequently. If the data is not ordered in any way, linear search is your only option. To understand this, imagine you are handed a normal phone book and asked to find the person who has a given phone number. Due to the fact that phone books are not ordered by phone numbers, your only recourse is to go through each and every line and check the numbers against what you are looking for. In any reasonably sized city this is something that no human would actually undertake.

If this were the only way to look through a telephone book people would not bother to keep them.[5] However, people do keep telephone books because they rarely look things up by number. Instead, people normally look things up by name and the telephone book is sorted by name. This ordering of the elements can lead to much more efficient searches. You might not be able to write a good algorithm for how you really look things up in a telephone book, but we can consider your first step and use that as a direction for writing an efficient algorithm.

Given a large phone book and a name, you will open it up and look at what is on the page you open to. Odds are good that you will not get exactly the right page. However, comparing what you are looking for to what is on the page gives you a significant piece of information. If what you are looking for comes earlier in the alphabet than what is on the page you will only look at other pages before that one. You basically throw everything after that page out of your search without even looking at it. Similarly, if what you are looking for comes after the page you have opened to you will only consider pages after the current one. You will generally repeat this process in a manner that is not easy to describe in an algorithm just because your method of picking pages might be impacted by things like the binding of the book and whether one page sticks out a bit further than another.

The idea of looking at a location and only considering things before or after it based on a sorted order can be used to create fast searching algorithms. The most general of which is the BINARY SEARCH. In a binary search, you keep track of a range of elements that you are considering by two integer indexes. We will call them `start` and `end`. At any given time, you know that the value you are looking for, if it is present, will be at an index in the range $i \in$ [`start`,`end`).[6] So if we are searching the whole `Array`, then initially start is 0 and end is the `length` of the `Array`. We consider the midpoint of the range, `mid=(start+end)/2`, and check if the element at `mid`. If it is, we return `mid`. Otherwise we check if what we are looking for is greater or less than the element at `mid` and cut down our range accordingly.

To begin with, we will present an imperative version of this algorithm that uses a `while` loop and several `var` declarations.

---

[5] Thanks to rapidly changing technology and remarkable changes in computer speed and usability, it is not clear people are bothering to keep phone books anyway.

[6] As a reminder, this notation implies that the range is inclusive for `start` and exclusive for `end`.

```
def binarySearch(a:Array[Int],value:Int):Int = {
  var start=0
  var end=a.length
  var mid=(end+start)/2
  while(end>start && a(mid)!=value) {
    if(value < a(mid)) {
      end=mid
    } else {
      start=mid+1
    }
    mid=(end+start)/2
  }
  if(end<=start) -1 else mid
}
```

The `while` loop continues as long as the range includes at least one element and the midpoint is not the value we want. Inside the loop, a check is performed to see if the midpoint is less than or greater than what we are looking for. If it is less, we set `end=mid`. This works because `end` is exclusive and we have just verified that the element is not at `mid`. Otherwise we set `start=mid+1`. The `start` is inclusive so we have to move it one element beyond the `mid`. When the loop is completed we return either the value of `mid` or -1 based on whether the element was found or not.

This version of the code is fairly straightforward, but there is a simpler approach. Binary search happens to be an algorithm that lends itself very nicely to implementation as a recursive algorithm. The following code shows what this might look like.

```
def binarySearchRecur(a:Array[Int],value:Int,start:Int,end:Int):Int = {
  if(end <= start) -1 else {
    val mid=(start+end)/2
    if(a(mid) == value) mid
    else if(a(mid) < value) binarySearchRecur(a,value,start,mid)
    else binarySearchRecur(a,value,mid+1,end)
  }
}
```

Clearly this code is shorter than what we had before. Most people would also find this code a bit easier to read than the imperative version. The only drawback is that the function has two extra arguments. The normal way to get around that is to provide a wrapper function that only has two arguments and have it call this version. An appropriate wrapper could be written this way.

```
def binarySearch(a:Array[Int],value:Int):Int =
  binarySearchRecur(a,value,0,a.length)
```

The logic of the recursive version is identical to the iterative version. Only the approach has changed.

Now that we have code to do a binary search, it is interesting to ask what order this function is. Again we say that the array has n elements in it. The worst case is the situation when the element is not found and we get down to one element in the range that is not what we are looking for. So we need to figure out how many comparisons happen to narrow the range down from n to 1. After one comparison the range is cut to $n/2$. After two comparisons it is $n/4$. In general, after t comparisons, there are roughly $n/2^t$ elements left in the range.[7]

[7]This is only exact if n is a power of two. Otherwise some rounding will occur, but that is a detail we can ignore when talking about the order.

n	$\sim \log_2 n$
1,000	10
1,000,000	20
1,000,000,000	30
1,000,000,000,000	40

TABLE 13.1: Table of approximate values of $\log_2 n$ as a function of n. We use the approximation that $2^{10} \approx 10^3$. The reality is that $2^{10} = 1024$, but this approximation is rather close and is a good one to keep in your head for quick approximations.

We now have enough information to find the maximum number of comparisons.

$$n/2^t = 1$$
$$n = 2^t$$
$$t = \log_2 n$$

This is typically called $O(\log n)$ as the difference between logs of different bases is simply a constant multiplier. This order is generally considered to be quite fast, as it goes slowly as the input gets bigger.

To get a feel for this, let us look at a few examples of how $\log_2 n$ scales with n. A list of approximate values for this are given in table 13.1. To really put this in perspective, consider the fact that the first number is the worst case for a sequential search and the second number is the worst case for a binary search. When n is small, the difference is not all that significant. However, as n gets large the cost savings of doing a binary search become quite apparent. The last two values of n are large enough that they pose a problem even for computers, despite their speed.

Of course, you can only use a binary search on sorted data and attempting an $O(n^2)$ sort on even a million items can be time consuming. So the real power of this $O(\log n)$ scaling is purely academic until we discuss some better ways to sort.

Linear Binary Search (Advanced)

The binary search is ideal for general sorted data. However, if you happen to know that your data is fairly evenly distributed you can do even better. Instead of having `mid` be the midpoint of the range, you can place it where you would expect the `value` to be based on data being linearly distributed between the values at `start` and `end-1`. Assuming the `Array` contains `Doubles`, this could be done with a line like the following.

```
val mid = start+(((value-a(start)/(a(end-1)-a(start))*(end-1-start)).toInt
```

This picks the value of `mid` based on a linear approximation. Some care would have to be taken to insure that this does not fall into an infinite loop. In addition, if the data is not uniformly distributed, this approach can wind up begin much slower than the normal binary search. In fact, it can degrade to $O(n)$.

Searching for Doubles (Advanced)

Careful readers might have noticed that between the section on sorts and the section on searches, a small change was made. The sorting section used an `Array[Double]` while the searching section used `Array[Int]`. The choice of the `Double` type for sorting was motivated by the fact that they are easy to generate and visualize. However, they are not ideal for searching. This is due to the nature of the `Double` type. In our discussions we have only presented the `Double` type as the type we use for numbers when a fractional part is required. We have glossed over the details of what is happening with the `Double` type. To understand the reason that `Double`s are not good for search algorithms requires us to dig a bit deeper into what they are.

The term `Double` stands for DOUBLE PRECISION FLOATING POINT NUMBER. The `Float` type, which we have generally ignored, is a SINGLE PRECISION FLOATING POINT NUMBER. The `Float` type can also represent fractional numbers, but it has a smaller range and lower precision. As with all numbers on a computer, floating point numbers, be they single or double precision, are really stored in binary. The best way to think about a floating point number is to think of numbers in normalized scientific notation. In decimal, you can think of a number in scientific notation as being in the following form,

$$(-1)^s * m * 10^e,$$

where $s \in [0, 1]$, $1 \leq m < 10$ or $m = 0$. We call s the sign, m the mantissa, and e the exponent. Not much changes when the number goes to binary except that s, m, and e are stored in bits instead of decimal digits, and we want powers of 2 instead of 10. So scientific notation in binary would be

$$(-1)^s * m * 2^e.$$

This still does not explain why it is hard to do searching with a floating point number. The key to that comes from the fact that we only have a finite number of bits to use to store m. To understand the implication of this, consider a situation where you only have 7 decimal digits to write m. Now try to write the decimal form of the fraction $1/3$. You would write $(-1)^0 * 3.333333 * 10^{-1}$. This is not exactly the same as $1/3$, but it is as close as you can get with only 7 digits. The reality is that to write $1/3$ in decimal you need an infinite number of digits. Lots of fractions require an infinite repeating representation in decimal.

This same thing happens in binary on the computer. Not having infinite binary digits can lead to some interesting results for factional numbers that can not be perfectly represented. To understand this, consider the following simple example from the Scala REPL.

```
scala> 0.1==1.0-0.9
res0: Boolean = false
```

Mathematically you expect the expression `0.1==1.0-0.9` to be `true`, but the decimal value `0.1` is an infinite repeating sequence in binary. As such, it is truncated at some point and we get an approximation. Similarly, `0.9` can not be represented perfectly either. The result is that subtracting `0.9` from `1` gives a value that is not exactly the same as the approximation to `0.1`. To see how different the two are we can subtract one from the other.

```
scala> 0.1-(1.0-0.9)
```

```
res1: Double = 2.7755575615628914E-17
```

This is an extremely small number, but it is not zero and because it is not zero, the two are not equal.

This is a well-known challenge with floating point numbers and people who do numeric work have learned not to do checks for equality on them as the results are generally unpredictable because of the rounding that is part of even the simplest arithmetic. For our discussion, what we have seen is evidence that using == in a search algorithm on the Double type is likely to produce unexpected results. So the next question is, how do we get around that.

The basic idea behind the solution is that we generally consider floating point numbers to be equivalent as long as they are close enough to one another. So how close is close enough? That depends on whether you are working with single or double precision numbers. In math the Greek symbol ϵ, epsilon, is typically used to represent a vanishingly small value. In computer numerics it is used to describe the smallest value that you can add to one and still get a number greater than one. If you go smaller than ϵ, the value will be rounded off in the sum and all you will get back is one. Here is code that declares and calculates this for both the Double and Float types.

```
scala> val doubleEpsilon = {
     |    var eps=1.0
     |    while(1.0+eps > 1.0) eps*=0.5
     |    eps*2.0
     | }

doubleEpsilon: Double = 2.220446049250313E-16

scala> val floatEpsilon = {
     |    var eps=1.0f
     |    while(1.0f+eps > 1.0f) eps*=0.5f
     |    eps*2.0
     | }
floatEpsilon: Double = 1.1920928955078125E-7
```

Any single operation can be expected to have errors on the order of ϵ. When you string many operations together the error grows. As such, it is standard practice to consider two values equal if the relative error is less than the square root of ϵ. This means you only trust half the bits of precision in the number. As such, the following values are what you really find important.

```
scala> val sqrtDoubleEpsilon=math.sqrt(doubleEpsilon)
sqrtDoubleEpsilon: Double = 1.4901161193847656E-8

scala> val sqrtFloatEpsilon=math.sqrt(floatEpsilon)
sqrtFloatEpsilon: Double = 3.4526698300124393E-4
```

For simple order of magnitude purposes, you might remember that for Double this is about 10^{-8} and for a Float it is 10^{-4}. We will use this approximate value in the code that follows.

So how could we modify our searches to work with the Double type? We simply need to replace the check for equality with a check of the relative difference between the two. For the linear search that would look like the following.

```
def linearSearch(a:Array[Double],value:Double):Int = {
    var i=0
    while(i < a.length && (a(i)-value).abs > 1e-8*value.abs) {
        i+=1
    }
    if(i > a.length) -1 else i
}
```

The second half of the condition in the while loop is performing the critical check. It takes the absolute value of the difference between the value in the `Array` and the value we are looking for. It compares to see if that is bigger than our approximate value for the square root of ϵ times the value we are looking for. You can not simply compare to 1e-8 because if both `value` and `a(i)` have a magnitude much smaller than unity, such a comparison could be erroneous. For example, imagine if the values were positions of atoms in an object measured in meters. A separation of 10^{-8} meters apart would likely be quite significant.

13.3 Performance and Timing

This chapter has introduced a number of different sorting and searching algorithms. We have discussed the performance of these algorithms in terms of order. This tells us how they scale as the number of inputs is changed and quite often this is all you really care about. You might not know or care about details of the hardware or the data sets you will be running on, or you know that such details will change over time. There are times though when you really do care exactly how fast an algorithm is on a particular machine and you need to compare it to other similar algorithms on that same machine. When you do this, you need to do timing tests on the algorithm.

There are a number of different ways to get timing information on a program. Linux has a command called `time` that will let you know how much time a program consumes when it is running. For some applications this is ideal. However, for our purposes we only want to measure how long a particular part of the program takes. We do not want to measure the time taken to start things up or to initialize the `Array`. We only want to measure the time spent sorting the `Array`. You can get very detailed information like this from a PROFILER,[8] but that is overkill for what we want to do. For our purposes, we just need the ability to determine a start time and a stop time and take the difference between the two. We can do this by calling `System.nanoTime()`. This is a call to the Java libraries that returns a `Long` measuring the current time in nanoseconds.[9]

Another thing that we want to do to make the tests even is sort the same numbers for each sort. To make this happen, we really need to sort a copy of the `Array` and keep the original so that we can make other copies of it. We need to do this for each of the sorts so it is nice to put the code into a function that we can easily call with different sorts. To make it work with different sorts, we need to pass in the sort function as an argument. The following function does this for us.

[8]You can invoke the Java profiler with the Java `-Xprof` option. To get Scala to run this option you set the `JAVA_OPTS` environment variable to include `-Xprof`.

[9]A nanosecond is 10^{-9} seconds.

```
def timeFunc(sortFunc:(Array[Double])=>Unit,a:Array[Double]) {
  val copy=Array.tabulate(a.length)(i => a(i))
  val start=System.nanoTime()
  sortFunc(copy)
  val end=System.nanoTime()
  println("Time:"+(end-start)/1e9)
  assert(isSorted(copy))
}
```

The print statement divides the time difference by `1e9`. This gives us back a value in seconds instead of nanoseconds, which is much easier for us to read and deal with. We can invoke this by putting the following code at the end of our script.

```
args(1) match {
  case "bubble" => timeFunc(bubbleSort,nums)
  case "flagged" => timeFunc(flaggedBubbleSort,nums)
  case "min" => timeFunc(minSort,nums)
  case "insert" => timeFunc(insertionSort,nums)
  case "shell" => timeFunc(shellSort,nums)
}
```

Before you use this, you will need to make sure you take out the calls that do the rendering as the renders intentionally slow things down so you can see what they are doing.

If you play around with this some you will notice a few things. First, you have to get up to at least 10000 numbers in the `Array` before the timing means much. If the `Array` is too small, the sort will be so fast that the resolution of the machine's clock will become a problem. Second, the amount of time spent doing the sort can vary on different invocations. For this reason, any true attempt to measure performance will run the code multiple times and take an average of the different values.

13.4 Classifying Bugs

By this point you have certainly learned that simply writing your code does not mean that it is correct and that it will work. Errors in programs are commonly referred to as BUGS. The term bug is historical in origin because one of the first ones was a moth. The earliest computers were huge. They took up rooms the size of gymnasiums. An early malfunction in one turned out to be a moth that had flown in and was causing a short circuit.

While the term "bug" has stuck around, the implications of this term no longer fit. The term and its history imply that it is something beyond the control of the programmer that just put itself into the code and now the programmer has to search for it. In reality, virtually all modern bugs are really mistakes on the part of the programmer. They are things that the programmer put into the code and now the programmer needs to correct.

Not all bugs are the same. There are three fundamentally different types of errors.

1. SYNTAX ERROR – Error in the structure of the code that is found by the compiler.

2. RUNTIME ERROR – Error that causes the program to crash while running.

3. LOGIC ERROR – Error that does not crash the code, but causes it to produce the wrong answer.

Each of these deserves a fair bit of discussion. We will also talk about how they compare to one another and the way in which they impact programmers.

When you are first learning to program, the errors that you likely run into the most are syntax errors. These can be as simple as typos or misspellings. They can also be more complex like type mismatch errors or calling functions or methods with the wrong number of arguments. The common element that makes something a syntax error is that it is discovered by the compiler when it is trying to translate the human written code into a format that is more computer friendly. Different programming languages do different amounts of checking for errors at compile time. This is often called STATIC CHECKING because it can be done without actually running the program.

Scala does a significant amount of static checking for errors.[10] The way we have been running our programs in the REPL or with scripts, the compile stage is not clearly separated from the running of the program. The Scala system is running a compiler in the background and then executing the results of the compile. The syntax errors display a message like the following:

```
timing.scala:5: error: not found: value This
This line will not compile
^
one error found
```

They tell you the name of the file along with a line number. Then they describe the error and show the line and where on the line the error occurred.

The second type of bug is a runtime error. This type of error occurs when everything is syntactically correct and the program compiles and runs. However, during the run this error causes the program to crash. In Scala a runtime error will produce a message that looks similar to the following:

```
java.lang.ArrayIndexOutOfBoundsException: -1
        at Main$$anon$1.<init>(timing.scala:5)
        at Main$.main(timing.scala:1)
        at Main$.main(timing.scala)
    ...
```

This message tells you what went wrong and then prints a stack trace that includes line numbers. In this simple example, the code failed on line 5.

There are many different reasons why runtime errors happen, and it might be dependent on user input. So a runtime error might not be found from running the program once or twice. To reduce the number of runtime errors in a program you have to run it with multiple different input test inputs as was discussed in section 8.5. In a general sense, it is impossible to prove that code has no runtime errors.

The last type of error is a logic error. An error is a logic error if the code compiles and runs to normal termination, but provides an incorrect output or behaves incorrectly in some other manner. These errors can come in all types of forms and there is no message that tells you there is a problem. You know there is a problem by checking the output or behavior of the program to see if it matches expectations. Like runtime errors, logic errors might not occur for all inputs. There might be only specific inputs that trigger the error.

If you have a choice, you want errors of types higher on the list. As a novice programmer you probably get tired of dealing with syntax errors and find them frustrating. However, the reality is that syntax errors are by far the best type of error. This is because syntax errors give you the most information on how to fix them and are detected by the compiler

[10]The significant static checking of types was a major influence in the selection of Scala for this textbook.

in a way that does not depend on inputs. Your second choice would be a runtime error for some of the same reasons, it provides you with some information related to what is wrong and, as a result, helps you fix the problem. By contrast, logic errors provide no information on how to fix them.

Different languages do more or less to help you with error detection. One of the significant advantages of Scala is that it is designed to maximize the number of errors that are syntax errors and reduce the number of runtime and logic errors. Part of this is the type checking system of Scala. The language does significant static type checking to make sure that all the values you are using are of a type that is appropriate for the usage.

Many of the higher-order functions and methods in Scala also help to prevent common errors that programmers face in other languages. This is also true of rules such as the requirement to initialize variables at declaration and the general preference of `val` declarations over `var` declarations. To understand this, simply consider the following line of code.

```
val dbl=nums.filter(_>0.5).map(_*2)
```

This line of code will give us a new collection of numbers that are twice the magnitude of the elements but only for the elements that were originally bigger than 0.5. There are not too many ways to mess up this line without having it be a syntax error. It is possible to get a logic error if you mistype the greater than or the multiplication, but that is something no programming language can really fix. The programmer has to correctly communicate the key logic, but this line does not have all that much other than that key logic. Despite this brevity, it is fairly easy to read.

To understand the real value of this line of code, you have to consider the alternative, which is what you would have to write in most other programming languages. We will write equivalent code that is specific for an `Array`. The Scala code works equally well for any sequence, but we will ignore that advantage for now. It will become more significant later on.

```
var cnt=0
for(i <- 0 until nums.length) {
  if(nums(i)>0.5) cnt += 1
}
val dbl=new Array[Double](cnt)
cnt=0
for(i <- 0 until nums.length) {
  if(nums(i)>0.5) {
    dbl(cnt)=nums(i){*}2
    cnt += 1
  }
}
```

The first loop counts how many elements are greater than 0.5. This is required because `Arrays` have to be given a size when they are created. Once we know how many there will be, we can make the `Array`. The second loop fills in the `Array` with values that are twice those in the original `Array`.

Clearly this second version is longer. More importantly, that are a lot more places where typos become runtime or logic errors. The reality is that the one line version is doing basically this same thing. However, most of the code is in the libraries and is not rewritten by each programmer every time. This works, in large part, because of the first-class functions and the ease with which function literals can be written in Scala.

13.5 Memory Layout

The memory of the computer is basically like a huge array of bytes. It is shared between the operating system and many different programs. Different parts of memory can be allocated for different things or associated with different devices. Scala hides most of the intricacies of memory from you. It is not a language designed for doing low-level system programming. At some point in your Computer Science training you should learn about the details of computer memory. For our purposes here, we will only care about the organization of memory inside of the allocation of a single program.

The memory for a program is broken into two broad pieces, the STACK and the HEAP. These terms were chosen intentionally and the images they invoke in your mind are probably fairly accurate. A stack is orderly with one item placed on top of the previous one. A heap is much less organized with items placed almost at random. Local variables and function arguments are allocated on the stack. As was discussed in section 7.8, the memory model in Scala is such that the variables are references and they refer to objects. In this memory model, the objects are allocated on the heap. Every time a new function is called, the memory for the arguments and the local variables, along with some memory for bookkeeping, is allocated in a block that is referred to as the STACK FRAME. If that function calls another, then another frame is allocated on top of it. When a function returns, the stack frame for that function is freed up. That same memory will be used later for another function.

This should help explain the output from a runtime error. The stack implicitly keeps track of where you are in each function when it calls the next. You can picture each of those functions stacked on top of the one that called it. That is what gets printed in the stack trace. Each line tells you what function has been called followed by the file name and line number.

The objects on the heap are allocated in free spaces. The memory for objects is freed up automatically when the object is no longer in use. The automatic freeing of heap memory is accomplished by a process called GARBAGE COLLECTION. An object can be collected if it can no longer be reached by following references that start on the stack. Not all languages include garbage collectors. In those that do not, the programmer is responsible for freeing memory that was allocated on the heap.

13.6 Sorting/Searching with `case classes`

In the sorts and searches that we have looked at, we were working with numeric types. More generally, the code that was written will work with any type that works with the comparison operators. As such, our code would have worked with `Array[Char]` or `Array[String]` if you simply altered the type that was passed into the sort. The `case classes` we have written do not meet this requirement. As such, we need to make other alterations to the code beyond the type if we want to sort a `case class`. Fortunately, these alterations are not all that significant.

We will work with the following case class.

```
case class Weather(id:String,year:Int,month:Int,precip:Double,tmax:Double,
  tmean:Double,tmin:Double)
```

This case class was created to store historical weather data. It was used to represent records for monthly data on temperature and precipitation. You would load an entire file of these records into an Array. If you want to see the hottest ten months you could sort the array by the high temperatures. Code for such a sort is shown here.

```
def bubbleSortWeatherHighTemps(a:Array[Weather]) {
  for(j <- 0 until a.length-1) {
    for(i <- 0 until a.length-1-j) {
      if(a(i).tmax > a(i+1).tmax) {
        val tmp=a(i)
        a(i)=a(i+1)
        a(i+1)=tmp
      }
    }
  }
}
```

A bubble sort was picked because of the simplicity of the sort. It is very easy to see what was changed. The only changes are the type in the parameter for the Array that is passed in and the fact that the comparison is done between fields of the Array elements. After applying this sort to the Array you can use take or takeRight to pull off the elements at one end or the other.

case classes also present another alternative that we did not really get with single values, the ability to have the comparison based on more than one field. For the weather data, it is likely to be stored in the data file in chronological order. Proper chronological order is a combination of both the year and month fields. If you wanted to have the ability to search for entries by time you might want to have a binary search that can look for a particular year and month. Code for doing that is listed here.

```
def binarySearchWeather(a:Array[Weather],year:Int,month:Int):Int = {
  var start=0
  var end=a.length
  var mid=(end+start)/2
  while(end>start && (a(mid).year!=year || a(mid).month!=month)) {
    if(year < a(mid).year || (year==a(mid).year && month < a(mid).month)) {
      end=mid
    } else {
      start=mid+1
    }
    mid=(end+start)/2
  }
  if(end <= start) -1 else mid
}
```

Note that the comparison has become significantly more complex. It has to compare the year first and then if there is a tie, break that tie with the month.

Unfortunately, when written in this way, we have to write a completely separate sort or search for each ordering we might want on the case class. For example, if you wanted wettest months instead of hottest months, you would need a separate sort. You might feel that copying a whole sort or search function only to make such small changes is not very efficient and that there should be a way to make a sort or search that works more generally. We can help improve these functions for case classes here. Later, in chapter 19, we will gain the ability to abstract these ideas so that the function works with multiple types, not just different sort orders on a single type.

Back in section 6.4 we saw how we could make some recursive functions more powerful by passing functions into them. This approach is exactly what is taken by the higher-order functions in the Scala collections libraries. As such, we have not had to go to it ourselves much since then. However, the desire to not write a completely new sort or search function for each and every possible ordering on a case class provides motivation to pull out these ideas again.

If you were to write a new version of bubble sort that sorts the `Weather` objects by precipitation, you would find that the only thing you change is the code related to the comparison of the elements. In order to get a sort that can sort by high temperature, precipitation, or anything else related to the `Weather` type, all we have to do is make it so that we can vary the comparison from the outside. This can be accomplished by passing in a function that does the comparison. For our sorts, we have been using less than and greater than for comparison so we just need to pass in a function that represents one of these. We pick less than here, though the code could easily be rewritten with greater than. The code for the sort after this change looks like the following.

```
def bubbleSortWeather(a:Array[Weather],lessThan:(Weather,Weather)=>Boolean) {
  for(j <- 0 until a.length-1) {
    for(i <- 0 until a.length-1-j) {
      if(lessThan(a(i+1),a(i))) {
        val tmp=a(i)
        a(i)=a(i+1)
        a(i+1)=tmp
      }
    }
  }
}
```

The comparison operator is represented as a function that takes two `Weather` objects and returns a `Boolean`. A call to this function is used in the `if` statement.

Using this modified version, we could sort by the high temperatures with a call like this.

```
bubbleSortWeather(weather,(w1,w2)=>{w1.tmax < w2.tmax})
```

Alternately, the same method could be used to sort by precipitation with a call like this.

```
bubbleSortWeather(weather,(w1,w2)=>{w1.precip < w2.precip})
```

What is more, using this version of the sort, it is easy to change it so that it sorts from greatest to least instead of the standard least to greatest. In the case of precipitation this is done by changing the call in the following way.

```
bubbleSortWeather(weather,(w1,w2)=>{w1.precip > w2.precip})
```

All that is changed in the direction of the comparison operator. As you can see, this abstracted version that uses a function instead of a hard-coded comparison operator is far more flexible.

Having seen the benefits we can get from using this in our sort, it would be nice to enable searching in the same way. If you start to write the search code with the same comparison operator you will find that there is a significant problem, the search requires more than just a less than or greater than. The search requires that we be able to tell if two things are equal. The standard way to deal with this is to have a function that returns an `Int` instead of a `Boolean`. The `Int` is negative, zero, or positive to represent less than, equal to, or greater than, respectively. The other change that makes search a bit different is that the function only takes one Weather object because we just want to know where the thing we are looking for is relative to this value we are searching for.

Translating all of this into code gives the following.

```
def binarySearchWeather(a:Array[Weather],comp:(Weather)=>Int):Int = {
  var start=0
  var end=a.length
  var mid=(end+start)/2
  var c=0
  while(end>start && {c=comp(a(mid)); c!=0}) {
    if(c<0) {
      end=mid
    } else {
      start=mid+1
    }
    mid=(end+start)/2
  }
  if(end <= start) -1 else mid
}
```

The function is called comp and it takes a Weather and returns an Int. In order to prevent having the code call comp more times than it needs to, we introduce a variable named c that stores the result of the most recent comparison. We assign this in a code block in the condition. The loop checks to make sure it is not zero. The if checks if it is negative. After we calculate a new mid we also make a new comparison.

Now the question is, how could we use this search to find different elements in the Array. The first example we will give is one that duplicates the search we had above searching for a specific year and month.

```
binarySearchWeather(data,w => {
  if(w.year>1896) -1
  else if(w.year<1896) 1
  else if(w.month>2) -1
  else if(w.month<2) 1
  else 0
})
```

In this case, we are searching for February of 1896. This version uses a sequence of if expressions to determine if the element w comes before or after that time. With a little math we can make a shorter version taking advantage of the fact that there are 12 months in each year.

```
binarySearchWeather(data,w => (1896*12+2)-(w.year*12+w.month))
```

Searching for data based on date is made more complex because it depends on two values. If we use our flexible sort to reorder the Array by precipitation we could use the following to search for a month in which there were 1.43 inches of rain.

```
binarySearchWeather(data,w => {
  val diff = 1.43-w.precip
  if(diff > 1e-8) 1
  else if(diff < -1e-8) -1
  else 0
})
```

The logic in this can be challenging to write. Thankfully, the behavior of returning negative, zero, or positive is a common standard and the Scala library contains code that does this with the built-in numeric types. As such, the code can be expressed more simply in this way.

```
binarySearchWeather(data, 1.43 compare _.precip)
```

The `compare` call is actually a method we can call on a `Double`. The fact that a period has other meaning for numbers led to us using operator notation for the method instead of the more standard notation with dot and parentheses.

Bucket/Radix Sort (Advanced)

The sorts discussed earlier in this chapter require nothing more than the ability to compare elements. There are situations where you have more information about the values and you can use that to sort them more efficiently. Examples of such non-general sorts include BUCKET SORT and RADIX SORT. Both use knowledge about the data being sorted to break them into groups.

To understand how this works, consider the situation where you are given a large number of folders to sort by name and you are in a big room. Many people would start off by going through the stack and breaking it into different piles. The piles would be for different parts of the alphabet. If the original stack were really large and you had a really big room, you might make one stack for each starting letter. You could continue to break things down in this way, again for each pile or switch to some other approach when the pile is smaller. This general approach is called a bucket sort and the piles would be called buckets. The bucket sort is really a general approach to breaking sorting problems into smaller, more manageable pieces.

The radix sort uses a similar approach, but it is a specific algorithm for sorting integer values. It organizes values, one digit at a time. In a counterintuitive way, it starts from the least significant digit. This only works because each pass through preserves the order of the elements. So if A has a lower digit than B in one pass, and the next pass puts them in the same bin, A will come before B in that bin.

By convention we will implement our radix sort using decimal digits using division and modulo with powers of ten. The following code is a reasonably efficient implementation that sorts an `Array[Int]` using an `Array[List[Int]]` for the bins. Note that moving items to the bins is not an in-place operation. The values are copied back to the `Array` after binning, but it does require at least twice the memory of an in-place sort.

```
def radixSort(a:Array[Int]) {
  var max = a.max max a.min.abs
  var powerOf10 = 1
  while(max>0) {
    val byDigit = Array.fill(19)(List[Int]())
    for(num <- a) {
      val digit = num/powerOf10%10+9
      byDigit(digit) ::= num
    }
    var i = 0
    for(bin <- byDigit; num <- bin.reverse) {
```

```
        a(i) = num
        i += 1
      }
      powerOf10 *= 10
      max /= 10
    }
  }
```

The `max` variable starts with the largest magnitude value and is divided by ten each time so that we know when to stop. The `powerOf10` variable keeps track of what digit we are currently binning. Each time through the loop, 19 empty bins are set up. You might wonder why there are 19 bins instead of just 10. The answer is that division and modulo preserve sign. If you only use 10 bins, this sort can only work on positive values. By going up to 19, it is able to handle negative numbers as well. To make that work, the bin value we get from division and modulo is incremented by 9 so that the values slide up to the domain of the `Array` index values.

The binning runs through all the numbers in the array and conses them onto the `List` for the proper digit. Consing to a `List` adds to the front. For that reason, the `for` loop that moves the valus back to the `Array` has to run through each bin in reverse.

This sorting algorithm is $O(kn)$ where $k = \lceil \log_{10}(max) \rceil$. In the case of an `Int`, the value of k can not be larger than 10. So the performance scales linearly for really large `Arrays`.

13.7 Putting it Together

To illustrate concepts in this chapter and link them together with earlier chapters we will return to the theme park example. Every month the theme park picks a top employee. This is based on performance relative to the average for that month. For every day of the month you have data that tells you what operators were working each ride, and how many people went on the ride. From this, we can calculate average ridership for each ride during the month as well as how many riders rode each ride for each day that a given operator was working it. Each operator can be given an efficiency for any particular ride as the average number of people who rode on days he/she was working divided by the average for all days. Averaging these efficiencies gives an overall rating to each operator. Those can be sorted and displayed to show relative performance. Code for doing that is shown here.

```
import scala.io.Source

case class DailyData(ride:String, operators:Array[String], numRiders:Int)
case class RideAverage(ride:String, avNum:Double)
case class OperatorDailyData(name:String, ride:String, numRiders:Int)
case class OperatorRideAverages(name:String, rideAvs:Array[RideAverage])
case class OperatorEfficiencyFactor(name:String,factor:Double)

def parseDailyData(line:String):DailyData = {
  val parts = line.split(" *; *")
  DailyData(parts(0), parts.slice(1, parts.length-1), parts.last.toInt)
```

```
}

def readData(fileName:String):Array[DailyData] = {
  val source = Source.fromFile(fileName)
  val lines = source.getLines
  val ret = (lines.map(parseDailyData)).toArray
  source.close
  ret
}

def insertionSortByEfficiency(a:Array[OperatorEfficiencyFactor]) {
  for(j <- 1 until a.length) {
    var i=j-1
    val tmp=a(j)
    while(i>=0 && a(i).factor>tmp.factor) {
      a(i+1) = a(i)
      i -= 1
    }
    a(i+1) = tmp
  }
}

val data = readData(args(0))
val rides = data.map(_.ride).distinct
val averages = for(ride <- rides) yield {
  val days = data.filter(_.ride==ride)
  RideAverage(ride, days.map(_.numRiders).sum.toDouble/days.length)
}
val dataByOperator = for(day <- data; op <- day.operators) yield {
  OperatorDailyData(op, day.ride, day.numRiders)
}
val operators = dataByOperator.map(_.name).distinct
val opRideAverages = for(op <- operators) yield {
  val opDays = dataByOperator.filter(_.name == op)
  val rideAvs = for(ride <- rides; if opDays.exists(_.ride==ride)) yield {
    val opRides = opDays.filter(_.ride == ride)
    RideAverage(ride, opRides.map(_.numRiders).sum.toDouble/opRides.length)
  }
  OperatorRideAverages(op, rideAvs)
}
val operatorFactors = for(OperatorRideAverages(op, rideAvs) <- opRideAverages)
    yield {
  val factors = for(RideAverage(ride,av) <- rideAvs) yield {
    av/averages.filter(_.ride==ride).head.avNum
  }
  OperatorEfficiencyFactor(op,factors.sum/factors.length)
}
insertionSortByEfficiency(operatorFactors)
operatorFactors.foreach(println)
```

This code assumes that the file has a line of data for each day that starts with the ride name, followed by operator names, with number of riders at the end. Each of these is separated by semicolons. That data is read and used to calculate the various values needed for efficiency. Once the efficiencies have been calculated, all the employees are sorted with an insertion sort and the results are printed.

13.8 End of Chapter Material

13.8.1 Summary of Concepts

- The act of ordering data according to some value is called sorting. It is a common operation on a computer as it benefits humans who view the data as well as programs when they process it. Several types of sorts were discussed in this chapter.
  - A bubble sort runs through the `Array` comparing adjacent elements and swapping them if they are out of order. This action is repeated until all the values are in proper order.
  - A selection sort picks specific elements and puts them into place. We demonstrated a `minSort`, which finds the smallest unsorted element and swaps it to the correct location. Selection sort does few swaps so it is most useful in a situation where moving data is an expensive operation.
  - An insertion sort takes each element and pushes it forward through the `Array` until it gets it into sorted order with the elements that came before it. Insertion sort is extremely efficient in situations where the data starts off close to the proper order.
  - The Shell sort is also called the diminishing gap sort. It does insertion sorts over partial data in such a way that things are moved toward proper ordering. It winds up being more efficient that any of the other three.
- When we talk about the performance of different algorithms in Computer Science, we typically use order analysis. This gives a rough idea of how the number of times an operation is performed will scale with the size of the input. The first three sorts above all do $O(n^2)$ comparisons.
- One of the most common tasks done on computers is looking for data, an activity we call searching.
  - If data is unorganized, that only approach is to go through all elements in a sequential/linear search. This type of search is $O(n)$.
  - When the data has a sorted order to it, you can use a binary search to find elements. A binary search is significantly faster that a sequential search because it effectively throws out half the data with each comparison. This provides $O(\log(n))$ performance.
- To compare the performance of different algorithms when you really care about how quickly a program runs, you can measure how much time it takes. To measure the runtime of a whole program you can use the Linux `time` command. To measure only specific parts of code call `System.nanoTime()` before and after the section you want to time and subtract the results. Profilers can also provide more detailed information on what is taking time in a program.
- Programmers often put errors into their code by accident. These errors are commonly called bugs. The process of fixing the errors is called debugging. Bugs can be categorized into different types.
  - Sytnax errors are errors where the programmer enters something that is not valid for the language. These are found when the program is compiled. The compiler

can generally give you helpful information about what is wrong and where the error is in the code.

  - Runtime errors occur when the code compiles fine, but crashes during the run. The crash itself can print helpful information for you. Unfortunately, runtime errors often only occur under certain situations, mkaing them harder to track down and fix.
  - Logic errors are the term we use to describe when the code compiles and runs without error, but the result is inaccurate. These are generally the worst of the three as the computer does not give you pointers to where the error is occurring or what it is caused by. Instead, the programmer has to track it down. This can be done using print statements at the level we are working at or using a debugger.

- The memory of the computer can be thought of as a really big array. Each program is given a different section of memory that is divided into the heap and the stack. The stack is well organized and is where local variables are allocated. Each function that is called gets a chunk of memory called a stack frame. When the function returns, the frame is released. The heap is where all objects are allocated in Scala. A garbage collector deals with objects that are no longer in use.
- Sort `case classes` requires minor alterations to the sort so that the appropriate fields are compared.

13.8.2 Exercises

1. Write a `minMaxSort` that finds both the minimum and maximum values in each pass and swaps them to the proper locations.
2. Do timing tests on `Arrays` of `Ints` and `Doubles` for the sorts presented in this chapter. Note that the radix sort that is presented only works with integer types. You want to have the number of elements in the `Array` or `List` grow exponentially so that you can see variation over a large range. Recommended sizes could be 100, 300, 1000, 3000, 10000, 30000, etc. Plot the data as a log-log scatter plot.[11]
3. Following onto the timing results, do a comparison of the number of comparisons done for different sorts presented in this chapter. Plot the results in the same way.
4. While most of the work for these sorts is in comparisons, it is also interesting to look at the number of memory moves. A standard swap is three assignments. Add code to count how many assignments are done in each of the sorts and plot the results.
5. Section 7.5.2 showed how you can abstract functions over types. This is something that will be dealt with in detail in the second half of the book. Review that section and see if you can write a version of insertion sort and min sort that is general with regards to type.
6. Do a little web searching to find some other type of sort not described in this chapter and write it.

[11]This means that the N value should be the x-axis and the time it takes should be the y-axis where both axes use a log scale. The advantage of this is that functions of the form $f(x) = x^n$ appear as straight lines in this type of plot with a slope equal to n

13.8.3 Projects

1. This project is intended for people who have been working on graphics and ray tracing, but it does not immediately link so you can do it even if you haven't been doing the previous ones. For this problem you will draw polygons to a `Graphics2D` using a "painter's algorithm." That is where you draw things from the back to the front so that the things in front appear on top of the things behind them. Doing this properly is a challenging problem. You should base your drawing on the point in the polygon that is closest to the viewer.

   To keep things simple, the viewer will be at the origin facing out the z-axis. That way the x and y coordinates are roughly what you expect. To make it so that things that are further away are smaller you divide the actual x and y by the z value to get the location you would draw them at. Given a point `(x,y,z)` you would want it drawn on an image or panel at `((x/z+1)*size.width/2,(1-y/z)*size.height/2)`. To represent a polygon for drawing to the `Graphics2D` you should use a `java.awt.geom.Path2D.Double`. You can `moveTo` the first point, then use `lineTo` to make lines. When you get to the last point you call `closePath` to close it off.

   Store your polygons in a file. You can decide the exact format. In addition to having the points in each polygon, each one should have a color that it will be drawn in. Remember to sort them by the z value so that the one with the smallest z value is drawn last.

2. The BASIC programming language was created to be a simple language for novice programmers. The original versions were organized by line number. Each statement of the program was a single line that has a number associated with it. The lines were ordered according to that number and flow control was implemented by allowing the program to jump to other line numbers. For this project you will create a simple GUI that lets you edit a simplified version of BASIC.

   For this simplified version you have a very limited set of possible commands. The GUI will also use a "line editor" style. The allowed commands including the following: `GOTO`, `IF-THEN`, `INPUT`, `LET`, and `PRINT`. Each of these must be preceded by a line number that is an integer. For our purposes, variable names are single characters and they will all be numbers. You can use a `Double`. The format of the commands is as follows.

   - The `GOTO` command should be followed by an integer number that is the line number the program should execute next. (Example: `100 GOTO 50`)
   - The `IF-THEN` command has the following syntax: `IF` *comp* `THEN` #. The *comp* is a comparison that can have a variable or a number on either side and either `=` or `<` between them. The # is a line number that the execution will jump to if the comparison in the condition is true. If the comparison is false, the execution continues on the next line. (Example: `110 IF a<21 GOTO 50`)
   - The `INPUT` command should be followed by a single variable name and when it is executed the program pauses and waits for the user to input a value that is stored in that variable. (Example: `120 INPUT b`)
   - The `LET` command has the keyword `LET` followed by a variable name with an equal sign and then either a single number/variable or two number/variable operands that separated by an operator. The operator can be `+`, `-`, `*`, or `/`. The result of the operation should be stored in the variable before the equal sign. (Example: `130 LET a=b+3`)

- The PRINT command can be followed either by a variable name or a string in double quotes. When executed, this will print either the value of the variable or the string to the output.

Variables do not have to be declared. They come into existence when first used and they should have a value of 0 to start with if no other value was given to them.

The GUI for this program should have three main elements. The program itself is displayed in a `ListView`. This makes it simple for users the select a line to edit without letting them type random text. There should also be a `TextField` where the user can enter/edit lines. If a line is selected in the `ListView`, the text from it should appear in the `TextField`. The user can edit that line or enter anything else. The number at the beginning of the line will be user to put it in place. If a number is used that duplicates an existing line, the new one replaces the old one. Lastly there is a `TextArea` that shows the output when the program is run.

When the user hits enter on the `TextField` your program should check if what was entered is a valid command. If it is, it should put it in the program and clear. If it is not, it should leave the text there and not alter the existing program.

There should be at least four menu items for this program: "Save", "Open", "Remove", and "Run". The "Save" option saves the program to a text file. The "Open" option allows the user to select a file and open it up as a program. The "Remove" option will remove the currently selected lines in the program. Nothing happens if nothing is selected in the `ListView`. The "Run" option runs the program. If starts running at the first line and continues until execution goes beyond the last line.

3. On the book, website, under this chapter, you will find files with historical weather data for some different cities in the United States along with a link to the source of the data. You should read this data into an `Array` of some case class that you create. The data is separated by commas. Note that the second line tells you what each column of data represents. You will skip the top two lines when reading the information. Write a script that will report the months with the five warmest average temperatures and those with the five lowest average temperatures.

   For a bit of an extra challenge make it so that the user can tell the program whether to report the top and bottom five months for any of the values in the file. You could do this with a lot of typing, but by passing in a function pointer you can cut down on the length of the code greatly.

   If you did any of the problems in the last chapter that included plotting points or data, you should consider sticking that functionality onto this project. That way you can bring up a GUI and show the plots of whatever field(s) the user selects.

4. If you did a game for one of the projects in chapter 12, for this one you can enhance it by adding a high scores list. That means having some way to score games. It also means saving scores to a file in a format of your choosing. Lastly, the high scores need to be sorted. Each score record needs to have at least a score and an identifying name or initials. All other details are up to the student.

5. If you have been working with the recipe projects, you now have the ability to order a shopping list according to where things are in the store. This follows most logically from project 11 (p.312), but it does not require the graphical functionality, only a data file listing what a isles different items are in and the ability for a user to make a shopping list.

Your script should have the ability to sort the grocery list they build to go through the store in either ascending or descending order by row. The sorted list should be displayable in a `TextArea` so that the user can cut and paste it for printing.

6. This project fits in with the sequence of schedule building projects. In particular, it makes sense to do if you have already done project 10 (p.282) where schedules were built and displayed in a GUI. That project showed the selected courses in the order the user selected them. However, it would often be helpful to have them displayed based on other criteria. For example, having them sorted by department and number or how much interest the user expressed in them could be helpful.

   For a bit of extra challenge, you could include some additional information with each course indicating things like whether it is required for graduation or what graduation requirement it fulfills. Allow the user to select from multiple orderings using a `ComboBox`.

7. If you did project 15 (p.313) or 12 (p.282) looking at box scores you might have noticed that one significant feature that was missing was the ability to change the player listing so that it orders the players based on a particular stat. Now that you know how to sort you can fix this. How you do this depends on how you chose to display the box score, but it can be as simple as adding a few `Buttons` for sorting by different statistical categories.

Additional exercises and projects, along with data files, are available on the book's website.

Chapter 14

XML

This chapter deals with XML, short for eXtensible Markup Language. XML is technically a completely language-independent topic. Scala was developed with XML in mind and makes it easy to work with.

In chapter 9, we learned how to read "from" and "write to" text files. The files that we used in that and the following chapters are what are called "flat" text files. They have the data in them with nothing that tells us about the nature of the data other than formatting. The advantages of this are that it is fairly simple to read and write and it can be both read and written with standard tools like a text editor. The disadvantages are that it can be slow and it lacks any inherent meaning so it is hard to move the information from one program to another. It is also somewhat error prone because so much information is in the formatting. XML addresses the latter disadvantage, without losing the advantages.

The eXtensible Markup Language is a standard for formatting text files to encode any type of information. It is a markup language, not a programming language. The standard simply defines a format for encoding information in a structured way. It is likely that you have heard of a different markup language called HTML, the HyperText Markup Language. HTML is used to encode web pages and has a format very similar to XML. Indeed, a newer standard called XHTML is basically HTML that conforms to the XML rules.

14.1 Description of XML

Everything in an XML file can be classified as either markup or content. Markup in XML is found between '<' and '>' or between '&' and ';'. The content is anything that is not markup. To help you understand XML we will look at an example XML file. This example is built on the idea of calculating grades for a course.

```
<course name="CSCI 1320">
   <student fname="Jason" lname="Hughes">
      <quiz grade="98"/>
      <quiz grade="100"/>
      <quiz grade="90"/>
      <test grade="94"/>
      <assignment grade="100">
         <!-- Feedback -->
         Code compiled and runs fine.
      </assignment>
   </student>
   <student fname="Kevin" lname="Peese">
      <quiz grade="85"/>
      <quiz grade="78"/>
      <test grade="67"/>
      <assignment grade="20">
         Code didn't compile.
      </assignment>
   </student>
</course>
```

Just reading this should tell you what information it contains. Now we want to go through the different pieces of XML to see how this was built.

14.1.1 Tags

Text between '<' and '>' characters are called TAGs. Nearly everything in our sample XML file is inside of a tag. The first word in the tag is the name of the tag and it is the only thing that is required in the tag. There are three types of tags that can appear in an XML file.

- Start tag - Begins with '<' and ends with '>',
- End tag - Begins with '</' and ends with '>',
- Empty-element tag - Begins with '<' and ends with '/>'.

Tags can also include attributes, which come after the name and before the close of the tag. These are discussed below.

14.1.2 Elements

The tags are used to define ELEMENTs which give structure to the XML document. An element is either a start tag and an end tag with everything in between or else it is an empty element tag. The empty element tag is simply a shorter version of a start tag followed by an end tag with nothing in between.

Elements can be nested inside of one another. In our sample file, there is a **course** element that encloses everything else. There are two **student** elements inside of that. Each of those includes elements for the different grades. Most of these are empty elements, but the **assignments** are not empty and have contents in them.

Any time there is a start tag, there should be a matching end tag. When elements are nested, they have to be nested properly. That is to say that if element-2 begins inside of element-1, then element-2 must also end before the end of element-1. In addition to other elements, you can place general text inside of elements.

14.1.3 Attributes

Additional information can be attached to both start tags and empty element tags in the form of ATTRIBUTEs. An attribute is a name value pair where the value is in quotes and the two are separated by an equal sign. The example XML is loaded with attributes. In fact, every start or empty element tag in the example has an attribute. It is not required that these tags have an attribute, but it is a good way to associate simple data with a tag. Some of the tags in the example show that you can also have multiple attributes associated with them.

14.1.4 Content

Between a start tag and an end tag you can not only put other tags, you can put plain text. This text is the contents of the element. In the example XML, the **assignment** element has text in it that serves as a comment on the grade. The text that you put inside of an element is unformatted and can include anything you want. Unlike the markup part of the XML document, there is no special formatting on content.

14.1.5 Special Characters

While the content does not have any special formatting, it is still embedded in an XML document. There are certain characters that are special in XML that you can not include directly in the content. For example, if you put a '<' in the content it will be interpreted as the beginning of a tag. For this reason, there is a special syntax that you can use to include symbols in the content of an XML file. This syntax uses the other form of markup that begins with & and ens with ;. There are five standard values defined for XML.

- & = &
- ' = '
- > = >
- < = <
- " = "

There are many more defined for other specific markup languages. If you have ever looked at HTML you have probably seen   used to represent spaces.

14.1.6 Comments

Just like with code, it is helpful to occasionally put comments in your XML. A comment begins with '<!--' and ends with '-->'. You can put whatever text you want between these as long as it does not include the sequence to end the comment.

14.1.7 Overall Format

There are two other rules that are significant for XML. First, the entire XML document must be inside of a single element. In the example above everything was in the course element. Had we wanted to have more than one course element, we would have had to create some higher-level element to hold them.

In addition, most XML files will begin with an XML declaration that comes right before the element containing the information. The declaration has a slightly different syntax and might look like the following line.

```
<?xml version="1.0" encoding="UTF-8" ?>
```

14.1.8 Comparison to Flat File

To better understand the benefit of XML, we will compare our XML file to a flat file that might be used to store basically the same information. Here is a flat file representation of the grade information from the XML above.

```
CSCI 1320
2
Jason Hughes
98 100 90
90
100
Kevin Peese
85 78
67
20
```

This is a lot shorter, but unless you happen to know what it is encoding, you can not figure much out about it. Namely, the numbers in the file are hard to distinguish. The 2 near the top you might be able to figure out, but without additional information, it is impossible to determine which lines of grades are assignments, quizzes, or tests.

The flat file also lacks some of the information that was in the XML. In particular, the comments on the assignments in the XML format are missing in this file. It would be possible to make the flat file contain such information, but doing so would cause the code required to parse the flat file to be much more complex.

14.1.8.1 Flexibility in XML

A significant "Catch 22" of XML is that there are lots of different ways to express the same information. For example, the comment could have been given as an attribute with the name comment instead of as contents of the element. Similarly, if you do not want to allow comments, you could shorten the XML to be more like the flat file by changing it to the following format.

```
<course name="CSCI 1320">
   <student fname="Jason" lname="Hughes">
      <quizzes>98 100 90</quizzes>
      <tests>94</tests>
      <assignments>100</assignments>
   </student>
   <student fname="Kevin" lname="Peese">
```

```
        <quizzes>85 78</quizzes>
        <tests>67</tests>
        <assignments>20</assignments>
    </student>
</course>
```

Here all the grades of the same type have been given as contents of elements with the proper names. This makes things much shorter and does not significantly increase the difficulty of parsing the grades. It does remove the flexibility of attaching additional information with each grade such as the comments.

There is not a rule that tells you if information should be stored in separate tags, as attributes in tags, or as text inside of elements. For specific applications there will be certain standards. XHTML is a perfect example of that. For your own data, it will be up to you as a developer or your team to determine how you want to encode the information. Whatever is decided, you need to hold to it consistently.

14.2 XML in Scala

XML is not part of the normal first semester topics in Computer Science, but Scala makes it so easy that there is no reason not too. To see this, simply go to the REPL and type in valid XML.

```
scala> <tag>Some XML.</tag>
res0: scala.xml.Elem = <tag>Some XML.</tag>
```

The Scala language has a built in XML parser that allows you to write XML directly into your code. You can see that the type of this expression is `scala.xml.Elem`.

The `scala.xml` package contains the types related XML. We will run through some of the more significant here.

- `Elem` - Represents a single element. This is a subtype of `Node`.
- `Node` - Represents a more general node in the XML document. This is a subtype of `NodeSeq`.
- `NodeSeq` - Represents a sequence of `Nodes`.
- `XML` - A helper object that has methods for reading and writing XML files.

There are quite a few other types that you can see in the API, but we will focus on these as they give us the functionality that we need for our purposes.

14.2.1 Loading XML

The XML object has methods we can use to either read from files or write to files. The `loadFile` method can be used to read in a file. If the first example XML that was shown is put in a file with the name 'grades.xml', then the following call would load it in.

```
scala> xml.XML.loadFile("grades.xml")
res4: scala.xml.Elem =
<course name="CSCI 1320">
```

```
        <student lname="Hughes" fname="Jason">
                <quiz grade="98"></quiz>
                <quiz grade="100"></quiz>
                <quiz grade="90"></quiz>
                <test grade="94"></test>
                <assignment grade="100">

                        Code compiled and runs fine.
                </assignment>
        </student>
        <student lname="Peese" fname="Kevin">
                <quiz grade="85"></quiz>
                <quiz grade="78"></quiz>
                <test grade="67"></test>
                <assignment grade="20">
                        Code did not compile.
                </assignment>
        </student>
</course>
```

Clearly this is the contents of the XML file that we had created. All that is missing is the comment, which was there for human purposes, not for other programs to worry about. In addition, the empty tags have also been converted to start and end tags with nothing in between. This illustrates that the empty tags were also just for human convenience and their meaning is the same as an empty pair of start and end tags.

14.2.2 Parsing XML

Once we have this `Elem` object stored in `res4`, the question becomes how you get the information out of it. The `NodeSeq` type, and hence the `Node` and `Elem` types which are subtypes of it, declare operators called \ and \\. These operators are used to search inside the contents of an object. Both operators take a second argument of a `String` that gives the name of what you want to look for. The difference is how far they search. The \ operator looks only for things at the top level, either `Nodes` in the current sequence if we have a true `NodeSeq`, or children of this node if we have a `Node` or `Elem`. The \\, on the other hand, finds anything that matches at any depth below the current level. To illustrate this, we will do three example searches.

```
scala> res4 \ "student"
res5: scala.xml.NodeSeq =
NodeSeq(<student lname="Hughes" fname="Jason">
                <quiz grade="98"></quiz>
                <quiz grade="100"></quiz>
                <quiz grade="90"></quiz>
                <test grade="94"></test>
                <assignment grade="100">

                        Code compiled and runs fine.
                </assignment>
        </student>, <student lname="Peese" fname="Kevin">
                <quiz grade="85"></quiz>
                <quiz grade="78"></quiz>
                <test grade="67"></test>
                <assignment grade="20">
```

```
                  Code did not compile.
              </assignment>
       </student>)

scala> res4 \ "test"
res6: scala.xml.NodeSeq = NodeSeq()

scala> res4 \\ "test"
res7: scala.xml.NodeSeq = NodeSeq(<test grade="94"></test>, <test
    grade="67"></test>)
```

The first two searches use the \ operator. The first one searches for elements that have the tag name "`student`". It finds two of them because they are at the top level and gives us back a `NodeSeq` with them in it. The second search looks for tags that have the name "`test`". This search returns an empty `NodeSeq`. This is because while there are tags with the name "`test`" in `res4`, they are nested more deeply inside of the "`student`" elements as as such, are not found by the \ operator. The last example searches for the same tag name, but does so with \\, which searches more deeply, and hence gives back a `NodeSeq` with two `Node`s inside of it.

The \ and \\ operators can also be used to get the attributes from elements. To get an attribute instead of a tag, simply put a '`@`' at the beginning of the string you are searching for. Here are three searches to illustrate this.

```
scala> res4 \ "@name"
res8: scala.xml.NodeSeq = CSCI 1320

scala> res4 \ "@grade"
res9: scala.xml.NodeSeq = NodeSeq()

scala> res4 \\ "@grade"
res10: scala.xml.NodeSeq = NodeSeq(98, 100, 90, 94, 100, 85, 78, 67, 20)
```

The first search uses \ to get the name of the top level node. Using \ to look for a `@grade` at the top level node does not give us anything, but using \\ will return the values of all the `@grade`s in the document.

Of course, what you really want to do is put the information from the XML file into a structure that can be used in the program. Given what we have learned, this would mean that we want to put things into `case classes`. The data in this XML file corresponds very closely to the student type that was created in chapter 10. That `case class` looked like this.

```
case class Student(name:String,assignments:List[Double],tests:List[Double],
   quizzes:List[Double])
```

In that chapter we parsed a flat file into an `Array` of `Students`. Now we will demonstrate how to do the same thing using the XML. We will start with a function that takes a `Node` that should be a `student Element` and returns a `Student` object. Such a function might look like the following.

```
def studentFromXML(elem:xml.Node):Student =
  Student((elem \ "@fname")+" "+(elem \ "@lname"),
    (elem \ "assignment").map(n =>
      (n \ "@grade").toString.toDouble).toList,
    (elem \ "test").map(n => (n \ "@grade").text.toDouble).toList,
```

```
(elem \ "quiz").map(n => (n \ "@grade").text.toDouble).toList)
```

This function builds a `Student` object and passes in the four required arguments. The first is the `name`, which is made from the `fname` and `lname` attributes of the element. After that are three `List`s of grades for the assignments, tests, and quizzes, respectively. These all have a similar form. They start by doing a search for to proper tag name and `mapping` the result to a function that converts the value of the `grade` attribute to a `Double`. The call to `text` is required because the result of \ here is a `Node`, not a `String`, and the `Node` type does not have a `toDouble` method. The last part of each grade type is a call to `toList`. This is required because the `map` is working on a `NodeSeq` and will give back a `Seq`, but a `List` is required for the `Student` type.

The use of `map` probably does not jump out to you at first. Hopefully at this point you have because quite comfortable with it and other higher-order methods. However, if you think a bit you will realize that is is a bit surprising here because the thing it is being called on is not a `List` or an `Array`. Instead, it is a `NodeSeq`. This works because the `NodeSeq` is itself a subtype of `Seq[Node]`, meaning that all the methods we have been using on other sequences work just fine on this as well.

This is useful for getting our array of students as well. The following line shows how we can use `map` and `toArray` to get the result that we want with the `studentFromXML` function.

```
scala> (res4 \ "student").map(studentFromXML).toArray
res15: Array[Student] = Array(Student(Jason
    Hughes,List(100.0),List(94.0),List(98.0, 100.0, 90.0)), Student(Kevin
    Peese,List(20.0),List(67.0),List(85.0, 78.0)))
```

Again, the call to `toArray` gives us back the desired `Array[Student]` instead of a more general `Seq[Student]`.

The `text` method applied to a full `Elem` will give you all of the text that appears inside of it and all subelements. So calling `text` on `res4` gives the two comments along with a lot of whitespace. To get just the comment on any particular grade, you would parse down to that specific element and call the `text` method on it.

14.2.3 Building XML

So now you know how to get the contents of an XML file into a useful form in Scala. What about going the other way? Assume that the code we just wrote were used in the menu-based application from chapter 10 and that changes were made and now we want to write the results back out to a file. The first step in this would be to build the `Node` that represents the data.

We saw above that we can put XML directly into a Scala program or the REPL and it will be parsed and understood. However, that alone does not give us the ability to put values from the program back into the XML file. Fortunately, this is not hard to do either. Inside of XML that is embedded in a Scala program you can embed Scala expressions inside of curly braces. We will start with a simple example.

```
scala> <tag>4+5 is {4+5}</tag>
res19: scala.xml.Elem = <tag>4+5 is 9</tag>
```

Here the expression `4+5` has been put in curly braces and as you can see it evaluates to the value `9` as it should. The code you put inside of the curly braces can be far more complex and built additional XML content or tags.

We will use this to write a function that packs a `Student` object into an XML node. This code looks like the following.

```
def studentToXML(stu:Student):xml.Node = {
  val nameParts=stu.name.split(" +")
  <student fname={nameParts(0)} lname={nameParts(1)}>
    {stu.quizzes.map(q => <quiz grade={q.toString}/>)}
    {stu.tests.map(t => <test grade={t.toString}/>)}
    {stu.assignments.map(a => <assignment grade={a.toString}/>)}
  </student>
}
```

The first line `splits` the student name into pieces around spaces. It is assumed that the first element is the first name and the second element is the second name. These are used as attribute values in the student tag. Inside of that element are three lines of code, one each for `quizzes`, `tests`, and `assignments`. Each of these `maps` corresponding `Lists` to a set of elements with grade attributes.

It is worth noting two things about using code for the attribute values. First, the quotes are not written anywhere. They are automatically provided when the value is Scala code. Second, the type of the Scala expression for the value has to be `String`. This is apparent with the grade values. They are `Doubles` and have to be explicitly converted to `Strings`.

14.2.4 Writing XML to File

Once you have the `Node` you want to write, the writing process is as easy as a call to the `save` method of the `XML` object.

```
xml.XML.save("grades.xml",node)
```

The first argument is the name of the file you want to write to. The second is the `Node` that you want to write.

Validating XML (Advanced)

When XML is used for large applications, it is important to be able to verify that the contents of a file match the format that is required by the applications. This process is called validation. When the XML standard was first released, validation was done with Document Type Definition files (DTDs)DTD). A DTD is a text file that has a fairly simple format that allows you to specify what types of elements should be in a file. For each element you can say what needs to be in it or what could be in it. This ability includes attributes as well at subelements.

DTDs were generally considered to have two problems. First, they were a bit limited and simplistic. For example, you could say that an element must contain an attribute, but you could not put any constraints on the nature of that attribute. You could not say that it had to be a number or a date. In addition, DTDs had their own syntax. The goal of XML was to be a general data storage format and some found it unfitting that you had to use some other format to specify what should go into an XML file.

For these reasons, XML schema were created. An XML schema is an XML document that uses certain tags to specify what can and can not go into a certain XML file. XML schema tend to be large and complex, but they provide great control over the format of an XML file.

There are tools for both DTDs and XML schema that will run through XML files and tell you whether or not they adhere to a given specification. There is also a package in the Scala standard libraries called scala.xml.dtd that can help with validation using DTDs.

14.2.5 XML Patterns

We have seen patterns in a number of places so far. With XML we can add another one. Not only can XML be written directly into the Scala language, it can be used to build patterns. The patterns look just like what you would use to build XML, except that what you put into curly braces should be names you want bound as variables. There is one significant limitation, you can not put attributes into your patterns. This usage can be seen with the following little example in the REPL.

```
scala> val personXML = <person><name>Mark</name><gender>M</gender></person>
personXML: scala.xml.Elem = <person><name>Mark</name><gender>M</gender></person>

scala> val <person><name>{name}</name><gender>{sex}</gender></person> = personXML
name: scala.xml.Node = Mark
sex: scala.xml.Node = M
```

14.3 Putting it Together

To illustrate the real power of XML we will make a more complete theme park program that includes the functionality of some of the earlier scripts along with editing abilities. All the information will be stored in a single XML file. This last part is something that was not highlighted before, but it implicitly comes with the ability to give meaning to data. In a flat text file, it is the position in the file that gives meaning to something. This makes it very hard to insert new information of different types. That is not a problem for XML as new tags can be added as desired. As long as the new tags have names that do not conflict with earlier tags, the earlier code will continue to work just fine.

In chapter 8, we wrote a script that would help with building schedules. In chapter 13 we wrote another script that could be used to determine the employee of the month. Both of these deal with information related to employees and rides, but there is not a 100% overlap between the required data and the file formats are very different. We want to write a script here that will include the functionality of both of those scripts, along with the ability to add ride and employee information while keeping all of the information stored in a single XML file.

Code for this is shown below. It starts with the definition of a number of case classes followed by functions that can build instances of those case classes from XML or build XML from them. After those functions are four lines that declare the main data for the program while reading it in from an XML file specified on the command-line. This is followed by slightly modified versions of the schedule builder and the employee ranker from previous chapters.

```
import scala.io.Source
import scala.xml._
```

```
case class DayData(ride:String, dayOfWeek:String, operators:Array[String],
    numRiders:Int)
case class MonthData(month:Int, days:List[DayData])
case class YearData(year:Int, months:List[MonthData])
case class RideData(name:String, numberOfOperators:Int, heavyCount:Int)
case class EmployeeData(name:String, rides:List[String])

def parseDay(node:Node):DayData = {
  val ride = (node \ "@ride").text
  val dow = (node \ "@dayOfWeek").text
  val num = (node \ "@numRiders").text.toInt
  val ops = (node \ "operator").map(_.text).toArray
  DayData(ride, dow, ops, num)
}

def dayToXML(day:DayData):Node = {
  <day ride={day.ride} dayOfWeek={day.dayOfWeek} numRiders={day.numRiders.toString}>
    {day.operators.map(op => <operator>{op}</operator>)}
  </day>
}

def parseMonth(node:Node):MonthData = {
  val month = (node \ "@month").text.toInt
  val days = (node \ "day").map(parseDay).toList
  MonthData(month, days)
}

def monthToXML(month:MonthData):Node = {
  <month month={month.month.toString}>
    {month.days.map(dayToXML)}
  </month>
}

def parseYear(node:Node):YearData = {
  val year = (node \ "@year").text.toInt
  val months = (node \ "month").map(parseMonth).toList
  YearData(year, months)
}

def yearToXML(year:YearData):Node = {
  <year year={year.year.toString}>
    {year.months.map(monthToXML)}
  </year>
}

def parseRideData(node:Node):RideData = {
  val name = (node \ "@name").text
  val numOps = (node \ "@numberOfOperators").text.toInt
  val heavy = (node \ "@heavyCount").text.toInt
  RideData(name, numOps, heavy)
}

def rideDataToXML(rd:RideData):Node = {
  <ride name={rd.name} numberOfOperators={rd.numberOfOperators.toString}
      heavyCount={rd.heavyCount.toString}/>
```

```
}

def parseEmployeeData(node:Node):EmployeeData = {
  val name = (node \ "@name").text
  val rides = (node \ "trainedRide").map(_.text).toList
  EmployeeData(name, rides)
}

def employeeToXML(ed:EmployeeData):Node = {
  <employee name={ed.name}>
    {ed.rides.map(r => <trainedRide>{r}</trainedRide>)}
  </employee>
}

val xmlData = XML.loadFile(args(0))
var years = (xmlData \ "year").map(parseYear).toList
var rideInfo = (xmlData \ "ride").map(parseRideData).toList
var employeeInfo = (xmlData \ "employee").map(parseEmployeeData).toList

def buildWeeklySchedules {
  val daysInfo = for(y <- years; m <- y.months; d <- m.days) yield d
  val days = daysInfo.map(_.dayOfWeek).distinct
  for(day <- days) {
    val thisDay = daysInfo.filter(_.dayOfWeek==day)
    val rides = thisDay.map(_.ride).distinct
    val operatorRides = rides.flatMap(ride => {
      val nums = thisDay.filter(_.ride==ride).map(_.numRiders)
      val avg = nums.sum/nums.length
      val rideData = rideInfo.find(_.name==ride).get
      Array.fill(rideData.numberOfOperators+(if(avg>=rideData.heavyCount) 1 else
          0))(ride)
    })
    val totalOps = operatorRides.length
    for(choice <- employeeInfo.combinations(totalOps)) {
      val perms = operatorRides.permutations
      var works = false
      while(!works && perms.hasNext) {
        val perm = perms.next
        if((perm,choice).zipped.forall((r,op) => op.rides.contains(r)))
          works = true
      }
      if(works) {
        println(day+" - "+choice.map(_.name).mkString(", "))
      }
    }
  }
}

case class RideAverage(ride:String, avNum:Double)
case class OperatorDailyData(name:String, ride:String, numRiders:Int)
case class OperatorRideAverages(name:String, rideAvs:List[RideAverage])
case class OperatorEfficiencyFactor(name:String,factor:Double)

def insertionSortByEfficiency(a:Array[OperatorEfficiencyFactor]) {
  for(j <- 1 until a.length) {
```

```
    var i=j-1
    val tmp=a(j)
    while(i>=0 && a(i).factor>tmp.factor) {
      a(i+1) = a(i)
      i -= 1
    }
    a(i+1) = tmp
  }
}

def rankEmployees(data:List[DayData]):Array[OperatorEfficiencyFactor] = {
  val rides = data.map(_.ride).distinct
  val averages = for(ride <- rides) yield {
    val days = data.filter(_.ride==ride)
    RideAverage(ride, days.map(_.numRiders).sum.toDouble/days.length)
  }
  val dataByOperator = for(day <- data; op <- day.operators) yield {
    OperatorDailyData(op, day.ride, day.numRiders)
  }
  val operators = dataByOperator.map(_.name).distinct
  val opRideAverages = for(op <- operators) yield {
    val opDays = dataByOperator.filter(_.name == op)
    val rideAvs = for(ride <- rides; if opDays.exists(_.ride==ride)) yield {
      val opRides = opDays.filter(_.ride == ride)
      RideAverage(ride, opRides.map(_.numRiders).sum.toDouble/opRides.length)
    }
    OperatorRideAverages(op, rideAvs)
  }
  val operatorFactors = (for(OperatorRideAverages(op, rideAvs) <- opRideAverages)
      yield {
    val factors = for(RideAverage(ride,av) <- rideAvs) yield {
      av/averages.filter(_.ride==ride).head.avNum
    }
    OperatorEfficiencyFactor(op,factors.sum/factors.length)
  }).toArray
  insertionSortByEfficiency(operatorFactors)
  operatorFactors
}

def rideInput(ri:RideData):Array[String] = {
  println(ri.name)
  println(employeeInfo.filter(_.rides.contains(ri.name)).map(_.name)
      .zipWithIndex.mkString(" "))
  readLine().split(" +")
}

def inputDay:List[DayData] = {
  println("What day of the week is this for?")
  val dow = readLine()
  println("For each ride displayed, enter the number of riders for the day followed
      by employee numbers from this list with spaces in between.")
  for(ri <- rideInfo;
      val input = rideInput(ri)
      if input.head.toInt>=0) yield {
```

```
    DayData(ri.name, dow, input.tail.map(_.toInt).map(employeeInfo).map(_.name),
        input.head.toInt)
  }
}

def inputRideDayData {
  println("What month/year do you want to enter data for?")
  readLine().trim.split("/") match {
    case Array(monthText, yearText) =>
      val (month, year) = (monthText.toInt, yearText.toInt)
      if(years.exists(_.year==year)) {
            years = for(y <- years) yield {
          if(y.year==year) {
            y.copy(months = {
              if(y.months.exists(_.month==month)) {
                for(m <- y.months) yield {
                  if(m.month==month) {
                    m.copy(days = inputDay ::: m.days)
                  } else m
                }
              } else MonthData(month, inputDay) :: y.months
            })
          } else y
        }
      } else {
        years ::= YearData(year,MonthData(month, inputDay)::Nil)
      }
    case _ =>
      println("Improper format. Needs to be numeric month followed by numeric year
          with a / between them.")
  }
}

def hireEmployee {
  println("What is the new employees name?")
  val name = readLine()
  employeeInfo ::= EmployeeData(name,Nil)
}

def trainEmployee {
  println("Which employee is training for a new ride?")
  println(employeeInfo.map(_.name).zipWithIndex.mkString(" "))
  val empNum = readInt()
  employeeInfo = for((e,i) <- employeeInfo.zipWithIndex) yield {
    if(i==empNum) {
      val avail = rideInfo.map(_.name).diff(e.rides)
      println("Which rides should be added? (Enter space separated numbers.)")
      println(avail.zipWithIndex.mkString(" "))
      e.copy(rides = (readLine().split(" +").map(_.toInt)).map(avail).toList :::
          e.rides)
    } else e
  }
}

def addRide {
```

```
  println("What is the name of the new ride?")
  val name = readLine()
  println("How many operators does it need?")
  val ops = readInt()
  println("At what rider count should another operator be added?")
  val heavy = readInt()
  rideInfo ::= RideData(name, ops, heavy)
}

var input = 0
do {
  println("""What would you like to do?
 1) Add ridership for a day.
 2) Add an Employee.
 3) Add training to an employee.
 4) Add a ride.
 5) Get schedule options for a week.
 6) Rank Employees.
 7) Quit.""")
  input = readInt()
  input match {
    case 1 => inputRideDayData
    case 2 => hireEmployee
    case 3 => trainEmployee
    case 4 => addRide
    case 5 => buildWeeklySchedules
    case 6 =>
      println("What month/year or year do you want to rank for?")
      println(readLine().trim.split("/") match {
        case Array(monthText,yearText) =>
          val year = yearText.toInt
          val month = monthText.toInt
          val y = years.filter(_.year==year)
          if(y.isEmpty) "Year not found."
          else {
            val m = y.head.months.filter(_.month==month)
            if(m.isEmpty) "Month not found."
            else {
              rankEmployees(m.head.days).mkString("\n")
            }
          }
        case Array(yearText) =>
          val year = yearText.toInt
          val y = years.filter(_.year==year)
          if(y.isEmpty) "Year not found."
          else {
            rankEmployees(y.head.months.flatMap(_.days)).mkString("\n")
          }
        case _ => "Invalid input"
      })
    case _ =>
  }
} while(input!=7)

XML.save(args(0), <themeParkData>
```

```
  {years.map(yearToXML)}
  {rideInfo.map(rideDataToXML)}
  {employeeInfo.map(employeeToXML)}
</themeParkData>)
```

There is completely new code at the bottom to allow for data entry that is added to the main variables. There is also a `do-while` loop that handles the menu functionality. The script ends by saving the main data elements back out to the same XML file.

14.4 End of Chapter Material

14.4.1 Summary of Concepts

- XML is a text markup language that can be used to encode arbitrary data. Being text means that it is as easy to work with as flat text files, but it allow you to attach meaning to the values in the file and the flexibility of the parsing makes it easier to extend XML files than flat text files.
- An XML file is composed of markup and content. The content is plain text. Markup has a number of options as must follow a certain format.
  - The primary markup is tags. A tag starts with `<` and ends with `>`. Each tag has a name.
  - The combination of a matching start and end tag defines an element. An element can contain content and other elements.
  - Start tags can be given attributes to store basic information.
  - Special characters that can not appear as plain text can be specified by markup tokens that begin with a `&` and end with a `;`.
  - You can put comments into an XML file by starting them with `<!--` and ending them with `-->`.
  - The entire contents of an XML file must be held inside of a single element.
- The Scala language has native support for XML. This makes it significantly easier to work with XML in Scala than in most other languages. XML elements can be written directly into Scala source code.
  - An XML file can be loaded using `XML.loadFile(fileName:String):Elem`.
  - The \ and \\ operators can be used to pull things out of XML.
  - You can build XML by typing in XML literals. You can put Scala code into the XML by surrounding it with curly braces.
  - XML can be written back out to file using `XML.save(fileName:String,node:Node)`.
  - Patterns can be made using XML with the limitation that attributes can not be part of the match. values can be bound by including names in curly braces.

14.4.2 Self-Directed Study

Enter the following statements into the REPL and see what they do. Some will produce errors. You should try to figure out why. Try some variations to make sure you understand what is going on.

```
scala> val xml1 = <tag>contents</tag>
scala> xml1.text
scala> val <tag>{str}</tag> = xml1
scala> val xml2 = <data type="simple">
     | <language>Scala</language>
     | <lesson>Programming is an art.</lesson>
     | <lesson>Software runs the world</lesson>
     | </data>
scala> xml2 \ "language"
scala> xml2 \ "lesson"
scala> val xml3 = <randPoints>
     | {(1 to 20).map(i => <point x={math.random.toString}
         y={math.random.toString}/>)}
     | </randPoints>
scala> (xml3 \ "point").map(p => {
     | val x = (p \ "@x").text.toDouble
     | val y = (p \ "@y").text.toDouble
     | math.sqrt(x*x+y*y)
     | })
scala> for(<point x={x} y={y}/> <- xml3) yield math.sqrt(x*x+y*y)
scala> val xml4 = <randPoints>
     | {(1 to 20).map(i => <point><x>{math.random}</x><y>{math.random}</y></point>)}
     | </randPoints>
scala> for(<point><x>{x}</x><y>{y}</y></point> <- xml4 \ "point") println(x+" "+y)
```

14.4.3 Exercises

1. Chapter 10 had several exercises where you were supposed to design `case classes` for a number of different types of data. For each of those, write code to convert to and from XML.

   - A transcript for a student.
   - Realtor information for a house.
   - Data for a sports team.

2. On the website for the book there are a number of XML data files. Write `case classes` to represent that data in each one and then write code to load the files and build objects from it.

3. Pick some other `case class` that you have written and create code to convert it to and from XML.

4. Find a Really Simple Syndication (RSS) feed for a website you visit and save the feed as a file. Use Scala to look through the XML.

14.4.4 Projects

1. If you have been working on the different graphics options for earlier projects, the material in this chapter gives you a clear extension, stores your geometry data in an XML file instead of a flat text file. You can use tags like "sphere" and "plane" to give meaning to the information. Add in a "light" tag as well in anticipation of adding lighting in a future project.

   After you have the XML format set up and some data to play with, alter the code from project 7 (p.311) to use this data format.

2. Project 3 (p.244) on the text adventure is very nicely extended with the use of XML data files. Using XML also makes it easier to extend the file to include additional information. For this project you should convert your map over to an XML format and have the code read in that format. In addition, you should add items. This will involve adding another `case class` for an item and putting some items into rooms in the XML data file.

   To make the items significant, you need to have it so that the `case class` for your room includes items in that room and the text description of the room lists what items are there. You should also give you player an inventory and implements commands for "get" and "drop". So if the player enters "get" followed by the name of an item in the room, that item will be taken out of the room and added to the players inventory. An "inv" command would be nice to let the player see what is in his/her inventory. If the player uses the "drop" command followed by an item in inventory, that item should be removed from inventory and placed in the current room. If "get" or "drop" are provided with an invalid item name, print an appropriate message for the user.

3. If you did project 4 (p.346) extending your game, you had a text file that specified the high scores for players of a particular game. For this project you should modify you code and the text file to use XML instead of a flat file.

4. If you have been doing the other recipe projects, you can change the format so that it uses XML to save the data. If you do this, you need to add in instructions for the recipes as well as the ability to add comments and other information like how much certain recipes are favored. You can merge what are currently separate text files into a single XML file if you want.

   The fact that your script keeps track of recipes and pantry contents points to one other piece of functionality you should be able to add in. Allow the user to see only the recipes that they have the ingredients to cook. If you have a preference level, sort them by preference.

5. If you have been doing the scheduling options, convert the data file for courses over to XML. In doing this, you can also add the ability to include comments on courses, instructors, or other things that make sense to you, but which did not fit as well in the limited formatting of a plain text file.

6. You can extend project 14 (p.312) on your music library by changing the format of the data file from a flat file to an XML file. Use the hierarchical nature of XML to simplify the file. You can have tags for `<artist>` at a higher level with `<album>` tags inside of those and `<song>` tags at the lowest level. With the XML you could add the ability for the user to insert notes about any element that will be displayed in a manner similar to the album cover when the appropriate item is selected in the GUI. Menu options could be used to allow editing of the notes.

7. This is a continuation of project 5 (p.310) on turtle graphics to draw fractal shapes generated with L-systems. You can find a full description of L-systems in "The Algorithmic Beauty of Plants", which can be found online at `http://algorithmicbotany.org/papers/#abop`. The first chapter has all the material that will be used for this project and a later one.

   In the last turtle project you made it so that you could use a turtle to draw figures from `String`s using the characters 'F', 'f', '+', and '-'. L-systems are formal grammars that we will use to generate strings that have interesting turtle representations. An L-system is defined by an initial `String` and a set of productions. Each production maps a character to a `String`. So the production `F -> F-F++F-F` will cause any `F` in a `String` to be replaced by `F-F++F-F`. The way L-systems work is that all productions are applied at the same time to all characters. Characters that do not have productions just stay the same.

   So with this example you might start with `F`. After one iteration you would have `F-F++F-F`. After the second iteration you would have `F-F++F-F-F-F++F-F++F-F++F-F-F-F++F-F`. The next iteration will be about five times longer than that. The `String` in an L-system grow exponentially in length. As a result, you probably want to have the length of the turtle move for an `F` or `f` get exponentially shorter. Start with a good value and divide by an appropriate value for each generation. For this example dividing by a factor of 3 is ideal. This one also works best with a turn angle of 60 degrees.

   The productions for an L-system can be implemented as a `List[(Char,String)]`. You can use the `find` method on the List and combine that with `flatMap` to run through generations of your `String`. You can decide how elaborate you want your GUI to be and if users should be able to enter productions or if they will be hard coded. Look in "The Algorithmic Beauty of Plants", chapter 1 for examples of other interesting production rules.

8. If you have done the box score options from any of the last three chapters you can extend to use XML encoding for the box score. You probably want to do this by having functionality to load in a plain text box score and add it to an XML file with all the box scores. To really take advantage of the XML formatting, you should allow the user to add comments to whatever elements you feel are appropriate.

Additional exercises and projects, along with data files, are available on the book's website.

Chapter 15

Recursion

Back in chapter 6 we got our first introduction to recursion. At that point we used recursion to provide iteration. In chapter 8 we learned how to produce iteration using loops and have used that more than recursion since that point. If the only capability of recursion was to produce iteration it would not be of much interest in Computer Science because loops would be a complete substitute that would have simpler syntax for most uses. However, that is not the case. Recursion allows us to express a lot more than just simple iteration, and because of this, recursion is a critical tool in writing concise solutions to many different problems.

15.1 Power of Recursion

To understand the real power of recursion and where it comes from, it might help to revisit some code from the end of chapter 6. Early in that chapter we used a recursive function to count down from a specified number using a single argument. The code for doing that looked like this.

```
def countDown(n:Int) {
  if(n>=0) {
    println(n)
    countDown(n-1)
  }
}
```

As long as the argument has not gotten below zero, this function prints the number and counts down. We also wrote code to count up using two arguments where one argument was incremented for each subsequent call. At the end of the chapter you were presented with the following code.

```
def count(n:Int) {
  if(n>=0) {
    count(n-1)
    println(n)
  }
}
```

You were told to enter this code and run it to see what it does. If you did so, you might have been rather surprised to see that this code counts up. That seems surprising because the argument is clearly decrementing. The reason this can count up has to do with the memory of the computer and in particular the call stack, which was introduced in section 13.5.

To understand how this works, consider figure 15.1. This shows a graphical representation for different frames on the call stack and what happens when you call `count(5)`. This call immediately calls `count(4)` before it does anything else. The call to `count(4)` gets a new stack frame that keeps track of where it was in the call to `count(5)` so that when it returns it can go back to that point. The call to `count(4)` also goes straight into a call to `count(3)`, which also gets a new stack frame. This continues until we get down to `count(0)` which does nothing at all, but return. When it returns, control returns to the call to `count(1)` and resumes right after the call to `count(0)`, which is the line with the print statement. So it prints a `1`. After the print, `count(1)` returns to `count(2)` which similarly does a print. Because the prints are happening as it pops back up the stack, the numbers get printed in ascending order even though the function only includes decrement. The memory of the stack is essential for this to work because each stack frame remembers its own value of `n`.

This is a very significant point to remember about recursive calls. While the variables in the recursive calls all have the same names, they are not the same variables. It is like multiple people with the same name. Just because two or three people are named "Pat", that does not mean they are the same person. In this case, there were six different versions of `n` that were created and they took on the values from `5` down to `0`. Each was distinct from the others and occupied different parts of memory.

This example shows how the stack can come into play with recursion, but it does not really show the power of recursion. To do that, we need to have a recursive function that can call itself more than once. In the following sections we will look at several different examples of this and see problems that really require the stack and are significantly harder to convert over to loops.

15.2 Fibonacci Numbers

The classic example of recursion is the Fibonacci numbers. This is a sequence of numbers where each number is defined as the sum of the previous two. So in mathematical notation this is written as $f(n) = f(n-1) + f(n-2)$. This is not a complete definition, however, because we need to know how the sequence starts. That is to say that we need a base case for the recursion. It is customary to have the first two elements be 1. So the sequence then is 1, 1, 2, 3, 5, 8, 13, 21, ...

We can write this function in Scala with one short function definition.

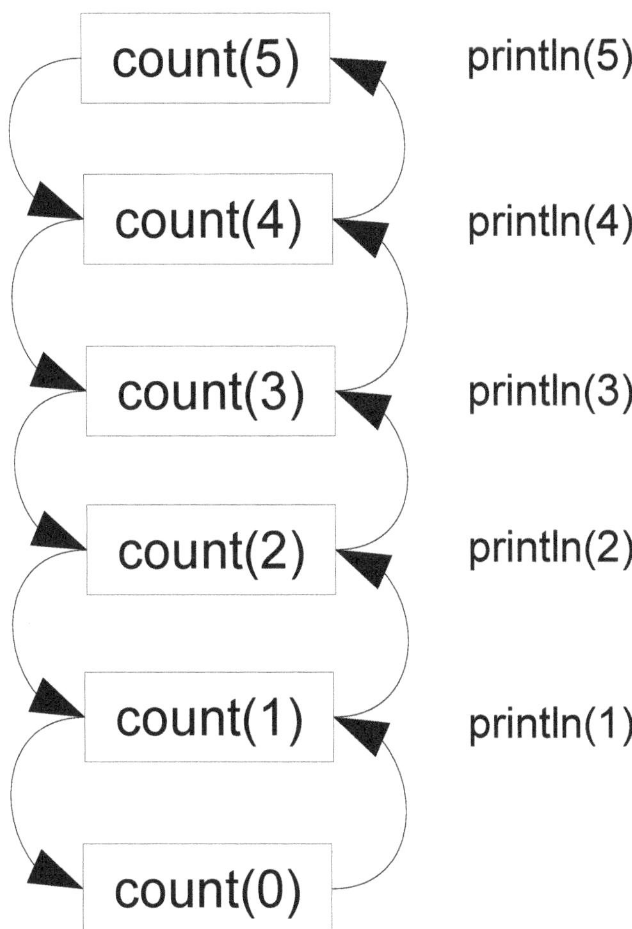

FIGURE 15.1: This shows the call stack for the function count which prints the numbers counting up. When you call the function with 5, it calls itself with 4 before doing anything else. This calls itself with 3 and so on down to 0. The call with 0 does nothing but return. It returns back into the call to `count(1)` that does the print and continues on back up the stack.

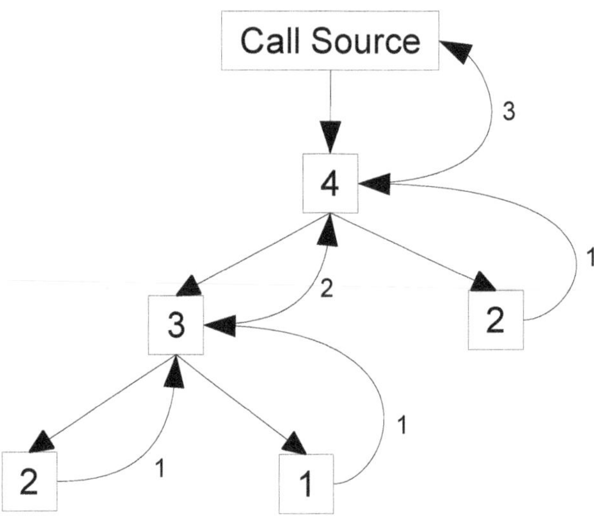

FIGURE 15.2: This figure shows a call tree for `fib(4)`. Boxes represent stack frames. Straight arrows point in the direction of function calls. Curved arrows show returns and are labeled with return values.

```
def fib(n:Int):Int = if(n<3) 1 else fib(n-1)+fib(n-2)
```

So for `n=1` and `n=2` we get back `1`. For `n=3` we get back `1+1=2`. For larger numbers, the process is more complex and it is instructive to try to visualize it. We do this by considering what happens on the call stack in a manner similar to figure 15.1. Unlike the code for figure 15.1, this function calls itself twice and that makes things a bit more complex.

For example, consider a call to `fib(4)`. This is equal to `fib(3)+fib(2)`. In order to know what that is, recursive calls have to be made to each of those functions. By convention, we will imagine that the calls are processed from left to right. So it calls `fib(3)`, which gets a stack frame. That in turn calls `fib(2)` which gets a stack frame. Unlike examples we have seen before, when `fib(3)` returns, there is another call to `fib(2)` still waiting. So immediately a new stack frame appears where the one for `fib(3)` had just been. This process of stack frames returning to be replaced with a new one happens a lot in this calculation of the Fibonacci numbers as well as other recursive functions that call themselves more than once. Instead of drawing this as a vertical stack as was done in figure 15.1, it is customary to draw it in a branching structure called a tree as is drawn in figure 15.2. Each new stack frame is still below the one that called it, but different stack frames at the same depth in the stack are drawn next to one another horizontally. Arrows are used to keep track of things.

In this representation, boxes represent the stack frames. We have dispensed with putting the function names and left only the argument value. The straight arrows show function calls. The curved arrows show returns and are labeled with return values. These types of diagrams can be very helpful to you when you are trying to figure out what happens in a complex recursive function. When you draw them yourself you can dispense with the boxes and arrows. Values with lines between them will typically suffice.

The Fibonacci numbers are a standard example of recursion, but they are not a great one outside of instructional purposes. You can fairly easily write a function that calculates Fibonacci numbers that uses a loop instead of recursion and it will be a lot faster. That is not true of the examples that follow.

15.3 Permutations

This next example is a bit more practical. We are given a `List` of values and we want to perform some function on all permutations of this `List`. A permutation of a collection contains all the same elements, but in potentially different orders. This type of behavior might be in order if you have different tasks and you need to find an optimal order in which to perform the tasks based on some rule.

So the next question is, how could we do this? We have seen previously that sequences in Scala have a method called `permutations` which will do for us, but we want to see how we could do it ourselves. To figure this out we will employ a common idiom when building recursive functions. It is one that we talked about in chapter 6 with our simpler recursive functions. The idea is to take a large problem and break it down so that we solve one step and then apply the same function to solve what is left. In this case, one step is picking a first element. Unlike what we did in chapter 6, there will generally be more than one option here. Any of the elements in the current `List` could be the first element in the permutation. What follows that is the permutations of everything else in the `List`. That gives us a recursive definition. We want to pick an element to go first, then make a recursive call on the rest. After that returns, we pick a different element to go first and recurse again. This should be repeated for all the elements in the `List`.

To convert this into a function we need to figure out what information has to be passed in. The obvious part is that we have to pass in the `List` of values we are permuting. We also need to pass in the elements in the permutation that has been built so far. This can be done as a second `List`. If we made the function such that it returns a `List` of `List`s with all the permutations, these would be the only arguments we would need. However, such a function is not useful for many situations because the number of permutations grows very quickly with `List` length and the cost in memory would become prohibitive. Instead, we will pass in a function that operates on a `List` that we will call each time a complete permutation is built. That way the caller can decide what is done with the different permutations.

So we are writing a function that takes three arguments, the `List` of numbers we want to permute, a function to apply to any finished permutations, and the current `List` for this permutation. This is done in the following code.

```
def permute(nums:List[Int],f:(List[Int])=>Unit,p:List[Int]) {
  if(nums.isEmpty) {
    f(p)
  } else {
    var before=List[Int]()
    var after=nums
    while(!after.isEmpty) {
      val perm = after.head :: p
      permute(before ::: after.tail,f,perm)
      before ::= after.head
      after = after.tail
    }
  }
```

```
}
```

When the List is empty, we have gotten a full permutation and we call the function on it. Otherwise we have a loop that runs through all the elements we want to permute and makes recursive calls assuming each one has been added to the head of the permutation. This is done using two variable Lists and a while loop. To start off with, one List is empty and the other is all the elements in the input List. In the loop we not only make the recursive call, we also move elements from one List to the other. We can check to see that this function works by calling it like this.

```
permute(List(1,2,3),println,Nil)
```

We are getting the six different permutations of a list with three elements and simply printing them. The first call should always use Nil for the last argument.

This function displays an interesting feature that is generally true of recursive functions. Because it is defined in terms of itself, we write it assuming that it works. When we are done writing it, it will work. This logic seems circular, but it is actually using a method called induction. First we make the function work for the simple base case. Every case above that is defined in terms of a smaller one. Eventually this gets down to the base case which generally works in a trivial way. If the base case works, the case right above it should work. When the one above the base case works, the one above that should work. This logic progresses upward to allow us to solve problems of arbitrary size.

We mentioned earlier that we do not return a List of the permutations because there can be a lot of them and that would require a lot of memory. It is worth taking a second to get a slightly better understanding of what "a lot" means here. Specifically, we would like to know the order of the number of permutations. This will not only tell us how long a list would be if we built one, it will give us a measure of how long this function will take to run for any task that we might give it as it will be the number of times that the function parameter, f, gets called.

As is normally the case for order analysis, we will think of things in terms of the size of our input. In this case, that is the size of the input List which we will call n. The first call to permute makes n calls to itself, each of which is passed a List with $n-1$ elements. Those them make $n-1$ calls with Lists of size $n-2$. This process continues until we get down to 0 elements in the List. If you picture this as a tree similar to that in figure 15.2, we have a top box with n branches off it. Those lead to n boxes that each have $n-1$ branches off of them. So at the third level we have $n*(n-1)$ boxes. At the next level we have $n*(n-1)*(n-2)$. In you continue this down to 1 we get the factorial function that we played with way back in chapter 6. Indeed, the order of the permutation function is $O(n!)$. If you recall from our earlier discussion, the factorial function grows very quickly. So quickly in fact, that we had to use the Long or BigInt types if we wanted to use even modestly large inputs. Clearly you do not want to try to get all the permutations of Lists with much more than 10 elements in them. Depending on what you are doing, even lists of 10 elements might take a while.

15.4 Towers of Hanoi

Our next example of these more powerful recursive functions is a bit more playful. You have likely seen and perhaps even played Towers of Hanoi. It is a "game" played with three pegs and a number of disks of different sizes that have holes in them so they can be put on the pegs. Your goal in playing is to move the disks from one peg to another following two rules:

1. You can only move one disk at a time.

2. No disk can be placed on a disk smaller than it.

We want to write a program that can solve this. To do so, we need to first analyze the problem a bit more.

If you start by picturing a peg with 7 disks on it stacked from largest at the bottom to smallest at the top and try to picture the solution, you will probably have a hard time. The first step is clearly to move the top disk to one of the other pegs, but which one? It gets more complex from there. Instead of trying to solve a hard problem, we should start with an easy problem and try to build up from there.

The easiest setup to solve is when you have one disk. Simply move the disk from where it started to the peg you want it to finish on. That was a bit trivial, so what about two disks? First you have to move the top disk to the peg you do not want to end on, then you move the second disk to the end peg and move the smaller one over on top of it. It might not be clear at this point, but that points toward a general solution. To see this, think of the situation with three disks. We could run through every single move, but we just saw how to move two disks so we can assume we know how to do that. Using that, we move two disks to the middle peg. Note that we do not literally move two disks at a time as that would violate the rules. Instead, we use the approach we just developed to move two disks in three separate one disk moves. Then the largest disk is moved to the end peg and the two disks are moved on top of it.

This points to a general solution. If we make the assumption that we know how to move $N-1$ disks, we can move N disks by first moving the top $N-1$ disks to the off peg, then moving the largest one to the destination, then moving the $N-1$ disks back on top of the largest one. Note that this is a recursive definition. The solution for N disks is relies on the solution to $N-1$ disks. The base case is the trivial one where we move a single disk across.

We now have an approach to solving this problem. Next we need to figure out how we want to represent the problem in the memory of the computer. We need to come up with a way to represent disks and pegs. For the disks, all we really care about is the size of the disk. This can nicely be represented by an `Int`. A peg is really just a stack of disks so we need something like an `Array` or a `List` of `Int`s. To decide which, we should consider how they are used. The number of disks on a peg changes over time, but they are always added or removed at one location, the top of the stack. This points to using a `List` where the `head` of the `List` is the top disk. We can use `::` to add a disk to a peg and the `tail` is what is left when a disk is removed. The script below shows code that implements the three pegs as an `Array[List[Int]]`. There is a function to move one disk that includes a check to make sure we are not breaking the second rule.

```
val num = if(args.length>0) args(0).toInt else 7
val pegs = Array(List.tabulate(num)(i => i+2), List[Int](), List[Int]())

def moveDisk(from:Int,to:Int) {
   require(pegs(to).isEmpty || pegs(from).head < pegs(to).head)
   pegs(to) = pegs(from).head :: pegs(to)
   pegs(from) = pegs(from).tail
}

def moveNDisks(n:Int,from:Int,to:Int) {
```

```
    val other = 3-from-to
    if(n==1) {
        moveDisk(from, to)
    } else {
        moveNDisks(n-1, from, other)
        moveDisk(from, to)
        moveNDisks(n-1, other, to)
    }
}

moveNDisks(pegs(0).size, 0, 2)

println(pegs.map(a => a.mkString(" ")).mkString("\n"))
```

The function `moveNDisks` is recursive and relies on the `moveDisk` function so we can feel certain the program is not cheating. The line `val other = 3-from-to` might seem a bit odd at first. The function is only passed the number of the peg to move from and the number of the peg to move to, but we have to refer to the third peg in the function as that is where the top `n-1` disks are to go first. That line calculates the number of the other peg using the simple observation that if you add the indexes of all the pegs you get 3. The script ends with a call to the `moveNDisks` function followed by a print of what is left at the end. This makes it simple to verify that the puzzle was actually solved.

Given the physical and visual nature of the Towers of Hanoi, the text print is a bit unsatisfying. This is a problem that really deserves to have a graphic display. The following script adds a `BufferedImage`, and `Panel`, and a `Frame` so that you can watch the disks move around as the recursion solves the puzzle.

```
import scala.swing._
import java.awt.image.BufferedImage
import java.awt.{Graphics2D,Color}

val num = if(args.length>0) args(0).toInt else 7
val pegs = Array(List.tabulate(num)(i => i+2), List[Int](), List[Int]())

val img = new BufferedImage(600, 300, BufferedImage.TYPE_INT_ARGB)
val g = img.createGraphics

val panel = new Panel {
        preferredSize = new Dimension(img.getWidth, img.getHeight)
        override def paint(g:Graphics2D) {
                g.drawImage(img, 0, 0, null)
        }
}

def renderPegs {
        g.clearRect(0, 0, img.getWidth, img.getHeight)
        g.setPaint(Color.red)
        for(i <- 0 until pegs.length) {
                for(j <- 0 until pegs(i).length) {
                        val s=pegs(i)(j)*10
                        g.fillRect(100+i*200-s/2,
                            img.getHeight-(pegs(i).length-j)*30, s, 20)
                }
        }
        panel.repaint()
```

```
        Thread.sleep(100)
}

def moveDisk(from:Int,to:Int) {
        require(pegs(to).isEmpty || pegs(from).head < pegs(to).head)
        pegs(to) = pegs(from).head :: pegs(to)
        pegs(from) = pegs(from).tail
        renderPegs
}

def moveNDisks(n:Int,from:Int,to:Int) {
        val other = 3-from-to
        if(n==1) {
                moveDisk(from, to)
        } else {
                moveNDisks(n-1, from, other)
                moveDisk(from, to)
                moveNDisks(n-1, other, to)
        }
}

val frame=new MainFrame {
        title="Hanoi"
        contents=panel
        resizable=false
}

frame.visible=true

moveNDisks(pegs(0).size, 0, 2)

println(pegs.map(a => a.mkString(" ")).mkString("\n"))
```

If you play with this code a bit you will notice that tall stacks of disks take a very long time to move. The leads to the question of just how many moves does it take to get N disks from one peg to another? The recursive code makes this question fairly easy to answer if we are happy with a recursive function for the answer.

$$f(N) = \begin{array}{ll} 1 & N = 1 \\ 1 + 2 * f(N-1) & otherwise \end{array}$$

This can be simplified to $f(N) = 2^N - 1$. So the number of moves grows exponentially in the number of disks. Stacks of 20 or 30 disks would indeed take a very long time to complete.

15.5 Mazes

Another class of problems that calls for recursion is those involving mazes. This might also seem to be more for recreation, but mazes are a simplified case of something called graphs which are very important in Computer Science and are used to represent all types of different problems. The same approaches that we will use here for mazes apply to graphs as well. There are also a number of applications where doing things like finding the optimal route through some type of restricted path like a maze is significant.

To keep things simple, we are going to use a rather basic approach to building our mazes. Instead of having a grid with walls between cells, we will have a grid where complete squares can either be open or be occupied by a wall. This representation means that we can use a 2-D `Array` of values that tell us if there is a wall or not. In theory we could use an `Array[Array[Boolean]]` for this purpose, but in practice we will have use of being able to put numeric values in the "rooms" so an `Array[Array[Int]]` will be more practical.

We will define the maze with code like the following.

```
val maze=Array(Array( 0, 0, 0, 0, 0, 0, 0, 0, 0, 0),
               Array(-1,-1,-1, 0,-1,-1,-1,-1,-1,-1),
               Array( 0, 0, 0, 0,-1, 0, 0, 0,-1, 0),
               Array( 0,-1, 0,-1,-1,-1,-1, 0, 0, 0),
               Array( 0,-1, 0,-1, 0, 0,-1,-1,-1, 0),
               Array( 0,-1,-1,-1,-1, 0, 0, 0, 0, 0),
               Array( 0,-1, 0, 0,-1, 0,-1,-1,-1, 0),
               Array( 0, 0, 0,-1,-1, 0,-1, 0, 0, 0),
               Array( 0,-1, 0,-1,-1, 0,-1, 0,-1,-1),
               Array( 0,-1, 0, 0, 0, 0,-1, 0, 0, 0))
```

This builds a 2-D `Array` of `Ints` that uses a 0 to represent an open square and a -1 to represent a wall. The use of -1 for a wall is intentional so that positive numbers can be used to represent other things later. This particular maze is only 10x10 in size. This is a good size to start off with for our purposes though it does not allow for a very complex maze.

Our first task is the thing that most people would probably want to do with a maze. We want to know how to get through it in the shortest number of steps. To start with, we will ask the simple question of how many steps long is the shortest path. This has a significant advantage over trying to actually return a path. Not only is it a simple numeric value, it is also unique. There can be multiple paths all with the same length. That produces ambiguity if we ask for the optimal path. It also happens that having a function to find the length of the shortest path is sufficient as multiple calls to this function can be used to construct the optimal path even though such a construction is less than optimal.

You can write a function to find the length of the shortest path using loops, but it is a fairly complex code. We can write a simpler solution using recursion. As with all recursive functions we can break the solution down into two broad parts, base cases and recursive cases. The base cases should be trivial to solve. There are two good base cases for the maze problem. One base case is when you reach the exit. If you are at the exit it takes zero steps to get out. Another base case is if you are checking a location which is out of the maze or in a wall. For that we want to return a value that could not possibly be a solution. We will come back to that after discussing the recursive cases.

Picture yourself standing at a location somewhere in the maze and you want to know how many steps it takes to get from your current location to the exit on the shortest path. The recursive way to do this is to imagine that you had an oracle that could tell you that answer from any of the neighboring squares and figure out how you would combine those values to determine the distance from where you are. As we said earlier in the chapter, we write a recursive function by assuming we have one that works and writing based on that assumption. Once we are done, assuming we have handled the cases properly, it will work. The oracle is our assumed working function. If you are in the maze and you can ask the oracle to tell you how many steps it takes to get out from one step to the north, south, east, and west, how would you use that information to determine the shortest distance from where you are?

To make this more concrete, picture yourself at a four-way intersection. You ask the oracle about each direction and get values of 7, 13, 4, and 19. So the shortest path lies in the direction that gave you back 4. However, that was the distance from one step in that direction. The minimum distance from where you are is one greater than that, or 5 steps. It is the minimum of the four, plus one step to get to that other location. The solution of taking the minimum also points us in the proper direction for the base case of hitting a wall or going out of bounds. If the best solution is the minimum, we should return a number that can not possibly be the minimum for a bad path. A large number such as one billion will suffice for virtually any maze. You might be tempted to use `Int.MaxValue`, but remember that we are adding 1 to it. Try doing `Int.MaxValue+1` in the REPL and you will see why that is not an ideal choice.

That defines our recursive case. There is only one other factor that needs to be considered. If you have ever been in a hedge maze or some other maze where you can not see over the walls, it can be quite difficult to tell one location from another, and unless you do something to mark your previous locations you can wind up running in circles and never finding your way out. A rather famous solution to this problem was described by the Brothers Grimm in Hansel and Gretel who left breadcrumbs to mark their path. We will choose the same approach. Assuming you have no birds in your computer this should work better for you than it did for the children. If you do have birds in your computer, you should probably deal with that before reading on.

So the last facet that we need for our algorithm is the ability to drop down breadcrumbs to mark our path. If we ever come back upon our breadcrumbs we will treat it just like being in a wall or out of bounds. We will also sweep up our breadcrumbs before we return from a path. This was not required by Hansel and Gretel who simply wanted to be able to find their path. We, however, want to find an optimal path and leaving breadcrumbs all over the place would often cause us to miss such a path. Converting this all to code produces the following.

```
def shortestPath(maze:Array[Array[Int]],x:Int,y:Int,endX:Int,endY:Int):Int = {
  if(x<0 || y<0 || y>=maze.length || x>=maze(y).length || maze(y)(x)!=0) {
    1000000000
  } else if(x==endX \&\& y==endY) {
    0
  } else {
    maze(y)(x)=1
    val dist=(shortestPath(maze,x+1,y,endX,endY) min
    shortestPath(maze,x-1,y,endX,endY) min
    shortestPath(maze,x,y+1,endX,endY) min
    shortestPath(maze,x,y-1,endX,endY)) + 1
    maze(y)(x)=0
    dist
  }
}
```

The whole function is built from `if` expressions. The first option is for going out of bounds, hitting a wall, or coming across one of our breadcrumbs, and it returns one billion, a number so large it can not possibly be a valid path, but one small enough that we also will not overflow it by adding 1 for other steps through the maze. The second case is where we have reached the end. These two can be reversed and would need to be if the exit were outside of the bounds or marked by a special value in the array. The last case in the recursive case which drops a breadcrumb, makes four recursive calls and combines them with `min`, then picks up the breadcrumb and returns. This code could be invoked with this call.

```
println(shortestPath(maze,0,0,9,9))
```

For the maze shown above, the return value from this call is 36. For this particular maze, that is the length of the only path from 0,0 to 9,9. You should play with the maze some to make sure that the function finds the shortest path for other configurations.

This function literally tests every single path through the maze and gives us back the shortest. There are strengths and weaknesses to this approach. The weakness is that if there are a lot of paths this could take a while. We will explore that more and consider ways to address it in chapter 28. The strength is that it takes only very minor modifications of this code to produce other similar functions such as one to find the longest path or to count up how many paths there are. These particular problems are left as exercises for the student, but we will tell you now that the only real changes are to alter the return values of the base cases and the way in which the recursive calls are combined.

15.6 Sorts

In chapter 13 we used loops to provide all of our iterations. We could have used recursion instead. In fact, some of the sorts are expressed very nicely as recursive sorts. Consider a `maxSort`. The two loops would be written as two recursive functions and one can be nested inside of the other. The inner one finds the index of the maximum value before some location. This is nested in a function that runs back through the array finding the max and swapping it into place before moving back one spot. The code for doing this looks like the following.

```
def maxSort(a:Array[Double],end:Int) {
  def maxIndex(end2:Int):Int =
    if(end2==0) 0 else {
      val m=maxIndex(end2-1)
      if(a(m) > a(end2)) m else end2
    }

  if(end>0) {
    val index=maxIndex(end)
    if(index!=end) {
      val tmp=a(index)
      a(index)=a(end)
      a(end)=tmp
    }
    maxSort(a,end-1)
  }
}
```

This is a bit longer than the version that uses a loop, but not by all that much. A wrapper function would be needed if we wanted to have a version we could call with only a single argument.

Perhaps an even better example is the insertion sort. The inner loop of the insertion sort is doing an insert so we write a recursive `insert` function. This gets called by an outer function that does the sort and simply walks down the `Array` putting each element in place.

```
def insertionSort(a:Array[Double],i:Int=1) {
  def insert(v:Double,i:Int) {
```

```
    if(i<=0 || v>=a(i-1)) a(i)=v else {
      a(i)=a(i-1)
      insert(v,i-1)
    }
  }

  if(i<a.length) {
    insert(a(i),i)
    insertionSort(a,i+1)
  }
}
```

The insertion sort is quite nice and neat as a recursive function. The fact that it does not have a full swap helps with that.

For the max sort and the insertion sort, the use of recursion is only a change in presentation. It does not really alter how they run or how efficient they are. The recursion in these functions is only used for iteration, nothing more. However, recursion also opens the door to some more efficient sorts. The next two subsections describe two of these.

15.6.1 Divide and Conquer Sorts

Recursion that calls itself multiple times opens up a new style of problem solving called DIVIDE AND CONQUER. The idea is exactly what the name sounds like. You take a big problem, divide it up, conquer/solve the pieces, then build a solution to the larger problem from the solutions to the pieces. The divide and conquer approach is very general and can be applied to many types of problems. In this section, we will look at two different sort algorithms that are built on this idea.

15.6.1.1 Merge Sort

The first sort we want to consider is the MERGE SORT. The general idea of a merge sort is that we are given a collection that we break into two even pieces. Each of those pieces is sorted individually and then the sorted results are merged back together.

This is a fast sort algorithm because the merge operation is $O(n)$. The reason for this is that when you want the lowest element from either collection, you do not have to look through all the elements, you only need to consider the elements at the low ends of the two collections. That means you only do one comparison. To find the next lowest element you again do one comparison. So you can make a sorted collection of n elements from two smaller sorted collections with only $n-1$ comparisons. The fact that you are repeatedly cutting the collection in half means that you get down to a single element in $\log_2(n)$ cuts. This gives an overall performance of $O(n\log(n))$. For large values of n, this is much better than the $O(n^2)$ performance of earlier sorts.

There is one minor problem with the merge sort, it can not be done in-place. Recall that this means that it takes additional memory at least proportional to n to to complete the sort. A well-crafted merge sort can get everything done with one additional `Array` of length n. Getting the memory usage down to that level is a bit complex so we will leave that for chapter 28. In this chapter, we will use the inability to do the sort in-place as an excuse for showing you the sort using `List`s, which require extra work space anyway because they are immutable.

The first solution that we will look at is all recursive. It has a recursive `mergeSort` function that takes a List[Int] and returns a List[Int].[1] This uses a recursive merge function to put the Lists together.

```
def merge(lst1:List[Int], lst2:List[Int]):List[Int] = (lst1,lst2) match {
  case (Nil,_) => lst2
  case (_,Nil) => lst1
  case (h1::t1, h2::t2) =>
    if(h1<h2) h1 :: merge(t1, lst2)
    else h2 :: merge(lst1, t2)
}

def mergeSort(lst:List[Int]):List[Int] = lst match {
  case Nil => lst
  case h::Nil => lst
  case _ =>
    val (l1, l2) = lst.splitAt(lst.length/2)
    merge(mergeSort(l1), mergeSort(l2))
}
```

Both of the functions use pattern matching for the different cases in the recursion. This makes them nice, short, and easy to read. Unfortunately, the recursive version of merge has to allocate a new stack frame for every element that is merged. That means that this version of the code can not scale up to large sizes.

To get around this limitation we need to use a more imperative merge that works with a while loop. such a version might look like the following.

```
def merge(lst1:List[Int], lst2:List[Int]):List[Int] = {
  var l1 = lst1
  var l2 = lst2
  var ret = List[Int]()
  while(l1.nonEmpty && l2.nonEmpty) {
    if(l1.head<l2.head) {
      ret ::= l1.head
      l1 = l1.tail
    } else {
      ret ::= l2.head
      l2 = l2.tail
    }
  }
  if(l1.nonEmpty) ret :::= l1.reverse
  else ret :::= l2.reverse
  ret.reverse
}
```

This version is not as pretty, and is about twice as long. However, it handles significantly longer Lists making it significantly more useful.

One last thing to note about the merge sort is the way in which it does work. As the recursive calls go down the stack, very little happens. The input List gets split in half and those two halves get passed down. This repeats until we reach a single element. The real work happens as the recursion pops back up the stack. Before a function can return, it calls merge, which does the real work. So if you picture in your head something like figure 15.1,

[1]Unlike version with Arrays, a sort using a List must return a new List as the original one can not be altered.

the down arrows are not associated with much work. It is the up arrows where the work is done.

15.6.1.2 Quicksort

Another divide and conquer sort is QUICKSORT. The idea of quicksort is to pick a special element called the pivot, then move it to the correct location with all the elements that are less than it before it and all those that are greater than it after it. Those elements to either side will not yet be sorted so recursive calls are made to sort each of them. The recursion terminates when it gets down to a single element.

Unlike merge sort, quicksort can be done in-place. However, that will be left as an exercise for chapter 28. For now we will make the quicksort like our merge sort and work on a `List`, for which the idea of being done in-place does not make sense.

The quality of a quicksort is largely dependent on the selection of a pivot. A good pivot will be in the middle of the collection to cut it in half. A really bad pivot would be the minimum or maximum element which does nothing to split up the data and only pulls out that one element from the next level of the recursion. Keeping with the idea of simplicity for the implementations in this chapter, we will use the first element as the pivot. This is not a good way to do this in general and can lead to very bad behavior on sequences that are already sorted. That too will be addressed in chapter 28.

With these simplifications in place, we can write a version of quicksort with the following code.

```
def quicksort(lst:List[Double]):List[Double] = lst match {
  case Nil => lst
  case h::Nil => lst
  case _ =>
    val pivot = lst.head
    val (less, greater) = lst.tail.partition(_<pivot)
    quicksort(less) ::: (pivot :: quicksort(greater))
}
```

The `pivot` is set to the `head` of the `List`. We then use `partition` to split the rest of the elements between those that are less than the `pivot` and those that are not. Finally, we call `quicksort` on those two sublists and put the whole thing together into a result. The result is a short and simple sort function that actually works fairly well on random data.

It is worth asking what the order of this sort really is. Unlike the merge sort, which is very stable in the amount of work that it does, the performance of our quicksort can vary dramatically depending on the input. Like the merge sort, each level of the recursion does $O(n)$ work. In this case that work is the `partition` and sticking `Lists` together. If the pivot is in the middle of the values, each level of the data is cut in half and we get $O(\log(n))$ levels for the recursion and $O(n \log(n))$ overall performance. For random data, this is the expected behavior. However, our simple pivot selection can lead to very poor behavior and this sort is $O(n^2)$ if the input is already sorted.

The last thing to note about quicksort is that unlike merge sort, it does most of its work going down the call stack. The call to `partition` is where all the comparisons happen. This `List` version does have to merge `Lists` on the way back up the call stack. The version we write with `Arrays` in chapter 28 will not have to do even that.

15.7 Putting it Together

To see another use of recursion we want to solve a problem that we have done before using a different approach. Twice now we have included code that makes suggestions for employee schedules based on data on ridership. The work of finding schedules was done using the `permutations` and `combinations` methods on sequences. As we saw earlier in this chapter, recursion can also be used to generate permutations. It can do combinations as well.

Instead of repeating exactly what we did before, we are going to use recursion to push it a bit further. Instead of showing possible groups of employees for each day, we want to show possible schedules for a full week that include a limit on how many days each week any given ride operator is willing/able to work.

We will not repeat the entire code from chapter 14 for the various menu options. This simply shows a function that can be used to build these more complete schedule types. For this to work a `daysPerWeek:Int` member was added to the `EmployeeData case class`. You can also modify menu option 5 to call this function:

```
def recursiveBuildWeeklySchedule {
  val daysInfo = for(y <- years; m <- y.months; d <- m.days) yield d
  val days = daysInfo.map(_.dayOfWeek).distinct

  case class WorkerDays(name:String, numDays:Int)
  case class WorkerAssigns(day:String, workerRide:List[(String, String)])

  def printSchedule(schedule:List[WorkerAssigns]) {
    println("Possible Schedule:")
    println(schedule.mkString("\n"))
  }

  def recurByWorker(daysLeft:List[String], workerAvail:List[WorkerDays],
      schedule:List[WorkerAssigns], workersLeft:List[String],
      ridesNeedingOps:List[String]) {
    if(ridesNeedingOps.isEmpty) {
      recurByDay(daysLeft, workerAvail, schedule)
    } else if(workersLeft.length>=ridesNeedingOps.length) {
      val worker = employeeInfo.filter(_.name == workersLeft.head).head
      for(ride <- worker.rides) {
        ridesNeedingOps.indexOf(ride) match {
          case -1 =>
          case i =>
            val newAvail = (for(w <- workerAvail) yield {
              if(w.name == worker.name) w.copy(numDays = w.numDays-1)
              else w
            }).filter(_.numDays>0)
            val newSchedule = schedule.head.copy(workerRide = (worker.name, ride) ::
                schedule.head.workerRide) :: schedule.tail
            recurByWorker(daysLeft, newAvail, newSchedule, workersLeft.tail,
                ridesNeedingOps.patch(i,Nil,1))
        }
      }
      recurByWorker(daysLeft, workerAvail, schedule, workersLeft.tail,
          ridesNeedingOps)
```

```
    }
  }

  def recurByDay(daysLeft:List[String], workerAvail:List[WorkerDays],
      schedule:List[WorkerAssigns]) {
    if(daysLeft.isEmpty) {
      printSchedule(schedule)
    } else {
      val day = daysLeft.head
      val thisDay = daysInfo.filter(_.dayOfWeek==day)
      val rides = thisDay.map(_.ride).distinct
      val operatorRides = rides.flatMap(ride => {
        val nums = thisDay.filter(_.ride==ride).map(_.numRiders)
        val avg = nums.sum/nums.length
        val rideData = rideInfo.find(_.name==ride).get
        Array.fill(rideData.numberOfOperators+(if(avg>=rideData.heavyCount) 1 else
            0))(ride)
      })
      recurByWorker(daysLeft.tail, workerAvail, WorkerAssigns(day, Nil)::schedule,
          workerAvail.map(_.name), operatorRides)
    }
  }

  recurByDay(days, employeeInfo.map(e => WorkerDays(e.name, e.daysPerWeek)),
      List[WorkerAssigns]())
}
```

This is a significantly more complex recursive function than what we looked at previously. It is worth taking some time to study what is going on. The recursion is broken into two separate functions that are nested in the primary function. The primary function does little more than pull together the information on the days and then make a call to `recurByDay`.

As the name implies, the `recurByDay` function is using days as the primary recursive argument. Each level down the call stack has one fewer elements on the `daysLeft List`. The base case is when `daysLeft` is empty. That means we have a full schedule for the week and we are ready to print it. This function borrows code from the earlier version to build up a `List` of ride names with multiple copies for the number of operators needed.

All of the information that comes into `recurByDay`, as well as other information that it figures out is passed into `recurByWorker`. The name here is again meant to give an image of what is going on. This function recurses through the workers, considering one worker on each call. The base case is when there are no more rides that need operators. In the recursive case the code runs through all the riders that worker is trained to operate and if any of them still need operators, one recursive branch is taken with that worker assigned to that ride. There is also a recursive branch at the end where the current worker is not assigned to work that day.

Note that the base case of `recurByWorker` contains a call back to `recurByDay` so these are MUTUALLY RECURSIVE FUNCTIONS. It is reasonably common in more complex situations, when the recursion needs to run through a space that has several different independent parameters, to break the problem up like this into different functions that handle changes in different options and depend on one another to get a complete solution.

15.8 End of Chapter Material

15.8.1 Summary of Concepts

- Recursion can be used for much more than iteration. The memory of the stack frames on the call stack allows recursive functions to call themselves more than once and try different possibilities for solving a problem.
- One of the most common examples of a recursive function that calls itself more than once is a function to generate the Fibonacci sequence. Each number in this sequence is the sum of the previous two, so this recursive function calls itself with arguments one less than the current value and two less than the current value.
- A more interesting example is the Towers of Hanoi. Here we saw that if we knew how to solve the problem with N disks, we could extend it to $N+1$ disks. This and a base case constitutes a complete solution for any value of N.
- Finding the shortest path through a maze works well as a recursive problem. At each point you need to find out the distance from the different options you have. Those can be combined to give you an appropriate value for the current location.
- Sorts can also be written recursively. Any of the sorts we looked at previously can be converted to recursion, but there is not a strong motivation to do so.
- Using a recursive divide and conquer approach to sorting leads us to two other sorts that do $O(n \log(n))$ comparisons instead of the $O(n^2)$ from our previous sorts.
  - A merge sort repeatedly breaks the collection in two going down the call stack, then merges the sorted results as it comes back up the call stack.
  - Quicksort works by picking a pivot and putting it in the right place, then recursing on the elements that are less than the pivot as well as those that are not.

15.8.2 Exercises

1. Write functions that calculate Fibonacci numbers using the following approaches.
   - Using a loop with an `Array` or `List`.
   - Using a loop with three `var` declarations.
   - Using a recursion function that takes three arguments, but only calls itself once.
2. Write a function that will calculate the longest non-self-intersecting path through a maze.
3. Write a function that will count the number of non-self-intersecting paths through a maze.
4. Find the size of the biggest completely empty maze one can solve with the recursive search function that finishes in a minute or less.

5. The following is a function called the Ackermann function, which is significant in theoretical Computer Science. Put this into Scala and play with it a bit.

$$f(0,n) = n+1$$
$$f(m+1,0) = f(m,1)$$
$$f(m+1,n+1) = f(m,f(m+1,n))$$

6. Write a recursive function that will build the power-set of a `List`. The power-set is the set of all subsets of a given set. The fact that it is a set means that we do not care about order, unlike with permutations. For example, the power set of `List(1,2,3)` is `List(Nil, List(1), List(2), List(3), List(1,2), List(1,3), List(2,3), List(1,2,3)`.

7. Write a recursive function that builds combinations of a `List` instead of permutations. The caller should be able to specify how many elements are desired in the combination. (Hint: All combinations of all sizes are part of the power-set.)

15.8.3 Projects

1. If you have been doing the ray-tracing and graphics options, this project continues with that. Ray tracers can implement both reflections and transparency using recursion. If you have a function that tells you what color to draw for a particular ray, you can do reflection by recursively calling that function with the reflected ray and combining that color with the color of the reflecting surface. How much of each goes into the color depends on the fractional reflectivity of the object.

   For this project you will give your objects colors, but you still will not worry about lighting. So when a ray hits an object, it gets the full color of that object unless the object is reflective. If it has reflectivity R then it gets $(1-R)$ of the color of that object and R of the color from the reflected ray. To add colors you need to be able to build your own colors and get the components out of colors. The `java.awt.Color` class has methods called `getRed`, `getGreen`, and `getBlue` that return the amount of each primary in a color. The values are `Ints` in the range of 0-255 inclusive. You can also make a new `Color` object by calling `new Color(red:Int,green:Int,blue:Int)`. The values passed in must be in the same range or you will get an error.

   To calculate reflection, you need the unit-normal vector for the surface you are reflecting off of. For a plane this is easy as the normal is the same everywhere. For a sphere, this is the vector from the center to the point of intersection. In both cases, you want it normalized. If you have that, the direction of the reflected ray is given by

$$\vec{r}_{reflected} = \vec{r} - 2\,(\vec{r}\cdot\vec{n})\,\vec{n},$$

   where $\vec{r}$ is the direction of the incoming ray and $\vec{n}$ is the unit normal. The unit normal will also be significant for calculating lighting in a later project.

2. For this project you will write a program that solves 9x9 Sudoku puzzles. It will do this through a recursive function.

   The input will be nine lines of text, each with nine characters on it. The characters will either be a space or a single digit number. You should output a board with all the blanks filled in and put spaces between the numbers to keep it readable. It turns out that most puzzles have more than one solution. You only need to provide one.

For an extra challenge, make it so your program has a Graphical User Interface (GUI). It should load puzzles from a file and use a `Table` or a `GridPanel` to display the puzzle. Users should be able to select a solve option to have other spaces filled in.

3. For this project, you will write a program that recursively parses a string for an arithmetic expression and returns the value of it. Examples would be `5+7` or `9*(5+7.5)/2`. Your parser should do proper order of operations (things in parentheses bind highest, `*` and `/` before `+` and `-`, and go from left to right for the same level of priority).

   The approach I want you to take for solving this problem is with a divide and conquer recursive algorithm that breaks the problem into pieces starting with the lowest priority operation. You will write a function `parse(exp:String):Double`. This function will return the value of the expression in the `String exp`. First it should find the lowest priority operator (it can not be in parentheses). If it finds one, it recurses twice with the substrings on either side of that operator (use the `substring` method of `String` or `take` and `drop`). If there is not an operator that is not in parentheses you can check if the `String` starts with '(' and pull the bounding parentheses off and recurse on what is left. If it does not start with '(' you know that part of the expression is just a number so use `toDouble` to get the value.

   The user will give you a formula, that does not include any spaces, at the command-line and you should simply print the value it evaluates to. So a potential invocation of your program might be as follows: `scala parser.scala 5+3*(70/5)`.

4. If you did project 7 (p.367), you can extend it in this project. If you go further into chapter 1 of "The Algorithmic Beauty of Plants", you will find that you can use L-systems to model grasses and trees. We will not do the 3-D implementation, but you can add handling for '[' and ']' to allow branching structures. This is best done using recursive calls that start on a '[' and return on a ']'. Instead of passing in a `String`, pass in the `Iterator[Char]` that you get by calling the `iterator` method on a `String`. The advantage of the `Iterator` is that the elements are consumed when you call next so that when you return from a function call, the code will automatically be working on what remains after the ']'.

5. Determining if a particular recipe can be made with items from your pantry is not all that hard and does not require using recursion. Planning an entire meal for a large dinner party is another matter. To do this, the recipes need to be annotated with what type of dish they are: entrée, dessert, side, etc. When planning a meal, the user must be able to specify how many of each type of dish they wish to make for the meal. You can then use recursion to find the meals that can be made fitting those requirements with ingredients that are on hand.

   You should attach information on how much the user likes certain recipes so that only the top 5 or so meals are shown. If you want a bit of an extra challenge, consider handling the situation where there are not any meals that can be made with what is on hand and then you list top meals that need few additional ingredients.

6. If you have been doing the schedule building problems, you can now extend the functionality using recursion. Annotate each course with how it fits into your curriculum. Then you can specify not only how many hours you want to take, but also what requirements you want to fulfill.

   This is a problem that you could solve using `permutations` and `combinations`, but as the number of course options grows, those options become less efficient. Using recursion, you can cut the recursion short for any combination if you determine that it can not possibly work. For example, if you hit the proper number of hours before you are done with courses, you do not need to consider variations in the remaining courses, just check if you have satisfied the other requirements.

7. Having the ability to do recursion opens up a lot of possibilities for the motion of computer-controlled characters in games that have barriers. Moving straight to a location is fine if the playing space is open and empty, but when there are obstacles, like in a maze, it is important to have a way to navigate them. Recursion gives you a way to do this. For this project you can implement your choice of simple game with enemies that use recursion to find ways around obstacles. Note that you probably do not want the enemy to have the ability to follow the shortest path to the player unless the player has a significant speed advantage. Instead, you can throw in some randomness to take choices that are somewhat less than optimal.

8. If you have been doing the text-adventure/text-map project options, you can write some utility functions that can be used to help check out the map. Recursion can let you see things like if you can get from one room to another, how many steps it takes, how many paths there are, or even the paths themselves. You can implement these as new commands when you run the program. For example a "canReach" command could be given a room number/identifier to see if the specified room can be reached from the current room.

9. If you have been working on the music library, you can throw some recursion into that as well if you give a few hints. You would need to annotate songs with hints as to what are good options for songs to follow it. You can make the recursion only follow from one song to another when the user has recommended that it is worth doing. Those recommendations can be given a "strength" as well. The user should select a starting song and the program should find the playlist with the highest total strength for all the connections.

   Note that this is a problem where you could run into problems if there are lots of songs with lots of connections. Doing this with a simple recursive algorithm could lead to extremely long runtimes. However, assuming that the user does not enter too many connections, recursion should work fine as long as you do not test song combos that have not been marked by the user. This is why you can not simply consider every ordering of songs.

Additional exercises and projects, along with data files, are available on the book's website.

Part II

Object-Orientation, Abstraction, and Data Structures

Chapter 16

Object-Orientation

From the beginning we have said that Scala is a completely object-oriented language. You have been using objects and calling methods on them since back in chapter 3 when we were taking our first steps in the language. Despite this, we have actually done very little with object-orientation. This has been intentional. So far we have been building up your logic skills and teaching you to break problems up into different functional units. Now that your problem-solving skills have reached a sufficient level, we will take the next step to have you doing full object-oriented decomposition of problems. Beginning with this chapter we will also start really revealing many of the details behind things that we have been using all along.

16.1 Basics of Object-Orientation

The basic idea of an object is that it is something that contains both data and the functionality that operates on that data. We grouped data originally with tuples and then in chapter 10 we started using `case classes` to group data together in a more meaningful way. What we did not do with either of these was to bind any functionality to them directly. We wrote functions that could operate on them, but it was independent of the data itself. In the case of tuples, we can not bind the functionality into objects. `case classes` do allow us to do this, we simply did not.

In this chapter we will learn how to do proper object-oriented programming in Scala. This is the model that we will use through the rest of the book as well. Scala is a CLASS-BASED object-oriented programming language. That means that programmers create `classes` in their programs and those `classes` define types that are used to make objects. We already saw how the `case classes` that we made defined new types. We did not put anything in them though except some data elements that we wanted any function to be able to access.

The `case` keyword in front of a `case class` worked well for our needs in chapter 10, but it is not required to define a `class`. We will be writing normal `classes` now and only using `case classes` when they are called for. The way that you should think of a `class` is as a blueprint for an object. The `class` tells Scala what goes into the objects of that type and what type of functionality those objects should have. The syntax for a `class` is as follows.

```
class TypeName(arg1:Type1, arg2:Type2, ...) {
  // Methods and Members
}
```

As you can see, the only things that are different between this and the `case classes` we have already worked with is the lack of the keyword `case` and the curly braces with stuff in them after the `class`. These curly braces, like all curly braces in Scala, define a new scope. This scope is associated with the `class` and anything you want to put in the class goes inside of them. There should be nothing too surprising about that.

16.1.1 Methods and Members

Inside of the `class` you can write code exactly the way you have anyplace else. There are small differences that will be discussed and the meaning and terminology changes a bit. When you write a variable declaration with `val` or `var` inside the body of a `class`, that is now a member declaration. It is data that is stored in the objects created from that `class`. When you use `def` to create a function, we call it a method because of its scope in the `class` and the fact that it is something that you will be able to call on objects created from this class. Any statements that you write that are not declarations of members (`val` or `var`) or methods (`def`) will simply be run when an object of this type is instantiated.

To see how this works, we will write a little script that includes a `class` declaration. For the `class` we will go back to the concept of grades in a course and write the `class` to represent a student. Here is a first draft of the `class` with some lines that use the class.

```
class Student(name:String,id:String) {
  var tests = List[Double]()
  var quizzes = List[Double]()
  var assignments = List[Double]()
```

```
  def testAverage = tests.sum/tests.size
  def quizAverage = quizzes.sum/quizzes.size
  def assignmentAverage = assignments.sum/assignments.size
}

val john = new Student("John Doe","0123456")
john.tests ::= 78
john.tests ::= 85
println(john.testAverage)
```

The `class` begins with a `class` declaration that looks a lot like a `case class` and really is not too different from a `def`. After the keyword `class` is the name of the type, in this case `Student`. This is followed by a list of parameters to the `class`. In this case we have two `Strings` called `name` and `id`. The body of the `class` is denoted with curly braces and inside of the curly braces we have three member variable declarations for grades and three method declarations for taking averages. More will need to be added, but we need to explore what this really means before we can do that.

After the class declaration are four lines of code. The first declares a variable called `john` that is set to be a `new Student` with the name `John Doe` and a student number of `0123456`. The second and third statements adds two test grades of 78 and 85 to this student. While this is simple, it is not clear that we really want the grades to be this accessible. We will talk more about that later. The last statement in the script prints the average for the tests by calling the `testAverage` method.

16.1.1.1 Parameters as Members

Everything in this example should seem fairly straightforward. It really does not look much different from things that we have done before. This would change though if we try to make the print statement a bit more informative.

```
println(john.name+" has a "+john.testAverage+" test average.")
```

If you change the `println` to this and run the script you will get an error message that might not make much sense to you.

```
ScalaBook/ch17.scala:15: error: value name is not a member of this.Student
println(john.name+" has a "+john.testAverage+" test average.")
             ^
one error found
```

You should definitely find this surprising. This would have worked for a case class. Indeed, if you put the `case` keyword at the beginning of the `class` declaration this works fine. It is worth exploring why it does not work without the `case` keyword and how we can fix it.

The error message tells us that `name` is not a member of `Student`. This is because, by default the arguments passed into a `class` are not turned into members. The logic behind this is that members have to be stored by each instance of the `class` that is created, so extra members consume memory. Even if they do have to be remembered, they will not be visible to any code outside of the `class`. For example, if you add the following method into the class you can call it to get the last name, but you still can not get the name itself.

```
  def lastName = name.split(" +").last
```

If you want an argument to be a member simply put `val` or `var` before the name. It will then become a member with the proper behavior for either a `val` or a `var`. For our `Student` class we likely want both to be constant so we should change the class declaration to the following.

```
class Student(val name:String,val id:String) {
```

If you make this change you can use the print statement shown above without an error.

16.1.1.2 Visibility

The possibility of things being visible or not is pretty much new to us, but it is a very important aspect of object-oriented programming because it allows you to hide details of the way an object works away from the code that uses the class. The value of this is that you can change those details without causing problems for the code that depends on the class to work. In the case of member data, you can also hide away details that you do not want other parts of code to have free access to.

A great demonstration of this last part comes in the form of a bank account. Let us consider the simplest possible class that we might write to represent a bank account.

```
class Account {
  var balance=0
}
```

This class does not take any arguments when it is constructed and, as a result, the parentheses have been left off. It has one piece of member data, an Int for the balance. As you might recall from chapter 3, money should be stored in Ints because the arithmetic of the Double type is inaccurate due to the limited precision. This representation is simple, but it has a significant drawback. The member balance is a var that can be set by any code that gets hold of an Account object. For a real bank program that would not be acceptable. The balance can not just be any value. For example, negative numbers generally are not allowed. In addition, changes to the balance are typically supposed to happen in specific operation, deposits and withdraws. Those operations can enforce certain rules and, for a real bank, would log any changes to the balance so that it is possible to go back and figure out what happened. While we could make functions for those operations, there would be nothing that forces programmers to use them. It would be possible for a programmer to "get lazy" at some point and access the balance directly. The results of this in banking software could be quite extreme.

The ability to force access to data to occur through certain methods is one of the most significant capabilities of object-orientation as a way of improving software construction and controlling complexity. The real benefit of this capability is to take responsibility for proper usage away from the users of the objects. If the only ways to use the object are through methods or data that is safe for them to use, then they know they can not mess it up and do not have to worry about that. Instead, they can focus on the logic that they are working on.

So how do we accomplish this in our classes? We do it by setting the visibility of members. There are three main levels of visibility in Scala and they are very similar to what you will find in other class-based object-oriented languages.

- Public - This means that something can be accessed by any code inside of outside of the class. In Scala this is the default visibility for elements so there is no keyword for it.
- private - This means that the member can only be accessed inside the class. Attempts to get to the member outside of the class will result in a syntax error. If

you prefix a declaration in a `class` with the `private` keyword, that member will be `private`.

- `protected` - This is like `private` except that `protected` members are also visible in subtypes. This will be discussed in detail in section 19.2.2.

In our example `Account` we really want the `balance` to be `private`. We can do this by simply adding it before the `var`.

```
class Account {
  private var balance = 0
}
```

Unfortunately, this leaves us with a `class` that is completely useless. Before making this change we could have done something like the following.

```
val myAccount=new Account
println(myAccount.balance)
myAccount.balance += 100
println(myAccount.balance)
```

If we do this now though we get a set of error messages.

```
/ScalaBook/ch17.scala:24: error: variable balance cannot be accessed in
     this.Account
myAccount.balance += 100
          ^
/ScalaBook/ch17.scala:24: error: reassignment to val
myAccount.balance += 100
                  ^
/ScalaBook/ch17.scala:25: error: variable balance cannot be accessed in
     this.Account
println(myAccount.balance)
                  ^
5 errors found
```

Note how the error message is different than what we saw previously for `name` on our `Student class`. Instead of telling us that `balance` is not a member, it tells us that `balance` can not be accessed. This is precisely because the `balance` is now `private` and the code is outside of the `class`.

To make this `class` useful, we would need to put some methods in it that are public that manipulate the `balance` in allowed ways. We will start with methods for `deposit` and `withdraw`. Here is what the `deposit` method might look like. Note that this is indented because it appears inside of the `Account class`.

```
  def deposit(amount:Int):Boolean = {
    if(amount>0) {
      balance += amount
      true
    } else false
  }
```

The method takes the `amount` to deposit as an `Int`. It returns a `Boolean` to tell us if the deposit went through. This method does not support logging because adding that functionality goes deeper than we want to at this point. It does check to make sure that the `amount` of the deposit is positive. If it is, then that amount is added to the balance and we give back true, otherwise nothing is done and we give back false.

We can add a very similar looking withdraw method.

```
def withdraw(amount:Int):Boolean = {
  if(amount>0 && amount<=balance) {
    balance -= amount
    true
  } else false
}
```

The only difference between deposit and withdraw is that you can not withdraw more money than you have and the amount is subtracted from the balance.

The last thing we need to add is a method to get the balance so that it can be checked. This method might look like the following.

```
def getBalance = balance
```

It simply gives back the balance in the account. This style of calling a method that retrieves a member using a method that begins with "get" is standard practice in Java. For reasons that we will see later, this is not the recommended style in Scala. Instead, in Scala we would use a different name for the private member and use a method with the normal name for getting the value. There are different approaches to how you name the private var. In this book we will typically use a shortened version of the word, in this case bal. You can pick whatever style you want, but be consistent with it. Having made this change, the class as a whole looks like this.

```
class Account {
  private var bal = 0
  def deposit(amount:Int):Boolean = {
    if(amount>0) {
      bal += amount
      true
    } else false
  }
  def withdraw(amount:Int):Boolean = {
    if(amount>0 && amount<=balance) {
      bal -= amount
      true
    } else false
  }
  def balance = bal
}
```

We can utilize this code by doing the following.

```
val myAccount=new Account
println(myAccount.balance)
myAccount.deposit(100)
println(myAccount.balance)
```

Note that because of the use of the name balance, the println statements appear the same way they did when balance was a public var. The difference is that you can not set balance now unless you go through either the deposit method or the withdraw method.

Advanced Visibility Options

Scala has more flexibility in visibility modifiers than what is described here. Either `private` or `protected` can be followed by square brackets and the name of an enclosing scope of a `package`, `class`, `trait`, or `object` type. (The `object` construct will be introduced later in this chapter. The `trait` construct will appear in a few chapters.) This will make that member publicly visible for everything at that scope and closer to the declaration. The primary benefit of this is that you can give special access to closely related bits of code without making a member public to the whole world.

The square brackets can also include the keyword `this`. Declaring a member to be `private[this]` means that is it `private` not only to other pieces of code, but also to any other instances of the enclosing `class`. Normally if a method gets hold of a reference to another object of the same type, they can access `private` members. This will not be the case if the member is `private[this]`.

16.1.2 `this` Keyword

By default, when you write code in a `class` that calls a method or uses a member of that `class`, the call is made on the current object or the value from the current object is used. All calls are implicitly made on the current instance. That is exactly what you want most of the time and given the scoping of variables it feels very natural. This implicit specification of an object prevents you from having to specify one. That is fine in general, but occasionally you will have a need to be able to put a name to the current object. For example, you might need to call a function/method that needs an instance of the object as an argument and you want to use the one that the code is currently executing on. When that situation arises you will use the `this` keyword. `this` is a name that always refers to the current instance in a `class`.

Scala is implicitly adding `this.` in front of your calls to methods and when you access local members. For example, the line

```
bal += ammount
```

is really short for

```
this.bal += ammount
```

Scala simply added the `this` for you.

16.1.3 Encapsulation/Separating Interface from Implementation

The binding of data and functions together is called ENCAPSULATION. It might not seem like much, but it can have a profound impact on the way that software is developed. Having functionality bound into them makes objects far more self-sufficient. It can also free the programmer from having to think about many types of details.

One of the goals we would like to strive for in our object-oriented programs is to have SEPARATION OF INTERFACE AND IMPLEMENTATION. The interface of an object or some other module of code is the set of constructs that you can interact with directly. For objects that are instances of a `class`, that would be the public members and methods. The implementation of that interface is what goes on to make the interface do what it is supposed to do. You want these things separate for a few reasons.

One reason is so that modifications to the details of the implementation do not break other code. If the interface and implementation are truly separated, then the code that makes calls to the interface does not depend at all on the details of the implementation. All that it depends on is that the elements of the interface adhere to a particular CONTRACT. That contract tells you what information is needed, what information you will get back, and what side effects, if any, will happen in between. It should not tell you how things are done in between. It should be possible to completely redo the details of how things happen, as long as the new approach does not violate the contract, and still have any code using the implementation work just fine.

Another reason for trying to keep interface and implementation separated is that the interface is typically much simpler than the implementation and it benefits coders to not have to keep too many details in their heads at once. Software has the ability to be arbitrarily complex. This is both a blessing and a curse. The flexibility that this provides is part of why software can do so many different things. However, humans can only keep so many details straight in their head at any given time. When the complexity of the software goes beyond what the programmer can deal with, it is essential that some parts of the complexity can be safely ignored.

The reality is that you have been taking advantage of this separation in everything you have done in this book and will continue to do so as long as you program computers. When you call `println`, you do not really care about all the details that go on between your call and having characters appear on a display. Thankfully you do not have to think about them either. All those details could change and as long as `println` keeps its contract of sending your information to the standard output you will not care that it has been changed. This is not just true of `println`, but of most of the Scala library code that we have been using.

16.1.4 Special Methods

There are a few method names that are special when you put them into `classes` in Scala. In a way, these special methods are rules that you have to learn as special cases. Such things often weigh upon a language. In this case there are not many of them, and the power they provide is quite remarkable.

16.1.4.1 Operator Methods

The first set of things that we might consider special are operator methods. It turns out that these are not really special in Scala. In reality they are just like any other methods in Scala. You can use any method call that takes a single argument using operator syntax as was discussed back in chapter 3. That means you can leave out the dot and the parentheses and just put the operator between the two operands. What makes this more interesting is that Scala accepts names that have symbols in them and most languages do not. This includes names for methods.

We have been using names that contain only letters and numbers and begin with a letter.[1] You can also include symbolic operator characters in certain ways. You can have names that include only operator characters, or you can have letters and numbers, beginning with a letter, followed by an underscore and operator characters. Note that once you get to operator characters in a name, all subsequent characters must be operator characters.

So while we have been using names like `timesTwo` or `times2`, the names `*`, `**`, `*+++*`, etc.,

[1]Names can also start with and include underscores though that has also been avoided in this book.

Precedence (highest at the top)
other special characters
* / %
+ -
:
= !
< >
&
^
|
all letters
assignment operators

TABLE 16.1: This table shows the precedence rules for operators in Scala. The precedence is based on the first character in the operator.

are also completely valid. As with so many things in life, just because you can use symbolic operator names for methods does not mean that you should. They have their place, but if you abuse them it can lead to code that is very difficult to read. As a general rule, you should only use symbolic names when they make sense. Changing from text names that people can read to arbitrary symbol sequences will make it nearly impossible for you, or others, to understand what is going on.

When you call a method using the normal method notation, the order in which things happen is obvious. This is because the argument to the method is found inside of parentheses and the parentheses determine the order. When you use operator notation, there have to be PRECEDENCE rules. You already know that `2+3*4` is `14`, not `20`. That is because multiplication has higher precedence than addition. This is true in Scala as it is in math. However, the `+` and `*` are simply names for methods in `Int` in this expression. So how does Scala decide that the `*` should happen before, or have higher precedence than, the `+`? When methods are used in operator notation, Scala uses the first character in the name to determine precedence. Table 16.1 lists the different levels of precedence that are used.

In the case of `+` and `*`, the first character happens to be the only character. The table tells us that `*` has higher precedence than `+` so it is done first. If you put methods called `*+*` and `+*+` in a class, the `*+*` would have higher precedence by virtue of starting with `*`. If you also put a method named `^*^` it would have lower precedence than either of the other two because of the low precedence of `^`. These types of precedence rules exist in most other programming languages as well. The only difference with Scala is that you have the capability of defining our own methods that can be used as operators and have the appropriate precedence.

What if two operators are at the same level of precedence? Which comes first? For example, is `5-3+2` equal to `4` or `0`? If the `-` happens first you get `4`. If the `+` happens first you get `0`. The order in which operators of the same precedence occur depends on ASSOCIATIVITY. You are used to operators being LEFT-ASSOCIATIVE. This means that the furthest left operator of a given precedence is done first. Such is the case of `+` and `-` and hence the value of the expression `5-3+2` is `4`. However, there are some operations that should be RIGHT-ASSOCIATIVE. An example of this from standard math is exponentiation. For example $3^{3^3} = 3^{(3^3)} \neq (3^3)^3$. The numbers are grouped from right to left here, not the normal left to right of other operations.

An operator in Scala is right-associative if it ends with a `:`. In addition to being right-associative, it is called on the object to the right side of it and the value on the left side is the argument to the method. We have actually seen this behavior before with the `::` operator.

```
3 :: 4 :: lst
```

This expression only makes sense if the `::` operator is right-associative. The `::` operator is not defined on `Int`, it is defined on `List`. So Scala sees this as the following.

```
3 :: (4 :: lst)
```

This happens to be the same as this.

```
lst.::(4).::(3)
```

There are quite a few other methods in the Application Programming Interface (API) that work in this way. We will see some more in chapter 20.

There is one other aspect of symbolic operator methods that is worth discussing before moving on and that is assignment methods. We have seen statements like `cnt += 1`. The operator in this case is an assignment operator. It performs an operation and does an assignment all in one step. You are allowed to write methods with names like `+=` or `-=`. If you do, Scala will use those in expressions like `cnt += 1`. However, if Scala can not find a match for an operator that ends in `=`, it will try to expand it out. So if there is no `+=` defined on the type of `cnt`, then Scala will try translating the `cnt += 1` into `cnt = cnt + 1`. If that still does not work, you will get a syntax error.

The significance of this is that unless you really need specific behavior from a particular assignment operator, you do not have to write one. Instead, you will get the behavior you want automatically by including just the binary operator. So if you define `+`, you will normally get `+=` for free. Note that this only works for symbolic operators. There is no assignment form for normal methods even though you can use them with operator syntax.

16.1.4.2 Unary Operators

Not all operators take two arguments. Operators that take only a single argument are called UNARY OPERATORS. You can write these in Scala by writing methods that do not take any arguments. Any method you write, whether it is made of operator characters or normal letters and numbers, can be used as an operator if placed after the object it is operating on. To see this, consider the `toInt` method of `String`.

```
scala> "5".toInt
res0: Int = 5

scala> "5" toInt
res1: Int = 5
```

The first version here uses the "normal" dot syntax. The second version leaves off the dot and simply specifies `toInt` as a POST-FIX operator. Post-fix simply means that the operator comes after the argument. The binary operators we considered in the last section were used IN-FIX. For some unary operators, it is standard to use PRE-FIX notation instead of post-fix notation. Consider the `-` to make a number negative. With a value like `-5` you could say it is just part of the number. However, with `-(a+7)` the role of the `-` is clearly that of an operation that is performed on the result of `a+7`. The `!` and `~` methods are other examples we have seen for pre-fix unary operators.

One of the major goals in allowing methods to be used in operator syntax is that you can create your own types that work in the same way you would expect language features to act. Indeed, the pre-fix unary operations for +, -, !, and ~ are part of the Scala libraries. You can write methods in a class that are interpreted as pre-fix unary operators. To do so, simply name the method unary_*op*, where *op* is replaced by +, -, !, or ~. So if you wanted your own negation you could create a method called unary_-.

16.1.4.3 Property Assignment Methods

Back in the bank account example we said that having a method that starts with the word get followed by the value you were getting was the Java style of doing things and that in Scala you simply name a method with the name you want people to see for the property. In our example this was balance. In Java style the partner to the get method is a set method. In Scala that counterpart to set is an assignment method. The advantage of the assignment method is that it allows you to set a property using normal assignment syntax. To create an assignment operator, write a method with the name *prop*_=, where *prop* is the name of the property.

To help you understand this, let us assume that you did want to be able to enter something like myAccount.balance = 700. We had this ability with the public var, but we decided that was too little control. We could get that back by adding the following method.

```
def balance_=(b:Int) = bal=b
```

Thanks to the way Scala handles assignment operators, this will let both of the following lines work.

```
myAccount.balance = 700
myAccount.balance += 40
```

So doing += will work like a deposit and -= will work like a withdraw. The downside is that neither returns a Boolean to let you know if it worked. As written, neither checks the value being deposited or withdrawn either.

This particular method leaves things too open. We might as well have the var because we are doing exactly what the var would do. However, the advantage of an assignment method is that it can be more complex and have more logic behind it. For example, the balance_= method could be altered to the following.

```
def balance_=(b:Int) = {
  if(b>=0) {
    if(b<bal) withdraw(bal-b) else deposit(b-bal)
  } else false
}
```

This version will reuse the code from withdraw or deposit include any error checking, logging, etc. It also throws in checking so that assignments to invalid balances will not go through.

Note that you only include a set method for properties that are mutable so there are many situations where they are left off. If you are programming in a more functional style, you will not have these assignment methods. It is only when objects specifically need to be able to mutate that these are helpful. The real advantage in that case is that the code can look like a normal assignment, but will call the method that can do various types of error checking.

16.1.4.4 The apply Method

Things get a bit more interesting in a much more subtle way with the apply method. You write apply just like any other method that you might put in a class. You can call it just like any other method as well. However, Scala will allow you to call the apply method without using a dot or the name apply. When you remove those, it makes it look like you are treating the object as a function. Indeed, this is how all functions work in Scala. As was said very early on, everything in Scala is an object. That includes functions. A function is just an object of a type that has an apply method that takes the appropriate methods.

The intelligent use of apply is how you have been indexing into all of the collection types. Consider the following code that you could write in the REPL.

```
scala> val arr=Array(5,7,4,6,3,2,8)
arr: Array[Int] = Array(5, 7, 4, 6, 3, 2, 8)

scala> arr(3)
res0: Int = 6

scala> arr.apply(3)
res1: Int = 6
```

The call to arr(3) is actually just a shortcut for arr.apply(3). Scala is doing nothing more here than assuming the presence of the call to apply.

How can we use this in our own code? Consider that we write a class called CourseSection that will keep track of students in a course for us. We could make it so that we can "address into" the course section using either a student name or a student ID.

```
class CourseSection(val subject:String,val number:Int,
    val title:String) {
  private var students=List[Student]()

  def addStudent(stu:Student) {
    students ::= stu
  }

  def apply(key:String):Option[Student] = {
    students.find(s => s.id==key || s.name==key)
  }
}
```

The apply method here returns an Option[Student] because it is possible that there is no student that matches. In that case this returns the value None. Otherwise, it will return a Some[Student] with the proper Student data. With this in place you could write the following code.

```
val course=new CourseSection("CSCI",1321,
  "Principles of Programming II")
course.addStudent(john)
course("0123456").get.quizzes += 88
```

This makes a new course then adds in the student we created above that is in the variable john. The last line looks up our student by his ID and adds a quiz grade. The call to get is required because of the fact that apply returns an Option. Note that if the course did not have a matching student, this code would result in an error.

16.1.4.5 The update Method

Having an apply method is sufficient for many situations. For example, it is all that is needed for immutable collections. It is also all we will want for our CourseSection class given how we have written it. However, it is not sufficient if you want to do something like an assignment into an Array. Basically, it looks things up, it does not change things. To do this, you use the update method. When Scala sees an expression like this

```
obj(args) = e
```

it converts it to a call of the form

```
obj.update(args,e)
```

So you can get assignments for indexing by simply implementing the proper update methods.

16.1.4.6 Overloading Methods

Methods can be overloaded. This means that you can have two different methods that have the same name as long as they take different numbers of arguments or different types of arguments. The ability to overload methods can come in handy, but it is something that you do not want to abuse as it can make it difficult for readers to understand what is going on or even make code ambiguous so that it does not compile. In the case of special methods, which have to have a particular name, overloading is the only way to make it so that you can get the desired functionality with different types.

To understand this, consider that you have a class that represents a grid of numbers, perhaps for a spreadsheet. This is a situation where it makes sense to use the apply method to get values out. However, there are several ways to refer to cells in a spreadsheet. The most obvious approach in a program would be to pass two Ints for the row and column. However, when you run a spreadsheet program it is common to refer to cells using letters for the column and numbers for the row. You might want to allow these to be passed in a String and an Int (e.g., "C", 13) or just a String (e.g., "C13"). With overloading you could write three different apply methods and update methods as shown here.

```
class Spreadsheet(rows:Int, columns:Int) {
  private val data = Array.fill(rows,columns)(0.0)

  def apply(c:Int, r:Int) : Double = data(r)(c)

  def update(c:Int, r:Int, v:Double) = data(r)(c)=v

  def apply(c:String, r:Int) : Double = data(r)(calcColumn(c))

  def update(c:String, r:Int, v:Double) = data(r)(calcColumn(c))=v

  def apply(cell:String) : Double = {
    val (c,r) = cell.partition(_.isLetter)
    data(r.toInt)(calcColumn(c))
  }

  def update(cell:String, v:Double) = {
    val (c,r) = cell.partition(_.isLetter)
    data(r.toInt)(calcColumn(c))=v
  }
```

```
  private def calcColumn(c:String) : Int = {
    c.toLowerCase.foldLeft(0)((v,c) => v*26+(c-'a'))
  }
}
```

Using this code you can do things like this.

```
val ss = new Spreadsheet(10,10)
ss("c5") = 14
ss(3,8) = 24
println(ss("c",5)+" "+ss("d8"))
```

This provides a very natural way to access elements using whatever style is convenient. In general you probably should not overload a method if you could use a slightly different name. Much of the time, different names will lead to code that is less error prone and easier to read.

16.1.4.7 Vector Example

To help demonstrate how these different features can work we will write a class that represents a 3-D vector. We will give it `x`, `y`, and `z` components then define the types of operations that we would want to be able to perform on it. One of the first major design decisions we have to make for this is whether we want the vector objects to be mutable or immutable. For this example we will go with the immutable version as it does a better job of demonstrating functional design and works better here. This approach is more memory intensive and can be slower in some applications because it does not mutate the vectors. However, it often produces better, more reliable code that, in some applications, will be faster because it prevents defensive copying. That is where you make a copy of an object to pass into other code when you do not want your version of the object to get changed.

We will start with the declaration of the `class` that includes operators for `+`, `-`, `*`, and `/`. It also has two versions of `apply`. Lastly there is a method called `toString` that we will learn more about in chapter 19. This method determines what we see when an object of the `Vect` type is printed out. It is very useful to have a form that helps you see what is going on to correct problems.

```
class Vect(val x:Double, val y:Double, val z:Double) {
  def +(v:Vect):Vect = new Vect(x+v.x,y+v.y,z+v.z)

  def -(v:Vect):Vect = new Vect(x-v.x,y-v.y,z-v.z)

  def *(c:Double):Vect = new Vect(c*x,c*y,c*z)

  def /(c:Double):Vect = new Vect(c/x,c/y,c/z)

  def apply(i:Int):Double = i match {
    case 0 => x
    case 1 => y
    case 2 => z
  }

  def apply(is:Seq[Int]):Seq[Double] = {
    is.map(Array(x,y,z))
  }
```

```
  override def toString = "<"+x+", "+y+", "+z+">"
}
```

The first version of `apply` lets you index into the different coordinates. It is written using a `match` expression though there are certainly other ways it could be done. The second version of `apply` is more unusual. It allows you get the `x`, `y`, and `z` values back as a sequence by passing in a sequence of indexes.[2]

With this code written, we can instantiate objects of the `Vect` type and perform the specified operations on them.

```
scala> val v1=new Vect(1,2,3)
v1: Vect = <1.0, 2.0, 3.0>

scala> val v2=new Vect(2,-4,7)
v2: Vect = <2.0, -4.0, 7.0>

scala> val v3=v1+v2
v3: Vect = <3.0, -2.0, 10.0>

scala> val v4=v3*8
v4: Vect = <24.0, -16.0, 80.0>
```

Note that with the code that we have here, we have to put the scalar on the right of the multiply. If you tried to do `8*v3` you would get an error. This is because Scala sees that as `(8).*(v3)`. The value `8` is an `Int`, it does not have a version of `*` that works on a `Vect`. It is possible to get around this, but it requires another more advanced concept called an implicit conversion which we do not want to consider at this time. For now we will be happy putting the scalar on the right.

There are two other common operations on 3-vectors: the dot product and the cross product. The dot product returns a scalar and the cross product returns a vector. When writing these, we have to make a significant decision, do we want them to be symbolic or words? We can not use just `*` for both because Scala could not figure out which one we meant to use at any given time. For that matter, people reading the code probably could not either. Scala does give us the ability to use longer symbol names with multiple operator symbols. So maybe `*+*` for a dot product and `*-*` for a cross product. You can not use `*.*` and `*x*` because `.` and `x` are not operator characters. The choice of `+` and `-` in these comes from knowledge of the formulas for dot and cross products, but neither really fits well or will read to most people as dot and cross. For that reason, we will stick with normal alphabetic method names.

```
  def dot(v:Vect):Double = x*v.x+y*v.y+z*v.z
  def cross(v:Vect):Vect =
    new Vect(y*v.z-z*v.y,z*v.x-x*v.z,x*v.y-y*v.x)
```

We can still use these in operator notation. The only problem is that characters have lower precedence than multiplication would so parentheses will have to be used to enforce order of operation.

There is no `update` method provided because these are immutable. You could, in theory, make them mutable and provide an `update`. You would probably also want to provide `+=`, `-=`, `*=`, and `/=` operators that mutate instead of making a new copy. Doing so is left as an exercise for the reader.

[2]This type of indexing is a standard operation in MATLAB® where dealing with arrays and matrices is the primary usage of the language.

16.1.5 `object` Declarations

Scala has another type declaration that is similar to a `class` in syntax, but which has a very significant difference in meaning. The `object` declaration creates a single object in the current scope with the members and methods that you put in it. Unlike a class, you can not pass an `object` any arguments. This is because you do not make the `object` with `new`. The declaration of the `object` is what makes it.

Even though an `object` declaration does not create a new type, we use a capital letter as the first letter in the name to distinguish it from normal instantiated objects. To use the members or methods of an `object`, use the `object` name and call it just like you would for an instance of a class. `object` declarations can be used in any scope where you want an object, but you only want one object of that type. At the top level they are typically used as a place to organize functions, but they have far more versatility than this overall.

16.1.5.1 Applications

The fact that you can use an `object` without instantiation is very significant. The `objects` effectively exist as soon as the program starts running. You do not have to have a line that makes a new one. This is most significant when we move beyond the REPL and scripts and into full-blown applications. A top-level `object` declaration defines an entry point to an APPLICATION if it contains a method called `main` that takes an argument of type `Array[String]` and returns `Unit`. Here is a simple example:

```
object FirstApp {
  def main(args:Array[String]) {
    println("My first application.")
  }
}
```

The scripts that we have been using can be converted to applications by doing nothing more than putting the code inside of a `main` that is in an `object`. If you use the variable name `args` for the command-line arguments, then references to them in a script will even work.

The main difference between a script and an application is how you organize the code in files and how you run them. By definition, a script is supposed to be a small program. All of the code for a script goes into a single file. When the amount of code that is required gets larger, it needs to be split apart and organized. The organizational structure you should follow in Scala in nearly all situations is to put each top-level `class` or `object` in a separate file that is the name of the `class` or `object` followed by ".scala".

Consider the example of a grade book application. We might use the `Student class` from above and put it in a file called "Student.scala". The `CourseSection class` from above would go in "CourseSection.scala". Finally, we would have an `object` declaration for a grade book that would go in "Gradebook.scala". In addition to having a `main` method, this would keep a collection of courses and provide whatever functionality we wanted for the grade book, whether it was a text or graphical interface.

Running an application like this is a two-step process. First you have to compile the code. This is done with the `scalac` command, which stands for "Scala compiler". You have to tell the Scala compiler what files it is you want to compile. If you organize your code so that all the code for an application is under a directory you can execute something like `scalac *.scala`. This will compile all of the files ending in ".scala" in the current directory.

When you do this, if you have any syntax errors they will be found and reported to you in much the same way they were when you ran programs as scripts. Once you have found and fixed all syntax errors, the compiler will produce a set of files in compiled BYTECODE that you can run. These files will end with the ".class" extension. To run the application you go back to using the `scala` command. Only now you follow it by the name of the `object` you want to run the `main` from. So in the case of our grade book we could run it by entering `scala Gradebook`. Other command-line arguments can be specified after the name of the `object`.

The application has to be an `object` because methods in a `class` can only be called on instances of that `class`, not the `class` itself. So in order to call a method you have to have executed `new` on that type. However, `main` has to be called right off the bat, before any other logic has been executed. This is not a problem with an `object` as the single instance exists without a call to `new`.

16.1.5.2 Introduction to Companion Objects

The most common use of `objects` is as COMPANION OBJECTS. A companion `object` is an `object` that has the same name as a `class`. Both the `class` and its companion `object` should be placed in the same file. The companion `object` has access to `private` members of the class it is a companion with. Similarly, the `class` can see `private` elements in the `object`.

You might have wondered why it is that when we are building objects in Scala, sometimes we use `new` and sometimes we do not. The reality is that making a new object always invokes `new`. When you do not type it, it means that you are calling other code that does. When you use the name of the type without `new` to build an object, you are calling the `apply` method on the companion `object` and that `apply` method is calling `new`.

Unfortunately, you can not easily write `classes` with companion `objects` in the REPL or in scripts. For us to create our own companion `objects`, we have to be working in the mode of writing applications, compiling with `scalac`, and running the compiled files with `scala`. We will come back to this next chapter.

16.1.6 Object Decomposition

At first, the use of object-orientation might not seem all that significant. Anything that is a method could be written as a function that takes the object as a first argument. However, over time you will come to see that bundling data and functionality actually has the ability to change a lot of how we approach solving problems. Instead of simply thinking of what functions you should break your problem down into and how to group the data as separate considerations, you take them together. You think of what data members and what functionality belong together in objects.

It is common to use the analogy when teaching object-orientation that objects are nouns and the methods are the verbs that act on those nouns. This analogy oversimplifies the situation though. Objects in programming do not always have to reflect things that we consider to be objects in the real world. They can represent much more abstract concepts. Indeed, functions in Scala are objects. A large portion of the second half of this book will be devoted to helping you see how we can use objects to express various ideas and to make our lives easier as we develop software.

16.2 Revisiting the API

Another significant change in the second half of the book will be our reliance on the API. In the next few chapters we will be uncovering many elements of Scala that will help you to understand different parts of the API and you will be expected to go into the API to get information on what you can do with the libraries.

We have already learned enough to help a bit with understanding what some things mean in the API. When you open up the API, the left side has a frame that looks like figure 16.1. We have used the generic word "type" to describe the things that are listed in this frame. Now we can be more specific. There are really three different declaration styles in Scala that appear over there and the API indicates which one you are dealing with.

The circles next to the names contain one of three letters in them. The meaning of these letters is as follows.

- c - For a `class`.
- o - For an `object`.
- t - For a `trait`, which we will learn about in chapter 19. For now you can think of a `trait` as being like a `class`.

When you click on one of these, you are shown the methods and members that are defined in them, whether it be a `class`, an `object`, or a `trait`. The API itself is built by running a program called `scaladoc` on Scala code to generate Hyper-Text Marup Language (HTML) descriptions of what is in the code.

When there is an "o" next to a "c" or a "t" it is a companion `object`. If you want to know if you can build objects without a direct use of `new`, look there and see if there is an `apply` method defined. If there is, it generally implies that is the preferred way to make instances of that type.

16.3 Meaning of `case classes`

The first usage of the term `class` that we encountered was back in chapter 10 when we looked at `case classes`. Now that you know a bit about regular `classes`, you might wonder what `case` keyword does for a `case` class. It does not do anything that you could not do without the use of `case`, it simply adds a lot of code for you.

First, all the data passed into the `case class` are turned into public `vals` instead of just being arguments. Second, it automatically creates a companion `object` that has an `apply` method that you can use to make instances of the `case class` without using `new`. It also adds a `copy` method to the `class` that has default arguments for all the values. There are a few other things added in that allow the pattern matching that we have not talked about yet.

16.4 `import` Options

The usage of `import` that we have seen so far with all the `import` statements at the top of the file giving full package names is the most basic usage in Scala. It mirrors the way `import` has to be used in Java. There are a lot of additional options in Scala that we have not really considered. Here is a list of some details of `import` starting with the general usage. Each one includes situations where you might want to use it.

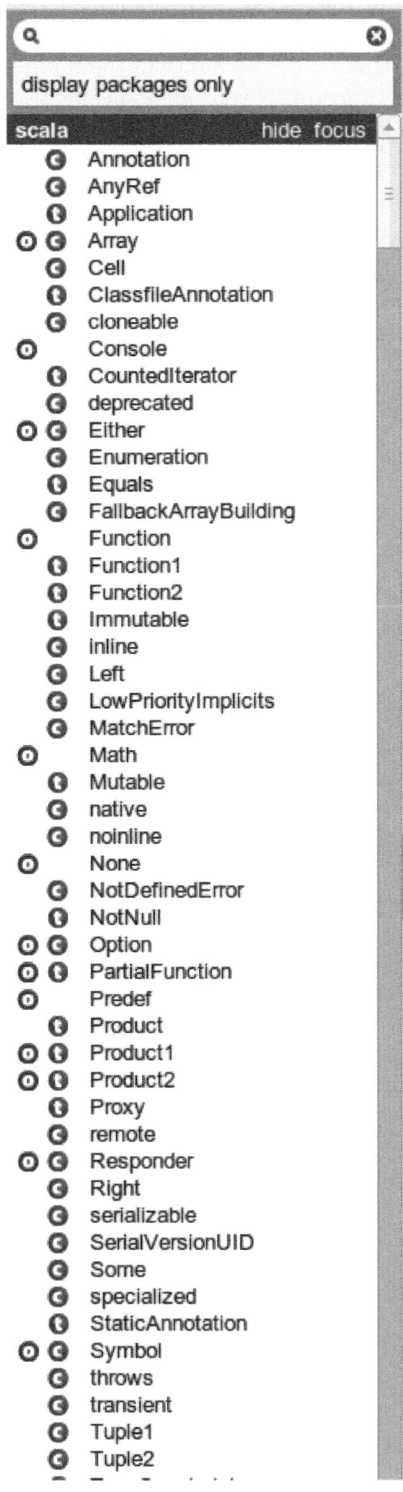

FIGURE 16.1: This is part of the right panel that you see when you first enter the API. The little circles with the letters c, o, and t indicate whether it is a `class`, `object`, or `trait`.

- So far, we have been putting `import` statements at the top of the file. This is not required though, they can go anywhere. `import` follows normal scoping rules. So if you put the `import` at the top of the file then the names brought in by it are visible through the whole file. However, if you put it inside of curly braces, the names that are `import`ed will only be visible through the end of the current scope.

- As we have seen before, you can use an underscore as a wildcard to represent everything in a certain scope.

- If you do not want to `import` everything, but you do want to `import` several things, you can put multiple names inside of curly braces.

- Names of `package`s and `class`es in Scala are truly nested. So if you have `import`ed all the contents of some `package`, and want to then `import` a sub`package`, you can leave off the first part of the name for the sub`package`. For example, after you do `import scala.swing._` you can do `import event._` instead of `import scala.swing.event._`. For that matter, because `scala._` is `import`ed by default, you can leave that off of everything.

- So far we have used `import` to bring in the contents of packages. It can be used to bring a lot more into the current scope. You can import package names or the members of objects. Things like the members of objects should definitely be done in limited scopes.

- You can give a different name to things that you `import` using a rocket, `=>` inside of the curly braces. Put the name the thing normally goes by on the left side and the name you want on the right side. For example, if you need to use the `java.util.List` you probably do not want it hiding the `List` in Scala. To get around this you could do `import java.util.{List => JUList}` so that the name `JUList` refers to the Java version. You can also hide a name by putting an underscore to the right of the rocket. So if you want to `import` all of the `java.util` package except the `List` you could do this: `import java.util.{List => _, _}`.

- Due to the nesting of packages, there are times when a particular name could refer to more than one package. To remove ambiguity, that name `_root_` can be used to refer to the base below all packages. So `import _root_.scala.io.Source` can be used to refer to the `Source` type even if there are conflicts with the names `io` or `Source`.

16.5 End of Chapter Material

16.5.1 Summary of Concepts

- Objects are constructs that combine data and the functionality that operates on that data.

- The standard way of making objects in Scala is to define a `class` and instantiate objects from it. You can think of a `class` as a blueprint for making objects.

  - A `def` declaration in a class defines a method. A `val` or `var` declaration defines member data.

  - classes can take parameters. By default they are not members. Adding val or var in front of the name makes it member data.
  - By default, constructs declared inside of a class are public and can be seen by any code. A big part of the power of object-orientation comes from the ability to hide things so that other code can not get to it. The private modifier makes it so no code outside of the class can see the declaration. There is also a protected modifier that makes declarations visible only to subtypes.
  - A big part of the benefit of making things private is that it can be used to hide implementation details. This facilitates the separation of interface and implementation.
  - To enhance the expressivity of the language, Scala allows a broader range of names than most languages and interprets some method names in special ways.
    * In addition to the "standard" names that start with characters or underscores and then have characters, numbers, or underscores, Scala allows two other types of names. One of those is names that include only operator characters. The other is a "normal" name followed by an underscore and one or more operator symbols. Operator syntax, dropping the dot and the parentheses works for any method with a single argument. Operator precedence is determined by the first character. Operators ending in : are right associative.
    * Unary prefix operators can be declared with a method name of unary_ followed by symbols including +, -, ~, and !.
    * Scala does not has a strong syntactic distinction between methods and members. You can mimic a property of the class by writing a method with no arguments or parentheses along with a special method for assignment. That method has the name of the property followed by _=.
    * Function call syntax is expanded out to calls to the apply method.
    * Indexed assignment is expanded out to a call to update.
  - You can have multiple methods with the same name as long as they have different arguments. This is called overloading.

- Scala also allows object declarations which create singleton objects. These can not take arguments and are not instantiated with new. A single instance just exists for using.
  - Scripts are good for small programs, but larger programs are written and run as applications. The entry point to an application is defined by a main method in an object. Top-level declarations are split to different files. Code is compiled with scalac and run using scala.
  - An object with the same name as a class can be declared in the same file and it becomes a companion object. Companions have access to private declarations.

- The case keyword on a case class effectively adds a bit of code to a class. It makes all parameters into val declarations. It also adds a copy method and code for pattern matching.

- You can do a lot of different things with import statements. They can go anywhere in the code, rename things or hide them.

16.5.2 Exercises

1. Code in the chapter developer an immutable vector `class`. For this exercise you should develop a mutable vector `class`. When you do this, keep the behavior of the existing methods the same. What makes it mutable is that you will also define `+=`, `-=`, `*=`, and `/=` methods that mutate the vector on which they are invoked.

2. You can turn any script that you wrote previously into an application by embedding everything into an `object` in the `main` method. The results of this should compile with `scalac` and the run with `scala`.[3]

   The result of this simple conversion is not generally ideal. Any method or `case class` declarations should generally be pulled out of `main`. The methods likely should go inside of the `object`, but outside of `main`. If those methods are not generally useful, they should be made `private`. The `case classes` could go inside of the `object` where they might or might not be `private`, depending on how generally useful they are. However, if they truly stand on their own and have meaning, they should go outside the `object` and into a separate file bearing their name.

   The results of this modification might not compile. That will depend on the quality of the original code and whether you used variables that were declared outside of in those methods. If you use variables defined outside of the methods in their body, you have one of two choices. Generally your first choice should be to pass those variables in by adding extra parameters to the method. If a variable really deserves to be a data member/property of the `object`, it can be moved up a level so it too is in the `object`, but not in `main`.[4]

   Your goal for this exercise is to run through this process on a number of scripts that you wrote earlier in the book. When you do this it is recommended that you make a subdirectory for each script you are converting, then copy files into there. This way you not only preserve the original script, you make it easier to compile that single application. Doing this on a number of scripts will really help to build your feel for how this new approach differs from what we had been doing previously.

3. Another thing that you can do to help you get used to some aspects of the object-oriented model is to convert `case classes` over to normal `classes`. This is generally a less significant modification, but it will require some changes in other parts of the code such as forcing you to use `new` when instantiating an object. You will also lose the `copy` method.

   While converting scripts to applications has some general use to it, keep in mind that not all `case classes` need to be changed. The `case classes` exist in Scala for a reason that goes beyond the educational purposes that have benefited us in this book. It is quite possible that a number of your `case classes` should ideally remain that way over the long term. Practice with changing them is good for you to see how that matters, but it is not always something you want to do.

4. For any of the things you code as applications, you can put in proper `scaladoc` comments and then run `scaladoc` on your source to see the HTML files that are generated.

[3]Remember that when you run an application with `scala` you give only the name of the `object`, not the name of the file. Most of the time the file name should start with the `object` name so you are just leaving off the `.scala`.

[4]Note that if a variable is set by user input, it almost certainly needs to stay in `main`. Having singleton `objects` that request users to enter input will lead to very odd and unexpected behaviors in larger programs.

16.5.3 Projects

All the projects in this chapter switch from running as scripts to running as applications. That means that you need to run `scalac` first to compile the application, then run `scala` on just the name of the `object` you put the `main` method in. This also allows you to split code up into separate files for different `classes` and `objects`. All code must be inside of a `class` or an `object` for this approach to work. If you put the files for the project in their own directory you can compile them all by running `scalac *.scala`.

1. One way to build on top of project 15.3 is to give the user the ability to plot functions. For this to work, the formula parser needs to support variables.

   In the formula parser, there were two possible cases where you did not find an operator: a number and parentheses. The variable becomes a third possibility. To keep things simple for now, the only variable allowed in a formula will be x. So if you do not find an operator and the value of the `String` you are parsing is "`x`", you will say it is a variable. To use variables, you need an extra argument to the evaluation, a `Double` that represents the value of the variable x.

   To fit in with the contents of this chapter, you are to make your code for this object-oriented by having it run as an application and including two separate `classes` called `Plot` and `Formula`. The `Formula class` is where your earlier parsing code will go. The `class` should take one argument that is a `String` for the formula. You will put the parsing code into a method in that `class`. The method should take a single argument of `x:Double`. It does not need to accept a `String` because the `class` as a whole knows what formula is being used. You could also add things like trigonometric function support so that you can plot more interesting curves.

   The `Plot class` will keep a `private var List` that tells it all the formulas that are to be plotted. This `List` could contain instances of `Formula` or of some `case class` that groups a `Formula` with a `Color` so that different functions are drawn in different colors. The `Plot` also needs to keep bounds for the x and y axes. You should put a `draw(g:Graphics2D)` method into `Plot` that can be called from the application itself. That method will loop through the different `Formula` objects, evaluating them at proper points inside the range to be plotted, and connecting the points to draw a line for the function.

   You can decide exactly how to create the application. Clearly the `object` will need to have a `main` and include a `MainFrame` and a `Panel` that calls `draw` on a `Plot` object. If you want an extra challenge to make the application more useful you can add the capability for the user to edit settings on the `Plot`. As with previous plotting work, it is recommended that you not try to support labels on axes. However, if the range or domain crosses zero, you might consider drawing in lines for the axes themselves.

2. If you did project 3 (p.387) and you did gravity integrations before that, then you can consider doing this option. The basic idea is that you will add variable support to your formula as described in the previous project and then have the user type in a force formula that is used instead of gravity.

   In the GUI for your integrator you should have a text field that lets users input a formula. You should use "`x`" for the distance between the bodies. You can also define standard variables that are used, like "`ma`" and "`mb`" for the masses of the first and second particles in an interaction. Using this, gravity would be input as "`-ma*mb/(x*x)`". A spring connection would be "-x". You can play with other options as well and see how they change the way particles move.

To make this work with the `Formula` type described above, you will need to build a `String` where the variables like `ma` and `mb` have been replaced by the proper numeric values for the masses of the particles being considered. You might even consider keeping those formulas in an `Array[Array[String]]` so that they do not have to be rebuilt each time. Then you have the `Formula` evaluate using the proper distance between the particles.

You could also extend the `Formula` type so that it accepts variables for `x`, `ma`, `mb`, and possibly others that you want to define. This second approach is less than ideal for the `Formula` type as it makes it less generic, but is probably the easiest way for you to get some flexibility user in the force-functions.

Note that this version of the integrator is going to be significantly slower than what you did before because every force calculation involves `String` parsing. In chapter 29 we will learn how to get back most of the speed while still allowing the force function to be determined by the user.

3. If you did the Tron in an earlier project, you could update it to be object-oriented. For Tron you can make a class the represents a light cycle with the position, direction, and color. It could have a method called `move` that alters the position appropriately for the direction. Other methods could include turning left or right that are called by reactions to key presses or a `draw` method that put a point at the appropriate location. You might also encapsulate whether a cycle is alive or not into the `class`.

   With this change it is now feasible to add in more players and have walls from one player disappear when that player dies. Extend the game to have three human players (or add a computer player) and make it so that when a player dies, the line for that player goes away.

4. If you worked on any other type of graphical game in earlier chapters, you should also be able to convert it to be an application with an object-oriented approach by making the different elements of the game into `classes` that store the relevant information and have appropriate methods.

5. If you implemented the Mandelbrot set calculation and display previously, this is the project for you. First, use what you have learned in this chapter to make a class called `Complex` that represents complex numbers and alter your original Mandelbrot calculations to use that. Once you have done that, you will take further advantage of it by adding the ability to display Julia sets of quadratic polynomials.

   Every point in the Mandelbrot corresponds to one of these Julia sets. The equation for a quadratic polynomial looks much like that for the Mandelbrot set, $z_{n+1} = z_n^2 + c$ with c being a fixed value and z_0 being the point in the plane you are calculating for. By contrast, the Mandelbrot set used a different c value for each point and $z_0 = 0$ so $z_1 = c$. Again, you iterate until either the value leaves the area around the origin or until some pre-specified number of iterations.

   If the user clicks on a location (use the `MouseClicked` event) then you should pop up another window (use `Frame`, not `MainFrame` for that window) with a different panel that will draw the Julia set associated with the point the user clicked on. You can decide what functionality you want to put into the window. It should start with a bounds of $r \in [-2, 2]$ and $i \in [-2, 2]$ for the real and imaginary axes, respectively. It should not be hard to add zooming, but that is not required.

6. This project continues project 5 (p.206), where you had to figure out how many

elements of each type were on the different sides of a chemical equation. You should recall that the output of that program was something like the following.

```
C: a*1=d*1
H: a*4=c*2
O: b*2=c*1+d*2
```

We want to treat this as a system of linear equations and solve it to find the proper values of a, b, c, and d. These particular formulas give you three equations for four unknowns. If you remember systems of equations, that is typically not a good thing as the system is underdetermined so there are an infinite number of solutions. The easy way to fix that is to assume that a certain number of the coefficients are 1 so that you are left with equal numbers of equations and unknowns.

The form given above would be fine for solving this equation by hand, but for solving it on a computer we will want to move things around a bit. The idea is that we want to get it into the form $Ax = y$, where A is a matrix giving the explicit numbers in the equations, x is a vector with the variables we are solving for, and y is a vector with whatever constants wind up being on the right side of the equal sign. Let us make this more explicit by rearranging the equations above into something more like the form we want.

```
1*a-1*d=0
4*a-2*c=0
2*b-1*c-2*d=0
```

Now we have to pick a coefficient to set to 1 so we get down to equal numbers of equations and coefficients. Any will do equally fine, but programmatically you will probably find it is easiest to set the last coefficient to a constant value (so everything is still zero indexed in your code). In this case that is d. If we do that and move constants to the right side we get the following equations.

```
1*a=1
4*a-2*c=0
2*b-1*c=2
```

This can be transformed into $Ax = y$ if

$$A = \begin{bmatrix} 1 & 0 & 0 \\ 4 & 0 & -1 \\ 0 & 2 & -1 \end{bmatrix}$$

and

$$y = \begin{bmatrix} 1 \\ 0 \\ 2 \end{bmatrix}.$$

Both x and y are column vectors here so they have a single column and multiple rows. Remember that

$$x = \begin{bmatrix} a \\ b \\ c \end{bmatrix}$$

and that is what we want to solve for. The way you will do this is through a process called Gaussian elimination. It turns out that there are many methods of doing this that have different numerical properties. Gaussian elimination is not the best, but it is the simplest to describe and sufficient for our purposes. Gaussian elimination is also exactly what you would do if you were solving this problem on paper so hopefully it will make sense.

What we do in Gaussian elimination is multiply and add together rows in A and y to remove coefficients and turn A into a triangular matrix. We might also have to swap rows at certain times. In fact, we will do that generally to improve numerical stability. To begin with, we want to remove the a term from everything but the first row. We could do this with A as it is, but for numerical reasons it is best if we keep the largest coefficient. You will see other reasons for this when we remove `b`. So the first thing we do is we note that the largest coefficient of a is 4 and it is in the second row so we swap the first and second rows. Note that we swap them in both A and y.[5] This gives the following values.

$$A = \begin{bmatrix} 4 & 0 & -1 \\ 1 & 0 & 0 \\ 0 & 2 & -1 \end{bmatrix}$$

and

$$y = \begin{bmatrix} 0 \\ 1 \\ 2 \end{bmatrix}$$

Now we eliminate the `a` terms in the second and third rows by multiplying their values appropriately so that when we subtract them from the top column we do not have anything left. If the `a` term is already zero in a row we can leave it alone. In this case we will remove the `a` term from the middle row by multiplying it by 4 and subtracting the top row from it. We will do nothing with the bottom row. This gives the following values.

$$A = \begin{bmatrix} 4 & 0 & -1 \\ 0 & 0 & 1 \\ 0 & 2 & -1 \end{bmatrix}$$

and

$$y = \begin{bmatrix} 0 \\ 4 \\ 2 \end{bmatrix}$$

Now the top row is set so we look at the smaller nested matrix ignoring the first row and column. We want to eliminate the `b` coefficients from that. Here it really matters that we swap up the row with the largest coefficient because what is there right now has a zero coefficient and that will cause division by zero if we do not move it. We do the swap and since the last row already has a zero in the `b` coefficient we will not do anything else. That leaves us with the following values.

$$A = \begin{bmatrix} 4 & 0 & -1 \\ 0 & 2 & -1 \\ 0 & 0 & 1 \end{bmatrix}$$

and

$$y = \begin{bmatrix} 0 \\ 2 \\ 4 \end{bmatrix}$$

[5]The fact that value in both A and y need to be swapped at the same time means that your implementation could represent both A and y in a single `Array[Array[Double]]` where each inner array stores a row of A followed by the corresponding y value.

This we can solve by working our way up the matrix. The bottom row tells us that c=4. We plug that into the next row and get 2*b-4=2 and find b=3. It is easy for us to say that, but we should probably examine how the computer will find these values. For the first one, the computer simply does c=y(2)/A(2)(2). Then as we move up we have to do some loops to subtract things off before we do the division. In this case b=(y(1)-c*A(1)(2))/A(1)(1). Note that we are always dividing by the component on the diagonal and we subtract off all the terms to the right of the diagonal. Now we are down to our top equation which is a=c/4 so a=1. In the program that will be a=(y(0)-c*A(0)(2)-b*A(0)(1))/A(0)(0). The values of y(0) and A(0)(1) are both zero, but the math in the program will include them and it will not matter.

Your program should take an input just like the earlier project where you type in a chemical equation. It should then print out the same output as in that assignment and follow that with a solution. In this case it would print out "a=1 b=3 c=4 d=1" as the last line of output. There might be some decimal points in there as well. This example was well behaved and gave us integer results. However, our procedure does not guarantee that. If you want an extra challenge you can scale things up so they are all integers.

Make this object-oriented by making an application and having a class that represents a chemical formula. You can add methods to access the information that is calculated using the approach described above.

7. It is time to make your map program into a true text adventure game. It does not matter how you do this, but you need to add in a goal to the game and have it inform the player if he/she wins or loses. Most of the original text adventure games had some type of puzzle solving where the items could interact with one another or with rooms in certain ways to make things happen. You probably do not want to go the Role-Playing Game (RPG) route and have equipment, combat, and the like though simple combat might be feasible. You must add a "help" command, which prints out the objective of the game and how to play. This should read and print the content of a file so you do not have to type in the full help in print statements.

8. If you did project 1 (p.387) for the ray tracing with recursion and the graphics projects before it, now it is time to add lighting and proper colors and make it so that you have a real image. You will also organize things into an application with objects and classes.

   You already have a case class scene that includes spheres, planes, and lights. So far you have not made use of the lights. Each of the geometry objects and lights should have a color attached to it. You need to write a function that takes a ray and the scene and tells you what color to use. If it never hits anything the color should be black. It is recommended for this that you make your own class called DColor that uses Doubles for the red, green, blue, and alpha values. You can put in methods to scale and combine these in meaningful ways. This might include a + method for adding DColors and a * method for scaling when multiplied by a Double.

   The reason for using your own DColor class is that you can represent colors with components that go outside the normal 0.0 to 1.0 range when you are doing calculations. You only have to bring things back to that range when you build a normal Color object. That operation makes another good method for the DColor type.

   You should also update the existing ray tracer code so that it uses classes for the various components including points/vectors, spheres, planes, lights, etc. The Point and Vect[6] classes in particular can have methods for doing mathematical operations that will make your code much simpler and easier to read. Methods like mag and normalize could also come in handy.

To color a point you need the color of the object that was hit, and the lighting on that point. To get the lighting, send a ray from the point of contact to the light source. If there is any geometry in the way (t value between 0.0 and 1.0), then that point is shadowed and the light does not get to it. If there is not anything in the way, that light contributed to the point. The color should be the component wise product of the colors of the light and the object hit, scaled by the cosine of the angle between the normal and the direction to the light. To find the cosine, keep in mind that $\overrightarrow{a} \cdot \overrightarrow{b} = |a|\,|b| \cos\theta$. To make this easier, you should probably store the colors of lights and geometry using DColor. This hopefully helps to demonstrate why putting proper methods in the DColor, Point, and Vect will help to make the code easier to write and to read.

If the geometry that is hit is reflective, you do the recursive call to find what it hits and add the resulting colors, scaled by the appropriate fractions, together to get the color value of the incoming ray.

9. You can make the scheduling script from project 6 (p.388) into an object-oriented application. To do this, you want to put a body on your Course case class. It is possible that you will leave it as an immutable case class and just add methods to it. The methods should probably include one that takes another Course and determines if the two overlap or not. There might also be methods designed to do specific types of copy operations that you find yourself doing frequently, such as making a copy of a course at a different time or with a different professor.

   The fact that the choice of professor often impacts how much you want to take a course, you could split this off and make a class called Professor that has the name of the Professor and a favorability rating. The Course could then hold a reference to Professor and have a rating that is independent of the professor. A method can be written to give you the combined favorability. This way you can easily copy course options with different professor selections and have that change automatically taken into account.

   You can also build a mutable Schedule class that keeps track of what the user wants in a schedule along with what has been met so far from the courses that have been put inside of the schedule. Put this all in an application and tie it together in a GUI.

10. To upgrade the recipe project from 15.5 to be more object-oriented, you will most convert case classes to have methods and possibly make them mutable classes. You will also move the main functionality of the script into an object with a main method.

    In addition to having classes for Pantry, Recipe, and Item, you might want to have a class that keeps track of all the known Item types that have been used in the program. That class should have methods that facilitate looking up items by names or other attributes so that users do not wind up creating duplicates of Item objects.

11. The music library script from project 9 (p.389) can be converted to an object-oriented application by filling out the different case classes with methods or turning them into mutable classes.

[6]There is a standard collection type called Vector. We suggest the shorter name Vect so as not to conflict with that.

You can also make types for playlists or other groupings of information that would benefit from encapsulation. Put the main script into an `object` with a `main` and you have an application.

12. There are a number of features of the L-Systems script from project 4 (p.388) that call for object-orientation. Starting at the top, an L-System should be represented by a `class` that includes multiple `Productions` as well as an initial value. The `Production` type should be a `class` that includes a `Char` for the character that is being changed and a `String` for what it should result in.

    One of the advantages of encapsulating a production in an L-System is that you can make probabilistic productions. These start with a single character, but can result in various different `Strings` depending on a random value. Whether a `Production` is deterministic or not does not really matter to the whole system as all the system does is ask for what a given character will be converted to.

    Using a `String` to represent the state of the L-System at any given time works fine, but there is some benefit to creating your own `class` that is wrapped around a `String`. This `class` might also include information on what generation you are at in applying the productions. The advantage of this is that the type makes it clear this is not any random `String`, instead it is the state of an L-System.

    You should make these changes and adjust the script to go into an `object` with a `main` to create a complete application.

Additional exercises and projects, along with data files, are available on the book's website.

Chapter 17

Bigger Programs/New Tools

We have been working in the command-line environment this far because it is a good thing to know and it works very well for both the REPL and scripts. It supported the little steps that we needed to take early on. However, it is now time to move beyond scripts and write larger programs as applications. This can be done using the command-line, as was described in the last chapter, but there are easier ways.

17.1 Eclipse IDE

Large scale software is generally developed with the assistance of many tools, including an Integrated Development Environment (IDE). At this point we want to move our own efforts from the command-line to an IDE. For this book we have chosen to work with the Eclipse IDE. There are others that you can certainly try, but Eclipse supports many languages and is one of the first platforms for which a Scala plug-in was created. It is also the IDE that is being officially supported for Scala by Typesafe.

The IDE makes it a lot easier to manage the multiple files that are typically created when writing larger applications. It also automates the process of compiling the code and can show you syntax errors as you are writing the code.

You can download Eclipse for free from `http://www.eclipse.org`. There are multiple different versions of Eclipse that have been customized for different types of usage. You should download the Classic version. Follow their instructions for installation. When you run Eclipse it will ask you for a workspace directory. The default should be perfectly fine to start with. The first time you run Eclipse you will get a welcome screen that looks something like figure 17.1. You can spend as much time as you like running through the tutorials and other information they have. When you are ready to start programming, click the bubble that takes you to the workspace.

Figure 17.2 shows what the Eclipse workspace will look like when you first run it. The

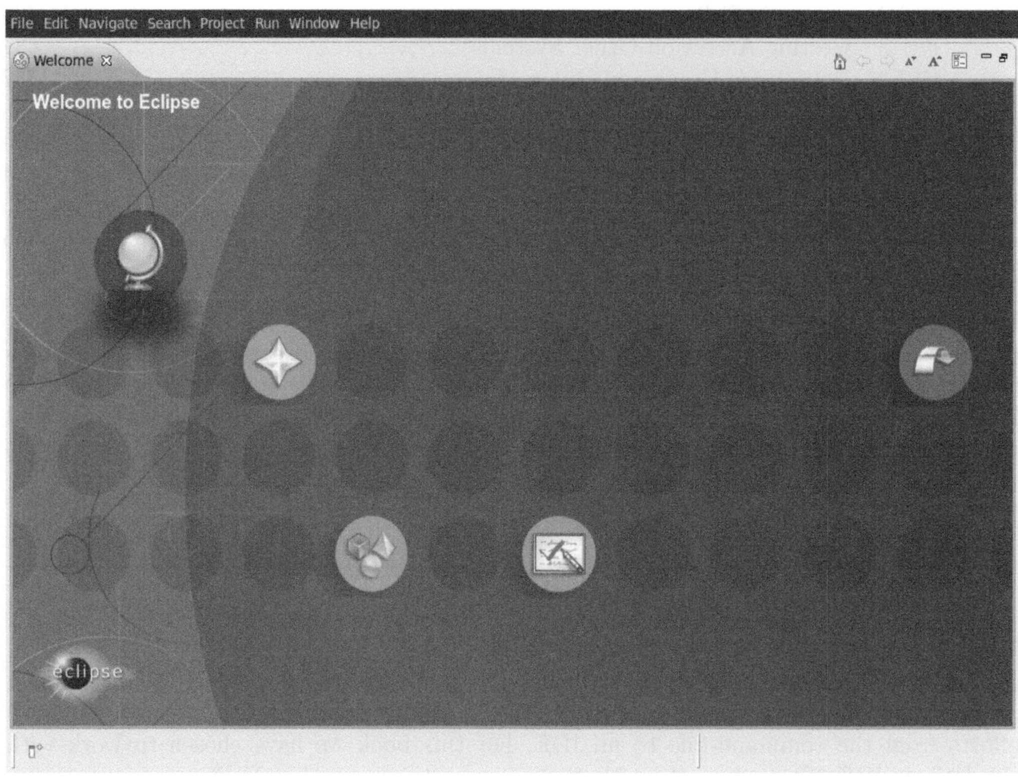

FIGURE 17.1: The welcome screen for Eclipse that you will see when you run it for the first time.

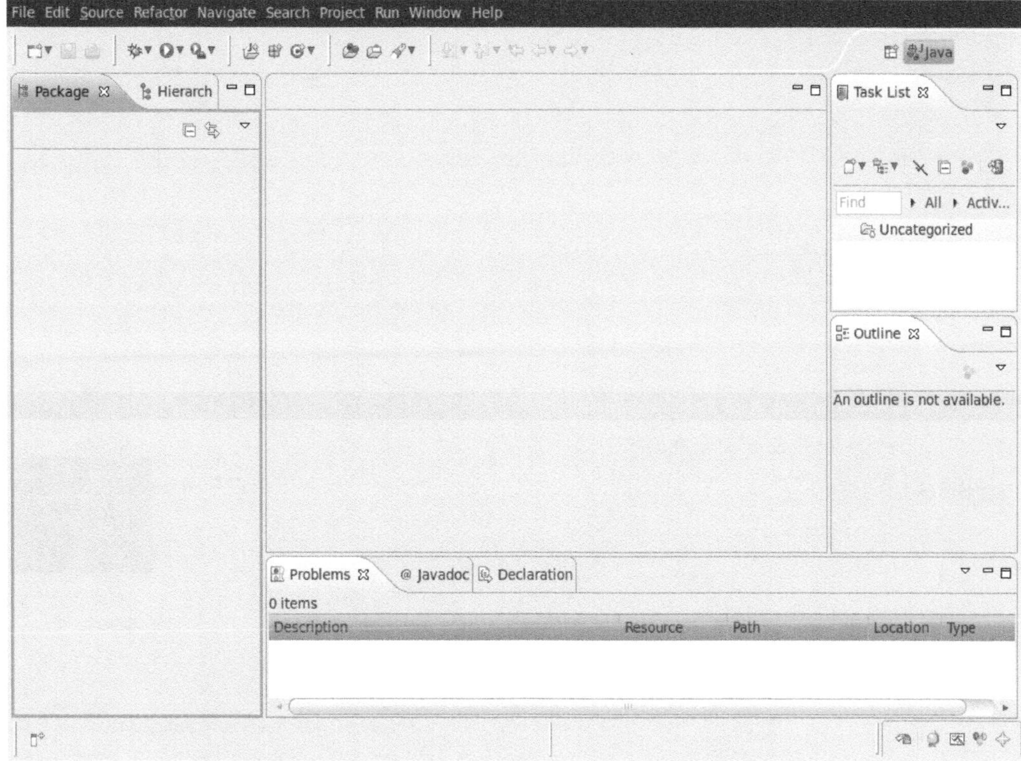

FIGURE 17.2: A blank Eclipse workspace. This is what you will start with when you are ready to begin programming in Eclipse.

window is divided into a number of different regions and has the normal set of menus you expect from a graphical application. If you look through the menus you will see that there are lots of different options. Do not let this scare you. We will only be using a small number of them in this book and you do not have to know what all the others do to make use of Eclipse.

By default, Eclipse Classic is set up to program in Java. You can tell this from the top-right corner where there is a little button that is selected with an icon and the word Java. In order to program in Scala we will have to add something to Eclipse. If you go to `http://www.scala-ide.org` they have instructions for adding the Scala plug-in to Eclipse. If you follow these correctly you will be asked to restart Eclipse. After the restart, things will look the same, but you will have the ability to write in Scala. To do this you will need to switch to the Scala perspective. You can do this by clicking the icon in the top-right corner of the workspace and selecting Scala as shown in figure 17.3. If Scala is not listed on the initial menu, pick "Other" and select Scala from the window that comes up.

In the command-line you could just start editing a file that ended in `.scala` and you had a program. Eclipse makes you go through some other steps to keep code organized. All code in Eclipse is placed in projects. Different projects can use different languages for development. We want to make a project that is associated with Scala. You can do this from the File menu or by right clicking on the "Package" region on the left side of the workspace. If you are in the Scala perspective the option to make a "Scala Project" should be right up at the top of the "New" submenu. If it is not, pick "Other" and find the "Scala Project" option. Doing this will bring up a dialog box that looks like figure 17.4. You can enter a

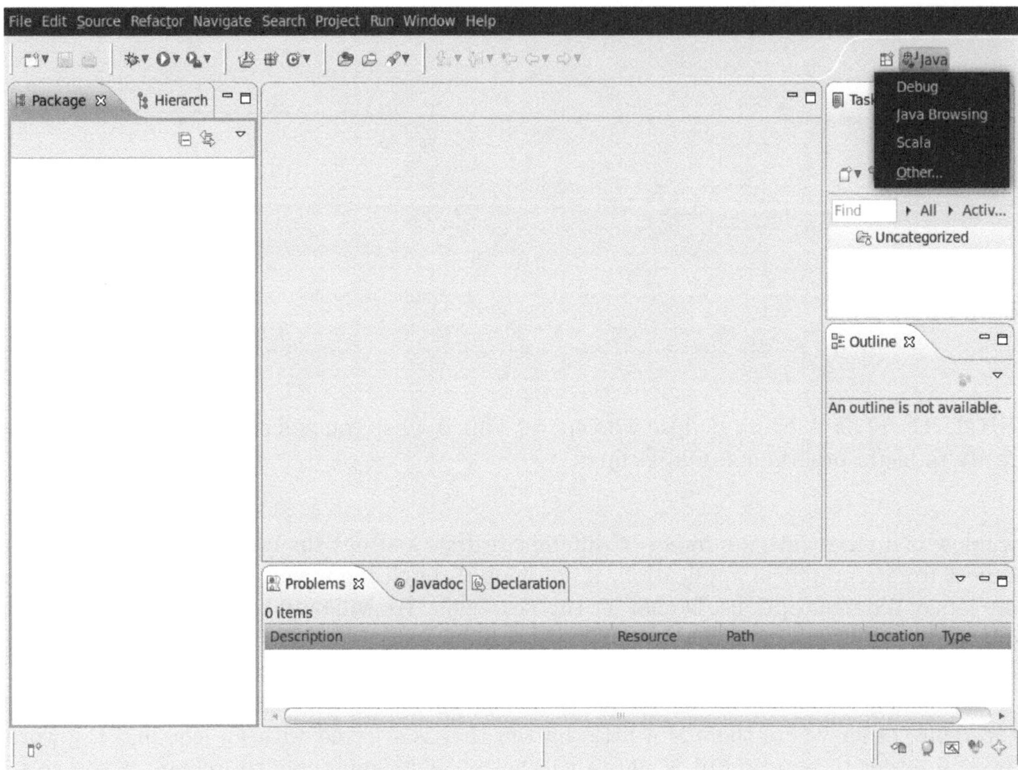

FIGURE 17.3: This figure shows the workspace with the perspective switching menu.

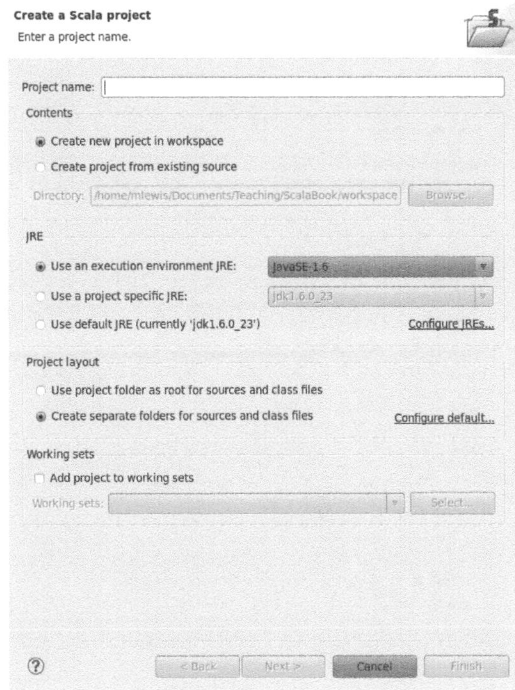

FIGURE 17.4: Dialog for making a new Scala project.

name and click "Finish". The default options should work well enough for our purposes at this point.

For this example we will make a project called "ScalaBook" that will be used for all the code written in the remainder of the book. You can pick a different name if you like. Just keep in mind that projects are intended to be groupings of fairly large groups of code. Do not make a new project for every little ten-line program you write. You make a new project when you truly have a project to be working on.

Inside of a project you need to organize your code into packages. We first ran into packages back in chapter 9 when we had to start importing code from the standard library packages. Now we are going to start making our own packages. When writing applications, it is considered poor style to not use packages. If you do not specify a package, your code is put into what is called the DEFAULT PACKAGE.

To make your code part of a package all you technically have to do is put in a `package` statement. This can be done in two ways. If you put a `package` statement at the top of the file, all the code that appears in the file will be part of that package. This style is adopted from Java. Scala packages have true nesting though and you can follow a `package` declaration with a code block and all the code inside the curly braces of that block will be in the package. In this book we will typically use the former approach, primarily for purposes of organizing code.

In Java, `packages` are also directories and when code goes in a `package` you have to have it in a directory with the same name as the `package`. While this is not required in Scala, it is recommended because it makes it easier to find the code you want. This is part of the reason that in this book we will adhere to the Java style. When you make `packages` in Eclipse the corresponding directories are automatically created for you.

We will go ahead and make our first `package` for putting our first bits of code into. To

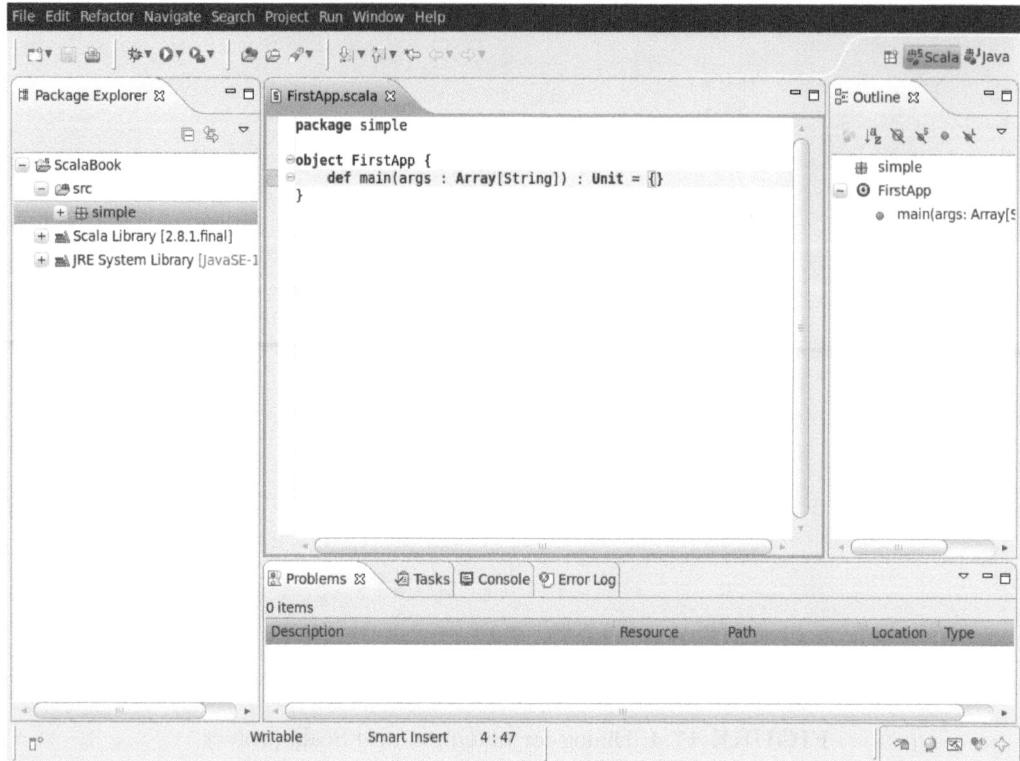

FIGURE 17.5: This figure shows what the workspace will look like after you have selected to create your first application.

make a `package` you can right click on the project and select "Package" from the "New" menu. This will bring up a dialog box where you can enter the package name. For now we will create a `package` called "simple". Note that the name is in all lowercase. This is the accepted style for `package`s. All letters should be lowercase with no spaces or special characters. We called this `package simple` because it is where we are going to put little code snippets that we will need to write occasionally. We will be creating other `packages` as well to organize other parts of the code that we write.

Now that we have a `package` we need to put something in it. If you expand the project and the `src` directory you can right click on the `package`. The "New" menu will have a number of different Scala related options. You will see that you can make a `class`, a `trait`, an `object`, or an application. We want to make an application, but we will do it by making an `object` and including an appropriate `main` method. You can give the `object` the name `FirstApp` and go with all the default settings. After you have done this, a file will be created and will be opened in the main editor area. After you enter the `main` method you should have a screen that looks like figure 17.5.[1]

The application is just a object with the proper `main` method defined. To make your first little program in Eclipse do something, simply fill in the body of the `main` method. A `println` will suffice to prove that it works. There are many ways to tell Eclipse to run a program. Perhaps the most direct is to right click on the editor window with the application and go down to the "Run As" submenu. Select to run it as a Scala application. The result

[1]The option to create an application directly uses a technique we have not covered yet.

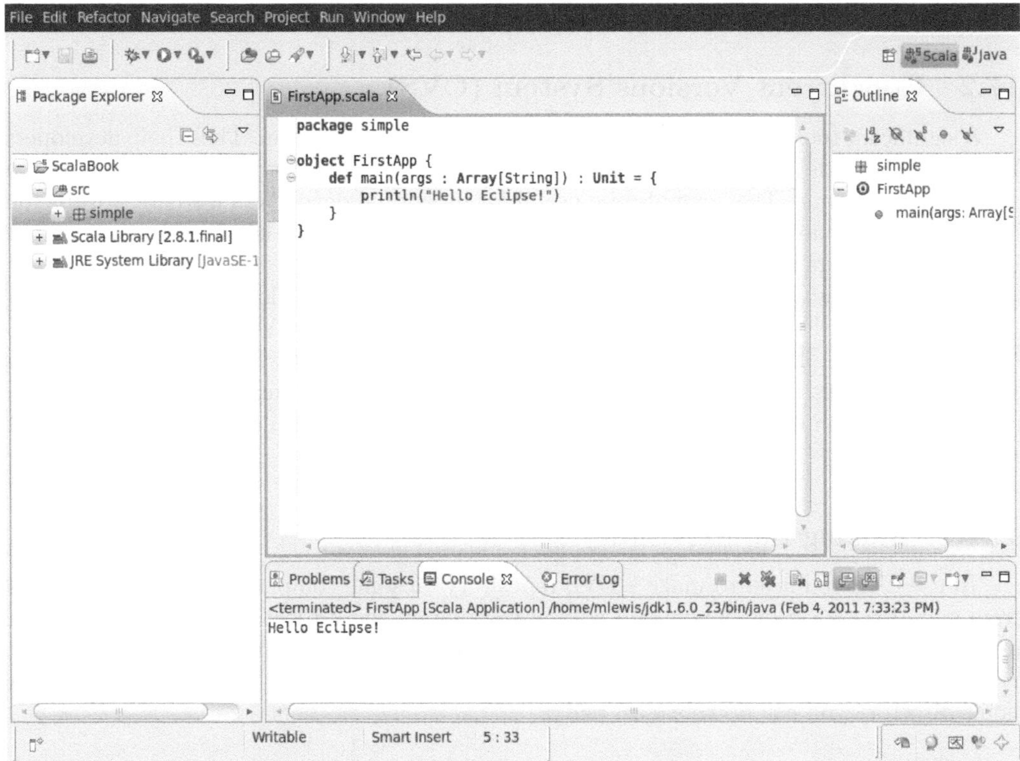

FIGURE 17.6: This shows the application with a single `println` and the result of running it with output in the Console.

you might see from doing this is shown in figure 17.6. Note that the output appears in the "Console" tab at the bottom.

When writing code in Eclipse you will write it in the object-oriented style putting all of your code in `class`es, `trait`s, or `object`s. You will not be writing scripts. Ideally each `class`, `trait`, or `object` that you declare at the top level will be in its own file that has the same name as the `class`, `trait`, or `object`. This is not required by Scala, but helps to organize the code and makes it easier to find things. When writing small programs alone it does not matter so much. However, as programs get larger or if you start writing code with other people, the ability to easily find things that you wrote long ago or that others wrote grows more important.

This does not mean that every five-line `class` you ever write needs its own file. Proper Scala style tends to lead to many smaller `class`es. Type declarations can be nested just like everything else in Scala. If you find a need for a quick little `class` in one part of your code and it will only be used in that part of the code, just write it where you need it, even if that is inside of some other `class` or `object`. You can even make new types inside of functions/methods. Proper scoping rules will apply.

To get yourself familiar with Eclipse you should consider re-solving some earlier problems writing them as applications instead of scripts. To keep things organized, put each problem is a separate `package` inside of the `simple package`. So you might have `simple.rational` or `simple.gui`. You can also bring up an REPL in Eclipse if you right click on a project and select "Scala > Create Scala Interpreter".

17.2 Concurrent Versions System (CVS)

A significant tool of software development is the versioning system. These help developers keep track of code and deal with the difficulties of having multiple people working on software at one time. They can also be useful for individual developers who are working on a piece of code from multiple different computers as it can help in keeping the code on the different computers in sync.

There are a number of different versioning systems that developers choose to use for different reasons. For this book we will be using the Concurrent Versions System (CVS). CVS is one of the older versioning systems, but it is very available. It is also the versioning system that comes by default with Eclipse. You could choose to use an alternate system, such as Subversion or Git, but you will have to install a different plug-in for Eclipse to get that functionality.

17.2.1 Making the Repository

To use CVS you need to make a CVS repository on some computer that you can access from any of the locations where you will be working on the project. If you are on a Linux system you can do this from the command-line. First you need to make a directory for the repository. A directory called `CVS` in your home directory will work fine. If you are in your home directory you can do this with `mkdir`.

```
mkdir CVS
```

Once you have the directory, you have to set it up as a CVS repository. The `cvs` command under Linux will do this, and any other functions you need to do with CVS. To see all the different options you have use `man`. If you just want to initialize the directory that you made you can use the following command.

```
cvs -d\$HOME/CVS init
```

The `$HOME` can be replaced with the fully specified path of your home directory if it is not defined on your system.

You can use the `cvs` command to commit code to the repository or to update it from the repository. We will be doing these functions from inside of Eclipse instead of from the command-line.

17.2.2 Set-up in Eclipse

To get started with CVS in Eclipse you should open the "CVS Repository Exploring" perspective in Eclipse. That will change the workspace to look like figure 17.7. If you right click on the "CVS Repositories" view in the top left you can choose to set up a new repository. That will bring up a window into which you need to enter the information for your system and repository. You might need to ask a system administrator for connection information if there are security restrictions. For example, if your system requires a secure connection, "extssh" could be required.

Once you have a repository set up you can go back to the Scala perspective and right click on the project you want to use with the repository. From that menu you will select "Team > Share Project". This will let you select the repository you want to place the code in. For sharing a repository, you can skip the previous step of setting up a repository an

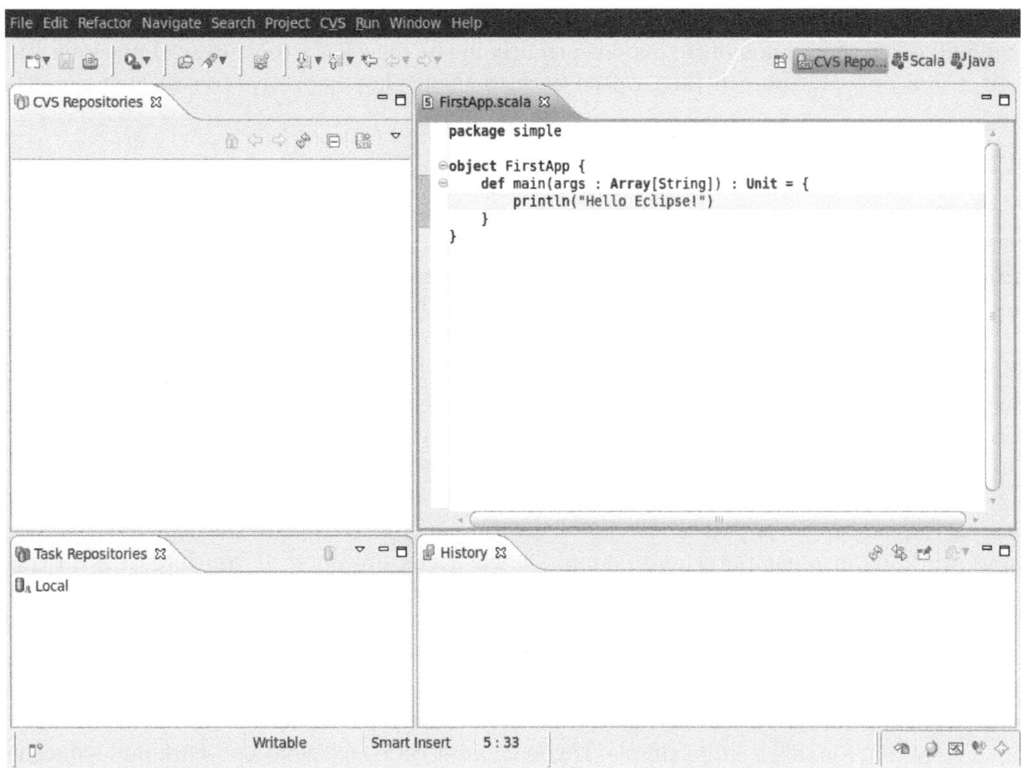

FIGURE 17.7: This figure shows the Eclipse CVS Repository Exploring perspective.

set one up when you choose to share. Going the other way requires that you set up the repository first though.

At this point you will be asked to commit your code to the repository. A dialog box will come up allowing you to select which files you want to commit and ask you for a comment associated with this commit. When you click that you are done, the current version of your code will be copied into the repository.

17.2.3 Getting Access on Other Systems

When you want to gain access to your project on another system you will bring up the "CVS Repository Exploring" perspective again and open the proper repository. Under the repository will be a selection called "HEAD". If you expand that it will tell you all the projects currently in that repository. Right click on the one you want and check it out. That will pull down the version of that code currently in the repository to your local machine and put it in a project. You can then switch back to the Scala perspective to edit that project.

17.2.4 Repository Usage and Rules

Editing the code on any machine does not automatically change what is in the repository. To do this, you have to commit your changes. You also do not automatically get changes from the repository. In order to do that you have to update. Once a project is connected to a repository, the "Team" menu will have options for both commit and update.

Often it is good to see what you are committing to or updating from the repository. You can do this by selecting the "Synchronize with Repository" option from the "Team" menu. This option will ask you to switch to a different perspective and you should let it. From there you can see what files you need to commit, what files you could update, and if any file needs both, it will show it as a conflict that you can look through and resolve.

When you are working by yourself and just using the repository to enable a single code base across multiple computers, you just have to remember to update the code anytime you start working on a computer and commit it when you are done. If you forget to do this you will not have the latest version when you go to work on a different machine. Working alone you might find you need to commit code that is not working because you need to go somewhere else, perhaps even to see someone who will be helping you to fix the problems. However, when you are working on a code base as a group, the primary rule of CVS or any other code repository system is that you never commit broken code.

The reason for this is quite simple. If you commit code that does not work and someone else updates before you fix it, now they have the broken code and it might prevent them from compiling or running the full project. Just fixing your problems before you commit is not quite enough though. After you have made your changes and made sure everything compiles and runs properly, you need to update from the repository again and check your code one more time before you commit. To understand why you need to do this, imagine you are in a work environment and you last updated your code an hour ago. There are multiple developers on your team, and some might have committed their own changes since you did an update. Just because your code works fine with what was in there an hour ago does not mean it will work fine with what is there now. If you commit without updating and your code does not work with other more recent commits, you will put the repository into a state where the code will be broken for anyone else who updates. So before you commit, you update and check for errors. If there are not any, you can commit. If there are errors, you have to fix them, update again and repeat the process. Basically, you can not commit until your code checks out with an update you have just done.

17.3 Scaladoc Comments

You were introduced to Scala comments back in chapter 3. Through most of the book we have not been all that particular about putting comments into the code as descriptions of the code were part of the text. With the move to doing larger, object-oriented projects there is another reason for including comments. The comments you have been using previously, with `//` for a single line comment and `/*` followed by `*/` for a multiline comment, were passive. You would put them in the code to help remind yourself what a piece of code was doing. If you had been working on code with a group of people, those comments could have been useful for letting other people understand your code. Such comments can be very helpful, but they are also prone to certain risks. The risk is that if programmers are not extremely diligent, the comments can get out of sync with the code. This can lead to significant confusion and cause real problems. For this reason, there are many programmers who try not to rely too much on comments to explain their code and instead try to make the code "self-documenting". This is done by choosing variable and method names that reflect what is happening and breaking problems down in ways that make them more readable.

Now that you are writing full `class`es and `object`s, there is another factor to consider with comments, external documentation. The point of comments up to this point has just been for the benefit of programmers looking at that source code. Hopefully people who have access to the code in this way know how to read it and can figure out many things from the code. `class`es and `object`s can be part of libraries that will be called upon by people who do not have access to your code or who really should not be expected to go through your code to figure out what it does. This type of thing is so common that there is special formatting for comments that support documentation that is intended to be used without the code. This is not a completely new concept. The Application Programming Interface (API) documentation that you have been using for quite a while now works in exactly this way. You need to know what types are in the libraries and what members those types have as well as how you can use them. The API tells you that without you having to go through the code to find it. The API is created by putting comments in the code that have a specific format, then running a program called `scaladoc` to generate web pages.

You can insert your own `scaladoc` style comments in your code and generate web pages from them. While these comments are just as likely to get out of sync with the code as normal comments, these serve a more active purpose for outside documentation and, as such, warrant the effort that is required to maintain them.

A `scaladoc` comment is just like a multiline comment except that the opening has an extra asterisk. So you open the comment with `/**`. This still starts with `/*` so the compiler will ignore it like any other multiline comment. However, his opening tells the `scaladoc` tool that the contents are related to the thing that follows the comment. You should put `scaladoc` comments directly above the `class/object/method/member` that the comment modifies. This code shows an example.

```
/**
 * This is an object where I am putting math functions.
 *
 * If I had more to say about this object I could put that here. You can
 * use HTML tags if you want. To show code use triple curly braces.
 * {{{ val a=5 }}}
 */
object MathFuncs {
  /**
```

```
   * This function can be used to get Fibonacci numbers.
   *
   * @param n index of the Fibonacci number oyu want to calculate.
   * @return that Fibonacci number.
   */
  def fib(n:Int):Int = if(n<2) 1 else fib(n-1)+fib(n-2)
}
```

Here we have a single `object` called `MathFuncs` with a single method called `fib`. Directly before the `object` declaration and the method declaration you see scaladoc comments. The first line in the comment should be the main thing you want people to know. In the case of methods, this is all that people will see unless they expand the view. After that you can put a full description. You can use HTML tags for formatting. You can also include code elements using triple curly braces. This is helpful if you want to demonstrate how something might be called.

The comment on the method includes two lines that start with `@param` and `@return`. These special tags are used to give extra information about calling that method. The `@param` tag is followed by the name of a parameter and a short description of that parameter. You should have one of these for each parameter the method takes. The `@return` tag should be followed by a description of what the method returns. Though not shown here, you can use `@tparam` to describe any type of parameters that are in square brackets for a method or `class`.

When you installed Scala, you got a number of other programs that are in the `bin` subdirectory of the install. One of these is called `scaladoc`. You can use `scaladoc -help` to see how this command is used. In the simplest usage, you follow it by all the files you want to generate web pages for. Doing this on the program above results in figure 17.8. As you can see, this looks just like the API for Scala.

There are many other tags and options for `scaladoc`, but this has covered the most significant ones. You should get into the habit of adding `scaladoc` comments for the types you declare and the non-private members and methods inside of them.

17.4 End of Chapter Material

17.4.1 Summary of Concepts

- Integrated Development Environments, IDEs, are tools that pull the different parts of software development together. They often make coding easier and faster.
- Eclipse is a free, open source, multilanguage IDE that we will be using for the rest of this book. There is a Scala plug-in you can install that will allow you to work with Scala.
  - All code in Eclipse is organized into projects. You should create a project for each large grouping of code you work on.
  - Code should be organized packages so that related pieces of code are grouped together. Packages are typically represented as directories.
- The Concurrent Versions System, CVS, is a tool for keeping code consistent across multiple workspaces and developers. It is an older tool that comes by default with Eclipse. There are also plug-ins for newer tools such as Subversion and Git.

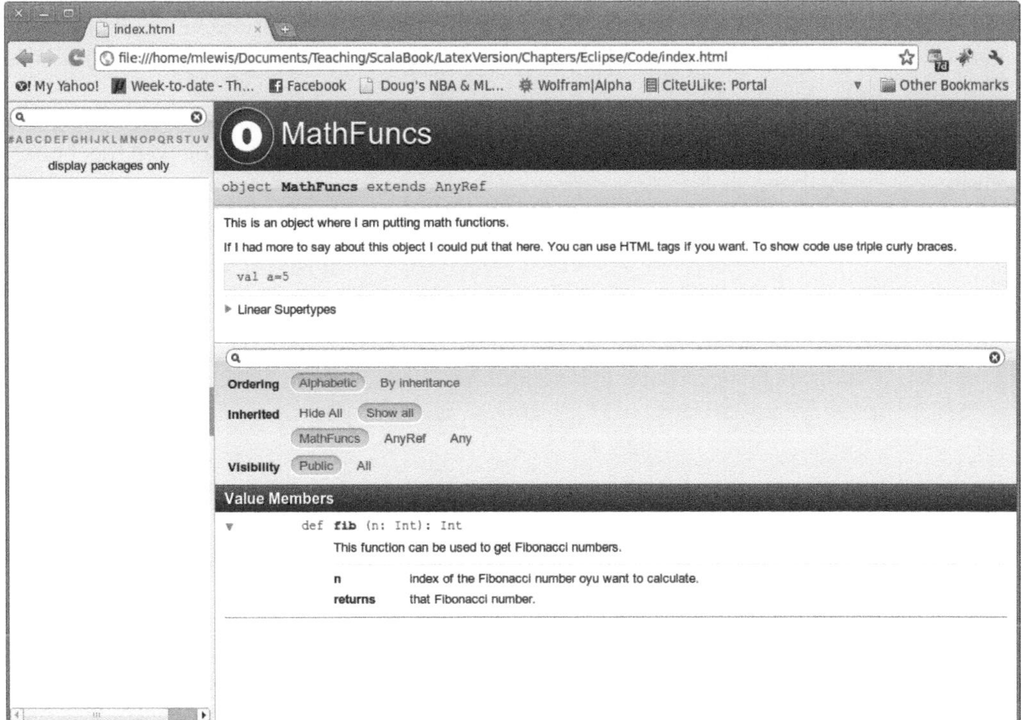

FIGURE 17.8: This figure shows the results of running `scaladoc` on the program above.

  - You can set up a CVS repository on a Linux box to keep track of your code.
  - When you want to pull down code you do an update. After you have made changes you commit the code to the repository. When working as a team you should never commit code that does not work.

- The `scaladoc` tool will convert commented code directly into API documentation. Comments for this should begin with `/**` and go directly before the thing they describe.

17.4.2 Exercises

1. Spend some time writing simple code samples in Eclipse. They do not need to be big. They just need to help you get familiar with how you work in Eclipse. They can all go into a single project for now. The fact that Eclipse does not work with scripts means that this will also give you some experience in doing things in a more object-oriented way.

2. Make a CVS repository and check in some code into it on one computer. Then go to another computer and check out that project.

17.4.3 Projects

Instead of listing individual project here, you should take any of the projects from the previous chapter and migrate them over to Eclipse. Do not put the code for this in the default package. Instead create at least one package with an appropriate name. If there are many classes and you feel that some belong together in their own package, you can organize them that way.

Additional exercises and projects, along with data files, are available on the book's website.

Chapter 18

A Project (Drawing Program)

Now that we are focusing on object-orientation and the development of larger programs, it will be helpful to have a larger program that we work on in this book. This chapter will introduce a project that we will develop incrementally in many of the following chapters. You will be building other large programs as part of the projects in the rest of this book as well, but it is helpful for you to see something besides the code that you are creating to give you a more complete view of how things work.

The program that we are going to develop is a drawing program. It is not going to be like Paint® or quite like any other "normal" drawing program. We will do a few things differently to make it so that this program can better illustrate key concepts through the rest of the book. A few things will be thrown in just for educational purposes, but in general features will be included to produce a final application that is capable of things that you do not normally see in drawing programs, including animation.

18.1 Software Development Stages

As was originally discussed in section 7.11. There are five stages involved in the production of any large software system. There are many different approaches to organizing these five stages, but the general nature of them is always the same, and they all have to be involved. The five stages are as follows.

Analysis This is where things begin. During analysis the developers figure out what it is that the program is supposed to do. This is a problem definition phase. The more detailed the analysis, the easier the other stages will be. In practice, it is not possible to get an initial analysis that covers everything, and inevitably, things change over time on real products. For this reason, while all software must start with analysis, it is not something you finish and leave, you generally have to come back to it. Analysis is very independent of programming and only looks at what you want to do, not how to do it.

Design After you have sufficient analysis so you know what you are trying to do, the development turns to figuring out how to produce the desired result in code. This is the design phase where you map out what will be happening in the code. Once again, a more complete design will make life easier later on, but realize it is basically impossible to make a perfect and complete design initially, so this is something that will also be revisited. The design phase discusses code, but does not actually involve writing code.

Implementation The middle of the five stages is where you actually write the code. This book, and much of your introduction to Computer Science (CS), focuses on developing your ability to write code. However, this is only one phase in the life cycle of software development and when things are done well it can actually be a small phase relative to others. This is the focus early on though because you can not really do any of the other phases, with the possible exception of analysis, unless you have a strong grasp of how the implementation will work.

Debugging At this point you should have already learned that just writing a program is not the biggest hurdle you have to clear. Finding the errors in the program can take far longer. The process of finding and fixing errors is called debugging. It is an ongoing process in software development and it often takes significantly longer than the implementation.

Maintenance After the software is deemed complete and sufficiently free of errors/bugs, it is released. This ushers in the maintenance phase. Maintenance is something that is hard to simulate in a classroom environment, but it is critical for most professional software. People who pay money for a piece of software often want to know that if they find something wrong, someone will be there to correct it. Often customers will think of additional features they would like to see as well. Maintenance is the general process of dealing with issues in the software after it has been released.

You will come back to these different steps in many later CS courses. The field of software engineering is all about how to do these things in a way that produces the best results with the minimal amount of effort.

18.2 Analysis

So what exactly is entailed in the analysis phase of software design? What do you have to do in order to be able to say that you know what problem you are solving? One key way to approach this question is to think about how the software is going to be used. Picture a user running the software. What different types of users are there? What are the different ways they can interact with the system? By answering these questions, you can come to truly understand the main functionality of the software.

There are many different approaches to doing software analysis (and design). Simply writing down prose that describes the system is an option, but not a very good one because it is dense and requires anyone viewing the document to do a significant amount of reading to get the needed information from it. For this reason, there are diagramatic approaches to analysis that can communicate key ideas quickly. These are backed up by prose that provide details which are not captured in the diagram. One of these is the Use-Case diagram. This

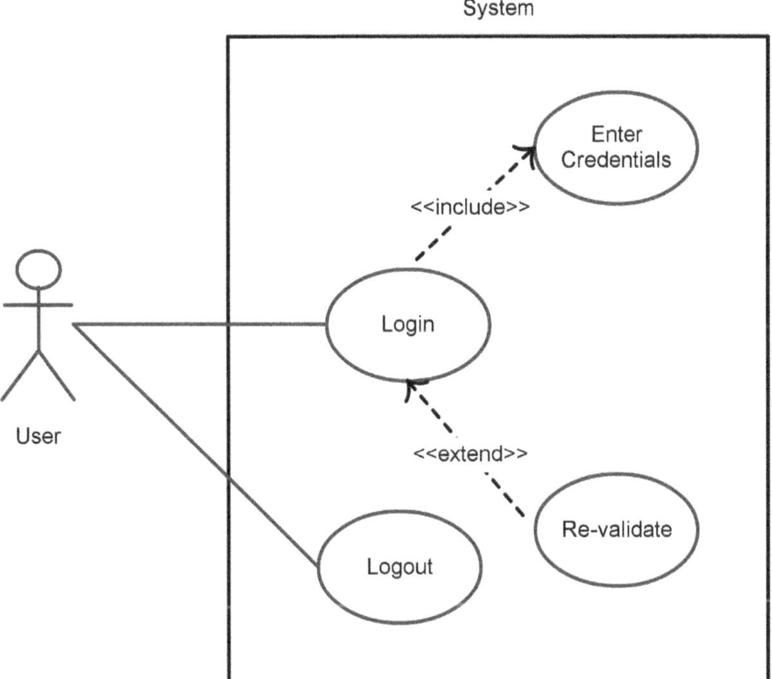

FIGURE 18.1: This is a simple example of a UML Use-Case diagram. Actors are drawn as stick figures. The system being analyzed is a box. The ways in which the actors use the system are drawn as circles.

is one of a number of different diagram specifications that constitute the Unified Modeling Language (UML).

A Use-Case diagram includes various actors and the use-cases they participate in. Actors often represent humans in different roles, but they can also represent software from outside of the current system that communicates with that system asking it to do things. Figure 18.1 shows a very simple use-case diagram with a single actor and a few basic use-cases. As you can see, actors are represented with stick figures. Ease of drawing is a significant feature of UML diagrams. One could say that they are whiteboard friendly. Indeed, whiteboards are a common feature of many software development companies and one of the main roles of use-case diagrams and the other UML diagrams is to provide an efficient, common language to communicate ideas to other people.

Figure 18.1 has a large box in it that represents the system being analyzed. This is the standard notation for use-case diagrams. Note that the actor is outside of the system. This makes sense, unless you happen to be putting people into your software. This is also why other pieces of software are represented as actors. They are not part of what you are focusing on. They are external and send in commands to make things happen.

Inside the system there are a few bubbles with text in them that represent use-cases. There are lines from the actor to the "Login" and "Logout" use-cases to indicate that these are actions that the "User" can initiate. The bubbles represent things that can be done in the software and the lines indicates who can do it. If your system had both regular users and administrators as two types of actors, the administrator would have lines to use-cases

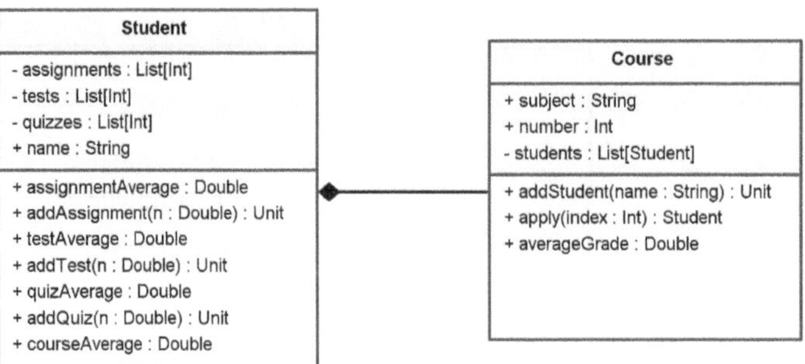

FIGURE 18.2: This is a simple UML class diagram that shows a course that can contain multiple students.

that were not available to the normal user. Actions like deleting a user account should not be part of the options given to normal users.

There are two other use-case bubbles shown with the text "Enter Credentials" and "Re-validate". These both have annotated arrows connecting them to the use-case for "Login". The arrow pointing from "Login" to "Enter Credentials" is annotated with the text "<<include>>". This indicates that when the "Login" use-case is invoked, it can include having the user enter his/her credentials. The arrow pointing from "Re-validate" to "Login" is annotated with "<<extends>>". This implies that the source of the arrow is a subtype of its destination. So having the user re-validate is shown to be a special case of logging in.

Use-case diagrams might seem a little silly, but they can force you to really think about all the different actions that a user should be able to do in your application and then to see how those different actions fit together. That can be very useful as you enter the design step of building your software.

18.3 Design

In the first half of this book, most of the work you did in the design process was to break the problem down into different functions and decide what data needed to be passed between those functions. When you are working in an object-oriented mode the design process involves figuring out the types of objects that you need in the programs and laying out `classes`, `objects`, and `traits` for them. Each `class`, `object`, and `trait` can have member data and and methods in it.

The way to represent this in UML is with a Class diagram. Figure 18.2 shows a simple class diagram showing two `classes`, one for a `Student` type and one for a `Course` type. Each `class` is represented with a box that has three different parts. The top is the name of the type. Below that are the data members of the type. The bottom section lists the methods for that type. The data members and methods are preceded by a symbol indicating their visibility. A '`+`' indicates that it is public, a '`-`' indicates that it is `private`, and a '`#`' indicates that it is `protected`.

This figure also shows an arrow between the `Course` and `Student classes`. There are multiple types of arrows the are used in class diagrams. We'll see more of them over time.

This particular style of arrow represents composition. It implies that a `Course` is composed of multiple `Student`s.

Like the use-case diagram, the class diagram does not show you everything about a system. However, it does communicate quite a bit of information in a format that people can read quickly and easily. The information that it imparts is particularly significant for helping them understand how the system works. The most obvious piece of information is the `class`es/types that are used in the system. It also gives you a quick reference to the capabilities of the different elements through the methods and member data.

What is less obvious is that the lines and arrows can be just as important. By showing you the links between different parts of a system, a class diagram lets you quickly see what parts are critical to the working of other parts. This lets you identify key points in the system that can lead to failures or vulnerabilities. Looking at a system at this higher level can also help you to see structures that might not otherwise jump out at you.

You will occasionally see an even less detailed version of the class diagram that represents each class by a simple box with arrows, leaving out the member data and methods. This view exists because often knowing the classes and the relationships between them is more important than knowing the details of any one class.

There is a lot more to UML class diagrams, including not only other types of arrows to represent different relationships, but also boxes for `package`s and annotations on the relationships between types. These different aspects will be introduced as needed in later chapters.

18.4 Making Packages

Large software projects that have a lot of code in them need to be organized. This is done by adding packages. We saw in the last chapter how we can create new packages in Eclipse. In Java, packages are always represented as directories and creating packages in Eclipse will mirror this arrangement. Scala does not require that source files be organized into a directory structure that mirrors the packages, but doing so is recommended to make it easier to find files.

In a source file, you specify a package using the `package` keyword. If everything in a file is in a single package, you can place a line at the top of the file that specifies the `package`. You can also follow a package declaration with curly braces and the code inside that block will be in the specified package. In this book we will use the first approach based on the idea that packages should be reasonably large groupings of code and single files should not get too long.

Note that in Eclipse, you need to be careful that the `object` and `package` names in files match those of files and directories if you want to run `main` method as an application. If the file has a `package` declaration that does not match the directory location, or the name of the file does not match the name of the `object`, Eclipse will tell you that it can not find the class when you try to run it.

18.5 Project Analysis and Design

Now it is time to start working on the project that we are going to carry forward through the rest of the book. This is going to be a fairly big piece of software, so as was just discussed, we should start with the analysis and then work on the design some before we try to implement anything. We are not going to try to figure out every possible option that could go into the software right now. We want to figure out enough that we have some direction to go in. We also want to identify where we might want to add functionality later so that as we move into the design phase, we can keep those in mind.

The program itself is a drawing program that has the flexibility of text input to issue commands. Figure 18.3 shows a use case diagram for our system. There are two actors in the diagram. The obvious one is the user who is running the program. This user has the ability to add elements to the drawing and edit elements of the drawing. They also have the ability to give text commands to the program. The less obvious parts of the diagram involves having two separate instances of the program talk to one another. This leads to use-cases where the local user makes connections to other instances of the program and does things such as sending them parts of drawings or other forms of communication.

This diagram shows some arrows connecting different use cases. The solid line with the open arrow indicates a generalization. This implies that one use-case is a specific type of another use-case. There is also a dotted arrow with an open head labeled <<include>>. That style of arrow indicates that whenever one use-case occurs, another will be included as part of it. This diagram is sufficient for our current needs. It lays out the basic operations of the program, and its creation forces us to think about what it is we want the program to be able to do. With that in place we can start to lay out a design.

Like the analysis, we will only take the design so far at this point. Unlike the analysis, part of this is because we need to learn a few more things to complete the design. There are a few things that we can put in place with our class diagram. These are shown in figure 18.4, which has four boxes representing three `classes` and one `object`. The `object` contains the `main` method for our application as well as some other aspects required for the Graphical User Interface (GUI). A `TabbedPane` is being used in the main window so that there can be multiple drawings open, each with its own tab. There is also a `private` method that is called when the user selects to create a new drawing and a `Buffer` of the drawings that are currently open. A <<uses>> arrow is drawn from the `DrawMain` object to the `Drawing` class to represent that `DrawMain` depends on the functionality of the `Drawing`.

The `Drawing` type keeps track of a root `Drawable` and two GUI components that are significant for interacting with the `Drawing`. It has a method for getting hold of a `Component` that lets the user interact with the properties of the `Drawing`. The `Drawing` class contains another class called `DrawPanel` which, as the name implies, is a custom-drawn panel in which the drawing will actually appear.

18.6 Implementing the GUI

At this point we have a sufficient understanding of what we want to do to start actually writing a little code. At least initially, the only thing we will set up is the GUI for the program and a little other code to get the program running.

The simplest `class` in this application is the `Drawable class`. This is largely because

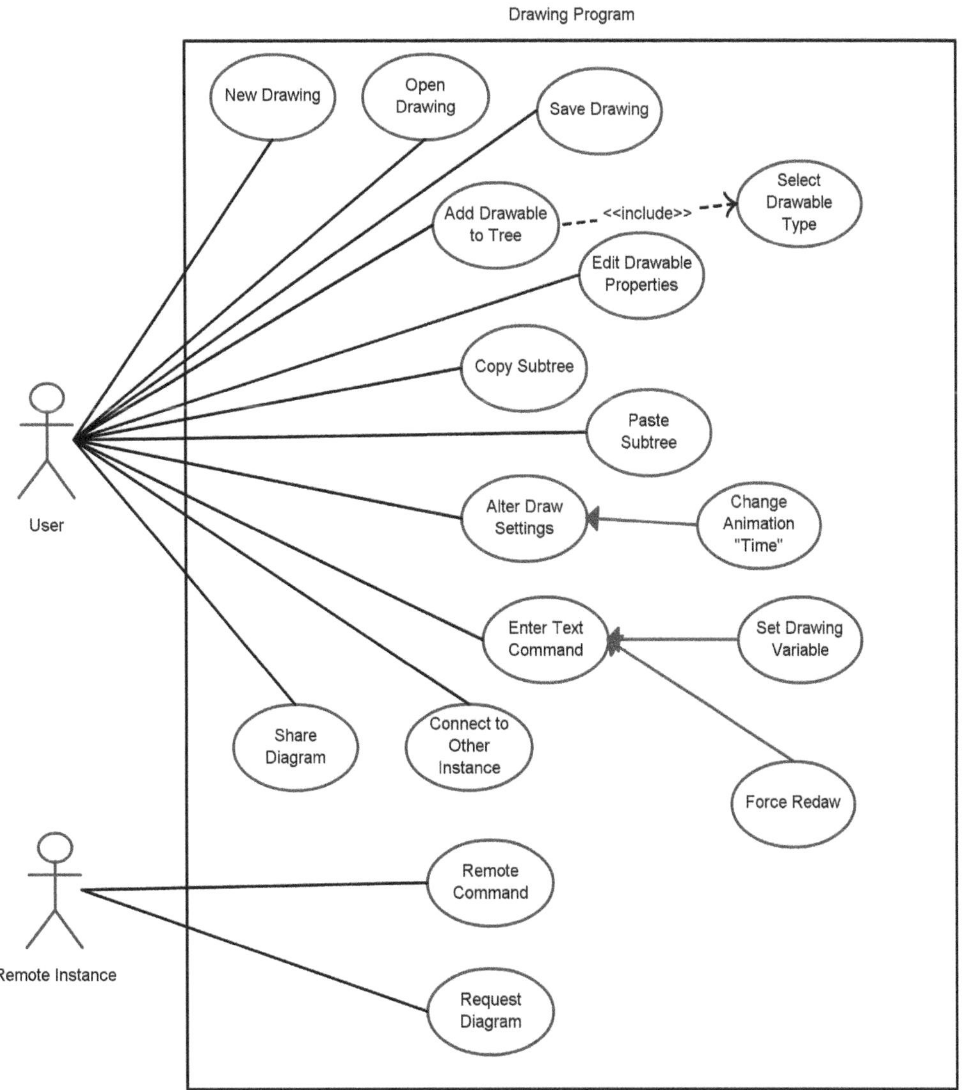

FIGURE 18.3: This is a use-case diagram for our drawing program.

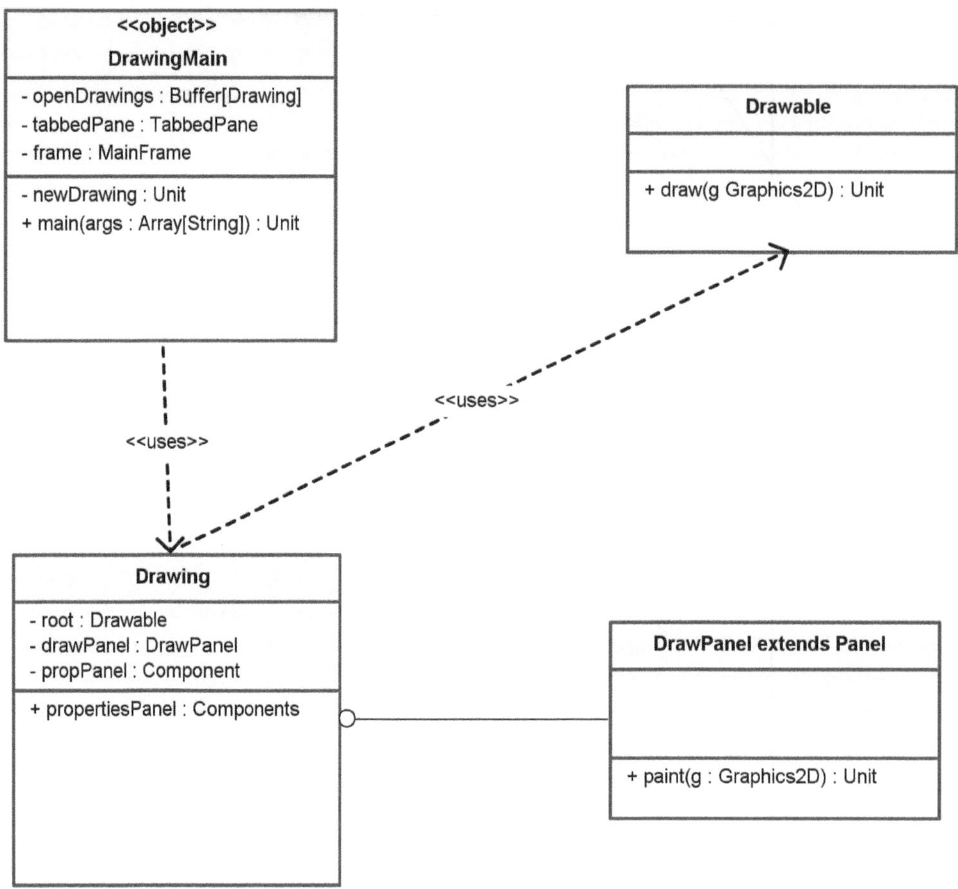

FIGURE 18.4: This is a first cut class diagram for our drawing project.

we do not yet have the knowledge of what should go inside of it. The current code, minus `package` statements and imports, is just this.

```
class Drawable {
  def draw(g:Graphics2D) {}
}
```

In many ways, this code is simply a placeholder for now so that the other parts of the code will compile. The next chapter will give us the means to put more into this.

The next step up in complexity is the `DrawingMain` object. At this point, this is a fairly basic GUI that opens a frame with a menu and a tabbed pane. Initially there are no tabs. Each drawing that the user creates or opens up will be displayed as a tab. The code for this is shown here.

```
object DrawingMain {
  private val tabbedPane = new TabbedPane

  private val openDrawings = mutable.Buffer[Drawing]()

  private def newDrawing {
    val nd = new Drawing
    openDrawings += nd
    tabbedPane.pages += new TabbedPane.Page("Unnamed",nd.propertiesPanel)
  }

  private val frame = new MainFrame {
    contents = tabbedPane
    menuBar = new MenuBar {
      contents += new Menu("File") {
        contents += new MenuItem(Action("New")(newDrawing))
        contents += new Separator()
        contents += new MenuItem(Action("Exit")(sys.exit(0)))
      }
    }
    size = new Dimension(800,600)
  }

  def main(args : Array[String]) : Unit = {
    frame.visible = true
  }
}
```

As you can see, there are currently only two menu options for adding a new drawing and exiting. The new drawing option makes a new drawing and adds a tab for it. Clearly this does not fully implement the use case diagram. That will take a bit more effort to build.

The last two `class`es are inside of a single file with the `DrawPanel` embedded inside of the `Drawing` type. This class is responsible for laying out the majority of the GUI. That is a significant task as most of the interaction of the program will take place in this environment. If it is too limiting it will not let us do the things that we want.

This drawing program is based on the concept of a SCENE GRAPH. This is a data structure that organizes different elements of a drawing in a hierarchy so that changes can be made in one place that will impact a whole group of geometry elements. There are four basic pieces of information we need displayed: the drawing itself, the hierarchy of `Drawable`s, the properties of a selected `Drawable`, and command input and output elements. It is hard to know in advance how much space a user will want to give to these elements so SplitPanes are a good choice of GUI component to use for the setup. Having things nested in ScrollPanes will also help the user see things if they get bigger than the space on screen. Here we see code for building such a GUI.

```
class Drawing {
  private val drawPanel = new DrawPanel
  private val root = new Drawable {}
  private var propPanel:Component = null

  def propertiesPanel = {
    if(propPanel == null) {
      val commandField = new TextField()
      val commandArea = new TextArea()
      commandArea.editable = false
      val commandPanel = new BorderPanel {
        layout += commandField -> BorderPanel.Position.North
        layout += new ScrollPane(commandArea) -> BorderPanel.Position.Center
        preferredSize = new Dimension(500,200)
      }
      val tree = new javax.swing.JTree
      val drawProps = new GridPanel(1,1)
      propPanel = new SplitPane(Orientation.Vertical,
        new GridPanel(2,1) {
          contents += new ScrollPane(new Component {
            override lazy val peer = tree
          })
          contents += new ScrollPane(drawProps)
        },new SplitPane(Orientation.Horizontal,
          new ScrollPane(drawPanel),
          commandPanel))
    }
    propPanel
  }

  class DrawPanel extends Panel {
    preferredSize = new Dimension(500,500)
    override def paint(g:Graphics2D) {
      g.setPaint(java.awt.Color.white)
      g.fillRect(0,0,size.width,size.height)
      root.draw(g)
    }
  }
}
```

There is a SplitPane with a vertical bar at the top of the GUI. On the left side of the bar is a GridPanel with a JTree on the top and an empty GridPanel on the bottom. The JTree will be used to display the hierarchy of Drawables and the empty panel, called drawProps, will display whatever information the selected Drawable requires for settings. On the right side of the top SplitPane is another SplitPane that has a horizontal bar. Above that bar is the drawing in the form of a DrawPanel, and below the bar is a panel that contains a TextField for inputing commands and a TextArea to display the output of those commands.

With these three files we can run the DrawingMain as an application and, after adding

FIGURE 18.5: This figure shows a screenshot of the basic GUI after one drawing has been added.

one new drawing, it produces the window shown in figure 18.5. As you can see, significant parts of this remain blank. The `JTree` has some things in it that you can interact with, but they have no relationship to what this program is doing. They are part of the default appearance and will go away later when we put in a proper implementation. The `TextField` will allow you to type in it, but it will not cause anything to happen. Many of these shortcomings will be fixed in the next chapter.

18.7 End of Chapter Material

18.7.1 Summary of Concepts

- The **software life cycle** is the different steps taken in creating a piece of software.
  - The process of figuring out what the program is actually supposed to do is called **analysis**.
  - Laying out how code is going to be written is called **design**.
  - **Implementation** is the actual act of writing the software.
  - After code has been written you have to go through **debugging** to correct any errors that have been put into the code.

- After software is released, the continued work on it for enhancements and further bug fixes is called **maintenance**.

- **UML** stands for Unified Modeling Language and it is a standard for a number of diagram types that are used for the software development process. This chapter introduced two of them.
  - The **Use-Case Diagram** is a style of diagram used during analysis to help consider the ways in which users and other outside entities interact with the software.
  - A **Class Diagram** is a diagram that shows a rough sketch of major code groupings like `classes`, `objects`, and `traits` as well as relationships between them.

- This chapter also introduced a project that will be used through the rest of the book. It also begins the analysis and design of that project.

18.7.2 Exercises

1. Describe the different phases of the software life cycle.

2. Imagine that you are writing software that has to model a car. Draw a UML diagram for how you might break down that problem.

3. Draw use-case and class diagrams for a project from the last chapter.

18.7.3 Projects

This chapter described a new project that will be followed through the rest of the text and there will be new lines of projects at the end of chapters too. There are several different projects that are developed through the rest of this book. Information on these and how different projects in each chapter are related to different end projects as well as some sample implementations can be found at the book's website. The projects in this chapter have you getting started by laying out some initial foundation and gaining some experience with what you are doing. Note that code you write this early is not likely to survive all the way to the end of the book. You are going to learn better ways of doing things and when you do, you will alter the code that you have written here.

1. One option has you writing a multiuser, networked text-based game, sometimes called a MUD. If you did the text-based game in the first half of the book then you already have some code you might be able to use for this. For this project at this time, you should make a number of `classes` to represent things like characters, rooms, and items in the game. Have it so that they can be read in from file and written out to file. Create an `object` with a `main` method that you can use to run the game and have a single player walk through the world.

   A full MUD implementation will also include computer-controlled characters. At this point only include one type of these. Make it so that character moves randomly around the map as the player executes commands. If the player enters the same room as that character, the description should show them much as it would items in the room.

2. Another option for the project is a web spider for data collection and processing. This option can be customized to fit lots of fields. The basic idea of a web spider is that it loads pages and follows links to get to other pages. Search engines like Google® use spiders to find data and index it so that people doing searches can find it. This project will be much more focused so that you do not require an entire farm of servers to make it run.

   The idea here is that you are spidering one site or a short list of sites for particular types of files. You might be looking for certain data sets, images, box scores, or anything else that you can find a way to recognize. You might want to have several possibilities. The program will compile that information and then provide useful ways for the user to view it.

   For this project option, you can currently build a GUI and think about what data you want to collect and set up some `class`es to store that information. Pull down one or two files manually (consider using `wget`) and have the GUI do a minimal display of the information from those files.

3. You can make a graphical game that is fairly simple, but includes networking so that multiple players can play with or against one another. The possibilities for this are extremely broad and range from puzzle games like Tetris where you have `class`es for `Board` and `Piece` to strategy or RPG games where you have `Units` and `Cities` or `Characters` and `Items`.

   At this time you should put together a GUI and get code up where you can have the player at the computer the application is running on move something around that is visible in the GUI.

4. Instead of doing networking, you can make a more complex graphical game that includes computer-controlled players with more significant Artificial Intelligence. The steps at this point will be about the same for this option as for the last one.

5. By the end of the semester you will have the ability to parse, evaluate, and manipulate functions. These are critical aspects of graphical mathematical software like Mathematica®. As such, one option you could do is to make a Mathematica-like worksheet program. In the end this program will have the ability to plot functions and do basic symbolic manipulations of those functions. It should also have some programmability in a simple language. All of this will be inside of a GUI.

   For now you should build a GUI, make a `class` for a `Worksheet`, and see if you can get it to plot some preprogrammed functions.

6. Another option you could work on is an image processing program somewhat similar in nature to Photoshop®. The goal is to have the ability to build an image from multiple layers as well as a number of different filters for processing images and some tools for creating images. The result will not be as complex as Photoshop itself, but it will be significantly more complex and functional than Paint.

   For now you can write something on the order of Paint. Give it the ability to load and save files as well as a number of different drawing tools for things like lines, rectangles, and similar primitives as well as color selections.

7. For the more numerically inclined you might consider the option of simulation workbench. There were some projects early in the book that had you do basic Newtonian mechanics on gravitational systems using the simplest possible integrator we could write. For this option, we will explore other systems and other integration techniques

including the ability for the user to specify a force law and see how such a system would behave. This will also look at some elements of discrete event simulations.

By the end of the semester, this project will evolve to the point where you can distribute work across multiple machines and view what a simulation is doing. You will also implement some spatial data structures to make integration of long-range forces more efficient. You might even be able to do things like simulate cloth using a particle mesh technique.

For now, make an application that does Newtonian gravity for a number of particles. Break this problem up into different `classes` and include a GUI so that you can watch the particles moving around. It should also have the ability to save to a file and load from a file.

8. Your instructor might have other ideas as well. Check with him/her about the other possibilities they have in mind and what they would entail.

Additional exercises and projects, along with data files, are available on the book's website.

Chapter 19

Abstraction and Polymorphism

Understanding the concept of abstraction is critical for improving your capabilities as a programmer. The ability to express our thoughts in more abstract ways is a significant part of what has allowed modern software systems to scale to the size they are today. We have dealt a little with abstraction already. Back in section 6.4 we saw how we could pass a function as an argument to allow us to accumulate multiple different types of values using a single recursive function. The basic idea of abstraction is that you want to write code where some part of what it does is not fully specified until it is used. We have used this idea significantly through the different methods in the Application Programming Interface (API). At this point it is time to explore the different mechanisms that allow us to create abstraction and write some code that uses them.

19.1 Polymorphism

One of the most powerful ways to add abstraction into code is through POLYMORPHISM. From the Greek roots, the word literally means "many shapes". In the context of programming though it means many types. So far the code that we have written, with the exception of a few lines of code in 7.5.2, has been monomorphic. This means that our code only worked with one type. Perhaps the best example of when this was a limitation comes from chapter 13 where we wrote sorts that used the `Double` type originally, but had to make a second version when we wanted to sort a `case class`. That was because our original sort was monomorphic. It only worked with the one type we wrote it for and nothing else.

If you go back and look at that code you will see how inefficient that can be. The difference between our sorts was very small. We did a copy and paste and then modified only two lines in the code. That type of duplication is wasteful and it leaves us with overly long code that is prone to errors and hard to work with. Plus, one would expect that with so few changes between the versions, it should be possible to make one version that handles them all. To put it a different way, we should be able to abstract out the differences so that we have one version that will work in the different situations we want. In this case, part of that abstraction is abstracting over types. To do that we need polymorphism.

There are multiple different styles of polymorphism, but we can group them into two broad categories: UNIVERSAL and *ad-hoc*. Universal polymorphism implies that code can work with an infinite number of types. By contrast, *ad-hoc* polymorphism only works with a finite number of types. We will primarily consider universal polymorphism in two different forms, INCLUSION POLYMORPHISM and PARAMETRIC POLYMORPHISM.

19.2 Inclusion Polymorphism (Inheritance and Subtyping)

Inclusion polymorphism is a form of universal polymorphism that we get from SUBTYPING. That is to say when all elements of one type are also part of another type. This is a topic that was touched on in section 7.5.3 and figure 7.1 showed a simplified Unified Modeling Language (UML) class diagram with some of the different types in Scala. Figure 19.1 shows an updated version of that diagram that includes direct subtyping relationships as well as some implicit conversions that are written into the Scala libraries to make things easier for you.

This figure shows you what subtyping relationships exist, but it does not explain how they get into the code or how we can use them to write polymorphic code. The answer to the first question is something called inheritance. Inheritance is the standard way we get subtyping and inclusion polymorphism in most object-oriented languages that are built on the concept of a class. When one class, `B`, inherits from another class, `A` it implies two things. One of those things is that `B` is a subtype of `A`. The other is that `B` gets all the data and methods that were part of `A`. The latter we will call CODE-REUSE. It helps us to not duplicate code, and might seem like the more important of the two aspects, however, the real power of polymorphism is the result of the subtyping.

So what does it mean for type `B` to be a subtype of type `A`? We can say that `B` is a subtype of `A` if any situation where an object of type `A` is needed, we can give it an object of type `B` and it will work. To put this into a more concrete form, consider the following example. You have a recipe that calls for three different types of fruit. The recipe gives you instructions

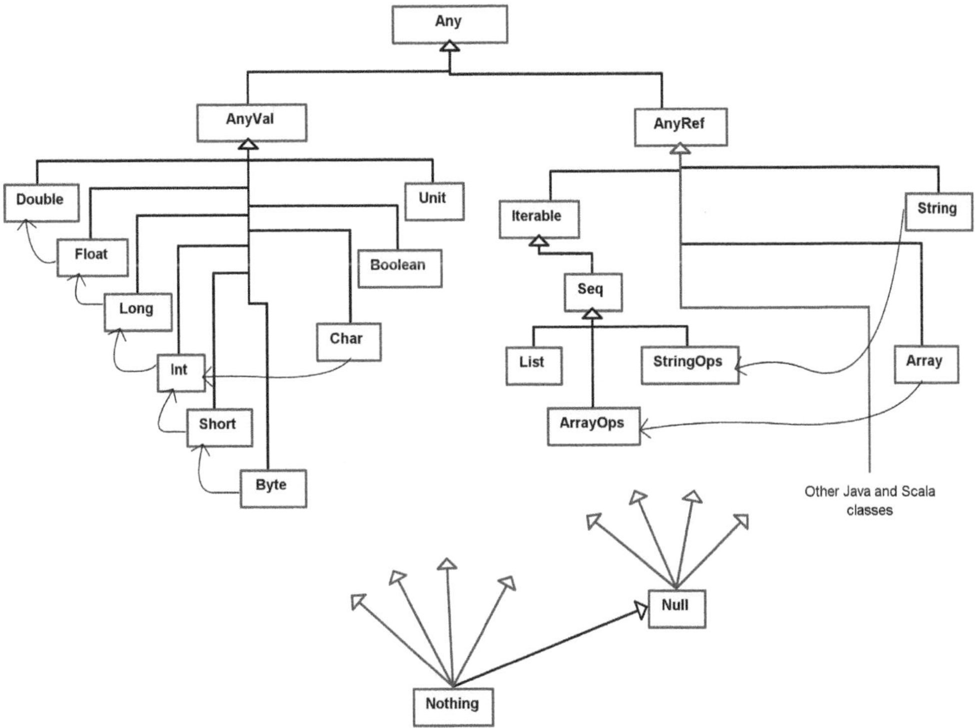

FIGURE 19.1: Diagram of general subtype relationships in Scala. This figure has been adapted from a similar figure in *Programming in Scala* by Odersky, Spoon, and Venners. The thick lines with full arrow heads indicate a subtyping relationship. The thin lines with an open arrow head indicate an implicit conversion.

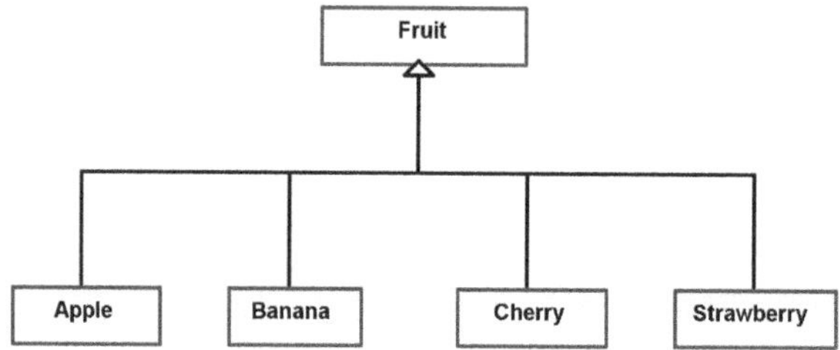

FIGURE 19.2: This is a simple class diagram showing subtypes of fruit.

for how to prepare the fruit and include it with the other ingredients. In this case, `Fruit` is the supertype and we have a description of how to do something with. There are many different subtypes of `Fruit` that we could choose to use including `Apple`, `Cherry`, `Banana`, and `Strawberry`. This particular example is illustrated using a simple UML diagram in figure 19.2. Here we see the type `Fruit` at the top with generalization arrows connecting it to the various subtypes. The direction of the arrows in UML is significant because while subtypes have to know about their supertype, a supertype generally does not, and should not, know about subtypes.

Why Supertypes Do Not Know about Subtypes

The reason that a supertype should not know about subtypes is that such knowledge limits the universality of the polymorphism. When the supertype is created there might only be a few subtypes. In our fruit example, the number of subtypes is four. For the polymorphism to be universal, it should be possible to create a new subtype at a later date without altering the supertype and have any code that already uses the supertype work well with the new subtype.

In code, this means that you can write a function like `makeBreakfastShake` that works with type `Fruit` and it should work with instances of `Apple`, `Banana`, `Cherry`, or `Strawberry`. Consider this code that might represent that function.

```
def makeBreakfastShake(fruit:Fruit) {
  if(!fruit.canEatSkin) {
    fruit.peel
  }
  blender += fruit
  blender += juice
  blender += ice
  blender.blend
}
```

The idea is that we could call this function using code like this.

```
makeBreakfastShake(new Banana)
makeBreakfastShake(new Strawberry)
```

We want this to work because `Banana` and `Strawberry` are both subtypes of `Fruit`. The question is, what is required for this to happen? For the type `Fruit` to work with

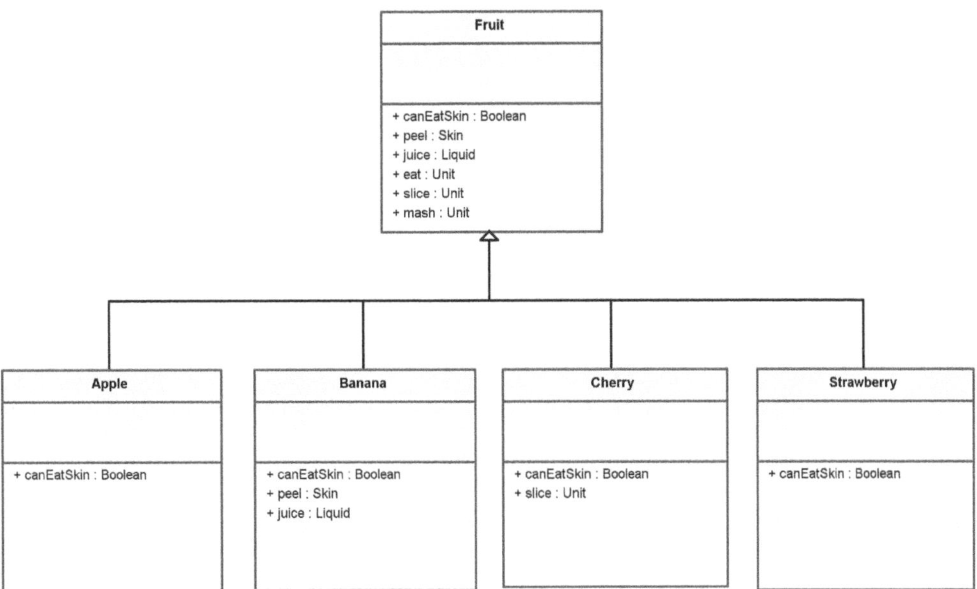

FIGURE 19.3: This class diagram for fruit adds some methods that we might want and how they might be overridden in the subclasses.

makeBreakfastShake it needs to have a method called canEatSkin that takes no arguments and returns a Boolean and a method peel that also takes no arguments. For objects of type Banana to work in place of Fruit, they have to have those same methods.

As you can see, it turns out that the code-reuse part of inheritance is not just there to reduce code duplication, it helps with subtyping. A subtype is guaranteed to have all the methods and data members of the supertype that might be called by outside code.

Of course, you might not always want those methods to do the same things. In this case, objects of the Banana type should return false for canEatPeel while objects of the Strawberry type should return true. Changing the implementation of a method in the supertype is called OVERRIDING. Indeed, this is why the override keyword was needed back in chapter 12 on the paint method. Figure 19.3 shows a UML class diagram using the more complete format for the classes so that you can see the methods that might be in them.

The Fruit type in this diagram has six different methods that pertain to all different subtypes of Fruit. Each of these can be implemented in some default way. The four subtypes only list the methods that they override. The other methods are present in their default implementation as a result of inheritance.

Going back to figure 19.1 you notice that there is a lot of inheritance that goes on in the basic structure of the type system in Scala. Every class that you create automatically inherits from a type called AnyRef, which inherits from Any. You can look in the API to see the types Any, AnyRef, and AnyVal right at the top of the list of classes. If you look in AnyRef you will see that there are a number of different methods that will be in any class you create. We can demonstrate this by making a little class in the REPL and calling one of those methods.

```
scala> class Vect3D(val x:Double,val y:Double,val z:Double)
defined class Vect3D
```

```
scala> val jHat = new Vect3D(0,1,0)
jHat: Vect3D = Vect3D@24753433

scala> jHat.toString
res0: java.lang.String = Vect3D@24753433
```

Here we have a `class` for an immutable 3-D vector type that has no methods, just public members for three components. The second line creates an instance on this class which is followed by a notification that the variable `jHat` is of type `Vect3D` and has a given value. The value is a bit odd, having the name of the class followed by an `@` symbol and a number. The third line shows a call to one of the methods inherited from `AnyRef`, the `toString` method. This call makes it clear where the text for the value came from. The `toString` method is actually defined on the type `Any` and, as such, is safe to call on any object in Scala. For that reason, it is used extensively to get a representation of objects. Here we see that it is called by the REPL.

Unfortunately, the default implementation of `toString` is not all that informative. For that reason, it is something that you often override. This code listing shows how we can do that.

```
scala> class Vect3D(val x:Double,val y:Double,val z:Double) {
     | override def toString():String = "Vect3d("+x+", "+y+", "+z+")"
     | }
defined class Vect3D

scala> val jHat = new Vect3D(0,1,0)
jHat: Vect3D = Vect3d(0.0, 1.0, 0.0)

scala> jHat.toString
res1: String = Vect3d(0.0, 1.0, 0.0)
```

The inclusion of `override` with a `def` that has the same signature as the inherited `toString` gives us a different implementation. You can see that this is now used in the REPL when the value of the object is printed as well as when the method is called directly.

This example shows an inherited method and how we can override it, but it does not actually demonstrate how to get inheritance in our code. It uses the inheritance that we get by default. If we want to inherit from something other than `AnyRef`, we use the `extends` keyword. This keyword is placed after the arguments to the class and before the open curly brace that starts the body of the class.

```
class Name(args) extends Supertype { body }
```

If you do not put in the `extends` keyword you get `extends AnyRef` by default.

To illustrate this, let us look at a different example that is a classic case of inheritance. We define a supertype called `Shape` that has some methods for things that we expect all shapes to be able to do.

```
class Shape {
  def area:Double = 0.0
  def circumference:Double = 0.0
  def draw(g:Graphics2D) {}
}
```

There are many different types of shape that we might want to have as subtypes of this class. Here are two possible examples.

```
class Rectangle(val width:Double,val height:Double) extends Shape {
  override def area:Double = width*height
  override def circumference:Double = 2.0*(width+height)
  override def draw(g:Graphics2D) {
    g.fill(new Rectangle2D.Double(0.0,0.0,width,height))
  }
}


class Circle(val radius:Double) extends Shape {
  override def area:Double = math.Pi*radius*radius
  override def circumference:Double = 2.0*math.Pi*radius
  override def draw(g:Graphics2D) {
    g.fill(new Ellipse2D.Double(0.0,0.0,2.0*radius,2.0*radius))
  }
}
```

Each of these classes takes some arguments and extends Shape, then overrides all of the methods of Shape.

We can now write code that takes an instance of Shape and pass it an instance of either Rectangle or Circle and it will work. The reason it will work is that inheritance guarantees us that any method we could call on an instance of Shape will be defined in the subtypes. Consider the following function and two calls to it.

```
def areaCircumferenceRatio(s:Shape):Double = {
  s.area/s.circumference
}

val circleACR = areaCircumferenceRatio(new Circle(5))
val rectACR = areaCircumferenceRatio(new Rectangle(4,5))
```

This is the heart of polymorphism, one piece of code that can work on multiple types. This is universal polymorphism because at any point in the future we could create a new subtype of Shape and it would work with this code as well.

As a general rule, inheritance should be used to represent an "*is-a*" relationship. In the examples we have seen so far, it seems quite natural to say that an apple *is-a@is-a* fruit or that a rectangle *is-a* shape. If you can not say that, then inheritance probably is not the construct you want to use. Instead, you should use composition. This is where you put an instance of the type as a member of some class instead of inheriting from it. To get the functionality, you make calls on that member. Composition is used to model a "*has-a*" relationship. We will see many examples of this type of construction later in the book. It should be emphasized that you do not want to abuse inheritance. The code-reuse aspect might seem great, but the subtyping should only be used when it makes sense.

The opposite is also true. There are times when you will have an *is-a* relationship and subtyping makes sense, but not all of the methods you would get through code-reuse fit. In those situations you should also refrain from using inheritance. An example of this is a square. Clearly a square *is-a* shape. In addition, a square *is-a* rectangle. That relationship can be expressed with the following one-line class.

```
class Square(length:Double) extends Rectangle(length,length)
```

Here the class Square takes one argument for the length of an edge. It extends Rectangle, and passes that length as both the height and the width arguments. As this example shows, when the subtype takes arguments, they are passed in an argument list that follows the name of the supertype after extends.

Given what we have written for Rectangle, this implementation of Square is perfectly acceptable and calls to all the methods of an instance of Square will work as desired. That is only because our Rectangle type happens to be immutable. Consider this alternative implementation of Rectangle.

```
class Rectangle(var width:Double,var height:Double) extends Shape {
  override def area:Double = width*height
  override def circumference:Double = 2.0*(width+height)
  override def draw(g:Graphics2D) {
    g.fill(new Rectangle2D.Double(0.0,0.0,width,height))
  }
}
```

If you use this with the above definition of Square everything compiles fine and it might even seem to run fine. However, there is something truly broken with this implementation. To see this, consider the following code.

```
val square = new Square(5)
square.width=20 // WARNING: Works in code, but not what we want.
println(square.area)
```

This does something that should not be possible. It makes a "square" that is 20 by 5. That certainly does not seem to fit the normal definition of a square.

The problem here is that when we make the Rectangle type mutable we effectively introduced two new methods: def width_=(w:Double) and def height_=(h:Double). These methods are not suitable for the Square, but by virtue of inheritance, our Square type gets them.

It is tempting to try to fix this by explicitly overriding the methods that set the width and height fields as is done in this code.

```
class Square(length:Double) extends Rectangle(length,length) {
  override def width_=(w:Double) { // ERROR: Can't override mutable variables.
    width = w
    height = w
  }
  override def height_=(h:Double) { // ERROR: Can't override mutable variables.
    height = h
    width = h
  }
}
```

This does not compile because you are not allowed to override methods associated with variables. Even if it were allowed, or if we used different methods to access and set the variables, it would not be a good idea. To understand why, consider what a programmer is expecting to happen when a line like rect.width = 20 is executed. Clearly the programmer who writes this line is expecting it to change the value stored in the width field. If the height field changes as well, that would be unexpected behavior and might well violate assumptions made elsewhere in the function. That could lead to bugs that are remarkably hard to find because the code would work fine when used with an instance of Rectangle, and reasoning about the program will not likely help the programmer find the error.

So be careful with how you use inheritance. It should only be used when you want both the subtyping *and* the code-reuse.

19.2.1 private Visibility and Inheritance

One thing that is lacking in our Shape example is a color for the shapes. Since there is a draw method, it would make sense for them to have a color to be drawn in, and because all shapes would have that color value, it would make sense for it to be part of the Shape class. Here are implementations of Shape, Rectangle, and Circle where the color field is a private, mutable value in Shape.

```
class Shape(private var color:Color) {
  def area:Double = 0.0
  def circumference:Double = 0.0
  def draw(g:Graphics2D) {}
}


class Rectangle(val width:Double,val height:Double,c:Color) extends Shape(c) {
  override def area:Double = width*height
  override def circumference:Double = 2.0*(width+height)
  override def draw(g:Graphics2D) {
    g.setPaint(color) // ERROR: Can't get to the private color data member.
    g.fill(new Rectangle2D.Double(0.0,0.0,width,height))
  }
}


class Circle(val radius:Double,c:Color) extends Shape(c) {
  override def area:Double = math.Pi*radius*radius
  override def circumference:Double = 2.0*math.Pi*radius
  override def draw(g:Graphics2D) {
    g.setPaint(color) // ERROR: Can't get to the private color data member.
    g.fill(new Ellipse2D.Double(0.0,0.0,2.0*radius,2.0*radius))
  }
}
```

The reason it is private is because it is mutable and we do not want to allow any piece of code that gets hold of a Shape object to be able to change that value. As the comments indicate, these classes will not compile. While subtypes get copies of everything in the supertype, they do not have direct access to private elements. So using this approach, the subtypes can not get to color unless we add a method for them to access it through.

This behavior is what we really desire. When something is made private, that means that other parts of code outside of that class should not be able to access it. Anyone can make a subtype of a class. If that gave them access to private data, then private would not really be all that safe. Granted, if color were declared as a val instead of a var, we would not need to hide it because the java.awt.Color class is immutable so other code could not change it, even if it could access it.

19.2.2 Protected Visibility

Of course, there are times, like our colored shape example, when you do want to have data or methods that are only accessible to the subtypes. This is the reason for the protected visibility. When a method or data member is modified with the protected keyword, it is visible in that class and its subclasses, but not to any other code. Using this, we could modify the Shape class to the following.

```
class Shape(protected var color:Color) {
```

```
  def area:Double = 0.0
  def circumference:Double = 0.0
  def draw(g:Graphics2D) {}
}
```

With color set to be protected, the previous code for Rectangle and Circle will work.

The protected visibility is something that is not used very often. If a value really needs to be hidden it should be private. Making it protected is a signal to programmers that it is needed by subclasses, but that they need to know what they are doing if they are going to use or alter it.

19.2.3 Calling Methods on the Supertype

Sometimes when you override a method in a subtype, you still want the code from the supertype to be run. There are actually situations where you are expected to call the method in the supertype before doing additional work. To support this type of behavior, you can use the super keyword like an object that references the part of the current object that is the supertype.

This can be used to solve the coloring problem in a way that keeps the color private. Consider the following implementation of Shape.

```
class Shape(private var color:Color) {
  def area:Double = 0.0
  def circumference:Double = 0.0
  def draw(g:Graphics2D) {
    g.setPaint(color)
  }
}
```

This has a private member color so it is not visible to the subtypes and is well encapsulated so we do not have to worry about other code changing it. Unlike the last version with a private color, this one has some code in the draw method that sets the paint on the Graphics2D object to color. Having this in the code allows us to do the following in Rectangle.

```
class Rectangle(val width:Double,val height:Double,c:Color) extends Shape(c) {
  override def area:Double = width*height
  override def circumference:Double = 2.0*(width+height)
  override def draw(g:Graphics2D) {
    super.draw(g)
    g.fill(new Rectangle2D.Double(0.0,0.0,width,height))
  }
}
```

By making a call to draw on super, this version can set the color when draw is called without actually having access to the value in the supertype. The same type of thing could be done in Circle.

One advantage to this approach is that as long as all the subtypes of Shape follow the pattern of making a call to super.draw before they draw their own geometry, it would be possible to add other settings such as different strokes or transformations to the Shape type without altering the draw methods in the subtypes. In large libraries, this same behavior can be problematic as it forces the supertype to stick with a certain behavior because all the subtypes, including those written by other authors, are expecting it to be maintained.

19.2.4 Anonymous Classes

At some point in this chapter you might have been thinking back to some things from earlier in the book and wondering how they were related to inheritance. A code segment like the following might specifically come to mind.

```
val panel = new Panel {
  override def paint(g:Graphics2D) {
    ...
  }
}
```

This was the first situation where we saw the keyword `override`. As you just learned, that signifies that you are creating a `class` that provides an alternate implementation for some method. Though you did not know it at the time, this syntax is creating a new `class`, an ANONYMOUS CLASS. In this case, that new `class` inherits from `Panel`. You can think of it as being like this code.

```
class AnonymousPanelClass extends Panel {
  override def paint(g:Graphics2D) {
    ...
  }
}

val panel = new AnonymousPanelClass
```

In the first version, the compiler actually creates a new `class` with a name that includes characters you are not allowed to use in your program for `class` names so it is impossible to refer to the type you have created. Instead, you have to treat the new object as an instance of the supertype.

You will see that this type of thing is done a lot in Scala. Fortunately, the syntax for anonymous `classes` is so natural in Scala that you rarely think of what you are doing as inheritance. Instead you think of it more like you are making a specialized instance of some type that has some additional functionality.

One place where the fact that you are in an anonymous `class` or some other nested `class` is when you try to use `this`. By default, `this` refers to the current instance of the most closely bound class. If you want to refer to the instance of a class that is wrapped around the code you are using, but is not the closest one, simply prefix `this` with the name of the bounding type you are interested in. So if the declaration of `panel` above was nested in some other `class` call `GUIProgram`, you could refer to the current instance of `GUIProgram` inside of the code you write for `panel` by using `GUIProgram.this`.

19.2.5 Abstract Classes

Let us now go back to the `Shape` example with its two subclasses. This version has a constant public `color`, but uses the call to `super` to provide uniform behavior.

```
class Shape(val color:Color) {
  def area:Double = 0.0
  def circumference:Double = 0.0
  def draw(g:Graphics2D) {
    g.setPaint(color)
  }
}
```

```
class Rectangle(val width:Double,val height:Double,c:Color) extends Shape(c) {
  override def area:Double = width*height
  override def circumference:Double = 2.0*(width+height)
  override def draw(g:Graphics2D) {
    super.draw(g)
    g.fill(new Rectangle2D.Double(0.0,0.0,width,height))
  }
}


class Circle(val radius:Double,c:Color) extends Shape(c) {
  override def area:Double = math.Pi*radius*radius
  override def circumference:Double = 2.0*math.Pi*radius
  override def draw(g:Graphics2D) {
    super.draw(g)
    g.fill(new Ellipse2D.Double(0.0,0.0,2.0*radius,2.0*radius))
  }
}
```

There are some aspects of this code that should feel less than ideal to you. They can really be summed up in the following line.

```
val s = new Shape(Color.red)
```

This line of code compiles and runs just fine. The question is, what does it really mean? If you were to call `area` or `circumference` on the object that `s` references, you would get back `0.0`. If you were to call `draw`, nothing would be drawn.

The problem is that while `Shape` is certainly a valid type, it is not a complete specification of an object. This is because while we certainly feel that all shapes should have areas and circumferences, we can not really define them in the completely general sense. We need to be able to say that a type has a method or some piece of member data, but not give it a definition. We also want `Shape` to represent a supertype only and objects should only be instantiated for the subtypes. This can be done with the following code.

```
abstract class Shape(val color:Color) {
  def area:Double
  def circumference:Double
  def draw(g:Graphics2D) {
    g.setPaint(color)
  }
}
```

Two things have been changed here. We have taken away the equal sign and what follows it for `area` and `circumference`. We have also added the keyword `abstract` to the `class` declaration.

Methods and member data that are not given a value in a class declaration are considered to be abstract. If anything in a `class` is abstract, then the `class` itself must be labeled as `abstract`. If you leave that off, you will get a syntax error. When a type inherits from an `abstract class`, it must either provide an implementation for the abstract members and methods or that subtype must also be labeled `abstract`.

Given this revised `Shape` class, the `Rectangle` and `Circle` classes shown above will work just fine and we will no longer have the conceptual difficulties produced by instantiating the `Shape` class directly. In addition, the `override` keyword is not required when you implement a method that is abstract in the supertype. There are two reasons for this. First, on the semantic side, you are not actually overriding an implementation in the supertype. Second, and more importantly, if the method is abstract and you do not implement it, that will generate a syntax error unless the class is labeled as `abstract`.

To help you understand this second reason, let us look at an example of why Scala requires the `override` keyword in the first place when you are overriding a method. Consider this version of `Rectangle`.

```
class Rectangle(val width:Double,val height:Double,c:Color) extends Shape(c) {
  def area:Double = width*height
  def circumference:Double = 2.0*(width+height)
  override def draw(g:Graphics) { // ERROR: Not really an override
    g.setPaint(color)
    g.fill(new Rectangle2D.Double(0.0,0.0,width,height))
  }
}
```

The `override` keyword has been dropped from `area` and `circumference` because they are abstract in the supertype. There is a comment on the `draw` method telling you that it will not compile. Can you tell what is wrong? There is a very subtle error here that most people would miss without the comment and it is quite possible that you are having a hard time identifying it even with the comment. The error is the `x` is listed as being type `Graphics` instead of `Graphics2D`. That kind of little typo is very easy to do when you are coding and, if it were not a syntax error it could produce a logic error that is very difficult to track down. Scala requires `override` in this usage so that it will be a syntax error if you change the type. In the situation where `override` is not required because the method in the supertype is abstract, a typo in the argument or method name will leave that method unimplemented, which will generate a different error indicating that the class has to be `abstract` because some method is abstract.

19.2.6 traits

The `class` and `object` constructs are not the only ways to create new types in Scala. A third option, which is significant to us in this chapter, is the `trait`. A `trait` is very similar to an `abstract class` in many ways. The two primary differences are that `traits` can not take arguments and you can inherit from more than one `trait`.

It was not specifically said above, but you might have noticed there was no description of how to list multiple `classes` after `extends`. This was not an oversight, you are not allowed to list multiple `classes` there. You are allowed to follow the `class` with multiple `traits` or just have one or more `traits`. The various types you inherit from are separated by the `with` keyword.

To understand why you would want to do this, consider the following example. Imagine you are writing software to simulate people moving through a building to help with designing the space. You have a type called `Person` that represents everyone that can be in the simulation. The company you are working with knows there are certain categories of people who behave in significantly different ways that the simulation must handle. For example, they need types for `Parent`, `Child`, and `GeneralAdult`. The need for facilities in the building means they also need types for `Male` and `Female`. This last part creates a problem for single inheritance because you want to be able to represent a type `Father` by inheriting from `Parent` and `Male`. The subtype relationship makes sense there, but if both of those type are written as `classes`, you will not be able to use them both.

There are two ways that you could approach this problem with traits. A somewhat standard inheritance scheme might use the following structure.

```
trait Person { ... }
trait Parent extends Person { ... }
trait Male extends Person { ... }
class Father extends Parent with Male { ... }
```

Here both `Parent` and `Male` are subtypes of `Person` and `Father` inherits from both. In this case `Person` has to be a `trait`. This is because a `trait` can only inherit from other `traits`, not `classes`. Below that, either `Parent` or `Male` could have been a `class`, but not both because `Father` could not inherit from both of them then. Without a pressing reason to make one a `class` the choice of a `trait` provides consistency.

An alternate approach to constructing these types could be to use the `traits` as mix-in types. Here is what the code could look like.

```
class Person { ... }
trait Parent { ... }
trait Male { ... }
class Father extends Person with Parent with Male { ... }
```

In this construction both the `Parent` and `Male` types are `traits` that do not inherit from the `Person`. Instead, they are mixed in with the `Person` to create the `Father`. This approach is perhaps a bit more advanced and if the `Parent` and `Male` types involve code that requires them to know they will be used with a `Person`, you will need to use self-types which are discussed in appendix B. What matters now is that you be aware that this option exists so that as we build different inheritance hierarchies you will understand what is going on.

`trait` or `abstract class`?

A standard question programmers deal with in writing Scala is whether to choose an `abstract class` or a `trait` when coding an abstract type that will be used with inheritance. The general rule of thumb here is to prefer a `trait` because it allows the flexibility of multiple inheritance. While it can not take arguments, any values that you would want to provide as arguments can be put into the `trait` and left undefined so that the `class` that implements them in the end will have to provide values for those.

19.2.7 `final`

When something is `abstract`, it basically means that it has to be "overridden" in a subclass.[1] There are times when it is also useful to be able to say that something can not be overridden, or changed in a subtype. This requirement can be enforced with the `final` keyword. You can use `final` to modify member data or methods that you do not want to be changed in subtypes. It can also be used to modify a whole class when you do not want to allow there to be any subtypes of that class.

You might wonder why you would want to make something `final`. One answer to this question is that you want to preserve some type of behavior in the current implementation that should not be altered in subtypes. It can also be a way of telling things that use a class that there is no possibility of getting different behavior from an object of that type than what they would get from that specific type.

[1]The term overridden is put in quotes here because technically something that is abstract is not overridden because it did not have a definition to start with.

The most common example of a place where you should use `final` to preserve a behavior in a type is when a class defines a type that is immutable. The strengths of immutable types, and how they can be used to simplify code is something that has been addressed quite a bit already in this book. Without inheritance, if you wrote a type to be immutable, you knew that any object of that type could be passed to other code without fear of it being changed.[2] With inheritance, that does not automatically hold true unless you make the `class` final. The reason for this is that subtypes could add mutable data and/or override methods such that the new implementations include some reference to a mutable state.

To help you understand this, consider the following example class.

```
class CharInStrComp(str:String) {
  def positionCompare(str2:String,index:Int) : Int = {
    if(index >= str.length) {
      if(index >= str2.length) 0 else 1
    } else if(index >= str2.length) -1 else {
      str(index).compareTo(str2(index))
    }
  }
}
```

This class is immutable. Technically it does not even have member data to mutate. So if you have an object created from this class it can be passed around freely. With inheritance though it would be possible to do this.

```
class CntCharInStrComp(str:String) extends CharInStrComp(str) {
  var cnt = 0
  override def positionCompare(str2:String,index:Int) : Int = {
    cnt += 1
    super.positionCompare(str2,index)
  }
}
```

If you make an instance of `CntCharInStrComp`, it can be used in any place that wants a `CharInStrComp`. The only problem is that if this is done in a place where the code relies on objects being immutable, that code can now break because it is being passed an object that is mutable. To prevent this from happening, the original version should be made `final`, as is shown here.

```
final class CharInStrComp(str:String) {
  def positionCompare(str2:String,index:Int) : Int = {
    if(index >= str.length) {
      if(index >= str2.length) 0 else 1
    } else if(index >= str2.length) -1 else {
      str(index).compareTo(str2(index))
    }
  }
}
```

Now any attempt to make a subtype, whether mutable or not, will produce an error.

The `final` keyword is often underused in programming. However, it is good to get into the habit of making things `final` when there are not supposed to be subtypes or when something should not be overridden. This is not only a safe practice to get into, code that is labeled as `final`, can sometimes be compiled to faster implementations as well.

---

[2]We will also see in chapter 21 that such objects can be used safely across multiple threads.

19.2.8 Method Resolution

When you are using inheritance, it is possible for you to create types for which there are multiple different method implementations. It is critical that you understand how Scala will determine which of the method implementations to use when the method is called on an object.

If the method is defined in the `class` or `object` the object was created from, that is the version that will be used. If it was not defined there, a version in one of the supertypes will be used. It will look at the supertypes beginning with the last one in the list. If it is not in the last type, it will try anything that type inherits from before going to the earlier elements of the list. This conversion from the full inheritance structure to a list of types to check through is called LINEARIZATION. One caveat is that if the type appears multiple times, the last one by order of appearance in the resolution list is used.

Let us consider the type `Father` defined above. The linearization for the two approaches is actually the same. In the first case, if you list all the types from the end of the list back, ignoring repeats you get `Father`, `Male`, `Person`, `Parent`, and `Person`. The first instance of `Person` is removed leaving us with `Father`, `Male`, `Parent`, and `Person`. In the second approach, starting from the end goes directly to `Father`, `Male`, `Parent`, and `Person`.[3]

Why Linearize?

Being able to inherit from many things is called MULTIPLE INHERITANCE and it is only allowed in some languages, C++ being the most notable case. However, it leads to many complications that many newer languages have chosen to avoid. In the case of Scala, you only get single inheritance from `class`es and multiple inheritance is only allowed with `trait`s.

The reason you are not allowed to do this with `class`es is that the semantics of inheriting from a `class` is to get a full copy of everything in that `class` in the subclass. With that in mind, consider the problems that are created by the UML diagram shown in figure 19.4. This situation is called the DIAMOND PROBLEM.

The C++ language follows the rule of full inclusion of superclasses in subclasses which illustrates how bad this problem can be. Using this rule, types `B` and `C` get full copies of `A` in them. Then `D` has a full copy of `B` and `C`, meaning it has two full copies of `A`, a situation that is clearly problematic. (C++ includes a construct called virtual inheritance that can take this down to one copy.) Imagine you make an instance of type `D` and try to use the value `a`, or call the methods `foo` or `bar`. Which one would it use? What code would be executed? `D` inherits those methods from two places and, in the case of `bar`, the code for the two can be different.

In C++, the programmer has to specify which of the supertypes to use in this situation. In Scala (and Perl and Python), the supertypes are "linearized" so that there is a specific order in which they are checked to resolve a method call. That removes any ambiguity. Other languages, such as Java, make it impossible to inherit multiple implementations or data values so this type of problem can not arise.

[3]In both of these cases, `AnyRef` and `Any` appear at the end of the list as every object implicitly inherits from `AnyRef`, which inherits from `Any`

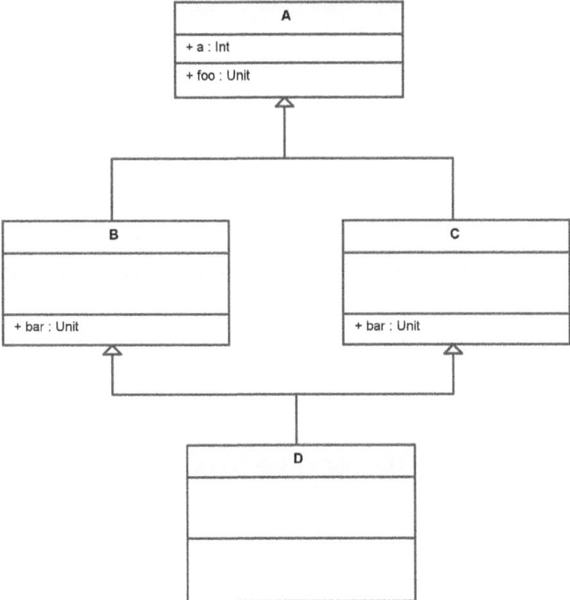

FIGURE 19.4: This UML class diagram shows what is commonly the diamond problem. The challenge with this construct is that it can lead to significant ambiguity. Different languages have different ways of dealing with this.

19.2.9 Inheriting from Function Types

We saw back in chapter 16 that when you treat an object like a function and pass it value arguments, Scala expands that to a call to the `apply` method on that object. When you enter the code `o(i)`, Scala sees `o.apply(i)`. As a result, you can easily make types that you treat like functions. In order for the type to really be a function type that you can use in places that expect functions, like calls to `map`, `filter`, `foreach`, etc., you need to have your type actually be a subtype of the function type.

To illustrate this, consider the simple example of a custom type that is a function that doubles integers. We could write such a type like this.

```
class G { def apply(i:Int) = 2*i }
```

If we instantiate an object of this type we can use it like a function.

```
scala> val g = new G
g: G = G@28f2e328

scala> g(5)
res0: Int = 10
```

However, if we try to use this same object in a call to `map`, we get an error.

```
scala> List(1,2,3) map g
<console>:10: error: type mismatch;
 found   : G
 required: (Int) => ?
      List(1,2,3) map g
```

As you can see from the error message, this does not work because the map method is expecting an argument that is a function of the form (Int) => ?. That is something that takes an Int as a single argument and outputs anything. We can make a type that fits this, but we need to use inheritance to do so.

```
scala> class F extends ((Int)=>Int) { def apply(i:Int) = 2*i }
defined class F

scala> val f=new F
f: F = <function1>

scala> List(1,2,3) map f
res1: List[Int] = List(2, 4, 6)
```

The only difference between type F and type G is that F explicitly extends (Int)=>Int. The first impact of this can be seen in the creation of the object f where we get an different version of the toString method.

19.3 Inheritance in the Project

You now know enough about inheritance that we can see how to use it in our project code. The most obvious place inheritance will appear is the different Drawable types that we will want to include.

19.3.1 Drawables

Each different type of thing that we might want to draw will be a different subtype of Drawable. At the very least, we can start with some basic shapes like rectangles and ellipses. We will also include another type that will group together multiple elements in the drawing and represent a transformation. Back in chapter 12 we learned about the java.awt.geom.AffineTransform type and used it with our drawings on a Graphics2D object. We want a Drawable element that can include that type of functionality in our drawings.

There is another aspect to the Drawable type that relates to how it is included in the Graphical User Interface (GUI). We want the different things that are being drawn to appear in the JTree element in the GUI. The javax.swing.JTree displays objects that have to be subtypes of the javax.swing.tree.TreeNode type.[4] To make this happen, we can just have the Drawable type extend TreeNode.

Given the description of Shape earlier in this chapter, you have probably also realized that the draw method in Drawable should be abstract because there really is not a good default implementation of how to draw. We will take the additional step of making Drawable into a trait. The resulting code looks like this.

```
/**
```

[4]The TreeNode type is what is called an interface in Java. This is like a Scala trait where all methods are left abstract and no member data is defined.

```
 * This represents the supertype for all the different types that can appear in
     our drawing.
 */
trait Drawable extends TreeNode {
  /**
   * Causes this object to be drawn to g.
   *
   * @param g a Graphics2D object to draw to.
   */
  def draw(g : Graphics2D) : Unit

  /**
   * Gives back a GUI component that allows the user to change drawing properties.
   *
   * @return A component that should be put in the GUI so the user can edit this
       object.
   */
  def propertiesPanel() : scala.swing.Component
}
```

Note that one additional method has been added. This is a method that gives us back a GUI component that can be displayed when the user selects a particular `Drawable`. That GUI component should allow the user to change the settings for that object.

We are going to start with the three subtypes of `Drawable` mentioned above. One possible way to arrange this is as is shown in figure 19.5. Looking at this figure you might wonder where all the methods came from in the `DrawTransform`, `DrawRectangle`, and `DrawEllipse` classes. All of the methods listed in these classes, with the exception of `draw`, are abstract methods from `TreeNode`. Because they are abstract, they have to be implemented in any concrete subtype, which is why they appear in this figure.

It turns out that the implementations of all the methods other than `draw` will be exactly the same in both `DrawRectangle` and `DrawEllipse`. In fact, the implementations of these methods will be exactly the same for all the `Drawable` subtypes other than `DrawTransform`. For that reason, it makes sense to put another type in the hierarchy as shown in figure 19.6.

The new `DrawLeaf` type can implement the methods that come from `TreeNode` in a way that will work for nearly all of the subtypes of `Drawable` that we will create in this book. While right now having this type only prevents us from making two copies of the code, before we are done, it will have saved us from making many separate copies of these methods. We are not doing this on the `DrawTransform` side because we do not expect to have other classes that have `Drawable` children. Were that decision to change, we would probably want to pull those methods out from that side of the hierarchy as well.

With this design work done, we can now turn to implementations of these three new `classes` and one new `trait`. We will start with the `DrawTransform` class because it stands on its own and might help us to see what we can do on the other side of the hierarchy in figure 19.6. Here is code for that.

```
package scalabook.drawing

import java.awt.Graphics2D
import javax.swing.tree.TreeNode
import scala.collection.mutable
import swing._
import event._
import java.awt.geom.AffineTransform
```

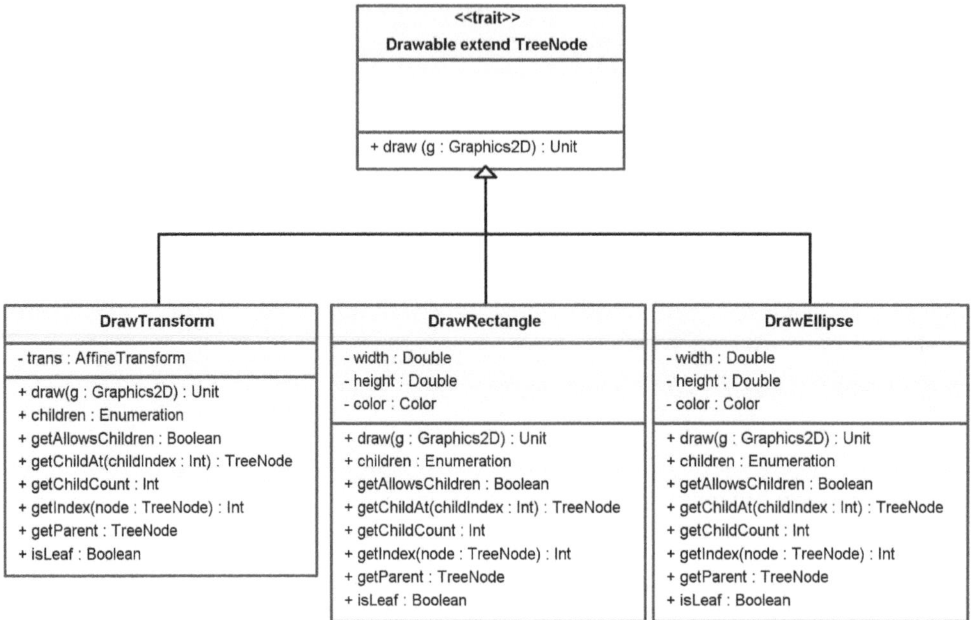

FIGURE 19.5: This UML class diagram shows a possible way to set up the hierarchy for our first three `Drawable` subtypes.

```
/**
 * This class represents a transformation with children.
 *
 * @param parent the DrawTransform that is the parent of this Drawable.
 */
class DrawTransform(parent:DrawTransform) extends Drawable {
  private val subnodes = mutable.Buffer[Drawable]()
  private var propPanel:Component = null

  private object TransformType extends Enumeration {
    val Translate,Rotate,Scale,Shear = Value
  }
  import TransformType._

  private var transformType = Translate
  private val transformValue = Array.fill(3)(0.0)

  /**
   * This method applies a transform and draws all the subnodes.
   *
   * @param g the Graphics2D object to draw to.
   */
  def draw(g:Graphics2D) {
    val oldTrans = g.getTransform
    g.transform(buildTransform)
    subnodes.foreach(_.draw(g))
```

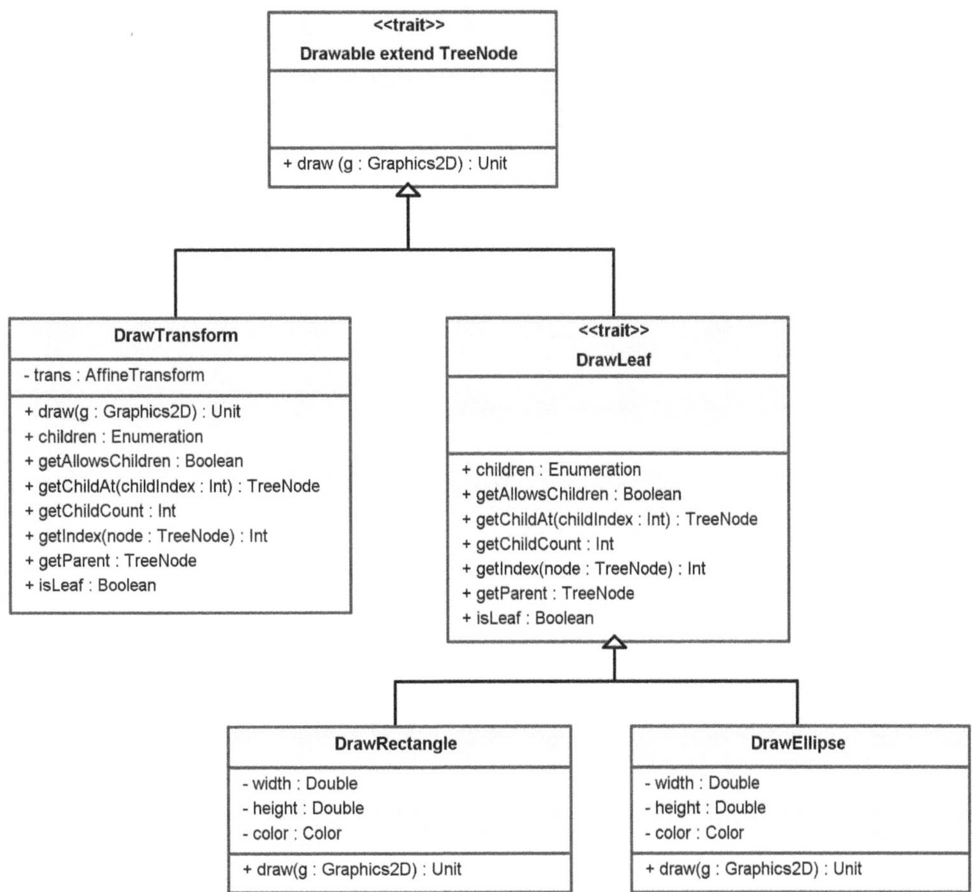

FIGURE 19.6: This UML class diagram shows a more useful hierarchy for our first three `Drawable` subtypes.

```
    g.setTransform(oldTrans)
  }

  private def buildTransform:AffineTransform = transformType match {
    case Translate =>
        AffineTransform.getTranslateInstance(transformValue(0),transformValue(1))
    case Rotate =>
        AffineTransform.getRotateInstance(transformValue(0),transformValue(1),transformValue
    case Scale =>
        AffineTransform.getScaleInstance(transformValue(0),transformValue(1))
    case Shear =>
        AffineTransform.getShearInstance(transformValue(0),transformValue(1))
  }

  /**
   * Returns a GUI element for setting parameters.
   * If this method is called multiple times, the same object will be returned.
   *
   * @return a Component that can be added to a Swing GUI.
   */
  def propertiesPanel() : Component = {
    if(propPanel == null) {
      propPanel = new BorderPanel {
        layout += new GridPanel(transformValue.length+1,1) {
          contents += new ComboBox(TransformType.values.toSeq) {
            listenTo(selection)
            reactions += { case e => if(selection.item!=null)
                transformType=selection.item }
          }
          for(i <- transformValue.indices) {
            val textField = new TextField(transformValue(i).toString) {
              listenTo(this)
              reactions += { case e:EditDone => transformValue(i)=text.toDouble }
            }
            contents += textField
          }
        } -> BorderPanel.Position.North
      }
    }
    propPanel
  }

  /**
   * Returns an object that can be used to run through the children.
   *
   * @return a java.util.Enumeration object for the subnodes.
   */
  def children : java.util.Enumeration[Drawable] = new
      java.util.Enumeration[Drawable] {
    val iter = subnodes.iterator
    def hasMoreElements() = iter.hasNext
    def nextElement() = iter.next
  }

  /**
```

```
 * Tells whether this can have children.
 *
 * @return always true for this type.
 */
def getAllowsChildren : Boolean = true

/**
 * Returns the child at a specified index.
 *
 * @param childIndex the index in the sequence of children.
 * @return the child at that index.
 */
def getChildAt(childIndex : Int) : Drawable = subnodes(childIndex)

/**
 * Returns how many children this node has.
 *
 * @return the number of children currently under this node.
 */
def getChildCount : Int = subnodes.length

/**
 * Returns the index of a particular object.
 *
 * @param node a TreeNode to search for in the sequence of children.
 * @return the index of that child of -1 if it isn't a child.
 */
def getIndex(node : TreeNode) : Int = subnodes.indexOf(node)

/**
 * Returns the parent of this node.
 *
 * @return the parent of this node. Will be null if this is the root.
 */
def getParent : DrawTransform = parent

/**
 * Tells is this node is a leaf in the tree.
 *
 * @return true if there are currently no children, otherwise false.
 */
def isLeaf : Boolean = subnodes.isEmpty

/**
 * Adds a child to this node.
 *
 * @param c the node to add.
 */
def addChild(c : Drawable) { subnodes += c }

/**
 * Removes the specified drawable from the sequence of children.
 *
 * @param c the node to remove.
 */
```

```
  def removeChild(c : Drawable) { subnodes -= c }

  override def toString() = "Transform"
}
```

This class has four private data members. The first, called `subnodes`, that is a `Buffer` which stores all of the children of the transform.[5] Those children are the things that the transform will be applied to. The next chapter will go into the details of the `Buffer` type. The second is a `Component` that keeps a reference to the properties panel. This starts off as `null` for reasons we will see below. The other two member data are an enumeration for the type of transform and an array of numeric values associated with the transform. The enumeration type is defined in the private object. See the aside for details on enumerations.

Enumerations

The `transformType` member of `DrawTransform` needs to be able to take on one of a small set of values. It would be possible to use an `Int` for this purpose, but the `Int` type allows over 4 billion different possible values. We only want to allow four. The need for a type that only has a small set of possible values is very common in programming. Types that satisfy this are called Enumerations. Many languages include a special language construct just for the purpose of creating these, often with the keyword `enum`. One of the goals of Scala was to be scalable and Enumerations are a perfect example of this.

The definition of `DrawTransform` shows an enumerated type that is created by having an object extend the `scala.Enumeration` class. The different values for the enumeration are created with a `val` declaration that has the four possible options we want set to be `Value`. The `Value` comes from `Enumeration`.

After this declaration, the type `TransformType.Value` only allows `Translate`, `Rotate`, `Scale`, and `Shear` as values. If you try to set a reference of that type, such as `transformType` to anything else, it will result in a syntax error.

After the members are the various methods, beginning with those that were in `Drawable`. First is the `draw` method. This method stores the current transformation from `g` in a temporary variable, then applies whatever transform it will use to `g`. The `buildTransform` method matches on the enumeration value and produces the appropriate type of transform passing it 2 or 3 of the values in the array. After this, each of the elements of `subnodes` is told to draw itself. After they are done, the transform on `g` is set back to what it had been originally.

After `draw` is the `propertiesPanel` method. As the scaladoc comment describes, this method should return the same object if it is called multiple times. For that reason, it always returns `propPanel`. However, if `propPanel` is `null`, it changes it to refer to a newly built panel that has a `ComboBox` with three `TextFields` with appropriate event handling to change `transformType` when the `ComboBox` selection is altered or the values in the `transformValue` array when a text field is edited.[6]

Next in the file are seven methods that are required by `TreeNode`. Most of these have straightforward implementations based on the contents of `subnodes`. There are two that are worth discussing further. The first of these is the `children` method. This method is notable

[5]The `Buffer` type will described in detail in the next chapter. The short version is that it is like an `Array` that has the ability to grow and shrink.

[6]This code does not do proper exception handling if the user types in a value that is not a valid `Double`. Fixing this is left as an exercise for the reader.

because it has to return an object of the type `java.util.Enumeration`. This type is not like the Enumerations we saw just above. This type is basically an iterator that was included in the Java libraries early on. Later versions of Java created their own `Iterator` type, also in `java.util`, but any library classes, like `TreeNode`, which already used `Enumeration` were stuck with that.

The Scala collections do not normally produce a `java.util.Enumeration` so we have to make our own. We do this by making an anonymous class that inherits from `java.util.Enumeration`. Fortunately, `Enumeration` only has two abstract methods that we have to implement: `hasMoreElements` and `newElement`. The implementations of these methods are very straightforward using an iterator that we get from `subnodes`.

The other method of interest is `getParent`. The implementation of this method is very simple, but it forces us to put a parameter at the top of the class so that when a `DrawTransform` is created, a parent has to be provided. The other point of interest on `getParent` is the return type. The `getParent` method in the `TreeNode` type specifies a return type of `TreeNode`. However, when you override a method, you are allowed to change the return type to a more specific subtype. In this case, we know that all parents will be of the type `DrawTransform`, which is a subtype of `TreeNode` so we are not violating what the supertype specifies. The `getChild` method is also returning a more specific type than required by `TreeNode`.

The next two methods in `DrawTransform` are `addChild` and `removeChild`. These methods were not inherited from any supertype. They exist here so that the `Drawing` type can do what it needs with the children without completely breaking the encapsulation on `subnodes`. At the very end is an override of the `toString` method so that transform objects will display nicely in GUIs.

With `DrawTransform` completed we can now turn to the other side of the `Drawable` hierarchy. We'll start with `DrawLeaf`. It is a `trait` that will only implement the methods that are required for `TreeNode`.

```
package scalabook.drawing

import javax.swing.tree.TreeNode
import java.util.Enumeration

/**
 * A supertype for Drawables that don't allow children.
 */
trait DrawLeaf extends Drawable {
  val parent:DrawTransform

  /**
   * Returns an empty enumeration.
   */
  def children : Enumeration[Drawable] = new Enumeration[Drawable] {
    def hasMoreElements() = false
    def nextElement() = null
  }

  /**
   * Returns false.
   */
  def getAllowsChildren : Boolean = false

  /**
```

```
   * Returns null for any input.
   */
  def getChildAt(childIndex : Int) : Drawable = null

  /**
   * Returns 0.
   */
  def getChildCount : Int = 0

  /**
   * Returns -1 for any input.
   */
  def getIndex(node : TreeNode) : Int = -1

  /**
   * Returns the parent of the current node.
   */
  def getParent : DrawTransform = parent

  /**
   * Returns true.
   */
  def isLeaf : Boolean = true
}
```

The methods are fairly straightforward. The only thing that might stand out in this code as being interesting is the abstract declaration of `parent`. The `getParent` method is defined here and has to be able to return something. However, a `trait` can not take arguments, so the way we can make sure a value is provided by the subtypes is to put in an abstract `val`. We can see how this plays out by looking at `DrawRectangle`.

```
package scalabook.drawing

import java.awt.Graphics2D
import java.awt.Color
import java.awt.geom.Rectangle2D
import swing._
import event._

/**
 * A Drawable that displays a filled colored rectangle.
 */
class DrawRectangle(p:DrawTransform) extends DrawLeaf {
  val parent = p
  private var propPanel:Component = null
  private var color = Color.black
  private var width = 100.0
  private var height = 100.0

  /**
   * Draws a rectangle.
   *
   * @param g the Graphics2D object that is drawn to.
   */
  def draw(g:Graphics2D) {
```

```
    g.setPaint(color)
    g.fill(new Rectangle2D.Double(0,0,width,height))
  }

  /**
   * Returns a Component for editing the settings of this object.
   *
   * @return a GUI component for editing the settings of the rectangle.
   */
  def propertiesPanel() : Component = {
    if(propPanel == null) {
      propPanel = new BorderPanel {
        layout += new BorderPanel {
          layout += new GridPanel(2,1) {
            contents += new Label("Width")
            contents += new Label("Height")
          } -> BorderPanel.Position.West
          layout += new GridPanel(2,1) {
            contents += new TextField(width.toString) {
              listenTo(this)
              reactions += { case e:EditDone => width = text.toDouble }
            }
            contents += new TextField(height.toString) {
              listenTo(this)
              reactions += { case e:EditDone => height = text.toDouble }
            }
          } -> BorderPanel.Position.Center
        } -> BorderPanel.Position.North
        layout += Button("Select Color") {
          val newColor = javax.swing.JColorChooser.showDialog(peer,"Select
              Color",color)
          if(newColor!=null) color=newColor
        } -> BorderPanel.Position.South
      }
    }
    propPanel
  }

  override def toString() = "Rectangle"
}
```

The fact that `DrawRectangle` is a class means that it can take an argument of the parent object. The first thing it does in the class is to define the abstract `parent` member and give it the value of what was passed in. After this are some member data declarations and the methods. First is a simple `draw` method, followed by `propertiesPanel` which is similar to what we saw in `DrawTransform` only it is build for the settings of the rectangle. At the end we override `toString` for the purposes of the GUI.

The DrawEllipse type is basically identical to the DrawRectangle type except the `draw` method calls `fill` on an `Ellipse2D`. With so much code in common, you might expect that it should be possible to abstract over this difference and use a single class for both `DrawRectangle` and `DrawEllipse`. Indeed this is possible. It was not done here because we want to illustrate inheritance. Doing this better is left as an exercise for the student.

19.3.2 Integration with Drawing

The code for these different types does not do anything for us until they have been used in the code for a Drawing. To do a simple test, you can change the declarations of `root` and `tree` and add a line after them in `Drawing` like this.

```
private val root = new DrawTransform(null)
private val tree = new javax.swing.JTree(root)
root.addChild(new DrawRectangle(root))
```

With this code you will get a `DrawTransform` with one rectangle under it for the drawing. If you run the program and make a new drawing, the tree should also display this instead of having the default and the drawing section should display the 100 by 100 pixel black rectangle. This still does not display the properties panel or let us change anything in the drawing. For that we need to add more code.

The following is a complete version of `Drawing` that includes buttons for adding and removing new `Drawable` objects as well as a reaction to clicking on the tree to that the properties panel for what is clicked on will be displayed.

```
package scalabook.drawing

import swing._
import javax.swing.event._

/**
 * This type represents a Drawing that you can have open in the program.
 */
class Drawing {
  private val drawPanel = new DrawPanel
  private var propPanel:Component = null
  private val root = new DrawTransform(null)
  private val tree = new javax.swing.JTree(root)

  /**
   * Returns a panel that displays the properties for this drawing.
   *
   * @return a Component that can be put into a GUI.
   */
  def propertiesPanel = {
    if(propPanel == null) {
      val commandArea = new TextArea()
      commandArea.editable = false
      val commandField = new TextField()
      val commandPanel = new BorderPanel {
        layout += commandField -> BorderPanel.Position.North
        layout += new ScrollPane(commandArea) -> BorderPanel.Position.Center
        preferredSize = new Dimension(500,200)
      }

      val drawProps = new GridPanel(1,1)

      propPanel = new SplitPane(Orientation.Vertical,
        new GridPanel(2,1) {
          contents += new BorderPanel {
            layout += new GridPanel(1,2) {
```

```
          contents += Button("Add"){
            executeOnSelection(d => addTo(d),addTo(root))
          }
          contents += Button("Remove"){
            executeOnSelection(d => remove(d))
          }
        } -> BorderPanel.Position.North
        layout+= new ScrollPane(new Component {
          override lazy val peer = tree
          tree.addTreeSelectionListener(new TreeSelectionListener {
            def valueChanged(e:TreeSelectionEvent) {
              executeOnSelection(d => {
                drawProps.contents.clear()
                drawProps.contents += d.propertiesPanel
                drawProps.revalidate
                drawProps.repaint
              })
            }
          })
        }) -> BorderPanel.Position.Center
      }
      contents += new ScrollPane(drawProps)
    },new SplitPane(Orientation.Horizontal,
      new ScrollPane(drawPanel),
      commandPanel))
  }
  propPanel
}

private def executeOnSelection(f: Drawable=>Unit,default: =>Unit = {}) {
  val path = tree.getSelectionPath
  if(path!=null) {
    val last = path.getLastPathComponent match {
      case drawable:Drawable if(drawable!=null)=>
        f(drawable)
      case _ =>
    }
  } else default
}

private def addTo(d:Drawable) = {
  assert(d!=null)
  val parent = d match {
    case dt:DrawTransform => dt
    case _ => d.getParent
  }
  val options = Seq("Ellipse","Rectangle","Transform")
  val choice = Dialog.showInput(propPanel,
     "What do you want to add?","Draw Type",Dialog.Message.Question,
     null,options,options(0))
  if(choice.nonEmpty) {
    parent.addChild(options.indexOf(choice.get) match {
      case 0 => new DrawEllipse(parent)
      case 1 => new DrawRectangle(parent)
      case _ => new DrawTransform(parent)
```

```
      })
      drawPanel.repaint
      tree.getModel match {
        case m:javax.swing.tree.DefaultTreeModel => m.reload
        case _ =>
      }
    }
  }

  private def remove(d:Drawable) {
    assert(d!=null)
    if(d.getParent!=null) {
      d.getParent.removeChild(d)
      drawPanel.repaint
      tree.getModel match {
        case m:javax.swing.tree.DefaultTreeModel => m.reload
        case _ =>
      }
    }
  }

  private class DrawPanel extends Panel {
    override def paint(g:Graphics2D) {
      g.setPaint(java.awt.Color.white)
      g.fillRect(0,0,size.width,size.height)
      root.draw(g)
    }
  }
}
```

The `executeOnSelection` method was added because adding, removing, and selecting all require finding the current selection and acting upon it. It was moved to a method to prevent duplication. The `addTo` and `remove` methods were added because they are longer and it improves the code structure to not have them nested deeply in function literals. In addition, `addTo` is called in two places so having the method prevents code duplication.

If you put this code into Eclipse or try to compile it, you will find that there are errors in both `addTo` and `remove`. These errors occur because the `getParent` method returns the type `TreeNode`, which does not have `removeChild` or `addChild` methods. This error might confuse you given that the `DrawRectangle` and `DrawLeaf` types both say they return `DrawTransform` from `getParent` and it does have those methods. The root of the problem is that the value passed into both of these methods is declared to be a `Drawable` and the `getParent` method has not been defined in `Drawable` so it automatically gets what was in `TreeNode`. To fix this, we need to add the following line of code to `Drawable`.

```
override def getParent : DrawTransform
```

Note that just because every subtype of `Drawable` that we had made returned a `DrawTransform` does not make it safe to assume that all subtypes will. Without this line of code, we could easily have created some other class that inherits directly from `Drawable` that returns some other type of `TreeNode` from `getParent`. This line of code is required to make it so that any attempt to do that results in a syntax error.

Enter the code that has been built to this point, or pull it down from the book's website, and run it. Play with it to see what it does and go through it closely to see how it works. There are things that are missing and behaviors that we probably do not want. This project will continue to grow through the rest of the book and you want to be certain you understand what is going on with it.

Scala Applications Revisited

If you have ever selected to option in Eclipse to make a Scala application, you have inevitably noticed that it does not create an `object` with a `main` method. Instead, it makes an `object` that `extends` a type called `App`. The `App` type provides a `main` method so your code simply goes into the body of the object. You can use the variable `args` to get hold of any command-line arguments. Here is a sample that does that.

```
object MyApp extends App {
  println(args mkString (" "))
}
```

Now that you know about inheritance, you can choose to use this method of making applications if you prefer.

19.4 Parametric Polymorphism

The inclusion polymorphism we get from inheritance and subtyping is not the only form of universal polymorphism. There is a second type called parametric polymorphism. This type of polymorphism was briefly discussed in section 7.5.2. It is also something that we have been using extensively since chapter 7. You use it every time you deal with a `List` or an `Array`. This is what we get when we write code that has type parameters.

You have become familiar with using parentheses to pass values to methods when you call them or classes when you construct a new object. In the same way, you can use square brackets to pass type parameters. Consider these two declarations.

```
scala> val lst1 = Array(1,2,3)
lst1: Array[Int] = Array(1, 2, 3)

scala> val lst2 = Array[Any](1,2,3)
lst2: Array[Any] = Array(1, 2, 3)
```

They both declare arrays that store the values 1, 2, and 3. Those values are all `Int`s so if we let Scala infer a type, as in the first case, we get `Array[Int]`. In the second case we explicitly pass in a parameter type of `Any` and we get back that type. The syntax here is to pass any type arguments in square brackets before the parentheses with the value arguments. You might feel like there is not any difference between these two lists, but the type parameter changes what you can do with them.

```
scala> lst1.sum
res0: Int = 6

scala> lst2.sum
<console>:9: error: could not find implicit value for parameter num: Numeric[Any]
```

```
      lst2.sum
          ^
```

You are not able to do sums with type Any. You might question why you can not sum the elements of lst2, but consider what happens here.

```
scala> lst2(1)=true

scala> lst2(2)=List('a','b','c')

scala> lst2
res3: Array[Any] = Array(1, true, List(a, b, c))
```

Type Any really does mean anything in Scala. Hopefully it is clear why trying to take the sum of this list should fail.

19.4.1 Parametric Types

What about using type parameters in our code to make it more flexible? You use a type parameter when you have code that you want to have work with a very broad set of types that possibly have nothing in relation to one another. The most common example of this is collections. Things like lists and arrays that should be able to hold anything you want, but where you do care about what they are holding because you need to know the type when you take things out. This is a usage that we will see a lot in chapters 20, 24, 25, 26, 29, 31, and 32. For now, we can provide some simple examples to give you a flavor for this use of parametric polymorphism.

We'll go back to the theme park example for this. As you write code for running the different aspects of the park, you realize that a lot of different things are recorded by time of day. This might be sales numbers, passenger counts, resource usage, or employees on duty. Without parametric polymorphism, you wind up duplicating a lot of code because while the code for the times aspect is the same, the types can be very different. Here is a first draft implementation of a class you might use to deal with this.

```
package scalabook.themepark

/**
 * This is a Time of Day Values collection to help reduce code usage
 * when dealing with values that are associated with the time of day.
 *
 * @tparam A the type of data being stored.
 */
class ToDValues[A] {
  private val values:Array[Option[A]] = Array.fill(24)(None:Option[A])

  /**
   * This allows you to get a value for a particular hour. If there isn't
   * a value, it will throw an exception.
   *
   * @param hour the hour of the day to get. Should be between 0 and 23 inclusive.
   * @return the value stored for that hour.
   */
  def apply(hour:Int) : A = values(hour).get

  /**
```

```
 * This allows you to get a value for a particular hour. If there isn't
 * a value, it will return None.
 *
 * @param hour the hour of the day to get. Should be between 0 and 23 inclusive.
 * @return an Option of the value stored for that hour.
 */
def get(hour:Int) : Option[A] = values(hour)

/**
 * Allows you to set the value in a particular hour.
 *
 * @param hour the hour of the day. Should be between 0 and 23 inclusive.
 * @param v the new value to set.
 */
def update(hour:Int, v:A) = values(hour) = Some(v)

/**
 * Allows you to set the value in a particular hour using a String for time.
 *
 * @param hour the hour of the day. Should be between 0 and 23 inclusive.
 * @param v the new value to set.
 */
def update(time:String, v:A) = {
  val hour = hourFromTime(time)
  values(hour) = Some(v)
}

/**
 * This method clears the value at a particular time.
 *
 *  @param hour the hour to clear.
 */
def clear(hour:Int) { values(hour) = None }

/**
 * This method clears the value at a particular time.
 *
 *  @param hour the hour to clear.
 */
def clear(time:String) {
  val hour = hourFromTime(time)
  values(hour) = None
}

/**
 * Allows you to combine two sets of data using a specified function.
 *
 * @param o the other set of data.
 * @param f The function to apply to the two data types.
 */
def combine(o:ToDValues[A],f:(Option[A],Option[A])=>Option[A]) : ToDValues[A] = {
  val ret = new ToDValues[A]
  for((v,i) <- (values,o.values).zipped.map((v1,v2) => f(v1,v2)).zipWithIndex) {
    ret.values(i)=v
  }
```

```
    ret
  }

  override def toString() : String = "ToD :\n"+
    (for((o,i) <- values.zipWithIndex) yield i+" : "+o).mkString("\n")

  private def hourFromTime(time:String) : Int = {
    time.substring(0,time.indexOf(':')).toInt +
      (if(time.endsWith("PM") && !time.startsWith("12")) 12 else 0)
  }
}
```

This class is parametric. The name, ToDValues is followed by a type argument [A]. The square brackets distinguish this from a value argument. It is customary to use single, uppercase letters, beginning with "A" for the names of type parameters. This code could be used like this.

```
val riders1 = new ToDValues[Int]
val riders2 = new ToDValues[Int]
val worker1 = new ToDValues[String]
val worker2 = new ToDValues[String]

riders1(12) = 5
riders1("8:24AM") = 10
riders1(14) = 7
riders2("2:13PM") = 8

worker1(12) = "Jason"

val totalRiders = riders1.combine(riders2,(o1,o2) => (o1,o2) match {
  case (None,None) => None
  case (Some(a),None) => Some(a)
  case (None,Some(b)) => Some(b)
  case (Some(a),Some(b)) => Some(a+b)
})

println(riders1)
println(totalRiders)
```

This code creates four different instances of the class, two that store Ints type and two that store Strings. It then adds some values into three of them and uses combine to add up the riders.

Having parametric polymorphism means that we can write one class and have it work with Int, String, or any other type we desire. This class still has some significant limitations. For one thing, it is hard to set up. If you had the data you want in some other sequence and you wanted to get it into here, you could not without writing code to manually copy it element by element. The combine method is also a bit limited because we can only combine data of the same type and get back that type. We can not, for example, combine workers with riders. That type of functionality could be useful, but it requires we put type parameters on methods.

19.4.2 Parametric Functions and Methods

The standard approach in Scala to letting user create object of a type in different ways is to create a companion object with versions of `apply` that take the different arguments you want. We then make it so that the class itself takes the proper data as an argument. Here is an example of what that might look like.

```
object ToDValues {
  // First attempt that lacks type parameter and does not compile.
  def apply() : ToDValues = new ToDValues(Array.fill(24)(None))
}
```

As the comment implies, this does not work. The reason is that `ToDValues` is not a valid type by itself. It has to have a type parameter to be fully specified. In order to get a type to use there, we need to give the method a type parameter. The syntax for this is straightforward. We simply put type parameters in square brackets between the name and the value parameters. Doing this, we can make two apply methods that look like the following.

```
object ToDValues {
  def apply[A]() = new ToDValues[A](Array.fill(24)(None))
  def apply[A](a:A*) = {
    val d = a.map(Option(_)).toArray
    new ToDValues[A](if(d.length<24) d.padTo(24,None) else
      if(d.length>24) d.take(24) else d)
  }
}
```

The second apply method allows us to have code that constructs a `ToDValues` object in the following ways.

```
    val riders1 = ToDValues[Int]()
    val riders2 = ToDValues(0,0,0,6,7,3,9)
```

What should jump out at you here is that the second usage does not specify the type parameter. This is because when they are used with methods, type parameters are generally inferred. As long at the type appears in the arguments to the method, you do not have to tell Scala what to use, it will figure it out. The line that creates `riders1` requires that we tell it `[Int]` because there are no arguments, but the second one can figure out that it is working with the `Int` type.

This code works because the class has been set up to take an argument. That means the declaration would have changed from what was shown above to something like this.

```
class ToDValues[A](private val values:Array[Option[A]]){
```

There is one very big problem with this, it is possible to make a `ToDValues` object with an array that does not have 24 elements. This is clearly bad behavior and we need to prevent it. One way to prevent it is to make it so that the class has a private constructor. This means that only code associated with the class or its companion object can make a call to new for this object. Syntactically this is done by putting the `private` keyword before the value arguments like this.

```
class ToDValues[A] private(private val values:Array[Option[A]]){
```

Now the only way that outside code can instantiate a ToDValues object is through the apply methods of the companion object and it is easy for us to make sure those always create instances using arrays with 24 elements.

The other addition we would like to make is to enhance the combine method so that it can work with different types. Ideally, we'd like to have the ability to make it so that the input types could be two distinct types and the output type is a third type. We can accomplish this by introducing two type parameters on the method like this.

```
def combine[B,C](o:ToDValues[B])(f:(Option[A],Option[B])=>Option[C]) :
    ToDValues[C] = {
  new ToDValues((values,o.values).zipped.map((v1,v2) => f(v1,v2)))
}
```

Now we could do things like combine workers with riders to keep track of correlations that might be used for bonuses.

If you look closely you will see that something other than the addition of the type parameters changed here. Instead of a single value parameter list with two parameters, there are two lists with one parameter each. You should recall this is a technique called currying. This is not required, but it makes using the method easier. Here is an example of usage.

```
val totalRiders = riders1.combine(riders2)((o1,o2) => (o1,o2) match {
  case (None,None) => None
  case (Some(a),None) => Some(a)
  case (None,Some(b)) => Some(b)
  case (Some(a),Some(b)) => Some(a+b)
})
```

If we do not curry this method, the local type inference in Scala will not automatically figure out the type of o2 and we would have to specify it in the function declaration with (o1,o2:Int).

There is another pair of tasks that we have done which could benefit from type parameters, sorting and searching. The sorts and searches we wrote in chapter 13 only worked on one type. We had to copy them and make changes when we wanted to work with a different type. Now we know that we'd like to have a method like sort[A] that could sort any type we want. A simple attempt at doing this will show us a detail of type parameters that we have not discussed that causes problems.

```
object Sorts {
  def bubbleSort[A](a:Array[A]) {
    for(i <- 0 until a.length) {
      for(j <- 0 until a.length-1-i) {
        if(a(j)>a(j+1)) { // Error: > undefined for Any
          val tmp = a(j)
          a(j) = a(j+1)
          a(j+1) = tmp
        }
      }
    }
  }
}
```

The comment in this code describes the error that we get if we try to compile this code. This might give you pause considering that the code does not use the type Any explicitly.

However, when we say that code must work with a type parameter, such as `A`, Scala has to make sure that it would work for any allowable `A`. Since there are no restrictions on this, that means that it must comply to the requirements of type `Any`.

One way to get around this is to put bounds on what we allow for type `A`. We'll see how to do that in the next section. For now we will look at an alternate approach. Instead of assuming the type will work with a particular method, like `<`, we will pass in an extra argument that is a function that plays the role of `<`. We'll call this function `lt`, short for less than.

```
object Sorts {
  def bubbleSort[A](a:Array[A])(lt:(A,A)=>Boolean) {
    for(i <- 0 until a.length) {
      for(j <- 0 until a.length-1-i) {
        if(lt(a(j+1),a(j))) {
          val tmp = a(j)
          a(j) = a(j+1)
          a(j+1) = tmp
        }
      }
    }
  }
}
```

This now provides us with a sort that can work for any type that we want, as long as we can define a function that represents less than for that type. As with the `combine` method, this works best if we curry the call to take two separate argument lists. If you load that into the REPL we can see it work like this.

```
scala> val rints = Array.fill(10)(util.Random.nextInt(30))
rints: Array[Int] = Array(29, 19, 15, 6, 14, 25, 1, 18, 10, 22)

scala> Sorts.bubbleSort(rints)(_<_)

scala> rints
res2: Array[Int] = Array(1, 6, 10, 14, 15, 18, 19, 22, 25, 29)

scala> val dints = Array.fill(10)(math.random)
dints: Array[Double] = Array(0.10388467739932095, 0.21220415385228875,
    0.8450116758102296, 0.5919780357660742, 0.9652457489710996,
    0.9401962629233398, 0.08314463374943748, 0.1502193866199757,
    0.7017577117339538, 0.9599077921736453)

scala> Sorts.bubbleSort(dints)(_<_)

scala> dints
res4: Array[Double] = Array(0.08314463374943748, 0.10388467739932095,
    0.1502193866199757, 0.21220415385228875, 0.5919780357660742,
    0.7017577117339538, 0.8450116758102296, 0.9401962629233398,
    0.9599077921736453, 0.9652457489710996)
```

Here the same method has been used to sort both `Int`s and `Double`s. Note that the `Array.fill` method is also curried. We first learned about that technique when we were learning about `fill` and `tabulate`.

You will often see this type of sorting done using a function that returns an `Int` instead of a `Boolean`. When you need to know if two values are possibly equal in addition to being less than or greater than, this approach is beneficial. If the values are equal the function will return 0. If the first one is less, it will return a negative value. If the first one is greater it will return a positive value. We will see this type of usage of comparison functions in later chapters.

The strength of this approach using a second function argument for comparison is that is makes the sort more flexible. For example, in either of the above examples we could have sorted the numbers from greatest to least instead of least to greatest by simply passing in `_>_` instead of `_<_`. In addition, it truly places no limitations on the type `A`. This allows you to sort things according to any sort order you want. For example, if you had a type for students you might want to sort by last name in some contexts or by grade in others. This approach would allow you to do that because the sort order is not attached to the object type. The downside of this approach is that the person calling the sort method has to write a comparison function each time they want to use it. In the examples above that was not a big deal because we had simply comparisons. However, there are times when that will not be the case.

19.4.3 Type Bounds

As we just saw, a normal type parameter, like `[A]`, is treated like `Any` when we use objects of that type in code. While this is a very common usage, there are certainly times when you want to limit what types can be used in a type parameter so that you can safely call methods other than the small set that are available with `Any`. The way you do this is to put bounds on the type.

The most common form of this is putting an upper bounds on a type. This is done with the `<:` symbol. To understand why you might want to do this, let's consider an example using code that we had before with fruit. The function, `makeBreakfastShake` that we wrote at the beginning of the chapter only used a single piece of fruit. This is not how it normally works. You would want to be able to include multiple pieces of fruit, perhaps of different types. A common example would be a shake with both bananas and strawberries. To do that, you might consider having code like this where you pass in an `Array[Fruit]`.

```
def makeBreakfastShake(fruits:Array[Fruit]) {
  for(fruit <- fruits) {
    if(!fruit.canEatSkin) {
      fruit.peel
    }
    blender += fruit
  }
  blender += juice
  blender += ice
  blender.blend
}
```

This code probably looks fine to you and in many situations it will work. However, it is not perfect. Here is an example of where it runs into problems.

```
scala> val berries = Array(new Strawberry,new Strawberry)
berries: Array[Strawberry] = Array(Strawberry@51d36f77, Strawberry@103b1799)

scala> makeBreakfastShake(berries)
<console>:16: error: type mismatch;
 found   : Array[Strawberry]
 required: Array[Fruit]
```

```
Note: Strawberry <: Fruit, but class Array is invariant in type T.
You may wish to investigate a wildcard type such as '_ <: Fruit'. (SLS 3.2.10)
      makeBreakfastShake(berries)
                         ^
```

The variable `berries` here has the type `Array[Strawberry]` and that does work. The way `makeBreakfastShake` is written here, the argument passed in has to actually be an `Array[Fruit]`. The type `Array[Strawberry]` is not a subtype of `Array[Fruit]`. To understand why this is the case, simply imagine what would happen if you if such a call were allowed in this code.

```
def subCherries(bowl:Array[Fruit]) {
  if(!bowl.isEmpty) bowl(0)=new Cherry
}
```

This is fine if you call it with an `Array[Fruit]`, but if you were allowed to call this function with an `Array[Strawberry]`, you would wind up with a `Cherry` in an array that is not allowed to hold one.

The solution to this problem is implied in the error message above. The `makeBreakfastShake` function needs to say that it can take an array of any type that is a subtype of `Fruit`. This is done by specifying a type bound using `<:` as seen here.

```
def makeBreakfastShake[A <: Fruit](fruits:Array[A]) {
  for(fruit <- fruits) {
    if(!fruit.canEatSkin) {
      fruit.peel
    }
    blender += fruit
  }
  blender += juice
  blender += ice
  blender.blend
}
```

Using the `<:` symbol lets you put a constraint on the type parameter `A`. In this way, we can say that it is safe to assume that objects of type `A` are something more specific than `Any`. This is required in this example because `Any` does not have methods for `canEatSkin` or `peel`.

You can also specify a lower bound on a type using `>:`. We will see uses of this type constraint in chapter 25 when we are building immutable data types.

Covariant and Contravariant

It is worth noting that the problem we had using an `Array[Strawberry]` would not have occurred had we used a `List` instead of an `Array`. The version of the code shown here would work for lists of any type that is a subtype of `Fruit` without the type bound.

```
def makeBreakfastShake(fruits:List[Fruit]) {
  for(fruit <- fruits) {
    if(!fruit.canEatSkin) {
      fruit.peel
```

```
    }
    blender += fruit
  }
  blender += juice
  blender += ice
  blender.blend
}
```

This works because List is a covariant type which means that the type List[Strawberry] is a subtype of List[Fruit], something that was not the case for Array.

If you look in the API, you will see that List has a type parameter of +A. This means that it is covariant and that subtype relationships on the whole type match the type parameters. The details of covariant, contravariant, and invariant are more advanced and appear in appendix B. A simple explanation of why this works is that because the List is immutable, our example of substituting a Cherry, which caused problems for the Array is not an allowed operation on a List.

One of the uses of traits is as mixins that add specific functionality to types. An example of this is the trait Ordered[A]. This trait has one abstract method, compare(that:A):Int. This method does a comparison to type A and returns a negative values if this comes before that in the ordering, a positive value if this comes after that in the ordering, and zero if they are equal. It has concrete methods to define the normal comparison operators based on the results of compare.

Any object that is a subtype of Ordered should work well in a sorting or searching algorithm. That fact that it is a subtype of Ordered means that it has a natural ordering that we want to use for the sort. This natural ordering means that we do not have to pass in a comparison function, it is built into the type. We simply have to limit what types can be used to subtypes of Ordered. A version of our bubble sort that takes this approach is shown here.

```
object Sorts {
  def bubbleSort[A <: Ordered[A]](a:Array[A]) {
    for(i <- 0 until a.length) {
      for(j <- 0 until a.length-1-i) {
        if(a(j)>a(j+1)) {
          val tmp = a(j)
          a(j) = a(j+1)
          a(j+1) = tmp
        }
      }
    }
  }
}
```

You should take special note of the type bound on this example: [A <: Ordered[A]]. This is a recursive type bound where the bound refers back to the type that it is constraining. In this context we require the ability to build a recursive type bound because inside the sort the code is doing a comparison between two objects of type A.

Subtype with Implicit Conversion

If you entered the sort that uses `Ordered` above into the computer and played with it, you probably discovered that it had some limitations you might not have been expecting. In particular, you could not use it to sort and array of `Ints`. The same is true of many of the other types that you might have tried to use it to sort.

This problem can be fixed by changing a single character in the code, though explaining what that character means and why it fixes the problem is a bit more involved. The fix is to replace the : with a % in the type bound.

```
object Sorts {
  def bubbleSort[A <% Ordered[A]](a:Array[A]) {
    for(i <- 0 until a.length) {
      for(j <- 0 until a.length-1-i) {
        if(a(j)>a(j+1)) {
          val tmp = a(j)
          a(j) = a(j+1)
          a(j+1) = tmp
        }
      }
    }
  }
}
```

The `<%` symbol says that you can use any type that is a subtype of the specified type, or that has an implicit conversion to a subtype. This sort will work for `Array[Int]` or `Array[Double]` because the standard libraries in Scala define implicit conversions from `Int` and `Double` to other types that are subtyped of `Ordered[Int]` and `Ordered[Double]`, respectively.

Implicit conversions are an advanced feature that you have been using all the time, but which are generally invisible to you. They make life easier by converting one type to another without forcing you to do any work. They can make it hard to tell what code is really doing in some places though, and for that reason, there are strict rules on what implicit conversions are allowed to happen. Full details are provided in Appendix B.

Structural Types

There are situations where you do not want to force the use of a subtype of any particular type. Instead, you want the code to be able to use any type as long as it has the methods you need to call. This is where structural types come into play. A structural type is written as curly braces with method signatures separated by semicolons. These are particularly useful when you want to be able to use code written by other people and you can not force that code to have a certain type, but it does use consistent naming of methods. In that situation, you can specify a type bound to a structural type.

A simple example of this would be the following code which will read all the integer values from any source that has the methods `hasNextInt` and `nextInt`.

```
def structRead[A <: { def hasNextInt():Boolean; def nextInt():Int
    }](source:A):mutable.Buffer[Int] = {
  val buf = mutable.Buffer[Int]()
  while(source.hasNextInt) buf += source.nextInt
  buf
}
```

An example of such a type is `java.util.Scanner`. The use of a structural type makes it so that this code will work with any type that has those two methods, regardless of what it inherits from.

It is worth noting that you can not normally use structural types with sorts. The reason is that structural types can not be recursive the way normal type bounds can. So you can not use a structural type to say that you want a type which can be compared to itself.

19.5 End of Chapter Material

19.5.1 Summary of Concepts

- Polymorphic code is code that can work with many types. Universal polymorphism implies that it can work with an infinite number of types.
- Inclusion polymorphism is a form of universal polymorphism that comes from subtyping. Scala, like most `class` based object-oriented languages, gets this through inheritance.
  - Subtypes can not get direct access to `private` members of the supertype. The `protected` visibility is meant to address this, but it is not as secure as `private` because you have little control over what code might inherit from any class you write.
  - The `super` keyword refers to the part of the current object that holds the supertype and can be used to make calls on on methods in the supertype that have been overridden in the current class.
  - Anonymous classes are created when a call to `new` on a type is followed by a block of code. This was done earlier in the book to make `Panels` and `Frames` without explicitly stating that is what you were doing.
  - Members and methods of a `class` can be left undefined. Such members are called `abstract`. If you have `abstract` members in a `class` the `class` itself must be declared abstract. You can not instantiate `abstract classes`.
  - A `class` can only inherit from one other `class`. There is a similar construct called a `trait` that allows for multiple inheritance. A `trait` is much like an `abstract class`.

  - Members and methods that should not be overridden can be declared `final`. You can also make whole `classes` `final` to prevent any other `classes` from extending them.

- Scala allows methods/functions, `classes`, and `traits` to type type parameters. These are passed in square brackets that come before any normal parameters in parentheses. This ability leads to parametric polymorphism, another form of universal polymorphism.

  - When a `class` takes a type parameter, the type must be specified at the point of instantiation with `new`. This is in contrast to parametric functions/methods which can typically infer the proper types of parameters.
  - You can place bounds on type parameters. Without bounds, the type parameters must be broadly assumed to be of the type `Any`.

19.5.2 Exercises

1. The code for the properties panels in the `Drawable` subtypes did not do a good job of handling exceptions if the user typed in something that was not a number. Edit the code to fix this.

2. The `DrawRectangle` and `DrawEllipse` classes shared a significant amount of code. Indeed, the two files only differed in that "`Rectangle`" was replaced by "`Ellipse`" in a few locations. Edit the code so that you have a single class called `DrawBoundedShape` that keeps the same member data as these classes, but abstracts over the exact shape that is created. There are several ways of doing this. One uses a function of `(Double,Double) => Shape`.

3. The `hourFromTime` method in the `ToDValues` class is not particularly robust. What are some situations where it could fail? Write a better replacement for it.

4. Draw out the inheritance hierarchies that you might design to represent the following types of objects.

   - Animal Life
   - School/University organization
   - Accounts for a large bank (savings, checking, etc.)
   - Types of Facebook friends
   - Types of students at your school
   - Whatever else you can think of that makes a good example of an "is-a" relationship.

5. Implement one of your hierarchies from above with a few simple methods.

19.5.3 Projects

There are several different options here that you could pick from based on which of the projects you started in the last chapter. The general idea though is that you will build a hierarchy of types for things that are related to your project. You can also put in parametric polymorphism if it fits your project.

When you do this, there is something very significant to keep in mind. Just because two things fit the "is-a" requirement does not mean they should be modeled with inheritance. One alternative is to use a single type and have the differences be data-driven. Whether you should use inheritance or just store values that provide the difference is best answered by whether the difference is truly a functional difference or a difference of magnitude. For example, in a game with different types of enemies or units, most of the different types should not be different subtypes because the only difference is one of things like strength or speed. However, when the difference would require one to use a very different implementation of a particular method, that is a reason to invoke inheritance.

1. In the first project on the MUD it was recommended that you create some character other than your player that wanders around the map. These other characters might play various roles in the final game. In a standard MUD, these would be the things that players do combat with. In many ways, these computer-controlled characters are just like players except that their movements and other actions are controlled by the computer instead of a human. This is a great potential use of inheritance. You might be able to come up with other uses as well.

2. For the two game projects, the possibilities depend a lot on what you are doing. As with the MUD, it is possible that you will have computer-controlled entities that mirror functionality of the player. Even if that does not fit, it is likely that there are other types of entities, active or not, which go into the game that share functionality in many ways, but not completely. (Note that for networked games the behavior of the local player and the remote player will be distinct, but that is something you do not need to deal with until chapter 23.)

3. One potential location for using inheritance in the web spider is with the types of data that you collect. You can make a supertype that represents data in general with abstract display capabilities. The details of this will depend on exactly what you are planning to do for your data collection.

4. If you are doing the mathematics workbench project, you will have a worksheet that can have multiple different types of things in it. That makes a great possibility for an inheritance hierarchy. At the very least, you should have commands and plots. However, you might have a lot more possibilities when you think about it more.

   Another element you could add to this project that uses inheritance is to implement the command pattern to provide undo and perhaps redo capabilities [6]. This works by creating a top-level command type that has two methods that might be called something like `execute` and `undo`. The the user does something that should change the worksheet in some way, instead of directly changing the worksheet, you build a subtype of the command type with an `execute` method that does the action and an `undo` method that undoes it. You pass that object to a main control object and it executes the command. You can keep track of the commands to provide undo and redo capabilities.

5. If you are working on the photoshop-like project, an obvious use of an inheritance hierarchy is the different things that can be drawn. Other elements that are options for the user to choose from can probably go into this as well. What is more, you can also implement the command pattern, described in the project above, to help make this work and to provide the undo and redo capabilities.

6. The simulation workbench project will be dealing with many different types of simulations. Each one will be a bit different, but they will all need to have the ability

to run, draw, provide analysis data, and perhaps be paused and restarted. Create a supertype that can provide this type of functionality, then write a simple N-body gravity simulation with the ability to draw itself. This is what was done for project 1 and the projects leading into it. Through inheritance, you will make that whole sequence of projects into a small piece of this project.

Chapter 20

Other Collection Types

Back in chapter 7 we talked about the most basic sequence collections in Scala, the `Array` and the `List`. Since then we have seen other similar types. These collections share a large number of different methods that we have used to great effect. The sequences are not the only type of collection though. The Scala collections library is far more extensive than what we have seen so far. In this chapter we will examine the inheritance hierarchy of the entire collections library and look at three other types of collections in detail.

20.1 The scala.collection Packages

At this point you have likely seen in the Application Programming Interface (API) that there are a number of packages that start with `scala.collection` including one called simply `scala.collection`. A class diagram showing the inheritance relationships in this top-level package is shown in figure 20.1. At the top of the hierarchy is the `Traversable[A]` type. This is a `trait` that underlies the rest of the hierarchy and can be used to represent anything that has the capability for running through it.

Below `Traversable[A]` is the `Iterable[A]` trait. As the name implies, this type, and its subtypes, have the ability to provide `Iterators`. Anything that is `Iterable[A]` will have a method called `iterator` that will give you an `Iterator[A]`. Though not shown in

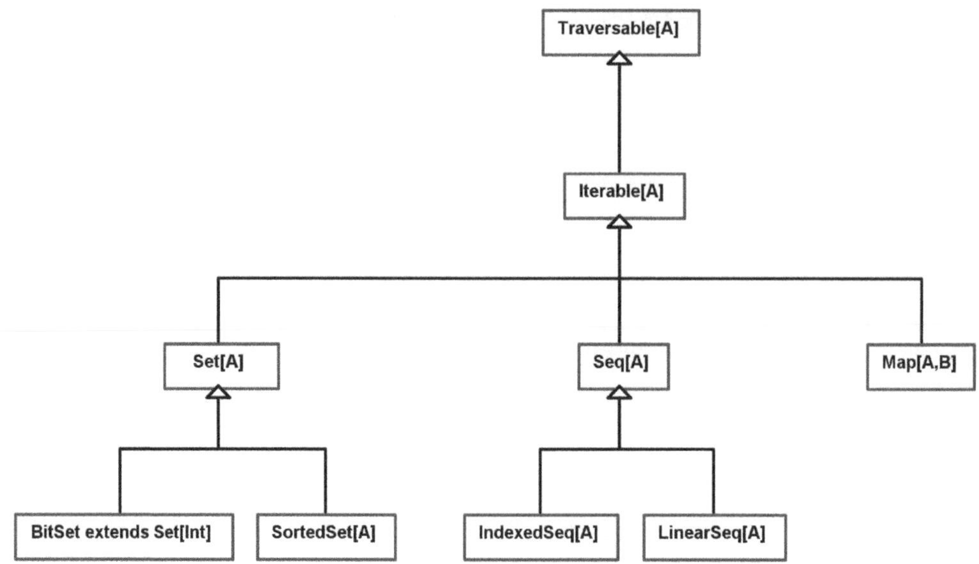

FIGURE 20.1: This is a UML diagram of the main types in the `scala.collection` package. The collection types we have dealt with, including `List[A]` and `Array[A]`, have been under the `Seq[A]` part of the hierarchy.

the figure, the `Iterator[A]` is also in `scala.collection` and it inherits from a type called `TraversableOnce[A]`. Both of these types represent collections that are consumed as they are read.

There are three main subtypes of `Iterable[A]` in the Scala libraries. These are `Seq[A]`, which we have worked with previously, `Set[A]`, and `Map[A]`. The latter two allow you to run through their contents, hence they are subtypes of `Iterable[A]`, but they handle their contents in a very different way than does a sequence. The sections below will go into the details of those types.

In this top package there are two subtypes of `Set` and `Seq`. The `BitSet` type is optimized to work with sets of `Int`. It stores one bit for each integer. If the value is in the set, the bit will be on, otherwise it will be off. The `SortedSet` type is a specialized version of `Set` that works when the contents have a natural ordering.

The two subtypes of `Seq` are `IndexedSeq` and `LinearSeq`. These two types differ in how the elements can be efficiently accessed. A `List` is an example of a `LinearSeq`. This type has the property that to get to an element, you have to run through all the ones before it. On the other hand, an `IndexedSeq` allows you to quickly get to elements at arbitrary indexes, like an `Array`. Each of these types implements their methods in a different way, depending on what is the most efficient way to get to things.

These types are implemented in two different way in the Scala libraries based on whether the implementations are immutable or mutable. To keep these separate, there are packages called `scala.collection.immutable` and `scala.collection.mutable`.

20.1.1 scala.collection.immutable

The `scala.collection.immutable` package stores various types of collections that are all immutable and which have been optimized so that common operations can be done

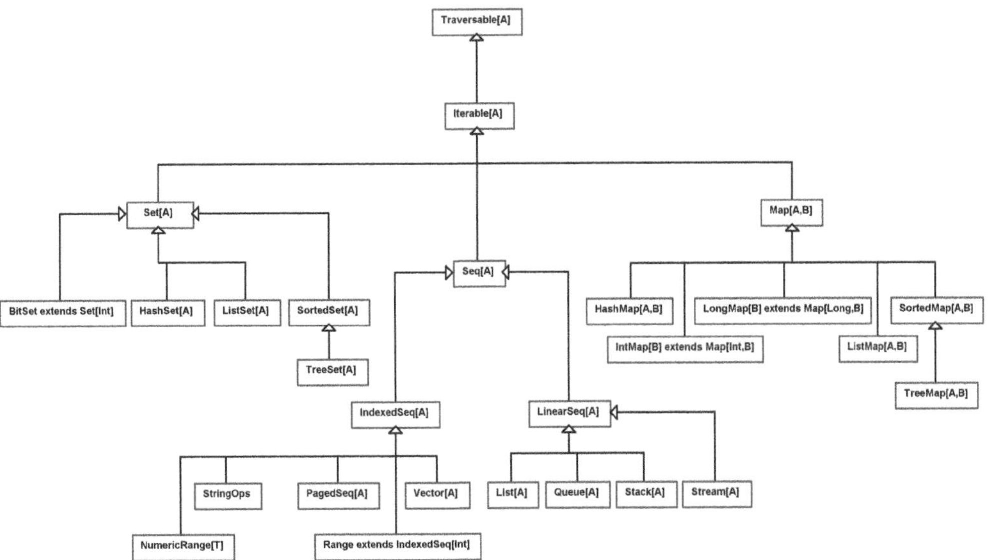

FIGURE 20.2: This is a UML diagram of the main types in the `scala.collection.immutable` package. This image is significantly simplified. If you look in the API you will see that most of the types inherit from multiple other types.

efficiently with them. This means that "updates" can be made without making complete copies. Of course, because these collections are immutable, no object is really updated, instead, a new, modified version is created. To be efficient, this new object will share as much memory as possible with the original. This type of sharing is safe because the objects are immutable. A simple class diagram showing the inheritance relationships between the main classes and traits in this package are shown in figure 20.2.

The top part of the hierarchy looks the same as that shown in figure 20.1 for the `scala.collection` package. It begins with `Traverable` and `Iterable` then has three main subtypes in the form of `Set`, `Seq`, and `Map`. The difference is that there are many more subtypes of those three listed here. We will not go into details on most of these, but it is worth noting a few. First there is `List`. This is the immutable sequence collection that we have been using for most of the book. You can see here that it is a `LinearSeq`. This means that you can't efficiently jump to random elements. We will see exactly why that is in chapter 25.

If you want an immutable collection that has efficient indexing, your default choice would generally be `Vector`. The other immutable subtypes of `IndexedSeq` include `Range`, which we saw in chapter 8, and a type called `StringOps`. The `StringOps` type provides Scala collection operations on strings. Many of the methods that you have been using with strings are not defined on the `java.lang.String` type. Instead, there is an implicit conversion from `java.lang.String` to `scala.collection.immutable.StringOps`. That is how you can do things like this.

```
val (upper,lower) = "This is a String.".filter(_.isLetter).partition(_.isUpper)
```

The `java.lang.String` type does not actually have methods called `filter` or `partition`. You can verify this looking in the Java API. When Scala sees this line of code, it sees that

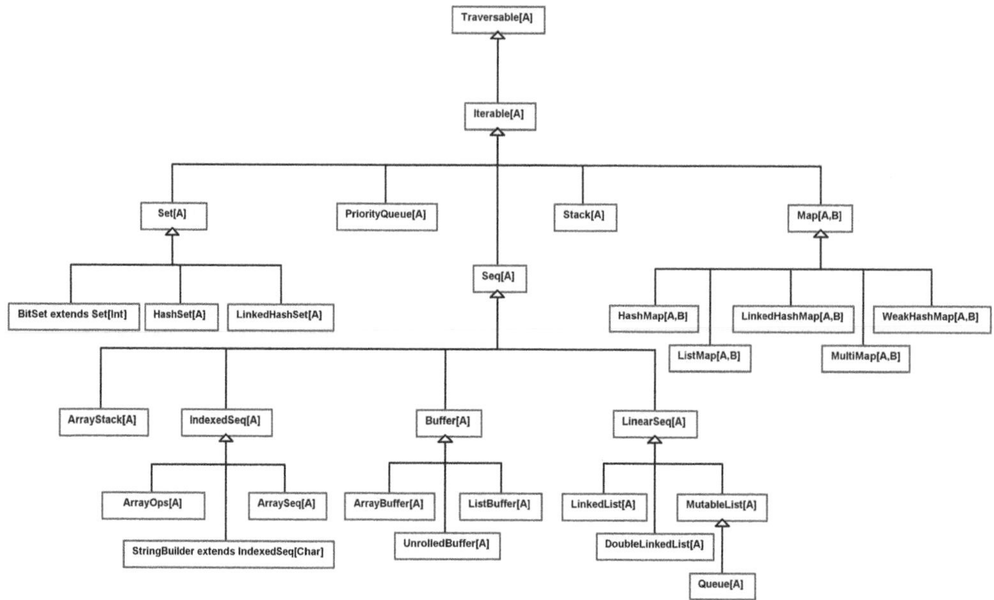

FIGURE 20.3: This is a UML diagram of the main types in the `scala.collection.mutable` package. This image is significantly simplified. If you look in the API you will see that most of the types inherit from multiple other types.

those methods are missing and looks for an implicit conversion that will give a type that has the needed methods. In this case, the conversion will be to `StringOps`.

There are a number of different types listed under `Map` as well. Two of these are optimized specifically to work with `Int` and `Long`. For most applications the `HashMap` will probably be ideal. The details of `HashMap` are beyond the scope of this book. You should cover those in a course on data structures. We will cover possible implementations of the `TreeMap` in chapter 29.

20.1.2 scala.collection.mutable

The last package under `scala.collection` that we will concern ourselves with right now is the `scala.collection.mutable` package. As the name implies, this package includes collections that are mutable. This is the most extensive of the three packages. It is worth scanning through the API to look at the various types in this package and to see what they are used for. Figure 20.3 shows a UML diagram with the main types in this package.

The fact that these types are mutable implies that they include methods that will change their values. When using them you need to try to remember that the "standard" methods, those we have been using regularly to this point, are still non-mutating methods. For example, `filter` does not change the contents of a collection, even if the collection is mutable. Instead, they introduce a few additional methods that do perform mutations. We'll talk about these a bit more below.

It is also worth noting that one of the mutable types is `ArrayOps`. This class is much like `StringOps`. In truth, the `Array` type that we have been using is `scala.Array` and it has very few methods defined on it. When you use the various sequence operations an implicit conversion is invoked to convert the `Array` object to an `ArrayOps` object.

Now we will look at the Set, Map, and Buffer types. These types can give you a significant boost in your programming and make it much easier to do certain tasks.

20.2 Sets

The concept of a set comes from math. There are two main differences between the Set type and the Seq type. The most important one is that a Set does not allow duplicate values. If you add something to a Set and there is already something there that is equal to it, it does not add a new value and the size of the resulting set is the same as what you had before. The other difference is that the order of elements in a Set is not guaranteed.[1] We can see both of these effects in the following sequence of REPL commands.

```
scala> Set(1,2,3)
res0: scala.collection.immutable.Set[Int] = Set(1, 2, 3)

scala> res0 + 4
res1: scala.collection.immutable.Set[Int] = Set(1, 2, 3, 4)

scala> res1 + 2
res2: scala.collection.immutable.Set[Int] = Set(1, 2, 3, 4)

scala> res1 + -10
res3: scala.collection.immutable.Set[Int] = Set(-10, 4, 1, 2, 3)
```

This begins with a set that has the values 1, 2, and 3. As with the List and Array types, Scala figures out that this is a Set[Int]. You can also see from the inferred type that by default we get an immutable set.

You can add to a set with +, you get back a new set that has the element after the plus sign added in. The second statement gives us back a set that includes 4 in addition to the 1, 2, and 3. In the third command we add a 2 to this new set. The result looks the same as the set before the 2 was added. This is because the set can not contain duplicate elements.

The last statement adds the value -10 into the set from res1. This value goes into the resulting set because there had not been a -10 previously. However, the result shows that order is not preserved in sets. Adding the -10 into the set not only put it in an odd place at the beginning, it also altered the location of the 4 in the set. From this you can see that you want to use sets when what you care about is the uniqueness of elements, not the order in which you get them.

The fact that the order can change means that "indexing" into a set in the manner we have done with an Array or a List do not have much meaning. You can perform that type of operation on a Set, but the meaning is a bit different. Consider the following examples.

```
scala> res3(2)
res4: Boolean = true

scala> res3(5)
res5: Boolean = false
```

[1] If you happen to be using a SortedSet, then the order of the objects will match the natural ordering of the type, but this is more the exception than the rule.

```
scala> res3(-10)
res6: Boolean = true
```

This type of operation on a set does not give you back a value at a particular location. Instead, it returns a Boolean that tells you whether or not the specified value is in the set. This works equally well if the set contains values other than `Ints`. Consider the following.

```
scala> val strSet=Set("Mark","Jason","Kevin")
strSet: scala.collection.immutable.Set[java.lang.String] = Set(Mark, Jason, Kevin)

scala> strSet("Jason")
res7: Boolean = true

scala> strSet("Bill")
res8: Boolean = false
```

In this example, the set contains Strings and the "indexing" is done by passing in Strings. The meaning is the same either way. The return value is true if the argument passed in is in the set and false if it is not. To see why this works this way, go look in the Scala API at one of the versions of `Set`. They all have a type parameter called `A` and they define the following method.

```
def apply (elem: A): Boolean
```

Remember that `apply` is a special method that is invoked when an object is used with function call notation. So `f(5)` in Scala gets expanded to `f.apply(5)`. Here the apply takes one argument, `elem`, which must have the type `A`. That is the type parameter of the set as a whole, so you have to pass in an object of the type that the set contains. This method then returns to you a `Boolean` telling you whether or not that value is in the set.

20.2.1 Running Through Sets

In addition to the standard higher order methods like `map`, `filter`, and `foreach`, that are available for both `Set` and the sequences that we have worked with previously, the `Set` type works perfectly well with `for` loops as well.

```
scala> for(v <- res3) println(v)
1
2
-10
3
4
```

In fact, you can not really use a `while` loop to run through a `Set`, because you can not index by position. Also, the order in which the loop runs through the set values is not something you can predict. It does not even have to match the order that the values are displayed when the Set is converted to a String for printing.

20.2.2 Mutable vs. Immutable

The examples above showed the default immutable Set type. Like the List type that we saw previously, an immutable set does not change after it is created. When we add elements to the set with the `+` operator, we get back a new set and the original set is unaltered. We can demonstrate this by calling back up the original set.

```
scala> res0
res9: scala.collection.immutable.Set[Int] = Set(1, 2, 3)
```

All of the operations that were done have not altered it in any way.

As we saw earlier, there is another `Set` type in the scala.collection.mutable package. You could get to this type using the fully specified name of scala.collection.mutable.Set. However, doing so not only requires a lot of typing, it also makes code difficult to read. You could import scala.collection.mutable.Set and then refer to it as just Set. The problem with this approach is that you can no longer easily refer to the immutable `Set` type. A recommended style for dealing with this is to import scala.collection.mutable. The packages in Scala are truly nested. As a result, this import will allow us to refer to a mutable set by the name `mutable.Set` and the immutable set as just `Set`.

All the operations that you can do with an immutable set are available on a mutable set as well. Above we used the + operator on the immutable set and it works equally well on the mutable set.

```
scala> import scala.collection.mutable
import scala.collection.mutable

scala> val mset=mutable.Set(1,2,3)
mset: scala.collection.mutable.Set[Int] = Set(1, 2, 3)

scala> mset + 4
res10: scala.collection.mutable.Set[Int] = Set(1, 4, 2, 3)

scala> mset
res11: scala.collection.mutable.Set[Int] = Set(1, 2, 3)
```

This code shows you the import statement, the declaration of a mutable set after the import, and the creation of a new set by adding an element to the old one. Just using + does not change the mutable set. The way in which the mutable set differs from the immutable version is that it includes additional operators and methods.

To see this we will call one of those extra operators. On a mutable set we can use `+=` the same way we might with an `Int` that was declared as a var.

```
scala> mset += 4
res12: mset.type = Set(1, 4, 2, 3)

scala> mset
res13: scala.collection.mutable.Set[Int] = Set(1, 4, 2, 3)
```

You can see from this that the use of `+=` actually changes the original set. The `+=` does not exist on the immutable set type. If you try to call it, you will get a syntax error. The list of methods/operators on the mutable set that mutate them includes the following for Set[A].

- `+=(elem : A)` - Adds the specified element to the set.
- `-=(elem : A)` - Removes the specified element from the set. Does nothing if it was not there initially.
- `++=(xs : TraversableOnce[A])` - Adds all the elements from a collection into the set. The `TraversableOnce` type is the most basic type for Scala collections so this can be used with any of the collections in Scala. `Traversable` inherits from `TraversableOnce`.

- `--=(xs : TraversableOnce[A])` - Removes all the elements in `xs` from the set.
- `add(elem : A) : Boolean` - Adds the specified element to the set. Returns true if it was not there before the call and false otherwise.
- `clear():Unit` - Removes all elements from the set.
- `remove(elem : A) : Boolean` - Removes the specified element from the set and returns true if it was there to remove or false otherwise.
- `retain(p : (A) => Boolean) : Unit` - Removes all elements that do not satisfy the predicate from the set.

Mutable sets can also be updated using assignment. This is basically the inverse operation of "indexing" that we saw before using `apply`. The index is the value you are adjusting in the set and the assignment is either to true or false. To see how this works, consider the following examples.

```
scala> mset(7)=true

scala> mset(2)=false

scala> mset
res14: scala.collection.mutable.Set[Int] = Set(1, 4, 7, 3)
```

The first line adds 7 to the set while the second removes 2. This works because the `scala.collection.mutable.Set` type defines an `update` method that looks like this.

```
def update (elem: A, included: Boolean): Unit
```

So the line `mset(7)=true` gets expanded to `mset.update(7,true)`.

Earlier in this section it was mentioned that the standard methods, like `map` and `filter`, will do the same thing on a mutable set as they do on an immutable set. This was very intentional. The same operations on subtypes types like this should always have the same outward appearance and significant effects. This reduces the chances of programmers making silly mistakes. If you want to mutate the set you have to call one of the specific methods for doing so. For example, you would call retain instead of filter. If you accidentally try to use one of these operations on an immutable type the code will not compile. If the operations with the same name had different behaviors, it would be very easy for the programmer to introduce logic errors because he/she believed an object was one type when it was actually another, or for bugs to be introduced later on by a programmer changing the specific type that is assigned to a variable.

20.2.3 Using a Set

To see how the `Set` type can be useful we will write a function that tells us all the unique words in a text file. The function will take the name of the file as an argument and return a `Set[String]` that has all the words in lowercase. As with any function of any significance, this can be done in many ways. We are going to do it making heavy use of the Scala collection methods. The fact that we are using the collection methods makes it more sensible to use a `Source` than a `Scanner`. Here is code that will do this.

```
def uniqueWords(fileName:String):Set[String] = {
  val source = io.Source.fromFile(fileName)
  val words = source.getLines.toSeq.flatMap(_.split(" +")).
```

```
    map(_.filter(_.isLetter).toLowerCase).toSet
  source.close()
  words
}
```

A function is defined using the signature that was described. This function includes only four statements. The first and third statements in the function open the file and close it. The last statement is just an expression for the set that was created from the source. The second statement is where all the work is done and is worth looking at in detail.

This line declares the variable words and makes it equal to a set of the words. To do that it gets the lines from the source and turns them into a sequence. The `toSeq` call makes types work out better with `flatMap`. As you might recall, `getLines` returns an `Iterator`. The `Iterator` type is a collection that you can only pass through once. It is consumed during that pass. By bumping it up to a sequence we get something that works better with the `Array[String]` that we get from the split method.

After the lines are converted to a sequence there is a call to `flatMap`. This is one of the standard higher-order methods on collections. We have not used it much so it is worth taking a minute to discuss. As you can tell from the name, `flatMap` is closely related to `map`. The `map` method takes each element of a collection, applies some function, and makes a new collection that has the results of that function application. The `flatMap` does basically the same thing only it expects that the function passed in is going to return collections. Instead of returning a collection of collections, as `map` would do, `flatMap` will flatten out all of the inner collections into one long collection. If you use map on the sequence of lines and use the same `map` function, you will get a `Seq[Array[String]]`. Each element of the sequence will contain an array of the words from that line. However, that is not quite what we want. We want a simple sequence of words, not a sequence of arrays of words. This is what `flatMap` does for us. It puts each of the arrays end to end in the new sequence so that we get a `Seq[String]`.

If we knew that the file only included simple words in lowercase we could stop there. However, most text files will includes capitalization and punctuation as well. This would mess up the set because "Hi" is not equal to "hi" and the both would appear in the result when we want only one. Similarly, the word "dog" would be different than "dog." or "dog," if the word happened to appear directly before a punctuation mark. The second line of the second statement includes logic to deal with this. Each of the words that the file was broken into is passed through a map. That map filters out any non-letter characters and converts the resulting strings to lower case. The filtering of non-letter characters is perhaps too aggressive because it removes things like apostrophes in contractions, but the result it provides is sufficiently correct for our purposes.

The last thing that happens in the second statement is that the sequence is converted to a `Set`. This gives us the type we desired and removes any non-unique elements by virtue of being a `Set`.

20.3 Maps

Another collection type that is extremely useful, probably more so than the `Set` in most programs, is the `Map`. Like the `Set`, the term `Map` has origins in math. A mapping in mathematics is something that takes values from one set and gives back values from another set. A function in math is technically a mapping. The `Map` collection type is not so different from this. It maps from value of one type, called the key type, to another type called the value type. The key and value types can be the same, but they are very often different.

To help you understand maps, let's look at some examples. We will start with a statement to create a new map.

```
scala> val imap=Map("one" -> 1, "two" -> 2, "three" -> 3)
imap: scala.collection.immutable.Map[java.lang.String,Int] = Map((one,1), (two,2),
    (three,3))
```

The `Map` type takes two parameters. In this example they are `String` and `Int`. The values passed in to build the `Map` are actually tuples where the key is the first element and the value is the second element. In this example we used the `->` syntax that was discussed in chapter 3. This syntax was added to the Scala libraries for this exact purpose. You can read this as making a map where the `String` "one" maps to the `Int` 1 and so on.

When you index into a map you do so with values of the key type. What you get in return is the corresponding value object.

```
scala> imap("two")
res0: Int = 2
```

If you like, you can think of a map as a collection that can be indexed by any type that you want instead of just the integer indexes that were used for sequences. If you try to pull out a key that is not in the map, you will get an exception. This works because the `Map` type defined an `apply` method like this.

```
def apply (key: A): B
```

The types `A` and `B` are the type parameters on `Map` with `A` being the key type and `B` being the value type.

In addition to allowing any type for the key, the map also has the benefit that it does not require any particular "starting" value for the indexes. For this reason, you might sometimes want to build a map that uses an `Int` as the index. Consider this example.

```
scala> val imap2=Map(10 -> "ten", 100 -> "hundred", 1000 -> "thousand")
imap2: scala.collection.immutable.Map[Int,java.lang.String] = Map((10,ten),
    (100,hundred), (1000,thousand))
```

Technically you could build a sequence that has the values "ten", "hundred", and "thousand" at the proper indexes. However, this is very wasteful because there are many indexes between 0 and 1000 that you will not be using and they would have null references in them.

As with the Set type, you can add new key, value pairs to the map with the `+` operator.

```
scala> imap + ("four" -> 4)
res1: scala.collection.immutable.Map[java.lang.String,Int] = Map((one,1), (two,2),
    (three,3), (four,4))
```

The key and value need to be in a tuple, just like when the original map was built. The parentheses in this sample are required for proper order of operation. Otherwise the `+` happens before the `->` and the result is a `(String,Int)` that will look rather unusual.

You can also use the `-` operator to get a map with fewer elements. For this operation, the second argument is just the key you want to remove. You do not have to have or even know the value associated with that key.

```
scala> imap - "two"
```

```
res2: scala.collection.immutable.Map[java.lang.String,Int] = Map((one,1),
    (three,3))
```

Here we remove the value 2 from the map by taking out the key "two".

As with the Set type, there are mutable and immutable flavors. As you can see from the examples above, the default for Map, like for Set, is immutable. To use the mutable version you should follow the same recommendation as was given above and import scala.collection.mutable and then refer to it as mutable.Map.

20.3.1 Looping Through a Map

You can use the higher-order methods or a for loop to run through the contents of a Map, but there is a significant difference between this and the other collections we have looked at. This is because the contents of the map are effectively tuples. To illustrate this, we'll look at a simple for loop.

```
scala> for(tup <- imap) println(tup)
(one,1)
(two,2)
(three,3)
```

Each element that is pulled off the map is a tuple. This has significant implications when using higher order methods like map. The argument to map is a function that takes a single argument, but that argument is a tuple with a key-value pair in it. Consider the following code.

```
scala> imap.map(tup => tup._2*2)
res3: scala.collection.immutable.Iterable[Int] = List(2, 4, 6)
```

Because the argument, tup, is a tuple, we have to use the _2 method to get a value out of it. The _1 method would give us the key. The alternative is to make the body of the function longer and use a pattern in an assignment to pull out the values. For many situations this is not significantly better.

Running through a map works much better with a for loop because the for loop does pattern matching automatically as was discussed in section 8.3.5. This syntax is shown here.

```
scala> for((k,v) <- imap) yield v*2
res4: scala.collection.immutable.Iterable[Int] = List(2, 4, 6)
```

This syntax is not significantly shorter for this example though it can be in others and it is easier for most people to read. The real advantage is that you can easily use meaningful variable names. This example used k and v for key and value, but in a full program you could use names that carry significantly more meaning.

You might have noticed that both the example with map and with the for loop gave back lists. This is also a bit different as we are used to seeing this type of operation give us back the same type that was passed in. The problem is that to build a map we would need key, value pairs. The way the libraries for Scala are written, you will automatically get back a map any time that you use a function that produces a tuple.

```
scala> imap.map(tup => tup._1 -> tup._2*2)
res5: scala.collection.immutable.Map[java.lang.String,Int] = Map((one,2), (two,4),
    (three,6))

scala> for((k,v) <- imap) yield k -> v*2
```

```
res6: scala.collection.immutable.Map[java.lang.String,Int] = Map((one,2), (two,4),
    (three,6))
```

These two examples show how that can work using either map or the `for` loop with a yield.

20.3.2 Using Maps

The example we will use for illustrating a use of a `Map` is closely related to the one we used for the `Set`. Instead of simply identifying all the words, we will count how many times each one occurs. This function will start off very similar to what we did before. We'll start with this template.

```
def wordCount(fileName:String):Map[String,Int] = {
  val source = io.Source.fromFile(fileName)
  val words = source.getLines.toSeq.flatMap(_.split(" +")).
    map(_.filter(_.isLetter).toLowerCase)
  val counts = // ???
  source.close()
  counts
}
```

Here `words` will just be a sequence of the words and we will use it to build the counts and return a `Map` with those counts. We will do this last part in two different ways.

The first approach will use a mutable map. The logic here is that we want to run through the sequence of words and if the word has not been seen before we add a count of 1 to the map. If it has been seen before, we add the same key back in, but with a value that is one larger than what we had before. The code for this looks like the following.

```
def wordCount(fileName:String):mutable.Map[String,Int] = {
  val source = io.Source.fromFile(fileName)
  val words = source.getLines.toSeq.flatMap(_.split(" +")).
    map(_.filter(_.isLetter).toLowerCase)
  val counts = mutable.Map[String,Int]()
  for(w <- words) {
    if(counts.contains(w)) counts += w -> (counts(w)+1)
    else counts += w -> 1
  }
  source.close()
  counts
}
```

Note that the `for` loop to run through the words here occurs after the declaration of counts. Because counts is mutable this approach works. You should also note that the return type for the function has been altered to be a `mutable.Map[String,Int]`. If you leave this off, you have to add a call to `.toMap` at the end, after counts, so that the counts will be converted to an immutable type. This distinction between mutable and immutable types is very significant for safety in programs. In general there could be other references to the `Map` you are given that are retained in other parts of the code. If it were possible for you to get a mutable map when you were expecting an immutable one, those other parts of the code could alter the map when you were not expecting it. This would lead to very hard to track down logic errors.

Using the mutable version is fairly logical and can have some efficiency advantages, but as was just alluded to, mutable values come with risks. For this reason, we will write a second version that uses an immutable map and the `foldLeft` method. The `foldLeft` method is a more complex method that we have not really dealt with before. However, it is ideal for what we are doing here as it effectively does what we wrote previously, but without the need for a `for` loop. The `foldLeft` method runs through the collection from left to right and applies a function to each element, compiling a result as it goes. The function is curried and the first argument list is a single value that tells what the compiled value should start off as. In this case, we are compiling a Map so we will start off with an empty one. The code we want to apply is very much like what was in the `for` loop previously. This is what it all looks like when put together.

```
def wordCount2(fileName:String):Map[String,Int] = {
  val source = io.Source.fromFile(fileName)
  val words = source.getLines.toSeq.flatMap(_.split(" +")).
    map(_.filter(_.isLetter).toLowerCase)
  val counts = words.foldLeft(Map[String,Int]())((m,w) => {
    if(m.contains(w)) m + (w -> (m(w)+1))
    else m + (w -> 1)
  })
  source.close()
  counts
}
```

The function that is passed into `foldLeft` takes two arguments. The first is the compiled value so far and the second is the value that is being operated on. In this case the compiled value is a `Map` and we use the variable `m` to represent it. The value being passed in is the next word that we want to count and is represented by the name `w` as it was in the `for` loop previously.

The body of the function passed into `foldLeft` as the second argument list differs from what was inside the for loop previously in two ways. First, instead of referring to the mutable map counts, we use the argument name `m`. Second, we do not use `+=`. Instead we just use `+`. The `+=` operator is not defined for the immutable map and what we are doing here is building a new immutable map that is passed forward to be used with the next word.

If you really want to push things a little harder you can get rid of the declaration of words all together. After all, the only thing it is used for is as the argument to `foldLeft` to build counts. Doing this and getting rid of some curly braces that are not technically needed produces the following code.

```
def wordCount3(fileName:String):Map[String,Int] = {
  val source = io.Source.fromFile(fileName)
  val counts = source.getLines.toSeq.flatMap(_.split(" +")).
    map(_.filter(_.isLetter).toLowerCase).
    foldLeft(Map[String,Int]())((m,w) =>
      if(m.contains(w)) m + (w -> (m(w)+1))
      else m + (w -> 1))
  source.close()
  counts
}
```

This code is compact, but whether you really want to do things this way or not is debatable. The introduction of the variable words did add extra length to the code, but for most readers it probably also made things more understandable.

Having any version of `wordCount` implicitly gives us a set of the words as well. This is because the `Map` type has a method called `keySet` that returns a `Set` of all of the keys used in the `Map`. Because each key can only occur once, nothing is lost when you get the keys as a set. For efficiency reasons you would probably want to keep the uniqueWords function around if you were going to be using it often instead of replacing calls to it with `wordCount("file.txt").keySet` because the counting of the words does extra work that is not needed to produce the set.

A more general use case for maps is using them as a fast way to look up groups of data like case classes by a particular key value. An example of this would be looking up a case class that represents a student by the student's ID or perhaps their unique login number. The real power of the `Map` type comes from the fact that the key and value types can be anything that you want.

20.4 Buffers

The last of the new collection types that we will look at in detail in this chapter is the `Buffer`. The `Buffer` type is like an `Array` that can change size. As a result, the `Buffer` type is implicitly mutable. There is no immutable `Buffer`. If you are using other mutable types, the rule of importing `scala.collection.mutable` and referring to a `Buffer` as `mutable.Buffer` will still work well and it makes it perfectly clear to anyone reading the code that you are using a mutable type. However, because the `Buffer` type is only mutable you could also consider doing an import of `scala.collection.mutable.Buffer` and simply calling it a `Buffer` in your code.

There are several subtypes of `Buffer` in the Scala libraries. The two most significant ones are `ArrayBuffer` and `ListBuffer`. One uses an array to store values. The other uses a structure called a linked list. The `List` type in Scala is also a linked list, though it is not directly used by the `ListBuffer` because the `ListBuffer` needs to be mutable. We will use the `ArrayBuffer` in this chapter. The nature of the `ListBuffer` and how it is different will become more apparent in chapter 25.

You can create a buffer in exactly the way that you would expect.

```
scala> val buf = mutable.Buffer(1,2,3,4,5)
buf: scala.collection.mutable.Buffer[Int] = ArrayBuffer(1, 2, 3, 4, 5)
```

You can also use some of the methods we learned about for `Array` and `List` like `fill` and `tabulate`.

```
scala> val rbuf=mutable.Buffer.fill(10)(math.random)
rbuf: scala.collection.mutable.Buffer[Double] = ArrayBuffer(0.061947605764430924,
    0.029870283928219443, 0.5457301708447658, 0.7098206843826819,
    0.8619215922836797, 0.5401420250956313, 0.6249953821782052,
    0.1376217145656472, 0.26995766937532295, 0.8716257556831167)
```

You can index into a buffer and change values the same way you did for an array.

```
scala> rbuf(3)
res0: Double = 0.7098206843826819

scala> buf(3)=99

scala> buf
res1: scala.collection.mutable.Buffer[Int] = ArrayBuffer(1, 2, 3, 99, 5)
```

The way in which the `Buffer` type differs from the `Array` type is that it can easily grow and shrink. There are a whole set of methods that take advantage of this. Here are some of them.

- `+=(elem: A): Buffer[A]` - Append the element to the buffer and return the same buffer.
- `+=:(elem: A): Buffer[A]` - Prepend the element to the buffer and return the same buffer.
- `++=(xs: TraversableOnce[A]): Buffer[A]` - Append the elements in `xs` to the buffer and return the same buffer.
- `++=:(xs: TraversableOnce[A]): Buffer[A]` - Prepend the elements in `xs` to the buffer and return the same buffer.
- `-=(elem: A): Buffer[A]` - Remove the element from the buffer and return the same buffer.
- `--=(xs: TraversableOnce[A]): Buffer[A]` - Remove all the elements in `xs` from the buffer and return the buffer.
- `append(elem: A): Unit` - Append the element to the buffer.
- `appendAll(xs: TraversableOnce[A]): Unit` - Append the elements in `xs` to the buffer.
- `clear(): Unit` - Remove all the elements from the buffer.
- `insert(n: Int, elems: A*): Unit` - Insert the specified elements at the specified index.
- `insertAll(n: Int, elems: Traversable[A]): Unit` - Insert all the elements in `elems` at index `n`.
- `prepend(elems: A*): Unit` - Prepend the elements to this buffer.
- `prependAll(xs: TraversableOnce[A]): Unit` - Prepend all the elements in `xs` to this buffer.
- `remove(n: Int, count: Int): Unit` - Remove `count` elements starting with the one at index `n`.
- `remove(n: Int): A` - Remove the one element at index `n` and return it.
- `trimEnd(n: Int): Unit` - Remove the last `n` elements from this buffer.
- `trimStart(n: Int): Unit` - Remove the first `n` elements from this buffer.

Most of these should be fairly self-explanatory. The only methods that you might question are the ones that involve symbol characters. There are two things of interest here. The first is that they all return the buffer that they are called on. The reason for this is that it allows you to string them together to append or prepend multiple elements in a row. Consider this example with appending.

```
scala> buf += 6 += 7
res2: buf.type = ArrayBuffer(1, 2, 3, 99, 5, 6, 7)
```

Here both 6 and 7 are appended in a single line.

The other operations that might seem odd are `+=:` and `++=:`. These operations say that they prepend. What is interesting about them is their usage. Here is an example.

```
scala> 0 +=: buf
res3: buf.type = ArrayBuffer(0, 1, 2, 3, 99, 5, 6, 7)
```

The value to be prepended is to the left of the operator and the buffer is on the right. In some ways that makes sense given where it is in the buffer after the prepending. We have seen something like this before as well. The cons operator on a list is `::` and it prepends the element to the list. That operator also ends with a colon. This is one of the more esoteric rules of Scala. Any symbolic operator method that ends in a colon is right associative instead of being left associative the way most operators are. This means that the object the operator works on is to the right of the operator and the operations are grouped going from the right to the left.

20.5 Collections as Functions

The different collection types all inherit from `PartialFunction`, which inherits from the standard function types. This means that you can use collections in places where you would normally be expected to use functions. For example, the `Seq[A]` type will work as a `Int => A`. Similarly, a `Map[A,B]` can be used as a `A => B`.

A simple example usage of this is to take every third elements of an array by mapping over a range with the desired indexes and using the array as the function argument.

```
val nums = Array.fill(30)(math.random)
val everyThird = (nums.indices by 3).map(nums)
```

In this way, applying map with a collection as the argument gives us the ability to do arbitrary indexing into a collection.

20.6 Project Integration

There are quite a few places where we can use these different collections. In the previous chapter the type `mutable.Buffer` was used in `DrawTransform` to hold children. This made it easy to include code that adds and removes children. In this section we are going to use the `Map` type to help with text commands. After that, we will make use of an `Array` to clean up adding `Drawables` to a `Drawing`. This is something we could have done before, but it fits better here when looking at different uses of collections.

20.6.1 Commands

The command processor is a part of the code that could be implemented with inheritance. It would be possible to make a type called `Command` that we have other types inherit from. That type of construct would make logical sense, but might not be what we really need. One of the aspects you want to learn about inheritance is to not use it when it is not required. To know if we need it, we have to do some analysis.

You might wonder why we would want to put a text-based command processor into a drawing program. Perhaps the best analogy for this is why you would want to include a command-line interface in an Operating System (OS) that has a Graphical User Interface (GUI). There are certain operations that simply work better when they are typed in than they would when you have to point and click them. For now we will start with some simple examples that we can use as use-cases for the command tool. There are three three commands that we will start with:

- echo - prints whatever follows the command,
- add - should be followed by a bunch of space separated numbers and prints the sum,
- refresh - takes no arguments and causes the drawing to refresh.

So the user might type in "echo Hello, World!", "add 1 2 3 4 5", or "refresh". The first two take arguments and the only result of calling them is that something is displayed in the output. The last one is very different in that it does not take arguments or display anything. Instead, it needs to cause the `Drawing` object that is displayed to repaint itself.

To make this happen in the code, we will need a listener on the `TextField` for the user input that reacts to an `EditDone` event. When that happens, the line that is typed in needs to have the first word identified and based on what it is, appropriate functionality needs to be invoked.

One way to implement this functionality is to have each command include a function of the form `(String,Drawing) => Any`. The first parameter to this function will be the rest of the line after the command. The second parameter is the `Drawing` object the command was invoked on. This is needed so that a command like `refresh` can cause something to happen associated with the drawing. The return value is anything we want to give back. For most things this object will be converted to a `String` and appended to the output.

Having these functions is not quite enough. We also need to have the ability to quickly identify which of these functions should be invoked based on the first word of the input. This is a perfect usage of a `Map`. More specifically, we want to use a `Map[String,(String,Drawing)=>Any]`. This is a map from each command `String` to the function that command should invoke.

Using this approach, we can declare an object that contains the command functionality. An implementation that includes the three commands described above might look like the following.

```
package scalabook.drawing

/**
 * This object defines the command processing features for the drawing.
 * Use apply to invoke a command.
 */
object Commands {
  private val commands = Map[String,(String,Drawing)=>Any](
      "add" -> ((rest,d) => rest.trim.split(" +").map(_.toInt).sum),
      "echo" -> ((rest,d) => rest.trim),
      "refresh" -> ((rest,d) => d.refresh)
  )

  def apply(input:String,drawing:Drawing):Any = {
    val spaceIndex = input.indexOf(' ')
```

```
    val (command,rest) = if(spaceIndex < 0) (input.toLowerCase(),"")
      else (input.take(spaceIndex).toLowerCase(),input.drop(spaceIndex))
    if(commands.contains(command)) commands(command)(rest,drawing)
      else "Not a valid command."
  }
}
```

This object contains an apply method that is used to invoke various commands. It takes two arguments. The first is the full input from the user and the second is the Drawing object for calls to be made back on. This method finds the location of the first space, assuming there is one, then breaks the input into a command part and the rest of the input around that space. If there is no space, the entire input is the command. Once it has split the input, it checks to see if the command is one that it knows.

The Map called commands is used to identify known commands and to invoke the command. A call to contains tells us if the command is a known command. If it is, then the apply method is called using function notation. If not, an appropriate String is returned.

To make this code function, two things must be done in the Drawing class. First, we need to add the refresh method that is called by the refresh command. It can be declared in the following way.

```
private[drawing] def refresh { drawPanel.repaint() }
```

This method is added to the class and does nothing more than tell the drawPanel to repaint. It does have one interesting feature, the visibility is modified with [drawing]. Putting square brackets after the private or protected visibility modifiers allows you to restrict that visibility to parts of the code. In this case, it says that it is private outside of the drawing package. Inside of that package it is effectively public.

The addition of refresh lets the code compile, but it still is not invoked. For it to be actively used, we need to modify the declaration of the commandField variable in Drawing in the following way.

```
val commandField = new TextField() {
  listenTo(this)
  reactions += {
    case e:EditDone =>
      if(!text.isEmpty) {
        commandArea append "> "+text+"\n"+Commands(text,Drawing.this)+"\n"
        text=""
      }
  }
}
```

This code tells the field to listen to itself and when an EditDone events occurs, it executes the command and appends the result to the command area then clears the text. This code has two interesting aspects. First, the call to the append method is done using operator syntax. This means that the dot and parentheses are left off. This is a completely stylistic choice here. The more significant piece of code is the second argument to Commands. You already know that the keyword this is used to refer to the current object. At this point in code, the current object is the TextField. We utilize that fact in the call to listenTo. The second argument to Commands needs to be of the type Drawing. This code is inside of the Drawing class so such an object is available. We just have to tell Scala that we want the this associated with one of the surrounding types. That is specified with dot notation so Drawing.this refers to the current instance of the Drawing class we are in.

The primary advantage of this design is that it makes it very easy to add new commands. Simply add a new member to the command map and you are done. For these commands, the code for the function could be done in line. Longer functions could easily be written as methods in the Commands object. If needed, the functionality for commands could even be placed in other parts of the code, as long as it is reachable from Commands or through the Drawing object that is passed in.

20.6.2 Adding Drawables

There is another place where we might use a collection in an interesting way to clean up the code so that it is more extensible and less error prone. This is in the code to add a new Drawable. The method we had in chapter 18 looked like this.

```
private def addTo(d:Drawable) = {
  assert(d!=null)
  val parent = d match {
    case dt:DrawTransform => dt
    case _ => d.getParent
  }
  val options = Seq("Ellipse","Rectangle","Transform")
  val choice = Dialog.showInput(propPanel,
     "What do you want to add?","Draw Type",Dialog.Message.Question,
     null,options,options(0))
  if(choice.nonEmpty) {
    parent.addChild(options.indexOf(choice.get) match {
      case 0 => new DrawEllipse(parent)
      case 1 => new DrawRectangle(parent)
      case _ => new DrawTransform(parent)
    })
    drawPanel.repaint
    tree.getModel match {
      case m:javax.swing.tree.DefaultTreeModel => m.reload
      case _ =>
    }
  }
}
```

This method is not all that long, but it contains one significant part that is particularly error prone and that could lead to irritating errors later in development as more Drawable types are added to the project. The weak point in this code is that the order of the match cases is correlated to the order of the strings in options. Imagine what happens over time as more options are added. You have to add a string to options, then add a new case in the same position. If you decide you want to keep the strings in alphabetical order this will get challenging as the numbers on all the cases need to be changed. Even if that were not the case, it is too easy to get things out of order using this mechanism and the checking of order between the two will be time consuming when the lists get long.

An alternate approach is to group the names with the functionality to make the Drawable objects. In this case, we will use an array of tuples for this because the showOptionDialog method needs an array to be passed in and gives us back an Int. If instead of giving back an Int, it gave back the value that was selected, then a Map might be preferred.

```
private def addTo(d:Drawable) = {
  assert(d!=null)
  val parent = d match {
```

```
      case dt:DrawTransform => dt
      case _ => d.getParent
    }
    val options = Array[(String, ()=>Drawable)](
        ("Ellipse", () => new DrawEllipse(parent)),
        ("Rectangle", () => new DrawRectangle(parent)),
        ("Transform", () => new DrawTransform(parent))
    ).toMap
    val choice = Dialog.showInput(propPanel,
        "What do you want to add?","Draw Type",Dialog.Message.Question,
        null,options.keys.toSeq,options.keys.head)
    choice match {
      case Some(ch) =>
        val newChild = options(ch)()
        parent.addChild(newChild)
        drawPanel.repaint
        tree.getModel match {
          case m:javax.swing.tree.DefaultTreeModel => m.reload
          case _ =>
        }
      case _ =>
    }
  }
```

The `options` array here is an `Array[(String, ()=>Drawable)]`. You might be tempted to use just an `Array[String, Drawable]` to get this. The problem with that approach is that any time an object is added, the code will create a new instance of every type of `Drawable` in the array. That is extremely wasteful since we only need one and some of the subtypes of `Drawable` that we create later might be fairly large objects with significant overhead in their creation.

Using `()=>Drawable` instead of just `Drawable` means that we do not actually create any of the objects unless we invoke the method. In this case, the code `options(ch)()` does that for us on the one function associated with the choice that was selected.

Of course, the real benefit of this approach is that the names and the code to create them are grouped together in such a way that they can be easily reordered or added to with very little chance of the programmer messing things up. Having the string and the type appear on the same line means that they can be moved around and a simple visual inspection can tell you whether or not they have been set up correctly.

Visibility Options

As was mentioned in the section above, you can put bounds on the visibility restrictions implied by `private` and `protected` using square brackets. Inside the square brackets you can put the name of a bounding scope that you want to have public access to that member. This can be the name of a `package`, `class`, or `trait` that the code is inside of. This allows you to make something visible to the closely related code in a package without opening it up to the whole world.

You can also specify the visibility of `private[this]`, which implies that the member can only be seen by the object it is in. This is more restrictive than plain `private` which will allow it to be seen by all instances of that `class` or `trait`.

20.7 End of Chapter Material

20.7.1 Summary of Concepts

- The Scala collections libraries go far beyond the sequences like `Array` and `List` that we have looked at previously. There is a significant type hierarchy in related different parts of the hierarchy.
- The `Set` type is based on the mathematical concept of a set. All values in the `Set` are unique and order does not matter. The advantage over a sequence is that the check for membership is fast. There are mutable and immutable `Set` types.
- A `Map` is a collection that allows you to store and look up values using an arbitrary key type. This allows you to store and index values by `Strings`, tuples, or anything else. The type used for keys should really be immutable. There are mutable and immutable `Map` types.
- The `Buffer` is the ultimate mutable sequence type. Values can be mutated like an `Array`, but you are also able to insert or remove elements in a way that alters the size.

20.7.2 Exercises

1. Implement the following function that will build a `Map` from any sequence of a type with a function that can make keys from values.

```
def buildMap[A,B](data:Seq[A], f: A => B):Map[B,A]
```

2. Write code to represent a deck of cards. Use the following as a starting point.

```
case class Card(suite:String,value:Char)
class Deck {
  private val cards = mutable.Buffer[Card]()

  def dealHand(numCards:Int):List[Card] = ...
  def shuffle() { ... }
  def split(where:Int) { ... }
}
```

   Note that `cards` is a `mutable.Buffer` so dealing a hand really moves cards and the other two methods should change the ordering of elements.

3. In math discussions of probability, it is common to use a "bag of marbles" with different colored marbles as a mental image for illustrating certain concepts. The "bag" is not a `Seq` because there is no ordering. In Scala, it is better modeled as a `Set`.

   Write a little script to make a `Set[Marble]` and try using these two definitions of `Marble`.

```
case class Marble(color:String)
class Marble(val color:String)
```

For each try making a set of three red and two blue marbles. How do those implementations differ? Why do you think that is?

4. This exercise is a continuation of project 2 (p.345) on the BASIC programming language from the first half of the book. In that project you will simply use a `Map` to store variable values. A `Map[String,Double]` will work nicely as an efficient way to store and lookup variables. The rest of the functionality for the project can remain the same.

5. A standard example of a `Map` is a telephone book. The key is the name. The value would be a `case class` with telephone number, address, etc. You should make a GUI to display the telephone book and store the data in an XML file. For the first cut you can make it so that there can only be one person with a given name and use a `Map[String,Person]`.

   If you want a bit of a challenge, make it so that there can be multiple people with the same name. For that you could use a `Map[String,mutable.Buffer[Person]]`.

6. If you did the graphics and ray-tracing projects in the first half of the book, you can alter your program to use a `Buffer` (or two) to store geometry. Put this into the GUI you wrote for project 4 (p.281) and keep whatever the most recent saving and loading options are. Also change your rendering code to work with a `Buffer`.

20.7.3 Projects

The project ideas for this chapter all have you adding functionality to what you had done before using one of the new types that was introduced.

1. If you are working on the MUD projects, one obvious use of the `Map` type is to look up commands that a user types in to get functions that execute the required actions. You could also consider using a `Map` for any other place where something will need to be indexed by a value other than a simple number. For example, items or characters in a room could be indexed by their name, with some care for the fact that you should allow multiples. The ability to have multiple values could be provided by a `Map[String,mutable.Buffer[Item]]`.

   Your program is also keeping track of things like all the rooms in the world and possibly all the players currently logged in. These are both things that would make good use of a `Buffer` because the number of them will change.

2. If you are working on the web spider you should probably keep a `Buffer` of the data that you have collected from the various pages. In addition, you definitely need a way to not repeat visiting pages and perhaps to associate data with the page that it came from. The former can be nicely done using a `Set` as it can efficiently tell you if a value is already a member or not. If you want to have the ability to see what data was associated with a particular URL, you could use a `Map` with the URL as the key. The fact that the keys of a `Map` are themselves a `Set`, means that you do not need a separate `Set` if you use a `Map` in this way.

3. The broad variety of possibilities for games, networked or not, makes it hard to predict the ways in which these new collections might be useful. However, the vast majority of possibilities will have entities in them whose numbers vary and could be implemented well with a `Buffer`. You might also find uses for `Map`s for efficient look-up or `Set`s to efficiently demonstrate uniqueness.

4. The math worksheet project will have commands, much like the MUD. At this point, that makes a clear case for using a `Map`. In addition, the worksheet itself is going to be a sequence of different elements that will grow as the user adds things, or possibly shrink when the user takes things away. After reading this chapter, that should sound like a `Buffer`.

5. The Photoshop® project is filled with different aspects that can be nicely implemented with `Buffers`. For example, one of the standard features of software like Photoshop is to include multiple layers that can be worked on independently. The elements that are drawn on the layers could also be stored in `Buffers`.

6. If you are working on the simulation workbench project, it makes sense to use a `Buffer` for different simulations that are running at one time. You could link that to a `TabbedPane` that shows the different simulations.

   Having a `Buffer` of particles could be interesting for some simulations, but probably not the gravity ones that you have right now.

   Another way to add more collections to this project is to add another simulation type all together. One that you can consider include adding a particle-mesh for doing cloth simulations. The way this works is that you have a number of particles that each have a small amount of mass. They are connected with "springs" so that if they move too close to one another or too far away, there is a restoring force. In many ways, this works just like the gravity simulation, the only difference is that the forces only occur between particles that are connected in the mesh, and the force is not a $1/r^2$ type of force.

   The most basic implementation of this would use a standard Hook's Law with $F_s = -kx$, where k is the spring constant and x is the offset from the rest length of the spring. The force is directed along the segment joining the particles. A grid of particles with springs between each particle and its neighbors in the same row and column provides a simple setup that you can test. To make it dynamic, consider fixing one or two particles so that forces can not move them. You might also add a somewhat random forcing to mimic wind.

   A more accurate simulation can be obtained by the addition of extra, weaker springs. These would go between each particle and the ones immediately diagonal from it on the grid, as well as those two down along the same row or column. These additional springs give the fabric some "strength" and will make it bend in a more natural way. The simple grid configuration can completely fold over with an unnatural crease.

Chapter 21

Multithreading and Concurrency

Our programs so far have been written based on the basic assumption that only one thing is happening at any given time. The instructions in the program execute one after another and control might jump from one method to another, but at any given time you can point to a single line of code and say that is what is happening at that time. This type of sequential processing of instructions is called a thread. Our programs so far have been written to utilize a single thread of execution. Such programs do not take full advantage of modern hardware. In this chapter we will learn how to make our programs use multiple threads. Multithreading can be used to simplify logic in some places, but most of the time we will use it to make our programs run faster.

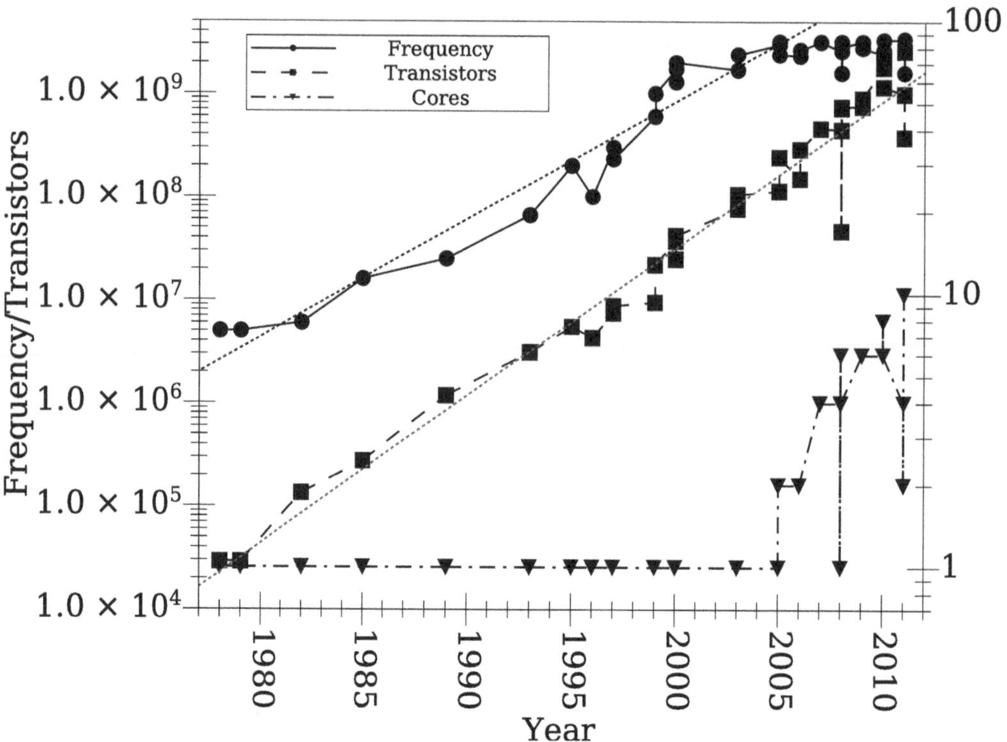

FIGURE 21.1: This figure shows frequencies, transistor counts, and core counts for x86 processors produced between 1978 and 2010. The log scale on the plot makes it so that exponential curves appear as straight lines. The dotted lines show fits to the full transistor count data and the frequency data through 2000. Clock speeds shown are the highest release for each chip.

21.1 The Multicore Future

For many decades after the development of the integrated circuit, computers got consistently faster because of a combination of two factors. The most significant was what is called *Moore's Law*. This was a prediction made by Gordon Moore at Intel that the number of transistors that could be etched on a chip would double roughly every two years. This growth was enabled by technological improvements that allowed the transistors to be made at smaller and smaller scales. Associated with this was the ability to push those transistors to higher frequencies.

Figure 21.1 shows clock frequencies, number of transistors, and core counts for x86 processors released by Intel and Advanced Micro Devices (AMD) between 1978 and 2011. The log scales on the vertical axes make it so that exponential growth displays as straight lines. The dotted line shows an exponential fit to the transistor count data. From this you can see that Moore's Law has held fairly well. Transistor counts have grown exponentially with a doubling time fairly close to the predicted two years.[1]

From the beginning of the figure until the 2000-2005 time frame, the clock frequencies of

[1]The low outlying points in 2008 and 2011 are chips specifically aimed at low-energy, mobile computing. This is why they have low transistor counts compared to other chips being made at the same time.

chips grew exponentially as well. The chips available in 1978 and 1979 ran at 5 MHz. By 2000 they had crossed the 1 GHz speed. The dotted line shows an exponential fit for the clock speeds through 2000. As you can see though, clock speeds stalled at a few GHz and have not gone up much since just after 2000. This was a result of problems with heat dissipation. Above these speeds, silicon-based chips need special cooling to keep them running stably.

Much of the speed boost that computers got until 2005 came from increases in clock frequency. When that ended, chip manufactures had to look for an alternate way to improve the speed of chips. After all, no one buys new computers so they can run programs slower than the old one. So beginning in 2005 commodity computer chips started going in a different direction.[2] The direction they went is shown by the dash-dot line with triangle points in figure 21.1. Instead of using additional transistors primarily to add complexity to the data paths for single instructions or just adding more cache, chip makers started making chips that had multiple cores.

When a chip has multiple cores, that means that it can be executing more than one instruction at any given time. We normally say that code can be executed in parallel. Unfortunately, a program written in the way we have done so far will not automatically take advantage of multiple cores. Instead, it has to be rewritten so that it will run faster on these newer computers. This is a significant change from what had been happening when processors were getting higher clock speeds. The significance of this change was described in an article titled "The Free Lunch is Over"[11]. The "free lunch" that is referred to is getting more speed from new processors without programmers having to do any work. To take full advantage of newer processors, programs have to be multithreaded. In this chapter we will begin to explore exactly what that means.

21.2 Basic Threads

When you imagine one of your programs running, you can picture the computer moving an arrow from one line to the next. The arrow points to the line that is currently being executed. When you call a function/method, the arrow jumps to that other part of the code, then jumps back when the function/method returns. The term we use for this behavior is a "thread". The programs we have written so far use a single thread of execution. We start this way because it is easier, but the current nature of computer hardware makes it important for you to understand the concept of multiple threads early on.

The idea of a multithreaded program is that you have two or more of those arrows pointing at different lines and moving through the program at the same time. These different threads can be run on different cores, giving us the ability to more completely utilize modern processors. Just like when multiple people cooperate on a task, there has to be some coordination between the individuals to get the maximum benefit. There are also typically limits to how many people can productively work together on a particular task. If you want to dig a small hole you probably can not efficiently use more than two or maybe three people. However, if you want to dig a large trench you can involve hundreds, if not thousands, of people productively. These same limitations can apply to programs.

There are many ways of using threads to produce parallelism in Scala. We will start off with the most basic pieces, the `java.lang.Thread` class and elements associated with it. This is not how you will likely do most of your parallel programming in Scala, but it

[2]High-end server chips had started doing this a few years earlier.

provides a good foundation and can show you the basic operations that are being used at the lowest level with all the different options we will look at.

We will start with a reasonably simple example. This is a script that will ask your age and count down from 10 to 1. If you do not answer before the countdown ends, it will terminate without you answering. If you answer, it prints a message based on whether or not you are 18 years of age and stops counting.

```
val countThread = new Thread {
  override def run {
    for(i <- 10 to 1 by -1) {
      println(i)
      Thread.sleep(1000)
    }
    println("Time's up.")
    sys.exit(0)
  }
}
countThread.start()
println("Enter your age.")
val age = readInt()
if(age<18) println("Sorry, you can't come here.")
else println("Welcome.")
sys.exit(0)
```

A new thread is created here using `new` and the type `Thread` and overriding the `run` method. You can also make a new `Thread` and pass it an instance of the `Runnable` type which must have a `run:Unit` method defined. In the `run` method here we place the code that we want the other thread to execute. In this case it has a `for` loop that counts down from 10 to 1. After each number is printed there is a call to `Thread.sleep(1000)`. The `sleep` method does what the name implies, it makes the thread pause its execution for a while. The one argument tells it how long to wait in milliseconds. The argument of 1000 here tells this thread to pause for a second between each number that is printed. After the loop, its print time is up and calls `sys.exit(0)`. This call forcibly causes a Scala program to terminate. The `0` is the exit code. By convention, a value of `0` indicates normal termination. Some other value would indicate an error.

Creating a new `Thread` object does not cause it to start working. For that to happen we have to call the `start` method. In this case, that is done right after it is defined. This is followed by a few simple lines of code reading input with a conditional print followed by another call to `sys.exit`. Enter this into a file and run it as a script to see how it works.

The idea behind this script is to demonstrate how two different things can be going on at the same time in a program. One thread is doing a countdown while another is doing some basic input and output to get information from a user. You can almost picture this as two different people doing things, though perhaps in that case the prompt should be printed in the thread code. What this code demonstrates is that the threads, like people, can operate independently. In this case, we are not using it to gain speed, instead we are simply aiming for concurrency.

The use of `sys.exit` is pretty heavy-handed. It works in this simple example, but it is easy to imagine situations where you do not want to terminate the entire execution of a program. The call to `sys.exit` in the thread is needed here because `readInt` is a blocking method that we can not get out of once we call it. The call to `sys.exit` at the end is needed because of the way we structured the loop in the thread. Let's try to see how we could get rid of both of these.

We will start with the call at the end of the script because it is easier to deal with.

There are two problems involved here. One is that we need some way for the code outside the thread to "talk" to the code inside the thread so that it can know when the user has answered. The second is that the `for` loop in Scala does not allow you to break out in the middle. If you want to have the possibility of terminating early, you need to use a `while` loop. The condition in the `while` loop should say that we keep counting while a counter variable is greater than zero and the user has not answered. This second part tells us how to get the two parts to "talk". We simply introduce a Boolean variable that keeps track of whether the user has answered. The `while` loop stops if they have. The line after reading the input should change this variable to say they have answered. A possible implementation of this looks like the following.

```
var answered = false
val countThread = new Thread {
  override def run {
    var i=10
    while(!answered && i>0) {
      println(i)
      Thread.sleep(1000)
      i -= 1
    }
    if(!answered) {
      println("Time's up.")
      sys.exit(0)
    }
  }
}
countThread.start()
println("Enter your age.")
val age = readInt()
answered = true
if(age<18) println("Sorry, you can't come here.")
else println("Welcome.")
```

Enter this code and see that it behaves the same as the earlier version.

This version still contains one call to `sys.exit`. This one is a little more difficult to get rid of. The problem is that `readInt` is what we call a *blocking method*. This is a method that stops the execution of the thread until some certain state is reached. In this case, it blocks until the user has typed in an appropriate input value. That is not acceptable in this situation because we want to make it so that the countdown thread can prevent the read from happening, and more importantly blocking, if it gets to the end before the user enters their value. An approach to this is shown here.

```
var answered = false
var timeUp = false
val countThread = new Thread {
  override def run {
    var i=10
    while(!answered && i>0) {
      println(i)
      Thread.sleep(1000)
      i -= 1
    }
    if(!answered) {
      println("Time's up.")
```

```
      timeUp = true
    }
  }
}
countThread.start()
println("Enter your age.")
while(!timeUp && !Console.in.ready) {
  Thread.sleep(10)
}
if(!timeUp) {
  val age = readInt()
  answered = true
  if(age<18) println("Sorry, you can't come here.")
  else println("Welcome.")
}
```

This code introduces another mutable Boolean variable that keeps track of whether time has expired. At the end of the countdown this variable is set to be true.

The more interesting part occurs right after the prompt for the users age. Here you find a `while` loop like this.

```
while(!timeUp && !Console.in.ready) {
  Thread.sleep(10)
}
```

This loop will stop under one of two conditions, either time must run out, or `Console.in` must be ready. To understand this second condition requires some digging through the Application Programming Interface. The object `scala.Console` is what you read from when you do calls like `readInt` or what you write to when you do `println`. The input comes through an object called `Console.in` which is a `java.io.BufferedReader`. We will talk more about the various classes in `java.io` in the next chapter, but for now you can know that this type has a method called `ready` that will tell you when you can read from it without having it block.

Inside the loop is another call to `Thread.sleep`. This is here simply to prevent the `while` loop from doing a "busy wait" and making the processor do a lot of extra work for no real reason. It will still check if the user has input something or if time has run out 100 times each second. This is more than sufficient for the type of user interaction that we have here. After the loop the same code from before appears inside an `if` so that it will not happen if the loop stopped as a result of time running out.

21.2.1 Problems with Threads

This example shows how simple using threads can be. Unfortunately, a lot of that apparent simplicity masks the many problems that can occur when you are using threads. The challenges of multithreading are the reason it is not done unless it is really needed. The two main problems that one runs into are called RACE CONDITIONS and DEADLOCK. We'll look at each of these in turn.

21.2.1.1 Race Conditions

The term race condition is used to describe a situation where the outcome of a program can depend on the order in which threads get to a different point in a program. Typically, if you run a program with the same inputs, you expect to have the same outputs. This deterministic nature is part of what makes computers so useful in the everyday world. Programmers do not have full control over what threads are running at any given time. You have the ability to do things like make a thread sleep for a period of time, but in a program with many threads, you do not get to explicitly say which ones will be running at any given time. Using the human analogy, you typically do not have the fine-grained control of telling each and every person exactly where to be and what to be doing. If they are digging a trench, you can not say exactly when each shovel will cut into the ground. If you tried to exert such control, it would probably slow things down in the end and defeat the purpose of having many people/threads working on the task.

Simply lacking detailed control does not lead to race conditions. Race conditions happen when you combine that lack of control with a shared mutable state. You are probably aware that most of the time your computer is running many different programs. These different programs are often called processes and they can utilize multiple cores as well. The difference between different processes and different threads is that each process gets its own chunk of memory that it can work with, but threads share access to the same memory. To see how this leads to problems, consider the classic example of a bank account.

```
/**
 * A class that represents a bank account. The balance is stored as an Int
 * representing a number of cents.
 */
class BankAccout(private var bal:Int = 0) {
  def balance:Int = bal

  def deposit(amount:Int) {
    assert(amount>=0)
    val oldBalance = balance
    val newBalance = oldBalance+amount
    bal = newBalance
  }

  def withdraw(amount:Int) {
    assert(amount>=0 && amount<=bal)
    val oldBalance = balance
    val newBalance = oldBalance-amount
    bal = newBalance
  }
}
```

This code has been written in a long way using local variables instead of using operators like `+=` and `-=` because this is what the code will really be expanded to when it is compiled and it makes it easier to see where the problem arises.

This is such a simple little class. What could possibly go wrong? To see the answer to that question, imagine that you make a withdrawal at an ATM at the same time as your paycheck is being deposited. Both of these activities read from and write to the `bal` member of the object that represents your account. The result at the end depends on the exact order in which these things happen. If one of them starts and finishes completely before the other, everything will be fine. The race condition occurs when both start before either finishes. In that situation, the one that executes `bal = newBalance` last will have its results stored. The other one will have the work that it did wipe out. You might not mind this if your withdrawal is not recorded and you get your money for free. However, you might have a problem with losing a whole month's salary because you happened to visit the ATM at the wrong time.

In fact, the script that we showed in the last section has the possibility for a race condition because both threads are accessing `answered` and `timeUp` with at least one writing to each. Though you could run the script many times and never see this, it is possible to answer the question, get a response, and be told that time is up in one run. To see how, picture the threads as arrows and imagine that you hit Enter at exactly the same time as the last second counts off. So the lines `if(!answered)` and `if(!timeUp)` execute at the same time. This means that both values will still be `false` and the program can get into both of those `if` statements.

Hopefully you can see how both of those examples could cause problems. However, they both require specific timing for the error to manifest. This does not make them unimportant, it just makes them hard to demonstrate. Indeed, that type of bug is one of the hardest to diagnose and fix because it is so rare and hard to reproduce. If you want to specifically see the problems that occur as a result of race conditions, the following code will do that for you.

```
var cnt = 0
val threads = Array.fill(100){
  new Thread {
    override def run {
      for(i <- 1 to 1000000) cnt += 1
    }
  }
}
threads.foreach(_.start)
threads.foreach(_.join)
println(cnt)
```

This code has a counter that starts at zero and then makes and starts 100 threads that each contain a `for` loop which counts to one million and as it does so it adds to the counter. At the end, the counter is printed.

Between the start and the print is a line that calls `join` on all the threads. The join method is designed to block until the thread it is called on finishes. After the thread finishes it continues. If this line were left out of the code, the print statement could, and likely would, happen while there were still threads adding to the counter. The use of join here simply makes sure that all the threads have finished before we do the print.

Just looking at this code it should be clear that we expect the final line to print out 100 million. However, if you run it, you will find that it never does.[3] Even if you run it on a machine with only a single core, the final printout will be less than 100 million. That is because on a machine with a single core, threads can not execute at exactly the same time so instead they swap from one to the other.

21.2.1.2 Deadlock

The way to prevent threads from having race conditions when there is a mutable state is to lock critical parts of code so that only one thread can be in those critical sections at a given time. There are also times when you need to have some threads wait for other threads to finish before they move on to further tasks. For example, you probably should not start filling the trench with water until after all the digging is done and everyone has gotten out.

Making it so that one thread can pause and wait for another to finish doing something is critical. However, it can also lead to a situation called deadlock where different threads are

[3]Technically it could, but the odds of that happening are vanishingly small.

waiting on one another and as a result, none of them can proceed. Consider the following example of pseudocode.[4]

```
var l1 = new Lock
var l2 = new Lock

def processA {
  l1.lock {
    do stuff A
    l2.lock {
      do more stuff A
    }
  }
}

def processB {
  l2.lock {
    do stuff B
    l1.lock {
      do more stuff B
    }
  }
}
```

Now imagine that one thread calls `processA` while a second thread calls `processB`. As soon as `processA` starts it, locks `l1` and does some work. Similarly, `processB` will lock `l2` then do some work. Before releasing the locks they have, they try to acquire the locks they do not have yet. This can cause a deadlock, because `processA` can not get the lock for `l2` until `processB` is done and `processB` can not get the lock for `l1` until `processA` is done. The result is that they sit around waiting for one another forever and the code simply stops.

You might argue that this is a contrived example and that you would never put such alternating locks into real code. In reality though, the inner lock would probably be located in some function/method that is called by the outer lock so it will not be at all obvious that such an alternating lock structure is present in the code.

21.2.2 Synchronization

The first approach to dealing with race conditions is synchronization. When a block of code is synchronized, it means that when that block executes it first checks a MONITOR on the specified object. If the monitor is unlocked, that thread acquires the lock on the monitor and executes the body. If the monitor is locked by another thread, the thread blocks until the lock is released by that thread. This is done by calling the `synchronized` method in `AnyRef`. That method has the following signature.

```
def synchronized[T](body: => T): T
```

Note that the body is passed by name and the return type is a parametric type. This makes it so that you can write whatever code you might have normally written as a body after the call to `synchronized`. The fact that it is defined in `AnyRef` means that you can call this inside of any `class`, `object`, or `trait` declaration, and you often do not have to specify the object it is being called on. You do have to be careful to keep in mind what object it

[4]The syntax shown here is not proper Scala code. We'll see how this type of thing should really be done in the following sections.

is being called on though as that will be the object the monitor is locked on and using the wrong one can lead to incorrect results.

As a first example, let us look at how we would fix the bank account class to be synchronized so that we do not have to worry about deposits and withdrawals happening at the same time.

```
/**
 * A class that represents a bank account. The balance is stored as an Int
 * representing a number of cents.
 */
class BankAccout(private var bal:Int = 0) {
  def balance:Int = bal

  def deposit(amount:Int) = synchronized {
    assert(amount>=0)
    val oldBalance = balance
    val newBalance = oldBalance+amount
    bal = newBalance
  }

  def withdraw(amount:Int) = synchronized {
    assert(amount>=0 && amount<=bal)
    val oldBalance = balance
    val newBalance = oldBalance-amount
    bal = newBalance
  }
}
```

As you see, all we did here was put a call to synchronize around the full method for `deposit` and `withdraw`. An equal sign was added because we are no longer using the short syntax for a method that returns `Unit`. That is all that is required for that example.

You might wonder why it is that synchronization works when the code we wrote before that used an `if` to check a variable then automatically set it does not. The reason behind this is that the checking and setting of the lock on the monitor is an atomic operation. What does it mean for an operation to be atomic? The Greek root of the word atom means indivisible. The name was given to the particles because it was believed at the time that they were the smallest pieces of matter and could not be split. While that turns out to not really be true of atoms in nature, it is true of atomic operations on a computer. In this alternate sense, when we say it can not be split, we mean it can not be interrupted. Once an atomic operation begins, nothing else can take over before it finishes. The problem with the `if` followed by a set was that it was not atomic and it was possible for the value of the variable to be changed by another thread between when this thread read the value, looked at it, and set it.

So what about the example that adds to a counter in multiple threads? Given what we have seen, your first attempt might look like this.

```
var cnt = 0
val threads = Array.fill(100){
  new Thread {
    override def run {
      for(i <- 1 to 1000000) synchronized {cnt += 1}
    }
  }
}
```

```
threads.foreach(_.start)
threads.foreach(_.join)
println(cnt)
```

This does nothing more than put the increment inside a synchronized block. Unfortunately, this does not work. This is one of those situations where we do have to specify what object is being synchronized on. The fact that we do not specify an object that the method is called on means it is being called on the "current" object. In this case, that is the `Thread` object that is being created. After all, this happens in the run method of the thread we are making. This is not good enough because each thread has its own monitor so this does not really prevent the different threads from accessing `cnt` at the same time.

In order to do that, we need to have all the threads synchronize on a single object. The most obvious candidate would be `cnt` itself. After all, what we want to prevent is having two threads attempt to increment that variable at the same time. However, this will not work because `Int` is not a subtype if `AnyRef`. If you look back at figure 19.1 on page 453, you will see that `Int` falls under `AnyVal`. As a result, it does not have a `synchronized` method to call. In the script form, the only other object we have is `threads` so that is what we will use.

```
var cnt = 0
val threads:Array[Thread] = Array.fill(100){
  new Thread {
    override def run {
      for(i <- 1 to 1000000) threads.synchronized {cnt += 1}
    }
  }
}
threads.foreach(_.start)
threads.foreach(_.join)
println(cnt)
```

Note that because this version uses `threads` inside of the declaration of `threads`, it becomes a recursive definition. For that reason, the type must be specified with the name. This is the same limitation that applies to methods and functions. We have simply never run into it before with a `val` or a `var`.

Normally you will not be writing multithreaded code inside of a script. After all, scripts are intended to be used with small programs and multithreading is most appropriate to speed up large programs. For that reason it is worth seeing what this would look like. The code below shows the same script moved into the main method of an object.

```
object ThreadCounting {
  def main(args:Array[String]) {
    var cnt = 0
    val threads = Array.fill(100){
      new Thread {
        override def run {
          for(i <- 1 to 1000000) ThreadCounting.synchronized {cnt += 1}
        }
      }
    }
    threads.foreach(_.start)
    threads.foreach(_.join)
    println(cnt)
  }
```

```
}
```

With this structure, the synchronization can be done on the ThreadCounting object. If ThreadCounting had been a class or a trait, you could have done ThreadCounting.this.synchronized.

At this point you might be wondering why you do not synchronize everything to avoid race conditions. The reason is that synchronization has a cost in execution speed. Checking and setting locks on monitors takes real time. In addition, it prevents multiple threads from operating. If you really did synchronize everything, you would likely make the program run very slowly with only a single thread active most of the time and have very good odds of deadlock. Instead, you have to be careful with how you synchronize things.

If you played with the different programs in this section you probably found that the programs that gave the right answer through synchronization were slower. In testing them for this book they were found to take about 10 times longer to complete than what it took with no synchronization. How do you get around this? Try not to share a mutable state. In this example, it would be much better to have one variable per thread that is incremented and at the end you add all the results together. Here is a solution that does that.

```
class CountThread extends Thread {
  var cnt = 0
  override def run {
    for(i <- 1 to 1000000) cnt += 1
  }
}
val threads = Array.fill(100)(new CountThread)
threads.foreach(_.start)
threads.foreach(_.join)
var cnt = threads.view.map(_.cnt).sum
println(cnt)
```

Notice that in this version we had to define a new class that is a subtype of Thread. This allows us to access the cnt field. The java.lang.Thread class does not have a member named cnt so we need a new type that does. This code runs very close to the same speed as the earlier version that had no synchronization and it gets the correct answer.

One last thing to note about this code is that it uses a view. This is a topic that was introduced as an advanced topic on page 197. It is used here because we are interested in timing results and it is slightly more efficient to not have map create a completely new collection.

21.2.3 Wait/Notify

The synchronized method can help you to prevent race conditions, but it is not generally sufficient to coordinate the behaviors of different threads. In the basic thread libraries this is done with wait, notify,[5] and notifyAll. More advanced tools have been introduced in libraries that we will discuss below, but there is some benefit to knowing how things work at the lowest level so we will discuss these briefly.

As with the synchronized method, wait, notify, and notifyAll are methods in AnyRef and can be called on any object of that type. The wait method does exactly what the name implies, it causes the current thread to stop executing and wait for something. That something is for it to be woken up by a call to notify or notifyAll in another thread.

[5] Use of notify is typically discouraged because you have no control over which thread it will wake up. This has a tendency to lead to deadlock.

All of these calls require that the current thread hold the lock to the monitor for the object the method is called on. In addition, `notify` and `notifyAll` only wake up threads that were set to wait on the same object. You will normally see these calls in synchronized blocks. The call to wait releases the lock on that monitor. It will be reacquired when the thread wakes back up.

To see how this can work, consider this example that has a number of different threads that use wait to count off in turns with no more than one thread active at any given time.

```
object WaitCounting {
  def main(args:Array[String]) {
    val numThreads = 3
    val threads = Array.tabulate(numThreads)(i => new Thread {
      override def run {
        println("Start "+i)
        for(j <- 1 to 5) {
          WaitCounting.synchronized {
            WaitCounting.wait()
            println(i+" : "+j)
            WaitCounting.notify()
          }
        }
      }
    })
    threads.foreach(_.start)
    Thread.sleep(1000)
    println("First notify.")
    synchronized { notify() }
  }
}
```

Each thread has a `for` loop that counts to 5. In the loop to synchronize on the object the code is in and calls `wait`. Once the thread is woken up by another thread, it will print out the thread number and the number in the loop, then notify some other thread. After that it immediately goes back to sleep as the next time through the loop leads to another call to wait. After the threads are built they are started and the main thread sleeps for one second before calling `notify`.

The call to sleep makes sure that all the threads have had time to start up and call `wait` before the first call to `notify` is made. If you leave this out, it is possible that the first call to notify will happen before the threads have started. More rigorous methods of accomplishing this would be to have a counter that increments as each thread starts.

Each call to `notify` wakes up a single thread that is waiting on the `WaitCounting` object so this works like dominoes with the prints running through the threads in whatever order `notify` happens to wake them up. The inability to predict what thread will be woken up is a problem. The real problem arises because it is possible that in later development some other thread might start waiting on this same object and that thread might not follow the same rule of printing then calling notify. That can lead to deadlock. In this example, if you simply add the following line to the bottom after the call to `notify`, you can create deadlock.

```
synchronized { wait() }
```

This causes a problem because now when one of the other threads calls `notify`, it wakes up the main thread instead of one of the printing threads. The main thread does not do anything after it wakes up though so we are left with printing threads that remain asleep forever.

The lack of predictability in notify has led to the standard rule that you should not use it. The recommended style is to have all calls to wait appear in while loops that check some condition and to always use notifyAll. The following code shows this rule in practice.

```
object WaitCountingSafe {
  val numThreads = 3
  def main(args:Array[String]) {
    var handOff = Array.fill(numThreads)(false)
    val threads = Array.tabulate(numThreads)(i => new Thread {
      override def run {
        println("Start "+i)
        for(j <- 1 to 5) {
          WaitCountingSafe.synchronized {
            while(!handOff(i)) {
              WaitCountingSafe.wait()
            }
            handOff(i) = false
            println(i+" : "+j)
            handOff((i+1)%numThreads) = true
            WaitCountingSafe.notifyAll()
          }
        }
      }
    })
    threads.foreach(_.start)
    Thread.sleep(1000)
    println("First notify.")
    handOff(0) = true
    synchronized { notifyAll() }
  }
}
```

Now each thread has its own Boolean flag telling it whether another thread has handed it control. If that flag is not true, the thread will go back to waiting. In this way we regain control over things and reduce the odds of deadlock. Indeed, adding the same call to wait at the end of this code will not cause any problems at all.

21.2.4 Other Thread Methods

There are a number of other methods Thread that are worth knowing about. We have already seen the sleep method which causes a thread to stop processing for a specified number of milliseconds. Similar to sleep is the yield method. Calling Thread.`yield` will make the current thread give up control so that if another thread is waiting it can take over. The backticks are required in this call because "yield" is a keyword in Scala. You use it with the for loop to make it into an expression. Using backticks allows you to have Scala interpret tokens as normal even if they are keywords that are part of the language.

The dumpStack method will print out the current stack trace. This can be helpful if you are running into problems in a certain method and you are not certain what sequence of calls got you there. You can also use Thread.currentThread() to get a reference to the Thread object that represents the current thread.

If you have a Thread object that you either created or got through a call to

`Thread.currentThread()`, you can call other methods on that thread. The `join` method is an example we saw previously. See the API for a complete listing. One pair that is worth mentioning are the `setPriority` and `getPriority` methods. Each thread has an integer priority associated with it. Higher-priority threads will be preferentially scheduled by the thread scheduler. If you have work that needs to happen in the background and is not critical or should not impact the function of the rest of the system you might consider making it a lower priority.

There are also a number of methods for `java.lang.Thread` that you can find in the API which are deprecated. This implies that they should no longer be used. This includes `stop`, `suspend`, and `resume`. These methods are fundamentally unsafe and should not be used. It might seem like a good idea to be able to tell a thread to stop. Indeed, the writers of the original Java library felt that was the case. However, experience showed them that this was actually a very bad thing to do. Forcibly stopping a thread from the outside will often leave things in unacceptable conditions. A simple example of this would be if you kill a thread while it is in the middle of writing to a file. This not only leaves the file open, it could very well leave the file in an inconsistent state that will make it unusable later.

Scala will give you warnings if you use deprecated library calls. You should pay attention to these, find them in your code, and look in the API to see how they should be handled. In the case of stopping threads, the proper technique is to have long running threads occasionally check Boolean flags that tell them if they should continue running. If the flag has been changed they should terminate in an appropriate manner.

21.3 Concurrency Library

Direct thread control gives you the power to do anything you want with multithreaded parallelism, but it is hard to do correctly and there can be a significant amount of code behind many common tasks. For this reason, the `java.util.concurrent` package was added in Java 5 to make many of the common tasks in parallel programming easier to do. In general, if you have detailed threading tasks that you can not do with the Scala libraries we will present later on, you will use elements from `java.util.concurrent` along with `synchronized` to accomplish what you want.

21.3.1 Executors and Executor Services

The process of creating new threads is expensive and is not abstract, meaning it can not be easily varied for different platforms. The `java.util.concurrent` package addresses this with the `Executor` and `ExecutorService` types. The `Executor` type is a very simple abstract type with a single method.[6]

```
def execute(command:Runnable):Unit
```

When this method is called, the `run` method of the `command` is called. What you do not know at this level is how that will be done. It is possible that a new thread will be created and it will run in that thread. However, it is also possible that it could be run in a thread that had been created earlier or even run in the current thread. Note that this last option implies that the current thread might not continue until after the `command` has finished.

[6]The signature of the method is really `void execute(Runnable command)` in Java form.

The abstract type `ExecutorService` extends `Executor` and adds significantly more functionality to it. The following methods are significant for us. Look in the Java API for a full list.

- `shutdown():Unit` - Tells the service to start an orderly shutdown allowing current tasks to finish. New tasks will not be accepted.
- `submit[T](task:Callable[T]):Future[T]` - Submits the specified task for execution and returns a `Future[T]` object representing the computation.
- `submit(task:Runnable):Future[_]` - Submits the specified task for execution and returns a `Future[_]` object representing the computation.

There are two new types presented in these methods: `Callable[T]` and `Future[T]`.

21.3.2 Callable and Futures

With normal threads, the computation can only be submitted as a `Runnable`. Unfortunately, the `run` method in `Runnable` does not return a value. This is a significant limitation as we often want to get results from the threads. The `Callable[T]` type contains one method that has a return value.

```
def call[T]():T
```

The submitted task might take a while to finish, so the `submit` method can not return the `T` object as that would require blocking. Instead, it returns a `Future[T]` object that can be used to check on the computation (use `isDone():Boolean`) and get the result when it is complete (use `get():T`).

Both `Executor` and `ExecutorService` are abstract. The standard way to get a concrete instance is to call one of several methods on `java.util.concurrent.Executors`. The methods you are likely to use are `newCachedThreadPool():ExecutorService` and `newFixedThreadPool(nThreads:Int):ExecutorService`. A cached thread pool will make as many threads as there are active tasks. Once a task has completed, the thread is stored in a wait state so that it can be reused for a later task. A fixed thread pool only has the specified number of threads. If a new task is submitted when all the threads are filled, it will wait until one of the active threads has completed.[7]

To demonstrate these things in action we will look at a version of factorial that uses `BigInt` and is multithreaded. Recall that a single threaded version of this can be written in the following way.

```
def fact(n:BigInt) = (BigInt(1) to n).product
```

If you call this for sufficiently large values of `n`, you will find that it can take a while to complete the calculation.[8] That makes it a reasonable function of us to consider parallelizing. How should we go about doing that? We want a function that breaks the multiplications up across a number of threads using an executor service. So the question that so often comes up when creating parallel programs is, how do we break up the work?

One approach to this is to break all the numbers into groups. If `n` were divisible by `nThreads` this could be viewed as n! = 1 * ... * n/nThreads * (n/nThreads+1) * ... *

[7]You must be careful when using a fixed thread pool if you add new tasks to the pool inside of other tasks. It is possible for the second task to be stuck waiting for the first to finish. If this happens and the first task has code to get the result of the second task you will get deadlock.

[8]Depending on your computer, sufficiently large will be at least several tens of thousands.

2*n/nThreads * (2*n/nThreads+1)*... When `n` is not divisible by `nThreads`, we need to make some of the groups bigger. Here is code that does that.

```
def parallelFactorial(n:BigInt,es:ExecutorService,nThreads:Int):BigInt = {
  val block = n/nThreads
  val rem = n%nThreads
  var i=BigInt(1)
  val futures = Array.tabulate(nThreads)(j => es.submit(new Callable[BigInt] {
    def call():BigInt = {
      val start = i
      val end = start + block + (if(BigInt(j)<rem) 1 else 0)
      i = end
      (start until end).product
    }
  }))
  futures.map(_.get).product
}
```

This function starts with calculations of the base size for each block as well as how many blocks need to be one bigger. It then creates a variable `i` that helps us keep track of where in the range we are. A call to `Array.tabulate` is used to submit the proper number of tasks to the `ExecutorService`. Each task uses `i` as the start value and adds an appropriate offset to get the end value. It then updates `i` and uses the standard `product` method to multiply the values in that chunk. The last thing this code does is map the various `Future[BigInt]` objects with `get`. This will block until they are done and give us a collection of the results. Another call to `product` produces the final result.

You can put this in a script or load it into the REPL to test it. When you do this, you probably want to start off with an import, then make your `Executor Service`.

```
import java.util.concurrent._
val es = Executors.newCachedThreadPool()
```

Now you can call the function and compare it to the non-parallel version above. Ideally you would specify the number of threads to be equal to the number of cores in your computer, but you can play with this to see how it impacts execution time. Either at the end of the script or before you exit the REPL you need to call `es.shutdown()`.

This approach to breaking up the work is not the only option we have. One potential problem with this approach is that the different threads we created really do not have equal workloads. The way the `BigInt` works, dealing with large numbers is slower than dealing with small numbers. Given the way the work was broken up here, the first thread winds up with a much smaller value than the last one and, as such, has a smaller work load. An alternate approach would be to have each thread work with numbers spread out through the whole range. Specifically, the first thread does 1*(nThreads+1)*(2*nThreads+1)*... while the second thread does 2*(nThreads+2)*(2*nThreads+2)*..., etc. The methods of `Range` types in Scala actually makes this version a lot shorter.

```
def parallelFactorial(n:BigInt,es:ExecutorService,nThreads:Int):BigInt = {
  val futures = Array.tabulate(nThreads)(j => es.submit(new Callable[BigInt] {
    def call():BigInt = {
      (BigInt(j+1) to n by nThreads).product
    }
  }))
  futures.map(_.get).product
}
```

The `by` method lets us get a range of values with the proper spacing. As long as it starts at the proper value, everything else will work the way we need.

You can use the `System.nanoTime()` method from Java to measure how long it actually takes to complete the calculation. Just call it before and after the calculation and take the difference to get the number of nanoseconds taken. You can divide that value by `1e9` to get a value in plain seconds.

If you play with the three versions of factorial that are presented here you should find that the parallel versions are faster than the sequential version, but it is not by all that much. If you have access to a machine with enough cores, you probably also noticed that there was a point of diminishing returns. Going from one thread to two shows a significant boost, perhaps even more than a factor of two. However, going up to four threads almost certainly does not return another factor of two improvement and further additions become less and less beneficial. There are a number of things going on here, but one of the big ones is what is know as Amdahl's law [1].

The basic idea of Amdahl's law is that the benefit you get from improving the speed of a section of your code depends on how much of the code you can make this improvement to. In the case of parallel processing, not all of the program will be done in parallel. Some sections are still done sequentially. If the faction of your code that is done in parallel is represented by P, and the fraction that remains sequential is $(1-P)$, then Amdahl's law states that the maximum speedup you can get with N processors is

$$\frac{1}{(1-P)+\frac{P}{N}}.$$

As N goes to infinity, this converges to

$$\frac{1}{1-P}.$$

So no matter how many processors you have, you are limited in the speed boost you can see by how much of the code you can actually do in parallel.

How much work are our functions here doing sequentially? In the second version it is pretty clear that the only sequential work is submitting the tasks and calculating the final product. For large values of `n`, submitting the tasks should be fairly negligible. Unfortunately, the final product will not be. Remember that operations with `BigInt` take longer for bigger numbers. The biggest numbers of the calculation will be the ones at the end that come up during this sequential section. If we really wanted to make this as fast as possible, we would need to make that last product happen in parallel as well. That task is left as an exercise for the reader.

Computer Performance (Advanced)

In chapter 13, when we were looking at sorting and searching algorithms, you should have gotten some experience with the fact that measuring the performance of a computer program can be challenging. The same program running with the same input on the same computer can take varying amounts of time to execute. To deal with that you need to take a number of timing tests and average them.

The performance can also vary in odd ways depending on the input size. This might have come out with the sorts as well. The amount of time spent in a bubble sort can increase by the expected factor of four when you double the input size for a range of inputs, but there will be points where it jumps dramatically. Instead of going up by a factor of four it might jump by as much as 100. This is generally due to the memory hierarchy of the computer. At some point in your studies you should learn about cache memory and how it works. In this context, once the data you are accessing gets bigger than one of the cache levels in your machine, the speed can drop dramatically as memory has to be fetched from a more distant memory store.

Doing parallel processing can make things even more complex. It would be nice if the runtime of programs would scale as the number of threads used up to the core count of the machine. We already saw that Amdahl's law precludes that behavior unless every line of code can be done in parallel. Things can be even odder though and not always in a bad way. On a test system, using two threads in the parallel factorial program reduced the runtime by significantly more than a factor of two. You might wonder how that is possible. Actually answering that requires detailed testing on any specific machine, but a common possibility deals with cache again. Multicore chips have some cache that is local to a core and some that is shared. Breaking up a problem across threads can lead to a situation where each thread uses only a fraction of the total memory. When this happens, the memory used by a thread might drop under one of the critical thresholds for the memory hierarchy of the chip. This can lead to abnormally good speed improvement. One the other hand, it is possible that multiple threads could be writing to adjacent pieces of memory regularly, which can slow things down because the cores have to talk to one another to make sure they have the latest version.

Modern processors are very complex and there are a lot of factors that go into how well they perform. The details of that go well beyond the scope of this book. For our purposes, we will try to find safe ways to break workloads across many threads without introducing unneeded overhead.

21.3.3 Parallel Data Structures

There are a number of different data structures provided in `java.util.concurrent` that help in performing common tasks. Instead of simply telling you what these do, we will consider scenarios where you want to do different things and then present the data structures that can help you to do those things.

21.3.3.1 Shared Barriers

Earlier we had the example of a large number of people working to dig a trench that would then be filled with water. This is the type of task where you have many workers, represented by threads on the computer, working on a task and there are certain critical points where no one can proceed until everyone has gotten to that point.

Consider a computer program that has to deal with collisions between moving objects in a scene. Collisions need to be handled at the exact time they occur for this to be accurate. In the program there are other things that need to be computed at certain intervals. We will call this interval the time step and it has a length of Δt. Collisions are processed in a more continuous way through the time step. So in any given step you would look for collisions that happen given the initial configuration, process collisions, with updates that happen as a result of earlier collisions, then do whatever processing is required at the end of the step.

To make things go fast, you want to break this work across as many threads as you can. You break the particles up into groups and have one thread run through and find all collisions during that time step given the initial trajectories of the particles. You can not start processing collisions until all the threads are done identifying the first round because it is possible that the last one found could be the first one you need to process. So all the threads have to finish finding collisions before the first one can be processed. There are schemes for processing collisions in parallel as well that make sure that you do not ever work on two collisions at the same time if they are too close. All collisions have to be processed and all threads have to get to the same point before the other processing can be done.

There are three different types in `java.util.concurrent` that can help give you this type of behavior. The first one is a type called `CountDownLatch`. You instantiate a `CountDownLatch` with `new CountDownLatch(count)`, where `count` is an integer value. The two main methods of this type are `countDown`, which decrements the count by one, and `await` which will block until the count reaches zero. If the count is already at zero, a call to `await` does nothing.

The second type that can provide this type of behavior is the `CyclicBarrier`. While the `CountDownLatch` is only good for one use, the `CyclicBarrier` can be cycled many times. You make a `CyclicBarrier` with `new CyclicBarrier(parties)`, where parties is an integer number specifying how many threads should be involved in the barrier. When each thread has finished doing its task, it makes a call to the `await` method on the `CyclicBarrier`. Once all "parties" have called `await`, they are all allowed to proceed forward.

The third type is the `Phaser` type. This type is more complex than the other two. It is also newer, having been added in Java 1.7. Interested readers should consult the API for usage of this type.

21.3.3.2 The Exchange

Imagine a situation where two parties work mostly independently, but they perform calculations that require some information from another party. You do not want to try reading the information at any old time as the value might not contain a valid, final result most of the time. Instead, you need to arrange a point when the parties pause and come together for an exchange.

An example might be a large simulation of a solid body like an airplane. These are often done with finite-element models. Each thread would have a chunk of the plane to work with and it would calculate the stresses on the elements in that chunk as they change over time. The stresses on one chunk will depend on what is happening with all the elements in adjacent chunks that are connected to elements in the first one. In order to do a full calculation, the threads need to meet up to exchange information about the boundary elements. This type of thing can be done with an `Exchanger`.

You instantiate an `Exchanger` with `new Exchanger[A]()`, where `A` is the type of information you want to exchange. When a thread is ready to exchange, it calls `exchange(x:A):A` on the `Exchanger`. This method will block for the first thread until the second thread also calls it. At that point, both calls return and each thread gets the value that was passed in by the other one.

21.3.3.3 Assembly Line

When you picture humans working in parallel on a task, one way to break things up is the assembly line. Here you have one group of people who do some part of the work and get products to a certain point. Then they pass their results on to others in a different group who do the next step. This continues until the final result is produced. This arrangement has a critical junction at the point where the product moves from one station to the next. Generally, one person in line can not start working on something until the person before them is done. If there are not any products in the right state, then that person needs to wait until one shows up.

There are two different types in `java.util.concurrent` that will allow you to achieve this behavior, `BlockingQueue[A]` and `BlockingDeque[A]`. The two main methods on both of these types are `put` and `take`. The `put` method will add a new value in. Some implementations of these types will have limited space. If there is not space to add the new value, the call will block until space is freed up. The `take` method will remove a value. If no value is available, then the call will block until one becomes available.

Both of these types are abstract. In the case of `BlockingDeque` there is one implementation provided called `LinkedBlockingDeque`. For the `BlockingQueue` type there are a number of different implementations: `ArrayBlockingQueue`, `DelayQueue`, `LinkedBlockingQueue`, `LinkedTransferQueue`, `PriorityBlockingQueue`, and `SynchronousQueue`. We will talk about queues in general and specific implementations using arrays and linked lists in chapters 24 and 25. After covering that material you might consider revisiting the concept of the `BlockingQueue`.

21.3.3.4 Ticketed Passengers

Some types of work are limited by resources. If you are building a house the number of people involved in driving nails will be limited by the number of hammers that you have. If you are transporting people on a plane, there are a limited number of seats. On a computer, there might be limits to how many threads you want involved in activities that deal with hardware, like reading from the network. There might be some types of processing that are particularly memory intensive for which you need to limit the number of threads doing them at any given time. For these types of situations, the `Semaphore` provides a way to limits how many threads can get to a certain region of code at a given time. Unlike the monitors described with `synchronized`, the `Semaphore` is more flexible than just allowing one thread in at a time.

You can instantiate a `Semaphore` with `new Semaphore(permits)`, where `permits` is an `Int` specifying how many of the resource you have. The main methods for the `Semaphore` are `acquire` and `release`. The `acquire` method will get one of the permits. If none are available, it will block until a different thread calls `release` to make one available. There is also a `tryAcquire` method that returns a `Boolean` that does not block. If a permit was available, it takes it and returns `true`. Otherwise it simply returns `false`.

21.3.3.5 Other Threadsafe Types

Various types in `java.util.concurrent` for working in applications with multiple threads. These types all begin with `Concurrent`. If you are using mutable data types from the Scala libraries, you can mix-in the proper synchronized traits, but because that gives you types where all methods are synchronized, the concurrent versions can provide better performance.

For example, if you are dealing with a mutable map, the Scala libraries would let you do the following.

```
val smmap = new mutable.HashMap[String,Data] with
    mutable.SynchronizedMap[String,Data]
```

This would give you a mutable map that uses a `String` for the key with some type called `Data` for the values where all the methods are synchronized so that it can be safely used across threads. Alternately, you could use the Java libraries and get a `ConcurrentHashMap` like this.

```
val cmmap = new ConcurrentHashMap[String,Data]
```

This version will likely have better performance, but it will lack most of the methods you are used to having on collections such as `map` and `filter`. It will also lack the ability to use indexing syntax to get and set values. Instead, you will have to explicitly call `get` and `put` methods.

21.3.4 Atomic (`java.util.concurrent.atomic`)

There are two packages under `java.util.concurrent`. The first is `java.util.concurrent.atomic`. This package contains a number of different classes that provide you with ways of storing and manipulating data in an atomic manner.

Back in subsection 21.2.2 we looked at a number of different programs that did counting across multiple threads. One of the solutions that we considered in that section synchronized the increments. This was needed because the seemingly simple statement `cnt += 1` is not atomic, it can be interrupted leading to race conditions. Indeed, without synchronization using this statement with a single mutable variable inevitably produced incorrect results.

The problem with using synchronization in that situation was that there is a lot of overhead in checking and locking monitors. The `atomic` package provides us with a slightly lighter weight solution. One of the classes in `java.util.concurrent.atomic` is `AtomicInteger`. Instances of this class store a single integer value. The advantage of using them is that they provide a number of different methods for dealing with that value that operates atomically without doing full synchronization. Here is code that does the counting using `AtomicInteger`.

```
import java.util.concurrent.atomic._
var cnt = new AtomicInteger(0)
val threads:Array[Thread] = Array.fill(100){
  new Thread {
    override def run {
      for(i <- 1 to 1000000) cnt.incrementAndGet()
    }
  }
}
threads.foreach(_.start)
threads.foreach(_.join)
println(cnt.get)
```

In limited testing, this code ran about twice as fast as a version with full synchronization. That is still significantly slower than a version that uses multiple counters and combines the results at the end, but there are some situations where you truly need a single mutable value. In those situations, types like `AtomicInteger` are the ideal solution.

21.3.5 Locks (`java.util.concurrent.locks`)

The other package under `java.util.concurrent` is `java.util.concurrent.locks`. Locks were added to address a shortcoming in using `synchronized`. When you use `synchronized` you are able to lock a monitor and restrict access from other threads for one block of code or one method. However, it is not uncommon to have code where you want to restrict access beginning with one method call and not release it back until some other method is called. There is not a concise way to do this with `synchronized`, but there is with a `Lock`. `Lock` is an abstract type with the methods `lock`, `unlock`, and `tryLock`.

It is much like a `Semaphore` with only one permit. The main concrete subclass is called `ReentrantLock`, which you can make by calling `new ReentrantLock`.

21.4 Parallel Collections

Even working with the concurrency library, parallelism can be hard to do well. For this reason, beginning with version 2.9, Scala has added another layer in the form of parallel collections. These are collections that have their basic operations performed in parallel. Not only do they do their operations in parallel, they have been written in such a way that they will automatically load balance. If one thread finishes its own work load, it will steal work from another thread that has not yet finished. For this reason, if your problem can be solved using parallel collections, you should probably use them. Not only will they be easier to write, they will probably be more efficient than anything you are going to take the time to write yourself.

The structure of the parallel collections largely mirrors the structure of the normal collections with a `scala.collection.parallel` package that has subpackages for `immutable` and `mutable` types. The parallel collections are not subtypes of the normal collections we have used to this point. Instead, they have mutual supertypes that begin with `Gen` for general. The reason for this is that the original collections had an implicit contract to go through their elements in order and in a single thread. A lot of the code we have written did not depend on this, but some of it did. The general types explicitly state that they do not have to preserve that behavior.

The parallel collections have names that begin with `Par`. Not all of the types we used previously have parallel equivalents. This is because some types, for example the `List`, are inherently sequential. If you think back to the last chapter, we saw that there were a whole set of types under the `LinearSeq` type that had linear access behavior. None of those types can be efficiently implemented in parallel because of that. Figure 21.2 shows a Unified Modeling Language (UML) class diagram of the parallel collections.

To convert from a normal collection to a parallel collection call the `par` method on the standard collection. To convert back, call the `seq` method on the parallel collection. These calls are efficient for types that have a parallel equivalent. You can convert a `Range`, a `Vector`, an `Array`, a `HashSet`, or a `HashMap` from sequential to parallel and back in O(1) time. For other types, the conversion will be less efficient as it has to build a completely new collection. Consider the following examples from the REPL.

```
scala> 1 to 10 par
res0: scala.collection.parallel.immutable.ParRange = ParRange(1, 2, 3, 4, 5, 6, 7,
    8, 9, 10)

scala> Array(1,2,3,4) par
res1: scala.collection.parallel.mutable.ParArray[Int] = ParArray(1, 2, 3, 4)

scala> List(1,2,3,4) par
res2: scala.collection.parallel.immutable.ParSeq[Int] = ParVector(1, 2, 3, 4)
```

The first two calls are able to simply wrap the sequential version in a parallel collection. The last one has to convert the `List` over the a `ParVector`. The choice of `ParVector` in this situation is made because preserving immutability is the primary consideration.

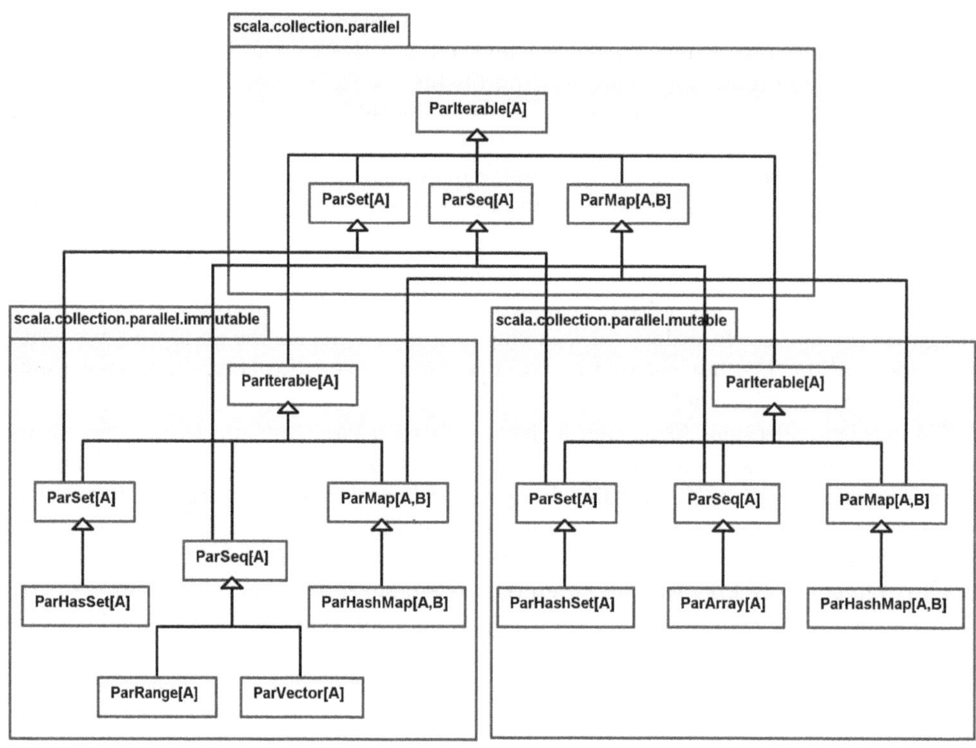

FIGURE 21.2: This UML diagram shows the main types in the three different parallel collection packages.

The parallel collections have all the normal methods that you are used to using. Given this, you can write a parallel version of factorial using `BigInt` with the following code.

```
def parFact(n:BigInt) = (BigInt(1) to n).par.product
```

This version is clearly shorter than the ones that we wrote earlier. What is more, if you test it, you will find that it is faster and scales better because it does not have as much inherently sequential code.

There is one method that is very significant for parallel collections called `aggregate`. In many ways, you can think of this method as being like a parallel version of the `fold` or `reduce` methods. The fact that you specifically call either `foldLeft`, `foldRight`, `reduceLeft`, or `reduceRight` tells you that those operations are sequential in their nature. A parallel version has to be able to run through the elements in whatever order it wants. Different threads will be running through different parts of the collection at the same time. For parallel, there are `fold` and `reduce` methods that will do those operations in an order-independent manner.

The `reduce` method is a full replacement for `reduceLeft` or `reduceRight` in situations where the order in which elements are processed does not matter. The same is not true of `fold`. To see why, compare the signatures of `fold` and `foldLeft` from `ParIterable[T]` in the API.

```
def fold [U >: T] (z: U)(op: (U, U)  U): U
def foldLeft [S] (z: S)(op: (S, T)  S): S
```

The `fold` method is more restricted in the types that it can use. While `foldLeft` could result in any type that you want, `fold` must result in a type that is a supertype of the type the collection contains. The reason for this restriction is that when a `fold` is done in parallel, you can have one thread apply `op` to one section of the collection while another thread does the same for a different section. To get your final answer, you have to be able to combine the results of the two threads. The combining operation only works with the same function if both parameters for the function have the same type as the output of that function.

The `aggregate` method addresses this shortcoming, but doing so requires two separate operators. Here is the signature of the method.

```
def aggregate [S] (z: S)(seqop: (S, T)  S, combop: (S, S)  S): S
```

The `seqop` is equivalent to the `op` that is passed into `foldLeft`. The `combop` is used to combine the results of processing the different pieces. Any time when you would use `foldLeft` or `foldRight` with a result type that is distinctly different from the type stored in the collection and the processing order is not critical to getting the correct answer, you can use aggregate instead.

As a simple example of this, consider the following code that generates a parallel array of 1000 random doubles then finds the min and max values using aggregate.

```
val data = ParArray.fill(1000)(math.random)
val (min,max) = data.aggregate((data(0),data(0)))(
  (mm,n) => (n min mm._1,n max mm._2),
  (mm1,mm2) => (mm1._1 min mm2._1,mm1._2 max mm2._2)
)
```

It would have been much shorter to simply write `data.min` and `data.max`, but each of those calls has to run through the full set of data. This approach only runs through the data once to find both the minimum and maximum values. Given the overhead in creating tuples it is not clear that this is a benefit when we are simply comparing Doubles, but if the values we were comparing were found via a complex calculation on the data, this approach might be beneficial. More significant is that it shows how you can use aggregate to do a fold type of operation in parallel with a result that is distinctly different from the type stored in the collection.

21.5 Introduction to Scala Actors

The creators of Scala are well aware of the challenges of multithreaded code as well as the significance of multithreading to the future of computing. For that reason, they have given a high priority to supporting multithreading in a way that is easy for programmers to use. This comes out clearly in the parallel collections. Even before the parallel collection libraries were added to Scala, there was another library supporting a different type of parallelism, the actors library in scala.actors.

Actors are an approach to parallelism that has been popularized by the Erlang programming language. In this approach, a system is built of multiple actors that communicate by sending messages to one another. You can imagine each actor doing its work in its own thread with the work being done in response to receiving a message.

The full details of actors are beyond the scope of this chapter and will be discussed in chapter 34. The actors package does provide two utility objects that are very helpful and easy to use, so we will describe them. In many situations, these objects will prevent us from having to make and use an ExecutorService. The scala.actors.Actor object has a method called actor with the following signature.

```
def actor (body: => Unit): Actor
```

This method provides a simple way to execute a block of code in a separate thread. The fact that this returns an Actor is not particularly relevant to us at this point. You can use this just like a call to submit on a cached thread executor service when we do not care about the return value. If you do want a return value, you would use the future method of the scala.actors.Futures object instead. This method has a signature like this.

```
def future [T] (body:  T): Future[T]
```

Note that the Future[T] here is a scala.actors.Future, not a java.util.concurrent.Future. When you want the value of this type of object, instead of calling get you simple invoke apply by using it as a function with no arguments.

Both of these functions take a single argument that is passed by name. Remember that Scala lets you use curly braces instead of parentheses for argument lists of a single parameter. The rule exists for just this type of situation. If you have imported scala.actors._ you can start some code in a separate thread just like this.

```
Actor.actor {
  statement1
  statement2
  ...
  statementN
}
```

The same will work with `Futures.future` though you likely want to assign the result to a `val`.

GPUs

Multicore chips are not the only example of computer hardware moving toward parallelism. Another significant trend is the use of the Graphics Processing Unit (GPU) to help with general purpose computing. This is commonly called GPGPU. Higher-end GPUs are typically found on separate graphics cards, though there are integrated graphics on motherboard and beginning around 2010, it became more common to find GPU elements etched onto the same silicon and CPUs.

The idea of doing general purpose calculations on a graphics card might seem odd at first. To see why it works, you need to know something about how graphics cards work. Graphics is basically a lot of math. Early GPUs were built specifically to do just the math that was needed with graphics. They did this in the fastest way possible and did not adhere to floating point number standards because you honestly do not need many digits of accuracy for building an image that is on the order of 1000 pixels across.

Over time, the desire for more realistic graphics capabilities led to the creation of pixel shaders. These allowed programmers to write limited logic that could be run for each pixel in an image or on the screen. This was a significant break from earlier designs as it allowed true programmable logic on the GPU. At first this logic was still very heavily focused on graphics-related calculations, but the fact that a GPU could do floating point math extremely fast tempted people into trying to use pixel shaders for other purposes.

Since then, GPUs have become more flexible and more programmable. The support for GPGPU became a selling point, so hardware manufacturers put in proper IEEE floating point arithmetic. First this was only in single precision, but later double precision was added as well. This has culminated in cards built specifically for GPGPU. These cards have lots of memory and one or more GPU chips in them, but they often do not have any graphics output. They are designed and sold just to support GPGPU style calculations.

While multicore chips have a few cores and there are plans to take them to tens or even a few hundreds, GPUs will support thousands or tens of thousands of threads. You have to break up a problem into a lot of pieces to keep a GPU busy. In addition, the threads on a GPU are not as fully capable as those on the cores of a CPU. Typically GPU threads need to be run in groups that are all doing very nearly the same thing to different data.

At this time there are two main interfaces to support GPGPU-style programming: CUDA® and OpenCL®. CUDA is an API created by NVIDIA®which only runs on NVIDIA GPUs. OpenCL, as the name implies, is an open standard that supports GPU and other high thread count computation models. There is currently preliminary support for OpenCL in Scala through the ScalaCL package. ScalaCL works in large part through an augmented set of collections that split the work up on the GPU instead of over multiple cores like the normal parallel collections do.

If you major in Computer Science, GPGPU is probably a topic that should be on your radar and all indications are that CPUs with integrated GPUs will become common. As was shown at the beginning of this chapter, it was not long ago that most programs did not need multithreading because most computers did not require it to run at full speed. A similar change could be in the works for GPUs and programmers will have to have some understanding of what is happening in the hardware to write programs that take full advantage of the resources they have.

21.6 Multithreaded Mandelbrot (Project Integration)

Now that you have seen many of the different possibilities for multithreading as well as the challenges that it poses, let's write a simple example that uses multithreading for our project. The code that we will write here will draw a Mandelbrot set in the drawing with the user able to specify the range to draw on the real and imaginary axes as well as the size of the region to draw. We will put this in a new class called **DrawMandelbrot** that extends DrawLeaf.

The Mandelbrot set is characterized by the mapping $z_{n+1} = z_n^2 + c$, where $z_0 = 0$ and c is a point in the plane. There is a different sequence for every point in the complex plane. For some points, the sequence will diverge to infinity. For others, it will stay bound. For our purposes here, it is not the math we are concerned with, but making the math happen in parallel while being translated to an image that we can draw.

Calculating the Mandelbrot set is an EMBARRASSINGLY PARALLEL problem. This means that it can be broken into separate pieces that can be solved completely independently. So you can put each piece in a different thread and they do not have to talk to one another. This makes it an ideal candidate for solving with the parallel collections library. This code shows just such a solution.

```
package scalabook.drawing

import java.awt.{Graphics2D,Color}
import java.awt.image.BufferedImage
import swing._
import event._

class DrawMandelbrot(p:DrawTransform) extends DrawLeaf {
  val parent = p
  private var propPanel:Component = null
  private var (rmin, rmax, imin, imax) = (-1.5, 0.5, -1.0, 1.0)
  private var (width, height) = (600, 600)
  private var maxCount = 100
  private var img:BufferedImage = null
  private var changed = false
  private val properties:Seq[(String,() => Any,String => Unit)] = Seq(
    ("Real Min", () => rmin, s => rmin = s.toDouble),
    ("Real Max", () => rmax, s => rmax = s.toDouble),
    ("Imaginary Min", () => imin, s => imin = s.toDouble),
    ("Imaginary Max", () => imax, s => imax = s.toDouble),
    ("Width", () => width, s => width = s.toInt),
    ("Height", () => height, s => height = s.toInt),
    ("Max Count", () => maxCount, s => maxCount = s.toInt)
  )

  def draw(g:Graphics2D) {
```

```
    if(img==null || changed) {
      if(img==null || img.getWidth!=width || img.getHeight!=height) {
        img = new BufferedImage(width,height,BufferedImage.TYPE_INT_ARGB)
      }
      for(i <- 0 until width par) {
        val cr = rmin + i*(rmax-rmin)/width
        for(j <- 0 until height) {
          val ci = imax - j*(imax-imin)/width
          val cnt = mandelCount(cr,ci)
          img.setRGB(i,j,if(cnt == maxCount) Color.black.getRGB else
            new Color(1.0f,0.0f,0.0f,cnt.toFloat/maxCount).getRGB)
        }
      }
    }
    g.drawImage(img,0,0,null)
  }

  def propertiesPanel() : Component = {
    if(propPanel==null) {
      propPanel = new BorderPanel {
        layout += new GridPanel(properties.length,1) {
          for((propName,value,setter) <- properties) {
            contents += new BorderPanel {
              layout += new Label(propName) -> BorderPanel.Position.West
              layout += new TextField(value().toString) {
                listenTo(this)
                reactions += { case e:EditDone => setter(text); changed = true }
              } -> BorderPanel.Position.Center
            }
          }
        } -> BorderPanel.Position.North
      }
    }
    propPanel
  }

  override def toString() = "Mandelbrot"

  private def mandelIter(zr:Double,zi:Double,cr:Double,ci:Double) =
       (zr*zr-zi*zi+cr, 2*zr*zi+ci)

  private def mandelCount(cr:Double,ci:Double):Int = {
    var ret = 0
    var (zr, zi) = (0.0, 0.0)
    while(ret<maxCount && zr*zr+zi*zi<4) {
      val (tr,ti) = mandelIter(zr,zi,cr,ci)
      zr = tr
      zi = ti
      ret += 1
    }
    ret
  }
}
```

The only part of this code that deals with making it parallel is the call to `par` in the outer

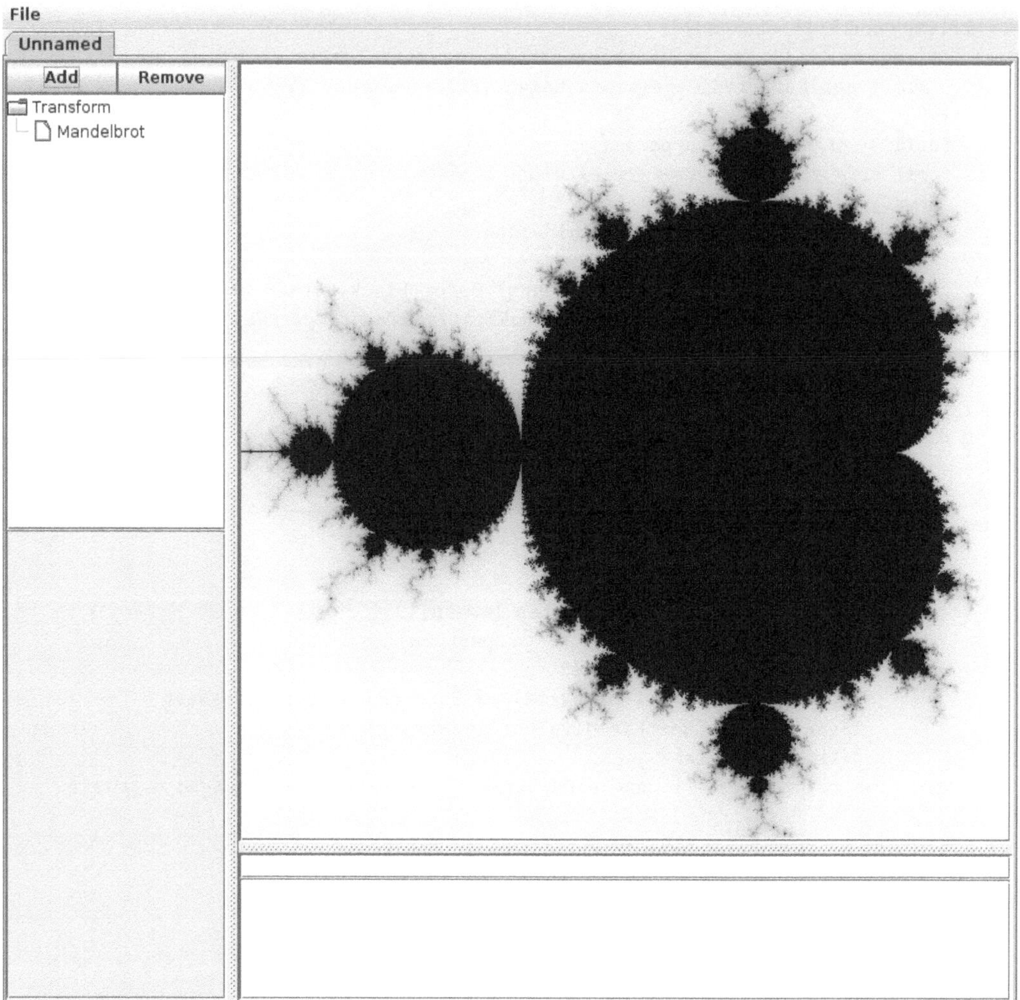

FIGURE 21.3: This is a screen capture of the drawing program with a Mandelbrot image drawn using the default values from the code.

`for` loop in the `draw` method. There is not a call to `par` in the inner loop based on the assumption that whole columns make a fairly good level at which to break things up for threads. To get this into a drawing you need to add the following line into the `addTo` method in `Drawing`.

```
("Mandebrot", () => new DrawMandelbrot(parent)),
```

Once you have done this you can run the program and add a Mandelbrot set. The result should look like figure 21.3.

While the parallelism of this code is not remarkably interesting, there is another part of this code that is worth taking a look at. This is the first `Drawable` type that we have written with a number of different user adjustable properties. To be specific, there are seven different properties for this. They all have a very similar style to them. The original version of this code had a much longer `propertiesPanel` method that contained seven different versions of the code that makes a `BorderPanel` with a `Label` and a `TextField`. The seven different versions were nearly identical except for one string and two variable names. This type of code hopefully strikes you as being less than ideal. It is a form of code duplication that tends to make code brittle and hard to deal with. In chapter 27 we will talk about "smells" associated with code. You might well say that the original code was "smelly". What you see here is one approach at fixing this.

The basic problem was that we had seven nearly identical pieces of code. What we want is one copy of the code that we can put in something like a loop, which abstracts the things that vary across the different versions. This goes back to the push we have in the second half of this book to emphasize abstraction as a tool to help make your programs more powerful and easier to maintain. In this case we introduced a sequence of three tuples. The choice of three values in the tuple came from the fact that there were three places in the original code that were different. The first was the `String` describing the name of the property. The second was where the value of that property was used to initialize the `TextField`. The third was the code that set the property when an `EditDone` event occurred. These became the three elements of the tuple. The type of the tuple is given at the declaration so that we do not have to specify that the argument `s` in the third element of each tuple is a `String`. The second element of each tuple is given as a function instead of just a value so that whenever the Graphical User Interface (GUI) is built, it will get the most up-to-date value.

21.7 Multithreading in GUIs

In reality, you have been writing multithreaded code since chapter 11. This is because code that includes GUIs is inherently multithreaded. You can demonstrate this with the following code.

```
val frame = new swing.MainFrame {
  title = "Thread Demo"
  contents = swing.Button("Click Me")(println("Button Clicked"))
}
frame.visible = true
for(i <- 1 to 20) {
  println(i)
  Thread.sleep(1000)
}
```

This will make a little frame with a single button that you can click. Any time you click the button it will print a message. After making the frame visible, the main thread goes into a `for` loop that counts to twenty with one-second pauses between values. When you click the button, the message is printed between numbers. This works because event handling in GUIs is done in its own thread. This other thread, called the EVENT THREAD, is responsible for handling all events that happen in the GUI and for repainting what is seen on the screen. When we call `repaint`, it does not actually do the painting then. Instead it schedules an event for the event thread so that a repaint will be done the next time it can.

The fact that a single thread is responsible for handling all events and repaints means that you have to be careful what you do in that thread or you can stop all responses from the GUI. The following code demonstrates that fact.

```
val frame = new swing.MainFrame {
  title = "Thread Blocking"
```

```
  contents = swing.Button("Click Me")(Thread.sleep(10000))
}
frame.visible = true
readLine
```

If you click the button, it blocks the event thread from doing anything else for ten seconds. If you run this program you can click the button and see the effects. One effect is that the button will "stay down". More interesting is seeing what happens if you try to close or resize the window. The close will not take effect until the end of the ten seconds. The window will not repaint itself during that time either.

To see this same effect on a larger scale, add the following line into the `commands` map in the `Commands` object in the project code.

```
"freeze" -> ((rest,d) => Thread.sleep(rest.trim.toInt*1000))
```

Now you can run the drawing program, add a drawing, and enter the command "freeze 10". After doing this, nothing you do in the GUI will take effect for ten seconds. This command will be useful in the next section as well to see how blocking the event thread impacts other things that happen in the GUI.

Much of this chapter has focused on the challenges associated with using threads. Those challenges remain for the threads associated with a GUI. As you just saw, keeping the event thread busy too long can lead to an unresponsive GUI. It is also risky to have other threads altering mutable data that is used by the event thread as that can lead to race conditions. This combination makes things a bit tricky because you need to have any significant workload happen in a separate thread, but you should not have that other thread mutate values that are used for painting or event handling. Instead, you should have the other thread notify the GUI using events. This is what happens with calls to `repaint`. Instead of doing the paint in the thread that makes the call, it fires off an event that tells the GUI to do a paint.

All of the different GUI components are `Publishers` so they have a method called `publish` that takes an event. To see how we can use this, we can make the command processing in `Drawing` so that it does the work in a separate thread and notified the GUI using an event. Consider the following alteration to the `commandField` declaration in `Drawing`.

```
val commandField = new TextField() {
  listenTo(this)
  reactions += {
    case e:EditDone =>
      if(!text.isEmpty) {
        val t = text
        Actor.actor {
          publish(new ResultEvent(t,Commands(t,Drawing.this)))
        }
        text=""
      }
    case e:ResultEvent =>
      commandArea append "> "+e.text+"\n"+e.result+"\n"
  }
}
```

This code uses `Actor.actor` to start the command executing in a separate thread. When the command is finished, the result is bundled into a `ResultEvent` along with the text that was executed and that is published. The `commandArea` is updated when the `ResultEvent` is received. To make this work, the following declaration of ResultEvent was added above the propertiesPanel method.

```
private class ResultEvent(val text:String,val result:Any) extends Event
```

The use of publish and having the change to commandArea happen in the event thread means that we can not have the thread created by actor put the GUI in an inconsistent state at a significant time, such as a redraw.

To really see the impact of this code you need to run DrawingMain and add a drawing, then try using the "freeze" command again. What you will find is that the GUI no longer freezes up. If you do "freeze 10" you can keep using the GUI and nothing will happen for ten seconds. At the end of the ten seconds, the output area will show the proper feedback to indicate that the command completed. You can even type in other commands while a freeze command is working. If the other commands finish before the end of the freeze time, their feedback in the output panel will come before that of freeze.

The other place where threading is significant for GUIs is during animation. The easiest, safe approach to doing animation in a GUI was introduced in chapter 12 where we created a javax.swing.Timer and gave it an ActionListener. The ActionListener is a type from the Java event handling libraries so this code will run the timed code in the GUI event thread and not cause any threading problems. We will use this approach again in the following section.

21.8 Animated Bouncing Balls (Project Integration)

To illustrate the use of a javax.swing.Timer in the drawing program we will make a new Drawable that shows balls bouncing around inside of a little box. While it is not really needed at this point, we will also make the code that moves the balls around execute in parallel using parallel collection functionality. Here is the code for this new class.

```
package scalabook.drawing

import java.awt.{Graphics2D,Color}
import java.awt.geom._
import swing._
import event._

class DrawBouncingBalls(p:DrawTransform,drawing:Drawing) extends DrawLeaf {
  val parent = p
  private var propPanel:Component = null
  private var (minSize, maxSize) = (0.01, 0.05)
  private var balls = Vector.fill(20){
    val size = minSize+math.random*(maxSize-minSize)
    new Ball(size+math.random*(1-2*size),size+math.random*(1-2*size),
        (math.random-0.5)*0.02,(math.random-0.5)*0.02,size)
  }

  private case class Ball(x:Double,y:Double,vx:Double,vy:Double,size:Double)

  private val timer = new javax.swing.Timer(100,Swing.ActionListener(e => {
    balls = (for(Ball(x,y,vx,vy,s) <- balls par) yield {
```

```
    var nvy = vy + 0.01 // gravity
    var nvx = vx
    var nx = x+nvx
    var ny = y+nvy
    if(nx-s<0.0) {
      nx = 2*s-nx
      nvx = nvx.abs
    } else if(nx+s>1.0) {
      nx = 2.0-(nx+2*s)
      nvx = -nvx.abs
    }
    if(ny-s<0.0) {
      ny = 2*s-ny
      nvy = nvy.abs
    } else if(ny+s>1.0) {
      ny = 2.0-(ny+2*s)
      nvy = -nvy.abs
    }
    Ball(nx,ny,nvx,nvy,s)
  }).seq
  drawing.refresh
}))

def draw(g:Graphics2D) {
  g.setPaint(Color.black)
  g.draw(new Rectangle2D.Double(0,0,100,100))
  g.setPaint(Color.green)
  for(Ball(x,y,_,_,s) <- balls) {
    g.fill(new Ellipse2D.Double((x-s)*100,(y-s)*100,s*200,s*200))
  }
}

def propertiesPanel() : Component = {
  if(propPanel==null) {
    propPanel = new BorderPanel {
      layout += new GridPanel(1,1) {
        val button = new Button("Start")
        button.action = Action("Start"){
          if(button.text == "Start") {
            timer.start()
            button.text="Stop"
          } else {
            timer.stop()
            button.text="Start"
          }
        }
        contents += button
      } -> BorderPanel.Position.North
    }
  }
  propPanel
}

override def toString = "Bouncing Balls"
}
```

The balls are represented by a case class called `Ball`. The box they are bouncing in goes from (0.0, 0.0) to (1.0, 1.0). The drawing scales that up by a factor of 100 so that it covers a reasonable range of pixels.

The properties panel includes a single button that controls the timer and alternates its text between "Start" and "Stop". The code to move the balls around is found in the timer itself, which is declared at the top of the class. Note the use of patterns on the case class to pull out values.

Currently this code does not have balls bouncing off of one another. That would add significant computation and potentially make the parallelism a real benefit. We will leave that as a task for a later chapter when we can talk about not just how to do it, but how to do it intelligently if the number of particles were to be very large.

21.9 End of Chapter Material

21.9.1 Summary of Concepts

- Multicore processors have become the norm and core counts are growing exponentially without increases in chip clock speed. This means that programs have to include parallelism to take full advantage of the hardware. Multithreading is on the primary ways to doing this.
- Through the Java libraries, Scala programs have access to basic thread operations. You can easily create instances of `java.lang.Thread` and start work running in the various threads.
  - One of the main problems with threads is that they all access the same memory. That means that if different threads are mutating parts of the memory, you can have race conditions where the outcome of the computation depends on the details of thread handling.
  - Another problem you run into with threads is that if you start pausing them to improve their cooperation, you run the risk of making a number of threads freeze up in a state called deadlock where they are all waiting around for the others to finish.
  - The primary way of dealing with race conditions is through synchronization. Critical blocks of code where values are mutated can be synchronized. Only one thread can be executing code that is synchronized on a particular object at any given time. Unfortunately, synchronization is a bit slow and too much of it can lead to deadlock.
  - Another way to control threads to help them work together is with `wait` and `notifyAll`. The `wait` method causes a thread to pause. It will stay paused until a notification wakes it up.
  - The `Thread` class has other methods like `sleep` and `yield` that can be quite useful.
- There are many tasks that are common to a lot of parallel programs. For this reason, a concurrency library was added to Java in the `java.util.concurrent` package.
  - `Executors` and `ExecutorServices` abstract away the task of picking a method

for running a piece of code. There are some useful methods in `Executors` that will provide `ExecutorService` objects which behave in generally useful ways. Many of these allow for the reuse of threads as there is significant overhead in creating `Thread` objects.

  - The `Callable` type allows you to run a task on an `ExecutorService` that gives you back a value. When such a task is submitted, you are given a `Future` object that will provide the result.
  - The concurrency library includes a number of parallel data structures as well.
    * The `CyclicBarrier` and `CountDownLatch` `classes` make it easy to have multiple threads pause at a point in the code until all threads have gotten there.
    * An `Exchanger` can be used to pass a value between threads at some particular point.
    * The `BlockingQueue` is a data structure that lets one group of threads exchange objects with another group of threads and blocks in situations when there is not space to put something or nothing is available.
    * A `Semaphore` provides a way to control how many threads are working in a particular part of the code at any given time.
  - The `java.util.concurrent.atomic` package provides `classes` for basic types that have operations which can not be interrupted. These often provide a simpler and more efficient approach to preventing race conditions than synchronization.
  - The `java.util.concurrent.locks` package defines locks that can be used for synchronization type behaviors that can span beyond a single method call or code block.

- Scala makes certain types of parallelism simple with parallel collections. These collections have methods that automatically split the work across multiple threads. Calling `par` on a regular collection gives you a parallel version. Calling `seq` on the parallel version gives you back a sequential one.
- There is another parallel library in Scala for Actor-based parallelism. For our current purposes, it provides easy, Scala-style methods for creating threads using `Actor.actor` and `Futures.future`.

21.9.2 Exercises

1. In the discussion of the parallel factorial and Amdahl's law we concluded that the primary limitation in the code we had written for scaling was the final product. Using techniques from `java.util.concurrent`, write code to make that operation happen in parallel.
2. A good way to make sure you understand parallelism is to work on implementing a number of different simple problems using each of the different techniques that were presented in this chapter. The problems you should solve are listed below. For each of them, try to write a solution using the following methods.
   - Normal `java.lang.Threads` or calls to `Actor.actor` with `synchronized` and `wait/notifyAll`.
   - Elements from `java.util.concurrent` such as `ExecutorServices`, thread-safe data structures, and atomic values.

- Scala parallel collections. (Note that this is is not flexible enough to handle all problems so only use it for those that work.)

For each solution that you write, you should do timing tests to see how the different solutions compare for speed using the highest core count machine you have access to. For speed comparisons, you should consider writing a sequential version as well. For most of these to be of interest for parallel, the inputs will need to be fairly large.

(a) Find the minimum value in a sequence of `Doubles`.

(b) Given a `Seq[Int]`, find the value for which the following function returns the largest value.

```
def goldbach(num:Int):Int = {
  var ret = 0
  var n = num
  while(n>1) {
    n = if(n%2==1) 3*n-1 else n/2
    ret += 1
  }
  ret
}
```

(c) Matrix addition using two matrices stored as `Array[Array[Double]]`.

(d) Matrix multiplication using two matrices stored as `Array[Array[Double]]`.

(e) Find all occurrences of a substring in a given `String`.

21.9.3 Projects

As you would expect, all of the projects for this chapter have you adding multithreading to your project. How you do this depends a lot on the project. There are two aspects of this that are worth noting in general. First, you can not generally multithread drawing to the screen. That is because the graphics is being done by a single device. Often, drawing to different images in parallel will not help either because drawing to images often involves the graphics pipeline on your graphics card so it is a bottleneck, even when you have separate threads doing separate work on separate images.

The other thing to note is not to overdo multithreading. Spawning many threads with little work for each to do will only slow your program down. Spawning too many can even consume enough resources to bring your machine to its knees. It is easy to spawn threads with things like `Actor.actor`, but you need to think about what you are doing when you use it. Overuse of threading can also lead to race conditions that are extremely difficult to debug. Appropriate usage of the thread-safe collections from `java.util.concurrent` can help dramatically with that problem.

1. Text adventures typically did not do anything unless the user entered a command. The code you have written at this point is probably the same. The call to `readLine` is blocking and nothing happens without that. This is not the case with a MUD. For one thing, there will be multiple players that could all be trying to do things at different rates. In addition, the other characters in a MUD have the ability to walk around even if the user is simply standing in one place doing nothing.

   Now that you know how to multithread, you can change this behavior. You will have the logic for the game run in one thread and the task of collecting input happen in another. The game thread will be like a GUIs timer. It will run through updating everything that needs to be updated, then sleep for a while before doing it again. The updates of the other characters can just be random walks at this point though you should consider how you will add combat or other features as well. The update of the player will interact with the input reading thread in some thread-safe manner. Some of the collections from `java.util.concurrent` could come in handy for this.

The thread the takes user input will still block on `readLine`. When the user inputs a line, that line will be put into whatever you are using to pass information across so that the main game thread can get it and execute it the next time through. The input thread then loops around and does another `readLine`.

2. Networking, like reading from and writing to a disk, can be a significant bottleneck. Unlike the disk though, it is possible that the bottleneck is not related to your machine. For this reason, there can be significant speed benefits to having multiple threads reading from the network all at once. If you are working on the web spider, this is exactly what you should do for this project. You can also set the stage for multiple threads processing data as well, in case that becomes a bottleneck.

   To do this, you can basically make two pools of threads, one that reads data and another that processes the data. Over time you can balance the number of threads in each to get maximum performance. Data needs to be passed between these threads in a thread-safe manner. One or more `java.util.concurrent.BlockingQueue`s would probably work well. As the data collecting threads pull down the data and get them into a form with minimal processing, they put their results on a queue. The analysis threads pull data off the queue and do the proper analysis on it. In the final version of this, the analysis will inevitably find more data that needs to be collected. That will be put on a different queue that the reading threads will pull from.

   You can stick with pulled data from files for this project, though it is unlikely the threading will help in that situation or that you have a large enough data set. In the next two chapters we will introduce concepts that will make it easier to pull the data from the web or other network sources.

3. If you are writing a networked game, you will be forced to put in multithreading to deal the networking. However, you can start preparing for that now. You can put in code to handle two players from the keyboard where the different players get messages from a single keyboard handler and a main thread runs the primary game logic. You might also be able to think of ways to split up the game logic across multiple threads.

4. For a non-networked game, the threading will primarily be done to help with the computer-controlled entities in the game. Care must be taken to avoid race conditions, but you should look for intelligent ways to split up the work across different threads. This could include grouping active entities if they have complex decision-making needs. It could also be just breaking things up by type. Note that the second option generally does not scale to high core counts as well.

5. If you are doing the Math worksheet problem, you have quite a few options for parallelism. Worksheets do not have to be linear. If one section of a worksheet does not depend on the results of another, they can be processed in parallel. In addition, complex calculations can often be done in parallel by working on different data values using different threads.

   If you have not done so already, you should consider integrating code for formula parsing and evaluation like that from project 16.1 into this program. The fact that the `String` parsing evaluator written there is remarkably slow provides a motivation for trying to do multiple evaluations in separate threads.

6. The Photoshop® project runs into some challenges with multithreading because most of the work that it is doing involves graphics and, as was described in the introduction to these projects, normal graphics does not generally benefit from multithreading. It is probably worth you spending a little bit of time to test this hypothesis though. You could render layers to separate `BufferedImage`s using separate threads. If you make the scenes complex enough, you can really see if multithreading provides any benefit.

   There are some other things that could be done which would definitely benefit from multithreading. The `getRGB` and `setRGB` methods on `BufferedImage` to not use normal graphics operations. They work with direct access to the memory behind the image. For this reason, any operations that you can do using only those methods will benefit from multithreading. An example of something that can be done this way is filters. Many effects that you might want to apply to an image can be done with only the Red, Green, Blue (RGB) values of a pixel and those around it.

   Depending on what projects you did in the first half of the book, you might want to add some more complex and interesting elements to the things you can add to drawings. That could include fractals like the Mandebrot shown in this chapter, Julia sets, L-Systems, or even ray-tracing elements. All of those could benefit from some or all parts being done in parallel.

7. The simulation workbench is also a plentiful area for parallelization because it is processor intensive and can be arranged in ways that do not conflict. This does not mean it is easy to do efficiently though. Using the simple gravity simulation, devise at least three different ways of calculating the forces in parallel, then do a speed comparison between those approaches and your original code.

Chapter 22

Stream I/O

In chapter 9 we saw how to use `scala.io.Source` and `java.util.Scanner` to read from files. We also saw how we could use `java.io.PrintWriter` to write out files. In this chapter we will explore the overall structure of the `java.io` package and see how streams can be used to represent a general approach to input and output. This will set us up in the following chapter to use streams for network communication.

22.1 The `java.io` Package

The original library for doing input and output in Java was the `java.io` package. While there is now a `java.nio` package that can provide higher performance, the original library is still used as a foundation for many other parts of the Java libraries, which makes it worth knowing and understanding.

The majority of the classes in the `java.io` package are subtypes of one of four different abstract types: `InputStream`, `OutputStream`, `Reader`, and `Writer`. The `InputStream` and `OutputStream` have basic operations that work with bytes. The `Reader` and `Writer` have almost the same set of methods, but they work with characters. Here are the signatures of the method in `InputStream`.

```
available():Int
close():Unit
```

```
mark(readlimit:Int):Unit
markSupported():Boolean
read():Int // This method is abstract
read(b:Array[Byte]):Int
read(b:Array[Byte], off:Int, len:Int):Int
reset():Unit
skip(long n):Long
```

The basic `read()` method is abstract. The other read methods have default implementations that depend on this one, though they can be overridden to provide more efficient implementations as well. The `read()` method returns the value of the next byte in the stream giving an `Int` in the range of 0-255 if the read succeeds. If the end of the stream has been reached, it returns -1. The methods for `Reader` are very similar. One major difference is that the read methods take an `Array[Char]` instead of an `Array[Byte]`. We will be focusing on the streams in this chapter. Refer to the Java Applications Programming Interface (API) for full details on `Reader` and `Writer`.

The term STREAM comes from the way that data is accessed. The data is in a flow and you can see what is there, but you can not really jump around to random places. Imagine water flowing through a stream. You can see the water as it goes by or you can look away and skip it. However, once it has passed you do not get the option to repeat seeing that same water. That is the idea of a stream. Note that the `InputStream` has methods called `markSupported`, `mark`, and `reset`. For certain streams, these methods will allow you to effectively cheat a bit and jump backward in a stream. However, not all streams will support this behavior and even in those that do, you are constrained to only being able to return to the point that was previously marked.

The `OutputStream` has a slightly shorter set of methods. Like the `InputStream`, it works with bytes. In this case, writing data to the stream is like throwing something into flowing water. It gets whisked away and you do not have the option of making alterations. Here are the methods for `OutputStream`.

```
close():Unit
flush():Unit
write(b:Array[Byte]):Unit
write(b:Array[Byte], off:Int, len:Int):Unit
write(b:Int):Unit // This method is abstract
```

Once again, there is one critical method, in this case it is `write(b:Int)`, that is `abstract`. While it takes an `Int`, only the lowest eight bits (the lowest byte) will be written out. The other 24 bits are ignored.

22.2 Streams for Files

The `InputStream` and `OutputStream` do not actually have any code to read from/write to anything. This is why they are `abstract`. Their sole purpose is to serve as base classes for inheritance. The subtypes are more specific about what source/sink of data they are attached to. The most obvious things to attach streams to are files. The `FileInputStream` and `FileOutputStream` are subtypes that implement the `read`/`write` methods so that they work with files. You instantiate these types using `new` and passing in the file you want either as a string or as an instance of `java.io.File`. Here is a simple example that illustrates using a `FileInputStream` to read a file. It prints each byte as a numeric value.

```
import java.io._

val fis = new FileInputStream(args(0))
var byte = fis.read()
while(byte>=0) {
  print(byte+" ")
  byte = fis.read()
}
println()
fis.close()
```

Running this program on itself produces the following output.

```
> scala ReadBytes.scala ReadBytes.scala
105 109 112 111 114 116 32 106 97 118 97 46 105 111 46 95 10 10 118 97 108 32 102
105 115 32 61 32 110 101 119 32 70 105 108 101 73 110 112 117 116 83 116 114 101
97 109 40 97 114 103 115 40 48 41 41 10 118 97 114 32 98 121 116 101 32 61 32 102
105 115 46 114 101 97 100 40 41 10 119 104 105 108 101 40 98 121 116 101 62 61 48
41 32 123 10 32 32 112 114 105 110 116 40 98 121 116 101 43 34 32 34 41 10 32 32
98 121 116 101 32 61 32 102 105 115 46 114 101 97 100 40 41 10 125 10 112 114 105
110 116 108 110 40 41 10 102 105 115 46 99 108 111 115 101 40 41 10
```

A slight modification will produce a script that can be used to copy files.

```
import java.io._

val fis = new FileInputStream(args(0))
val fos = new FileOutputStream(args(1))
var byte = fis.read()
while(byte>=0) {
  fos.write(byte)
  byte = fis.read()
}
fis.close()
fos.close()
```

This script takes two arguments, just like the Linux `cp` command. It uses the first as the file name for the `FileInputStream` and the second for the name of the `FileOutputStream`. Instead of printing the bytes as numbers to standard output, they are written to the output stream. When everything is done, both files are closed.

Wrapping an `InputStream` in a `Source`

In chapter 9 we argued that one advantage of the `scala.io.Source` type was that it was a Scala collection. This meant that all the methods we have gotten comfortable with when using other collections will work on it as well. If you have an `InputStream` that is reading text you can take advantage of this benefit. Instead of calling `Source.fromFile(name:String)`, you can call `Source.fromInputStream(is:InputStream)` to get a `Source` that will read from the specified InputStream. Much of the code we will use with streams will be reading binary data instead of text data, but you should know that this option exists.

22.3 Exceptions

If you run one of the scripts and provide a name for the input file that is not valid you will get something that looks somewhat like the following.

```
java.io.FileNotFoundException: notthere.txt (No such file or directory)
        at java.io.FileInputStream.open(Native Method)
        at java.io.FileInputStream.<init>(FileInputStream.java:138)
        at java.io.FileInputStream.<init>(FileInputStream.java:97)
        at Main$$anon$1.<init>(ReadBytes.scala:3)
        at Main.main(ReadBytes.scala:1)
        at Main.main(ReadBytes.scala)
        ...
```

You have probably seen quite a few outputs like this in working with Scala. This is a stack trace and it was printed out because something bad happened that caused the program to throw an exception. To this point we have generally ignored exceptions and expected the user or the computer to behave in a way that would allow the program to run without errors. It is time to learn how to deal with these situations instead of ignoring them and hoping they will not occur.

Exceptions are meant to occur in exceptional circumstances, such as when there is an error. In the example above, you can see that it was a `java.io.FileNotFoundException` that occurred. This should make sense given that we tried to open a file that was not there. Real applications need to be able to deal with this. For example, if a user specified a file that did not exist, the program should tell the user that and perhaps ask for an alternate file. It should not crash. In order to prevent the crash, we need a construct that has the ability to execute some alternate code when an exception occurs. The construct for doing this is the `try`/`catch`.

To understand this construct, it helps to have some terminology. The act of raising an exception is described by the term "throw". You can use the `throw` keyword to indicate when an exception occurs in your own code. Here we demonstrate this in the REPL.

```
scala> throw new Exception("Testing")
java.lang.Exception: Testing
        at .<init>(<console>:8)
        at .<clinit>(<console>)
        at .<init>(<console>:11)
        at .<clinit>(<console>)
        ...
```

You can throw any subtype of the `java.lang.Throwable` class. The two immediate subtypes of this are `java.lang.Error` and `java.lang.Exception`. Typically an `Error` represents something that you will not be able to recover from. An `Exception` is something that you should consider dealing with. While this example throws an instance of `Exception` itself, that is not something you should do normally. Instead, you should throw a subtype that matches what has happened. If such a subtype exists in the libraries, use it. If not, create your own subtype.

You can also see here that the `Exception` can be passed a string argument. One of the main advantages of exceptions over other ways of dealing with errors is that exceptions can be informative. The ideal exception gives the programmer all the information they need to fix the error. All exceptions give you a stack trace so that you can see what line caused the exception and how that was reached. Often you need additional information to know what went wrong. In the case of the `FileNotFoundException`, the name of the file is critical information. If you try to access a value that is out of bounds on an `Array`, knowing what index was being accessed and how big the `Array` really was are critical pieces of information. So when you throw an exception or make a new subtype of `Exception`, you should make sure that it has all the information that will be needed to fix it.

22.3.1 try-catch-finally

Any time you open a file that was provided to you by the user, you know that something might go wrong and it might throw an exception. To stop that exception from crashing the thread that it is running in, you tell Scala that you are going to "try" to execute some code. If an exception is thrown that you want to deal with, you should "catch" it. The `try` keyword is followed by the block of code that you want to try to run. After the block of code you put the `catch` keyword followed by a block of cases that include the handlers for different exceptions that might arise. In the case of opening a file and printing the contents, code might look like this.

```
import java.io._

try {
  val fis = new FileInputStream(args(0))
  var byte = fis.read()
  while(byte>=0) {
    print(byte+" ")
    byte = fis.read()
  }
  println()
  fis.close()
} catch {
  case ex:FileNotFoundException =>
    println("The file "+args(0)+" could not be opened.")
  case ex:IOException =>
    println("There was a problem working with the file.")
}
```

This has two different `case`s in the `catch`. The first deals with the exception we saw before where the file could not be opened. This can happen even if the file exists if you do not have permission to read that file. The second `case` catches the more general `IOException`. This is a supertype of `FileNotFoundException` and represents any type of error that can occur with input and output. In this case, it would imply that something goes wrong with the reading of the file. Other than hardware errors it is hard to picture anything going wrong with this particular code because it makes no assumptions about the structure of the data in the file. It simply reads bytes until it runs out of data to read.

This code has a problem with it that would be even more significant if we were doing more interesting things with the reading. The problem arises if the file opens, but an exception occurs while it is being read. Imagine that an `IOException` is thrown by the `fis.read()`

call in the `while` loop because something happens to the file. The exception causes the `try` block to stop and control jumps down to the proper `case` in the `catch`. After the code in that `case` is executed, the program continues with whatever follows the `catch` block. This means that the code in the `try` block after the location where the exception occurs is never executed. This is a problem, because that code included the closing of the file. Failure to close files can be a significant problem for large, long-running programs as any given program is only allowed to have so many files open at a given time.

What we need to fix this problem is a way where we can specify a block of code that will run after a `try` block regardless of what happens in the `try` block. This is done with a `finally` block. You can put a `finally` block after a `catch` block or directly after a `try` block. To do our file reading problem correctly we could use the following code.

```
import java.io._

try {
  val fis = new FileInputStream(args(0))
  try {
    var byte = fis.read()
    while(byte>=0) {
      print(byte+" ")
      byte = fis.read()
    }
    println()
  } catch {
    case ex:IOException =>
      println("There was a problem working with the file.")
  } finally {
    fis.close()
  }
} catch {
  case ex:FileNotFoundException =>
    println("The file "+args(0)+" could not be opened.")
}
```

This has one `try` block that contains everything, and inside of that is another `try` block that reads the file with a `finally` for closing the file. This is a bit verbose. We will see shortly how we can set things up so that we do not have to repeat this.

As with nearly every other construct in Scala, the `try-catch-finally` has a value and can be used as an expression. If the `try` block executes without throwing an exception then the value will be that of the `try` block. If an exception is thrown and caught, the value of the whole expression is the value of the `catch` for that `case`. Any value for `finally` is ignored as `finally` is typically used for clean-up, not part of the normal computation.

There is also the possibility that the `try` block executes and throws an exception that is not caught. In this situation the value of the full `try` expression is `Nothing`. You might remember this type from figure 19.1. It is at the very bottom of the type diagram. The `Nothing` type is a subtype of everything else. There are no instances of `Nothing`. It is mostly used by the type inference system. When trying to figure out the type of a complex expression with multiple possible return values, Scala picks the lowest type in the hierarchy that is a supertype of all the possibilities. Since a `try` can throw an exception that is not caught, it is almost always possible for it to produce a `Nothing`. The fact that `Nothing` is at the base of the type hierarchy means that it does not impact the inferred type.

If you use a `try-catch-finally` as an expression, you have to be careful of the types of the last expressions in each of the different cases of the `catch`. In particular, you probably do not want the last thing in any of the cases to be a print statement. Calls like `println` have a return type of `Unit`. Unless all the options are `Unit`, the inferred type will almost certainly be `AnyVal` or `Any`. Those are types that do not provide you with significant functionality. The `catch` `cases` of a `try` that is used as an expression probably should not print anything. If it should be an expression, then it should communicate results with a return value instead of through side effects.

22.3.2 Effect of Exceptions

So far we have treated an uncaught exception as causing the thread to terminate. This is not really the case. When an exception is thrown in a method/function, it causes execution of the current block to stop and control jumps either to the `catch` of an enclosing `try` block, or out of that method/function all together. If there is not an enclosing `try`, or if the `catch` on that `try` does not have a `case` for the exception that was thrown, the exception will pop up the stack to the method that called the current function. If the call point in that function is in a `try` block, the `cases` for that `try` block are checked against the exception. This process continues up the call stack until either a `case` is found that matches the exception or the top of the stack is reached and the thread terminates.

The fact that exceptions propagate up the call stack means that you often do not catch them at the place where they originate. Instead, you should catch exceptions at a place in the code where you know how to deal with them. Often this is several calls above the one that caused the exception to occur.

22.3.3 Loan Pattern

The need to double nest `try` expressions with a `finally` to properly handle closing a file is enough of a pain that it would prevent most programmers from putting in the effort to do it. This same problem existed in Java so they altered the language with an enhanced `try-catch` in Java 7. The syntax of Scala lets you streamline this code without altering the language by making it easy to implement the *loan pattern* [9]. Here is code that does the reading we have been doing in this way.

```
import java.io._

def useFileInputStream[A](fileName:String)(body:FileInputStream=>A):A = {
  val fis = new FileInputStream(fileName)
  try {
    body(fis)
  } finally {
    fis.close()
  }
}

try {
  useFileInputStream(args(0))( fis => {
    var byte = fis.read()
    while(byte>=0) {
      print(byte+" ")
      byte = fis.read()
    }
    println()
  } )
} catch {
```

```
  case ex:FileNotFoundException =>
    println("The file "+args(0)+" could not be opened.")
  case ex:IOException =>
    println("There was a problem working with the file.")
}
```

Here the `try` with the `finally` has been moved into a separate function called `useFileInputStream`. This function takes a file name in one argument list and a function in a second. The function is used after that with a `try` that is only responsible for handling the errors. There are still two `try` statements, but the one in the `useFileInputStream` only has to be written once. The rest of the code can call that function and safely include only a single `try-catch` without a `finally`.

While it is not used in this example, the `useFileInputStream` function takes a type argument that allows it to be used as an expression. The type is inferred from the code in the `body` function.

22.4 Decorating Streams

The `InputStream` and `OutputStream` types only allow reading/writing of single bytes or `Arrays` of bytes. This is technically sufficient to do anything you want as all the data on a computer is stored as sequence of bytes when you get down to a certain level. Being sufficient though does not mean that it is ideal. What do we do if we are reading a stream that is supposed to contain a mixture of different data that includes things like `Int` or `Double` and not just `Byte`. You can build an `Int` from four bytes, but you do not want to have to write the code to do so yourself, especially not every time you need that functionality. This limitation of the basic streams, and others including the file streams, are addressed by decorating streams.[1]

Types like `FileInputStream` know what they are pulling data from, but they are not very flexible or efficient. This shortcoming is addressed by having stream types that are constructed by wrapping them around other streams. This wrapping of one stream around another to provide additional functionality is the decorating.

22.4.1 Buffering

One of the shortcomings of `FileInputStream` and `FileOutputStream` deals with performance. Accessing files is a slow process, it is one of the slower things that you can do on a computer. Reading one byte at a time is particularly slow. It is much better to read a large block of data than to read all the bytes, one at a time. This is the reason why `BufferedInputStream` and `BufferedOutputStream` were created. These types have the same methods as the standard `InputStream` and `OutputStream`. The only difference is how the `read`/`write` method is implemented.

Both types keep a buffer in memory. In the `BufferedInputBuffer` if that buffer has unread bytes in it, the read pulls from the buffer. When the buffer runs out of data, a full buffer is read at once. For the `BufferedOutputStream`, the bytes that are written go into the buffer until it is full, then all of them are dumped out at once. If you need to force things to go out before the buffer is full you can call the `flush` method.

[1]This terminology comes about because the `java.io` library uses the *decorator pattern*.[6]

The buffered types do not actually read from or write to any particular source or sink. In order to use one of the buffered types you need to give it a stream to pull data from or push data to. The syntax for doing that could look like this.

```
val bis = new BufferedInputStream(new FileInputStream(fileName))
```

The fact that disk access has really high latency means that you should always wrap file streams with buffered streams. The speed boost of doing so for large files can be remarkable.

Latency and Bandwidth

To understand why it is better to read/write a big block of data than to read separate bytes, you need to know something about how we characterize communication speeds. When we talk about how fast data can move inside of a computer or between computers there are two values that are significant, latency and bandwidth. Most of the time you hear people refer to bandwidth. This is how much data can be moved in a particular period of time. Bandwidths often fall in the range from a few Mb/s to a few Gb/s.[2]

The bandwidth value is only relevant in the middle of a block of communication. Every time a new communication starts, there is a pause called latency. The amount of time it takes to read a block of N bits is roughly given by $time = latency + N/bandwidth$. If you read your data in one byte/eight bit increments the second term is small and you spend almost all your time waiting for the latency. Reading larger blocks minimizes the latency overhead and gives you an actual speed closer to the full bandwidth.

22.4.2 Binary Data

The other drawback of the standard `InputStream` and `OutputStream` is that they only work with bytes. Pretty much all applications will need to work with data in some more complex format. We have seen how to do this using flat text files and using XML. When we introduced XML in chapter 14, we looked at several of the shortcomings of flat text files. The XML format put more information in the file so that anyone could pick it up and have a chance of figuring out what was in it. It also increased flexibility. XML did not address the problems of speed and using a significant amount of memory. If anything, it made these worse. In order to get speed and to use a small amount of space we have to sacrifice the benefits of XML and write the data in a format that matches how it is stored in the memory of the computer. We need to store the data in binary format.

The first way to read/write data in binary format is with the `DataInputStream` and `DataOutputStream`. These types get wrapped around other streams like the `BufferedInputStream` and `BufferedOutputStream`. What they give us is a set of additional methods. For the `DataInputStream` these methods include the following.

```
readBoolean():Boolean
readByte():Byte
readChar():Char
readDouble():Double
readFloat():Float
readInt():Int
```

[2]These are short for megabits per second and gigabits per second.

```
readLong():Long
readShort():Short
readUTF():String
```

Each of these reads the specified data value from the stream that is wrapped. The `DataOutputStream` has matching methods for writing data. It also has a few extras that can write strings in other ways.

```
writeBoolean(v:Boolean):Unit
writeByte(v:Int):Unit
writeBytes(s:String):Unit
writeChar(v:Int):Unit
writeChars(s:String):Unit
writeDouble(v:Double):Unit
writeFloat(v:Float):Unit
writeInt(v:Int):Unit
writeLong(v:Long):Unit
writeShort(v:Int):Unit
writeUTF(str:String):Unit
```

The combination of these methods gives you the ability the write data to a file and read it back in.

If you are working with files for your binary data you still really need to have buffering for performance reasons. The beauty of the way the `java.io` streams library works is that you can decorate streams however you want. In this situation you want to wrap the data stream around the buffered stream, which is wrapped around the file stream. Code for that looks like the following.

```
val dis = new DataInputStream(new BufferedInputStream(new FileInputStream(file)))
```

The order is significant here, mainly because the methods we want to be able to call are part of the `DataInputStream`. The general rule is that the outermost type needs to have the methods that you want to call. The ones between that and the actual source/sink stream should implement the basic methods in an altered fashion, such as buffering. Those should be stacked in an order that makes sense for the application.

The challenge in working with binary data files is that they can not be easily edited with any standard programs. To understand this, consider the following code.

```
import java.io._

def withDOS[A](fileName:String)(body:DataOutputStream=>A):A = {
  val dos = new DataOutputStream(new BufferedOutputStream(new
      FileOutputStream(fileName)))
  try {
    body(dos)
  } finally {
    dos.close()
  }
}

def withDIS[A](fileName:String)(body:DataInputStream=>A):A = {
  val dis = new DataInputStream(new BufferedInputStream(new
      FileInputStream(fileName)))
  try {
    body(dis)
```

```
    } finally {
      dis.close()
    }
  }

  def writeDoubleArray(fileName:String,data:Array[Double]) {
    withDOS(fileName)(dos => {
      dos.writeInt(data.size)
      data.foreach(x => dos.writeDouble(x))
    } )
  }

  def readDoubleArray(fileName:String):Array[Double] = {
    withDIS(fileName)(dis => {
      Array.fill(dis.readInt)(dis.readDouble)
    } )
  }
```

This code contains two functions that can be used to generally work with `DataOutputStreams` and `DataInputStreams`. In an application, these should probably be methods in an object that contains a number of such utility methods. The other two methods use those first two and write out or read back in an `Array` of `Doubles`. If you load that into the REPL, you can test it with the following.

```
scala> writeDoubleArray("data.bin",Array.fill(10)(math.random))

scala> readDoubleArray("data.bin")
res3: Array[Double] = Array(0.6609985904587437, 0.49319578039338174,
    0.5859229163803784, 0.965200016272522, 0.4832664731158809, 0.7969665973550756,
    0.7572733105097633, 0.1578906904549643, 0.08416227543386434,
    0.6824199253206102)
```

After you have done this you should go look at the contents of the "data.txt" file. What you will find using `less`, `cat`, or `vi` is that the contents look like random garbage characters. If you were to edit the file using `vi` then try to read it in with the `readDoubleArray` the data will almost certainly be messed up if it manages to read at all.

The problem with looking at a file like "data.bin" is that normal characters only account for a fairly small fraction of the possible values that each byte can take. Binary data tends to use all possible values, including many that do not print well. There are command-line tools like `hexdump` and `xxd` that can be used to view binary files. The following shows the output of `xxd`.

```
> xxd data.bin
0000000: 0000 000a 3fe5 26e6 8417 1dea 3fdf 9085 ....?.&.....?...
0000010: 08d4 253c 3fe2 bfe1 6a7a 94d5 3fee e2eb ..%<?...jz..?...
0000020: 24ff a71e 3fde edd6 8052 4d5c 3fe9 8024 $...?....RM\?..$
0000030: 17f4 9f55 3fe8 3b95 3cd8 bd20 3fc4 35c3 ...U?.;.<.. ?.5.
0000040: 1bec 683c 3fb5 8ba8 ac8b 9ec8 3fe5 d662 ..h<?.......?..b
0000050: 4fac 814c                                O..L
```

The first column shows the position in the file. The next eight columns show hexadecimal values for the contents of the file. There are two characters for each byte because $256 = 16^2$. So each line shows 16 bytes. The last column shows the ASCII characters with any non-printable characters appearing as a dot. The `xxd` tool also does a reverse encoding. You can have it output to a file, edit the hex section of the file, and run the reverse encoding to get back a binary file. Doing so is a somewhat delicate operation because it is easy to mess up.

Big Endian vs. Little Endian

The data in a file produced by a `DataOutputStream` probably does not exactly match what was in the memory of your computer because Java libraries write the data out in a platform-independent way. The discussion of binary data from chapter 3 is a generally accurate description of binary arithmetic, and it is perfectly correct at the byte level. However, when computer makers start laying out bytes in memory for larger groups like `Int` or `Double`, different computer makers picked different orders. For the x86 chips, Intel put the least significant byte first. This order is called Little Endian. Most other chip makers used the opposite ordering, called Big Endian. (The terms Big Endian and Little Endian are a reference to Gulliver's Travels where the Lilliputians were fighting a bitter war over which end of an egg one should crack when eating a soft-boiled egg.)

Looking closely at the hex dump above, you can see that Java writes files out in Big Endian order. You can tell this because we know that the first thing in the file is the `Int` value 10. An `Int` is stored in four bytes so the hex is the eight characters "0000000a". If the file were written using Little Endian this would be "0a000000". The reason Java uses this format, even though your computer inevitably does not (as you are most likely using an x86 based machine to run Scala), is inevitably related to the fact that when Sun®created Java, they had their own SPARC®architecture that was Big Endian.

22.4.3 Serialization

Often when you are saving data to a file, you want to write out whole objects. The process of converting an object to some format that you can write out and read back in later is called serialization. Java and a number of other modern platforms have built in serialization methods. The Java serialization method uses a binary output format. You can also write your own code to serialize objects in other formats. XML happens to be a useful serialization format. We will look at each of these separately.

22.4.3.1 Binary Serialization

The Java platform has a rather powerful form of serialization that is built into the system. Other systems have different ways of supporting this. To make a type that can be serialized in Scala have it inherit from `Serializable`.[3] The reason you would want to have things that are serializable is that they can be used with `ObjectInputStream` and `ObjectOutputStream`. These types have the same methods as `DataInputStream` and `DataOutputStream` plus `readObject():AnyRef` and `writeObject(obj:AnyRef):Unit`, respectively.

To help illustrate this, consider the following little application that includes a serializable class and some code to either write an instance of it to a file or read one in from a file and print it.

[3]Case classes automatically inherit from `Serializable`.

```
import java.io._

class Student(val name:String,val grades:Array[Int]) extends Serializable

object Main {
  def main(args:Array[String]) {
    args(0) match {
      case "-r" =>
        val ois = new ObjectInputStream(new BufferedInputStream(new
            FileInputStream(args(1))))
        ois.readObject() match {
          case s:Student => println(s.name+" "+s.grades.mkString(", "))
          case _ => println("Unidentified type.")
        }
        ois.close()
      case "-w" =>
        val oos = new ObjectOutputStream(new BufferedOutputStream(new
            FileOutputStream(args(1))))
        val s = new Student(args(2),args.drop(3).map(_.toInt))
        oos.writeObject(s)
        oos.close()
      case _ =>
        println("Usage: -r filename | -w filename name g1 g2 ...")
    }
  }
}
```

Put this in a file and compile it with `scalac`. From the command-line you can invoke this first step using this command.[4]

```
scala Main -w obj.bin John 98 78 88 93 100 83
```

After running this there will be a file called "obj.bin". You can look at it with `cat` or `xxd`. You will see that this is clearly a binary file, but there are some parts that are human readable strings. One of these is the name itself, but there are others that gives type names like `Student` and `java/lang/String`. These have to be in the file because when you *deserialize* the file it has to know what types of objects to create. You can verify that this process works with the following command.

```
scala Main -r obj.bin
```

This will read back in the file you just created and print the `Student` object that was read.

One critical aspect to note about the code for the `-r` option is that it includes a `match`. If you leave out the `match` and assign the result of `ois.readObject()` to a variable, the variable will have type `AnyRef`. That is because `readObject` has a return type of `AnyRef`. You will not be able to get `name` or `grades` from an `AnyRef` because that type does not have those. The `match` allows you to check the type of the object that was read in and do the print statement we want if it is a `Student` or print an error message if it is not.[5]

When an object is serialized, some indication of its type is written out, followed by a

---

[4]You can do this in Eclipse as well. You need to open the `Run` settings and change the arguments. Using command-line arguments is easier in the terminal. If you want to test this in Eclipse you might consider changing from using `args` to standard input and output.

[5]This same type of operation can be done with `isInstanceOf[A]` and `asInstanceOf[A]` methods. However, the use of those methods is strongly frowned upon in Scala. Using a `match` is the appropriate Scala style for determining the type of an object and getting a reference to that object of the proper type.

serialization of its contents. This only works if all the contents are serializable. If you try running this code here, you will find that it throws an exception.

```
import java.io._

class OtherData(val id:String,val course:String)
class Student(val name:String,val grades:Array[Int],val od:OtherData) extends
    Serializable

object Main {
  def main(args:Array[String]) {
    val oos = new ObjectOutputStream(new FileOutputStream("fail.bin"))
    val s = new Student("John",Array(98,90),new OtherData("0123","CS2"))
    oos.writeObject(s)
    oos.close()
  }
}
```

The details of the exception are shown here.

```
java.io.NotSerializableException: OtherData
```

The problem with this code is that `od:OtherData` is a member of `Student`, but it is not serializable. So when the serialization process gets to the `od` member, it fails. In this case it is simple enough to fix that by making it so that `OtherData` extends `Serializable`. In other situations you will not have that type of control because the type that you are dealing with might have been written by someone else and not be `Serializable`.

One way to deal with information that is not serializable is to simply not write it out. There are other times when this approach is valid as well. For example, the `DrawRectangle` type keeps the `propPanel` variable so that it does not make a new Graphical User Interface (GUI) component every time the user looks at the settings. However, if you save off a `DrawRectangle`, there is no reason to save all the information associated with the GUI component. That could be easily recreated from the other information. In order to do this in Scala we use an ANNOTATION.

An annotation is specified by a normal name that is preceded by an `@`. Annotations provide meta-information about a program. That is information that is used by higher-level tools and is not really part of the normal program code. There are a number of standard annotations that are part of the Scala compiler. You can identify them in the API because they start with lowercase letters. The two associated with serialization are `@transient` and `@SerialVersionUID`. If you have a member in a class that you do not want to have serialized, simply annotate it with `@transient`. Here we have code where this has been done with `od`.

```
import java.io._

class OtherData(val id:String,val course:String)
class Student(val name:String,
  val grades:Array[Int],
  @transient val od:OtherData) extends Serializable

object Main {
  def main(args:Array[String]) {
    val oos = new ObjectOutputStream(new FileOutputStream("pass.bin"))
    val s = new Student("John",Array(98,90),new OtherData("0123","CS2"))
```

```
    oos.writeObject(s)
    oos.close()

    val ois = new ObjectInputStream(new FileInputStream("pass.bin"))
    ois.readObject() match {
      case s2:Student => println(s2.name+" "+s2.grades+" "+s2.od)
      case _ => println("Unknown type read.")
    }
    ois.close()
  }
}
```

This code also reads the object back in and prints the different fields. This is done to illustrate what happens when you read a serialized object that that has a transient field. That field is not part of the serialization, so when the object is read back in, that field is given a default value. For any subtype of `AnyRef`, the default value is `null`. Running this code will show you that the last value printed is indeed `null`.

The other annotation, `@SerialVersionUID`, is used to attach a version number to a type. This is done so that when you load back in an object, the version number of the saved version can be compared to that of the loaded version to make sure they are the same. Java will automatically generate these for you, but they change nearly every time the code for the type changes. This will cause saved files to break, even if they would still work fine. To prevent this, you might consider putting this annotation, followed by an argument of a numeric ID, before the type declaration.

The implication this has for your programming is that anything that is transient generally must be a `var` instead of a `val`. This is due to the fact that when you deserialize such an object, that member will have a default value that almost certainly needs to be changed. The other side of this is that places that use any transient members likely need to have conditional code that will initialize the values if they are not set properly. This is exactly what was done with the `propPanel` in the `Drawable` types that we have made. If the field is accessed a lot, you should consider having a `private` local `var` that is only accessed through a method. Here is an alternate version of `Student` which follows that rule.

```
class Student(val name:String,
    val grades:Array[Int],
    @transient private var lod:OtherData) extends Serializable {
  assert(lod!=null)
  def od : OtherData = {
    if(lod==null) {
      lod = new OtherData("012345","Default")
    }
    lod
  }
}
```

You can substitute this version in the code above and you will no longer get `null` in the print out. This is because now the reference to `s2.od` is a call to the method that will create a default value if the member `lod` has not been assigned.

Custom Serialization

Default serialization is wonderful in many situations. However, it can be very inefficient for some types. In addition, there might be situations where you really do need to store information for a member even though it is not serializable. For this reason, it is possible to override the default and use your own custom serialization.

To override the default implementation, you implement the private methods `writeObjects(oos:ObjectOutputStream)` and `readObjects (ois:ObjectInputStream)`. The first thing you will do in these methods is call `oos.defaultWriteObject()` and `ois.defaultReadObject()` respectively. This will write out or read in all of the information that should be part of the default serialization. Even if there are no fields to be serialized you still need to call this. After that you write to or read from the stream. Here is an example of `Student` using this approach.

```
class Student(val name:String,
    val grades:Array[Int],
    @transient private var lod:OtherData) extends Serializable {
  assert(lod!=null)
  def od = lod
  private def writeObject(oos:ObjectOutputStream) {
    oos.defaultWriteObject()
    oos.writeUTF(od.id)
    oos.writeUTF(od.course)
  }
  private def readObject(ois:ObjectInputStream) {
    ois.defaultReadObject()
    lod = new OtherData(ois.readUTF,ois.readUTF)
  }
}
```

The `lod` member is still a `private var`. It needs to be `private` because it is a `var`. It needs to be a `var` because of the line in `readObject` that makes an assignment to it. Outside code gets access to the value using the `od` method. Note that the method is simpler in this case because it does not have to check the value of `lod`.

This might seem like a silly example, but it is not hard to find a real use for custom serialization. It happens that the `java.awt.image.BufferedImage` class that we introduced in chapter 12 is not serializable. This type is used quite frequently and you are likely to find that it would be helpful if it were written with an object. The following demonstrates how you might do that.

```
class Painting(val name:String,
    // ...
    @transient private var lImg:BufferedImage) extends Serializable {
  assert(lImg!=null)
  def img = lImg
  private def writeObject(oos:ObjectOutputStream) {
    oos.defaultWriteObject()
    oos.writeInt(img.getWidth)
    oos.writeInt(img.getHeight)
    oos.writeInt(img.getType)
    for(i <- 0 until img.getWidth; j <- 0 until img.getHeight) {
      oos.writeInt(img.getRGB(i,j))
```

```
    }
  }
  private def readObject(ois:ObjectInputStream) {
    ois.defaultReadObject()
    val (w,h,t) = (ois.readInt(),ois.readInt(),ois.readInt())
    lImg = new BufferedImage(w,h,t)
    for(i <- 0 until w; j <- 0 until h) {
      lImg.setRGB(i,j,ois.readInt())
    }
  }
}
```

This code writes the width, height, and type of the image first, then follows it with all the Red, Green, Blue (RGB) pixel values. When reading an object back in, the first three values are read, then the image is created, then the rest of the image is read.

22.4.3.2 XML Serialization

Java's default serialization comes with the standard advantages and limitations of a binary format. In addition, it only works for programs running under the Java Virtual Machine (JVM). That includes not only Java and Scala, but a number of other languages as well. It does not include all languages. C and C++ stand out as languages that do not have a JVM implementation. There are also some situations where you do not need the benefits of binary format and you would rather have the portability and readability of XML.

There is not a built-in method of converting objects to XML or getting them back, but you can add this type of functionality to your own code fairly easily. To do this we will follow three fairly simple rules with each class we want to be able to serialize.

- Put a `toXML : scala.xml.Node` method in the class that returns an XML element with all the information you want to save for those objects.
- Include a companion object with an `apply(node:scala.xml.Node) : Type` method that deserializes the XML node and returns an object of the type you have created.
- Make all the fields that are serialized be arguments to the class.

If you do this consistently across your types, when one type includes a reference to another, you can simply call `toXML` on that type.[6] You can use the normal methods in `scala.xml.XML` to read the XML from a file or write it back out to a file.

Compressed Streams

If you have spent much time doing things on a computer or downloading files from the Internet, odds are good that at some point you have come across a ZIP file. Compressed files in Windows®use this format. It allows you to combine many files into one, and for some files, particularly those that store text data, it makes them much

[6]The only shortcoming of this approach is that if there are two or more references to any given object, it will be duplicated. The Java serialization code caches objects so this does not happen.

smaller. If you have large text files, large XML files would qualify, zipping them can save a significant amount of space.

When your program creates those large files it would be more efficient to have them written out directly to the ZIP format and then read back in from that as well. Fortunately, there are some wrapper streams in `java.util.zip` which can do exactly that for you. The `ZippedInputStream` and `ZippedOutputStream` can be used to decorate other streams in exactly the same fashion as a `BufferedInputStream` or `BufferedOutputStream`. These streams significantly alter the contents of what is written or read back in so that everything uses the ZIP format.

The ZIP format has the ability to store multiple files inside of a single ZIP file. There is a type called `ZipEntry` that represents a single file. If you are working with ZIP files, you have to make some extra calls to position the stream in a particular entry.

22.5 Saving Drawings (Project Integration)

Now it is time to integrate these different concepts into our project. We will do this by giving our program the ability to save drawings. For completeness, there will be three different save options: default serialized binary, zipped serialized binary, and XML. These three options will allow us to cover nearly everything that was introduced in the chapter in a manner that is more significant than the small samples shown with each topic.

There are three separate pieces to adding this code. We need to have GUI elements for the user to interact with. Those GUI elements need to call code that deals with the files. Then changes have to be made to the Drawing and Drawables so that they can be serialized and deserialized using both the default method and XML. We will start with the GUI code for two reasons. First, it is fairly easy to add menu items. Second, in order to test the other parts we need to have the ability to make the program call that code and that is what the GUI elements do. This is as easy to do as adding the following four lines to `DrawingMain`.

```
contents += new MenuItem(Action("Save Binary")(saveBinary))
contents += new MenuItem(Action("Save Zip")(saveZip))
contents += new MenuItem(Action("Save XML")(saveXML))
contents += new MenuItem(Action("Load Drawing")(load))
```

If you also add empty stubs for the methods called in the actions into `DrawingMain`, the code will compile and let you run it to see the menus.

At this point things get more interesting because we have to implement those four methods. The `saveBinary` and `saveZip` methods have a fair bit of code in common. They both let the user select a file to write to, open a `FileInputStream`, wrap it in some way, and then write the currently selected drawing out to the stream. The `saveXML` method is going to have the user select a file, then use `scala.xml.XML.save` to write a `Node` to disk. The node needs to come from the currently selected drawing and given what was said above, this will probably be done by adding `toXML` methods to `Drawing` and other places. The load method needs to allow the user to select a file, then identify the file type by the extension and execute the appropriate deserialization code.

We will start with `saveBinary` and `saveZip` and pull any code that would be duplicated out into other methods. This is what they look like.

```
private def saveBinary {
  withSaveFile(file => {
    val oos = new ObjectOutputStream(new BufferedOutputStream(new
        FileOutputStream(file)))
    withOutputStream(oos)(strm => {
      serializeDrawingToStream(strm,file.getName())
    })
  })
}

private def saveZip {
  withSaveFile(file => {
    val zos = new ZipOutputStream(new BufferedOutputStream(new
        FileOutputStream(file)))
    zos.putNextEntry(new ZipEntry(file.getName().dropRight(3)+"bin"))
    val oos = new ObjectOutputStream(zos)
    withOutputStream(oos)(strm => {
      serializeDrawingToStream(strm,file.getName())
    })
  })
}
```

The `withSaveFile` method opens a dialog box and allows the user to select a file, then executes the function that is passed on that file. The `withOutputStream` separates out the `try/finally` code for making sure the streams are closed off.[7] Lastly, the `serializeDrawingToStream` method takes an `ObjectOutputStream` and a name for the file then serializes the current drawing to that stream and changes the text on the tab to the file name.

We can reuse the `withSaveFile` method in the `saveXML` method using the following code.

```
private def saveXML {
  withSaveFile(file => {
    val drawing = openDrawings(tabbedPane.selection.index)
    xml.XML.save(file.getAbsolutePath(),drawing.toXML)
    tabbedPane.selection.page.title = file.getName()
  })
}
```

In order to get this code to compile, you have to add a `toXML : xml.Node` method to the `Drawing` class. At this point, that function can just return `<drawing></drawing>`.

The `load` method is a bit longer because it can handle any of the three types of files based on the extension. It would be a better overall design for an application to be symmetric in the load and save options, but mixing them in this situation allows us to demonstrate how you might do it with one menu option or several. This sacrifices proper UI design for other pedagogical purposes.

```
private def load {
  val chooser = new FileChooser()
  if(chooser.showSaveDialog(tabbedPane)==FileChooser.Result.Approve) {
    if(chooser.selectedFile.getName().endsWith(".bin")) {
```

[7]The way that object streams work makes it so you need to close the object stream, not one of the streams it is wrapped around, like the file stream.

```
      val ois = new ObjectInputStream(new BufferedInputStream(new
          FileInputStream(chooser.selectedFile)))
      withInputStream(ois)(strm => {
        deserializeDrawingFromStream(strm,chooser.selectedFile.getName())
      })
    } else if(chooser.selectedFile.getName().endsWith(".zip")) {
      val zis = new ZipInputStream(new BufferedInputStream(new
          FileInputStream(chooser.selectedFile)))
      zis.getNextEntry
      val ois = new ObjectInputStream(zis)
      withInputStream(ois)(strm => {
        deserializeDrawingFromStream(strm,chooser.selectedFile.getName())
      })
    } else if(chooser.selectedFile.getName().endsWith(".xml")) {
      val nd = Drawing(xml.XML.loadFile(chooser.selectedFile))
      openDrawings += nd
      tabbedPane.pages += new
          TabbedPane.Page(chooser.selectedFile.getName(),nd.propertiesPanel)
    }
  }
}
```

Similar helper files are used here to those that were used for saving. For the XML option to work, a companion object must be added for `Drawing` that has an apply method that takes an `xml.Node` and returns a `Drawing`. For now this can simply return a `new Drawing`.

A complete listing of the revised `DrawingMain` is shown here.

```
package scalabook.drawing

import scala.swing._
import scala.collection.mutable
import java.io._
import java.util.zip._

object DrawingMain {
  private val tabbedPane = new TabbedPane

  private val openDrawings = mutable.Buffer[Drawing]()

  private def newDrawing {
    val nd = Drawing()
    openDrawings += nd
    tabbedPane.pages += new TabbedPane.Page("Unnamed",nd.propertiesPanel)
  }

  private def withOutputStream[A,B <: OutputStream](os:B)(body:B=>A):A = {
    try {
      body(os)
    } finally {
      os.close()
    }
  }

  private def serializeDrawingToStream(oos:ObjectOutputStream,name:String) {
    val drawing = openDrawings(tabbedPane.selection.index)
```

```
  oos.writeObject(drawing)
  tabbedPane.selection.page.title = name
}

private def withSaveFile(body: File=>Unit) {
  val chooser = new FileChooser
  if(chooser.showSaveDialog(tabbedPane)==FileChooser.Result.Approve) {
    try {
      body(chooser.selectedFile)
    } catch {
      case ex:FileNotFoundException => ex.printStackTrace()
      case ex:IOException => ex.printStackTrace()
    }
  }
}

private def saveBinary {
  withSaveFile(file => {
    val oos = new ObjectOutputStream(new BufferedOutputStream(new
        FileOutputStream(file)))
    withOutputStream(oos)(strm => {
      serializeDrawingToStream(strm,file.getName())
    })
  })
}

private def saveZip {
  withSaveFile(file => {
    val zos = new ZipOutputStream(new BufferedOutputStream(new
        FileOutputStream(file)))
    zos.putNextEntry(new ZipEntry(file.getName().dropRight(3)+"bin"))
    val oos = new ObjectOutputStream(zos)
    withOutputStream(oos)(strm => {
      serializeDrawingToStream(strm,file.getName())
    })
  })
}

private def saveXML {
  withSaveFile(file => {
    val drawing = openDrawings(tabbedPane.selection.index)
    xml.XML.save(file.getAbsolutePath(),drawing.toXML)
    tabbedPane.selection.page.title = file.getName()
  })
}

private def withInputStream[A,B <: InputStream](is:B)(body:B=>A):A = {
  try {
    body(is)
  } finally {
    is.close()
  }
}

private def deserializeDrawingFromStream(ois:ObjectInputStream,name:String) {
```

```
    val obj = ois.readObject()
    obj match {
      case nd : Drawing =>
        openDrawings += nd
        tabbedPane.pages += new TabbedPane.Page(name,nd.propertiesPanel)
      case _ =>
    }
  }

  private def load {
    val chooser = new FileChooser()
    if(chooser.showSaveDialog(tabbedPane)==FileChooser.Result.Approve) {
      if(chooser.selectedFile.getName().endsWith(".bin")) {
        val ois = new ObjectInputStream(new BufferedInputStream(new
            FileInputStream(chooser.selectedFile)))
        withInputStream(ois)(strm => {
          deserializeDrawingFromStream(strm,chooser.selectedFile.getName())
        })
      } else if(chooser.selectedFile.getName().endsWith(".zip")) {
        val zis = new ZipInputStream(new BufferedInputStream(new
            FileInputStream(chooser.selectedFile)))
        zis.getNextEntry
        val ois = new ObjectInputStream(zis)
        withInputStream(ois)(strm => {
          deserializeDrawingFromStream(strm,chooser.selectedFile.getName())
        })
      } else if(chooser.selectedFile.getName().endsWith(".xml")) {
        val nd = Drawing(xml.XML.loadFile(chooser.selectedFile))
        openDrawings += nd
        tabbedPane.pages += new
            TabbedPane.Page(chooser.selectedFile.getName(),nd.propertiesPanel)
      }
    }
  }

  private val frame = new MainFrame {
    contents = tabbedPane
    menuBar = new MenuBar {
      contents += new Menu("File") {
        contents += new MenuItem(Action("New")(newDrawing))
        contents += new MenuItem(Action("Save Binary")(saveBinary))
        contents += new MenuItem(Action("Save Zip")(saveZip))
        contents += new MenuItem(Action("Save XML")(saveXML))
        contents += new MenuItem(Action("Load Drawing")(load))
        contents += new Separator()
        contents += new MenuItem(Action("Exit")(sys.exit(0)))
      }
    }
    size = new Dimension(800,600)
  }

  def main(args : Array[String]) : Unit = {
    frame.visible=true
  }
}
```

The withInputStream and withOutputStream methods are curried to facilitate the type inference. Without the currying, it would be necessary to put a type on the strm argument for the functions that are passed in.

The error handling in this code is minimal, at best. Having a print of a stack trace in the code can be very helpful for programmers. However, it tells the user nothing. For a GUI based application like this one, it is likely that the user will not even see a console where the stack trace would print out. This code should pop up a window letting the user know what has happened. It has been left out here to not bloat the code and is left as an exercise for the reader.

If you try to run this code at this point, none of the options will really work. To make them work we have to alter Drawing and the Drawable hierarchy to support default and XML serialization. We will start with Drawing because that is what we are actually serializing. For the default serialization this means that the Drawing class needs to extend Serializable and that members that are not saved should be made transient. In addition, code needs to be altered so that transient fields do not cause problems when they are deserialized with a value of null.

For the XML-based serialization, that means implementing the toXML method to put any values we want saved into an XML element and implementing a deserializing apply method in a companion object. To make this work, values that are saved should also be moved up to be arguments of the class. After doing this to Drawing, we get code that looks like the following.

```
package scalabook.drawing

import swing._
import event._
import javax.swing.event._
import actors.Actor

/**
 * This type represents a Drawing that you can have open in the program.
 */
class Drawing(private val root:DrawTransform) extends Serializable {
  @transient private var lDrawPanel = new DrawPanel
  @transient private var propPanel:Component = null
  @transient private var lTree = new javax.swing.JTree(root)

  root.drawing=this

  private def drawPanel = {
    if(lDrawPanel==null) lDrawPanel = new DrawPanel
    lDrawPanel
  }

  private def tree = {
    if(lTree==null) lTree = new javax.swing.JTree(root)
    lTree
  }

  private[drawing] def refresh { drawPanel.repaint() }

  private class ResultEvent(val text:String,val result:Any) extends Event

  /**
```

```
 * Returns a panel that displays the properties for this drawing.
 *
 * @return a Component that can be put into a GUI.
 */
def propertiesPanel = {
  if(propPanel == null) {
    val commandArea = new TextArea()
    commandArea.editable = false
    val commandField = new TextField() {
      listenTo(this)
      reactions += {
        case e:EditDone =>
          if(!text.isEmpty) {
            val t = text
            Actor.actor {
              publish(new ResultEvent(t,Commands(t,Drawing.this)))
            }
            text=""
          }
        case e:ResultEvent =>
          commandArea append "> "+e.text+"\n"+e.result+"\n"
      }
    }
    val commandPanel = new BorderPanel {
      layout += commandField -> BorderPanel.Position.North
      layout += new ScrollPane(commandArea) -> BorderPanel.Position.Center
      preferredSize = new Dimension(500,200)
    }

    val drawProps = new GridPanel(1,1)

    propPanel = new DrawingPane(new SplitPane(Orientation.Vertical,
      new GridPanel(2,1) {
        contents += new BorderPanel {
          layout += new GridPanel(1,2) {
            contents += Button("Add"){
              executeOnSelection(d => addTo(d),addTo(root))
            }
            contents += Button("Remove"){
              executeOnSelection(d => remove(d))
            }
          } -> BorderPanel.Position.North
          layout+= new ScrollPane(new Component {
            override lazy val peer = tree
            tree.addTreeSelectionListener(new TreeSelectionListener {
              def valueChanged(e:TreeSelectionEvent) {
                executeOnSelection(d => {
                  drawProps.contents.clear()
                  drawProps.contents += d.propertiesPanel
                  drawProps.revalidate
                  drawProps.repaint
                })
              }
            })
          }) -> BorderPanel.Position.Center
```

```
        }
        contents += new ScrollPane(drawProps)
      },new SplitPane(Orientation.Horizontal,
        new ScrollPane(drawPanel),
        commandPanel))
    )
  }
  propPanel
}

def toXML : xml.Node =
<drawing>
  {root.toXML}
</drawing>

private def executeOnSelection(f: Drawable=>Unit,default: =>Unit = {}) {
  val path = tree.getSelectionPath
  if(path!=null) {
    val last = path.getLastPathComponent match {
      case drawable:Drawable if(drawable!=null)=>
        f(drawable)
      case _ =>
    }
  } else default
}

private def addTo(d:Drawable) = {
  assert(d!=null)
  val parent = d match {
    case dt:DrawTransform => dt
    case _ => d.getParent
  }
  val options = Array[(String, ()=>Drawable)](
      ("Bouncing Balls", () => DrawBouncingBalls(parent)),
      ("Ellipse", () => new DrawEllipse(parent)),
      ("Mandebrot", () => new DrawMandelbrot(parent)),
      ("Rectangle", () => new DrawRectangle(parent)),
      ("Transform", () => new DrawTransform(parent))
  ).toMap
  val choice = Dialog.showInput(propPanel,
      "What do you want to add?","Draw Type",Dialog.Message.Question,
      null,options.keys.toSeq,options.keys.head)
  choice match {
    case Some(ch) =>
      val newChild = options(ch)()
      parent.addChild(newChild)
      drawPanel.repaint
      tree.getModel match {
        case m:javax.swing.tree.DefaultTreeModel => m.reload
        case _ =>
      }
    case _ =>
  }
}
```

```
  private def remove(d:Drawable) {
    assert(d!=null)
    if(d.getParent!=null) {
      d.getParent.removeChild(d)
      drawPanel.repaint
      tree.getModel match {
        case m:javax.swing.tree.DefaultTreeModel => m.reload
        case _ =>
      }
    }
  }

  private class DrawPanel extends Panel {
    override def paint(g:Graphics2D) {
      g.setPaint(java.awt.Color.white)
      g.fillRect(0,0,size.width,size.height)
      root.draw(g)
    }
  }

  class DrawingPane(c:Component) extends GridPanel(1,1) {
    contents += c
    def drawing = Drawing.this
  }
}

object Drawing {
  def apply(data:xml.Node):Drawing = {
    new Drawing(DrawTransform(null,(data \ "drawTransform")(0)))
  }

  def apply():Drawing = {
    new Drawing(new DrawTransform(null))
  }
}
```

The primary alterations are at the beginning and the end. In order for this code to compile, you must add a `toXML` method and a companion object with an `apply` method to `DrawTransform`. There are some other changes in here that are required because of alterations that we will run into later.

The fact that `DrawTransform` is part of an inheritance hierarchy, combined with the fact that default serialization is done through inheritance means that we will want to start at the top. So before looking at `DrawTransform`, we will look at `Drawable`. Here is the code for that trait.

```
package scalabook.drawing

import java.awt.Graphics2D
import javax.swing.tree.TreeNode

/**
 * This represents the supertype for all the different types that can appear in
     our drawing.
 */
trait Drawable extends TreeNode with Serializable {
```

```
  /**
   * Stores the Drawing this Drawable is part of.
   */
  @transient protected var lDrawing:Drawing = null

  /**
   * Causes this object to be drawn to g.
   *
   * @param g a Graphics2D object to draw to.
   */
  def draw(g : Graphics2D) : Unit

  /**
   * Gives back a GUI component that allows the user to change drawing properties.
   *
   * @return A component that should be put in the GUI so the user can edit this
       object.
   */
  def propertiesPanel() : scala.swing.Component

  /**
   * Return an XML serialization of this object.
   */
  def toXML : xml.Node

  /**
   * Make it so that the getParent inherited from TreeNode returns a DrawTransform.
   */
  override def getParent : DrawTransform

  /**
   * Returns the drawing this Drawable is part of.
   */
  def drawing = lDrawing

  /**
   * Sets the Drawing this is part of. Implementations in subtypes
   * should recursively descend through Drawables.
   */
  def drawing_=(d:Drawing):Unit
}
```

We have added `Serializable` to the list of types being inherited from and also put in an abstract `toXML` method. The former now makes all subtypes serializable. The latter will force us to put in concrete `toXML` methods in the subtypes. For the subtypes then we need to make the proper values transient, write `toXML`, and add a companion object with the appropriate `apply` method. This code also has a protected member called `lDrawing` with public methods that can be used to access or set it. This is included to deal with a problem that we will encounter with `DrawBouncingBalls`. Having it there will also allow us to do some things that we did not previously.

For the subtypes, we will start with `DrawRectangle` because it is fairly simple. Here is code for the revision to that class, leaving out the `draw`, `propertiesPanel`, and `toString` methods which are not changed.

```
class DrawRectangle(p:DrawTransform,
```

```
    private var width:Double = 100.0,
    private var height:Double = 100.0,
    private var color:Color = Color.black) extends DrawLeaf {
  val parent = p
  @transient private var propPanel:Component = null

  ...

  def toXML : xml.Node = {
    val colStr = color.getRGB.toHexString
    <drawRectangle width={width.toString} height={height.toString}
        color={"0"*(6-colStr.length)+colStr}/>
  }
}

object DrawRectangle {
  def apply(p:DrawTransform,data:xml.Node) = {
    new
        DrawRectangle(p,(data\"@width").text.toDouble,(data\"@height").text.toDouble,
        new Color(Integer.parseInt((data\"@color").text,16)))
  }
}
```

The simplicity of DrawRectangle means that the XML element can be written with the short form and given three properties. To make it easy to store the color, the RGB value is converted to hexadecimal padded to eight characters. The eight characters represent AARRGGBB for alpha, red, green, and blue respectively. To get that back to an Int, we use the Integer.parseInt method from the Java libraries because it can take a second parameter that specifies the base to use. The changes to DrawEllipse are the same as those to DrawRectangle.

You might note that the code for DrawRectangle is lacking the drawing_= method. That has been added to DrawLeaf so that it does not have to be added to all the subtypes. The method in DrawLeaf looks like this.

```
def drawing_=(d:Drawing) { lDrawing = d }
```

The fact that the leaf elements do not have children means that all they need to do is set their locally stored value of lDrawing.

The other Drawable subtypes each represent a more significant challenge. In the case of DrawMandelbrot there are significantly more data members that need to be moved around to be arguments with default values. You can see the changes here.

```
package scalabook.drawing

import java.awt.{Graphics2D,Color}
import java.awt.image.BufferedImage
import swing._
import event._

class DrawMandelbrot(p:DrawTransform,
    private var rmin:Double = -1.5,
    private var rmax:Double = 0.5,
    private var imin:Double = -1.0,
    private var imax:Double = 1.0,
    private var width:Int = 600,
```

```
  private var height:Int = 600,
  private var maxCount:Int = 100) extends DrawLeaf {
val parent = p
@transient private var propPanel:Component = null
@transient private var img:BufferedImage = null
@transient private var changed = false

private def properties:Seq[(String,() => Any,String => Unit)] = Seq(
  ("Real Min", () => rmin, s => rmin = s.toDouble),
  ("Real Max", () => rmax, s => rmax = s.toDouble),
  ("Imaginary Min", () => imin, s => imin = s.toDouble),
  ("Imaginary Max", () => imax, s => imax = s.toDouble),
  ("Width", () => width, s => width = s.toInt),
  ("Height", () => height, s => height = s.toInt),
  ("Max Count", () => maxCount, s => maxCount = s.toInt)
)

def draw(g:Graphics2D) {
  if(img==null || changed) {
    if(img==null || img.getWidth!=width || img.getHeight!=height) {
      img = new BufferedImage(width,height,BufferedImage.TYPE_INT_ARGB)
    }
    for(i <- 0 until width par) {
      val cr = rmin + i*(rmax-rmin)/width
      for(j <- 0 until height) {
        val ci = imax - j*(imax-imin)/width
        val cnt = mandelCount(cr,ci)
        img.setRGB(i,j,if(cnt == maxCount) Color.black.getRGB else
          new Color(1.0f,0.0f,0.0f,cnt.toFloat/maxCount).getRGB)
      }
    }
  }
  g.drawImage(img,0,0,null)
}

def propertiesPanel() : Component = {
  if(propPanel==null) {
    propPanel = new BorderPanel {
      val props = properties
      layout += new GridPanel(props.length,1) {
        for((propName,value,setter) <- props) {
          contents += new BorderPanel {
            layout += new Label(propName) -> BorderPanel.Position.West
            layout += new TextField(value().toString) {
              listenTo(this)
              reactions += { case e:EditDone => setter(text); changed = true }
            } -> BorderPanel.Position.Center
          }
        }
      } -> BorderPanel.Position.North
    }
  }
  propPanel
}
```

```
  override def toString() = "Mandelbrot"

  def toXML : xml.Node = {
    <drawMandelbrot rmin={rmin.toString} rmax={rmax.toString}
      imin={imin.toString} imax={imax.toString}
      width={width.toString} height={height.toString} maxCount={maxCount.toString}/>
  }

  private def mandelIter(zr:Double,zi:Double,cr:Double,ci:Double) =
      (zr*zr-zi*zi+cr, 2*zr*zi+ci)

  private def mandelCount(cr:Double,ci:Double):Int = {
    var ret = 0
    var (zr, zi) = (0.0, 0.0)
    while(ret<maxCount && zr*zr+zi*zi<4) {
      val (tr,ti) = mandelIter(zr,zi,cr,ci)
      zr = tr
      zi = ti
      ret += 1
    }
    ret
  }
}

object DrawMandelbrot {
  def apply(p:DrawTransform,data:xml.Node) = {
    new DrawMandelbrot(p,
        (data\"@rmin").text.toDouble,
        (data\"@rmax").text.toDouble,
        (data\"@imin").text.toDouble,
        (data\"@imax").text.toDouble,
        (data\"@width").text.toInt,
        (data\"@height").text.toInt,
        (data\"@maxCount").text.toInt)
  }
}
```

In addition to adding the code that handles XML, the other significant change is that `properties` has been converted from a `val` to a `def` and the point where it is used in the `propertiesPanel` method now introduces a temporary variable called `props`. This is done because data members interact with serialization while methods do not. The `properties` data member does not need to be serialized. It can be rebuilt whenever it is needed. Previously a `val` was slightly easier to use. With serialization, that is no longer the case.

The only other subtype of `DrawLeaf` that we have created so far is `DrawBouncingBalls`. The challenges here center around the fact that it contains a full collection so the XML has to have the ability to deal with that.

```
package scalabook.drawing

import java.awt.{Graphics2D,Color}
import java.awt.geom._
import swing._
import event._
import javax.swing.Timer
```

```
class DrawBouncingBalls(p:DrawTransform,
    private var balls:Vector[DrawBouncingBalls.Ball]) extends DrawLeaf {
  val parent = p
  @transient private var propPanel:Component = null
  @transient private var lTimer:Timer = null

  private def timer = {
    if(lTimer==null) {
      lTimer = new Timer(100,Swing.ActionListener(e => {
        balls = (for(DrawBouncingBalls.Ball(x,y,vx,vy,s) <- balls par) yield {
          var nvy = vy + 0.01 // gravity
          var nvx = vx
          var nx = x+nvx
          var ny = y+nvy
          if(nx-s<0.0) {
            nx = 2*s-nx
            nvx = nvx.abs
          } else if(nx+s>1.0) {
            nx = 2.0-(nx+2*s)
            nvx = -nvx.abs
          }
          if(ny-s<0.0) {
            ny = 2*s-ny
            nvy = nvy.abs
          } else if(ny+s>1.0) {
            ny = 2.0-(ny+2*s)
            nvy = -nvy.abs
          }
          DrawBouncingBalls.Ball(nx,ny,nvx,nvy,s)
        }).seq
        drawing.refresh
      }))
    }
    lTimer
  }

  def draw(g:Graphics2D) {
    g.setPaint(Color.black)
    g.draw(new Rectangle2D.Double(0,0,100,100))
    g.setPaint(Color.green)
    for(DrawBouncingBalls.Ball(x,y,_,_,s) <- balls) {
      g.fill(new Ellipse2D.Double((x-s)*100,(y-s)*100,s*200,s*200))
    }
  }

  def propertiesPanel() : Component = {
    if(propPanel==null) {
      propPanel = new BorderPanel {
        layout += new GridPanel(1,1) {
          val button = new Button("Start")
          button.action = Action("Start"){
            if(button.text == "Start") {
              timer.start()
              button.text="Stop"
            } else {
```

```
            timer.stop()
            button.text="Start"
          }
        }
        contents += button
      } -> BorderPanel.Position.North
    }
  }
  propPanel
}

override def toString = "Bouncing Balls"

def toXML : xml.Node =
  <drawBouncingBalls>
    {balls.map(_.toXML)}
  </drawBouncingBalls>
}

object DrawBouncingBalls {
  def apply(p:DrawTransform,data:xml.Node) = {
    new DrawBouncingBalls(p,Vector((data\"ball").map(bxml => {
      val x=(bxml\"@x").text.toDouble
      val y=(bxml\"@y").text.toDouble
      val vx=(bxml\"@vx").text.toDouble
      val vy=(bxml\"@vy").text.toDouble
      val size=(bxml\"@size").text.toDouble
      Ball(x,y,vx,vy,size)
    }):_*))
  }

  def apply(p:DrawTransform,minSize:Double = 0.01,maxSize:Double = 0.05) = {
    new DrawBouncingBalls(p,Vector.fill(20) {
      val size = minSize+math.random*(maxSize-minSize)
      Ball(size+math.random*(1-2*size),size+math.random*(1-2*size),
        (math.random-0.5)*0.02,(math.random-0.5)*0.02,size)
    })
  }

  case class Ball(x:Double,y:Double,vx:Double,vy:Double,size:Double) {
    def toXML = <ball x={x.toString} y={y.toString}
        vx={vx.toString} vy={vy.toString} size={size.toString}/>
  }
}
```

The need for the collection to be properly serialized in both manners requires moving the code that does the creation of the balls down into apply methods of the companion object. In addition, the timer needs to be transient so that the button on the properties panel will continue to work after a file is loaded in.

The only type that remains to convert is the DrawTransform. The modifications to this type to support default serialization are fairly minimal. Those for the XML serialization are a bit more significant.

```
class DrawTransform(val parent:DrawTransform,
```

```
    private var transformType:DrawTransform.TransformType.Value =
        DrawTransform.TransformType.Translate,
    private val transformValue:Array[Double] = Array.fill(3)(0.0)
    ) extends Drawable {
  private val subnodes = mutable.Buffer[Drawable]()
  @transient private var propPanel:Component = null

  import DrawTransform.TransformType._

  ...

  def toXML : xml.Node =
    <drawTransform type={transformType.id.toString}
        values={transformValue.mkString(":")}>
      {subnodes.map(_.toXML)}
    </drawTransform>

  private def buildTransform:AffineTransform = transformType match {
    case Translate =>
        AffineTransform.getTranslateInstance(transformValue(0),transformValue(1))
    case Rotate =>
        AffineTransform.getRotateInstance(transformValue(0),transformValue(1),
                    transformValue(2))
    case Scale =>
        AffineTransform.getScaleInstance(transformValue(0),transformValue(1))
    case Shear =>
        AffineTransform.getShearInstance(transformValue(0),transformValue(1))
  }

  ...

  def drawing_=(d:Drawing) {
    lDrawing = d
    subnodes.foreach(_.drawing=d)
  }

  ...
}

object DrawTransform {
  object TransformType extends Enumeration {
    val Translate,Rotate,Scale,Shear = Value
  }

  def makeDrawable(p:DrawTransform,cxml:xml.Node):Drawable = {
    cxml.label match {
      case "drawTransform" => DrawTransform(p,cxml)
      case "drawRectangle" => DrawRectangle(p,cxml)
      case "drawEllipse" => DrawEllipse(p,cxml)
      case "drawMandelbrot" => DrawMandelbrot(p,cxml)
      case "drawBouncingBalls" => DrawBouncingBalls(p,cxml)
      case _ => throw new IllegalArgumentException("XML contains unknown type:
          "+cxml)
    }
  }
```

```
  def apply(p:DrawTransform,data:xml.Node) = {
    val ret = new DrawTransform(p, TransformType((data\"@type").text.toInt),
        (data\"@values").text.split(":").map(_.toDouble))
    data.child.filter(cxml => cxml match {
      case e:xml.Elem => true
      case _ => false
    }).map(cxml => makeDrawable(ret,cxml)).foreach(ret.addChild)
    ret
  }
}
```

The `Enumeration` for the type of transformation was moved into the companion object to eliminate problems it might cause with serialization. The `subnodes` member is left out of the argument list because when we deserialize a transform, the `Drawables` under it need to have it as their parent. For this reason, we have to instantiate a complete transform that does not yet have children, then make the children and add them to the transform. This is all handled by methods in the companion object. Lastly, the `drawing_=` method implemented here not only sets the local value, it also recursively sets the values on all the subnodes.

With these changes in place, you now have the ability to make whatever drawings you wish and save them off in one of three styles, then load them back in. In practice you would likely choose between the binary serialization and XML serialization methods as few applications would need to support both.

22.6 End of Chapter Material

22.6.1 Summary of Concepts

- The `java.io` package contains quite a few useful types. There are four extended type hierarchies rooted in `InputStream`, `OutputStream`, `Reader`, and `Writer`. The first two have operations for reading and writing bytes. The second two operate on characters. We previously saw how to use `scala.io.Source` and `java.io.PrintWriter` for text access so this chapter focuses mainly on the two stream hierarchies.
- The `InputStream` and `OutputStream` `classes` are both abstract. Their subtypes come in two general forms, those that actually specify a source or sink for reading or writing, and those that modify the manner or functionality of reading and writing. The `FileInputStream` and `FileOutputStream` are of the former types. They provide stream behaviors attached to files.
- Exceptions are a way of dealing with unexpected conditions or other errors in code. When something goes wrong, an exception can be `thrown`. It is then up to the code that called the function where the error occurred to figure out what to do about it.
  - Exception handling is done through the `try-catch` expression. Code that might fail is put in a `try` block. The `catch` partial function has cases for the different exceptions the coder knows how to handle at that point.
  - When an exception is thrown, it immediately begins to pop up the call stack

until it comes to a `catch` that can handle it. If none exists, it will go all the way to the top of the call stack and crash that thread.

  - The loan pattern is a handy approach to use in Scala to make it easy to write code that can throw exceptions.

- There are other subtypes of the primary stream types that do not inherently attach to a source or sink. These `class`es alter behavior of the basic methods or add additional methods to provide new functionality. You use them to "decorate" an existing stream.
  - One of the most common forms of decoration is buffering. This makes sure that slow operations like disk access are done in large chunks.
  - The `DataInputStream` and `DataOutputStream` provide additional methods that allow you to read or write data in binary. This is generally faster and more compact than text, but it loses the convenience of human readability.
  - The `ObjectInputStream` and `ObjectOutputStream` go a step beyond providing basic binary data reading and writing capabilities, they allow you to serialize whole objects assuming that the objects adhere to certain rules.

22.6.2 Exercises

1. Make the save for a drawing be a single option, and give the dialog different file types and extensions.
2. Add code to what was written in the chapter where you pop up error messages using `Dialog.showMessage` if something goes wrong.
3. The code shown for saving drawings has a significant bug that occurs if you try to save when there are no open drawings. Figure out what this is, and edit the code to fix it.
4. Make a `class` called `Matrix` that has inside of it an `Array[Array[Double]]`. Write code so that your class can be serialized and deserialized using Java serialization, XML, and custom binary formats.
5. Make a binary file with 1,000,000 integer values written with a `DataOutputStream` using `writeInt`. Try reading that back in with and without a `BufferedInputStream` between the `DataInputStream` and the `FileInputStream`. Time the results to see how much it matters on your hardware.
6. Write a binary file of integers with ten values in it. Make the values all different, perhaps random. Using `xxd` attempt to increment every value by one and put the result in a different file. Write Scala code to read back in that file and see if what you did worked.
7. If you did the last exercise and you are looking for a significant challenge, try to do the same thing with `Double`s instead of `Int`s.
8. Write code to save some XML data out to a file. The challenge here is that it should be written to a ZIP file, not as plane text. Use an unzip utility to see if what you did worked. Also try to read it back in using Scala.

22.6.3 Projects

The projects for this chapter involve saving and loading data through serializing objects. You can decide whether you want to take the route of default binary serialization, XML based serialization, or a custom binary serialization. Which is best depends a lot on the project you are doing.

Something to note in general about the default binary serialization is that if you go this route, you almost certainly want to label your serializable classes is `@SerialVersionUID` so that the saved files do not break quite so often. They will still break, just not as often. The extra control you have over the XML gives it something of an advantage in this regard.

1. The main thing that will need to be saved in a MUD is the players. Over time they will collect different items and should probably gain in abilities. At this point, you can only have one player playing the game at a time. However, you can add the ability for a player to log in and log back out. That way the game can shift from one player to another. When the player logs out, their information should be saved. You can decide the format and method you want to use for saving the players.

2. If you are building the web spider, you inevitably want to store the data that you are interested locally in a format that you can easily work with. You might not want to store images locally, but at least keep a local copy of the URL. For text data that you process, storing it in the processed form will make it much easier to work with.

   So for this project you need to start building an internal representation of the data you care about, then work on serializing that out. Using the default binary serialization is probably ideal for numeric or tabular data. Things that are really text and images could work with XML. Choose the format that matches what you are interested in.

3. Like the MUD, a networked, multiplayer game will inevitably need a way to store off information about how various players are doing. Implement that for this project along with menu options for saving.

4. Even games that are not networked likely have a state that needs to be remembered. If you are working on that type of project, add menu options for saving and loading games.

5. If you are working on the math worksheet, that needs to be saved. You can pick between default serialization and XML for that.

6. Obviously, a Photoshop® type application needs to be able to save what you are drawing. This is one option where the default serialization is probably the easiest route. You will have to do some customized serialization code for images, but images do not store well in XML either.

7. Storing simulation data in plain text can make it easy to load in with simple plotting tools. However, large simulations need the space efficiency and speed of binary. For this, the ideal is neither XML, nor default serialization. Instead, you want to write custom binary serialization. For example, if you have N particles with x, y, z, v_x, v_y, and v_z coordinates, as well as a mass, the efficient binary serialization starts with writing N as a binary `Int`, then following it with $7N$ binary `Doubles` for the data values of the particles.

Chapter 23

Networking

In the 1990s Sun Microsystems®, a server hardware company and the creator of Java, had an advertising campaign with the slogan, "The network is the computer". Given the date, their proclamation has had remarkable foresight. Today there are few things that you do on a computer that do not use network access and many computing devices are built specifically for use with a network connection of some form. You might often find that your devices become far less useful to you when they are not connected.

At a basic level, networking is just having computers talk to one another. This allows them to share information and for programs on one computer to react to things that happen on another computer. So far we have written programs that only run locally and only deal with data they can read as files.[1] In this chapter we will branch out beyond that and learn how to make programs that can talk to other computers.

23.1 TCP and UDP

When computers talk via a network they basically send bits either over a wire or through wireless signals. In order for other computers to understand those bits and for the bits to get to the computer they are supposed to go to, there have to be agreed-upon ways of interpreting things. This agreement for how to interpret the binary data is what we call a network PROTOCOL. There are many different protocols that have been developed and there are different ones for different levels of communication. For moving information across a network the Java libraries provide support for Transmission Control Protocol (TCP) and User Datagram Protocol (UDP)/indexUDP.

Both of these protocols break the information you send up into packets. Each packet

[1] It is quite possible that some of the work you have done has happened on a networked file system. In that case you have been using a network connection, even if you did not realize it.

contains a certain amount of your data as well as protocol data with information such as where that data should go. Most network communication uses TCP, and we will follow along with that. The reason is that TCP is "safe". You can think of TCP as being something like certified mail. When a TCP packet is sent out, it goes through the network until it reaches its destination. The receiver sends back a verification that it got the packet. If something is lost, TCP will try to resend it. If it fails too many times, the program sending the data will be notified with an error message. In Scala, that notification will come in the form of an exception.

UDP is more like standard mail and lacks the safety mechanisms of TCP. You put your information in a UDP packet and send it out and just hope it arrives. If you are on a good network, most of the packets will. However, UDP will not actually tell you if they do or do not. It is your responsibility as a program to write the code in such a way that packets can be dropped and the program can deal with it. The advantage of UDP being "unsafe" is that it can be much faster. The system does not ever sit around waiting for verification that data got across, it just sends things out. This makes it the preferred protocol for things like games where the network speed can significantly impact user experience. It also places an extra development cost on game companies because effort does have to go into making the programs deal with dropped packets.

23.2 Sockets

Communication between machines is done through SOCKETS. The program on one machine, often called the "server", will wait for connections from the outside. programs running on other computers, called "clients", then make a connection to the one that is waiting. In order to support multiple different types of connections, each computer has many "ports" that sockets can connect to. The server specifies which port it will listen to and clients should connect to the appropriate port for the server they want to talk to.

23.2.1 TCP Sockets

The libraries for using sockets can be found in `java.io.net`. For TCP connections, the program on the server side creates a `ServerSocket` and gives it a port number to use. A call to `accept` on the `ServerSocket` will block until a connection is made. The `accept` method returns an instance of `Socket`. If you wanted to accept a single connection and do something with it, the code might look like this.

```
import java.net._

val ss = new ServerSocket(8000)
val serverSock = ss.accept()
// Stuff to communicate across serverSock
```

The counterpart to this code on the client side just uses the `Socket` class to make a connection to a remote computer. You pass in the machine to connect it and a port number like this.

```
val clientSock = new Socket("localhost",8000)
// Stuff to communicate across clientSock
```

The port numbers on the client and server need to agree for them to connect. In this example, the client would be run on the same machine as the server. The name "localhost" is used to represent the local machine. You could also provide a remote machine by name or IP address.

Details of Port Numbers

Port numbers are 16-bit unsigned values so they can range from 0 to 65535. The first 1024 (from 0 to 1023) of these can only be used by programs run with administrator privileges. Most of these are assigned to well-known system level programs. For example, ssh uses port 22 by default while the SMTP mail service uses 25.

The ports from 1024 to 49151 are registered ports that have assigned uses based on applications to the IANA. On most systems these are open for users, though you can not use a port that is already in use by some other application. The code in this chapter uses port 8000. Technically, this port has been registered for the Intel®Remote Desktop Management Interface (iRDMI). However, unless you happen to be running that on your system, it should work just fine for your programs.

The ports from 49152 to 65535 can not be registered and are used for custom applications or temporary purposes. You should feel free to have your application use a port in that range to make sure that you do not conflict with anything on whatever system the application might be run on.

23.2.2 UDP Sockets

With UDP, the role of client and server are less obvious. There is a single `DatagramSocket` class that is used for both sending and receiving data. It does this through a `DatagramPacket`. The two simplest ways to build a `DatagramPacket` are the following.

```
new DatagramPacket(buf:Array[Byte], len:Int) // Receiving packet
new DatagramPacket(buf:Array[Byte], len:Int, address:InetAddress, port:Int) //
    Sending packet
```

In the first case you are building a packet to receive data. The data goes into the specified array and the specified number of bytes are supposed to be read. The length of the array does not have to match `len`, but for simple usage it probably should.

The second one takes additional arguments that tell where the data should be sent. The `InetAddress` type is also in `java.net`. There are several different methods in `InetAddress` that can be used to get an instance of the type. The two you are most likely to use are the following.

```
getByAddress(addr:Array[Byte]):InetAddress
getByName(host:String):InetAddress
```

You are probably most familiar with using names to represent computers and will likely use the second option.

The `DatagramSocket` has methods called `send` and `receive` that each take a `DatagramPacket` and do the appropriate action with them. The `receive` method will block until a message comes in on the proper port. The real challenge of working with UDP is that you have to pack whatever data you want to sent into the arrays of bytes in the packets yourself. For this reason, we will not be making much use of UDP in this book. The following example is as far as we will go.

```
import java.net._
import scala.collection.mutable

object DatagramDoubles {
  def packDouble(d:Double,v:mutable.IndexedSeq[Byte]) {
    var dlong = java.lang.Double.doubleToLongBits(d)
    for(i <- v.indices) {
      v(i) = dlong.toByte
      dlong >>= 8
    }
  }

  def unpackDouble(v:mutable.IndexedSeq[Byte]):Double = {
    val dlong = v.foldRight(0L)((b,dl) => (dl << 8) | (b & 0xff))
    java.lang.Double.longBitsToDouble(dlong)
  }

  def packArray(ds:Array[Double]):Array[Byte] = {
    val ret = new Array[Byte](8*ds.length)
    for(i <- ds.indices) packDouble(ds(i),ret.view(8*i,8*(i+1)))
    ret
  }

  def main(args:Array[String]) {
    if(args.length>1) {
      val socket = new DatagramSocket()
      val data = packArray(args.map(_.toDouble))
      val packet = new
          DatagramPacket(data,data.length,InetAddress.getByName("localhost"),8000)
      socket.send(packet)
    } else if(args.length==1) {
      val socket = new DatagramSocket(8000)
      val num = args(0).toInt
      val data = new Array[Byte](8*num)
      val packet = new DatagramPacket(data,data.length)
      socket.receive(packet)
      for(i <- 0 until 8*num by 8) println(unpackDouble(data.view(i,i+8)))
    } else println("Specify one int for number to read or multiple doubles to
        send.")
  }
}
```

This code has the ability to send an `Array` of `Doubles` using UDP. Note that in order to pack the `Doubles` into an `Array` of bytes we use some calls in the Java library such as `doubleToLongBits` as well as bitwise operations like >>, <<, |, and &.

To run this code you first start the receiving end, as the `receive` method will block. If we want to send three numbers you could do that with `scala DatagramPoint 3`. This will just sit there until we run the sender with a call like this, `scala DatagramPoint 2.3 7.8 1.8775`. After this call the other instance will print out the three values on different lines.

Clearly this is code that you would only go through the effort to write if you really need the speed of UDP. Instead of hand packing the data you can also use the `java.io.ByteArrayInputStream` wrapped in either a `java.io.DataInputStream` or a

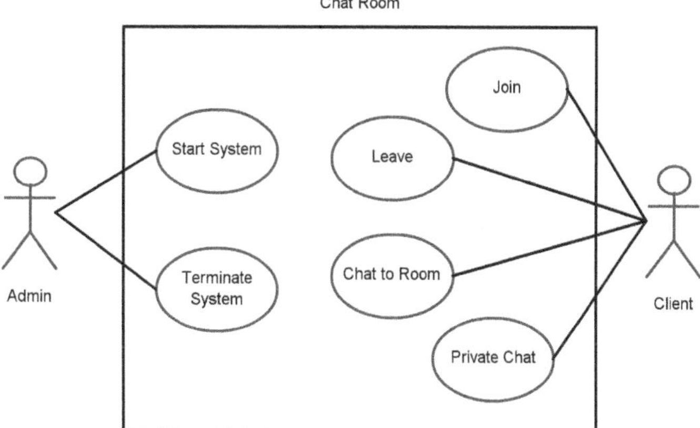

FIGURE 23.1: This figure shows the different use-cases for the socket/stream-based chat program that we want to write.

`java.io.ObjectInputStream`. If you use this approach, you can get the data out with a `java.io.ByteArrayOutputStream` wrapped in the appropriate data or object output stream. This approach is not all that much simpler.

23.2.3 Streams from Sockets

For UDP we saw that data has to be packed into the `DatagramPackets` to be sent to other machines and then unpacked when they arrive. We did not discuss how data is sent using the plain `Socket` class for TCP. To do this, we use streams. The `Socket` class has the following methods.

```
getInputStream() : InputStream
getOutputStream() : OutputStream
```

These methods do pretty much exactly what you would expect them to given their names. If you call `getInputStream` on a socket, it will return an implementation of `InputStream` that connects to the socket so that when data is read from it, data sent through the socket from the other machine will be read. You can wrap these streams in the same manner you would have wrapped streams attached to files. This is part of the flexibility of the stream library. You can write code that uses an `InputStream` or an `OutputStream` and it can work regardless of the type of stream that is passed to it. So the same code can be used to work with sockets or files.

To see how this works, we will create a simple chat room with a server and the ability to include multiple clients. To see what we really want, we should do some analysis. Figure 23.1 shows a use-case diagram for what we want to do. There are two types of users of the system, one will start of the chat room as a server on a particular computer. Other users will be clients of the server and will connect into the room and have options for chatting and leaving. When a person connects they have to give a handle they will go by so that others will know who has chatted certain messages and so that private messages can be sent. For this to work, only a single user at any given time can have a particular handle.

We will start with the client code because it is simpler. It only needs to make a socket to the server, get the user's handle, ask the user for chat lines that are sent to the server, and display chat text coming from the server. Here is code that does that with a console interface.

```
package scalabook.chat

import java.net._
import java.io._
import actors.Actor.actor
import annotation.tailrec

/**
 * This application runs a server for a chat room that demonstrates the use of
 * socket based networking.
 */
object ChatClient {
  def main(args : Array[String]) {
    if(args.isEmpty) {
      println("Usage: scala ChatClient host [port]")
    } else {
      val port = if(args.length>1) args(1).toInt else 8000
      val sock = new Socket(args(0),port)
      val dis = new DataInputStream(new BufferedInputStream(sock.getInputStream()))
      val dos = new DataOutputStream(new
          BufferedOutputStream(sock.getOutputStream()))
      println("What is your handle for the chat room?")
      val name = readLine()
      dos.writeUTF(name)
      dos.flush()
      val response = dis.readUTF()
      if(response!=":quit") {
        actor { incoming(dis) }
        println("Welcome! Begin chatting.")
        outgoing(dos)
      } else {
        println("The server has rejected you.")
      }
      sock.close()
    }
    sys.exit(0)
  }

  @tailrec private def incoming(dis:DataInputStream) {
    val line = dis.readUTF()
    println(line)
    incoming(dis)
  }

  @tailrec private def outgoing(dos:DataOutputStream) {
    print("> ")
    val input = readLine().trim
    dos.writeUTF(input)
    dos.flush()
    if(input!=":quit") outgoing(dos)
  }
}
```

Typing in ":quit" as a chat line will terminate the chat. Note that the incoming messages are handled in a separate thread. This thread is spawned using the `actor` shortcut. The reason for doing this is that both incoming and outgoing message handling contain blocking methods in the form of `readUTF` and `readLine`. To handle both in a single thread we would have to write more elaborate code that only reads when there are things to be read to prevent blocking.

The `incoming` and `outgoing` methods include something new that we have not seen before, an annotation called `@tailrec`. This annotation tells the Scala compiler that a particular recursive method must be optimized for tail recursion. If it can not, the compiler should generate an error. What does this mean? Tail recursion is a term used to describe a recursive function that contains only one recursive call and does not do significant work after the call returns. These types of recursive functions can often be transformed into `while` loops. In this way you can get the efficiency of a `while` loop while getting the programming benefits of recursion. The `@tailrec` annotation will force an error if the Scala compiler can not convert the function to a loop. This is important, because you might recall that recursive calls consume memory on the stack, and if you recurse too deeply, you run out of memory and the program crashes from a stack overflow.

The code for the server is a bit more complex because it has more work to do. It needs to have a `ServerSocket` that is accepting new connections, code for each socket that takes in chat messages, and code that lets the person who runs the server enter a command to stop it. Like the client, some of these things involve blocking calls so multithreading can be helpful. Here is one way of writing the code.

```
package scalabook.chat

import java.net._
import java.io._
import actors.Actor._
import annotation.tailrec
import collection.mutable

/**
 * This application runs a server for a chat room that demonstrates the use of
 * socket based networking.
 */
object ChatServer {
  def main(args : Array[String]) {
    val port = if(args.isEmpty) 8000 else args(0).toInt
    actor { startServer(port) }
    checkQuit()
    sys.exit(0)
  }

  private case class User(name:String, sock:Socket, dis:DataInputStream,
      dos:DataOutputStream)
  private val users = new mutable.HashMap[String,User] with
      mutable.SynchronizedMap[String,User] {}

  private def startServer(port:Int) {
    val ss = new ServerSocket(port)
    while(true) {
      val sock = ss.accept()
      actor { addUser(sock) }
    }
  }
```

```
  private def addUser(sock:Socket) {
    val dis = new DataInputStream(new BufferedInputStream(sock.getInputStream()))
    val dos = new DataOutputStream(new BufferedOutputStream(sock.getOutputStream()))
    val name = dis.readUTF()
    if(users.contains(name)) {
      dos.writeUTF(":quit")
      dos.flush()
      sock.close()
    } else {
      dos.writeUTF(":accept")
      dos.flush()
      val user = User(name,sock,dis,dos)
      users += name -> user
      doChat(user)
    }
  }

  @tailrec private def doChat(user:User) {
    val input = user.dis.readUTF().trim
    if(input!=":quit") {
      val index = input.indexOf("<-")
      if(index>0) {
        val oname = input.take(index).trim
        if(users.contains(oname)) {
          users(oname).dos.writeUTF(user.name+" -> "+input.drop(index+2).trim)
          users(oname).dos.flush()
        } else {
          user.dos.writeUTF(oname+" is not a valid handle.")
          user.dos.flush()
        }
      } else {
        val output = user.name+" : "+input
        for((name,ouser) <- users; if name!=user.name) {
          ouser.dos.writeUTF(output)
          ouser.dos.flush()
        }
      }
      doChat(user)
    } else {
      users.remove(user.name)
    }
  }

  @tailrec private def checkQuit() {
    if(readLine() != ":quit") checkQuit()
  }
}
```

The main method gets a port number, then starts the server in a separate thread. The main thread waits for input from the user.

The function that creates the `ServerSocket` goes into an infinite loop where it accepts connections and calls `addUser` in new threads created with `actor`. This means that each client could be in a separate thread. In reality, the actor library handles threads and we can not be certain of how many threads it will create to do the processing.

The `addUser` method gets streams from the socket and wraps them in data streams. It then reads a string which should be the name/handle of that client. Note that the server defined a case class called `User` and keeps a mutable map associating names with users. This mutable map is synchronized so that it does not run into problems with the multiple threads involved. This map allows us to easily insure that there are never two users with the same handle. It also makes it easier to do private messages. The name that is read in is checked against the map. If the map already has that as a key, it tells the client to terminate before closing the socket. Otherwise, it tells the client that it was accepted, adds a new `User` to the map, and moves on to the `doChat` method.

The `doChat` method is another tail recursive method. It reads from the input stream of that user and checks to see if they have disconnected. If they have not, it either sends out a private message to the correct user or runs through all users sending everyone a message. If they quit, that user name is removed from the map and there is no recursive call. You will also see that private messages are handled in this method. It looks for `<-` in the text and assumes that what is before that is a user handle and what is after it is the message to send.

There are several places in both the client and the server where you see output streams getting flushed. In fact, there is a flush after every write. The reason is that most of the writes will be small and the whole thing will sit in the buffer if we do not flush it.

Enter these files and check that they work. If you use "localhost" as the machine name you can have conversations with yourself on a single computer. After playing with this for a while you might notice some areas in which it is lacking. First, it is not very pretty. It is plain text in the console. Chat messages that come in will print in the middle of what you are typing in a way that can make it not only a bit ugly, but can also make it hard to use as you can not keep track of what you are writing.

There are some other problems that are more than cosmetic. Currently anything is allowed for a handle. That will not cause a problem as long as your users are not trying to break the system. However, there are names that could lead to less than ideal behavior. Strings that are only whitespace or that include whitespace at the beginning or end will prevent private messages from being sent. More checking should be put into the server when it receives a name to reject anything that would cause a problem.

The other problem you might have noticed is that if a client disconnects without entering ":quit" there is an exception in the server. Networking code throws even more exceptions than code using files. This code never includes a `try/catch`, but they could be useful around every call to `readUTF` or `writeUTF`. The ones that will cause exceptions if a client disconnects improperly are the first line in `doChat` and the `writeUTF` calls for sending messages. The cure for this is having a try/catch that will detect when a link fails. At that point it should stop the `doChat` for that user and remove it from the `users` map. Doing this is left as an exercise.

Network Security

The topics of security and information assurance in computing are extremely broad in scope and of critical importance in our modern, computer-driven world. Programs that deal with confidential information need to take efforts to protect that information so that outside entities can not get to it. Programs that run critical systems need to take precautions to make sure that they can not be taken down by individuals without the authorization to do so.

The chat program we just wrote failed on all accounts in regards to security. Adding the proper error handling would be a major step in the direction to make sure that only the person running the server can take it down. However, every message coming in to and out of the server is sent pretty much as plain text that could be easily read at any number of points between the client and the server if sent across a network. Granted, this was not intended to be a highly secure system. It does nothing to even verify who is connecting to it. This does not mean that it would never be used to discuss things that people might want to keep private. Someone might feel that if they set up the server on a machine just temporarily, then no one they do not invite would have a chance to come in and snoop on what is being said. The problem with this assumption is that another party could be watching communication without even knowing about the server.

The standard way to circumvent someone from listening to network communications is to encrypt messages. The person could still see the message going by, but because it is encrypted they can not read it. The simplest way to do this with socket communication in Scala is to use a `javax.net.ssl.SSLServerSocket` on the server and connect with `javax.net.ssl.SSLSockets`. These make sockets that encrypt all communications using the Secure Socket Layer (SSL) protocol. This is the same protocol used by `ssh` and `scp`.

23.3 URLs

As was mentioned at the beginning of the chapter, you have likely been using networked programs for years. Most notable of these would be your web browser. Web browsers were originally written as little more than a program to pull information down from servers and display that information on your screen. Many web applications now have some 2-way communication, but there is still a lot on the web that can be accessed by simply pulling down information.

When you use a web browser to access information you have to tell it where you want to pull the information from and what information you want to pull down. This is done with a Uniform Resource Locator or URL. Most of the URLs you enter into a browser begin with "http". This tells the browser that it is supposed to communicate with the server using the Hypertext Transport Protocol (HTTP). This is a standard for requesting information across sockets using fairly simple text commands. By default, HTTP servers sit on port 80 though other port numbers can be specified as part of the URL. You could pull down information from an HTTP server by opening a socket and sending the appropriate commands. However, because HTTP is so ubiquitous, there is a `java.net.URL` class that will help to do this for you.

There are a number of ways to build a URL where you can specify different parts of the URL, but the simplest to describe just takes a `String` with the same text that you would enter into a browser giving a complete specification of what information you want. For example, you could use the following line of code to get a URL for the main page of Scala.

```
import java.net._
val scalaURL = new URL("http://www.scala-lang.org/")
```

The URL class has a method called openStream that will return an InputStream that can be used to read the contents of the URL. We could use this in code like the following to read the complete contents of that URL and convert it to a String.

```
import java.io._
import collection.mutable
val urlis = new BufferedInputStream(scalaURL.openStream())
val buffer = mutable.Buffer[Byte]()
var res = urlis.read()
while(res>=0) {
  buffer += res.toByte
  res = urlis.read()
}
val contents = new String(buffer.toArray)
```

This code opens the stream and wraps it in a BufferedInputStream to improve performance. It then creates a Buffer[Byte] to hold the values that are read, then enters a loop where bytes are read one at a time and added to the buffer. This process ends when the value read is less than zero, indicating that the end of the stream has been reached. After the end has been reached, the buffer is converted to an array and a new String is build from that array.

The use of a while loop here is called for because we do not know how much data we are loading in advance and other approaches would require that. Instead of a while loop this could easily be converted to use a recursive function as well. Just make sure to specify @tailrec so that you know what you write will not overflow the stack. Knowing the length of the content would allow this alternate version, which is significantly simpler. The challenge is knowing what goes in place of the question marks.

```
import java.io._
val urlis = new BufferedInputStream(scalaURL.openStream())
val buffer = new Array[Byte](???) // Needs the size of the contents in bytes.
urlis.read(buffer)
val contents = new String(buffer)
```

It is tempting to call available on the stream. Unfortunately, this is not a good solution. That method returns the number of bytes that can be read before blocking. That number is often different from the total size. If the stream was just opened it will be small. Waiting for the full contents to become available is not an appropriate solution.

One way to do this that will work from some content is using a java.net.URLConnection. The URL class as a method called openConnection:URLConnection that can return one of these objects. The URLConnection has a getContentLength:Int method that will tell you the size of the contents of that URL. There is also a getContentLengthLong:Long method that can give you the size for content that exceeds the maximum size of an Int. A call to getInputStream:IntputStream on the URLConnection will provide an input stream that you can read from.

```
import java.io._
val connection = scalaURL.openConnection()
val urlis = new BufferedInputStream(connection.getInputStream())
val buffer = new Array[Byte](connection.getContentLength)
urlis.read(buffer)
```

```
val contents = new String(buffer)
```

Unfortunately, the getContentLength method does not work with all types of content. If the length is not available, a value of -1 is returned which will cause this code to crash. For this reason, the while loop is probably the most robust solution.

If the URL pointed to something other than a text file, the data that was read would have to be handled in a different way. For example, it might point to an image. The URLConnection has a method called getContentType:String that can help you determine what type of data the URL references. You would then have to write code to load in the data and process it appropriately. There is also a method called getContent:AnyRef that is part of both URL and URLConnection. This method can return various things depending on the type of the contents. For example, getContent will return an image object if that is what the URL refers to. The fact that it returns AnyRef means that you should do a match on the value and handle it appropriately depending on what it is. The code might look something like this.

```
url.getContent() match {
  case img:java.awt.Image =>
    // Code for handling an image.
  case is:InputStream =>
    // This is what you get for text so you need code to read it.
  case _ =>
    // Some other type.
}
```

23.4 Remote Method Invocation (RMI)

In the next section we will be adding networking to the project in a way that will let users collaborate on a drawing while working on different programs. In order for this to work, they need to be able to share a drawing, chat with one another, and modify elements of that shared drawing. We could also consider adding other features to the sharing such as a sketch pad where they can doodle ideas. Now picture the code to implement these features using sockets.

The challenge that such an implementation will face is that there are many different types of messages being sent between the two users and those different messages can have an impact on very different parts of the code on either end. The code to handle this type of situation using sockets gets very messy. It might be tempting to use a different port for each type of message and keep many different sockets open, but such a solution is sloppy and does not scale. The operating system puts limits on the number of resources a single application can use. Instead, you have to send additional information with each message that identifies what type of message it is, and what objects it should be directed toward.

Dealing with that extra bookkeeping is challenging and time consuming to code. It is also difficult to debug. For these reasons, and the fact that this is something that happens in many applications, the Java libraries provide a simplified alternative called Remote Method Invocation (RMI). As the name implies, RMI lets you make method calls on objects that are located on other machines. It automatically handles the bookkeeping so that information is sent to the right place and the correct code is invoked on the opposite side. The downsides of RMI are that it does have a bit of a learning/setup curve, it only works between languages running on the Java Virtual Machine (JVM) , and it runs slower than well- done specialized code for the same task. The rest of this section will talk about the setup and help you climb the learning curve. Our application is running on the JVM so that is not a problem for us. Our application is not highly dependent on performance either so the costs involved will not be significant and will be more than make up for with the convenience that RMI provides.

There are a number of steps that are required to get an application to use RMI. Here they are listed out.

1. Set up abstract types that specify the methods that can be called remotely. This is done by making traits for any types that you want to have remote access to. These traits need to inherit from `java.rmi.Remote` and they should only include methods that are abstract. So the only things in the trait should be declarations using `def` that do not specify any code to run.[2] Each method needs to say it can throw a `java.rmi.RemoteException`. More on that below.

2. Write a class that extends `java.rmi.server.UnicastRemoteObject` with the trait you just created. You have to implement all the methods from the trait. Provide others as needed to make a complete implementation for your program.[3]

3. Create a server application that binds the server object to an RMI registry. You do this by calling `bind` or `rebind` on `java.rmi.Naming`. Those methods take two arguments. The first is the name the registry will know the object by and the second is the object to bind. The `bind` method will throw an exception if that name is already in use. The `rebind` method will unbind the old object and bind this new one in its place.

4. Create a client application that uses the `lookup` method of `java.rmi.Naming` to get hold of the server object. The `lookup` method takes a single argument that is a URL for the server and object to look up. The protocol for the URL should be "rmi" and the name should be separated from the server by a slash. An example would be "rmi://computer.college.edu/ServerObjectName". The name used here must match what the server used when it was bound to the registry.

5. Lastly you need to run `rmiregistry` then start up the server and clients. The `rmiregistry` program is part of the Java Development Kit (JDK) installation. For a Java program, the registry needs to have access to the compiled ".class" files for the application. This can be accomplished either by setting a `CLASSPATH` environment variable or by running `rmiregistry` in the based directory that the program compiles to.[4] For a Scala program, odds are good that it will also need to be able to find certain Scala jar files. These are in the `lib` file of the Scala install directory. More details on this below in the example.

This might seem like a daunting list, but it really is only five steps and once you have done that it becomes much easier to extend your networked application. We will go back through the steps shortly as part of an example.

Before the example, there is one other detail of RMI that needs to be covered, the passing of arguments. Back in sections 7.8-7.9 we talked about how arguments are passed in Scala. The normal style is to pass a reference to the object with support for pass-by-name sematics

[2]Java has a construct called an interface that is purely abstract. That is what RMI actually requires. Scala traits with no data that have only abstract methods get compiled to Java interfaces.

[3]There are other servers you can use instead of `UnicastRemoteObject`. That one is specified because it is the most straightforward.

[4]If you are using Eclipse, there is a `bin` directory in the project that is the base directory for the compiled files.

as well. With RMI there are two different ways to pass things: pass-by-value and pass-by-reference. Which happens depends on the type of the object being passed. Pass-by-value implies that the object is copied and the other machine gets its own copy that is independent of the first one. This happens for Serializable data and includes the standard types you are used to working with in Scala. Objects with a type that implements `Remote` are passed as a remote reference. Instead of getting a copy of the object, the remote machine gets an object that implements the remote interface and which has the ability to communicate calls across the network. Methods called on these objects go back to the original value. The fact that there is network communication involved means that they are much slower than calls on local objects.

If you try to pass a value that is not serializable or remote, you will get an exception. The reason for this should be fairly clear to you, RMI is utilizing object streams to perform communication. Only objects that can be serialized will go through those. So the discussion of how to make objects serializable in 22.4.3 is relevant when using RMI as well.

To help you see how this all goes together, we will write another chat application that uses RMI instead of making direct use of sockets. We will run through the steps from above and point out what code is written for each one. Before we can write it though, we need to figure out what we are writing. This chat application is going to be a bit different from the last one. In particular, it will have a GUI for the client. The GUI will include not only standard chat elements, but a list of everyone logged onto the server. The idea is that users can send private messages by selecting one or more users from the list when they do their chats. Only the selected users will get their messages.

Starting with step 1 from the RMI steps, we lay out the types we want to have remote access to and put in the methods that we need to be able to call remotely. We start with the server.

```
package scalabook.chat

import java.rmi._

trait RMIServer extends Remote {
  @throws(classOf[RemoteException]) def connect(client:RMIClient):String
  @throws(classOf[RemoteException]) def disconnect(client:RMIClient):Unit
  @throws(classOf[RemoteException]) def getClients:Seq[RMIClient]
  @throws(classOf[RemoteException]) def publicMessage(client:RMIClient,
      text:String):Unit
}
```

The fundamental structure is that we have a trait that extends `java.rmi.Remote`. Inside of that trait there are four methods that we want to be able to call on a server remotely. These allow the client to connect, disconnect, and get a list of the current clients. There is also a method called `publicMessage` that should be called whenever a client sends out a message publicly. This allows the server to keep a brief history so that someone new logging in can see what has been discussed recently.

This code uses a new annotation that we have not seen before, `@throws`. This annotation allows us to specifically state exceptions that can be thrown by methods. It takes a single argument for the class of exception that might be thrown. This is not something that we have to do very often in Scala. The equivalent code is much more common in Java and in this case it is required for Scala because we need our code to work smoothly with the Java RMI libraries.

A second trait is set up for the methods that we want to be able to call remotely on the client. It looks like the following.

```
package scalabook.chat

import java.rmi._

trait RMIClient extends Remote {
  @throws(classOf[RemoteException]) def name:String
  @throws(classOf[RemoteException]) def message(sender:RMIClient, text:String):Unit
  @throws(classOf[RemoteException]) def clientUpdate(clients:Seq[RMIClient]):Unit
}
```

There are only three methods on the client. These get the name of the client, deliver a message to this client, and tell the client to update its list of clients logged into the server.

Looking at these two interfaces you might feel like there is something missing, methods on the server for sending out public or private chats. In the last chat, when one person wrote a chat message, that was sent to the server as just a String, and the server figured out what to do with it then sent it to the various chat participants. This uses the standard client-server network application structure where all communication goes through the server. This worked well for sockets and made our life easier. However, it is not very efficient as the server becomes a significant bottleneck. What is more, if you really wanted your messages to be private, you lack some security because they all have to go through the server for processing. Basically, the server sees everything and does a lot of work sending out all the messages.

RMI makes it easy for us to deviate from the client-server structure and use more of a peer-to-peer approach. The clients won't just have a remote reference to the server, they will have remote references to all the other clients. This means that each client can call the `message` method on the other clients it wants messages to go to. This reduces pressure on the server and makes it less of a bottleneck. If we did not want to keep a history of what was being said on the server, chat traffic would not be sent to the server at all.

This next piece of code includes steps 2 and 3 for the server. Step 2 is in the class `RMIServerImpl` while step 3 is in the companion object.

```
package scalabook.chat

import java.rmi._
import java.rmi.server.UnicastRemoteObject
import collection.mutable

class RMIServerImpl extends UnicastRemoteObject with RMIServer {
  private val clients = mutable.Buffer[RMIClient]()
  private var history = mutable.ListBuffer("Server Started\n")

  @throws(classOf[RemoteException]) def connect(client:RMIClient):String = {
    clients += client
    sendUpdate
    history.mkString("\n")+"\n"
  }

  @throws(classOf[RemoteException]) def disconnect(client:RMIClient) {
    clients -= client
    sendUpdate
  }

  @throws(classOf[RemoteException]) def getClients:Seq[RMIClient] = clients
```

```
  @throws(classOf[RemoteException]) def publicMessage(client:RMIClient,text:String)
      {
    history += client.name+" : "+text
    if(history.length>10) history.remove(0)
  }

  private def sendUpdate {
    val deadClients = clients.filter(c =>
      try {
        c.name
        false
      } catch {
        case ex:RemoteException => true
      }
    )
    clients --= deadClients
    clients.foreach(_.clientUpdate(clients))
  }
}

object RemoteServerImpl {
  def main(args:Array[String]) {
    val server = new RMIServerImpl
    Naming.rebind("ChatServer",server)
  }
}
```

The class defines two fields for storing the clients and history, then implements the four methods from the interface with rather short methods. You can see that the `connect` method returns the history. Both `connect` and `disconnect` make use of the one private method that is also included, `sendUpdate`. This method sends out a message to all the clients letting them know that the client list has been modified. This method is a bit longer because it first runs through all the clients and calls their `name` method. That value is not used for anything, but it serves as a check that the client is still there. This is part of a filter operation that is checking for dead clients. The clients that can not be reached are removed from the list before that list is sent out to all the active clients.

The main method for running the server is remarkably simple. It makes an instance of `RemoteServerImpl` then uses rebind to bind it to the `rmiregistry`.

This next piece of code shows steps 2 and 4 for the client. This code is a bit longer because it includes support for the GUI. Even that does not make it all that long.

```
package scalabook.chat

import java.rmi._
import java.rmi.server.UnicastRemoteObject
import collection.mutable
import swing._

class RMIClientImpl(myName:String,server:RMIServer) extends UnicastRemoteObject
    with RMIClient {
  private val chatText = new TextArea(server.connect(this))
  chatText.editable = false
  private var clients = server.getClients
  private val userList = new ListView(clients.map(_.name))
```

```
private val chatField = new TextField("") {
  listenTo(this)
  reactions += {
    case e:event.EditDone => if(text.trim.nonEmpty) {
      val recipients = if(userList.selection.items.isEmpty) {
        server.publicMessage(RMIClientImpl.this,text)
        clients
      } else {
        userList.selection.indices.map(clients).toSeq
      }
      recipients.foreach(r => try {
        r.message(RMIClientImpl.this,text)
      } catch {
        case ex:RemoteException => chatText.append("Couldn't send to one
            recipient.")
      } )
      text = ""
    }
  }
}

private val frame = new MainFrame {
  title = "Chat"
  contents = new BorderPanel {
    import BorderPanel.Position._
    val scrollList = new ScrollPane(userList)
    scrollList.preferredSize = new Dimension(200,500)
    layout += scrollList -> West
    layout += new BorderPanel {
      val scrollChat = new ScrollPane(chatText)
      scrollChat.preferredSize = new Dimension(500,200)
      layout += scrollChat -> Center
      layout += chatField -> South
    } -> Center
  }
  listenTo(this)
  reactions += {
    case e:event.WindowClosing => server.disconnect(RMIClientImpl.this)
    case _ =>
  }
}
frame.visible = true

@throws(classOf[RemoteException]) def name:String = myName

@throws(classOf[RemoteException]) def message(sender:RMIClient,text:String) {
  chatText.append(sender.name+" : "+text+"\n")
}

@throws(classOf[RemoteException]) def clientUpdate(cls:Seq[RMIClient]) {
  clients = cls
  if(userList!=null) userList.listData = cls.map(c =>
    try {
      c.name
    } catch {
```

```
        case ex:RemoteException => "Error"
      } )
  }
}

object RMIClientImpl {
  def main(args:Array[String]) {
    val mName = Dialog.showInput(null, "What server do you want to connect to?",
        "Server Name", Dialog.Message.Question, null, Nil, "")
    mName match {
      case Some(machineName) =>
        Naming.lookup("rmi://"+machineName+"/ChatServer") match {
          case server:RMIServer =>
            val name = Dialog.showInput(null, "What name do you want to go by?",
                "User Name", Dialog.Message.Question, null, Nil, "")
                    if(name.nonEmpty) new RMIClientImpl(name.get, server)
          case _ => println("That machine does not have a registered server.")
        }
      case None =>
    }
  }
}
```

The GUI includes a `TextArea` to display chat text, a `TextField` to enter chat text into, and a `ListView` to show the names of the clients so that private messages can be sent. The text field has event-handling code to send out messages when the user stops editing the field. This is where the peer-to-peer aspect of the code comes into play as each client can directly call `message` on the other clients.

Those elements are set in a frame using nested `BorderPanels`. Both the list and the main area are embedded in `ScrollPanes` that have their preferred size set so that the GUI will look reasonable. The frame also includes event code so that when the window is closed, this client will disconnect from the server. The GUI as a whole looks like figure 23.2.

The implementations of the remote methods are quite short for the client as well. Only the updating of the client list has any length and that is due to the fact that the call to `name` is a remote call that needs some error handling so that the code does not die on the first client that can not be reached.

That is all the code we need. All that remains is step 5 where we start of the `rmiregistry` and run the server and clients. The use of Scala does add one complication to that first part because `rmiregistry` is a Java program, not a Scala program and by default it will know about the Java libraries, but not the Scala ones. For this reason, we need to tell it where to find the Scala libraries.

When you install Scala there are a number of JAR files that are put in a directory called `lib`. JAR stands for Java archive and is a standard way of packaging files for Java. Not all programs will need all of these files, but the one called "scala-library.jar" will be needed by the majority of Scala programs, including this one. On a Linux or Mac, if you are in the root directory for the compiled .class files, you could do the following: `CLASSPATH=$(SCALA_HOME)/lib/scala-library.jar:. rmiregistry`. If `SCALA_HOME` has not been set up on your system just replace that with the path to your Scala install. Once that is running start up the server on that same machine and then start clients on that machine or other machines that can reach it. Each client will ask you what machine you want to connect to and what name you want to be known by. If the client is on the same machine as the server, you can use "localhost" as the machine name.

There is one aspect of this application that could be seen as a weakness. Calls to the

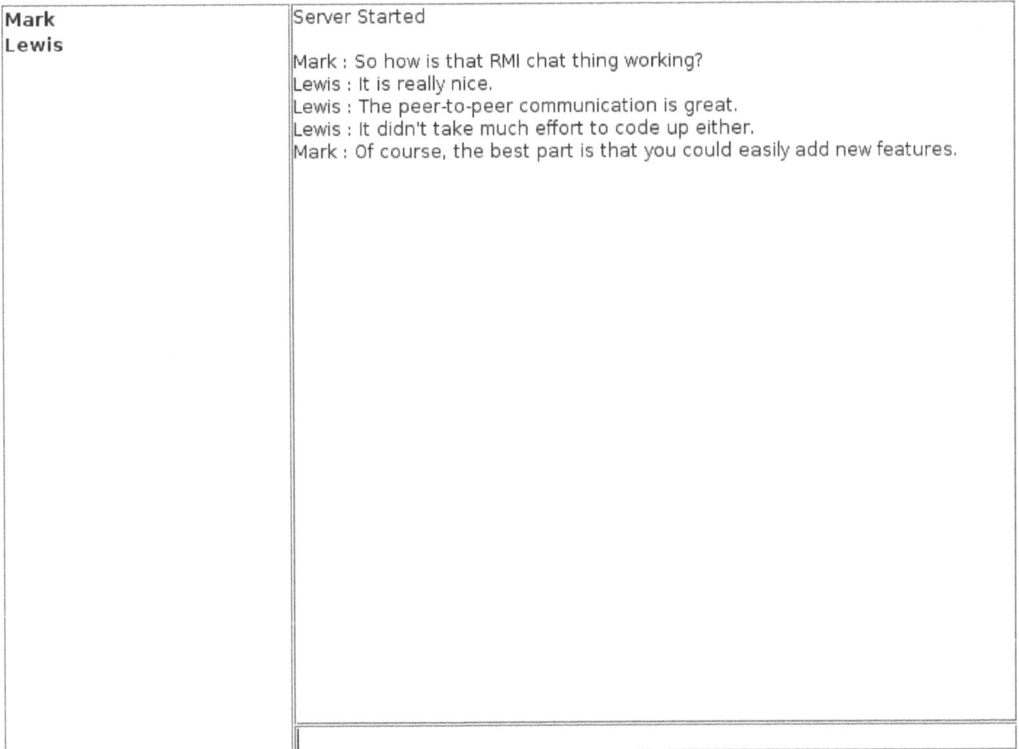

FIGURE 23.2: This is a screen shot of the RMI chat client with two users connected.

client `name` method are made more often than they probably should be. The fact that `name` is a method in a remote trait means that every call to this will incur network overhead. In this case, "frequent" really just means when chats go through or when client lists are updated. Given that those things are driven by human interactions this likely is not a problem. Frequently updating names likely make it harder to "spoof" the system and act like someone else. In general, you do need to be cognizant of when you are making remote calls because they do have a significant overhead. There might be times when you pass things by value to allow a client to do fast processing then pass a modified copy back just to avoid the overhead of many remote calls.

To really understand the benefits of RMI, consider what it takes to add features. Perhaps you want to make a menu option for things like sharing images or having a shared sketchpad. Using socketing, those additions would make the code significantly more complex. With RMI, you would just add one or two methods into the `RMIClient` trait and the `RMIClientImpl` class. You implement those methods and put in code to call them and you are done. Setting up the RMI is a bit more hassle, but once in place, you can add new features to the RMI-based application just about as easily as you would a non-networked application.

23.5 Collaborative Drawing (Project Integration)

Now it is time to put things together into the drawing program. We will do this in two ways. To demonstrate socketing, we will add a simple feature that allows a copy of a drawing to be sent from one user to another. This will be simple to implement because we already have code that sends a drawing to a stream that writes to a file. The other aspect of networking that we want to add is more significant and we will use RMI. This will be a collaboration panel where multiple users can post drawings for others to pull down, chat with one another, and even draw basic sketches.

We will add a second menu with the title "Collaborate" onto the menu bar for the frame in `DrawingMain`. This menu will have four options. The first two will work with the basic sending of drawings over the network. The second two will work with the more sophisticated RMI-based collaboration scheme. We'll start with just the first two and add the following code into the `DrawingMain` class after the creation of the "File" menu.

```
contents += new Menu("Collaborate") {
  contents += new MenuItem("Start Server") {
    action = Action("Start Server"){
      startServer
      enabled = false
    }
  }
  contents += new MenuItem(Action("Send Drawing")(sendDrawing))
}
```

The first option is intended to set up a `ServerSocket` on the current machine that will sit there until the program is stopped and accept connections from other machines that want to share drawings. The second option allows us to send the drawing in the tab that is currently selected to another machine. For this code to compile, we need to implement the `startServer` and `sendDrawing` methods.

These two methods are fairly simple and they can be put directly in `DrawingMain` because we have already made it so that the `Drawing` type can be serialized. If this had not been done for saving through streams earlier, we would have to add that functionality now. A possible implementation of the `startServer` method is shown here.

```
private def startServer {
  actors.Actor.actor {
    val ss = new ServerSocket(8080)
    while(true) {
      val sock = ss.accept
      val ois = new ObjectInputStream(new
          BufferedInputStream(sock.getInputStream()))
      val sender = ois.readUTF()
      val title = ois.readUTF()
      ois.readObject() match {
        case drawing:Drawing =>
          val response = Dialog.showConfirmation(tabbedPane,"Accept "+title+" from
              "+sender+"?","Accept Drawing?")
          if(response==Dialog.Result.Yes) {
            addDrawing(title,drawing)
          }
        case _ =>
```

```
      }
      ois.close()
      sock.close()
    }
  }
}
```

This makes a new ServerSocket and then enters into an infinite loop of accepting connections. This code needs to go into a separate thread because the loop will never exit. Recall from chapter 21 that all events in a GUI are handled in the same thread. If you do not put the server code in a separate thread, the GUI will become completely unresponsive because this method never ends.

That infinite loop accepts connections, opens the stream with wrapping to read objects, then it reads two strings and an object. The strings tell us who the sender is and the title of the drawing. The last read is for the drawing itself. A match statement is used to make sure the object that is read really is a Drawing. If it is, the user is asked if they want to accept that drawing from that sender. If they answer yes, it is added as a new tab using a call to the following method.

```
def addDrawing(title:String,drawing:Drawing) {
  openDrawings += drawing
  tabbedPane.pages += new TabbedPane.Page(title,drawing.propertiesPanel)
}
```

Once that is all done, the stream and the socket are closed.

The other side of this communication is sending a drawing to a different machine. This method is invoked when we select the "Collaborate → Send Drawing" menu option.

```
private def sendDrawing {
  val host = Dialog.showInput(null, "What machine do you want to send to?",
      "Machine Name", Dialog.Message.Question, null, Nil, "")
  if(host.nonEmpty) {
    val sock = new Socket(host.get,8080)
    val oos = new ObjectOutputStream(new
        BufferedOutputStream(sock.getOutputStream()))
    val name = Dialog.showInput(null, "Who do you want to say is sending?",
        "Name", Dialog.Message.Question, null, Nil, "")
    if(name.nonEmpty) {
      val title = Dialog.showInput(null, "What is the title of this drawing?",
          "Title", Dialog.Message.Question, null, Nil, "")
      if(title.nonEmpty) {
        oos.writeUTF(name.get)
        oos.writeUTF(title.get)
        oos.writeObject(openDrawings(tabbedPane.selection.index))
        oos.close()
        sock.close()
      }
    }
  }
}
```

This method asks for the machine we want to send to, then opens a socket to that machine and wraps the OutputStream of the socket for sending objects. Once those are open, it asks for the name of the sender and the title of the drawing. Those are sent through the socket, followed by the drawing in the currently selected tab. Once that is sent, everything is closed.

Neither of these methods includes proper error checking. That helps to significantly reduce the length of the code, but it is something that production network code really can not ignore as networking is a task with the ability to produce many types of exceptions.

This basic socketing approach is technically sufficient to let us share drawing, but if you use it for a while you will notice that it really does not feel very collaborative. There is not any real interaction between the different users. Putting in real interaction requires that there are many types of different messages sent between the machines. This is the type of thing that leans towards the strengths of RMI.

The ideal feature to add would be a true collaboration on a drawing where two or more users could view and edit a single drawing at the same time. This path was considered for this chapter, but eventually dismissed because the required code was too involved. It is made more challenging due to the fact that we are already using standard serialization for saving drawings and RMI features interact with serialization. In addition, users on both sides of the network connection need to interact with elements through a GUI. It happens that GUI elements are serializable, but the event handling is a bit more complex to do properly. In the end it was decided to go with this simpler form of collaboration that demonstrates real-time interactions without including many other details that would obscure the educational objectives.

The use of RMI means that we need to follow the basic steps of RMI. That begins with determining what remote interfaces we want to have and what methods should go into them. Our application here is still simple enough that we can make due with one server type and one client type. In this case, the server type will be fairly simple. Both the remote trait and the implementation are shown here.

```
package scalabook.drawing

import java.rmi._
import server.UnicastRemoteObject
import collection.mutable

trait CollaborationServer extends Remote {
  @throws(classOf[RemoteException]) def joinCollaboration(col:Collaborator) :
      (Array[Collaborator], Array[(String,Drawing)])
  @throws(classOf[RemoteException]) def addDrawing(title:String, drawing:Drawing) :
      Unit
}

class CollaborationServerImpl extends UnicastRemoteObject with CollaborationServer
    {
  private val collaborators = mutable.Buffer[Collaborator]()
  private val drawings = mutable.Buffer[(String,Drawing)]()

  @throws(classOf[RemoteException]) def joinCollaboration(col:Collaborator) :
      (Array[Collaborator], Array[(String,Drawing)]) = {
    collaborators += col
    (collaborators.toArray,drawings.toArray)
  }

  @throws(classOf[RemoteException]) def addDrawing(title:String, drawing:Drawing) {
    drawings += title -> drawing
    for(c <- collaborators) {
      try {
        c.addDrawing(title,drawing)
```

```
      } catch {
        case ex:RemoteException =>
      }
    }
  }
}
```

The trait has only two methods, one for joining and one for sharing a drawing. The implementation includes those two methods, which are both fairly short, along with two data members for storing who is currently participating in the collaboration and the drawings that have been shared, along with their titles. The `joinCollaboration` method returns all of the collaborators as well as the drawings that have been shared. It is interesting to point out that these things are passed differently. The drawings and their titles are all serializable but not remote, so they are passed by value and the new collaborator gets a copy of them. On the other hand, the collaborators are remote so the array that goes across will be filled with remote references that allow the new user to communicate with everyone else.

It is worth noting that we have used a slightly different approach to arranging files for these types. Instead of having only a single class or trait in the file, this file has both the remote trait and the implementing class.[5] This was done because the two are intimately related and it cuts down on additional imports of things like `java.rmi`. The downside of doing this is that someone looking for `CollaborationServerImpl` might take a long time to find it as there is no file with that name. This is very much a style issue and something that different instructors and employers might request different rules for.

The majority of the code for implementing the collaboration feature goes into the client code. This includes a remote interface for the `Collaborator` as well as an implementation. Here again, the two are put into a single file. This time the file is longer because the implementation is longer. If both the trait and the implementation were long we would certainly split them into two files to avoid long files. However, moving the trait out does not significantly reduce the length of this file. The code gives us the ability to produce a window that looks like figure 23.3.

```
package scalabook.drawing

import java.rmi._
import server.UnicastRemoteObject
import java.awt.image.BufferedImage
import java.awt.{Color,Font}
import java.awt.geom._
import swing._
import event._
import collection.mutable

trait Collaborator extends Remote {
  @throws(classOf[RemoteException]) def name:String
  @throws(classOf[RemoteException]) def post(text:String):Unit
  @throws(classOf[RemoteException]) def requestSketch:Array[Byte]
  @throws(classOf[RemoteException]) def updateSketch(who:Collaborator,
      img:Array[Byte]):Unit
  @throws(classOf[RemoteException]) def addDrawing(title:String,
      drawing:Drawing):Unit
}
```

[5]This is something that you can do in Scala, but not in other languages, like Java. In Java you can only have one public type per file and it must have the same name as the file.

```
/**
 * This implementation will run on the clients and bring up a window that will show
 * thumb nails of the shared drawings, a simple sketch pad, and chat window.
 */
class CollaboratorImpl(server:CollaborationServer) extends UnicastRemoteObject
    with Collaborator {
  private[drawing] val sketch = {
    val img = new BufferedImage(CollaboratorImpl.sketchSize,
        CollaboratorImpl.sketchSize, BufferedImage.TYPE_INT_ARGB)
    for(i <- 0 until img.getWidth; j <- 0 until img.getHeight)
      img.setRGB(i,j,0)
    img
  }

  private[drawing] val (sketches,drawings) = {
    val (cols,draws) = server.joinCollaboration(this)
    mutable.Map(cols.map(c => {
      try {
        Some(c -> ImageHandling.imageFromBytes(c.requestSketch))
      } catch {
        case ex:RemoteException => None
      }
    }).filter(!_.isEmpty).map(_.get):_*) -> mutable.Buffer(draws:_*)
  }

  private val nameField = new TextField("Name")
  private val chatField = new TextField
  private val chatArea = new TextArea
  chatArea.editable = false

  object SharedPanel extends Panel {
    preferredSize = new Dimension(150,200)
    val bigFont = new Font(Font.SERIF,Font.PLAIN,75)

    override def paint(g:Graphics2D) {
      g.setPaint(Color.white)
      g.fillRect(0,0,size.width,size.height)
      g.scale(0.15,0.15)
      g.setPaint(Color.black)
      for(i <- drawings.indices) {
        drawings(i)._2.renderTo(g)
        g.setPaint(Color.black)
        g.setFont(bigFont)
        g.drawString(drawings(i)._1,0,100/0.15f)
        g.translate(0,200/0.15)
      }
    }

    def setSize() {
      preferredSize = new Dimension(150,200*drawings.size)
    }

    listenTo(mouse.clicks)
    reactions += {
```

```
      case mc:MouseClicked =>
        if(mc.clicks==2 && mc.point.y<drawings.size*200) {
          val index = mc.point.y/200
          DrawingMain.addDrawing(drawings(index)._1,drawings(index)._2)
        }
    }
  }

  val sketchPanel = new SketchPanel(this)

  sketchUpdated()

  import BorderPanel.Position._

  val collabSplit = new SplitPane(Orientation.Horizontal,
    sketchPanel,
    new BorderPanel {
      layout += chatField -> North
      layout += chatArea -> Center
    }
  )

  private val frame = new Frame {
    contents = new BorderPanel {
      layout += new BorderPanel {
        layout += new Label("Name") -> West
        layout += nameField -> Center
      } -> North
      layout += new ScrollPane(SharedPanel) -> West
      layout += collabSplit -> Center
    }
    listenTo(chatField)
    reactions += {
      case ed:EditDone =>
        if(chatField.text.trim.nonEmpty) {
          foreachCollaborator(_.post(name+" : "+chatField.text))
          chatField.text = ""
        }
    }
    size = new Dimension(800,600)
  }

  def showFrame {
    frame.visible = true
    collabSplit.dividerLocation = 0.7
  }

  private def forCollaborators[A](f:Collaborator=>A):Seq[A] = {
    (for(c <- sketches.keys) yield try {
      Some(f(c))
    } catch {
      case ex:RemoteException => None
    }).toSeq.filter(!_.isEmpty).map(_.get)
  }
```

```
  private def foreachCollaborator(f:Collaborator=>Unit) {
    for(c <- sketches.keys) try {
      f(c)
    } catch {
      case ex:RemoteException =>
    }
  }

  @throws(classOf[RemoteException]) def name:String = nameField.text

  @throws(classOf[RemoteException]) def post(text:String) {
    chatArea.append(text+"\n")
  }

  @throws(classOf[RemoteException]) def requestSketch:Array[Byte] = {
    ImageHandling.imageAsBytes(sketch)
  }

  @throws(classOf[RemoteException]) def
      updateSketch(who:Collaborator,img:Array[Byte]) {
    sketches(who)=ImageHandling.imageFromBytes(img)
    sketchPanel.sketchSlider.max = sketches.size-1
    sketchPanel.repaint()
  }

  private[drawing] def sketchUpdated() {
    val bytes = ImageHandling.imageAsBytes(sketch)
    foreachCollaborator(c => {
      c.updateSketch(this,bytes)
    })
  }

  @throws(classOf[RemoteException]) def addDrawing(title:String,drawing:Drawing) {
    drawings += title->drawing
    SharedPanel.setSize()
    SharedPanel.repaint()
  }
}

object CollaboratorImpl {
  val sketchSize = 1000
}
```

The class starts with the creation of a sketch image that is filled in with transparent color. After that, is code that joins the collaboration, then uses the return from that to get the sketches of all the other collaborators as well as all shared drawings. This is followed by code to set up the GUI in a frame and two helper methods that make it easier to perform actions on the collaborators without worrying about exceptions. The implementation class ends with definitions of the five remote methods that were put into the trait. A companion object includes a constant that can be used to size the images used in the sketches.

The `CollaboratorImpl` depends on code in two other files. The fact that `BufferedImage` is not serializable presents a problem for sending sketches between the collaborators. If you look in the remote interface, sketches are passed around using the type `Array[Byte]`.

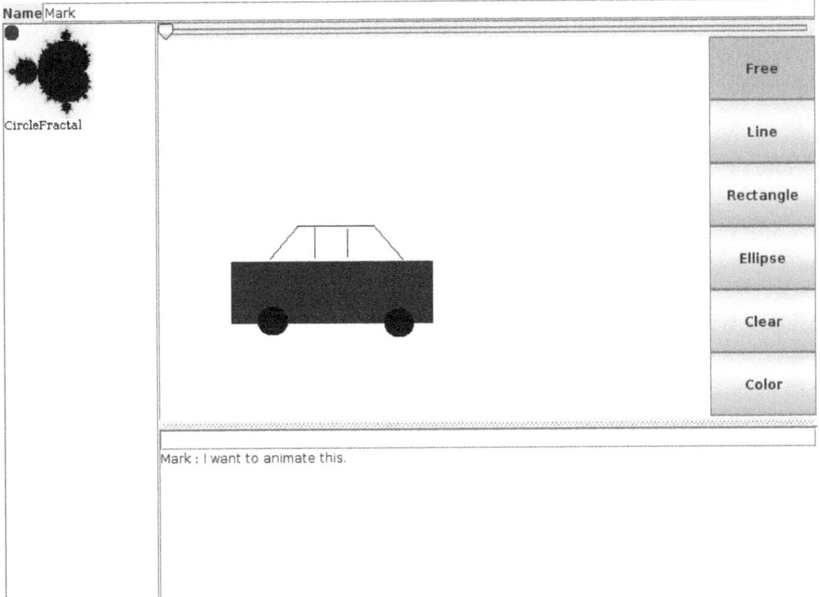

FIGURE 23.3: This is a screen shot of the collaboration window. On the left is a column showing shared drawings. On the bottom is a chat area. The main portion is a shared sketch region where different collaborators can draw using some simple types.

To convert from `BufferedImage` to `Array[Byte]` and back, we have an object called `ImageHandling`, which has two methods for performing these conversions.

```
package scalabook.drawing

import javax.imageio.ImageIO
import java.awt.image.BufferedImage
import java.io._

object ImageHandling {
  def imageAsBytes(img:BufferedImage):Array[Byte] = {
    val baos = new ByteArrayOutputStream()
    ImageIO.write(img,"png",baos)
    baos.toByteArray()
  }

  def imageFromBytes(bytes:Array[Byte]):BufferedImage = {
    val bais = new ByteArrayInputStream(bytes)
    ImageIO.read(bais)
  }
}
```

There are quite a few different ways that you could convert a `BufferedImage` into something that is serializable. One way would be to simply make a big `Array[Array[Int]]` and store Red, Green, Blue (RGB) values of pixels in that. This approach will work, but is very inefficient on the networking side because that sends a completely uncompressed version of the image. This class uses `javax.imageio.ImageIO` to compress the image using PNG format. A `java.io.ByteArrayOutputStream` is used to get `ImageIO.write` to put the data into an array that we can send without having to write to, and read from disk. A ByteArrayInputStream is used on the other side with ImageIO.read to get back to an image. The PNG format is fairly efficient compression which is non-lossy so it works well for our purposes.

The CollaboratorImpl also depends on a type called SketchPanel. This is a class that holds the GUI component for the sketch. This is a large enough piece of code that it was worth pulling it completely out. Here is what it looks like.

```
package scalabook.drawing

import swing._
import event._
import java.awt.{Color,Graphics2D}
import java.awt.geom._
import java.awt.image.BufferedImage

class SketchPanel(collab:CollaboratorImpl) extends BorderPanel {
  class NamedSketch(val name:String,val col:Collaborator,val img:BufferedImage) {
    override def toString = name
  }
  val sketchSlider= new Slider
  sketchSlider.min = 0
  sketchSlider.max = collab.sketches.size-1
  var curColor = Color.black
  def drawCurrent {
    val g=collab.sketch.createGraphics()
    g.setPaint(curColor)
    curDrawer.draw(g)
    collab.sketchUpdated()
  }
  trait Drawer {
    var mouseDown = false
    def pressed(x:Int,y:Int)
    def released(x:Int,y:Int)
    def dragged(x:Int,y:Int)
    def draw(g:Graphics2D)
  }
  object FreeDrawer extends Drawer {
    var path:GeneralPath = new GeneralPath
    def pressed(x:Int,y:Int) {
      mouseDown=true
      path=new GeneralPath
      path.moveTo(x,y)
    }
    def released(x:Int,y:Int) {
      path.lineTo(x,y)
      drawCurrent
      mouseDown=false
    }
    def dragged(x:Int,y:Int) { path.lineTo(x,y) }
    def draw(g:Graphics2D) { if(mouseDown) g.draw(path) }
  }
  trait PositionDrawer extends Drawer {
    var (sx,sy,ex,ey) = (0,0,0,0)
    def pressed(x:Int,y:Int) {
```

```
    mouseDown=true
    sx=x
    sy=y
    ex=x
    ey=y
  }
  def released(x:Int,y:Int) {
    ex=x
    ey=y
    drawCurrent
    mouseDown=false
  }
  def dragged(x:Int,y:Int) {
    ex=x
    ey=y
  }
}
object LineDrawer extends PositionDrawer {
  def draw(g:Graphics2D) { if(mouseDown) g.draw(new Line2D.Double(sx,sy,ex,ey))
      }
}
object RectangleDrawer extends PositionDrawer {
  def draw(g:Graphics2D) { if(mouseDown) g.fill(new Rectangle2D.Double(sx min
      ex, sy min ey, (ex-sx).abs, (ey-sy).abs)) }
}
object EllipseDrawer extends PositionDrawer {
  def draw(g:Graphics2D) { if(mouseDown) g.fill(new Ellipse2D.Double(sx min ex,
      sy min ey, (ex-sx).abs, (ey-sy).abs)) }
}
var curDrawer:Drawer = FreeDrawer
val buttons = new GridPanel(6,1) {
  val b = Array(
    new ToggleButton {
      selected = true
      action = Action("Free")(curDrawer = FreeDrawer)
    },
    new ToggleButton { action = Action("Line")(curDrawer = LineDrawer) },
    new ToggleButton { action = Action("Rectangle")(curDrawer =
        RectangleDrawer) },
    new ToggleButton { action = Action("Ellipse")(curDrawer = EllipseDrawer) }
  )
  val group = new ButtonGroup(b:_*)
  b.foreach(contents += _)
  contents += Button("Clear") {
    for(i <- 0 until collab.sketch.getWidth; j <- 0 until
        collab.sketch.getHeight)
      collab.sketch.setRGB(i,j,0)
    collab.sketchUpdated()
  }
  contents += Button("Color") {
    val c = javax.swing.JColorChooser.showDialog(peer,"Select Color",curColor)
    if(c!=null) curColor = c
  }
}
val sketchDraw = new Panel {
```

```
      override def paint(g:Graphics2D) {
        g.setPaint(Color.white)
        g.fillRect(0,0,size.width,size.height)
        val sketchSeq = collab.sketches.toSeq
        for(index <- 0 until collab.sketches.size) {
          val (c,i) = sketchSeq((index+sketchSlider.value)%sketchSeq.size)
          g.drawImage(i,0,0,null)
        }
        g.setPaint(curColor)
        curDrawer.draw(g)
      }
    }
    import BorderPanel.Position._
    layout += sketchDraw -> Center
    layout += buttons -> East
    layout += sketchSlider -> North
    listenTo(sketchDraw.mouse.clicks,sketchDraw.mouse.moves,sketchSlider)
    reactions += {
      case mp:MousePressed =>
        curDrawer.pressed(mp.point.x,mp.point.y)
        sketchDraw.repaint()
      case mp:MouseDragged =>
        curDrawer.dragged(mp.point.x,mp.point.y)
        sketchDraw.repaint()
      case mp:MouseReleased =>
        curDrawer.released(mp.point.x,mp.point.y)
        sketchDraw.repaint()
      case vc:ValueChanged =>
        sketchDraw.repaint()
    }
}
```

The class gets passed the `CollaboratorImpl` because it needs to have access to the sketch and some other parts of that object. Much of this class is focused on dealing with the user drawing to his/her sketch. There is a group of `ToggleButtons` that allow the user to select how they are drawing as well as other buttons to clear the sketch and select a color. The `ToggleButtons` are added to a `ButtonGroup` so that only one can be selected at a time.

The State Pattern is used to make it so that mouse input is interpreted properly for each different type of thing that the user can draw [6]. The way this works is that we have a trait called `Drawer` that has four abstract methods in it. These methods deal with the mouse being pressed, released, and dragged as well as how to draw what would be drawn if the mouse were released. Each style of drawing is represented by an object that extends this interface. A mutable variable named `curDrawer` stores a reference to one of these objects, the one associated with the currently selected drawing mode. Events are passed through to that object, and are placed where the proper shape needs to be drawn to screen of an image.

You might be tempted to write code like this by keeping an integer value telling you what style you are drawing in and then having multiple if/else conditions to take the right action in different situations. While that approach can work, it is less flexible and more bug prone. The real advantage of the abstraction shown here is how easy it is to add new drawing modes. You simply add a new button to the GUI and a new `Drawer` object for handling that style and it works. This is an example of using abstraction to make code more flexible.

The last thing we need is to add this functionality to the GUI. First we add the following two lines to DrawingMain so that it can keep track of a server and a client.

```
private var server:CollaborationServer = null
private var client:CollaboratorImpl = null
```

After that, the following code can be added to the "Collaborate" menu below the two options for simple sharing.

```
contents += new Separator
val joinItem = new MenuItem("Join") {
  action = Action("Join") {
    text = "View"
    if(server==null) {
      val host = Dialog.showInput(null, "What server host do you want to
          use?", "Server Name", Dialog.Message.Question, null, Nil, "")
      if(host.nonEmpty && host.get.trim.nonEmpty) {
        try {
          Naming.lookup("rmi://"+host.get+"/Collaborate") match {

            case svr:CollaborationServer =>
              server = svr
              client = new CollaboratorImpl(server)
            case _ => Dialog.showMessage(tabbedPane,"The wrong type of
                object was found on that server.")
          }
        } catch {
          case ex:NotBoundException =>
            server = new CollaborationServerImpl()
            Naming.bind("Collaborate",server)
            client = new CollaboratorImpl(server)
        }
      }
    }
    if(client!=null) client.showFrame
  }
}
contents += joinItem
contents += new MenuItem(Action("Share"){
  val title = Dialog.showInput(null, "What do you want to call this
      drawing?", "Drawing Title", Dialog.Message.Question, null, Nil, "")
  if(title.nonEmpty) {
    server.addDrawing(title.get, openDrawings(tabbedPane.selection.index))
  }
})
```

This gives a "Join" option that asks for a host, then uses RMI to try to look up a server on that host. If one is found, it connects to it. Otherwise one is created and bound to that host. The share option simply asks for a title for the drawing and then sends the currently selected drawing to the server.

This section contains a fair bit of code that really requires some exploration. It is highly recommended that you enter this all in and play with it to see how it works. It is worth noting that while the RMI code might not look like it involves threads, RMI calls do come across on a separate thread so some applications will have to take that into account.

23.6 End of Chapter Material

23.6.1 Summary of Concepts

- Networking has become an essential aspect of computers. When computers talk to one another over a network they have to go about it in an agreed-upon manner known as a protocol.
  - The most commonly used networking protocol in the Transmission Control Protocol, TCP. A TCP connection guarantees transmission or throws an exception so that the program knows it has failed. This makes it slower, but safe.
  - Applications needing higher speed that are willing to live with the occasional loss of a packet typically use the User Datagram Protocol, UDP. This protocol throws out packets of information and does not care if they get where they are going. As a result, some will be lost. The program will have to have mechanisms in place for dealing with packets that do not get through.
- Computers communicate through sockets. There are different socket types in the `java.net` package.
  - For TCP connections you want to create a `ServerSocket` on one machine that will accept connections on a particular port. Other machines can create `Sockets` that go to the port on the machine with the `ServerSocket`.
  - The `Socket` class has the ability to give you an `InputStream` or an `OutputStream`. These can be used just like any other streams to send or receive information.
  - A machine that wants to receive connections over UDP can open a `DatagramSocket`. The information has to be packed up by a client in a `DatagramPacket`.
- The `java.net.URL` class provides a simple way to read data information from web servers.
- Java includes an approach to network communication called Remote Method Invocation, RMI. Using RMI, you can make method calls to objects that reside on other machines using the standard syntax of a method call. There are a few steps you have to take to get an application set up with RMI.

23.6.2 Exercises

1. Set up a server socket and have it accept one connection. Have it open both the input and output streams and run through a loop where it parrots back to output what was written in input with minor modifications of your choosing. If the input is "quit" have it close the socket and terminate. Run this on a machine, then use `telnet` to connect to that port and see what happens.
2. Write a simple chat server that works using `telnet` as the client.
3. Fix the error handling in the first chat program example in the chapter.
4. Clean up the allowed names code in the first chat program example in the chapter.

5. Take the chat we have and add some extra commands. For example, `:who` should display everyone in the room. Other options might include private interactions other than normal chat.

6. Make a chat using sockets that has a GUI client with a list of all people in the chat room that is kept updated at all times.

7. Write a text-based chat client and server with RMI as the mode of communicating between the two sides.

8. Write a GUI-based chat client and server with RMI as the mode of communication.

9. Using the serialization concepts from the last chapter, add functionality to the GUI-based chats that you have written to send other types of information as serialized objects.

23.6.3 Projects

For this chapter, the obvious addition to the projects is to put in networking. For those projects that are based on networking, it is easy to see how this will happen. For the other projects we are going to force it in a bit, but only in ways that are potentially useful to you.

1. If you are working on a MUD, it is customary for players to be able to access the game through `telnet`. If you want to take that approach, then you simply add a `ServerSocket` to your map game and set it up so that many people can connect and give commands. You can also choose to create a more advanced viewer, but keep in mind that this is supposed to be text based.

   To do this, you will have a separate thread handling the `ServerSocket` with calls to `accept`. The new connections, once the user is logged in, will need to be handed off to the main game thread. It is suggested that you use one of the types in `java.util.concurrent` for doing this. The main game code need not change too much for this addition. Instead of reading from `readLine`, you will be pulling from the `InputStream` of the users `Socket`. You can also support multiple players so that on each iteration, the program checks for input from all users and executes them in order. Keeping the main processing sequential is one way to prevent race conditions.

2. If you are writing the webspider it is finally time to add real functionality. Use the `URL class` and give your program the ability to download pages. You could add a very simple search for "`href=`" so that you can find links to other pages. Use the ability of the `URL class` to build relative URLs to make this easier. Have your program pull down a few hundred pages and collect the data that you want from them. If you had the code setup as described in the previous chapter, this should be as simple as making the threads that are acquiring data pull from a URL instead of from files.

   If you do not already have a graphical display of information, you should add that in. That way you can display to the user what pages have been downloaded then let them look at the data you have collected in some manner. Remember that you need to have a `Set` or `Map` to make sure that you do not visit the same page multiple times.

3. For the multiplayer graphical game you have to make it so that two or more instances of the game can be run on separate computers that talk to one another. Exactly how you do this depends on the style of game that you are creating. If your game allows many users to play in a networked fashion, you should probably use a client-server

model where there is a separate application running as the server and then various clients connect.

In the client-server model, much of the game logic will likely be processed on the server. The clients become graphical displays and the medium for user interactions. The commands given to the clients are likely sent to the server with little interpretation.

If your game is only a two-player networked game, you might skip having a server and instead have a single application where the user can set their instance up as something of a local server for a second instance to connect to. In this situation, the game logic will happen primarily on that "main" instance and the second client is primarily just sending and displaying information.

One thing that you will probably realize fairly quickly when you work on this is that distributed parallel applications, those running on multiple machines across a network, can have their own types of race conditions as the different clients might not always be seeing the same world.

4. A game that is not supposed to involve networking is probably the most difficult project to tie networking into. One reason that you might want to include networking is to be able to keep track of things going on in the game outside of the normal display or provide other types of basic interaction. This is largely useful for debugging.

   Large and complex software often needs to have the ability to do "logging." This is where different steps in the code write out what they are doing to a file or something similar. When things go wrong, you can consult the log files to see what happened. Instead of logging to a file, you can make your game log to the network. Have your game create a `ServerSocket` that accepts connections. All logging should go to all the connections. If there are no connections, the logging is basically skipped.

   While logging to the network lacks some of the advantages of logging to a file, like an easy way to go back over things, it provides the flexibility that you can use commands written to the network connection. These commands could be to query information for use in debugging, or they could be like a superuser mode to set up situations that would be hard to create when you are testing your program.

5. The math workbook program could involve networking in at least two different ways. One would be that addition of collaborative features, like those added to the drawing program in this chapter, so that users can send a complete or partial worksheet from one computer to another.

   It is also possible to add logging features, like those described in project 4. The logging features for a math worksheet are most likely going to be for debugging purposes to inspect what is happening in the background.

   You can decide whether you want to implement one or both of these options into your code. Pick what interests you most and fits with your objectives for the project.

6. For the Photoshop® project you could add networking in the form of collaboration, logging, and/or using a URL to pull down images from the web. The collaboration could be much like what was shown in the chapter, though it should be altered to fit the flow of the program you are developing. Logging would be most useful for debugging as described in project 4. Both of those options are fairly significant, and you only need to pick one.

   The addition of loading images from URLs across the web is very straightforward and really is something that should be in this type of application. Whichever of the other two networking options you choose, you should probably add this possibility on top of it.

7. The simulation workbench provides a very different feature to add networking to the application. The fact that simulations can be highly processor intensive means that there can be benefits to breaking the work-load not only across threads, but also across multiple machines. This form of distributed computing requires networking.

   You should make a second application for a helper node that you can run on other computers. That helper application will make a network connection to the main application and wait to be given work. The simplest way to divide the work is to have separate simulations on each machine with information passed back occasionally for the purposes of displaying data to the user and perhaps saving. If the number of particles in a simulation is really large, it can make sense to distribute that one simulation across multiple machines. This is a more challenging problem as the different machines have to communicate the state of the particles they are controlling for force calculations.

   It is left to the student to decide how involved you wish to get with distributing your simulations.

Chapter 24

Stacks and Queues

The last few chapters have extended the depth of your knowledge in libraries so that you have the capability to do more inside of your programs. Now we are going to switch gears for a while and start looking at what goes into constructing data structures and writing some basic libraries. You started using simple data structures in the default libraries in chapter 7. We saw other types of data structures in chapter 20. Most of the time you are writing programs, you will use these types of pre-existing libraries. After all, they have been written by experienced people who put a lot of time and effort into tuning them so that they are flexible and efficient.

The fact that you will typically use libraries written by others might tempt you to believe that you really do not need to know what is going on inside of these data structures. On the contrary, understanding the inner working of data structures is an essential skill for any competent developer. One reason for this is that you often have many different options to choose from when you select a data structure and picking the right one requires knowing certain details. For example, so far we have treated the `List` and `Array` types almost interchangeably when we need a `Seq` with the exception that the `List` type is immutable. In chapter 25 we will see that the two are much more different than that and that even when we just focus on lists, there are many options.

Knowing how different data structures work and how to implement them is also required if you ever have to implement your own. This might happen for one of several reasons. One is that you are developing for a system where you are required to use a lower-level language that does not come with advanced collections libraries. Another is that you have a problem with very specific requirements for which none of the provided structures are really appropriate. If you do not understand and have experience writing data structures, you will almost certainly lack the write an efficient and correct one for a specific task. These are the types of skills that separate passable developers from the really good ones.

In this chapter, we will start our exploration with some of the simplest possible data structures, the stack and the queue. We will also write basic implementations of a stack and a queue that are based on an array for storage and which use parametric polymorphism so that they can be used with any type. Particular attention will be paid to how well these implementations perform and what we have to do to make sure they do not slow down when larger data sets are being used.

24.1 Abstract Data Types (ADTs)

Implementations of the stack and the queue are found in Scala's collections library because they hold or collect data. They could also be referred to using the term data structure. However, the most accurate term for them is ABSTRACT DATA TYPE (ADT). The "abstract" part has the same meaning as the usage we first saw in chapter 19. An ADT specifies certain operations that can be performed on it, but it does not tell us how they are done. The *how* part can vary across implementations, just as the implementation of abstract methods varies in different subtypes.

Different implementations can have different strengths and weaknesses. One of the learning goals of this second half of the book is for you to know the capabilities of different ADTs as well as the strengths and weaknesses of different implementations so that you can make appropriate choices. In some situations, the hardest part might be knowing exactly the usage in a particular application.

24.2 Operations on Stacks and Queues

What defines a specific ADT is the set of methods that have to be included in it along with descriptions of what they do. Most implementations will go beyond the bare minimum and add extra methods to improve usability, but it is those required methods that matter most. In the case of both the stack and the queue ADT there are only two methods that are absolutely required. One adds elements and the other removes them.

In the case of the stack these methods are commonly called `push` and `pop`. For the queue they are called `enqueue` and `dequeue`. These ADTs are so remarkably simple that the removal methods, `pop` and `dequeue`, do not take arguments. You can not specify what you are taking out. What makes the stack and the queue different from one another is how each selects the item to give you when something is removed. In a stack, the object that is removed by `pop` is the one that was most recently added. We say that it follows a "Last In, First Out" (LIFO) ordering. Called to `dequeue`, conversely, it remove the item that was least recently added with `enqueue`. The queue has a "First In, First Out" (FIFO) ordering.

The names of the ADTs and the methods on them are meant to invoke an image. In the case of a stack, you should picture a spring-loaded plate holder like you might see in a cafeteria. When a new plate is added, its weight pushes down the rest of the stack. When a plate is removed, the stack pops up. Following this image, you only have access to the top plate on the stack. Generally these devices prevent you from randomly pulling out any plate that you want.

In many English-speaking parts of the world, the term queue is used where American's use the term line. When you go to bủy tickets for an event, you do not "get in line", instead you "get on queue". The queue ADT is intended to function in the same way that a service line should. When people get into such a line, they expect to be serviced in the order they arrive. It is a first-come, first-serve type of structure. The queue should work in this same way with the `dequeue` method pulling off the item that has been waiting in the queue the longest.

24.3 Real Meaning of O

Back in chapter 13, you were introduced to the concept of O order. This was used to describe how fast the searches or sorts were. We want to explore this concept a bit more deeply here because we are going to use it a lot in discussing the performance of different ADTs and we will also use it to put constraints on what we will accept as implementations of certain ADTs.

While we often care about how long something will take to run in units of time, this really is not a very good way to characterize the performance of programs or parts of programs. The runtime can vary from one machine to another and do so in odd ways that are very particular to that hardware. A more stable definition is to consider one or more operations of interest, and look at how the number of times those operations are executed varies with the size of the input set we are working with. Imagine that we have some part of a program that is of interest to us, it might be a single function/method or the running of an entire program. We would run the program on various inputs of different sizes and count how many times the operations of interest occur. We could call the variation of input size $g(n)$. We say that $g(n)$ is $O(f(n))$, for some function $f(n)$, if $\exists m, c$ such that $c \times f(n) > n(n)$ $\forall n > m$. The best way to see what this means is graphically. Figure 24.1 shows how you might imagine $g(n)$ and $c \times f(n)$ looking in a plot.

There are some significant things to notice about this definition. First, O is only an upper bound. If $g(n)$ is $O(n)$, then it is also $O(n^2)$ and $O(n^3)$. In addition, the O behavior of a piece of code only really matters when n gets large. Technically, it is an asymptotic analysis. It deals with the situation where $n \to \infty$. In general this is very significant because so much of what we do on computers deals with large sets of data and we are always trying to push them to work with more information. However, if you have a problem where you know the size will be small, it is quite possible that using a method with an inferior O behavior could be faster.

The factor c in the definition of O means that you generally ignore all constant factors for performance. So $2n$ and $100n$ are both $O(n)$, even though the former is clearly better than the latter. Similarly, while $100n$ has a better order than $3n^2$ ($O(n)$ vs. $O(n^2)$), the latter is likely to be more efficient for small inputs.

The O notation also ignores all lower-order terms. The functions $n^3 + 2n^2 + 100$ and $n^3 + \log n$ are both $O(n^3)$. It is only the fastest growing term that really matters when n gets large. Despite all of the things that technically do not apply in O, it is not uncommon for people publishing new algorithms to include coefficients and lower-order terms as they can be significant in the range of input size that individuals might be concerned with.

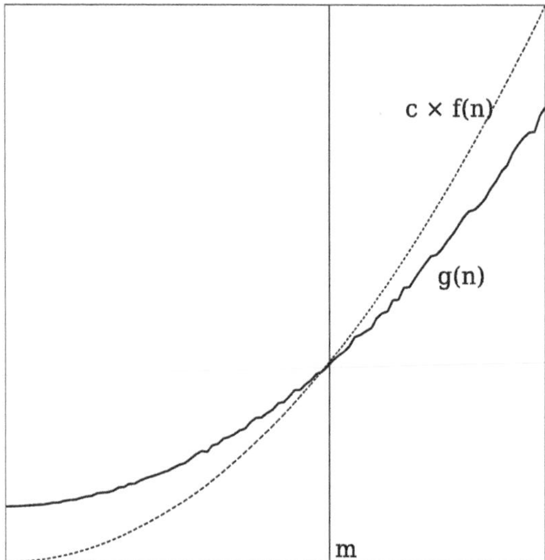

FIGURE 24.1: This figure shows a graphical representation of the meaning of O. The solid line represents $g(n)$, which would be a measure of the amount of computation done for an input size n. The dotted line is a function, $c \times f(n)$ where we want to say that $g(n)$ is $O(f(n))$. This plot labels a position m, above which the dotted curve is always above the solid curve.

24.4 $O(1)$ Requirement

The fact that the stack and queue are so very simple allows us to place a strong limit on the methods that we put into them. We are going to require that all methods we put into our implementations be $O(1)$. To understand why we want such a stringent limit, consider what would happen to the speed of a full program if a stack or queue grew to having n elements. In order to get there, and then make that useful, you have to have at least $O(n)$ calls to both the adding and removing methods. Combine the fact that orders can be multiplied with a slower implementation that is $O(n)$ and we would get whole program performance that is $O(n^2)$. This is prohibitively expensive for many applications.

The requirement of $O(1)$ performance will only be for the amortized cost. This is to say that we are willing to have an occasional call take longer and scale with the size of the data structure as long as this happens infrequently enough that it averages to a constant cost per call. For example, we can accept a call with an $O(n)$ cost as long as it happens only once every $O(n)$ times.

For most ADTs, it is not possible to impose such a strict performance requirement. It is only the simplicity of the stack and the queue that make this possible. For others we will have to accept $O(n)$ behavior for some of the operations or we will be happy when we can make the most expensive operation have a cost of $O(\log n)$.

24.5 Array-Based Stack

Having laid out the groundwork, it is now time to write some code. We will start with the stack ADT, and our implementations in this chapter will use arrays to store the data. For usability, we will include `peek` and `isEmpty` methods on both our stack and queue in addition to the two required methods. The `peek` method simply returns the value that would be taken off next without taking it off. The `isEmpty` method returns `true` if the stack or queue has nothing in it and `false` otherwise.

We start by defining a trait that includes our methods. This trait has a type parameter so that it can work with whatever type the user wants and provide type safety so that objects of the wrong type can not be added.

```
package scalabook.adt

/**
 * This trait defines a mutable Stack ADT.
 * @tparam A the type of data stored
 */
trait Stack[A] {
  /**
   * Add an item to the stack.
   * @param obj the item to add
   */
  def push(obj:A)

  /**
   * Remove the most recently added item.
   * @return the item that was removed
   */
  def pop():A

  /**
   * Return the most recently added item without removing it.
   * @return the most recently added item
   */
  def peek:A

  /**
   * Tells whether this stack is empty.
   * @return true if there are no items on the stack, otherwise false.
   */
  def isEmpty:Boolean
}
```

Scaladoc comments are added here for the trait. They will be left off of implementations `to conserve space`, but your own code should generally include appropriate Scaladoc comments throughout. This code should be fairly straightforward. If there is anything worthy of questioning, it is the fact that `pop` is followed by parentheses while `peek` and `isEmpty` are not. This was done fitting with style recommendations that any method that causes a mutation should be followed by parentheses, while methods that simply return a value, and which could potentially be implemented with a `val` or `var` should not. This makes it clear to others what methods have side effects and allows flexibility in subtypes.

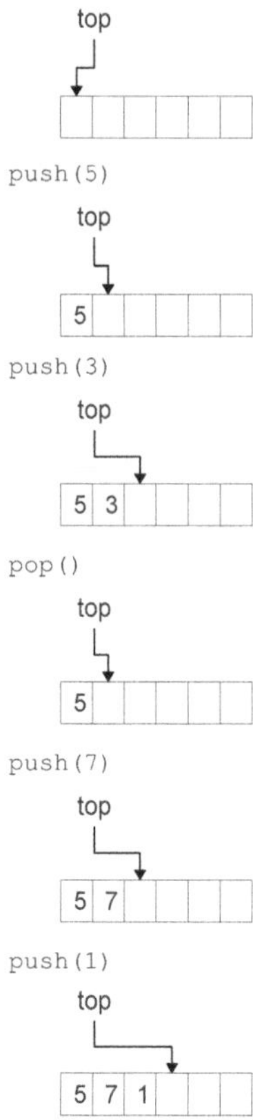

FIGURE 24.2: This figure shows a graphical representation of an array-based stack and how it changes over several operations. The stack stores data in the array and uses an integer index to keep track of the position of the top.

This trait effectively defines the ADT in Scala. Once we have it, we can work on a class that extends the trait and provides concrete implementations of the methods. Before we show the code for such an implementation, we should pause to think about how it might work and what it would look like in memory. A possible view of this is shown in figure 24.2. For illustration purposes, this stack stores `Ints`. We have an array that stores the various values as well as an integer value called top that stores the index where the next element should be placed in the stack. Each time a value is pushed, it is placed in the proper location and top is incremented. When a value is popped, top is decremented and the value at that location is returned.

Though this figure does not show it, we also need to handle the situation where enough elements have been pushed that the array fills. In this situation, a larger array needs to be allocated and the values should be copied over so that we have room for more elements.

Code that implements this view of a stack can be seen here.

```
package scalabook.adt

class ArrayStack[A : Manifest] extends Stack[A]{
  private var top = 0
  private var data = new Array[A](10)
```

```
  def push(obj:A) {
    if(top >= data.length) {
      val tmp = new Array[A](data.length*2)
      Array.copy(data,0,tmp,0,data.length)
      data = tmp
    }
    data(top) = obj
    top += 1
  }

  def pop():A = {
    assert(!isEmpty,"Pop called on an empty stack.")
    top -= 1
    data(top)
  }

  def peek:A = data(top-1)

  def isEmpty:Boolean = top==0
}
```

We will run through this beginning at the top. The class is called `ArrayStack` and it extends `Stack`. The `ArrayStack` takes a type parameter called `A` that is passed through to `Stack`. This type parameter is modified in a way we have not seen before, it is followed with `: Manifest`. This is required for us to be able to make an array of `A` using `new Array[A](...)`. The details of why this is needed are explained in the advanced section below.

Inside the class we have two private var declarations for `top` and the array called `data`. The implementations of `pop():A` decrements top and returns the data that is there. Both `peek` and `isEmpty` are even simpler. All of these methods are clearly $O(1)$ because they do math on an integer and direct access on an array, both operations that occur in constant time.

The only complex method is `push(obj:A)`. This method ends with two lines that actually place the object in the array and increment `top`. The first part of the method is a check to make sure that there is space to place the object in the array. If there is not, a new array is made and the contents of the current array are copied into the new one, then the reference `data` is changed to point to the new array. Everything here is $O(1)$ except the call to `Array.copy`. That call takes $O(n)$ time because it has to move $O(n)$ values from one array to another.

The obvious question is, will this average out to give us an amortized cost of $O(1)$? For this code, the answer is yes, but it is very important to see why, as it would have been very easy to write a version that has an amortized cost of $O(n)$. The key is in how we grow the array. In this code, every time we need more space, we make the array twice as big. This might seem like overkill, and many would be tempted to instead add a certain amount each time. Perhaps make the array 10 slots bigger each time. The table below runs through the first six copy events using these two approaches. At the bottom it shows generalized formulas for the mth copy event.

	data.length*2		data.length+10	
Resize Event	Size	Total Work	Size	Total Work
1	10	10	10	10
2	20	30	20	30
3	40	70	30	60
4	80	150	40	100
5	160	310	50	150
6	320	630	60	210
...	...	...	...	...
m	10×2^m	$10 \times (2^{m+1} - 1)$	$10m$	$5m(m+1)$

To get the amortized cost, we divide the total work by the size because we had to call `push` that number of times to get the stack to that size. For the first approach we get

$$\frac{10 \times (2^{m+1} - 1)}{10 \times 2^m} = \frac{2^{m+1} - 1}{2^m} = 2 - \frac{1}{2^m} \approx 2 \in O(1).$$

For the second approach we get

$$\frac{5m(m+1)}{10m} = \frac{m+1}{2} \in O(m).$$

Neither the multiple of 2, nor the addition of 10 is in any way magical. Growing the array by a fixed multiple greater than one will lead to amortized constant cost. Growing the array by any fixed value will lead to an amortized linear cost. The advantage of the fixed multiple is that while the copy events gets exponentially larger, they happen inversely exponentially often.

Manifests (Advanced)

To understand Manifests, it helps to have some understanding of the history of Java as the nature of the Java Virtual Machine (JVM) plays a role in their existence. The parallel to Scala type parameters in Java is generics. Generics did not exist in the original Java language. In fact, they were not added until Java 5. At that point, there were a huge number of JVMs installed around the globe and it was deemed infeasible to make changes that would invalidate all of them. For that reason, as well as a motivation to keep the virtual machine fairly simple, generics were implemented using erasure. What this means is that the information for generics was used at compile time, but was erased in the compiled version so that at runtime it was impossible to tell what the generic type had been. In this way, the code that was produced looked like what would have been produced before generics were introduced.

This has quite a few effects on how Java programs are written. It comes through in Scala as a requirement for having a manifest on certain type parameters. The manifest basically holds some of the type information so that it is available at runtime. To understand why we need this, we should go back to the expression that was responsible for us needing a manifest: `new Array[A](10)`. This makes a new `Array[A]` with 10 elements in it and sets the 10 elements to appropriate default values. It is the last part that leads to the requirement of having a manifest. For any type that is a subtype of `AnyRef`, the default value is `null`. For the subtypes of `AnyVal`, the default varies. For `Int`, it is 0. For `Double` it is 0.0. For `Boolean` it is `false`. In order to execute that expression, you have to know enough at least enough about the type to put in proper default values.[1] Type erasure takes away that information from the runtime. The manifest allows Scala to get back access to it.

It is also worth noting that our implementation of the stack technically does not actually remove the value from the array when `pop()` is called. This remove is not really required for the stack to execute properly. However, it can improve memory usage and performance, especially if the type `A` is large. Unfortunately, that would require having a default value to put into the array in that place, which we have just discussed it is hard from a completely general type parameter.

24.6 Array-Based Queue

Like the stack, the code for the array-based queue begins by setting up the ADT as a completely abstract `trait`. This trait, which takes a type parameter so that it can work with any type, is shown here.

```
package scalabook.adt

/**
 * This trait defines a mutable Queue ADT.
 * @tparam A the type of data stored
 */
trait Queue[A] {
  /**
   * Add an item to the queue.
   * @param obj the item to add
   */
  def enqueue(obj:A)

  /**
   * Remove the item that has been on the longest.
   * @return the item that was removed
   */
  def dequeue():A

  /**
   * Return the item that has been on the longest without removing it.
   * @return the most recently added item
   */
  def peek:A

  /**
   * Tells whether this queue is empty.
   * @return true if there are no items on the queue, otherwise false.
   */
  def isEmpty:Boolean
}
```

[1]You also need other information such as how big a memory block should be reserved. For example, `Int` requires 4-bytes while `Double` requires 8.

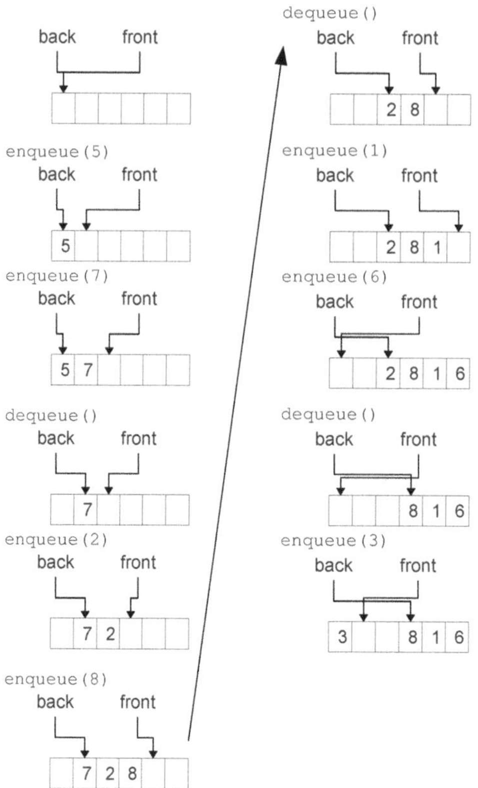

FIGURE 24.3: This figure shows a graphical representation of an array-based queue and how it changes over several operations. The queue stores data in the array and uses integer indexes to keep track of the front and back.

As with the stack, this has only the signatures of the methods along with comments describing what they do.

The queue ADT can also be implemented using an array. Figure 24.3 shows how this might look for several operations. Unlike the stack, the queue needs to keep track of two locations in the array: a front and a back. The back is where new elements go when something is added to the queue and the front is where values are taken from when they are removed. Both indexes are incremented during the proper operations. When an index moves beyond the end of the array, it is wrapped back around to the beginning. This allows us to continually add and remove items without having to grow the array or copy elements around. Both of those would require $O(n)$ operations and prevent our total performance from being $O(1)$.

This figure does not show what happens when the queue gets full. At that point a bigger array is needed and values need to be copied from the old array into the new one. The details for the queue are not quite as straightforward as they were for the stack. Figure 24.4 demonstrates the problem and a possible solution. Doing a direct copy of elements is not appropriate because the wrapping could have resulted in a situation where the front and back are inverted. Instead, the copy needs to shift elements around so that they are contiguous. The values of front and back need to be adjusted appropriately as well. There

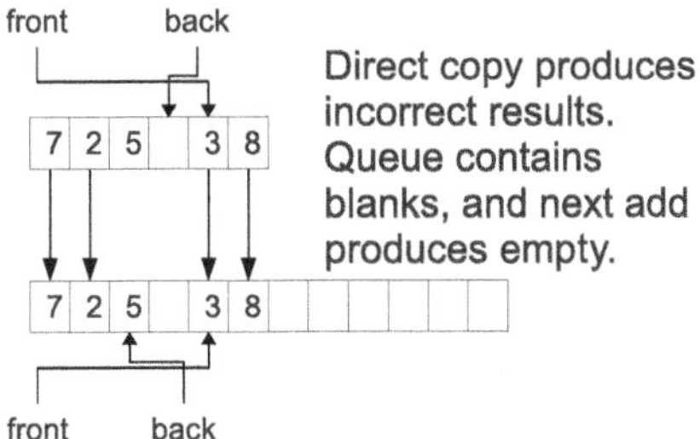

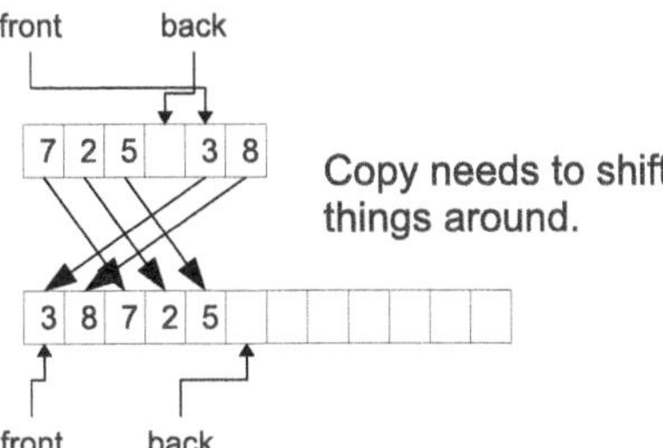

FIGURE 24.4: This figure shows a graphical representation of an array-based queue when it needs to grow on an enqueue. The queue is seen as empty if `front==back`. For that reason, the grow operation should happen when the last slot will be filled. In addition to growing a bigger array, the copy operation needs to move elements of the array around and alter the location of front and back.

are many ways that this could be done. Figure 24.4 demonstrates the approach that we will take in the code here where the front is moved back to the beginning of the array.

This graphical description of implementing an array-based queue can be implemented in code like the following.

```
package scalabook.adt

class ArrayQueue[A : Manifest] extends Queue[A] {
  private var front,back = 0
  private var data = new Array[A](10)

  def enqueue(obj:A) {
    if((back+1)%data.length==front) {
      val tmp = new Array[A](data.length*2)
      for(i <- 0 until data.length-1) tmp(i) = data((i+front)%data.length)
      front = 0
      back = data.length-1
      data = tmp
    }
    data(back) = obj
    back = (back+1)%data.length
  }

  def dequeue():A = {
    assert(!isEmpty,"Dequeue called on an empty queue.")
```

```
    val ret = data(front)
    front = (front+1)%data.length
    ret
  }

  def peek:A = data(front)

  def isEmpty:Boolean = front==back
}
```

This code is very similar to what we had for the stack. The only element that you might find odd is that the wrapping from the end of the array back to the beginning is done with a modulo. The wrap in dequeue and the one at the end of enqueue could easily be done with an `if`. For the other two, the wrapping needs to be part of an expression. This makes the modulo far more effective.

All the operations have an amortized cost of $O(1)$. As with `push` on the stack, the `dequeue` method will be $O(n)$ on the calls where it has to grow. However, those are rare enough that they average out to constant time.

24.7 Unit Tests

As you are well aware by this time, writing code that compiles does mean that you have code that is correct. There can still be runtime and logic errors present in the code. In order to find these you need to test the code. The topic of testing was first addressed back in section 8.5. The tests described at that point were manual tests. Each test had to be run explicitly by the developer if we wanted to test the code. In addition, some of the tests would print values that had to be verified for correctness by the programmer. In large-scale development, these approaches fall short of helping us to produce reliable code. In this section we will address this by introducing the concept of automated UNIT TESTS.

A unit test is a test that checks the functionality of some particular unit of a program. Typically it is something like a single class or even some methods in a class. The unit test is intended to be run automatically and, as such, does not rely on programmer interaction to determine if the test passed or failed. That logic should be built into the test. This allows unit tests to be run frequently. The idea is that every time you make changes to the code, you rerun all the unit tests to make sure that the changes did not break anything.

There are many types of unit-testing frameworks. Some, such as ScalaTest, ScalaCheck, and Specs, have been written specifically for Scala. In this book we will use JUnit, the primary unit-testing framework for Java. The reasons for this choice include the fact that JUnit is broadly used, it is easy to use, and it has support built into Eclipse and other IDEs.

24.7.1 Setup

There are a few steps that we need to take to get JUnit running with our project in Eclipse. First, go to `http://www.junit.org/` and click on "Download JUnit". That will take you to a page with many different downloads. You should get the newest release of the "Basic jar". Save this file in your Eclipse workspace or some other location where it will be easy to find.

Now go into Eclipse and right click on the project you want to add testing to. From the pop-up menu, select "Build Path" > "Configure Build Path...". Go to the "Libraries" tab and click on "Add External JARs..." button. After you have navigated to the file and selected it you are done setting up that project to run JUnit tests.

24.7.2 Writing Tests

Unlike tests that we would put in a `main` method in a companion object, unit tests are put in separate files and even in separate packages or projects. We will take the approach of putting all test code in a parallel package hierarchy that begins with `test`. As the `ArrayStack` and `ArrayQueue` were in the `scalabook.adt` package, our tests will be in `test.scalabook.adt`. To make the tests we simply write classes with methods and annotate our test methods with the `@Test` annotation from JUnit. For that to compile you should do an import for `org.junit._`. Every method that you put the `@Test` annotation on will be run at part of the unit testing.

Inside of the test methods, you need to have calls to JUnit `assert` methods. There are methods such as `assertTrue`, `assertFalse`, and `assertEquals`. Typically, tests are written building up from those that test very simple requirements to more significant ones. To understand how simple, consider the following examples.

```
@Test def emptyOnCreate {
  val stack = new ArrayStack[Int]
  assertTrue(stack.isEmpty)
}

@Test def nonEmptyOnPush {
  val stack = new ArrayStack[Int]
  stack.push(5)
  assertFalse(stack.isEmpty)
}

@Test def pushPop {
  val stack = new ArrayStack[Int]
  stack.push(5)
  assertEquals(5,stack.pop)
}
```

The first two only check that a new stack is empty and that after something has been pushed it is not empty. It is not until the third test that we even pop the stack and check that we get back the correct value.

You should notice that all of these methods begin with the same line. Code duplication like this is frowned on in testing as much as it is in normal code development. However, because JUnit runs every test separately, you have to use a tag to make sure that the code you want to run happens before each and every test. The tag for this purpose is the `@Before` tag. If you have code that needs to follow every test to clean up after something that happens in the "before" method, there is also a `@After` tag. You can see these used, along with a more complete set of tests for the `ArrayStack` in this code.

```
package test.scalabook.adt

import org.junit._
import org.junit.Assert._
import scalabook.adt._
```

```
class TestArrayStack {
  var stack:Stack[Int] = null

  @Before def initStack {
    stack = new ArrayStack[Int]
  }

  @Test def emptyOnCreate {
    assertTrue(stack.isEmpty)
  }

  @Test def nonEmptyOnPush {
    stack.push(5)
    assertFalse(stack.isEmpty)
  }

  @Test def pushPop {
    stack.push(5)
    assertEquals(5,stack.pop)
  }

  @Test def pushPopPushPop {
    stack.push(5)
    assertEquals(5,stack.pop)
    stack.push(3)
    assertEquals(3,stack.pop)
  }

  @Test def pushPushPopPop {
    stack.push(5)
    stack.push(3)
    assertEquals(3,stack.pop)
    assertEquals(5,stack.pop)
  }

  @Test def push100Pop100 {
    val nums = Array.fill(100)(util.Random.nextInt)
    nums.foreach(stack.push(_))
    nums.reverse.foreach(assertEquals(_,stack.pop))
  }
}
```

These tests are incomplete. It is left as an exercise for the student to come up with and implement additional tests.

With these tests written now we just need to run them and make certain that our implementation passes all the tests. If you select a run option for that file in Eclipse, you should have an option for "JUnit Test". If you select that, you should see something similar to what is shown in figure 24.5. This shows the six tests from above all passing. When everything passes, we get a green bar. If any tests fail there would be a red bar and the tests that fail would be shown in red. You can click on failed tests to find out what assert caused them to fail.

Similar tests can be written for our queue. They might look like the following.

```
package test.scalabook.adt
```

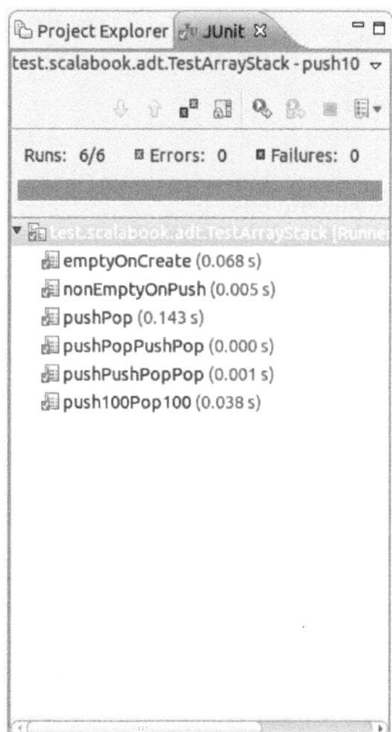

FIGURE 24.5: This is what Eclipse shows when a JUnit test is run and all tests are passed. A green bar means that everything is good. If you get a red bar then one or more tests have failed and they will be listed for you.

```
import org.junit._
import org.junit.Assert._
import scalabook.adt._

class TestArrayQueue {
  var queue:Queue[Int] = null

  @Before def initQueue {
    queue = new ArrayQueue[Int]
  }

  @Test def emptyOnCreate {
    assertTrue(queue.isEmpty)
  }

  @Test def nonEmptyOnEnqueue {
    queue.enqueue(5)
    assertFalse(queue.isEmpty)
  }

  @Test def enqueueDequeue {
    queue.enqueue(5)
    assertEquals(5,queue.dequeue)
  }

  @Test def enqueueDequeueEnqueueDequeue {
    queue.enqueue(5)
```

```
    assertEquals(5,queue.dequeue)
    queue.enqueue(3)
    assertEquals(3,queue.dequeue)
  }

  @Test def enqueueEnqueueDequeueDequeue {
    queue.enqueue(5)
    queue.enqueue(3)
    assertEquals(5,queue.dequeue)
    assertEquals(3,queue.dequeue)
  }

  @Test def enqueue100Dequeue100 {
    val nums = Array.fill(100)(util.Random.nextInt)
    nums.foreach(queue.enqueue(_))
    nums.foreach(assertEquals(_,queue.dequeue))
  }
}
```

You can check these as well and see that our array-based implementation of a queue passes all of these tests.

24.7.3 Test Suites

The purpose of automated testing is to make it easy to run all the tests regularly during development. Having the ability to run all of your tests on a regular basis makes you more confident when you go to make changes to your code. If you run the tests and they pass, then make a change and one or more fail, you know exactly what caused the failure and it will be easier to locate and diagnose.

Unfortunately, going through every test class and running it would be just as much of a pain as going through a large number of objects that have a main defined for testing. To consolidate your test runs, you can make a test suite. This lets you specify a group of different classes that contain test code that should be run together. To build a test suite you make a class and annotate it with the `@RunWith` and `@Suite.SuiteClasses` annotations. Here is an example of a suite that contains the two tests we just wrote.

```
package test

import org.junit.runner.RunWith
import org.junit.runners.Suite

@RunWith(classOf[Suite])
@Suite.SuiteClasses(Array(
    classOf[scalabook.adt.TestArrayStack],
    classOf[scalabook.adt.TestArrayQueue],
    classOf[scalabook.util.TestRPNCalc]
))
class AllTests
```

Note that the class does not actually contain anything at all. It is the annotations that tell JUnit all the work that should be done here. As we add more test classes, we will also add them to our suite.

24.7.4 Test-Driven Development

Unit tests are a central part of an approach to development called "Test Driven Development" (TDD). This style of development is highlighted by the idea that you write tests for code before you write the actual code. In a TDD environment, a programmer who wants to write a new class would create the class and put in blank methods which have just enough code to get the class to compile. Methods that return a value are set to return default values such as `null`, 0, or `false`. No logic or data is put in.

After that, the developer writes a basic test for the class and runs the test to see that it fails. Then just enough code is added to make that test pass. Once all existing tests pass, another test is written which fails and this process is repeated until a sufficient set of tests have been written to demonstrate that the class must meet the desired requirements in order to pass them all.

While the textbook does not follow a TDD approach, there are some significant benefits to it. The main one is that it forces a developer to take small steps and verify that everything works on a regular basis. It also gives you a feeling of security when you are making changes because generally feeling confident that if you make a change which breaks something, it will immediately manifest as a red bar in testing. This can be important for novice programmers who often leave large sections of code in comments because they are afraid to delete them in case something goes wrong. For this reason, the literature on TDD refers to tests as giving you the "courage" to make changes.

There is a lot more to unit testing that we have not covered here. Interested readers are encouraged to look at the JUnit website or look into Scalabased testing frameworks such as ScalaCheck and ScalaTest. Full development needs more than just unit testing as well. Unit tests, by their nature, insure that small pieces of code work alone. To feel confident in a whole application, you need to test the parts working together. This is called integration testing and it too is done on a fairly regular basis in TDD.

24.8 RPN Calculator

A classic example usage of a stack is the Reverse Polish Notation (RPN) calculator. In RPN mode, operations are entered after the values that they operate on. This is also called post-fix notation. The RPN representation of "2 + 3" is "2 3 +" While this might seem like an odd way to do things, it has one very significant advantage, there is no need for parentheses to help specify order of operation. Using standard notation, also called in-fix notation, the expressions "2 + 3 * 5" and "5 * 3 + 2" are both 17 because multiplication occurs before addition, regardless of the order things are written. If you want the addition to happen first, you must specify that with parentheses, "(2 + 3) * 5". Using post-fix notation, we can say "3 5 * 2 +" to get 17 or "2 3 + 5 *" if we want the addition to happen first so we get 25. This lack of parentheses can significantly reduce the length of complex expressions which can be appealing when working on such things.

The evaluation of a post-fix expression can be done quite easily with a stack that holds the numbers being worked with. Simply run through the expression and for each element you do one of the following. If it is a number, you push that number onto the stack. If it is an operator, you pop two elements front the stack, perform the proper operation on them, and then push the result back on the stack. At the end there should be one value on the stack which represents the final result. We could implement this basic functionality using the following object.

```
object RPNCalc {
  def apply(args:Seq[String]):Double = {
    val stack = new ArrayStack[Double]
    for(arg <- args; if(arg.nonEmpty)) arg match {
      case "+" => stack.push(stack.pop+stack.pop)
      case "*" => stack.push(stack.pop*stack.pop)
      case "-" => val tmp = stack.pop; stack.push(stack.pop-tmp)
      case "/" => val tmp = stack.pop; stack.push(stack.pop/tmp)
      case x => stack.push(x.toDouble)
    }
    stack.pop
  }
}
```

These arguments are passed in as a sequence that is traversed using a `for` loop. The subtract and divide operators are slightly different from those for addition and multiplication simply because they are not commutative, the order of their operands matters. This code also assumes that anything that is not one of the four operators is a number.

There are a few options that would be nice to add to this basic foundation. For example, it would be nice to have a few other functions beyond that standard four arithmetic operations. It would also be nice to have support for variables so that not every value has to be a literal numeric value. A final object that supports these additions is shown here.

```
package scalabook.util

import scalabook.adt.ArrayStack

object RPNCalc {
  def apply(args:Seq[String],vars:collection.Map[String,Double]):Double = {
    val stack = new ArrayStack[Double]
    for(arg <- args; if(arg.nonEmpty)) arg match {
      case "+" => stack.push(stack.pop+stack.pop)
      case "*" => stack.push(stack.pop*stack.pop)
      case "-" => val tmp = stack.pop; stack.push(stack.pop-tmp)
      case "/" => val tmp = stack.pop; stack.push(stack.pop/tmp)
      case "sin" => stack.push(math.sin(stack.pop))
      case "cos" => stack.push(math.cos(stack.pop))
      case "tan" => stack.push(math.tan(stack.pop))
      case "sqrt" => stack.push(math.sqrt(stack.pop))
      case v if(v(0).isLetter) => stack.push(try { vars(v) } catch { case ex => 0.0
          })
      case x => stack.push(try { x.toDouble } catch { case ex => 0.0 })
    }
    stack.pop
  }
}
```

This has added the three main trigonometric functions along with a square root function. These operations pop only a single value, apply the function, and push back the result. This code also includes an extra argument of type `Map[String,Double]` to store variables by name. In this case, the `scala.collection.Map` type is used. This is a supertype to both the mutable and immutable maps. This code does not care about mutability so this provides additional flexibility. Lastly, some error handling was added so that if a value comes in that is neither a number, nor a valid variable, we push 0.0 to the stack.

Before we can proclaim that this code works, we need to test it.[2] Here is a possible test configuration.

```
package test.scalabook.util

import org.junit._
import org.junit.Assert._
import scalabook.util.RPNCalc

class TestRPNCalc {
  @Test def basicOps {
    assertEquals(5,RPNCalc("2 3 +".split(" "),null),0.0)
    assertEquals(6,RPNCalc("2 3 *".split(" "),null),0.0)
    assertEquals(3,RPNCalc("6 2 /".split(" "),null),0.0)
    assertEquals(1,RPNCalc("3 2 -".split(" "),null),0.0)
  }

  @Test def twoOps {
    assertEquals(20,RPNCalc("2 3 + 4 *".split(" "),null),0.0)
    assertEquals(3,RPNCalc("2 3 * 3 -".split(" "),null),0.0)
    assertEquals(5,RPNCalc("6 2 / 2 +".split(" "),null),0.0)
    assertEquals(0.25,RPNCalc("3 2 - 4 /".split(" "),null),0.0)
  }

  @Test def vars {
    val v = Map("x" -> 3.0, "y" -> 2.0)
    assertEquals(20,RPNCalc("2 x + 4 *".split(" "),v),0.0)
    assertEquals(3,RPNCalc("y 3 * x -".split(" "),v),0.0)
    assertEquals(5,RPNCalc("6 2 / y +".split(" "),v),0.0)
    assertEquals(0.25,RPNCalc("x y - 4 /".split(" "),v),0.0)
  }

  @Test def specials {
    val v = Map("pi" -> 3.14159, "x" -> 3.0, "y" -> 2.0)
    assertEquals(0,RPNCalc("pi cos 1 +".split(" "),v),1e-8)
    assertEquals(math.sqrt(2)+3,RPNCalc("y sqrt x +".split(" "),v),1e-8)
    assertEquals(1,RPNCalc("pi 2 / sin".split(" "),v),1e-8)
    assertEquals(0.0,RPNCalc("x y y 2 / + - tan".split(" "),v),1e-8)
  }
}
```

As with the earlier tests, these are fairly minimal. There are literally an infinite number of tests that could be run to verify the RPN calculator. This set simply picks a few different tests related to the different types of functionality the calculator is supposed to support.

True unit testing adherents would prefer that each `assertEquals` be split into a separate method because none depends on the results before it. This is not just a style issue either, there are valid functionality benefits to that approach. JUnit will run all the test methods independently, but each test method will stop when an assert fails. So if the processing of "2 3 +" were to fail, we would not get results from the other three lines of that test as it is written now. That makes it harder to know why it failed. If "2 3 +" fails and all the other single operator calls pass, we can feel fairly confident the problem is with the handling of "+". As it is written now, we will not get that extra information. The only reason for using the format shown here is that it takes less than half as many lines to make this readable.

[2]It is worth noting that there was a bug in this code originally that was caught by these tests.

This RPNCalc object can also be integrated into the drawing program as a command. To make use of the variables, we add an extra val to the Drawing class called vars that is a mutable.Map[String,Double]. Then we add two commands called "rpn" and "set". The first invokes that RPNCalc object with the specified arguments. The second allows you to set variables in the Map for that drawing. Here is code for the modified version of Commands

```
package scalabook.drawing

/**
 * This object defines the command processing features for the drawing.
 * Use apply to invoke a command.
 */
object Commands {
  private val commands = Map[String,(String,Drawing)=>Any](
      "add" -> ((rest,d) => rest.trim.split(" +").map(_.toInt).sum),
      "echo" -> ((rest,d) => rest.trim),
      "refresh" -> ((rest,d) => d.refresh),
      "freeze" -> ((rest,d) => Thread.sleep(rest.trim.toInt*1000)),
      "rpn" -> ((rest,d) => scalabook.util.RPNCalc(rest.trim.split(" +"),d.vars)),
      "set" -> ((rest,d) => {
         val parts = rest.trim.split(" +")
         d.vars(parts(0)) = parts(1).toDouble
         parts(0)+" = "+parts(1)
      })
  )

  def apply(input:String,drawing:Drawing):Any = {
    val spaceIndex = input.indexOf(' ')
    val (command,rest) = if(spaceIndex < 0) (input.toLowerCase(),"")
      else (input.take(spaceIndex).toLowerCase(),input.drop(spaceIndex))
    if(commands.contains(command)) commands(command)(rest,drawing)
      else "Not a valid command."
  }
}
```

Thanks to the flexibility of the original design, there is very little effort required for adding two additional commands. More effort was required to add vars to Drawing. The obvious addition is in the top line where arguments are passed in.

```
class Drawing(private val root:DrawTransform,private[drawing] val
    vars:mutable.Map[String,Double]) extends Serializable {
```

The greater challenge comes from the fact that the apply method in the companion object that deserializes XML needs to be given an appropriate map. Here is code for the two versions of apply and the one method that builds a map from XML.

```
  def apply(data:xml.Node):Drawing = {
    new Drawing(DrawTransform(null,(data \ "drawTransform")(0)),buildVars(data \
        "variable"))
  }

  def apply():Drawing = {
    new Drawing(new DrawTransform(null),mutable.Map[String,Double]())
  }
```

```
def buildVars(nodes:NodeSeq):mutable.Map[String,Double] = {
  mutable.Map((for(n <- nodes) yield (n \ "@name").text -> (n \
      "@value").text.toDouble):_*)
}
```

To pull the data out of the XML means that it has to get there in the first place. For this reason, we must also alter the `toXML` method in `Drawing` to include extra elements for the variables. This can be done with the following code.

```
def toXML : xml.Node =
<drawing>
  {root.toXML}
  { for((n,v) <- vars) yield (<variable name={n} value={v.toString}/>) }
</drawing>
```

A `for` loop is used here to simplify the syntax of dealing with the tuples that come out of the map.

24.9 End of Chapter Material

24.9.1 Summary of Concepts

- The term Abstract Data Type, ADT, is used to describe a programmatic structure that has well-defined operations, but that does not specify how those operations are carried out. This concept links back to the ideas of separation of interface and implementation.
- A stack is a simple ADT with two basic operations, `push` and `pop`. The `push` operation adds an element to the top of the stack. The `pop` operation removes the most recently added element. This order of content removal is called Last In First Out, LIFO.
- A queue is another simple ADT with `enqueue` and `dequeue` operations. The `enqueue` adds an element to a queue while the `dequeue` takes out the element that has been in the queue the longest. This ordering scheme for removal of elements is called First In First Out, FIFO.
- Order analysis helps us understand the way in which algorithms behave as input sizes get larger. The O notation ignores constant multiples and lower-order terms. It is only really relevant when the input sizes get large.
- Due to their simplicity, any reasonable implementation of a stack or a queue should have operations that have an $O(1)$ amortized cost.
- One way to implement that stack ADT is using an `Array` to store the data. Simply keep an index to the `top` of the stack in the `Array` and build up from 0. When the `Array` is full, it should be grown by a multiplicative factor to maintain $O(1)$ amortized cost over copying operations.
- It is also possible to implement a queue with an `Array`. The implementation is a bit more complex because data is added at one end of the queue and removed at the other. The `front` and `back` are wrapped around when they hit the end so the `Array`

stays small and no copies are needed for most operations. Like the stack, when the `Array` does have to grow, it should do so by a multiplicative factor.

- Testing is a very important part of software development. Unit Tests are tests of the functionality of small units of code, such as single classes. The `JUnit` framework provides a simple way of building unit tests that can be run regularly during software development.

24.9.2 Exercises

1. Add extra functions to the RPN calculator.
2. Write more tests for the JUnit tests that are shown in the book.
3. Write some code so that requests that are sent across a network are queued up and handled in the order they arrive. You can simulate a long-running calculation with `Thread.sleep`.
4. For this exercise you should implement a few possible versions of both the queue and the stack and then perform timing tests on them to see how well they scale. You will include the implementation that we developed in this chapter plus some slightly modified implementations to see how minor choices impact the final speed of the implementation.
5. Build your own version of `BlockingQueue` using `synchronized`, `wait`, and `notifyAll` to reproduce the behavior of the version in `java.util.concurrent`.

24.9.3 Projects

As you can probably guess, the projects for this chapter have you adding stacks and/or queues to your code. A number of them have you making thread-safe queues to put into your code in places you might have been using a `java.util.concurrent.BlockingQueue`.

1. If you have been working on the MUD project, for this iteration you should make your own thread-safe queue, specifically for handing players who have logged in from the server socket thread to the main game thread. You can decide exactly what functionality to include, but the two critical aspects are that it needs to be a FIFO and thread-safe.
2. If you have been working on the web spider project, make and use your own thread-safe queue for handing information between the socket reading threads and the parsing threads and back. There will be at least two instances of queues for these functions with at least one for each direction of data flow.
3. If you are working on a networked graphical game you need to write a thread-safe queue for the same purposes as that needed for the MUD.
4. Back in project 4 (p.494) on the math worksheet it was suggested that you could integrate polymorphism into your project by using the command pattern. One of the main uses of the command pattern is to implement undo and redo capabilities. A stack is an ideal data structure for doing this. You have the main application keep track of two stacks. One is the undo stack, which you push commands onto as they

are executed. When the user selects to undo something, you pop the last command on the undo stack, call its undo method, then push it to a second stack, which is a redo stack. When a user selects redo, things move in the opposite direction. Note that any normal activity should probably clear the redo stack.

The stack that you use for this will probably be a bit different from the default implementation. For one thing, you need the redo stack to have a clear method. In addition, you might not want unlimited undo so you might want the stack to "lose" elements after a certain number have been pushed. It is left as an exercise for the student to consider how this might be implemented in a way that preserves $O(1)$ behavior.

5. Project 5 (p.494) suggested that you add the command pattern to the Photoshop® project in order to implement undo and redo capabilities. As was described in the project above, this is best done with two stacks. For this iteration you should do for Photoshop what is described above for the math worksheet.

6. So far the projects for the simulation workbench have had you working with N-body simulations like gravity or a cloth particle-mesh. Now we want to add a fundamentally different type of simulation to the mix, discrete event simulations. These are discussed in more depth in chapter 26, but you can create your first simple one here, a single-server queue.

   The idea of a single-server queue is that you have people arriving and getting in line to be receiving service from a single server. The people arrive at random intervals and the server takes a random amount of time to process requests. When people arrive and the server is busy, they enter a queue. The simulation can keep track of things like how long people have to wait and the average length of the queue.

   The key to a discrete event simulation is that all you care about are events, not the time in between. In this system, there are only two types of events, a customer arriving and a service being completed. Only one of each can be scheduled at any given time. To model arrivals you will use an exponential distribution and to model service times you will use a Weibull distribution. The following formulas can be used to generate numbers from these distributions, where r is the result of a call to `math.random`.

   - $a(\lambda)_{exponential} = -\frac{\ln(1-r)}{\lambda}$
   - $a(k,\lambda)_{Weibull} = \lambda(-\ln(1-r))^{\frac{1}{k}}$

   For the Weibull, use $k = 2$. The λ values determine the average time between arrivals or services. You should let the user alter those.

   The simulation should start at $t = 0$ with no one in the queue. You generate a value from the exponential distribution, $a_{exponential,1}$ and instantly advance to that time, when the first person enters. As no one is being served at that time, they go straight to the server. You also generate another number from the exponential distribution, $a_{exponential,2}$ and keep that as the time the next person will enter. The first person going to the server causes you to generate a service time, $a_{Weibull,1}$ for when that person will finish the service, and you store that.

   You then instantly advance to the lower of the two event times. So either a second person arrives and enters the queue, or the first person is serviced and the server becomes free with no one present. Each time one person arrives, you schedule the next arrival. Each time a person goes to the server, you schedule when they will be done. Note that the generated times are intervals so you add them to the current time to get when the event happens.

Let the user determine how long to run the simulation. Calculate and display the average wait time, the average line length, and the fraction of time the server is busy.

7. For non-networked games, it is hard to predict what will fit well. For some you can implement undo and redo functionality as described in project 4. That applies best to turn-based games. It is also possible that you can find places where queues can be used to pass information between threads or to queue up commands from the user if there are delays on how long actions take.

Chapter 25

Linked Lists

So far in this book we have been very happy to use the collections libraries without having a full understanding of how they work. That is fine for much of what we have done and in a first year of Computer Science (CS) study, but real developers need to have knowledge of what is really happening so that they can make informed opinions about what collection types to use in different situations. This chapter will begin to reveal how we can build our own structures for storing data by linking objects that store small pieces of data together.

25.1 The List/Seq ADT

The topic of this chapter is the linked list. This is an approach to storing data in a way that can be applied to some different Abstract Data Types (ADTs). Later in the chapter we will see how a linked list can be used as the storage for stacks and queues. Before that, we want to introduce a new ADT that is a step up on the ladder of capabilities, the list ADT. This ADT provides the ability to store a sequence of data elements that are indexed by integers with functionality for random retrieval, insertion, and removal. In a language-independent way one might expect these to be implemented in methods like these.

```
def get(index:Int) : A
def set(index:Int, data:A)
def insert(index:Int, data:A)
def remove(index:Int) : A
```

Here `A` is the type of data being stored in the list. The insert and remove methods shift things around so that all elements after the specified index are moved back on index on an insert or forward one index for a remove.

This set of methods is most appropriate for a mutable list implementation. It is also possible to make immutable list implementations. These would return new collections when operations such as inserting or removing are performed. They also typically avoid doing general insertions or removals as those operations can be very inefficient.

In the Scala collections, the behaviors of this ADT are represented without `insert` and `remove` functionality by the `Seq` trait and more complete sense by the `Buffer` trait. They have slightly different methods than what was listed above. We will see these in later sections where we build implementations that inherit from those traits.

25.2 Nature of Arrays

To understand linked lists, it is helpful have something to compare them to. Arrays are the simplest type of the collection. An array is a block of memory that is big enough to hold everything that needs to be stored. Memory in the computer is laid out with numeric addresses and each element of an array has to be the same size. This means that the location of any element of an array can be found with a few mathematical operations. So if you know where to find the first element of an array, you can find any other with $O(1)$ work.

The list ADT can be implemented using an array for the storage. This is what an `ArrayBuffer` does. When you build a list with an array, the `get` and `set` methods can happen in $O(1)$ operations because they simply do a direct access into the array. This makes them very fast and if you are going to be calling these a lot on the list, then an array-based list could be ideal. On the other hand, the insert and remove methods in an array-based list have to copy $O(n)$ elements in order to either make room for a new element or fill in a whole left when an element is removed. More specifically, all the elements after the specified index have to be moved. This is not only $O(n)$ operations, the writing to memory part of the copy tends to be particularly slow on most modern hardware so it is $O(n)$ in a fairly costly operation. If you are always adding to or removing from the end of the list, then the access is much like the stack we wrote previously and everything is $O(1)$ or close to it. If your program will spend a lot of time doing random insertions and removes, or worse, doing them near the beginning of the list, then the array based version probably is not ideal.

Like the array-based version of the stack and the queue, the list should allow for the array to be larger than the current list with the true size stored in an `Int`. In addition, the array should be grown in multiplicative steps instead of additive ones so that the amortized cost is $O(1)$. The actual construction of such a list is left as an exercise for the reader.

25.3 Nature of Linked Lists

In some ways, the linked list can be viewed as an alternative to the array which is strong in the areas the array is weak, but weak in the areas where the array is strong. Instead of having a single large block of memory such that the computer can efficiently access any

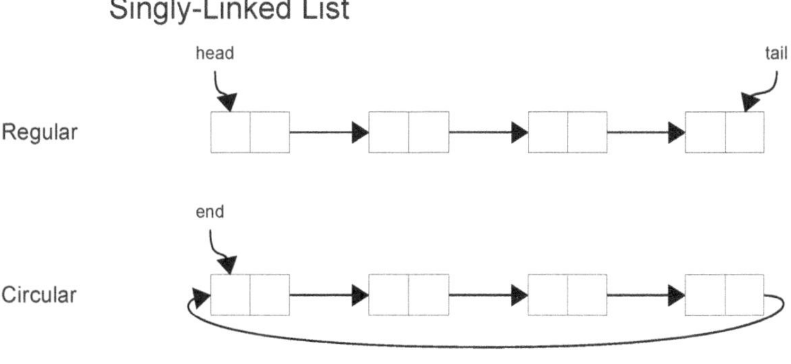

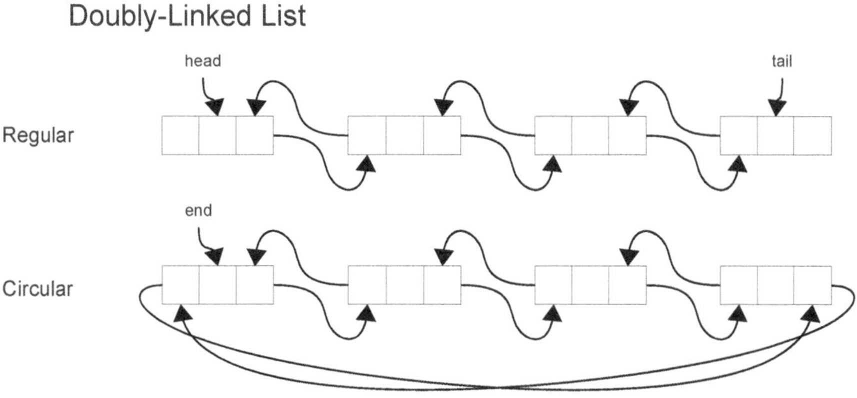

FIGURE 25.1: This figure shows the basic structure of some different forms of linked lists.

element if it simply knows the location of the first one, linked lists are composed of nodes that store individual elements which are linked together. The organization in memory is effectively random so the only nodes you can directly reach from one node are the others it links too. There are many different types of linked data structures that have this type of form. A linked list has a sequential structure, and each node will link to the node after it and possibly the node before it. When there is only a link to the next node it is called a singly-linked list. When there is a link to both the previous and the next it is called a doubly-linked list.

Figure 25.1 shows a graphical representation of a number of different flavors of linked lists. In this figure, each node is represented by a set of boxes. For a singly-linked list, the first box represents a reference the to data stored in that node while the second is a reference to the next node. Arrows show where the references link to. No arrows are shown for the data as that is typically independent of the functionality. In a sense, lists do not care about what they contain. This type of graphical representation can be extremely helpful for understanding the code used to manipulate linked lists.

From the figure you can see that both singly-and doubly-linked lists can be set up in regular or circular configurations. In the regular configuration, the end (first or last element) of the list is represented by a `null` reference. In the circular configuration these point back to the other end of the list.

The program must always keep track of at least one element of a linked list. For a regular, singly-linked list, this must be the head. All elements of the list are reachable from the head, but because references are one way, if you lose your reference to the head, any nodes that had been in front of the one you keep will be lost. The figure shows another reference at the other end of the list to a tail. Storing the tail is not generally required, but it can make certain operations significantly faster. In particular, adding to the end of a mutable list is $O(n)$ if you do not keep track of the tail, but can be done in $O(1)$ if you do keep track of the tail. For this reason, we will keep track of the tail in our final mutable, singly-linked list despite the fact that it will add a fair bit of complexity to the code.

25.4 Mutable Singly-Linked List

The first form of linked list that we will implement is a very basic mutable, singly-linked list. We will include only the four methods that were listed above even though this minimal an implementation is not very useful. It is instructive to start here so that the details of other methods do not get in the way of your understanding.

The class below shows what such a linked list might look like. The `get` and `set` methods have been changed to `apply` and `update` to take advantage of the Scala syntax rules for these special methods, but the logic inside of them has not changed.

```
package scalabook.adt

/**
 * This is a very simple implementation of a basic linked list.
 * @tparam A the type stored in the list.
 */
class SinglyLinkedList[A] {
  private class Node(var data:A,var next:Node)
  private var head:Node = null

  /**
   * This methods gets the value at a specified index. If the index is beyond the
   * length of the list a NullPointerException will be thrown.
   * @param index the index to get.
   * @return the value at that index.
   */
  def apply(index:Int) : A = {
    assert(index>=0)
    var rover = head
    for(i <- 0 until index) rover = rover.next
    rover.data
  }

  /**
   * Sets the value at a particular index in the list. If the index is beyond the
   * length of the list a NullPointerException will be thrown.
   * @param index the index to set the value at.
```

```
   * @param data the value to store at that index.
   */
  def update(index:Int, data:A) {
    assert(index>=0)
    var rover = head
    for(i <- 0 until index) rover = rover.next
    val ret = rover.data
    rover.data = data
    ret
  }

  /**
   * Inserts a new value at a particular index in the list. If the index is beyond
   * the length of the list a NullPointerException will be thrown.
   * @param index the index to set the value at.
   * @param data the value to insert at that index.
   */
  def insert(index:Int, data:A) {
    assert(index>=0)
    if(index==0) {
      head = new Node(data,head)
    } else {
      var rover = head
      for(i <- 0 until index-1) rover = rover.next
      rover.next = new Node(data,rover.next)
    }
  }

  /**
   * Removes a particular index from the list. If the index is beyond the length
   * of the list a NullPointerException will be thrown.
   * @param index the index to remove.
   * @return the data value that had been stored at that index.
   */
  def remove(index:Int) : A = {
    assert(index>=0)
    if(index==0) {
      val ret = head.data
      head = head.next
      ret
    } else {
      var rover = head
      for(i <- 0 until index-1) rover = rover.next
      val ret = rover.data
      rover.next = rover.next.next
      ret
    }
  }
}
```

The class takes a type parameter named A which represents the type of data the list contains. This class contains a little private class called Node that stores the references to the data and a reference to the next node. The linked list class also keeps track of the head of the list in a variable called head.

The four methods in the class are all fairly simple and they contain very similar structures. Let us look first at the apply method. It begins with an assert to insure that the index is not less than zero. The comments also say that if an index that is beyond the length of the list is provided that a NullPointerException will be thrown. While this is sufficient for our current needs, this is not ideal behavior. It would be far better if asking for an index that is out of bounds threw an appropriate exception with information on what was requested and how many there were. This shortcoming will be corrected in the more complete implementation that comes later.

All of the methods in this class declare a variable called rover that is used to run through the list. With an array, we move through the elements by counting with an index. Nodes do not have to be consecutive in memory, so we must run through the elements by following references from one node to another. This is what the assignment rover = rover.next inside of the for loop does.

Question: How could you remove duplicate code from this simple implementation of a linked list?

The insert and remove methods have an element that is missing from both apply and update, a special case for dealing with the beginning of the list. Both insert and remove alter references in the list and have to include a special case for when that reference is the first element. The normal case alters a next reference, while the special case alters head.

The reader should also notice that insert and remove have for loops that only run $n-1$ times. This is because in a singly-linked list, adding or removing a node requires altering the next reference of the node before it. If you want to remove the node at index 3, you must have a reference to the node at index 2 and change its next reference to point to the node currently at index 4. This is what happens in the line rover.next = rover.next.next. If rover were to go all the way to the node we wanted to remove, we would have no access to the reference that needs to be modified because it is not possible to move backwards through a singly-linked list.

To convince ourselves that this code works, we could implement the following tests that do some minimal checking.

```
package test.scalabook.adt

import org.junit._
import org.junit.Assert._
import scalabook.adt._

class TestSinglyLinkedList {
  var list:SinglyLinkedList[Int] = null

  @Before def setup {
    list = new SinglyLinkedList[Int]
  }

  @Test def appendOne {
    list.insert(0,10)
    assertEquals(10,list(0))
  }

  @Test def appendTwo {
    list.insert(0,10)
    list.insert(0,20)
```

```
    assertEquals(20,list(0))
    assertEquals(10,list(1))
  }

  @Test def appendTwoB {
    list.insert(0,10)
    list.insert(1,20)
    assertEquals(10,list(0))
    assertEquals(20,list(1))
  }

  @Test def appendTwoUpdate {
    list.insert(0,10)
    list.insert(0,20)
    list(0)=5
    list(1)=8
    assertEquals(5,list(0))
    assertEquals(8,list(1))
  }

  @Test def appendRemove {
    list.insert(0,5)
    list.insert(1,10)
    list.insert(2,15)
    list.insert(3,20)
    assertEquals(5,list(0))
    assertEquals(10,list(1))
    assertEquals(15,list(2))
    assertEquals(20,list(3))
    list.remove(2)
    assertEquals(5,list(0))
    assertEquals(10,list(1))
    assertEquals(20,list(2))
    list.remove(0)
    assertEquals(10,list(0))
    assertEquals(20,list(1))
  }
}
```

As you should be coming to expect in this text, the tests shown here are not really sufficient to give us confidence in the code. One test that has been left out which is of particular significance for collections like this is a test that adds to the list, empties it completely using remove, then begins adding again. Such boundary cases are particularly common sources of errors in collections.

25.4.1 Implementing `mutable.Buffer`

The previous example of a linked list was extremely minimal and lacked a lot of the functionality that is really needed to make a collection useful. One way for use to access significantly more functionality without all the much additional effort is to build on top of the foundation provided by the `collection` packages in the Scala libraries. For a mutable linked list, the obvious choice for a foundation type is the `mutable.Buffer` type.

If you look in the Application Programming Interface (API) you will find that `mutable.Buffer` lists the following nine methods as being abstract.

```
def +=(elem: A): Buffer.this.type
def +=:(elem: A): Buffer.this.type
def apply(n: Int): A
def clear(): Unit
def insertAll(n: Int, elems: Traversable[A]): Unit
def iterator: Iterator[A]
def length: Int
def remove(n: Int): A
def update(n: Int, newelem: A): Unit
```

This means that if we want to extend the `mutable.Buffer` type we must implement these nine methods. The reward for doing this is that the large number of other methods that are part of `mutable.Buffer` will automatically work for us. This happens because those other methods are implemented in terms of the those nine abstract methods. So for the cost of nine you get many, many others.

The following code shows an implementation of just such a class. This implementation keeps track of both a head and a tail for the list. They are called `hd` and `tl` so that they do not conflict with the `head` and `tail` members that are part of the general Scala sequence. The class also keeps track of an `Int` named `num` so that calls to `length` and bounds checking can both be done efficiently.

```
package scalabook.adt

import collection.mutable

class MutableSLList[A] extends mutable.Buffer[A] {
  private var hd:Node = null
  private var tl:Node = null
  private var num = 0
  private class Node(var data:A,var next:Node)

  def +=(elem: A) = {
    if(tl==null) {
      tl = new Node(elem,null)
      hd = tl
    } else {
      tl.next = new Node(elem,null)
      tl = tl.next
    }
    num += 1
    this
  }

  def +=:(elem: A) = {
    hd = new Node(elem,hd)
    if(tl==null) {
      tl = hd
    }
    num += 1
    this
  }

  def apply(n: Int): A = {
    if(n<0 || n>=num) throw new IndexOutOfBoundsException(n+" of "+num)
    var rover = hd
```

```
  for(i <- 0 until n) rover = rover.next
  rover.data
}

def clear() {
  hd = null
  tl = null
  num = 0
}

def insertAll(n: Int, elems: Traversable[A]) {
  if(n<0 || n>=num+1) throw new IndexOutOfBoundsException(n+" of "+num)
  if(elems.nonEmpty) {
    val first = new Node(elems.head,null)
    var last = first
    num += 1
    for(e <- elems.tail) {
      last.next = new Node(e,null)
      last = last.next
      num += 1
    }
    if(n==0) {
      last.next = hd
      hd = first
      if(tl==null) tl=last
    } else {
      var rover = hd
      for(i <- 1 until n) rover = rover.next
      last.next = rover.next
      rover.next = first
      if(last.next==null) tl = last
    }
  }
}

def iterator = new Iterator[A] {
  var rover = hd
  def hasNext = rover!=null
  def next:A = {
    val ret = rover.data
    rover = rover.next
    ret
  }
}

def length : Int = num

def remove(n: Int): A = {
  if(n<0 || n>=num) throw new IndexOutOfBoundsException(n+" of "+num)
  num -= 1
  if(n==0) {
    val ret = hd.data
    hd = hd.next
    if(hd==null) tl = null
    ret
```

```
    } else {
      var rover = hd
      for(i <- 1 until n) rover = rover.next
      val ret = rover.next.data
      rover.next = rover.next.next
      if(rover.next==null) tl = rover
      ret
    }
  }

  def update(n: Int, newelem: A) {
    if(n<0 || n>=num) throw new IndexOutOfBoundsException(n+" of "+num)
    var rover = hd
    for(i <- 0 until n) rover = rover.next
    rover.data = newelem
  }

  override def toString = mkString("MutableSLList(",", ",")")
}
```

Comments on public methods have been omitted as they would mirror those from the main API.

As with the previous list, `insert` and `remove` have special cases in them for handling the end cases. These are situations where either the head or the tail of the list are modified. There is more special case code here, because a reference to the tail is included. The primary motivation for including this reference is that the `+=` method for appending to the list is $O(1)$ when we have a tail and we should expect this to be called fairly regularly.

25.4.2 Iterators

This code also includes an implementation for the `iterator` method. This method is supposed to give us back an `Iterator` that can be used to run through the list efficiently. Using a loop with an index and `apply` to go through a linked list is not efficient. To be more specific, running through a linked list in that way requires $O(n^2)$ operations. This is because `apply` runs through the entire list up to the requested index. So the first request has to take 0 steps, but the next takes 1 and the third takes 2. This continues up to the last call which has to run through $n-1$ links. The sum of these is $n(n-1)/2$ which is $O(n^2)$. Using an `Iterator` provides a way for any outside code to walk through the elements of the list in $O(n)$ operations. As we have seen before, an `Iterator` has methods called `next` and `hasNext` that let us walk through the elements. These are the only things we have to write to create our own `Iterator`. In the implementation of our list, they look much like the code we use for the loops that are internal to the class. We have a `rover` that begins at the head. The `hasNext` method simply checks if `rover` has gotten to the `null` at the end of the list. The `next` method moves `rover` to be `rover.next` and returns the data `rover` had been referencing.

Fitting the Scala Collection Libraries

If you play a bit in the REPL with the code that is shown here for the mutable singly-linked list, you will find that while it works correctly, there are a few behaviors that are less than ideal. In particular, if you call methods like `map` or `filter`, the resulting collection will be an `ArrayBuffer`. This is not the behavior you have come to expect from Scala collections where these methods tend to return the same type they operate on. This happens because we did not write those methods and we are relying on the implementations in the `Buffer` supertype to provide them. The `Buffer` type defaults to giving back an `ArrayBuffer` so that is what we see here.

If the solution to this required rewriting all of the methods that return collections, it would greatly reduce the benefit of extending the `Buffer` trait. The Scala collections library has been designed so that this is not required, but getting these methods to return a different type is still a bit challenging. To make your collection really fit in with the main collections library, there are a few things that need to be done. Instead of just extending the `Buffer` type, we also extend the `BufferLike` type. The `BufferLike` trait is one of a number of traits that end in `Like` that can be used to add onto the libraries.

The `BufferLike` trait takes two parameters. One is the type of the collection and the other is the collection type. In this case, `MutableSLList[A]`. That second type of parameter tells it what should be returned for calls that create a new collection, such as `filter` and `map`. Beyond this, things get a bit more complex. We need to specify a way for these types to be built. For our code, this leads to inheriting from `GenericTraversableTemplate`, making a companion object that is a `SeqFactory`, and making our class inherit from `Builder` as well. Putting it all together, the changes look like this.

```
import collection.generic._

class MutableSLList[A] extends mutable.Buffer[A]
    with GenericTraversableTemplate[A,MutableSLList]
    with mutable.BufferLike[A,MutableSLList[A]]
    with mutable.Builder[A,MutableSLList[A]] {

  ...

  override def companion: GenericCompanion[MutableSLList] = MutableSLList

  def result() = this

  ...
}

object MutableSLList extends SeqFactory[MutableSLList] {
  implicit def canBuildFrom[A]: CanBuildFrom[MutableSLList[_], A,
      MutableSLList[A]] =
    new GenericCanBuildFrom[A]
  def newBuilder[A]: mutable.Builder[A, MutableSLList[A]] =
    new MutableSLList[A]
}
```

This begins to pull in aspects of the Scala language that are mainly important for library writers. As the goal of this book is not to turn you into a proficient Scala library writer, we will not be putting this type of functionality into future collections as it obfuscates the concepts that you should be learning.

25.5 Mutable Doubly-Linked List

Now we want to consider a doubly-linked list. The obvious advantage of a doubly-linked list over a singly-linked list is that it can be traversed in either direction. This comes with the cost of an extra reference per node and extra code to handle it. However, as we saw in the last section, the main source of additional code was not handling the next references, but dealing with the special cases for head and tail references. Fortunately, there is a way we can remove that special case code when we are using a doubly-linked list so that handling the previous reference is more than compensated for. The way we do this is through the introduction of a special node called a sentinel.

A sentinel is an extra node that does not store data, but which functions as both the beginning and end of the list. To make this possible the list needs to be circular. If we call the sentinel `end` then `end.next` is the head of the list and `end.prev` is the tail of the list. This alteration saves a significant amount of code because how the head and tail are just `next` and `prev` references, so adding and removing at those locations is no different from doing so at other locations.

Code for this implementing the `Buffer` trait is shown here. The declarations at the top of the class are very similar to those in the singly-linked list, except that there is one one node reference called `end` and initially the `next` and `prev` of that node are set to point back to it. There are two other points to note about the top part of this class. The `end` node has to be given a data value. This causes a bit of a problem because there is not any single value that can be used as the default for the type parameter `A`. To get around this we use the approach seen with our array-based stacks and queues of requiring the type have a manifest and then the data in `end is` set by making an array with one element and using that first element. There are other solutions to this that use inheritance, but they require more code and potentially obfuscate the purpose of this example.

```
package scalabook.adt

import collection.mutable

class MutableDLList[A : Manifest] extends mutable.Buffer[A] {
  private var num = 0
  private class Node(var data:A, var prev:Node, var next:Node)
  private val end = new Node(new Array[A](1)(0),null,null)
  end.next = end
  end.prev = end

  def +=(elem: A) = {
    val newNode = new Node(elem,end.prev,end)
    end.prev.next = newNode
    end.prev = newNode
    num += 1
    this
  }

  def +=:(elem: A) = {
    val newNode = new Node(elem,end,end.next)
    end.next.prev = newNode
    end.next = newNode
    num += 1
```

```
  this
}

def apply(n: Int): A = {
  if(n<0 || n>=num) throw new IndexOutOfBoundsException(n+" of "+num)
  var rover = end.next
  for(i <- 0 until n) rover = rover.next
  rover.data
}

def clear() {
  end.prev = end
  end.next = end
  num = 0
}

def insertAll(n: Int, elems: Traversable[A]) {
  if(n<0 || n>=num+1) throw new IndexOutOfBoundsException(n+" of "+num)
  if(elems.nonEmpty) {
    var rover = end.next
    for(i <- 0 until n) rover = rover.next
    for(e <- elems) {
      val newNode = new Node(e,rover.prev,rover)
      rover.prev.next = newNode
      rover.prev = newNode
      num += 1
    }
  }
}

def iterator = new Iterator[A] {
  var rover = end.next
  def hasNext = rover!=end
  def next:A = {
    val ret = rover.data
    rover = rover.next
    ret
  }
}

def length : Int = num

def remove(n: Int): A = {
  if(n<0 || n>=num) throw new IndexOutOfBoundsException(n+" of "+num)
  num -= 1
  var rover = end.next
  for(i <- 0 until n) rover = rover.next
  val ret = rover.data
  rover.prev.next = rover.next
  rover.next.prev = rover.prev
  ret
}

def update(n: Int, newelem: A) {
  if(n<0 || n>=num) throw new IndexOutOfBoundsException(n+" of "+num)
```

```
    var rover = end.next
    for(i <- 0 until n) rover = rover.next
    rover.data = newelem
  }

  override def toString = mkString("MutableDLList(",", ", ")")
}
```

The other methods should be fairly straightforward after you have seen the singly-linked list. The biggest changes are that there are two pointers that have to be assigned for each node-node link and that all the special case code is gone. On the whole, this version is shorter, less complex, and less error prone due to the removal of the special case code. The reader should draw pictures to see the effect of a line like `rover.prev.next = rover.next`. This particular line in `remove` takes advantage of the fact that doubly-linked lists can be traversed forward and backward so a reference to the node that we are removing is sufficient. We do not need to have a direct reference to the prior node as was required for the singly-linked list.

Sentinels through Inheritance

Forcing the type to have a manifest and creating a little array simply to get a data value to store in the sentinel is less than ideal from a stylistic standpoint. There is an alternative. Using inheritance, we can make a special node type to represent the sentinel which does not have data in it. Alternate code that uses that approach is shown here.

```
class MutableDLList[A] extends mutable.Buffer[A] {
  private var num = 0
  private trait Node {
    var data:A
    var prev:Node
    var next:Node
  }
  private class DNode(var data:A, var prev:Node, var next:Node) extends Node
  private class ENode extends Node {
    def data:A = throw new IllegalArgumentException("Requested value of
         sentinel node.")
    def data_=(d:A) = throw new IllegalArgumentException("Value assigned to
         sentinel node.")
    var next:Node = this
    var prev:Node = this
  }
  private val end = new ENode
```

Using this approach, the `Node` type is now a trait with subtypes called `DNode` and `ENode`. The node defines abstract `var`s for the different fields we had in `Node` earlier. The `DNode` type inherits from this and otherwise looks just like our original `Node` class. Where this approach differs is in the `ENode` for the sentinel. This class uses regular `var` declarations for `next` and `prev`, but it handles `data` quite differently. Instead of declaring a `var`, there are two `def` declarations for getting and setting `data`. Both of these throw exceptions instead of returning a value. This works because a `var` declaration actually compiles to these getter and setter methods and the `throw` expression has the type `Nothing`, which is a subtype of every other type in Scala.

The choice between these approaches is mostly one of style, but it is possible that the use of inheritance here could produce slightly slower code when it is compiled. Inherited methods use virtual function calls, which have more overhead than calling a `final` method that nothing overrides. Whether the compiler produces code that is slower in this case depends on the compiler. Testing this on your own platform is left as an exercise for the reader.

25.6 Immutable Singly-Linked List

The two linked list implementations we looked at earlier were both for mutable linked lists. So we could instantiate a list and then modify the contents in various ways. This worked well with the `Buffer` trait, but it turns out that the first linked list we worked with in Scala was the `List` class, which represents an immutable linked list.

Immutable linked lists only make sense as singly-linked lists because you are not allowed to change data or links. Once a list is created, it must remain completely intact and unchanged. Adding to the front of the list, an operation that we have done with `::`, appends a new node without altering the ones before it in any way. Unlike the mutable linked lists, where the list wraps around the nodes and hides them from view, in an immutable linked list, the nodes are the list. There is nothing wrapped around them.

The code below gives an implementation of an immutable, singly-linked list. This uses a design similar to that for the `List` class itself. Instead of being a single class, it is a small hierarchy with the primary class we normally use to refer to the list being abstract and two subtypes to represent empty and non-empty nodes.

```
package scalabook.adt

import collection.immutable.LinearSeq

sealed abstract class ImmutableSLList[+A] extends LinearSeq[A] {
  def ::[B >:A](elem:B):ImmutableSLList[B] = new Cons(elem,this)
  override def iterator = new Iterator[A] {
    var rover:LinearSeq[A] = ImmutableSLList.this
    def hasNext = !rover.isEmpty
    def next = {
      val ret = rover.head
      rover = rover.tail
      ret
    }
  }
}

final class Cons[A](hd:A,tl:ImmutableSLList[A]) extends ImmutableSLList[A] {
  def length = 1+tl.length
  def apply(index:Int):A = if(index==0) hd else tl.apply(index-1)
  override def isEmpty = false
  override def head = hd
```

```
  override def tail = tl
}

object MyNil extends ImmutableSLList[Nothing] {
  def length = 0
  def apply(index:Int) = throw new IllegalArgumentException("Nil has no contents.")
  override def isEmpty = true
  override def head = throw new IllegalArgumentException("Nil has no head.")
  override def tail = throw new IllegalArgumentException("Nil has no tail.")
}

object ImmutableSLList {
  def apply[A](data:A*):ImmutableSLList[A] = {
    var ret:ImmutableSLList[A] = MyNil
    for(d <- data.reverse) {
      ret = new Cons(d,ret)
    }
    ret
  }
}
```

There are a few elements of this code that we have not seen before that could use some explanation. Starting at the top, the main class is labeled as `sealed` and `abstract`. We have seen `abstract` before, but `sealed` is new. What this keyword implies is that all the subtypes of this class have to appear in the current file. This can be used to provide better type checking for patterns in `match`. In this case it provides us with the security of knowing that other programmers can not produce undesired subtypes that behave in ways different from that desired for our list. Often we can enforce that using `final` to make sure that nothing inherits from a class. In this case that is not an option as we need to have some inheritance.

Next there is the type parameter for `ImmutableSLList`. It is `+A` instead of just `A`. The plus sign indicates that the list is covariant in the type parameter. This is a topic that is covered in more detail in appendix B. For now it is sufficient to say that this means that if `B` is a subtype of `A` then `ImmutableSLList[B]` is a subtype of `ImmutableSLList[A]`. Without the plus sign, the type parameters are invariant and the type any different type parameters produce and the type of the class for different type parameters are effectively unrelated.

The first method in the class is called `::` to mimic that used in the normal `List` class. The odd aspect of this is the bounds on the type parameter. This is the first usage we have seen for the `>:` type bounds. This indicates that we can use any type, `B`, as long as it the same as `A` or a supertype of `A`. This notation was mentioned in subsection 19.4.3, but we did not have a use for it until now. To see the effect that this can have, consider the following code using the standard `List` in the REPL.

```
scala> val nums = List(1,2,3)
nums: List[Int] = List(1, 2, 3)

scala> "hi" :: nums
res0: List[Any] = List(hi, 1, 2, 3)
```

Note the type of `res0`. It is a `List[Any]`. Our list will behave in this same way. The new element, `"hi"` is not an `Int`, so Scala has to figure out a type `B` to use. The only type that is a supertype of both `String` and `Int` is `Any`, so that is what is used. If there were a common supertype lower in the inheritance hierarchy, that would be used instead.

In the `iterator` method the declaration of `rover` needs an explicit type because rover.tail is a method of the supertype. It is also interesting to note that the rover is initialized to be the current node. This is a side effect of the fact that in this implementation, the nodes are not hidden in a list, they truly represent a list. In fact, each node represents a list either as the empty list or the head of some list.

Following the definition of the main class are the two subtypes. The first represents a non-empty node. It is called Cons here and it keeps track of the data element in that node as well as a node for the rest of the list. To complete the LinearSeq, the class has to implement length and apply. To make the implementation more efficient for standard usage, we also overload isEmpty, head, and tail. These same methods are put in the object MyNil, which represents an empty list. This is an object because we only need one of them to exist. It extends ImmutableSLList[Nothing] so that it can be a subtype of any list type. This is where the covariance of ImmutableSLList with respect to A becomes important. Without that, a separate empty list object would be needed for every type of list.

The code ends with a companion object that has an apply method. This makes it possible to create our new list type using the same syntax that we are accustomed to with List.

Some test code works well to demonstrate the usage of our list. You can see that it is very much like the normal List type in how it functions.

```
package test.scalabook.adt

import org.junit._
import org.junit.Assert._
import scalabook.adt._

class TestImmutableSLList {
  @Test def nilIsEmpty {
    assertTrue(MyNil.isEmpty)
  }

  @Test def buildCons {
    val lst = 1 :: 2 :: 3 :: MyNil
    assertEquals(3,lst.length)
    assertEquals(1,lst.head)
    assertEquals(2,lst(1))
    assertEquals(3,lst(2))
  }

  @Test def buildApply {
    val lst = ImmutableSLList(1,2,3)
    assertEquals(3,lst.length)
    assertEquals(1,lst.head)
    assertEquals(2,lst(1))
    assertEquals(3,lst(2))
  }
}
```

One thing that our code will not do is pattern matching on ::. Implementing this would require further advanced techniques that are left for an appendix.

The actual scala.collection.immutable.List code is more complex than what we have written here, but mainly to introduce further efficiencies. This class is completely immutable in that no node can every change after it is created. This forces us to build them by appending to the head if we want to have any efficiency at all. This appears in our code at the end when the input to the apply method is reversed in the for loop. This is one type of overhead which the true library version manages to get around.

25.7 Linked List-Based Stacks and Queues

Recall that the primary characteristic of an ADT is that it tells us what different methods do, not how they are done. The linked list is a real implementation and, as we have seen, it can be used to implement the list ADT. It can also be used as the storage mechanism for other ADTs. In particular, it works well for the only other ADTs that have been considered so far, the stack and the queue. As with the array-based implementation, we will require that all the operations on these collections be $O(1)$ because they are fundamentally simple.

25.7.1 Linked List-Based Stack

To implement a stack using a linked list we could put an instance of one of the lists that we wrote in a previous section into this class as a member and perform operations on that. Such an approach would have value from a software engineering standpoint where code reuse can significantly boost productivity. However, the focus of this chapter is on the nature and structure of linked lists. For that purpose, there is value to building things from scratch and including a node type and code to manipulate it directly in this class. Such an approach also helps us to have greater control over the efficiency of what we write.

An implementation of a stack using a singly-linked list is shown here. The Node type is made using a case class because the values in it do not change. That provides an easy way to make an immutable node.

```
package scalabook.adt

class ListStack[A] extends Stack[A] {
  private case class Node(data:A,next:Node)
  private var top:Node = null

  def push(obj:A) {
    top = Node(obj,top)
  }

  def pop():A = {
    assert(!isEmpty,"Pop called on an empty stack.")
    val ret = top.data
    top = top.next
    ret
  }

  def peek:A = top.data

  def isEmpty:Boolean = top==null
}
```

The top member is a var that serves as the head of the list. The implementation of the methods are remarkably straightforward. As you can see, this is a very short class, even though we are re-implementing a singly-linked list here. The simplicity is allowed because of the simplicity of the stack ADT. No loops or complex logic is required because all access, both adding and removing, happens at the top.

Test code for this class is identical to that for `ArrayStack` with the word `Array` changed to `List`. Only one of those substitutions is functional instead of cosmetic, the instantiation of the stack in the `@Before` method.

25.7.2 Linked List-Based Queue

The same procedure can be used to create a queue that uses a linked list for the data storage. For the stack we only needed to keep one end of the list that we could add to and remove from. In the case of a queue, we need to keep track of both ends of the list. One of them will be the front and the other will be the back. Which is the head and which is the tail is determined by the functionality that we need. New elements should be added at the back and removed at the front. We just saw that the head can do both of these operations efficiently. What about the tail?

In the case of a singly-linked list, we can only add efficiently at the tail, we can not easily remove from there. This goes back to that problem that in order to remove from a singly-linked list, you need to know the node in front of the one you are removing and we can not walk backwards up the list. For this reason, the back of the queue must be the tail of the list if we want to maintain $O(1)$ performance.

Code for a list-based queue is shown here. There are a few more special cases that have to be dealt with which make this code longer than that for the stack. In particular, there are adjustments to the back that are required when an element is added to an empty queue or a remove takes out the last element from the queue.

```
package scalabook.adt

class ListQueue[A] extends Queue[A] {
  private class Node(val data:A,var next:Node)
  private var front:Node = null
  private var back:Node = null

  def enqueue(obj:A) {
    if(front==null) {
      front = new Node(obj,null)
      back = front
    } else {
      back.next = new Node(obj,null)
      back = back.next
    }
  }

  def dequeue():A = {
    assert(!isEmpty,"Dequeue called on an empty queue.")
    val ret = front.data
    front = front.next
    if(front==null) back = null
    ret
  }

  def peek:A = front.data

  def isEmpty:Boolean = front==null
```

```
}
```

25.8 End of Chapter Material

25.8.1 Summary of Concepts

- A slightly more complex ADT than the stack or queue is that List/Seq ADT. This ADT provides methods for random access of elements.
- The `Arrays` that we have been working with are stored as single blocks of memory that can hold the required number of references. This means when you know where the block is, you know where every element is. However, inserting and removing is slow.
- Linked lists are an alternative approach to storing values in a sequential manner. Instead of having a large block, there are nodes that link to one or two other nodes. The list keeps track of at least one node and then others can be reached from that. This makes random access slow, but inserting and removing can be done with $O(1)$ memory moves.
- In a singly-linked list, each node keeps a reference to the one node in front of it and the list must keep track of the first node. It can also keep track of the last. These boundaries become special cases that make the code more complex.
- A doubly-linked list has nodes that keep references to both the next element and the previous element. They are best coded with a sentinel node that stores no data and is effectively the first and last node in the list. The sentinel keeps the code for a doubly-linked list short without special cases.
- While random access is slow for linked lists, iterators allow efficient linear access.
- A singly-linked list can also be written to be immutable. This is what the `List` type does. In this code, the node itself represents a list and there is a special `object` for an empty list.
- The fact that an ADT does not specify how things are done can be illustrated by reimplementing the stack and queue ADTs using linked lists as the mechanism for storing data.

25.8.2 Exercises

1. Write an array-based implementation of the list ADT. You can choose to implement just the four methods listed in the first section of this chapter or implement the `Buffer` trait.
2. Compare speed of your array-based list and a linked list using the following set of manipulations.
3. Compare speed of doubly-linked lists using the two different node types.

- Add at the end, randomly read elements, randomly set elements, remove from the end.
- Add at random locations, randomly read elements, randomly set elements, remove from random locations.
- Add at the beginning, randomly read elements, randomly set elements, remove from the beginning.
- Add at random locations, read sequential elements with an iterator, remove from random locations.

4. Alter the code for a linked list so that you pull out the node walking using `rover` into a single function and call that function in places where that behavior is needed.
5. Add more tests to the linked list test code. Make sure you include the case where the list is completely emptied using remove and the refilled.
6. Compare speed of list-based stack to the array-based stack for different sizes and usage modes.
7. Compare speed of list-based queue to the array-based queue for different sizes and usage modes.

25.8.3 Projects

For this chapter there are not separate descriptions for each of the projects. In all of them, you should already be using some form of `Seq` or `Buffer` for storing significant data. For this project, you need to convert your code to use a linked-list class of your construction instead of the Scala library classes you have been using.

Chapter 26

Priority Queues

The queue Abstract Data Type (ADT) that we have looked at previously can be thought of as modeling a service line. All the entities that are added in are considered equal, and they are taken off in the order they arrive. Not every system works in such an equitable way. In real life there are things like VIP lines. Some systems have a serious reason to treat entities differently. An example of such a system is an emergency room. If you show up to an ER with a paper cut, you can expect to wait for a very long time. This is because the order in which people are served in an ER is based much more on the seriousness of the person's condition than on the time of arrival. Situations where people set up appointments are another example. Arriving long before your appointment time generally does not lead to you being seen significantly earlier.

These types of systems call for a new ADT, the priority queue. A priority queue has the same methods as a normal queue, but there is a different behavior for the dequeue and peek methods in that they give the highest priority object on the queue and only consider length of time in the queue if the highest priority items have equal priority.

26.1 Two Approaches

We will consider two different approaches to writing a priority queue in this chapter, both use a linked list for storage. The difference between the two is whether extra work is done when new items are added to the queue or if that work is put off until the items are removed. We will discuss both and compare their efficiency, but only code the one that is better. Unlike the normal queue, it is not possible to get $O(1)$ performance for all the methods in a priority queue. For the implementations we will talk about here, some methods will be $O(n)$. We will discuss another implementation that can improve on that in chapter 32.

All implementations of the priority queue will extend the following trait.

```
package scalabook.adt

/**
 * This trait defines a mutable Priority Queue ADT.
 * @tparam A the type of data stored
 */
trait PriorityQueue[A] {
  /**
   * Add an item to the priority queue.
   * @param obj the item to add
   */
  def enqueue(obj:A)

  /**
   * Remove the item that has the highest priority or, in the case of a tie,
   * has been on the longest.
   * @return the item that was removed
   */
  def dequeue():A

  /**
   * Return the item that has the highest priority or, in the case of a tie,
   * has been on the longest without removing it.
   * @return the most recently added item
   */
  def peek:A

  /**
   * Tells whether this priority queue is empty.
   * @return true if there are no items on the priority queue, otherwise false.
   */
  def isEmpty:Boolean
}
```

26.1.1 Searching by Priority

One way to create a priority queue using a linked list is to add elements to tail just like a regular queue, but then alter the code in `dequeue` and `peek` so that they run through the elements to find the first one with the highest priority. In the case of `dequeue`, that one is then removed.

We can evaluate the quality of this approach by looking at the amount of work done by each of the methods. The `enqueue` and `isEmpty` methods will be $O(1)$. However, both `dequeue` and `peek` have to do a search through all the elements to find the one with the highest priority. This requires $O(n)$ time. Looking closer than just order, this search is particularly costly because it is impossible to know what we are looking for early on so every element must be considered for each search.

26.1.2 Sorted Linked List

A second approach is to do the additional work when elements are added, and keep the list in sorted order so that we always know where the highest priority item is and we can remove it efficiently. Using this approach, the `dequeue`, `peek`, and `isEmpty` methods are all

$O(1)$. The enqueue method, however, must perform a search to find the correct insertion locations, and is therefore $O(n)$.

Both of these approaches contain at least one method which is $O(n)$. As most applications will have roughly equal numbers of calls to enqueue and dequeue, and the number of calls will be at least $O(n)$ as well, we can expect either of these implementations to have an overall performance of $O(n^2)$. This second approach is better for most applications for a few reasons. The first is that calls to peek are $O(1)$. While there is a logical reason why calls to enqueue and dequeue should roughly balance, there could easily be more calls to peek than either of the other two. In addition, placing items into a sorted list does not require walking the full list. This procedure is like what happens in the inner loop of an insertion sort. You only have to walk through the list until you find the right place to put the new element. For many applications, new elements will have a naturally lower priority than many previous elements. This means that the code will do comparisons to fewer than half the elements, on average, for each enqueue, compared to comparisons of all elements for dequeue and peek in the other approach.

Here is a listing of a sample implementation of a priority queue using a sorted linked list.

```
package scalabook.adt

class SortedListPriorityQueue[A : Manifest](comp: (A,A)=>Int) extends
    PriorityQueue[A] {
  private class Node(var data:A, var prev:Node, var next:Node)
  private val end = new Node(new Array[A](1)(0),null,null)
  end.next = end
  end.prev = end

  def enqueue(obj:A) {
    var rover = end.prev
    while(rover!=end && comp(obj,rover.data)>0) rover = rover.prev
    rover.next.prev = new Node(obj,rover,rover.next)
    rover.next = rover.next.prev
  }

  def dequeue():A = {
    val ret = end.next.data
    end.next = end.next.next
    end.next.prev = end
    ret
  }

  def peek:A = end.next.data

  def isEmpty:Boolean = end.next==end

  def removeMatches(f: A => Boolean) {
    var rover = end.next
    while(rover!=end) {
      if(f(rover.data)) {
        rover.prev.next = rover.next
        rover.next.prev = rover.prev
      }
      rover=rover.next
    }
```

```
  }
}
```

Note that a comparison function is passed in at creation so that any type can be used. This implementation uses a doubly-linked list with a sentinel to reduce special cases and keep the code compact. The highest priority items are kept at the head of the list, `end.next`, and this is where `dequeue` pulls from. The `enqueue` method runs from the tail of the list backwards until it gets to the sentinel or finds a node with higher priority data. Note that if each new item had a lower priority than the one before it, this version of `enqueue` would be $O(1)$.

The test code for this class can look just like that for the regular queue with a few minor changes. In addition to changing the type, there is a comparison function required for the priority queue at creation. Here we can use the `compareTo` method that is part of `Int` and most other built-in types in the library that have a natural ordering. This makes higher values have a higher priority. If you were in a situation where lower values should have the higher priority then you could use `b.compareTo(a)` instead.

```
var queue:PriorityQueue[Int] = null

@Before def initQueue {
  queue = new SortedListPriorityQueue[Int]((a,b) => a.compareTo(b))
}

@Test def enqueue100Dequeue100 {
  val nums = Array.fill(100)(util.Random.nextInt)
  nums.foreach(queue.enqueue(_))
  nums.sorted.reverse.foreach(assertEquals(_,queue.dequeue))
}
```

Using the priority queue also requires that code checking the dequeue order needs to be altered. The last method shown here was one that used a queue of random values. The checks can be done against a reversed, sorted version of the array because the comparison made higher values have higher priority and they will come off first.

26.1.3 Problems with Arrays

For both the regular stack and the regular queue, it was possible to write an implementation using an array instead of a linked list utilizing a similar amount of code and having a very similar performance. This is not the case for the priority queue using either of these approaches, because they involve either insertion or deletion from random locations in the array, which requires doing $O(n)$ copies to move elements around.[1]

In one sense, the array-based implementations still $O(n)$ in the same methods. However, the type of operations they are $O(n)$ in has changed. The linked list approaches are $O(n)$ in comparisons and memory reads, but they are only $O(1)$ in memory writes. Proper array-based implementations add $O(n)$ write overhead on top of the comparisons and memory reads.

This overhead is completely unavoidable for the version that keeps the array sorted all the time as it is not possible to maintain a sorted array without doing inserts that shift elements around. The sort and remove approach can get around this if we relax the requirement that items with tied priority go in the order of arrival. This is not as bad as it

[1]In chapter 32, we will see that an array is exactly what we want to use for a priority queue implementation that is more efficient than either of the options considered here.

might sound because many applications will have very few, if any, ties. With that relaxation, when the highest priority item is found, it can be removed and the last item in the array can be swapped into its place instead of copying all the items down.

26.2 Project Integration: Discrete Event Simulation

One area that makes significant use of priority queues is the field of discrete event simulation. This is a style of simulation in which all changes to the system are modeled as events that happen at specific times and the simulation jumps instantly from one event to the next. The way this is normally done is to have a priority queue that events are placed onto. Calls to dequeue pulls events off in order by the time they occur. Most events have the ability to schedule one or more other events. The simulation continues until either a certain amount of time has passed, or until the queue becomes empty.

A very standard example of a discrete event simulation is a line of people at a teller. The events in this system are people arriving and getting in line and tellers finishing with one person and taking another. Simulations of this type allowed businesses to determine how many people need to be working during different hours. A minor variation on these can be used to set schedules for traffic lights.

26.2.1 Cell Splitting

A very simple example of a discrete event simulation is a simple model of cells dividing in a dish. Part of the simplicity of this simulation is that there is only one type of event, a cell dividing. The dish starts off with a single cell. After a certain period of time, that cell divides into two cells. Each of those later divides into two more. This process continues until we stop the simulation. We ignore cell death to keep things simpler.

The delay between when a cell is first "born" and when it splits is not a fixed value. Instead, it has a certain randomness. It is possible to spend a whole semester covering random numbers, how to pull random numbers from various statistical distributions, and how to find appropriate distributions for different data sets. As these details do not concern us, we will say that each cell splits somewhere between a minimum and a maximum number of time units after the split that created it with an equal probability of any value in that range. This uses a uniform distribution, which is what we get from `math.random`.

The events, in this case, are simply splitting times and can be represented with nothing more than a `Double`. We begin by placing one value on the queue, which is the time the first cells splits. We also need to start a counter that keeps track of how many cells there are. It begins at one and it gets incremented by one for each event as one cell splits into two. Basic code for this looks like the following.

```
def runSimulation:Double = {
  val pq = new SortedListPriorityQueue[Double]((a,b) => b.compareTo(a))
  pq.enqueue(minSplitTime+math.random*(maxSplitTime-minSplitTime))
  var pop = 1
  while(pop<maximumPopulation-1) {
    val time = pq.dequeue()
    pop += 1
    pq.enqueue(time+minSplitTime+math.random*(maxSplitTime-minSplitTime))
    pq.enqueue(time+minSplitTime+math.random*(maxSplitTime-minSplitTime))
```

```
    }
    pq.dequeue()
  }
```

This code will run the simulation, but it does not really do anything useful for us because it does not do much for us. It returns the amount of time it took to reach the stop population. We can produce something that is a bit more functional by putting this inside of a class that is `Drawable` and having it plot the population as a function of time. Code that does that is shown here.

```
package scalabook.drawing

import swing._
import event._
import java.awt.{Color}
import java.awt.geom._
import java.awt.image.BufferedImage
import collection.mutable
import scalabook.adt._

class DrawCellSim(p:DrawTransform,
    private var width:Int,
    private var height:Int,
    private var maximumPopulation:Int,
    private var minSplitTime:Double,
    private var maxSplitTime:Double,
    private var numSims:Int) extends DrawLeaf {
  val parent = p
  @transient private var propPanel:Component = null
  @transient private var img:BufferedImage = null

  private def runSimulation(splits:mutable.Buffer[Double]) {
    val pq = new SortedListPriorityQueue[Double]((a,b) => b.compareTo(a))
    pq.enqueue(minSplitTime+math.random*(maxSplitTime-minSplitTime))
    while(splits.length+1<maximumPopulation) {
      val time = pq.dequeue()
      splits += time
      pq.enqueue(time+minSplitTime+math.random*(maxSplitTime-minSplitTime))
      pq.enqueue(time+minSplitTime+math.random*(maxSplitTime-minSplitTime))
    }
  }

  private def redoImage {
    if(img==null || width!=img.getWidth || height!=img.getHeight) {
      img = new BufferedImage(width,height,BufferedImage.TYPE_INT_ARGB)
    }
    for(i <- 0 until width; j <- 0 until height) img.setRGB(i,j,0) // transparent
    val data = mutable.Buffer[Double]()
    var maxTime = -1.0
    val g = img.createGraphics()
    g.setPaint(Color.black)
    for(i <- 0 until numSims) {
      runSimulation(data)
      if(maxTime < 0) {
        maxTime = data.last*1.2
```

```
    }
    var (lastX,lastY) = (0.0,height-height/maximumPopulation)
    for(i <- data.indices) {
      val (x,y) = (data(i)*width/maxTime,height-(i+1)*height/maximumPopulation)
      g.draw(new Line2D.Double(lastX,lastY,x,y))
      lastX = x
      lastY = y
    }
    data.clear
  }
  drawing.refresh
}

def draw(g:Graphics2D) {
  if(img==null) redoImage
  g.drawImage(img,0,0,null)
}

def propertiesPanel() : Component = {
  if(propPanel == null) {
    import BorderPanel.Position._
    propPanel = new BorderPanel {
      layout += new BorderPanel {
        val properties = Seq[(String, () => String, (String) => Unit)](
            ("Width", () => width.toString, str => width = str.toInt),
            ("Height", () => height.toString, str => height = str.toInt),
            ("Max Pop.", () => maximumPopulation.toString, str =>
                maximumPopulation = str.toInt),
            ("Min Split Time", () => minSplitTime.toString, str => minSplitTime =
                str.toDouble),
            ("Max Split Time", () => maxSplitTime.toString, str => maxSplitTime =
                str.toDouble),
            ("Num Sims", () => numSims.toString, str => numSims = str.toInt)
            )
        layout += new GridPanel(6,1) {
          for((name,_,_) <- properties) {
            contents += new Label(name)
          }
        } -> West
        layout += new GridPanel(6,1) {
          for((_,get,set) <- properties) {
            contents += new TextField(get()) {
              listenTo(this)
              reactions += { case e:EditDone =>
                try {
                  set(text)
                } catch {
                  case ex:NumberFormatException =>
                }
              }
            }
          }
        } -> Center
      } -> North
      layout += Button("Run Sims"){ redoImage } -> South
```

```
      }
    }
    propPanel
  }

  override def toString() = "Cell Simulation"

  def toXML : xml.Node = {
    <drawCellSim width={width.toString} height={height.toString}
      maxPop={maximumPopulation.toString} minSplitTime={minSplitTime.toString}
      maxSplitTime={maxSplitTime.toString} numSims={numSims.toString}/>
  }
}

object DrawCellSim {
  def apply(p:DrawTransform,data:xml.Node) = {
    new DrawCellSim(p,(data \ "@width").text.toInt,
        (data \ "@height").text.toInt,
        (data \ "@maxPop").text.toInt,
        (data \ "@minSplitTime").text.toDouble,
        (data \ "@maxSplitTime").text.toDouble,
        (data \ "@numSims").text.toInt)
  }

  def apply(p:DrawTransform) = {
    new DrawCellSim(p,200,200,100,1.0,5.0,10)
  }
}
```

This lets the user set a number of different parameters for running the simulation and plot lines for many different simulations. The plots are scaled to fit the range in the specified size. No attempt is made to show scale values as that is a significantly more challenging problem. Figure 26.1 shows an example using this class in a drawing.

26.2.2 Collision Handling

Another example where discrete event handling makes sense is in realistic hard-sphere collision handling. The events in this system are times when balls run into one another or into other obstacles such as walls. They have to be handled in order because one collision can alter or completely prevent a later one. It is possible to do this constantly for the whole simulation, though there are reasons to bring all the particles up to a particular time every so often. For example, if you are going to render the simulation or save off the configuration to a disk, it helps if all the particles have a location for the same time instead of having them at the last position for which they had an event.

Using events for collisions is more accurate and potentially significantly faster than letting particles move forward for a whole timestep, then checking for overlap. The latter approach requires small time steps and it potentially misses details. The obvious example of this is if two small, fast-moving particles should collide in the middle of a timestep. Unless the actual times of collisions are found, they can pass through one another and have no overlap at the end of the step. Similar types of errors can occur when one particle should collide with two or more other particles during a step.

In a purely event-driven simulation, the "timesteps" can also be handled as events. So when these events are reached, all the particle positions are updated and any special

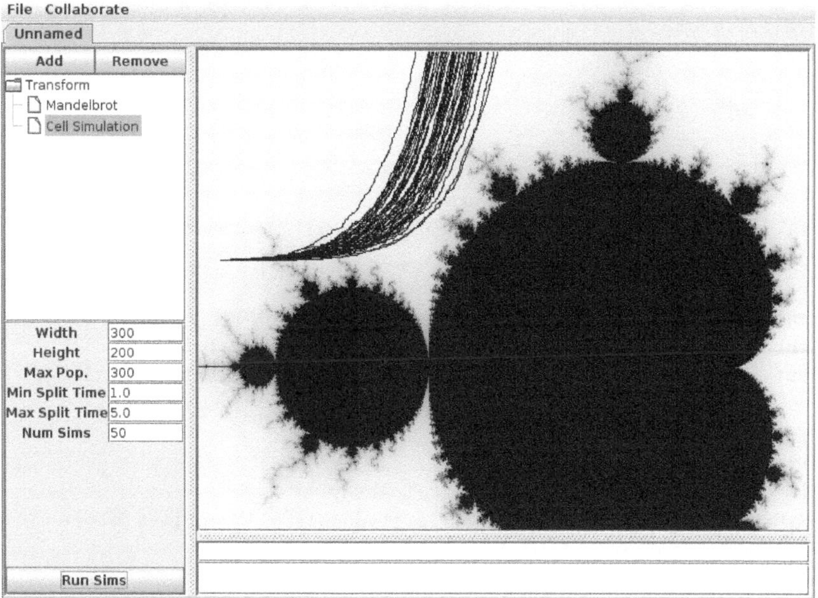

FIGURE 26.1: This screen shot shows the output of the cell simulation drawable. The properties that are used are visible on the left side of the window.

handling, such as rendering the particles or the application of non-collisional forces like gravity, can be performed. This type of handling can also be performed by breaking out of the event-handling loop occasionally. With some minor modifications, we can make it so that the `DrawBouncingBalls` class we wrote previously works in this way.

The first modification we have to make is not to the bouncing balls, but to our queue. Unlike the other examples that have been mentioned, this example requires that we are able to remove events from the queue without processing them. Each time we process an event that modifies the velocity of a particle, all future events involving that particle must be removed. The most general way for us to add that functionality is with a method like the following.

```
def removeMatches(f: A => Boolean) {
  var rover = end.next
  while(rover!=end) {
    if(f(rover.data)) {
      rover.prev.next = rover.next
      rover.next.prev = rover.prev
    }
    rover=rover.next
  }
}
```

This takes a predicate function and removes all the elements for which the function is true.

With this added in, we can make modifications to `DrawBouncingBalls` itself. We will show the entire class here as more has been modified or added than was left alone.

```
package scalabook.drawing

import java.awt.{Graphics2D,Color}
```

```
import java.awt.geom._
import swing._
import event._
import javax.swing.Timer
import scalabook.adt._

class DrawBouncingBalls(p:DrawTransform,
    private var balls:Vector[DrawBouncingBalls.Ball]) extends DrawLeaf {
  val parent = p
  @transient private var propPanel:Component = null
  @transient private var lTimer:Timer = null
  @transient private var workCopy:Array[DrawBouncingBalls.Ball] = null
  @transient private var pq:SortedListPriorityQueue[CollEvent] = null

  private def timer = {
    if(lTimer==null) {
      if(workCopy==null) workCopy = balls.toArray
      if(pq==null) pq = new SortedListPriorityQueue[CollEvent]((a,b) =>
          b.time.compareTo(a.time))
      lTimer = new Timer(100,Swing.ActionListener(e => {
        for(i <- balls.indices) workCopy(i) = balls(i).copy(vy = balls(i).vy+0.01,
            time = 0.0)
        for(i <- balls.indices) {
          findEventsFor(i,i+1 until balls.length,0.0)
        }
        while(!pq.isEmpty) {
          val event = pq.dequeue()
          event.handle
        }
        balls = Vector(workCopy.map(_.advanceTo(1.0)):_*)
        drawing.refresh
      }))
    }
    lTimer
  }

  def draw(g:Graphics2D) {
    g.setPaint(Color.black)
    g.draw(new Rectangle2D.Double(0,0,100,100))
    g.setPaint(Color.green)
    for(DrawBouncingBalls.Ball(x,y,_,_,s,_) <- balls) {
      g.fill(new Ellipse2D.Double((x-s)*100,(y-s)*100,s*200,s*200))
    }
  }

  def propertiesPanel() : Component = {
    if(propPanel==null) {
      propPanel = new BorderPanel {
        layout += new GridPanel(1,1) {
          val button = new Button("Start")
          button.action = Action("Start"){
            if(button.text == "Start") {
              timer.start()
              button.text="Stop"
            } else {
```

```
          timer.stop()
          button.text="Start"
        }
      }
      contents += button
    } -> BorderPanel.Position.North
  }
}
propPanel
}

override def toString = "Bouncing Balls"

def toXML : xml.Node =
  <drawBouncingBalls>
    {balls.map(_.toXML)}
  </drawBouncingBalls>

private def collisionTime(b1:DrawBouncingBalls.Ball,
    b2:DrawBouncingBalls.Ball):Double = {
  val (sx1,sy1) = (b1.x-b1.vx*b1.time, b1.y-b1.vy*b1.time)
  val (sx2,sy2) = (b2.x-b2.vx*b2.time, b2.y-b2.vy*b2.time)
  val radSum = b1.size+b2.size
  val (dx,dy) = (sx1-sx2, sy1-sy2)
  val (dvx,dvy) = (b1.vx-b2.vx, b1.vy-b2.vy)
  val c = dx*dx+dy*dy-radSum*radSum
  val b = 2*(dx*dvx+dy*dvy)
  val a = dvx*dvx+dvy*dvy
  val root = b*b-4*a*c
  if(root<0) {
    -1.0
  } else {
    (-b-math.sqrt(root))/(2*a)
  }
}

private def findEventsFor(i:Int,against:Seq[Int],curTime:Double) {
  for(j <- against) {
    val t = collisionTime(workCopy(i),workCopy(j))
    if(t>=curTime && t<1.0) {
      pq.enqueue(new BallBallColl(t,i,j))
    }
  }
  for((tfunc,bfunc) <- DrawBouncingBalls.wallInfo) {
    val t = tfunc(workCopy(i))
    if(t>=curTime && t<1.0) {
      pq.enqueue(new BallWallColl(t,i,bfunc))
    }
  }
}

private trait CollEvent {
  def time:Double
  def handle:Unit
}
```

```
private class BallBallColl(val time:Double,val b1:Int,val b2:Int) extends
    CollEvent {
  def handle {
    val ball1 = workCopy(b1).advanceTo(time)
    val ball2 = workCopy(b2).advanceTo(time)
    val m1 = ball1.size*ball1.size*ball1.size
    val m2 = ball2.size*ball2.size*ball2.size
    val cmvx = (ball1.vx*m1+ball2.vx*m2)/(m1+m2)
    val cmvy = (ball1.vy*m1+ball2.vy*m2)/(m1+m2)
    val dx = ball1.x-ball2.x
    val dy = ball1.y-ball2.y
    val dist = math.sqrt(dx*dx+dy*dy)
    if(dist>1.01*(ball1.size+ball2.size)) {
      println("Warning: collision with big separation. "+b1+" "+b2+" "+dist)
    }
    if(dist<0.99*(ball1.size+ball2.size)) {
      println("Warning: collision with little separation. "+b1+" "+b2+" "+dist)
    }
    val vx1 = ball1.vx-cmvx
    val vy1 = ball1.vy-cmvy
    val nx = dx/dist
    val ny = dy/dist
    val mag = nx*vx1+ny*vy1
    workCopy(b1) = ball1.copy(vx = ball1.vx-1.9*mag*nx, vy = ball1.vy-1.9*mag*ny)
    workCopy(b2) = ball2.copy(vx = ball2.vx+1.9*mag*nx*m1/m2, vy =
        ball2.vy+1.9*mag*ny*m1/m2)
    pq.removeMatches(_ match {
      case bbc:BallBallColl => bbc.b1==b1 || bbc.b2==b1 || bbc.b1==b2 ||
          bbc.b2==b2
      case bwc:BallWallColl => bwc.b==b1 || bwc.b==b2
      case _ => false
    })
    val others = workCopy.indices.filter(b => b!=b1 && b!=b2)
    findEventsFor(b1,others,time)
    findEventsFor(b2,others,time)
  }
}

private class BallWallColl(val time:Double,val b:Int,val
    newDir:(Double,Double,Double,Double) => (Double,Double)) extends CollEvent {
  def handle {
    val ball = workCopy(b)
    val nx = ball.x+(time-ball.time)*ball.vx
    val ny = ball.y+(time-ball.time)*ball.vy
    val (nvx,nvy) = newDir(ball.x,ball.y,ball.vx,ball.vy)
    workCopy(b) = DrawBouncingBalls.Ball(nx,ny,nvx,nvy,ball.size,time)
    pq.removeMatches(_ match {
      case bbc:BallBallColl => bbc.b1==b || bbc.b2==b
      case bwc:BallWallColl => bwc.b==b
      case _ => false
    })
    findEventsFor(b,workCopy.indices.filter(_!=b),time)
  }
}
```

```
}

object DrawBouncingBalls {
  def apply(p:DrawTransform,data:xml.Node) = {
    new DrawBouncingBalls(p,Vector((data\"ball").map(bxml => {
      val x=(bxml\"@x").text.toDouble
      val y=(bxml\"@y").text.toDouble
      val vx=(bxml\"@vx").text.toDouble
      val vy=(bxml\"@vy").text.toDouble
      val size=(bxml\"@size").text.toDouble
      Ball(x,y,vx,vy,size,0.0)
    }):_*))
  }

  def apply(p:DrawTransform,minSize:Double = 0.01,maxSize:Double = 0.05) = {
    new DrawBouncingBalls(p,Vector.fill(20) {
      val size = minSize+math.random*(maxSize-minSize)
      Ball(size+math.random*(1-2*size),size+math.random*(1-2*size),
        (math.random-0.5)*0.02,(math.random-0.5)*0.02,size,0.0)
    })
  }

  case class Ball(x:Double,y:Double,vx:Double,vy:Double,size:Double,time:Double) {
    def toXML = <ball x={x.toString} y={y.toString}
        vx={vx.toString} vy={vy.toString} size={size.toString}/>

    def advanceTo(t:Double) = {
      val dt = t-time
      copy(x = x+vx*dt, y = y+vy*dt, time = t)
    }
  }

  private val wallInfo=Seq[(Ball => Double, (Double,Double,Double,Double) =>
      (Double,Double))](
      (b => if(b.vx<0) b.time-(b.x-b.size)/b.vx else -1, (x,y,vx,vy) =>
          (vx.abs,vy)),
      (b => if(b.vx>0) b.time+(1-b.x-b.size)/b.vx else -1, (x,y,vx,vy) =>
          (-vx.abs,vy)),
      (b => if(b.vy<0) b.time-(b.y-b.size)/b.vy else -1, (x,y,vx,vy) =>
          (vx,vy.abs)),
      (b => if(b.vy>0) b.time+(1-b.y-b.size)/b.vy else -1, (x,y,vx,vy) =>
          (vx,-vy.abs)))
}
```

The most significant changes are reflected at the top with two new class-level variables and modifications to the code in the timer. The new class-level variables are an array to mirror the vector while the processing is happening and a priority queue that holds a type called `CollEvent`, which is a trait declared later in the code.

The timer code checks to see if either of those is `null` and creates them if so. This has to be done for the purposes of serialization. The main changes to the logic occur in the action for the timer. The `workCopy` variable is initialized to have a copy of the balls with gravity applied. Then for each ball, all events that happen during the next time unit are found relative to all the balls after it in the array. The condition of only checking against particles later in the list prevents us from double finding collisions. Then as long as the priority queue is not empty, we pull off the next event and handle it. Then the balls are copied over the a `Vector` that will be used for the purposes of drawing. The code for drawing, making the properties, and producing XML remain the same.

After the `toXML` method there are a number of new methods and classes that have been added to the main class. First is a method that takes two `Ball` objects and returns a collision time for the two. This method assumes the balls move on straight lines, so the distance between them squared is a quadratic in time. We simply have to solve for the point in time when they get to a distance equal to the sum of the radii. Each particle keeps track of the "time" it is at during a step so that has to be accounted for. If there is no collision, this method returns -1.0. Otherwise, it returns the lesser root as that is when the particles hit. The other root would be when they separate if they moved through one another. Here you can see the math for finding the coefficients in the quadratic equation.

$$\begin{aligned} d^2(t) &= ((x_1 + vx_1 * t) - (x_2 + vx_2 * t))^2 + ((y_1 + vy_1 * t) - (y_2 + vy_2 * t))^2 \\ &= (\Delta x + \Delta vx * t)^2 + (\Delta y + \Delta vy * t)^2 \\ &= \Delta x^2 + \Delta y^2 + 2(\Delta x \Delta vx + \Delta y \Delta vy)t + (\Delta vx^2 + \Delta vy^2)t^2 \end{aligned}$$

When particle motion is more complex than a straight line, more complex code has to be involved to find the collision time as normally there is not a closed-form solution like there is in this case.

After the method that finds the time of a collision is a method that finds all the events for a particular particle after the current time against some set of other particles and the walls. The way the code has been set up, the structure for handling the particles and the walls is very similar. We loop over the objects we want to check against, calculate an event time, and if that event time falls into the valid range for the current step, we add it to the queue. We will come back to how the wall code is written to work in this way.

This simulation has two types of collisions that are subtypes of the `CollEvent` trait. They are `BallBallColl` and `BallWallColl`. Each of these knows a time when the event happens and contains code for handling the event. The wall class is simpler and is probably worth looking at to see what is going on. First the position of the ball at the time of the event is found. Then we call a function that was passed in at the construction of the event which takes the position and velocity of the ball and returns a new velocity. That new velocity and position are used to build a new ball object which is stored in the array in place of the old one. Then future events involving that ball are removed and new events are found.

The code for ball-to-ball collisions is doing the same thing, but for two balls and it involves more math for handling that type of collision. It is not significant that you understand this math unless you want to. It starts by calculating relative masses of the balls, then finding the center of mass velocity of the two balls.[2] This code checks the distance against the sum of the radii and prints an error message if it is not close enough to the expected value. This code can be very helpful for making sure that everything is working correctly, and while it is not technically needed in the final code, it can be helpful to keep in there to prevent problems with future modifications. The modification in velocity for each particle is calculated as a multiple of the projection of the velocity along the line separating the particles. In this code, the constant 1.9 is used so that there is some energy loss. Perfect collisions would use a value of 2.0. Unfortunately, the gravity integration for this code does not conserve energy, so it is helpful to have the collisions damp the system a bit.[3]

[2]Collisions are much easier to calculate in the center of mass frame. In this moving frame of motion, it is as if each ball ran into a stationary wall.

[3]This code applies a full downward acceleration to each particle at the beginning of the step. If the particle reverses vertical direction part way through the step, the influence of gravity should have been reversed. That does not happen here so particles can gain energy with each bounce off the bottom.

While the math in this code might be a bit confusing to some, the most interesting piece of the code it probably the way in which collisions with the walls are handled. The event finding code uses something called `wallInfo` and functions from that are passed into the `BallWallColl` objects. The magic is contained in the declaration of `wallInfo` at the bottom of the code. This is a sequence of tuples where each element has two functions. These functions represent a wall. The first one tells when a given ball would collide with that wall and the second one gives the modified velocity of the ball after it collides with the wall.

The beauty of this approach is that it makes it fairly simple to add additional barriers. For example, the following tuple defines a barrier that runs diagonally across the middle of the cell that bounces particles toward the top-left corner.

```
(b => if(b.vx>0 || b.vy>0) b.time+(1-b.x-b.y)/(b.vx+b.vy) else -1,
    (x,y,vx,vy) => (-vy.abs,-vx.abs))
```

This is not technically a wall because it only considers the center of the particle, not the edges. It also allows particles to pass up through it, just not down. However, it demonstrates the flexibility of the approach. With a bit more math, it would be possible to throw in barriers like a round wall that is a cup at the bottom. As soon as the functions are added to the `wallInfo` sequence, it automatically works with the rest of the code.

If you enter the code that is given above and run the drawing program, then add a bouncing balls drawable, you will see 20 balls of random sizes added in, just like before. You can select that drawable from the tree, and click the button to start the timer going. If you watch for a while, you will see that the large balls bunch up near the bottom and the small ones occasionally get shot higher up into the box. This type of behavior is called equipartition of energy and is a standard feature of physical systems where bodies of different masses interact. The fact that we see it in this system is a good indication that our code is behaving properly.

26.3 End of Chapter Material

26.3.1 Summary of Concepts

- A priority queue has the same methods as a queue, but the `dequeue` method functions differently.
  - Instead of always taking the object that has been present the longest, it takes the object with the highest priority.
  - Tied priority can go to the one that has been present longest.
  - We can create a reasonably efficient priority queue using a sorted linked list.
- Discrete event simulations use priority queues.
  - Events are prioritized by the time they occur.
  - We looked at a cell splitting simulation and collision handling.
  - Collision handling required an extra method to remove certain events.

26.3.2 Exercises

1. Code the search-based priority queue and do speed tests to compare it to the version that maintains a sorted linkedlist. For the speed tests, insert elements with random times and dequeue them. Vary the number of elements used across an exponential range to measure how that impacts performance.

2. Code `Array`-based implementations of a priority queue using both a sorted `Array` and an unsorted `Array` that you search. Do speed tests to compare them to each other and to the linked list-based implementations. For the search-based method with an `Array`, do both shifting elements down and and swapping the last element into the hole left by the one that is removed.

3. Implement priority queues using `ArrayBuffer` and `ListBuffer` to see how efficient they are relative to our "custom" implementations.

4. Implement some extra wall types for the collision simulation. Consider trying walls that are not lines or writing a method that creates wall segments.

5. Look up some different distributions for random variables and try the cell splitting simulation using those.

26.3.3 Projects

1. There are a few ways that you could use a priority queue in your MUD to good effect. All of them involve having a single priority queue that actions can be added to so that they will run at a later time. This will allow things like spells to have delayed effects or to wear off. It can also be used by computer controlled characters and to implement random spawning of characters. The advantage of a priority queue is that you do not have to have checks in every single character/room/item for whether something should happen. Instead, all those things could place events on the priority queue with a delay, and the main thread only pulls off things as needed.

   The idea here uses a sorted list or `Array` for the priority queue so that the `peek` operation is fast. The idea is that in most "ticks" there will not be any events on the priority queue to process. In that case, the code only does a `peek`, sees that the time of the first element is after the current time, then goes on to whatever else needs to be done.

2. If you are working on the web spider project, you might have noticed that many websites are extremely large. Even if you limit your spider to a single domain, say only pages coming from domains ending with cnn.com, espn.go.com, or google.com, there can still be an extremely large number of pages. If you let the processing proceed in the order that links are found, you are likely to never get to a lot of the data that you are interested in because of resource limitations. One way to deal with this is to make the queue of pages to be read work as a priority queue.

   If you know the types of pages you are looking for and have an idea of what paths they are at on the site you are running through, you can give those pages a higher priority. That way you will visit those pages first. As long as the program keeps finding pages that have a structure you have coded to be interesting, it will work on those pages and not go to other pages that are less likely to have interesting information.

3. If you are working on either a networked or non-networked graphical game, you can use a priority queue in the main loop of the game in much the same way as is described above for the MUD. It is possible that players will be able to do things that should not have effects for a while. It is also possible that you will have many units in the game that do not need to update every single tick. In either of these cases, you can make a priority queue of events/commands that are only executed when it comes time for them to happen.

4. This chapter gave a reasonable introduction to the topic of discrete event simulation. Add some capabilities to your simulation workbench that support general discrete event simulations as well as collisional simulations. Make a few specific instances of discrete event simulations that users can create and run. Allow them to adjust certain significant parameters and track significant aspects of the status of the simulation.

5. The math worksheet does not provide any obvious places to include a priority queue. Two possibilities would be to give the ability for the user to build simple discrete event simulations in a worksheet or to have elements of a worksheet prioritized for processing. The latter might be useful if you either want things that are currently on screen to be calculated before things off screen or if you decide that processing of graphical elements should be a lower priority than processing numeric elements.

   If neither of these appeals to you or fits into what you want to do with your project, you could do the first three exercises in this chapter instead.

6. The Photoshop® project also does not automatically lend itself to including a priority queue. If you can think of an application, you could write one. Prioritizing draw elements could work, but you have to make sure you do not break the user control over what covers what in a painters algorithm.

   If this option does not fit into what you want to do with your project, you could do the first three exercises in this chapter instead.

Chapter 27

Refactoring

In the second half of this book, as you have worked with longer pieces of code, there are a few things that you should have realized about working on bigger programs. For example, the first time that you write a piece of code, you probably do not do it in an optimal way. Also, when you first lay out how to put something together, you likely miss details that have to be added on. What is more, sticking pieces onto code in a haphazard way often leads to really brittle code that is hard to work with. If you were writing software in a professional environment, there would be another rule that could be added on here, the customer often does not really know what he/she wants.

When you put all of these together, it leads to the general rule that code needs to be dynamic and adaptable. You have to be willing and able to modify how things are done over time. It is also ideal for this to be done in a good way so that the results do not just pile mistaken approach upon mistaken approach in a crudely rigged solution to a final goal.

In this chapter, we will look at one of the primary tools in the toolkit of the software developer for effectively managing code over time as old approaches are replaced by newer ones and as features get piled up. This tool is called refactoring and it is the alternation of existing code without changing functionality. At first the idea of changing code without altering the functionality might seem odd, but there are good reasons for why it should be done.

27.1 Smells

Not all code is created equal. While there are effectively an infinite number of ways to build correct solutions to a problem, you have hopefully learned that many solutions that are technically correct, are less than ideal. In some situations this is obvious. In others, it depends on the application or even who you ask. Code is very much an art and, as such, beauty is often in the eye of the beholder. Even when code is not quite ideal, that does not mean it is worth the effort to rewrite to a better form. The technical term for code that has a property that makes it less than ideal is that it smells.

The term smell carries significant imagery. Smells can be strong or mild. Even if something smells bad, if it is really mild then you probably will not go to any lengths to correct it unless there is a good reason to. In addition, not all programmers will be equally sensitive to different smells. In his book *Refactoring: Improving the Design of Existing Code*, Martin Fowler, along with Kent Beck, list 22 different smells that can afflict code [5]. We will list them all here with brief explanations along with possible examples of where they have potentially popped up in code we have written.

Alternative Classes with Different Interfaces - When you have methods in different classes that do the same thing, but call it different names. This makes code very hard to work with. You should pick a name that fits and rename everything to match that.

Comments - Some comments are definitely good. Too many, especially when they do not really add extra information, can be smelly. Comments that are out of date with the code are also horribly smelly. Programmers will read comments to help them figure out what is happening in the code. If the comments have not been updated with the code, this can lead to significant confusion and waste a lot of time. You are better off with no comments than bad comments.

Data Class - These are classes that do nothing but store data and have no real functionality. Littering your code with individual files that contain these makes it very hard for people to navigate the code. You might consider having these nested inside of objects that have a meaningful relationship to the case class.

Data Clumps - Pieces of data that are almost always found together and should be united into a class. A simple example of this is information for a data point. Instead of keeping x, y, and z separate and passing them as three separate arguments into many different methods, use a class that can store the three values. One benefit of this is that calling code explicitly knows the data is a point, not three unrelated `Doubles`.

Divergent Change - When you change the same piece of code (often a class) at different times for different reasons. This can imply that code is taking on overly broad responsibilities in the code. Try to break it into smaller pieces that are more focused in their tasks.

Duplicated Code - The name says it all here. We have been talking about ways to avoid this smell since very early in the book. The reader should also by this point be well aware of the pitfalls of duplicate code, such as duplication of errors and the need to modify all copies for certain changes. Multiple approaches have been introduced to reduce code duplication, from function to collections and loops, then finally to abstraction and polymorphism, you should strive to write any particular piece of logic as few times as possible.

Feature Envy - This is when a method in one class makes lots of calls to another class, especially if it does so just to get data for doing a calculation. This likely implies that the method is in the wrong class and should be moved closer to the data it is using.

Inappropriate Intimacy - An object is supposed to encapsulate an idea. If you have two classes that are constantly referring to elements of one another, it implies that things have not been broken up appropriately. Consider moving things around so that each class stands more on its own.

Incomplete Library Class - This smell occurs when a library that you are using lacks functionality that you need. The significance of it being in a library is that you typically can not modify library code. In Scala, you can often get around this using implicit conversions. This is a more advanced technique that is discussed in appendix B.

Large Class - Classes that get too large become hard to manage. You want to have the functionality of any given class be specific to a particular task. If a class gets too large it is an indication that it should probably be broken into parts.

Lazy Class - If you have a class that does not do much of anything it is lazy and you should consider getting rid of it. At the very least, put classes like this inside of an appropriate other class or object so they do not take up their own file.

Long Method - As was discussed in the first half of this book, long functions are error prone. The same goes for methods. If a method starts to get too long it should be broken up into pieces.

Long Parameter List - Methods that have extremely long parameter lists are hard to use.[1] Often, long parameter lists go along with the "Data Clumps" smell. The parameter list can be shortened by passing in a smaller number of objects containing the needed values.

Message Chains - This smell occurs when one class goes through many objects to get to what it needs. For example, `val stu = school.teacher.course.student`. The problem with this is that whatever class you write this in new depends on the exact name and general contract for the methods `teacher`, `course`, and `student` in three different classes. If any of those things change, this code will break. That is bad.

Middle Man - This smell arises when too many methods in one class do nothing but forward calls to another class. The one doing the forwarding is the "middle man." If this happens too much, and is not providing other benefits, you should consider going straight to the object that does the work.

Parallel Inheritance Hierarchies - This is a special case of "Shutgun Surgery." In this case, you have two inheritance hierarchies that mirror one another. When a class is added to one, it must be added to the other as well.

Primitive Obsession - In Java and many other languages, types like `Int`, `Double`, `Char`, and `Boolean` are called PRIMITIVE types and they are treated differently than objects.[2] Scala does not maintain this distinction, but this smell can still apply. Code has this

[1] Having default values and named parameters in Scala does minimize this problem a bit.

[2] Primitive types are built into the hardware of the machine so they execute much more quickly and take up less space than object. When you use an `Int` in Scala, the compiler will try to compile it to a Java primitive `int` for better performance.

smell when you use basic or built-in types too much instead of creating your own types. Before you learned about case classes, all of your data grouping was done with tuples. This made for smelly code because references to _1 and _2 tell you nothing about the meaning of the value. Similarly, an (Int, Int) could be a month/year combination or coordinates on a planar grid. Using types specific to the task not only makes code easier to read, it allows the type checker to help you determine when you have made an error by producing a syntax error on a type mismatch.

Refused Bequest - This smell occurs when a class inherits from another class or trait, but does not "want" some of the methods in that supertype. This is typically a sign of a poor design choice. The motivation behind inheritance should be to become a subtype of the supertype. That means that any method calls that could be made on an instance of the supertype should work on the subtype. If this is not the case, then inheritance is probably the wrong approach.

Shotgun Surgery - This is when a particular change in your code has a tendency to force you to make edits in several different places. This is smelly because it is too easy to forget one of those edits, which leads to bugs. Adding a new Drawable has something of a shotgun surgery feel to it. We could fix this by moving the makeDrawable method from the DrawTransform object into the Drawing object next to the makeOptionsFor method. Then adding a new drawable would only require adding code to those two methods that are adjacent to one another instead of altering code in two separate files.

Speculative Generality - We often want to have a lot of flexibility in our code, but code that is far more general than the usage can be artificially hard to understand, use, and maintain. This happens when a code writer expects to need certain flexibility that the final usage does not take advantage of.

Switch Statements - Scala does not have switch statements. The closest approximation is the match, which is far more powerful than a normal switch statement. This smell can still occur in Scala code if matches are used in particular ways. This smell occurs when you have multiple match statements/expressions that have exactly the same cases and have whenever a case has to be added to one, it must be added to all. This turns the adding of a case into a shotgun surgery. Generally, this should be dealt with making a trait that has one method for each of the different match occurrences and a different concrete subtype for each case. Each match in the code then becomes a call to the appropriate method of an object that keeps track of the state.

Temporary Field - This is when you have a field of a class that is only needed at certain times. This makes code hard to understand because the field will have a value at other times, but that value might not really have any meaning for the object.

The goal of refactoring, and what you should be thinking about when considering smells in code, is that you want to keep your code readable and maintainable as well as keeping it correct and sufficiently optimized for your needs. Some of these smells almost compete with one another, pushing you in opposite directions. Before you make changes, consider the motivation. Will the change improve the code and make it easier to work with? Would the side-effects of a cure for a particular smell be worse than the original smell? Just because you see something in your code that seems like it matches one of these smells does not mean the code is really smelly. That judgment requires really thinking about how the code behaves and how the programmer interacts with it.

27.2 Refactorings

Fowler's book on refactoring lists 72 different refactoring methods. Only a few will be discussed here. Not only should interested readers turn to the original book for a full description, many of the refactoring methods presented in the book involve concepts that are beyond the scope of this book or what the reader is expected to understand at this point in his/her Computer Science education.

27.2.1 Built-in Refactoring Methods

Many of the refactoring methods are algorithmic in nature. For this reason, Integrated Development Environments (IDEs), such as Eclipse, have a number of them written in. Using built-in refactoring tools can make things go much faster. They are also set up so that they have no impact on the behavior of the code. They change the structure without changing the functionality.

As of this writing, the following options are available under the refactoring menu when you right click on code in Eclipse under Scala.[3]

- Rename - This can be applied to any programmer given name such as variable names, method names, or class names. Just as comments that no longer fit the code are smelly, using names in your programs that are hard to understand or, worst yet, are misleading, should have you detecting a rank stench. Of course, you can change the name of something on your own, but that means finding all the places it is used and changing them as well. For something like the name of a public method in a large project, this can mean tracking down invocations in many different files. This refactoring option does all that for you. So if you find that a name you chose early in the development of a project no longer fits, select this option and change it to something that does.

- Inline Local - This will remove a local `val` and change all references to it to the value it is initialized to. This should only be done with immutable values as this operation can change behavior if the object is mutated in some of the usages. You would use this if you put something in a variable expecting it would be long or used many times and then come to find out neither is true.

- Organize Imports - This very handy option, which has the keystroke Ctrl-Alt-O, will automatically add imports for any types that are lacking them. It also organizes the import statements at the top of the file. It should be noted that this uses Java-style imports that go at the top of the file and have one class per line. If you need an import to be local, you should add that yourself. The fact that it uses a longer format is not that important as you do not have to type it.

- Extract Local - This is the inverse of Inline Local. It can be used if there is an expression in a method that either appears multiple times, or which would make more sense if a name were associated with it. This is sometimes also called Introduce Explaining Variable.

- Extract Method - This remarkably convenient refactoring should be used to break up

[3]Given the rate at which work is being done on Scala and the Eclipse plug-in, it is inevitable that this list will likely be longer for you.

large methods or to extract code into a method when you find that you need to use it more than you originally believed. Simply highlight the code you want in the method and select this option. It will figure out the parameters that are needed and you just pick a name.

The built-in refactoring tools in IDEs like Eclipse can make certain types of changes remarkably easy. The most important aspect of this is that they reduce the barrier so that you really have no excuse to not use them. If you leave a poorly named variable or method in your code, or if you decide to live with a method that is 200 lines long, you are simply being lazy.

27.2.2 Introduce Null Object

Not all refactoring methods are quite so easy to perform. Some will require a bit of effort on your part, but when they are called for, they are definitely worth it. To see this, we can consider one of the exercises from chapter 25 where you were asked to write a doubly-linked list without using a sentinel. If you actually went through that exercise you found that the code wound up being a nest of conditionals set up to deal with the boundary cases where a `prev` or `next` reference was `null`. This type of special case coding to deal with `null` references is fairly common. So common, in fact, that there is a refactoring method specifically for dealing with it called, "Introduce Null Object." As the name implies, the goal is to introduce an object that is used in place of `null` references. This object should have methods on it that are basically stubs with limited functionality. You want to do this if you find that your code has many lines that look something like this.

```
if(ref==null) ...
```

The sentinel played the role of a Null Object in the doubly-linked list. An even better example is `Nil` for the immutable singly-linked list. You could theoretically make an immutable singly-linked list without making a `Nil` object. Instead, the lists would be terminated with `null` references as they were for our mutable singly-linked list. Such code tends to be far messier to write and to work with, making it harder to debug and to maintain. We will see another usage of a Null Object in chapter 29 when we make our immutable binary search tree.

27.2.3 Add and Remove Parameter

The code we wrote back in chapter 25 used the Null Objects initially, we did not have to refactor for it. Describing such a refactoring would be challenging. To make it more clear how this might work, consider the "Add Parameter" refactoring. The idea here is that you have written a method that takes a certain list of parameters. As an example, consider a method that logs employee pay. This might be part of some larger application used by a company. The current version looks like this.

```
def logHours(employeeID:ID, hoursWorked:Int, wagesPaid:Int) = {
  // Stuff to do logging
}
```

It works and everything is fine until it is determined that the logs need to include additional information to comply with some new government regulation. For example, maybe this is a large, multinational corporation and they have to also keep records of what location the employee was working at for those hours. Clearly this is going to require changes to the logHours method. The idea of refactoring is that you break those changes into two parts. The first part is to perform the Add Parameter refactoring. That gives you this code.

```
def logHours(employeeID:ID, hoursWorked:Int, wagesPaid:Int,
    location:OfficeLocation) = {
  // Same stuff as before
}
```

Note that the code in the method is unchanged. Adding this parameter forces changes at the call sites, but it does not actually change the behavior of the code. You should make this change, get the code to compile again, then run all existing tests and verify that things work. Once you have confirmed that the code is still functional, you can proceed to the second step where you actually alter the functionality of the method.

```
def logHours(employeeID:ID, hoursWorked:Int, wagesPaid:Int,
    location:OfficeLocation) = {
  // Modified stuff to include the location.
}
```

The whole goal of refactoring is to formalize your approach to change so that when code needs to change, you are more willing to do it and more confident that it will work. Changing existing code can be a very scary thing to do. Often students will wind up with large segments of their code commented out, but not deleted, because they are afraid to really commit to changes. Combining the techniques of refactoring and automated unit testing can help to give you the courage to make needed changes. Code should be dynamic and flexible. That only happens if programmers are willing to change things.

The counterpart to Add Parameter is "Remove Parameter". The motivation for this refactoring would be that alterations to a method or the code around it have made it so that parameters, which at one point provided needed information, are no longer being used. Perhaps alterations in the system have made it so that the value of the wages paid should be calculated using the employee information and how many hours were worked instead of being passed in. In that case, you might have code that still has wagesPaid as a parameter, but that parameter is never used. That is smelly as it is not only overhead for the code, it is likely to confuse people looking at it. Such a parameter should be removed to make the code easier to understand and work with.

27.2.4 Cures for Switch Statements

As was mentioned above, Scala does not have a switch statement in the traditional sense. However, the smell that is generally associated with switch statements can be produced in any conditional execution including not only match, but also if. Running with the example from above, consider that we need to calculate a number of different values associated with employees and the nature of the calculation depends on the type of employee under consideration. This could be done with code that looks something like this.

```
val monthlyPay = employeeType match {
  case Wage => ...
  case Salary => ...
  case Temp => ...
}
```

This same pattern might be repeated in different parts of the code for calculating benefits and leave time as well. The values Wage, Salary, and Temp could be vals defined in some object, or objects themselves. They could also be enums, a topic discussed in appendix B. This could also appear in code using `if` expressions.

```
val monthlyPay = if(employeeType == Wage) ...
  else if(employeeType == Salary) ...
  else ...
}
```

The key to this being a smell is that the pattern above is repeated in multiple locations. If a new category or employee were added, all of those would need to be modified. Missing one or more occurrences will lead to difficult to track errors.

There are two refactoring methods that can be employed to remove this smell. They are "Replace Conditional with Polymorphism" and "Replace Type Code with Subclasses". This would appear in the code as an abstract supertype that has methods for each of the different types of actions/calculations that are required. That might look like the following.

```
trait EmployeeType {
  def monthlyPay(hours:Int, wage:Int):Int
  def logBenefits(employee:Employee, hours:Int):Unit
  def leaveTime(hours:Int):Double
}
```

Using this, the `match` or `if` constructs shown above would be replaced by this simple line.

```
val monthlyPay = employeeType.monthlyPay(hours,wage)
```

This works because the `employeeType` will be a reference to an appropriate subtype of `EmployeeType`. That subtype will have the appropriate code in it for that type. The code might look like the following with proper implementations of all the methods. Then the parts of the code that give a value to `employeeType` would provide the appropriate subtype.

```
class WageEmployee extends EmployeeType {
  def monthlyPay(hours:Int, wage:Int):Int = ...
  def logBenefits(employee:Employee, hours:Int):Unit = ...
  def leaveTime(hours:Int):Double = ...
}

class SalaryEmployee extends EmployeeType {
  def monthlyPay(hours:Int, wage:Int):Int = ...
  def logBenefits(employee:Employee, hours:Int):Unit = ...
  def leaveTime(hours:Int):Double = ...
}

class TempEmployee extends EmployeeType {
  def monthlyPay(hours:Int, wage:Int):Int = ...
  def logBenefits(employee:Employee, hours:Int):Unit = ...
  def leaveTime(hours:Int):Double = ...
}
```

Note that if each of these had no state, fields that differ between different instances, they could be declared as objects instead of classes.

There are two main benefits to this change that improve upon the switch statement smell. The first is that this new structure makes it significantly easier to add new types of employees. If it is decided that any of these categories needs to be resolved into different classifications, that can be done easily. It also localizes code from any given type of employee and makes it much easier to find the code when changes need to be made.

27.2.5 Consolidate Conditional Expression

In the calculation of how much you should pay wage employees, it is likely that you will have to include expressions for normal overtime and perhaps even double overtime. These might look something like `hours>=40 && hours<60` and `hours>=60`. The first problem that you notice with these expressions is that they include "magic numbers" in the form of 40 and 60. As a general rule, you do not want to have these magic numbers spread through your code. The other potential problem is that these expressions do not have obvious meaning and it is possible that the requirements might need to be changed in the future. All of these can be fixed, to one degree or another, by pulling the Boolean expression out into a method like this.

```
def overtime(hours:Int):Boolean = hours>=40 && hours<60
```

Then your condition in other parts of the code becomes `overtime(hours)`. This is easy to read, highly informative, and makes it easy to change the nature of the condition. However, if the expression is not used multiple times, this approach is overkill and only adds to the complexity of the code.

27.2.6 Convert Procedural Design to Objects

In part I of this book, the focus was functional decomposition of problems. We did not get into constructing classes or proper object-oriented design. This was for educational purposes. Object-orientation introduces additional concepts that are not essential for solving small problems. This approach where data and functionality are viewed as being independent is called the procedural approach. Bigger programs, however, gain a lot from being built in a proper object-oriented manner. There is also a benefit to consistent design across a project. For this reason, if you have aspects of a large program that contains occasional segments that are more procedural, you should refactor them to use an object-oriented approach. This is done by grouping the functionality with the data that it should operate on and putting the two together into objects. This can be done in an incremental way by extracting methods from long procedures and moving them into the appropriate `class`, `trait`, or `object`.

27.2.7 Encapsulate Collection

Picture a class that has a data member that is a mutable collection. This might be something like an `Array` or a `Buffer`. For example, in our company you have a type for a `Department` that has a collection filled with the `Employee` type. Now imagine that the class has a method like the following.

```
def employees:mutable.Buffer[Employee] = lEmployees
```

Note that this method is public and gives any outside code access to the local employee collection. This is a risky thing to do with a mutable collection. The reason being that any code can call this method and then start making whatever changes it desires to the employees in that department without the `Department` class having any control over it. That is not something one would expect to allow in real life and it is best avoided in code as well. Even if it seems like a reasonable thing to do when you write it, even if the collection is not that critical to the working of the class, giving up that much control is likely to cause problems down the line.

The alternative is to return an immutable copy from this method and then write other methods for adding or removing elements as needed. This could be implemented in the simplest terms with the following code.

```
def employees:Seq[Employee] = lEmployees.toSeq
def addEmployee(e:Employee) = lEmployee += e
def removeEmployee(index:Int) = lEmployee.remove(index,1)
```

With just this code there is not much of a benefit. However, if at some point you decide that you need to check conditions on new employees or to log when employees are removed, that will be easy to do. Using the original implementation, that would have been nearly impossible.

The general rule of thumb here is that mutable data should not be publicly viewable and any changes to mutable data should go through methods that have a chance to limit access or do something in response to it. While the ability to define special member assignment methods in Scala makes it reasonable to have public `var` members, care should still be taken in regards to how mutable members are shared with outside code.

27.2.8 Push Down or Pull Up Field or Method

Image for a moment that the drawing program were originally constructed without writing a `DrawLeaf` class. In that situation, the types such as `DrawRectangle` and `DrawEllipse` would include all the methods currently located in `DrawLeaf`. This would be a significant amount of code duplication. Somewhere between the second and third time that the many methods of `TreeNode` were all implemented in exactly the same way, we should have started to look for a better way to lay things out. That better way would have been to create and insert the `DrawLeaf` type between `Drawable` and all of its subtypes other than `DrawTransform`. After that class was created, the common method and fields should have been pulled up into that class. This action would have utilized the "Pull Up Field" and "Pull Up Method" refactoring methods.

These methods provide a simple way of dealing with any duplication of code that you find between different types inheriting from a common subtype. If all the subtypes have the same implementation, it is clear that the shared method implementation should be moved up to the common supertype. If only some share those methods, you should consider making a new intermediate type and moving the methods into that type.

There are also times when you want to do the opposite. You originally feel that a method will be implemented the same way for an entire hierarchy, only to realize later that most subtypes wind up overriding that method. This starts to have the smell of a refused bequest. In that situation, you should perform "Push Down Field" or "Push Down Method" to minimize the odor. If there are some common implementations, introduced intermediate types to hold that code. If a large number of the subtypes are going to override a method, you should strongly consider having it be abstract in the highest level type that contains the method. The advantage of this is that because you know many implementations are going to be different, you force anyone who inherits from your type to do something with that method or member. This forces them to think about it and reduces the odds that they will accidentally inherit an inappropriate implementation that leads to challenging to find bugs.

27.2.9 Substitute Algorithm

The last refactoring to be considered in this chapter is "Substitute Algorithm." There are many different reasons why your first choice of an algorithm for some part of a problem might not wind up being ideal. A simple example that the author dealt with was a sort algorithm for a small number of data values. By default, the built-in sort from the libraries was selected. It was easy to use, and it was a well-tuned, general purpose merge sort algorithm. One of the language library writers had certainly worked hard to make it efficient. However, when the full application was completed, performance turned out to be quite poor. Profiling indicated that an inordinate amount of time was being spent in the sort. A simple insertion sort written with the specific comparison required for that code was put in place instead. This switch of algorithm did not alter the final results of the program at all. However, it led to a remarkable improvement in speed.

When you first solve a problem you are likely to use whatever is easiest to write correctly. That is exactly what you should do because most of the time that will work best and be the easiest to maintain. There are situations where your initial guesses about the nature of the data will be incorrect and different approaches turn out to be significantly faster. Optimizing for speed is one of the last things you should do in developing a program and even then it should only be done when you know that it is required. A well-known quote from Donald Knuth is that "We should forget about small efficiencies, say about 97% of the time: premature optimization is the root of all evil". [7] When you do come to realize that an early choice of algorithm is slowing things down in an unacceptable way, do not be afraid to refactor the code and put in a different algorithm that is better suited for the specific task at hand.

27.3 End of Chapter Material

27.3.1 Summary of Concepts

- Smells
  - Something in the code that makes it less than ideal.
  - Sometimes related to a specific piece of code, but often deals with structure and includes many distributed elements.
  - They often build up over time as code is modified.
  - Make code harder to work with and maintain.
- Refactoring
  - The act of changing the structure of code or how it works without changing the end behavior.
  - Can be done to remove smells.
  - Also done to help change the structure of code to allow new features to be implemented.
  - Many IDEs include certain options for automatic refactoring.

27.3.2 Exercises

1. Test each of the different built-in refactoring options in Eclipse.

2. The website for the book has some examples of smelly code. Go through each one, figure out why it is smelly, and refactor it.

27.3.3 Projects

Instead of giving individual descriptions for each of the projects, this chapter has more of a general description. Go through the code you have written and find things that smell. When you find them, refactor them so that they do not smell.

Going forward, there will likely be projects that will force you to change some of the design aspects of your project. When those things come up, make sure that you refactor the code first, then add the new features. Strive to not mix the steps. Do the refactoring and verify that the code works as it did before, then add to it.

Chapter 28

Recursion

Recursion first appeared in this book in chapter 6 when it was used to provide basic iteration. This was extended upon in chapter 15 where the memory of the stack was utilized with recursive functions that call themselves more than once. This chapter will revisit the concept of recursion to help solidify it in your mind.[1] The next three chapters will utilize recursion, making this an ideal time to provide a refresher and integrate it into our project.

28.1 Refresher

A recursive function is nothing more than a function that calls itself. This means that it is defined in terms of itself. This seemingly circular type of definition works because all recursive functions need some form of base case where they do not call themselves. The other cases should progress toward the base case. Recursive functions that only call themselves once provide iteration and are typically replaced with loops in most programming languages.[2]

The real power of recursion comes out when a function calls itself more than once. This allows a recursive function to test multiple alternatives. The memory of the stack is essential to this as what really happens is that one call is executed, and after it has finished, the stack keeps track of where the code was and it is done again. Functions like this can be converted into loops, but it typically requires significantly more effort. We saw in chapter

[1] Recursion is a topic many students struggle with, but which is very powerful, so there is value in returning to it several times.

[2] It is worth noting that even if a function only calls itself once, it can still benefit from being recursive if it does some work as it pops back up the stack. We will see examples of this in chapter 29.

15 how certain problems that would be quite challenging to solve using normal loops, could be dealt with very easily by means of recursive code.

28.2 Project Integration: A Maze

One of the problems that we considered in chapter 15 was that of completing a maze. In particular, we wanted to find the length of the shortest path through the maze. The recursive approach that was taken also lent itself well to similar problems like finding the longest path through the maze or finding how many paths there were through the maze. Mazes are something that humans like to represent graphically. As such, they lend themselves to being put into our drawing program.

To do this we will create a drawable for the maze. The final code put together is shown here. We will go through different parts of it in the subsections that follow.

```
package scalabook.drawing

import java.awt.{Graphics2D,Color}
import java.awt.geom._
import scala.swing._
import scala.swing.event._
import actors.Actor

class DrawMaze(p:DrawTransform,m:Seq[Seq[Int]],
    private var ex:Int,
    private var ey:Int) extends DrawLeaf with Clickable {
  val parent = p
  private var maze = m.map(_.toArray).toArray
  @transient private var propPanel:Component = null
  @transient private var clickAction:(Int,Int)=>Unit = null

  def draw(g:Graphics2D) {
    lastTransform = g.getTransform()
    g.setPaint(Color.black)
    g.drawRect(-1,-1,20*maze.length+1,20*maze(0).length+1)
    for(i <- 0 until maze.length; j <- 0 until maze(i).length) {
      g.setPaint(maze(i)(j) match {
        case -1 => Color.black
        case -2 => Color.yellow
        case 0 => Color.white
        case s => new Color(0,(255-s) max 0,0)
      })
      g.fillRect(i*20,j*20,20,20)
    }
    g.setPaint(Color.green)
    g.fillOval(ex*20,ey*20,20,20)
  }

  def propertiesPanel() : Component = {
    if(propPanel == null) {
      import BorderPanel.Position._
      propPanel = new BorderPanel {
```

```
layout += new BorderPanel {
  layout += new GridPanel(2,1) {
    contents += new Label("Width")
    contents += new Label("Height")
  } -> West
  layout += new GridPanel(2,1) {
    contents += new TextField(maze.length.toString) {
      listenTo(this)
      reactions += { case e:EditDone =>
        val size = text.toInt
        if(size>5) resize(size,maze(0).length)
      }
    }
    contents += new TextField(maze(0).length.toString) {
      listenTo(this)
      reactions += { case e:EditDone =>
        val size = text.toInt
        if(size>5) resize(maze.length,size)
      }
    }
  } -> Center
  val radioButtons = Seq(
      new RadioButton { action = Action("Draw Walls") {
          clickAction = (x,y) => {
            maze(x)(y) = if(maze(x)(y) == -1) 0 else -1
            drawing.refresh
          }
      } },
      new RadioButton { action = Action("Place Exit") {
          clickAction = (x,y) => {
            ex = x
            ey = y
            drawing.refresh
          }
      } },
      new RadioButton { action = Action("Shortest Path") {
          clickAction = (x,y) => Actor.actor {
            Dialog.showMessage(propPanel,"The shortest path has length " +
                 shortestPath(x,y) + ".")
            drawing.refresh
          }
      } },
      new RadioButton { action = Action("Shortest Path Fast") {
          clickAction = (x,y) => Actor.actor {
            Dialog.showMessage(propPanel,"The shortest path has length " +
                 shortestPathFast(x,y,0) + ".")
            drawing.refresh
          }
      } }
      )
  val group = new ButtonGroup(radioButtons:_*)
  layout += new GridPanel(radioButtons.length+1,1) {
    radioButtons.foreach(rb => contents += rb)
    contents += Button("Clear Breadcrumbs") {
      for(i <- maze.indices; j <- maze(i).indices)
```

```
          if(maze(i)(j)>0) maze(i)(j) = 0
          drawing.refresh
        }
      } -> South
    } -> North
  }
}
propPanel
}

override def toString() = "Maze"

def toXML : xml.Node = {
  <drawMaze ex={ex.toString} ey={ey.toString}>
    {maze.map(_.mkString(",")).mkString("\n")}
  </drawMaze>
}

def react(evt:Event,cx:Double,cy:Double) {
  evt match {
    case mc:MouseClicked =>
      val x = cx.toInt/20
      val y = cy.toInt/20
      if(x>=0 && x<maze.length && y>=0 && y<maze(x).length && clickAction!=null) {
        clickAction(x,y)
      }
    case _ =>
  }
}

private def resize(w:Int,h:Int) {
  maze = Array.tabulate(w,h)((i,j) => {
    if(i<maze.length && j<maze(i).length) maze(i)(j) else 0
  })
  drawing.refresh
}

private def shortestPath(x:Int,y:Int):Int = {
  if(x==ex && y==ey) 0
  else if(x<0 || x>=maze.length || y<0 || y>=maze(x).length || maze(x)(y)<0) {
    1000000000
  } else {
    maze(x)(y) = -2
    drawing.refresh
    Thread.sleep(10)
    val ret = 1+(shortestPath(x+1,y) min
                 shortestPath(x-1,y) min
                 shortestPath(x,y+1) min
                 shortestPath(x,y-1))
    maze(x)(y) = 0
    ret
  }
}

private def shortestPathFast(x:Int,y:Int,steps:Int):Int = {
```

```
    if(x==ex && y==ey) 0
    else if(x<0 || x>=maze.length || y<0 || y>=maze(x).length || maze(x)(y)<0) {
      1000000000
    } else if(maze(x)(y)>0 && maze(x)(y)<=steps) {
      1000000000
    } else {
      maze(x)(y) = steps
      drawing.refresh
      Thread.sleep(10)
      val ret = 1+(shortestPathFast(x+1,y,steps+1) min
                   shortestPathFast(x-1,y,steps+1) min
                   shortestPathFast(x,y+1,steps+1) min
                   shortestPathFast(x,y-1,steps+1))
      ret
    }
  }
}

object DrawMaze {
  def apply(p:DrawTransform,data:xml.Node) = {
    new DrawMaze(p,
        data.text.trim.split("\n").map(_.split(",").map(_.toInt).toSeq).toSeq,
        (data\"@ex").text.toInt,
        (data\"@ey").text.toInt)
  }

  def apply(p:DrawTransform) = {
    new DrawMaze(p,
        Seq(Seq( 0,-1, 0, 0, 0, 0, 0, 0, 0, 0),
            Seq( 0,-1, 0,-1,-1,-1,-1,-1,-1, 0),
            Seq( 0,-1, 0,-1, 0, 0, 0, 0, 0, 0),
            Seq( 0,-1, 0,-1, 0,-1,-1,-1,-1,-1),
            Seq( 0, 0, 0,-1, 0, 0, 0, 0, 0, 0),
            Seq( 0,-1,-1,-1,-1,-1,-1,-1,-1, 0),
            Seq( 0,-1, 0, 0, 0, 0, 0, 0,-1, 0),
            Seq( 0,-1, 0,-1, 0, 0,-1, 0, 0, 0),
            Seq( 0,-1, 0,-1,-1, 0,-1,-1,-1,-1),
            Seq( 0, 0, 0,-1, 0, 0, 0, 0, 0, 0)),
        9,9)
  }
}
```

Clicking on the maze in the drawing activates different functionality depending on what the user selects. This includes finding the shortest path through the maze.

28.2.1 Revisiting the Basic Approach

The `shortestPath` method uses the approach that we covered previously. This tests all paths through the maze and returns the length of the shortest one. The code for this method is shown here.

```
  private def shortestPath(x:Int,y:Int):Int = {
    if(x==ex && y==ey) 0
    else if(x<0 || x>=maze.length || y<0 || y>=maze(x).length || maze(x)(y)<0) {
      1000000000
```

```
    } else {
      maze(x)(y) = -2
      drawing.refresh
      Thread.sleep(10)
      val ret = 1+(shortestPath(x+1,y) min
                   shortestPath(x-1,y) min
                   shortestPath(x,y+1) min
                   shortestPath(x,y-1))
      maze(x)(y) = 0
      ret
    }
  }
```

The base cases are getting to the exit, which returns 0 as the number of steps that need to be taken to get to the exit, and going out of bounds, which returns 1000000000 as a number that can not be the solution and will not ever be the minimum. That is significant because the recursive case makes recursive calls in the four different directions and takes the minimum of them.

The line `maze(x)(y) = -2` is dropping a "breadcrumb" on the maze so that the computer does not run in loops. You can think of this as moving the recursion toward a base case. Negative values in squares are bases cases and this line makes another square negative before calling the recursion. That can only happen a finite number of times before the recursion runs out of empty squares. This action is reversed when the four recursive calls are done so that a square can be revisited on a future path that gets to it in a different way. The lines of code to refresh the drawing and sleep the thread are there simply to animate the search process. You can reduce the sleep length or remove that line completely to speed things up for mazes that take too long to solve.

28.2.2 Graphical Editing

This is the first drawable that truly needs to have the user interact with it using the mouse. Solving the maze requires a starting location that would be nice to be able to adjust. A proper maze should also be editable, allowing us to change the locations of walls as well as where the exit is. It would be possible to put this type of editing into the properties panel, but that is a bit silly considering that the maze is already being drawn in the drawing. There is no reason we should not be able to make it so that the user can interact with the maze through mouse activity on the maze in the drawing.

Of course, there are some challenges. These arise from three different sources. The most obvious challenge is that the panel the user will be clicking on is in the `Drawing` class, not in the `DrawMaze` or other `Drawable` types that we might want to have performing the reaction. Another challenge is that the elements in the drawing can be transformed so that the location of the click on the panel is very indirectly linked to the location of the click relative to the drawn object. The most obscure challenge, and perhaps the one that is hardest to figure out how to solve, is that we do not want to implement changes that break the serialization. All three of these problems can be partially addressed by adding the following trait and making some changes to the `Drawing` class.

```
package scalabook.drawing

import scala.swing._
import scala.swing.event._
import java.awt.geom._
```

```
trait Clickable {
  @transient private var lReactor:Reactor = null
  @transient protected var lastTransform:AffineTransform = null

  def reactor = {
    if(lReactor==null) {
      lReactor = new Reactor {}
      lReactor.reactions += {
        case me:MouseEvent =>
          val dest = new Point2D.Double
          lastTransform.inverseTransform(me.point,dest)
          react(me,dest.x,dest.y)
      }
      lastTransform = new AffineTransform
    }
    lReactor
  }

  def react(evt:Event,cx:Double,cy:Double)
}
```

The idea is that this trait should be inherited by any drawable that we want the user to be able to interact with using the mouse. The call to `inverseTransform` takes a click location from the screen coordinates to the coordinates of that drawable. In order to make this work, the line `lastTransform = g.getTransform()` has to be included in the `draw` method of that drawable.

The trait itself includes code for making a `Reactor` that deals well with serialization and keeps track of the transform the drawable was last drawn with. It also provides a special type that we can look for in the `Drawing` to determine if mouse information should be sent somewhere. To make this work, the following code has to be added to `Drawing`.

```
  // This was added in the declarations at the top.
  @transient private var currentReactor:Reactor = null

  // It is used in this code that was added at the end of the executeOnSelection
  // code in the TreeSelectionListener.
                if(currentReactor!=null) {
                  currentReactor.deafTo(drawPanel.mouse.moves,
                      drawPanel.mouse.clicks,
                      drawPanel.mouse.wheel)
                }
                d match {
                  case r:Clickable =>
                    currentReactor = r.reactor
                    currentReactor.listenTo(drawPanel.mouse.moves,
                      drawPanel.mouse.clicks,
                      drawPanel.mouse.wheel)
                  case _ =>
                }
```

The type information is significant for the `match` expression that checks if the selected `Drawable` is a subtype of `Clickable`. If it is, it remembers the reactor and sets it to listen for mouse events on the panel. When something else is selected, this reactor will be made deaf to those events.

Instead of creating a `Clickable` type, it is tempting to simply have the drawable inherit from Reactor directly. Indeed, this is what the author did originally. However, that breaks serialization as the Reactor type includes a link to non-serializable data. The code was reworked with the newly created Clickable trait to fix serialization, and it had the added benefit of allowing the transformation code to be consolidated. The full Event is passed through to the react method to allow matching as well as to provide screen coordinates in case those are desired.

Inside of the DrawMaze class, the react method uses integer division to find the block that was clicked, then uses a function variable called clickAction to determine what happens on a click. The value of clickAction is set when the user selects a radio button from the properties panel. This approach abstracts the behavior away from the react method and puts it directly into the radio button code so that the names shown to the user and the code for the behaviors are adjacent. Note that the path finding calls use Actor.actor to run in a separate thread because those actions can take a significant amount of time, and we want the event handling thread free to repaint for the animations.

28.2.3 Optimizing the Maze

If you played around with the maze some back in chapter 15 or with this code here, you might have noticed something. If you keep taking out wall after wall and run the shortest path function, as you remove walls, the function will start to run more and more slowly. At a certain point, you will make the maze empty enough that the function will take longer than you have the patience to let it run. This happens because the basic algorithm tests all possible paths to find the shortest. The number of possible paths is extremely large for an empty maze. The DrawMaze allows you to change the size as well. A 10 by 10 maze is sufficient to take many times longer than a human life if left empty, going to 20 by 20 or larger makes things far worse.

There are ways around this. The way we will discuss here makes the breadcrumbs smarter so that they can provide additional information so that paths which have no chance of being ideal are eliminated from consideration before they have been fully explored. The code for doing this is shown here.

```
private def shortestPathFast(x:Int,y:Int,steps:Int):Int = {
  if(x==ex && y==ey) 0
  else if(x<0 || x>=maze.length || y<0 || y>=maze(x).length || maze(x)(y)<0) {
    1000000000
  } else if(maze(x)(y)>0 && maze(x)(y)<=steps) {
    1000000000
  } else {
    maze(x)(y) = steps
    drawing.refresh
    Thread.sleep(10)
    val ret = 1+(shortestPathFast(x+1,y,steps+1) min
                 shortestPathFast(x-1,y,steps+1) min
                 shortestPathFast(x,y+1,steps+1) min
                 shortestPathFast(x,y-1,steps+1))
    ret
  }
}
```

This looks very similar to the earlier version, but there is an extra argument to the function, one extra base case, and the breadcrumbs are not picked up. The extra argument keeps track of how many steps have been taken along the current path. The breadcrumbs, instead of just marking that you have been to a square, keep track of how many steps it took to get there on the shortest path so far. The extra base case is for a room that is reached taking more steps than the current step count. In that situation, the path can not possibly be the shortest and the recursion is terminated.

If you make the maze big enough, even this approach will be insufficient. If you test an empty 30 by 30 maze you will discover this for yourself. An alternate approach is to abandon recursion completely and do a *breadth-first* search of the maze. By its nature, recursion is *depth-first*. This means that it goes as far down one path as possible before trying another path. For a shortest-path algorithm, the first solution reached is the winner. For that reason, it is more efficient to try all paths at the same time, taking one step forward on each. That approach works very well of a maze, but there are some problems for which the number of independent paths is so large that you can not keep track of all the different paths without running out of memory. The breadth-first approach also does not help with problems such as finding the longest path or counting the number of paths.

How Many Paths Are There?

You can empirically check to see that an empty maze takes a long time to solve. With a little work we can estimate more accurately how long it would take. Imagine a 10 by 10 empty maze. In the recursive case, the function calls itself four times. Some of those will lead to base cases and immediately return. For our estimate we will say that two of the calls are not base cases. So the initial call splits to two, which split to four, and so on. After n steps in the recursion, there are roughly 2^n paths being worked on. The longest path in a 10 by 10 maze is 100 steps. To keep things simple we will stick with that number for how deep the recursion goes. It is an overestimate, but it is somewhat offset by assuming only two branches. This implies on the order of 2^{100} paths.

Hopefully, it is clear that 2^{100} is a very big number. Such big numbers can be hard for people to really comprehend. After all, computers are really fast. Modern computers run at clock frequencies of several gigahertz. To help us put this number into perspective, assume that your computer can do 10^9 recursive calls each second.[3] We then use two other rough estimates: $2^{10} \sim 10^3$ and $1yr \sim 3 \times 10^7 sec$. This gives us the following:

$$2^{100} ops \sim 10^{30} ops \sim 10^{21} sec \sim 3 \times 10^{13} yr.$$

That is a few times longer than the best estimates for the current age of the Universe.

This type of exponential growth is common for recursive algorithms. If each non-base case calls the function m times and this goes n calls deep, then there will be m^n total calls. For some problems, this seems to be unavoidable. A lot of effort goes into trying to find ways of getting around this. You should learn about approaches to dealing with this later in your studies.

[3]It is very unlikely your computer can do this as going through our function requires hundreds of clock cycles, but it is a reasonable assumption for the level of detail being considered here.

28.3 Graph Traversals

A maze is a special case of a very general data structure called a graph. If you draw a bunch of dots and put lines between them, you have made a graph. If you are following the end-of-chapter projects, you might well have a form of a graph in the code that you are writing. For example, the multiuser text game has rooms that are connected by exits. That is an example of a graph. For a graphical game you might have different rooms/levels that have doors or other connections between them. This is also a graph.

Using the foundation of code that we created for the maze, it is not to hard to make a DrawGraph type which the user can interact with by clicking on the drawing. The full code is shown here. It is fairly long in large part because of the ability to interactively edit the graph. There is a text field that can be used to enter an integer weight for edges that are added to the graph.

```
package scalabook.drawing

import java.awt.{Graphics2D,Color}
import java.awt.geom._
import scala.swing._
import scala.swing.event._
import actors.Actor
import collection.mutable

class DrawGraph(p:DrawTransform,nloc:Seq[Point2D],edges:Seq[(Int,Int,Int)],en:Int)
    extends DrawLeaf with Clickable {
  val parent = p
  import DrawGraph.{Node,Edge}
  private val nodes = mutable.Buffer[Node]()
  for(p <- nloc) nodes += new Node(p.getX(),p.getY())
  for((f,t,w) <- edges) nodes(f).edges ::= new Edge(nodes(f),nodes(t),w)
  private var endNode = if(en<0) {
    if(nodes.nonEmpty) nodes(0) else null
  } else nodes(en)
  @transient private var propPanel:Component = null
  @transient private var weight:TextField = null
  @transient private var clickAction:(MouseEvent,Double,Double)=>Unit = null
  @transient private var hoverNode:Node = null
  @transient private var hoverEdge:Edge = null
  @transient private var startNode:Node = null
  @transient private var (dx,dy) = (0.0, 0.0)
  @transient private var pathSet:Set[Node] = null


  def draw(g:Graphics2D) {
    def drawNode(n:Node) {
      g.setColor(if(n == endNode) Color.green else Color.black)
      g.fill(new Ellipse2D.Double(n.x-5,n.y-5,10,10))
      if(n==hoverNode) {
        g.setColor(Color.red)
        g.fill(new Ellipse2D.Double(n.x-4,n.y-4,8,8))
      }
    }
```

```
    def drawEdge(e:Edge) {
      g.setColor(if(e==hoverEdge) Color.red else Color.black)
      g.draw(new Line2D.Double(e.from.x,e.from.y,e.to.x,e.to.y))
      g.drawString(e.weight.toString,0.5f*(e.from.x+e.to.x).toFloat,
        0.5f*(e.from.y+e.to.y).toFloat)
    }

    for(n <- nodes) {
      drawNode(n)
      for(e <- n.edges; if(nodes.indexOf(e.from)<nodes.indexOf(e.to))) {
        drawEdge(e)
      }
      g.setPaint(Color.black)
      if(startNode!=null) g.draw(new Line2D.Double(startNode.x,startNode.y,dx,dy))
    }
    if(pathSet!=null) {
      g.setPaint(Color.blue)
      for(n <- pathSet) {
        g.fill(new Ellipse2D.Double(n.x-2,n.y-2,4,4))
      }
    }
  }

  def propertiesPanel() : Component = {
    if(propPanel == null) {
      import BorderPanel.Position._
      propPanel = new BorderPanel {
        val radioButtons = Seq(
          new RadioButton { action = Action("Add Node") {
              clickAction = (e,x,y) => e match {
                case mc:MouseClicked =>
                  nodes += new Node(x,y)
                  if(nodes.length==1) endNode = nodes(0)
                case _ =>
              }
          } },
          new RadioButton { action = Action("Add Edge") {
              clickAction = (e,x,y) => e match {
                case mp:MousePressed =>
                  if(hoverNode!=null) {
                    startNode = hoverNode
                    dx = startNode.x
                    dy = startNode.y
                  }
                case mp:MouseDragged =>
                  if(startNode!=null) {
                    dx = x
                    dy = y
                  }
                case mp:MouseReleased =>
                  if(startNode!=null && hoverNode!=null) {
                    startNode.edges ::= new
                        Edge(startNode,hoverNode,weight.text.toInt)
```

```
          hoverNode.edges ::= new
              Edge(hoverNode,startNode,weight.text.toInt)
        }
        startNode = null
      case _ =>
    }
  } },
  new RadioButton { action = Action("Move Node") {
      clickAction = (e,x,y) => e match {
        case mp:MousePressed =>
          if(hoverNode!=null) {
            startNode = hoverNode
            dx = startNode.x
            dy = startNode.y
          }
        case mp:MouseDragged =>
          if(startNode!=null) {
            startNode.x = x
            startNode.y = y
            dx = x
            dy = y
          }
        case mp:MouseReleased =>
          if(startNode!=null && hoverNode!=null) {
            startNode.x = x
            startNode.y = y
          }
          startNode = null
        case _ =>
      }
  } },
  new RadioButton { action = Action("Remove") {
      clickAction = (e,x,y) => e match {
        case mc:MouseClicked =>
          if(hoverNode!=null) {
            nodes -= hoverNode
            for(n <- nodes) n.edges = n.edges.filter(_.to != hoverNode)
            if(hoverNode==endNode && nodes.nonEmpty) endNode = nodes(0)
            pathSet = null
          } else if(hoverEdge!=null) {
            hoverEdge.from.edges = hoverEdge.from.edges.filterNot(_ sameAs
                hoverEdge)
            hoverEdge.to.edges = hoverEdge.to.edges.filterNot(_ sameAs
                hoverEdge)
          }
        case _ =>
      }
  } },
  new RadioButton { action = Action("Set End") {
      clickAction = (e,x,y) => e match {
        case mc:MouseClicked =>
          if(hoverNode!=null) {
            endNode = hoverNode
          }
        case _ =>
```

```
            }
        } },
        new RadioButton { action = Action("Reachable") {
            clickAction = (e,x,y) => e match {
              case mc:MouseClicked =>
                if(hoverNode!=null) Actor.actor {
                  if(endNode!=null) {
                    Dialog.showMessage(propPanel,"The end node is"+
                        (if(canReach(hoverNode,mutable.Set())) "" else " not")+"
                            reachable.")
                  } else Dialog.showMessage(propPanel,"There must be an end node.")
                }
              case _ =>
            }
        } },
        new RadioButton { action = Action("Shortest Path") {
            clickAction = (e,x,y) => e match {
              case mc:MouseClicked =>
                if(hoverNode!=null) Actor.actor {
                  if(endNode!=null) {
                    shortestPath(hoverNode,Set()) match {
                      case None =>
                        Dialog.showMessage(propPanel,"There is no path.")
                      case Some((len,ps)) =>
                        Dialog.showMessage(propPanel,"There is a path of length
                            "+len+".")
                        pathSet = ps
                        drawing.refresh
                    }
                  } else Dialog.showMessage(propPanel,"There must be an end node.")
                }
              case _ =>
            }
        } }
      )
      val group = new ButtonGroup(radioButtons:_*)
      layout += new GridPanel(radioButtons.length+1,1) {
        radioButtons.foreach(rb => contents += rb)
        weight = new TextField("1")
        contents += weight
      } -> North
    }
  }
  propPanel
}

override def toString() = "Graph"

def toXML : xml.Node = {
  <drawGraph en={nodes.indexOf(endNode).toString}>
    {nodes.map(n => <node x={n.x.toString} y={n.y.toString}/>)}
    {for(n <- nodes; e <- n.edges) yield
      <edge from={nodes.indexOf(e.from).toString}
          to={nodes.indexOf(e.to).toString} weight={e.weight.toString}/>
    }
```

```
    </drawGraph>
  }

  def react(evt:Event,cx:Double,cy:Double) {
    evt match {
      case me:MouseEvent =>
        hoverNode = null
        hoverEdge = null
        var lastDist = 1e100
        for(n <- nodes) {
          val dx = cx-n.x
          val dy = cy-n.y
          val dist = math.sqrt(dx*dx+dy*dy)
          if(dist<10 && dist<lastDist) {
            hoverNode = n
            lastDist = dist
          }
          if(lastDist>3) for(e <- n.edges;
              if(nodes.indexOf(e.from)<nodes.indexOf(e.to))) {
            val line = new Line2D.Double(e.from.x,e.from.y,e.to.x,e.to.y)
            val edist = line.ptSegDist(me.point).abs
            if(edist<3 && edist<lastDist) {
              hoverEdge = e
            }
          }
        }
        if(hoverNode!=null) hoverEdge = null
        if(clickAction!=null) clickAction(me,cx,cy)
        drawing.refresh
      case _ =>
    }
  }

  private def canReach(n:Node,visited:mutable.Set[Node]):Boolean = {
    if(n==endNode) true
    else if(visited(n)) false
    else {
      visited += n
      n.edges.exists(e => canReach(e.to,visited))
    }
  }

  private def shortestPath(n:Node,visited:Set[Node]):Option[(Int,Set[Node])] = {
    if(n==endNode) Some(0 -> visited)
    else if(visited(n)) None
    else {
      val newVisited = visited+n
      n.edges.foldLeft(None:Option[(Int,Set[Node])])((last,e) => {
        (last,shortestPath(e.to,newVisited)) match {
          case (None,Some((len,v))) => Some((len+e.weight,v))
          case (_,None) => last
          case (Some((len1,_)), Some((len2,v))) => if(len1<=len2+e.weight) last
              else Some(len2+e.weight,v)
        }
      })
```

```
    }
  }
}

object DrawGraph {
  def apply(p:DrawTransform) = new DrawGraph(p,List(new
      Point2D.Double(100,100)),List(),-1)

  def apply(p:DrawTransform,n:xml.Node) = {
    val end = (n \ "@en").text.toInt
    val pnts = (n \ "node").map(pn => {
      val x = (pn \ "@x").text.toDouble
      val y = (pn \ "@y").text.toDouble
      new Point2D.Double(x,y)
    })
    val edges = (n \ "edge").map(en => {
      val from = (en \ "@from").text.toInt
      val to = (en \ "@to").text.toInt
      val weight = (en \ "@weight").text.toInt
      (from,to,weight)
    })
    new DrawGraph(p,pnts,edges,end)
  }

  private class Node(var x:Double,var y:Double) extends Serializable {
    var edges = List[Edge]()
  }

  private class Edge(val from:Node,val to:Node,val weight:Int) extends Serializable
      {
    def sameAs(e:Edge):Boolean = {
      weight==e.weight && ((from==e.from && to==e.to) || (from==e.to && to==e.from))
    }
  }
}
```

In general, the edges in graphs can be directed so that they only go from one node to another, and not back. The way we store the nodes in the example would allow that flexibility, but the code maintains two-way linkages so that the user interface does not have to be able to distinguish two connections between the same two nodes.

The similarity to the `DrawMaze` type should enable you to read most of this code fairly easily. You should put this into your project, add appropriate code to `Drawing` and `DrawTransform` to handle adding this type and loading it from XML, then play with it for a while. The main aspect that we want to focus on here are the two methods that recursively run through the graph as some of the project options include adding similar functionality to your code.

The first method simply checks if the end node can be reached from the one that you click on. This method uses a mutable set to keep track of the nodes that have been visited. This set serves the purpose of "breadcrumbs" in this algorithm. The use of a `Set` type is for the purposes of efficiency. Having the set be mutable gives us a behavior like an algorithm that does not pick back up the breadcrumbs. This is fine here because we only want to know if a node can be reached. We are completely unconcerned with the path that is taken.

```
private def canReach(n:Node,visited:mutable.Set[Node]):Boolean = {
  if(n==endNode) true
  else if(visited(n)) false
  else {
    visited += n
    n.edges.exists(e => canReach(e.to,visited))
  }
}
```

The base cases are reaching the end node and hitting a node that has already been visited. Each time a node is visited, that node is added into the set. The recursion occurs in a call to `exists` on the edges coming out of the current node. The `exists` method is ideal for us because it will stop execution if any of the elements that are run through produce a value of true. So if the first edge from a node manages to reach the end, none of the other edges need to be taken.

The second method is another shortest path algorithm, but this time on a weighted graph instead of a maze. Here again a set is passed in to keep track of the visited elements. In this case, it is immutable because the path is significant and we need to pick up the breadcrumbs. This method also has a significantly more complex return type. It is `Option[(Int,Set[Node])]`. The `Option` part indicates that it is possible that there is no solution. This would be the case if the end were not reachable from the selected start node. This approach is generally preferred over using `null` to indicate that nothing worked. The `Int` is the length of the path, and the `Set[Node]` is all the nodes that were visited on the path that was found. This technically is not the same as returning a path, but it provides enough information that the user can easily figure out the path taken in most situations. It was also easy to build as we are keeping track of the visited nodes using immutable sets anyway.

```
private def shortestPath(n:Node,visited:Set[Node]):Option[(Int,Set[Node])] = {
  if(n==endNode) Some(0 -> visited)
  else if(visited(n)) None
  else {
    val newVisited = visited+n
    n.edges.foldLeft(None:Option[(Int,Set[Node])])((last,e) => {
      (last,shortestPath(e.to,newVisited)) match {
        case (None,Some((len,v))) => Some((len+e.weight,v))
        case (_,None) => last
        case (Some((len1,_)), Some((len2,v))) => if(len1<=len2+e.weight) last
            else Some(len2+e.weight,v)
      }
    })
  }
}
```

The base cases here are the same as for the reachable method. They simply return slightly different values. The recursive case is also a bit longer to deal with the different options. It uses a fold to compile an answer, running through the different edges out of the current node. The fold starts with a value of `None` specifying the appropriate type. The function in the fold does a match on the incoming value and the recursive call of the current edge. When there is a `None`, the other value is used. If both of the values are a `Some`, a pattern is used to pull out the length so the shorter one can be returned. The patterns for the match here display the use of _ to produce names for the minimum number of values.

This shortest path solution for the graph is just like the original solution to the maze and has the same pitfalls in that it can take exponentially long for large graphs. Here again, there are alternate approaches that can produce superior performance. Graphs can be used to represent many different systems and algorithms on them are a significant concept that you should see much more of during your computing career.

28.4 Divide and Conquer

One standard problem-solving technique that uses recursion is the divide and conquer approach. The idea is to take a big problem, and cut it into two or more smaller pieces, solve those pieces, then combine the results into a complete solution. This is similar to the idea behind problem decomposition that we first talked about in chapter 5. The standard approach to solving hard problems is that they should be broken down into easier problems. It is an essential aspect of programming, as programs can be arbitrarily complex. A similar approach can be used in programs to solve problems.

Divide and conquer is recursive, because the function that solves the whole problem is used to solve the pieces as well. The base case is whenever you get down to a size that you can solve without breaking it down. Often this has you go all the way down to the point of being trivial. The recursion moves you toward the base case as each recursive call is on a smaller problem than what you started with. We will explore several different examples of divide and conquer algorithms to see how this approach works.

The first examples of divide and conquer that we will consider are sorting algorithms that we looked at briefly in chapter 15, `mergesort` and `quicksort`. Both of these sorts work by taking a sequence of numbers and breaking it into two pieces that are handled recursively. The default base case is a single element as an array with one element is always properly sorted. However, it can be more efficient to stop the recursion at some larger number of elements and switch to one of the sorts that we have explored previously to finish off the smaller segments.

28.4.1 Merge Sort

Merge sort works by recursively dividing the array in two equal parts[4] and recursively calls on those two parts. The real work in a merge sort happens as the code pops back up the call stack and the sorted subsections are merged together. One of the key advantages of a merge sort is that it is remarkably stable in how much work it does. It always divides down the middle so the recursion always goes $\log_2(n)$ levels deep. The merge operation for two lists of $n/2$ elements requires $n - 1$ comparisons and n memory moves. Given this, a merge sort will always scale as $O(n \log_2(n))$, regardless of the form of the input.

The primary disadvantage of a merge sort is that it can not be done in-place. This is because you can not merge an array with two halves that are independently sorted into a single sorted array without using extra space. This make a merge sort rather well suited for lists, which merge easily, but are immutable so the idea of in-place operations does not make sense. Writing a merge sort that works with arrays and does not consume more than $O(n)$ memory is a bit trickier.

We begin with the simplest form of merge sort. This is a sort of a `List` that uses a recursive merge. The code for that functions is shown here.

---

[4] Or within one of being equal in the case of odd numbers of elements.

```
def mergeSort[A](lst:List[A])(comp:(A,A)=>Int):List[A] = {
  def merge(l1:List[A],l2:List[A]):List[A] = (l1,l2) match {
    case (_,Nil) => l1
    case (Nil,_) => l2
    case (_,_) =>
      if(comp(l1.head,l2.head)<=0) l1.head :: merge(l1.tail,l2)
      else l2.head :: merge(l1,l2.tail)
  }

  val len = lst.length
  if(len<2) lst
  else {
    val (front,back) = lst.splitAt(len/2)
    merge(mergeSort(front)(comp),mergeSort(back)(comp))
  }
}
```

Note that the `merge` function is nested inside of `mergeSort`. This works well as it prevents it from being accessed by outside code and gives it implicit access to the comparison function. The body of the primary method appears at the bottom which has a base case for lists with a length of less than two. Otherwise it splits the list at the midpoint, recursively calls sort on the halves, and merges the result. The length of the list is stored in a variable because it is an $O(n)$ call that we do not want to repeat more than is required.

The merge here is recursive with three cases. If either list is empty, the result is the other list. When neither is empty, a comparison of the two heads is performed and the lesser is consed on the front of the merge of what remains. This is fairly simple code for a sort that promises $O(n\log(n))$ performance, but unfortunately, it will not work well for long lists. The recursive merge function is not tail recursive and, as a result, it can overflow the stack if the lists being merged are too long. This limitation can be removed by using a `merge` function that employs a `while` loop.

```
def merge(l1:List[A],l2:List[A]):List[A] = {
  var (lst1,lst2,ret) = (l1,l2,List[A]())
  while(lst1.nonEmpty || lst2.nonEmpty) {
    if(lst2.isEmpty || (lst1.nonEmpty && comp(lst1.head,lst2.head)<=0)) {
      ret ::= lst1.head
      lst1 = lst1.tail
    } else {
      ret ::= lst2.head
      lst2 = lst2.tail
    }
  }
  ret.reverse
}
```

This function builds the merged list in reverse order, then calls the `reverse` method to turn it around. This might seem odd, but remember that adding to the head of a `List` is $O(1)$ while adding to the tail is $O(n)$. For this reason, it is significantly more efficient to take the approach shown here than to append and not reverse.

This gives us a reasonably good merge sort for lists. The use of an immutable type for sorting forces us to create a separate storage for the result. When dealing with an array, however, the mutability means that many sorts can use no more memory than the provided array and one temporary space holder. As was mentioned above, that is not an option for the merge operation. However, using the style we used for the list when dealing with an array would be extremely inefficient, requiring $O(n\log(n))$ memory. It turns out that we can fairly easily create an array-based version that uses $2n \in O(n)$ memory by making a second array and doing merges from the original array to the second array and back. The following code demonstrates that.

```
def mergeSort[A : Manifest](a:Array[A])(comp:(A,A)=>Int) {
  val data = Array(a,new Array[A](a.length))

  def mergeSortRecur[A](start:Int,end:Int,dest:Int) {
    val src = 1-dest
    if(start==end-1) {
      if(dest==1) data(dest)(start) = data(src)(start)
    } else {
      val mid = (start+end)/2 // Can fail for arrays over 2^30 in length
      mergeSortRecur(start,mid,src)
      mergeSortRecur(mid,end,src)
      var (p1,p2,pdest) = (start,mid,start)
      while(pdest<end) {
        if((p2>=end || comp(data(src)(p1),data(src)(p2))<=0) && p1<mid) {
          data(dest)(pdest) = data(src)(p1)
          p1 += 1
        } else {
          data(dest)(pdest) = data(src)(p2)
          p2 += 1
        }
        pdest += 1
      }
    }
  }

  mergeSortRecur(0,a.length,0)
}
```

Here the first line creates an `Array[Array[A]]` that holds the original array and our second array which is allocated to be the same size. There is a nested recursive function declared that takes the start and end of the range we are sorting as well as the index of the destination in `data`. That destination alternates between 0 and 1 as we go down the call stack. The merge is done from `src` into `dest`. The initial call uses a destination of 0 so the final sorted version winds up being in the original array.

The comment on the calculation of `mid` is worth noting. For arrays with a bit more than a billion elements, the quantity `start+end` can overflow and `Int`. This will lead to a negative value for `mid`. You can get around this by using `mid = start+(end-start)/2` instead. This expression will work for all legal array sizes.

28.4.2 Quicksort

A second sorting example of divide and conquer is the quicksort. Unlike merge sort, the quicksort can be done in-place and, when written that way, it does all of its work going down the stack instead of on the way back up. Unlike merge sort, the quicksort is less stable and can degrade to $O(n^2)$ performance. This happens due to poor selection of a pivot. Recall from chapter 15 that a quicksort works by selecting one element to be the "pivot" and then moves elements around so that the pivot is in the correct location. If pivots are selected which belong near the middle of the array, this produces $O(n\log(n))$ performance. However, if you consistently pick elements near the edges, the recursion has to go $O(n)$ levels deep and the resulting performance is truly horrible. The version written back in chapter 15 will do exactly that for sorted data, in addition to the overhead of not occurring in-place. Just selecting a pivot at random makes this type of behavior unlikely. With a bit more logic, we can make it almost impossible.

We will start with a general version of an in-place quicksort that can be easily modified in regards to picking a pivot. This first version always uses the first element, but it has that pulled off in a function so that it is easy to modify.

```
def quicksort[A](a:Array[A])(comp:(A,A)=>Int) {
  def pickPivot(start:Int,end:Int) = start

  def qsRecur(start:Int,end:Int) {
    if(start<end-1) {
      val pivot = pickPivot(start,end)
      val p = a(pivot)
      a(pivot) = a(start)
      a(start) = p
      var (low,high) = (start+1,end-1)
      while(low<=high) {
        if(comp(a(low),p)<=0) {
          low += 1
        } else {
          val tmp = a(low)
          a(low) = a(high)
          a(high) = tmp
          high -= 1
        }
      }
      a(start) = a(high)
      a(high) = p
      qsRecur(start,high)
      qsRecur(low,end)
    }
  }
  qsRecur(0,a.length)
}
```

This code retains that unfortunate characteristic that is used on a sorted array, it will have $O(n^2)$ performance. However, that can be corrected by making the following change.

```
  def pickPivot(start:Int,end:Int) = start + util.Random.nextInt(end-start)
```

This modified `pickPivot` function picks a random value for the pivot. It can still have $O(n^2)$ behavior, but the odds are very low, especially for large arrays, as it requires randomly picking elements that are near the smallest or the largest in a range repeatedly.

The behavior of quicksort can be improved further by incorporating a bit more logic into the pivot selection and by using insertion sort when the segment we are sorting gets small enough. It might seem counterintuitive to rely on a sort that we know is $O(n^2)$ to help improve a sort that is supposed to be $O(n \log(n))$, however, this can be understood by remembering that order notation is really only relevant for large values of n. When n is small, the sorts that we normally think of as being inferior can actually demonstrate significantly better performance. Code that incorporates both of these improvements is shown here.

```
def quicksort[A](a:Array[A])(comp:(A,A)=>Int) {
```

```
  def insertionSort(start:Int,end:Int) {
    for(i <- start+1 until end) {
      val tmp = a(i)
      var j = i-1
      while(j>=0 && comp(a(j),tmp)>0) {
        a(j+1) = a(j)
        j -= 1
      }
      a(j+1) = tmp
    }
  }

  def pickPivot(start:Int,end:Int) = {
    val mid = start + (end-start)/2
    val sm = comp(a(start),a(mid))
    val se = comp(a(start),a(end-1))
    if(sm<=0 && se>=0 || sm>=0 && se<=0) start
    else {
      val me = comp(a(mid),a(end-1))
      if(sm<=0 && me<=0 || sm>=0 && me>=0) mid else end-1
    }
  }

  def qsRecur(start:Int,end:Int) {
    if(start<end-7) {
      val pivot = pickPivot(start,end)
      val p = a(pivot)
      a(pivot) = a(start)
      a(start) = p
      var (low,high) = (start+1,end-1)
      while(low<=high) {
        if(comp(a(low),p)<=0) {
          low += 1
        } else {
          val tmp = a(low)
          a(low) = a(high)
          a(high) = tmp
          high -= 1
        }
      }
      a(start) = a(high)
      a(high) = p
      qsRecur(start,high)
      qsRecur(low,end)
    } else {
      insertionSort(start,end)
    }
  }
  qsRecur(0,a.length)
}
```

The pivot selection works by finding the median of the first, middle, and last elements in the range. While technically this only eliminates selecting the smallest or largest element, it improves the probability of picking an element near the middle dramatically. It also provides ideal performance for arrays that were already sorted.

This method of picking a pivot also calls for using something other than quicksort below three elements. After all, the pivot selection process does not make sense if the first, middle, and last elements are not different. In this code, we have switched over to an insertion sort for any array with fewer than seven elements. This value was selected using some empirical testing to count the number of comparisons that were performed on random data sets.

28.4.3 Formula Parser

The last divide and conquer problem we will consider in this chapter is the problem of formula parsing. Consider an application where you want the user to be able to type in basic math formulas for evaluation. This would give you applications somewhat like graphing calculators, which can evaluate expressions the way you would write them on paper. So if the user were to enter "3+5*2" you want a function that can take that as a `String` and return the value 13.[5]

There are a number of different approaches to this. One approach divides the formula at the lowest precedence operator, recurses on the two sides of that operator, and then applies the operator to the two return values at get an answer. If no operator is found, we have a base case for a number or we could write code to support parentheses, in which case the function should recurse on the contents of the parentheses.

An object with a method called `eval` for doing just this is shown here. It simply calls the recursive function, passing the formula with all spaces removed. The parse defines some variables, then has a loop that finds the lowest precedence operator. It ends with a set of ifs to deal with different possibilities.

```
package scalabook.util

object Formula {
  val ops = "+-*/".toSet

  def eval(form:String):Double = evalParse(form.filter(_!=' '))

  private def evalParse(f:String):Double = {
    var opLoc = -1
    var parensCount = 0
    var i = f.length-1
    while(i>0) {
      if(f(i)=='(') parensCount += 1
      else if(f(i)==')') parensCount -= 1
      else if(parensCount==0 && (f(i)=='+' || f(i)=='-' && !ops.contains(f(i-1)))) {
        opLoc = i
        i = -1
      } else if(parensCount==0 && opLoc == -1 && (f(i)=='*' || f(i)=='/')) {
        opLoc = i
      }
      i -= 1
    }
    if(opLoc<0) {
      if(f(0)=='(') {
        evalParse(f.substring(1,f.length-1))
      } else f.toDouble
    } else {
```

[5]This example was chosen because it illustrates that we want to have proper order of operations so that multiplication is done before addition.

```
      f(opLoc) match {
        case '+' => evalParse(f.take(opLoc))+evalParse(f.drop(opLoc+1))
        case '-' => evalParse(f.take(opLoc))-evalParse(f.drop(opLoc+1))
        case '*' => evalParse(f.take(opLoc))*evalParse(f.drop(opLoc+1))
        case '/' => evalParse(f.take(opLoc))/evalParse(f.drop(opLoc+1))
      }
    }
  }
}
```

The first variable is opLoc, which should store the location of the operator. It is initialized to -1 and if no operator is found, it will still have that value at the end. After that is parensCount, which keeps track of how deeply nested we are in parentheses. The function is written so that no operator that is in parentheses can be considered to be top level. Last is a loop variable, i, which starts at the end of the string. The operators we are using are all left-associative, so the lowest precedence operator will be the furthest right.

The while loop starts off with checks to see if there is a parentheses. After that are checks to see if we have found different operators. There are some details to these that are worth noting. Both of the operator checks require that the parentheses count be zero. The check for a minus sign also requires that the character in front of it not be an operator. This is to make sure that formulas like "5+-3" work. In this situation, the '-' is negation, not a binary operator. When either '+' or '-' are found, the location is stored in opLoc, and the value of i is set to be negative so the loop will stop. This is because the first addition or subtraction that is found will always be the lowest precedence operator.

For multiplication and division, the behavior is a bit different. That characters only matter if opLoc is -1 and they do not change the value of i. This is because finding a '*' or '/' does not automatically mean we have found the lowest precedence operator. There could still be a '+' or '-' further left that would have lower precedence. For that reason, the loop must continue working. However, if another '*' or '/' is found to the left of the first one seen, it should be skipped over because of the associativity.

We will continue playing with the formula parser in later chapters. In the next chapter we will add the ability to handle variables. That will let us include formulas in our project. We do not want to do it with this code, because string parsing is fundamentally slow and we do not want to use code that parses the string every time a formula is evaluated.

28.5 End of Chapter Material

28.5.1 Summary of Concepts

This chapter was intended to give you another look at recursion and to explore some algorithms that use it in a bit more detail.

- Algorithms for optimal path finding through a maze.
  - The brute force method checks all paths and picks the shortest. There can be a huge number of paths making this impractical for some situations.
  - Using "smart breadcrumbs" allows for certain paths to be ignored as soon as it is found that they can not be correct.
- A drawable can be made to respond to mouse clicks on the drawing.

  - The drawable needs to inherit from a specific type so that the drawing knows if it should be listening for things.
  - Using `Reactor` as the supertype breaks serialization.
  - Transforms can be considered by remembering the last ones used and using `inverseTransform`.

- We took our first look at graphs. These combine nodes and edges and can be used to represent many different types of problems.
  - Code for a reachability algorithm does recursive edge following with a mutable set of visited nodes.
  - Code for a shortest-path algorithm does recursive edge following with an immutable set of visited nodes.

- Divide and conquer algorithms. These take a large problem and break it into pieces, solve the pieces, and combine the solutions to get a final answer. The base case it typically when you get down to a size that the solution is trivial.
  - Revisited merge sort and quicksort and explored a number of implementations of them.
  - Wrote a divide and conquer solution to formula parsing that allows basic operators and parentheses.

28.5.2 Exercises

1. Add a longest path function and Graphical User Interface (GUI) option to the `DrawMaze`.
2. Add a path count function and GUI option to the `DrawMaze`.
3. Make the animations thread safe by preventing clicks from going through while a maze solution is being calculated.
4. Compare the speed of the different mergesort and quicksort versions.
5. Extend the formula parser so that it can also do square roots and the three basic trigonometric functions.

28.5.3 Projects

1. It is now time to implement a number of additional commands in your MUD. Hopefully you have been adding things to make it playable and working on having a real map. In a real MUD, map editing is something that can be done by certain users using standard commands from inside the MUD. You should not have to go out and edit XML by hand or anything like that. If you have not yet added this functionality to your MUD, you should now.

   There are some other commands that will be very handy for anyone trying to build a map, mainly to check that the result looks the way you want. In particular, you need to be able to jump to random rooms, check if you can get from one room to another, and see how many paths there are between rooms. These are not normal commands.

They should only be available to users who have the rights for creating rooms. The last two can be written as recursive algorithms that run from room to room following the exits that connect them.

2. A good recursive problem for the web spider is to determine how many clicks it takes to get from one page to another. Closely related to that, you can measure distances, in clicks, between different data elements in a data set.

3. Recursion can be used for a number of different tasks in graphical games. Which you pick really depends on what your game is like. All are at least roughly related to the maze algorithm. If your game has multiple screens that are connected (picture classic Zelda) it could be useful to have a tool that tells you how different screens are connected.

   If your game has computer-controlled characters and obstacles that can block movement, then a recursive algorithm can be used to find a route from the character to the player. Note that in general, you do not want to allow an Artificial Intelligence (AI) to follow the shortest route to the player. The reason being that it becomes impossible for the player to evade the enemy.

4. If you are doing the math worksheet project, you should add the ability for the user to define functions with names. Then make it so that the definition of one function can include a call to another function. This will use recursion in your code. You do not have to go all the way of including conditionals in the functions so that the functions themselves can be recursive. You are certainly allowed to try, but that is functionality that you will be able to add in later.

5. Every good paint program needs a flood fill option. Your Photoshop® project is no exception. Flood fill can be written in a very compact form using recursion. You should implement this and try it. What you will find if you have a large image is that you can easily overflow the stack by clicking in an open area.

   To avoid this, you have to switch from using recursion to using a loop with a manual stack or queue for keeping track of things. If you use a stack, the order of the fill will match what happens in recursion. If you use a queue, you get a breadth-first ordering instead. For this application, the difference will not be significant.

6. In chapter 26 you added a collisional simulation to the simulation workbench using a priority queue. That code has two different aspects that force every step to take $O(n^2)$ time, the comparison of particles to determine collision times, and the inserting of items into a sorted linked list priority queue. We will look at one way to improve on the first of these.

   Currently your code finds collisions by doing comparisons between every pair of particles in the simulation. In reality, collisions are localized. During the fairly short period of time represented by a timestep in the simulation, it generally is not possible for a particle to collide with other particles on the opposite side of the simulation area. Technically, you can only collide with particles where $d < \Delta t|\vec{v_1} - \vec{v_2}|$, where d is the distance between the particles, Δt is the length of the timestep, and $\vec{v_1}$ and $\vec{v_2}$ are the velocities of the two particles. For this reason, you really only need to compare to things that are close by. The problem is, you do not know in advance what is close by.

   One way to deal with this is to maintain in each particle knowledge of its nearest neighbors. Have each particle keep track of the closest m other particles. The value of m should probably be at least four for a 2-D simulation and six to eight for a

3-D simulation. Building those connections initially requires a full search and $O(n^2)$ checks. After it has been built, searches for particles to do collision checks against can be done with a recursive search running through the connections. The search can stop anytime it gets to a particle that is too far away.

This same method can also be used to maintain the closest m particles. If the recursive search ever hits a particle that is closer than one of the one currently being tracked, the new one can replace it. Every so often another $O(n^2)$ search can be done to make sure the connections are all good, but this allows most of the steps to be done with each particle compared to a significant smaller number of other particles.

Chapter 29

Trees

"I think that I shall never see
A poem as lovely as a tree." - Joyce Kilmer

In this chapter you will learn about the concept of a tree as they are known in Computer Science. While we will not be writing poems about them, we will construct some lovely code to provide efficient implementations of some Abstract Data Type (ADT).

29.1 General Trees

Chapter 25 introduced the concept of the linked structure using the simplest form of a linked list. The most general expression of a linked structure is the graph that was briefly presented in chapter 28. A tree is a linked structure that falls somewhere between these two extremes. Like the other two, a tree is formed of nodes with edges between them. We consider these edges to have a direction, pointing from one node to another. Unlike a graph, where we can do pretty much anything we want with with the edges between the nodes, to qualify as a tree, there are certain limitations. We will run through these and introduce the basic terminology, which is largely a mixture of metaphors from biology and genealogy.

A tree can include any number of nodes. One node is the ROOT of the tree. Consider that you have a node C and another node P such that there is an edge from P to C. See figure 29.1 for an example. We call P the PARENT of C and C is the CHILD of P. To qualify as a tree, every node has one incoming edge except the root, which has zero. As a result, there is one and only one path from the root to any given node. Other nodes that have an edge going from P to them are the SIBLINGs of C. Going further with the genealogy metaphor, the nodes can can be reached going down edges from a node are called the DESCENDANTs of that node. Similarly, the nodes that are on the path from a node to the root, including the root, are the ANCESTORs of the node.

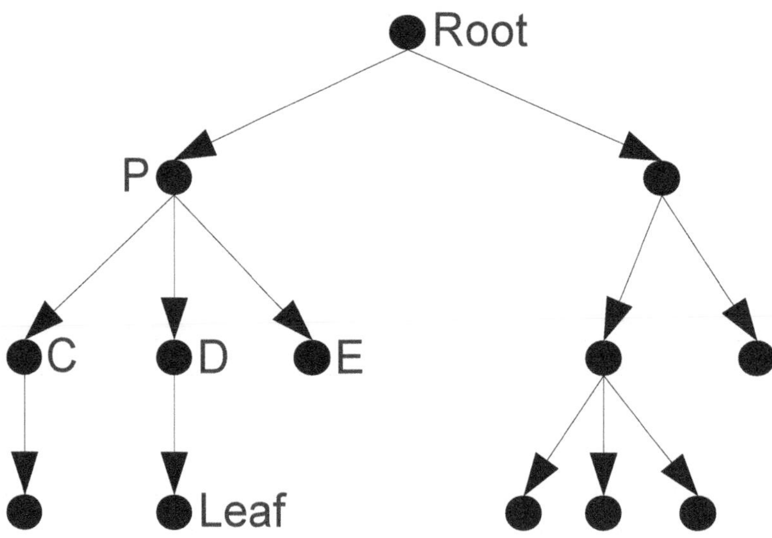

FIGURE 29.1: This figure shows a sample tree with some of the nodes labeled. Node P is the parent of nodes C, D, and E. D and E are siblings of C. Node P has a depth of 1 and a height of 2, while node C has a depth of 2 and a height of 1. The size of node P is six, while the size of node C is 2.

Following the biology metaphor, any node that does not have outgoing edges is called a LEAF. The number of edges between the root and a node is called the DEPTH of the node. The number of edges between a node and the deepest leaf is called the HEIGHT of that node. The term SIZE refers to the number of descendants of a node, plus itself.

We can also define a tree in a recursive way. Using this definition, a tree is either an empty node or a node with edges to zero or more non-empty subtrees under it. This definition will be directly reflected in some of the implementations.

Note that for all these definitions, when you draw out a tree, there are no loops of any kind. This is probably the key image to have in your mind when you think about trees, they do not allow loops. You should also note from this description that a linked list is actually a perfectly happy tree. The root of the tree is equivalent to the head of the list, and each node has only one outgoing edge to a child. It might be a boring tree, but the standard, singly-linked list definitely qualifies as a tree.

29.1.1 Implementations

A general tree allows each node to have an arbitrary number of children. This leads to a number of different implementation strategies. One approach is to have the node store whatever data is needed along with a sequence of children. Our `Drawable` type defines a node in a tree and we used this approach. Recall that `Drawable` actually inherits from `javax.swing.tree.TreeNode`. This was done so that we could use it with the `javax.swing.JTree` in our Graphical User Interface (GUI). In that tree, the internal nodes are all of the `DrawTransform` type while the leaves extend `DrawLeaf`. The `DrawTransform` contains the following line of code.

```
private val subnodes = mutable.Buffer[Drawable]()
```

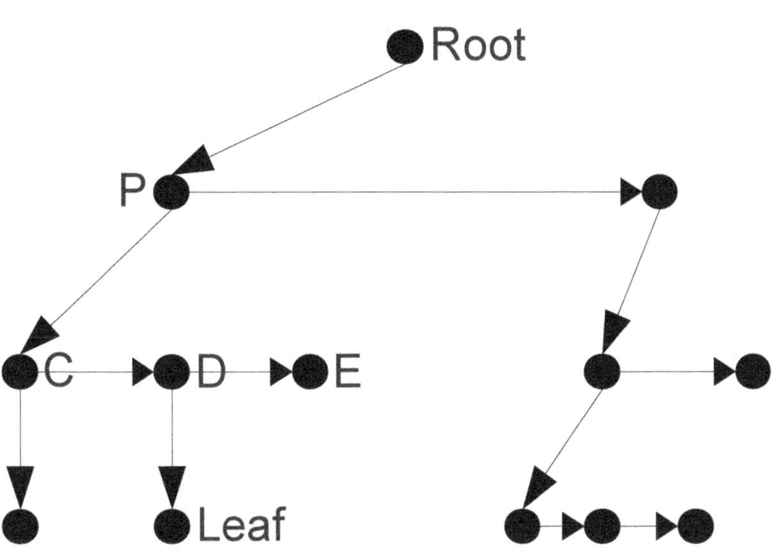

FIGURE 29.2: This figure shows the same tree that was displayed in figure 29.1, but draws arrows for the first-child, next-sibling approach to building the tree.

This is the sequence of children. For this example, a mutable buffer was used, but other sequence implementations could work just as well for other trees.

Another example of a general tree is the file system on a computer. Depending on your operating system, you will have a drive or a root directory. Under that will be some combination of files and directories. The normal files are all leaves, while the directories can be internal nodes with children.

Instead of using a sequence in the node, a general tree can be built by explicitly utilizing a linked list. In this approach, each node keeps the data it needs along with references to the first child and the next sibling. The links for the next sibling provide a linked list of the children for a given node. A graphical representation of what this looks like in memory is shown in figure 29.2. Given this implementation, no node has to store more than two references to other nodes. This generally makes it a bit more efficient in memory, if somewhat more challenging to deal with in code.

29.1.2 Traversals

As with linked lists, trees are ways for us to store information. The tree simply has more structure to it. One of the most significant activities we want to perform on any collection of data is a traversal through the elements. For an array we do this by counting indexes. For a linked list this was accomplished by having a reference, which we called `rover`, start at the head and follow links down the list. For a tree, things are a bit more complex. There is no single natural ordering for a tree. It makes sense for children to be handled from left to right, but there are questions about whether parents should come before or after children or whether children should be handled before siblings. This leads to the possibility for multiple different traversals of trees.

The family of traversals we will consider first are the DEPTH FIRST traversals. These all handle children before following siblings. They are a family of options, because even with

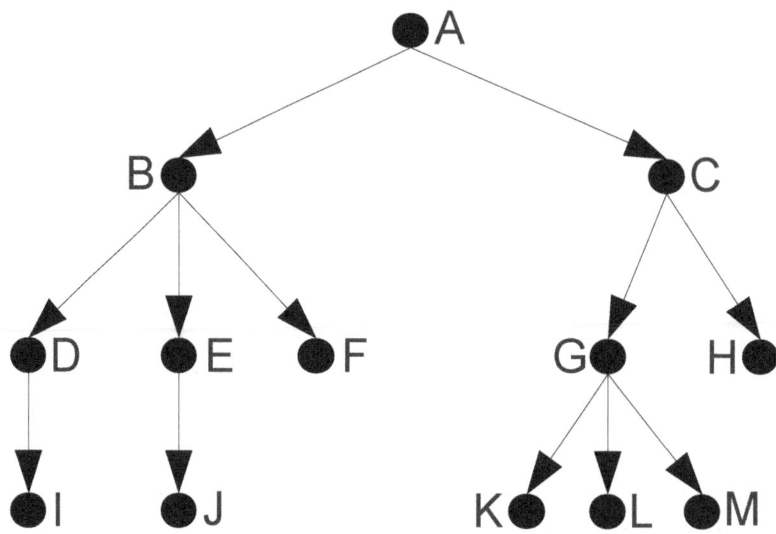

FIGURE 29.3: This figure shows the same tree that was displayed in figures 29.1 and 29.2. Here all of the nodes are labeled with letters to aid the discussion of traversal.

that rule, there is the question of when the parent is handled relative to children. If the parent is handled before the children, it is called a PRE-ORDER traversal. If the parent is handled after the children it is called a POST-ORDER traversal. When the parent is handled between children, it is an IN-ORDER traversal. We will go through these one at a time and consider the order of handling for the tree shown in figure 29.3. We will also consider the code we would write for such traversals.

All of the depth-first searches can be traced through the tree in the same way. You go from a node to its first child and everything below it, then pop back up to the node and go to the next child. Nodes are not handled as soon as they are hit or every time they are hit, but instead when the desired traversal specifies.

A pre-order traversal handles parents before children and children before following siblings. With our sample tree, this gives the following order for the nodes: A, B, D, I, E, J, F, C, G, K, L, M, H. In the pre-order, parents always come before their children. As a result, the root is handled first. The lowest child on the far right side will be handled last. Assume that our `Node` class looks something like this,

```
case class Node(data: A, children: Seq[Node])
```

where `A` is the type of the data being stored in the tree. Then the code for a pre-order traversal might look like the following.

```
def preorder(n: Node,handler: A => Unit) {
  handler(n.data)
  n.children.foreach(n2 => preorder(n2,handler))
}
```

The `handler` is a function that is passed in that should be called at the appropriate time. Note that for a pre-order traversal, it is called before the recursion handles all of the children. A full traversal can be initiated by calling this function on the `root` node of the tree.

It is also possible to handle the parent after the children. This give a post-order traversal. The post-order traversal for our sample tree has the following order for handling nodes: I, D, J, E, F, B, K, L, M, G, H, C, A. In this ordering, each node comes after all of its descendants. As a result, the root has to come last. The lowest element on the left side comes first. Using the `Node` class defined above, a post-order traversal could be written this way.

```
\begin{center}
def postorder(n: Node,handler: A => Unit) {
  n.children.foreach(n2 => postorder(n2,handler))
  handler(n.data)
}

\end{center}
```

Note that the only change is that the call to `handler` has moved below the processing of the children.

It is also possible to put the handling of the current node between some of the children. This produces an in-order traversal. For a general tree where each node can have an arbitrary number of children, this traversal is not particularly well defined. We will revisit this traversal later on when we have trees with somewhat more limited structures.

The depth-first traversals, where children are handled before following siblings, works well with recursion because the memory of the stack keeps track of what is happening for nodes where only some of the children have been processed. It turns out we could write a depth-first traversal without recursion using the `Stack` types we wrote back in chapters 24 and 25. This code would push the root node, then loop while the stack is not empty, popping elements off, handling them, and pushing the children. The code for this looks like the following.

```
def loopPreorder(r: Node, handler: A => Unit) {
  val stack = new ListStack[Node]
  stack.push(r)
  while(!stack.isEmpty) {
    val n = stack.pop
    handler(n.data)
    n.children.reverse.foreach(stack.push)
  }
}
```

The call to `reverse` on the children insures that children are handled from left to right as the last one pushed to the stack will be the first one processed.

Handling the siblings, and cousins, before the children produces a fundamentally different type of traversal called a BREADTH-FIRST traversal. This runs across each level, or generation, of the tree from left to right before going to the next level down. This type of traversal does not work with recursion. Instead of repeatedly trying different paths down from a node, which recursion does well, this simply runs across nodes of the same depth. This approach does not have the same branching type of behavior that characterizes recursion. Indeed, to implement this type of traversal, one uses a queue instead of a stack. Here is an example implementation.

```
def breadthFirst(r: Node,handler: A => Unit) {
  val queue = new ListQueue[Node]
  queue.enqueue(r)
  while(!queue.isEmpty) {
```

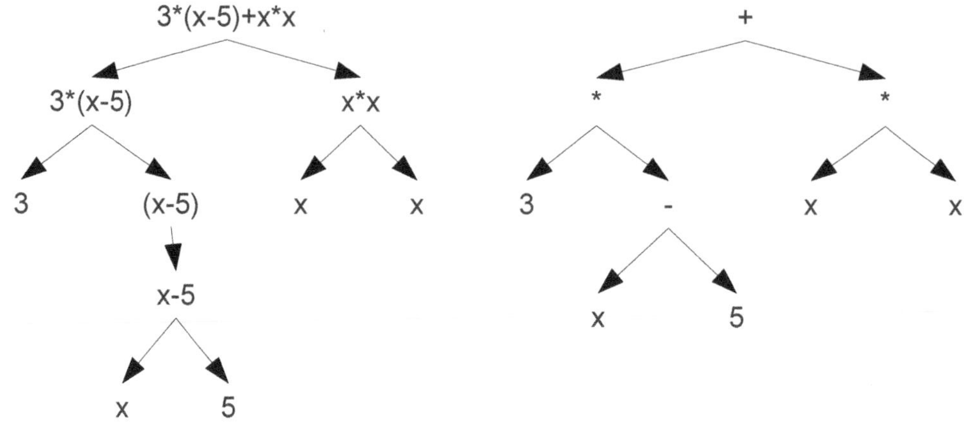

FIGURE 29.4: This figure shows the recursion that happens when we parse the formula 3*(x-5)+x*x on the left side. On the right is the tree that we would want to build from this formula.

```
      val n = queue.dequeue()
      handler(n.data)
      n.children.foreach(queue.enqueue)
    }
  }
```

Note how this looks nearly the same as the loop based pre-order traversal except with a queue instead of a stack. If you run this on our sample tree the order of handling is A, B, C, D, E, F, G, H, I, J, K, L, M.

29.2 Project Integration: Formula Parsing

As was already noted, there is a tree in the current project code that keeps track of, and gives order to the drawables. In this section we will add a somewhat different type of tree to our formula parsing code that will make formula evaluation faster so that we can introduce variables and evaluate the same variable many times using different values for the variables. We can then add an extra element to the GUI that adjusts a time variable so that our drawings can have animation.

Recall that the formula parser found the lowest priority operator, and split on that. It recursively evaluated the two sides and combined their answers using the operator that was found. This involves a lot of string manipulation and running through the string to find things. However, this same process can generate a nice tree structure that is fast to evaluate.

Figure 29.4 shows two different trees. On the left is a tree showing the recursion, assuming the function were augmented to handle variables. Each node is a function call with the string that would be passed into that call. On the right is a tree with the significant operator, number, or variable from each call. This tree effectively represents the original formula and can be used for fast evaluation.

To do this, we make a `trait` that represents a node in the tree, and give it a method so that it can be evaluated. We then make classes that inherit from that trait, which define all the different types of nodes that we want to have. The `evalParse` function that we wrote previously can be changed to just do a parse and return a `Node` instead of a `Double`. The following code shows an implementation of that. In addition to parsing to a tree, this code adds support for some trigonometric functions and a square root function.

```
package scalabook.util

import scala.collection.Map

class Formula(val formula:String) extends Serializable {
  private val root = Formula.parse(formula)

  def eval(vars:Map[String,Double]):Double = root eval vars

  override def toString = formula
}

object Formula {
  val ops = "+-*/".toSet

  def eval(form:String, vars:Map[String,Double] = null):Double = {
    val root = parse(form.filter(_!=' '))
    root.eval(vars)
  }

  private def parse(f:String):Node = {
    var opLoc = -1
    var parensCount = 0
    var i = f.length-1
    while(i>0) {
      if(f(i)=='(') parensCount += 1
      else if(f(i)==')') parensCount -= 1
      else if(parensCount==0 && (f(i)=='+' || f(i)=='-' && !ops.contains(f(i-1)))) {
        opLoc = i
        i = -1
      } else if(parensCount==0 && opLoc == -1 && (f(i)=='*' || f(i)=='/')) {
        opLoc = i
      }
      i -= 1
    }
    if(opLoc<0) {
      if(f(0)=='(') {
        parse(f.substring(1,f.length-1))
      } else if(f.startsWith("sin(")) {
        new SingleOpNode(parse(f.substring(4,f.length-1)),math.sin)
      } else if(f.startsWith("cos(")) {
        new SingleOpNode(parse(f.substring(4,f.length-1)),math.cos)
      } else if(f.startsWith("tan(")) {
        new SingleOpNode(parse(f.substring(4,f.length-1)),math.tan)
      } else if(f.startsWith("sqrt(")) {
        new SingleOpNode(parse(f.substring(5,f.length-1)),math.sqrt)
      } else try {
        new NumberNode(f.toDouble)
```

```
      } catch {
        case ex:NumberFormatException => new VariableNode(f)
      }
    } else {
      f(opLoc) match {
        case '+' => new BinaryOpNode(parse(f.take(opLoc)), parse(f.drop(opLoc+1)),
            _+_)
        case '-' => new BinaryOpNode(parse(f.take(opLoc)), parse(f.drop(opLoc+1)),
            _-_)
        case '*' => new BinaryOpNode(parse(f.take(opLoc)), parse(f.drop(opLoc+1)),
            _*_)
        case '/' => new BinaryOpNode(parse(f.take(opLoc)), parse(f.drop(opLoc+1)),
            _/_)
      }
    }
  }

  private trait Node extends Serializable {
    def eval(vars:Map[String,Double]):Double
  }

  private case class NumberNode(num:Double) extends Node {
    def eval(vars:Map[String,Double]):Double = num
  }

  private case class VariableNode(name:String) extends Node {
    def eval(vars:Map[String,Double]):Double = vars get name getOrElse 0.0
  }

  private case class BinaryOpNode(left:Node, right:Node, op:(Double,Double) =>
      Double) extends Node {
    def eval(vars:Map[String,Double]):Double = op(left eval vars,right eval vars)
  }

  private case class SingleOpNode(arg:Node, op:Double => Double) extends Node {
    def eval(vars:Map[String,Double]):Double = op(arg eval vars)
  }
}
```

This code includes a class called `Formula`, in addition to the object, so that you can instantiate an object of the `Formula` type and that will parse the string once, then remember the tree for later evaluations. Calling `Formula.eval` on the companion object will be even slower than what was done before, but it has been left in for simple "one-off" calls. The import at the top makes it so that the `Map` type we are using is general and can be mutable or immutable.

Going through this code closely, you will see that there are a few places where we use a fundamentally different style for our Scala code. There are a number of places where methods are used with operator notation, even when they have normal names. For example, `root eval vars` near the top. This is the same as `root.eval(vars)`. The `eval` method is just used with an operator syntax so there is no dot and no parentheses are required. This style is quite common among professional Scala developers and it will appear more as we go through the book.

With this code written, we can add significant functionality to our drawing. First, we need to make a few additions and minor alterations to the `Drawing` type to put a `Slider` at the top of the GUI and have it update a value in the vars map that is associated with the name "t" for use in formulas.

```
// Add this.
@transient private var lTimeSlider:Slider = null

...

// Add this.
private def timeSlider = {
  if(lTimeSlider==null) {
    lTimeSlider = new Slider {
      min = 0
      max = 800
      listenTo(this)
      reactions += {
        case vc:ValueChanged =>
          vars("t") = value
          drawPanel.repaint()
      }
    }
    if(!vars.contains("t")) vars("t") = 0
    lTimeSlider.value = vars("t").toInt
  }
  lTimeSlider
}

...

// Change to this.
      },new SplitPane(Orientation.Horizontal,
        new BorderPanel {
          layout += new ScrollPane(drawPanel) -> Center
          layout += timeSlider -> North
        },
        commandPanel))
```

The vars map was put in place when we added the Reverse Polish Notation (RPN) calculator to the commands. It can be used again here in connection with the Formula type. This is not just reuse, it adds power. Now the commands can be used to set variables that are referenced in formulas. All we need now is to make something that we draw utilize the formulas as well.

This is not hard to do either. For example, the width and height of the DrawRectangle type can be changed to use formulas with the following alterations.

```
class DrawRectangle(p:DrawTransform,
    private var width:Formula = new Formula("100.0"),
    private var height:Formula = new Formula("100.0"),
    private var color:Color = Color.black) extends DrawLeaf {

  ...

  def draw(g:Graphics2D) {
    g.setPaint(color)
```

```
    g.fill(new
        Rectangle2D.Double(0,0,width.eval(drawing.vars),height.eval(drawing.vars)))
  }

  ...

          contents += new TextField(width.toString) {
            listenTo(this)
            reactions += { case e:EditDone => width = new Formula(text) }
          }
          contents += new TextField(height.toString) {
            listenTo(this)
            reactions += { case e:EditDone => height = new Formula(text) }
          }

  ...

}

object DrawRectangle {
  def apply(p:DrawTransform,data:xml.Node) = {
    new DrawRectangle(p,new Formula((data\"@width").text),new
        Formula((data\"@height").text),
       new Color(java.lang.Long.parseLong((data\"@color").text,16).toInt))
  }
}
```

The changes can be summarized as making `width` and `height` be of type `Formula` instead of `Double`, then altering the few other parts of the code that are required to make those changes compile. You can make similar changes to `DrawTransform` by changing the `transformValue` from an `Array[Double]` to an `Array[Formula]` and then making minor alterations to fix the syntax errors that creates. This same procedure of changing `Double` to `Formula` and fixing syntax errors can be applied to any other drawables that you want. The benefit of doing this is that you can then make it so that your drawing changes as you move the slider. This allows simple animation or other effects to be produced.

29.2.1 Formula Tree Traversals and In-order Traversal

For the formula tree, the in-order traversal is well defined if we make the rule that for binary operator nodes, the left child is handled first, followed by the current node, and then the right child. It is interesting to consider the three different traversals, pre-order, post-order, and in-order, on the tree shown in figure 29.4.

- Pre-order: + * 3 - x 5 * x x
- Post-order: 3 x 5 - * x x * +
- In-order: 3 * x - 5 + x * x

The in-order traversal is the original formula as we are used to seeing it, but without the parentheses. The parentheses are required for in-order expressions to get the proper order of operations. The pre-order version is similar to the syntax of languages like LISP and Scheme. In Scheme, this expression would be written as (+ (* 3 (- x 5)) (* x x)). Parentheses are again needed to specify order of operation, but it is easy to extend the syntax so that more than two operands can be specified for any operator. The post-order traversal is exactly what you would enter into a RPN calculator and needs no parentheses to specify order of operation. While you might not normally think of equations as trees, it turns out that form contains their meaning very nicely and translates easily into different useful versions of the expression.

29.3 Binary Search Trees: Binary Trees as Maps

The example above is a fun use of a tree and an interesting illustration of the power of trees. However, the most use of these structures is for storing data. In this section we will explore the details of the binary search tree, BST, as a means of implementing the Map ADT.

The nature of the map ADT was described in chapter 20 when we looked at the ones that are built into the Scala collections libraries. A binary search tree is first a binary tree. This is a tree that allows 0, 1, or 2 children per node. In the case of the BST, the children are typically called left and right, and it is possible to have a right child without having a left. At least part of the data stored in nodes includes a key that has a complete ordering. That means that for any two key values you can ask whether one is less than, equal to, or greater than the other. The primary criteria for a BST is that descendants to the left have letter key values while descendants to the right have greater key values.

The name binary search tree should bring to mind the binary search that was discussed in chapter 13. This is not by accident. A well-balanced BST will behave much like a binary search when the user looks for elements. To understand this, we will consider an example. Figure 29.5 shows an example BST that uses basic integers for keys to keep things simple. This tree contains the values 0 - 9 in a proper BST organization.[1] The bold arrows indicate the path taken by a search for the value 2.

Adding new elements to a basic BST is fairly straightforward. If the tree is empty, there is no root, then we make a node with the current element and make that the root. For any other situation, we start at the root and walk down in the same way we would for a search, going to the left when the new value is less than what is stored in the node and the right when it is greater. Eventually this process reaches a point where there is no child in the indicated direction. A new node is created to store the new value and it is placed at that location in the tree. Figure 29.6 goes through this process adding the values 5, 3, 1, 8, 7, 4, 2, 9, 0, 6 in that order. The result is the tree shown in figure 29.5. Note that the order of adding the numbers is significant. While some other orderings would produce the same tree, most will not.

The process of removing from a BST is a bit more complex than adding. Consider the situation where we want to remove the node with the value 5 from our sample tree. This happens to be the root node. We can not simply take it away as that would leave us with two separate trees. We have to do something that modifies the tree in a way so that the result is still a valid BST. One approach to this would be to put the right subtree at the bottom right of the left subtree or vice versa. In this case that would put either the 8 right of 4 or the 3 left of 7. While these technically work, they are far from ideal because they have a tendency to unbalance the tree. We will see shortly how that can make a big difference.

The other approach is to move the data from a lower node, which is easier to remove, up to replace the 5 and then remove that lower node. Nodes are easy to remove if they

[1]As we will see shortly, there are many other ways that this could be drawn. The only requirement is that lesser values are to the left and greater values are to the right.

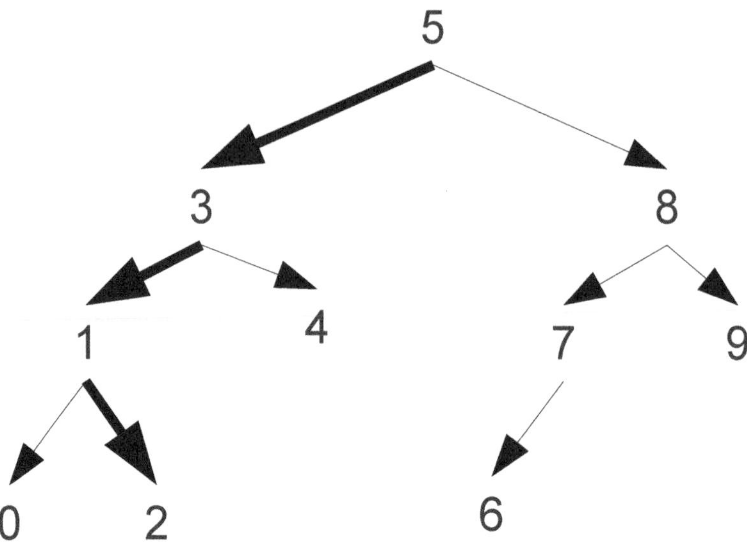

FIGURE 29.5: This is an example of a binary search tree using integer value keys. Searching for an element in the tree follows steps very much like those a binary search would perform. The bold arrows show the path that would be taken in a search for the value 2.

have zero or one child. Of course, the data that is moved has to be able to take the place of the node being removed without breaking the proper ordering of the tree. There are only two nodes that can do that, the greatest element on the left or the smallest element on the right. In this case, those are the values 4 and 6. If 5 were replaced by either of those, the rest of the tree structure remains perfectly intact and retains proper ordering. While this procedure is fairly easy to describe, unfortunately it has many special cases that have to be dealt with.

The following is code that uses a BST to implement a mutable map from the Scala libraries. This has type parameters for both a key and value type, called K and V in this code. The `TreeMap` must be passed a comparison function at construction. This provides the complete ordering that is needed for the keys. The `mutable.Map` type has four abstract methods that must be implemented, `+=`, `-=`, `get`, and `iterator`. A complete implementation is shown here.

```
package scalabook.adt

import collection.mutable
import annotation._

class TreeMap[K,V](comp:(K,K)=>Int) extends mutable.Map[K,V] {
  private class Node(var key:K, var data:V) {
    var left:Node = null
    var right:Node = null
  }

  private var root:Node = null

  def +=(kv:(K,V)) = {
```

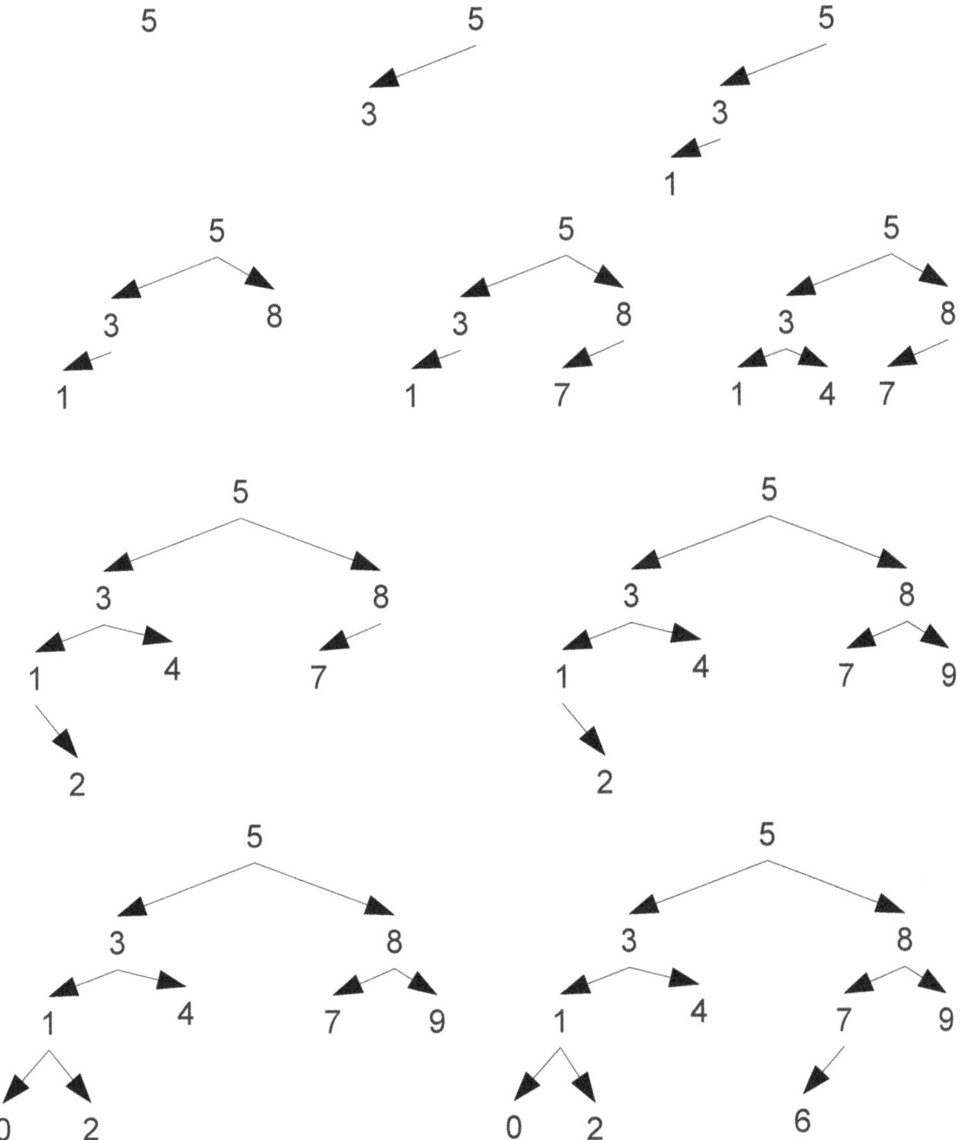

FIGURE 29.6: This figure shows the creation of the BST from figure 29.5. Each element is placed by running down the tree in the same manner as a search until we hit a location where there is not a child. A new node with the value is placed in that location.

```
    if(root==null) {
      root = new Node(kv._1,kv._2)
    } else {
      var rover = root
      var parent:Node = null
      var c = comp(kv._1,rover.key)
      while(c!=0 && rover!=null) {
        parent = rover
        rover = if(c<0) rover.left else rover.right
        if(rover!=null) c = comp(kv._1,rover.key)
      }
      if(c==0) {
        rover.key = kv._1
        rover.data = kv._2
      } else if(c<0) {
        parent.left = new Node(kv._1,kv._2)
      } else {
        parent.right = new Node(kv._1,kv._2)
      }
    }
    this
  }

  def -=(key:K) = {
    def findVictim(n:Node):Node = {
      if(n==null) null
      else {
        val c = comp(key,n.key)
        if(c==0) {
          if(n.left==null) n.right
          else if(n.right==null) n.left
          else {
            val (key,data,node) = deleteMaxChild(n.left)
            n.left = node
            n.key = key
            n.data = data
            n
          }
        } else if(c<0) {
          n.left = findVictim(n.left)
          n
        } else {
          n.right = findVictim(n.right)
          n
        }
      }
    }

    def deleteMaxChild(n:Node):(K,V,Node) = {
      if(n.right==null) {
        (n.key,n.data,n.left)
      } else {
        val (key,data,node) = deleteMaxChild(n.right)
        n.right = node
        (key,data,n)
```

```
      }
    }

    root = findVictim(root)
    this
  }

  def get(key:K):Option[V] = {
    var rover = root
    var c = comp(key,rover.key)
    while(rover!=null && c!=0) {
        rover = if(c<0) rover.left else rover.right
        if(rover!=null) c = comp(key,rover.key)
    }
    if(rover==null) None else Some(rover.data)
  }

  def iterator = new Iterator[(K,V)] {
    val stack = new ListStack[Node]
    pushRunLeft(root)
    def hasNext:Boolean = !stack.isEmpty
    def next:(K,V) = {
      val n = stack.pop()
      pushRunLeft(n.right)
      n.key -> n.data
    }
    @tailrec def pushRunLeft(n:Node) {
      if(n!=null) {
        stack.push(n)
        pushRunLeft(n.left)
      }
    }
  }
}

object TreeMap {
  def apply[K,V](data:(K,V)*)(comp:(K,K)=>Int):TreeMap[K,V] = {
    val tm = new TreeMap[K,V](comp)
    val d = data.sortWith((a,b) => comp(a._1,b._1)<0).toIndexedSeq
    def binaryAdd(start:Int,end:Int) {
      if(start<end) {
        val mid = (start+end)/2
        tm += d(mid)
        binaryAdd(start,mid)
        binaryAdd(mid+1,end)
      }
    }
    binaryAdd(0,data.length)
    tm
  }
}
```

The simplest method in this code is the `get` method. It makes a rover and a variable to store comparison results, then has a `while` loop that will go until the right key is found or the rover falls off the bottom of the tree. The variable `c` is used to store comparison results because the comp function might be expensive and we need to check for equality and at least one form of inequality each time through the loop. Storing the result in a variable reduces the number of calls by two and does not really hurt code readability.

The += method is the next simplest and is heavily based on the get method. It has a special case for adding to an empty tree, as that has to change the value of root instead of a left or right in a node. The other case contains a while loop that is much like the one in get with the exception that it keeps a reference called parent that follows the rover down the tree. This happens because in the case of a new key value, the rover will fall off the bottom of the tree. At that point, we need to know the parent to be able to add a node. This is much like removing from a singly-linked list, where we needed a reference to the node before the one we were removing. That could also be done with a trailing node equivalent to parent, but it was easier to simply check the next reference in the node. In this case, there are both left and right references so it is easier to keep use the trailing reference approach. After the loop breaks, there are three cases for the key being less than, greater than, or equal to the last node checked.

The condition where the key passed to += is the same as one currently in the tree brings up something that was avoided above, what happens when there are duplicate keys? The map ADT is designed to only hold one value for any given key. For that reason, the code here replaces what is stored in the node with what is passed in. If a BST is used in a situation where duplicate keys are allowed, such as a multimap, a different approach must be taken. One option would be to have the node store a list of values instead of a single one, other options, such as having equal values go left or right, can cause problems for some applications.

The iterator method is the next simplest. Any traversal would work, but there is a certain logic to using an in-order traversal as that traversal on a BST handles the nodes in sorted order by their keys. While it is simple enough to write a recursive traversal of a BST, that does not work well for the iterator, which needs to be able to provide elements one at a time instead of handling all elements as part of a single call. For this reason, our iterator needs to keep a stack to remember elements that have been passed over and not yet visited.

To get an in-order traversal, we repeatedly push all elements running down the left branch of any subtree. This is done by the pushRunLeft method. It is called on the root initially to set up the stack. Then each time a node is popped, that method is called on the left child of the node before the data element is returned. As long as the stack contains items, the iterator is not empty.

That leaves us with the -= method. The implementation shown here is not the most efficient, but it is far easier to understand than more efficient implementations. This version breaks the problem down a bit further and includes two nested functions that actually do the work. There are many possible special cases for removing from a BST. The overhead of these in the code is reduced by having the recursion set up so that values are assigned at each level.

The first recursive method is called findVictim and it is repeated here. It returns the node that the parent, or root, should reference. As the name implies, the primary purpose of this method is to find the key value that needs to be removed. The easy base case is when the function is called with a null node. That implies that the walk fell off the tree without finding the element so nothing needs to be deleted.

```
def findVictim(n:Node):Node = {
  if(n==null) null
  else {
    val c = comp(key,n.key)
```

```
    if(c==0) {
      if(n.left==null) n.right
      else if(n.right==null) n.left
      else {
        val (key,data,node) = deleteMaxChild(n.left)
        n.left = node
        n.key = key
        n.data = data
        n
      }
    } else if(c<0) {
      n.left = findVictim(n.left)
      n
    } else {
      n.right = findVictim(n.right)
      n
    }
  }
}
```

The second base case is when a matching key is found. If either the left or the right is `null`, the other one is returned so that it becomes the new child and the matching key is unlinked from the tree. The more challenging case is when the matching node has two children. In that case, the other method is invoked to find replacement data and remove the node it is in. The other two cases simply recurse to the left or right.

The second helper method also recurses on nodes going down the tree. In addition to returning the node that should be connected to by the parent, it also returns the key and data for the maximum key in that subtree. Note that the `findVictim` method calls this method on the left child.

```
def deleteMaxChild(n:Node):(K,V,Node) = {
  if(n.right==null) {
    (n.key,n.data,n.left)
  } else {
    val (key,data,node) = deleteMaxChild(n.right)
    n.right = node
    (key,data,n)
  }
}
```

The recursion here is much simpler because it simply runs down until it finds a node with no right child. The base case returns the key and data of that node as well as the left child which the parent of the node being removed should link to. As the stack pops back up, links are put back into place. The approach of making links while popping back up the stack does do many extra reference assignments as often the returned node was already the child. Without this though, the times when the child should be different becomes a special case and there are other challenges to building that code as well. Developing and testing such code is left as an exercise.

29.3.1 Order Analysis

So what is the order of the methods for searching, adding to, or removing from a BST? It turns out that depends a lot on the structure of the tree. To understand this, consider what happens if you add the values 0, 1, 2, 3, 4, 5, 6, 7, 8, and 9 to a BST in that order.

These are the same values that were used in the previous example, only now they are sorted. It should be clear that doing this produces a tree with 0 and the root and the other values running down the right. This is a tree that has the structure of a singly-linked list and the operations on it have the same performance as a singly-linked list. That is to say, the three main operations are all $O(n)$ for this tree.

The structure of a tree is often described as being balanced or unbalanced based on how similar the heights of left and right subtrees are. The original example tree shown in figure 29.5 is balanced. There is no configuration of nodes for those key values that has a lower total height and if you pick and node in the tree, the difference in height between the left and right subtrees is never more than one. On the other hand, the tree you get if those same key values are added in sorted order is completely unbalanced.

A balanced BST has $O(\log(n))$ performance for searching, insertion, and removal. This is a result of the fact that all of these operations run from the root down a single path through the tree. In a balanced tree, the number of nodes at a depth, d, scales as $O(2^d)$ so the total number depth for n nodes goes as $O(\log_2(n))$. This should not be surprising given the similarity to the binary search that was discussed earlier.

So an ideal tree gives $O(\log(n))$ behavior while the worst possible tree configuration is $O(n)$. These are called the BEST-CASE and WORST-CASE orders. Often we care more about the AVERAGE-CASE behavior of a system. If keys are added to a BST in random order, the resulting tree is fairly well balanced giving an average-case behavior of $O(\log(n))$. The average case has a higher coefficient than the best case, but it is still the same order. Unfortunately, the scenarios that produce worst-case performance in a BST are not all that uncommon. We have already seen that sorted, or nearly sorted data will lead to worst-case behavior. It is also possible to start with a fairly well-balanced tree and have it become unbalanced. For example, if the usage of a BST is such that old nodes are removed and new ones added, but new ones tend to be above or below the average of the old ones, the tree might start off balanced, but it will skew over time. A possible solution to this is described in chapter 35.

This discussion should make it clear why the file ends with a companion object that has an `apply` method to build a tree using a recursive decent of the sorted data. This insures that the tree that is produced is perfectly balanced. Of course, this comes with the overhead of having to sort the sequence and convert it to an indexed sequence. That conversion is needed to prevent the random access in the `binaryAdd` method from producing $O(n^2)$ behavior.

29.3.2 Immutable BSTs

Now you have seen how a BST can be used to implement a mutable map, what about an immutable one? To do this, one can use an approach very similar to what was done for the immutable singly-linked list. The type for the immutable map itself is abstract with redefinitions of methods that are required to make it so that the return a more specific type. There are subtypes for a node as well as for an empty tree. The latter is the equivalent of `Nil` for the `List` type. Here is a possible implementation of this type.

```
package scalabook.adt

abstract sealed class ImmutableTreeMap[K <% Ordered[K],+V] extends Map[K,V] {
  def +[B >: V](kv:(K,B)):ImmutableTreeMap[K,B]
  def -(k:K):ImmutableTreeMap[K,V]
}

private class Node[K <% Ordered[K],+V](private val key:K,
```

```
  private val data:V, private val left:ImmutableTreeMap[K,V],
  private val right:ImmutableTreeMap[K,V]) extends ImmutableTreeMap[K,V] {

def +[B >: V](kv:(K,B)) = {
  if(kv._1 == key) new Node(kv._1, kv._2, left,right)
  else if(kv._1 < key) new Node(key, data, left+kv, right)
  else new Node(key, data, left, right+kv)
}

def -(k:K) = {
  if(k == key) {
    left match {
      case _:Empty[K] => right
      case i:Node[K,V] => {
        right match {
          case _:Empty[K] => left
          case _ => {
            val (k, d, newLeft) = i.removeMax
            new Node(k, d, newLeft, right)
          }
        }
      }
    }
  } else if(k < key) {
    new Node(key, data, left-k, right)
  } else {
    new Node(key, data, left, right-k)
  }
}

def get(k:K):Option[V] = {
  if(k==key) Some(data)
  else if(k<key) left.get(k)
  else right.get(k)
}

def iterator = new Iterator[(K,V)] {
  val stack = new ListStack[Node[K,V]]
  pushRunLeft(Node.this)
  def hasNext:Boolean = !stack.isEmpty
  def next:(K,V) = {
    val n = stack.pop()
    pushRunLeft(n.right)
    n.key -> n.data
  }
  def pushRunLeft(n:ImmutableTreeMap[K,V]) {
    n match {
      case e:Empty[K] =>
      case i:Node[K,V] =>
        stack.push(i)
        pushRunLeft(i.left)
    }
  }
}
```

```
  private def removeMax:(K,V,ImmutableTreeMap[K,V]) = {
    right match {
      case e:Empty[K] => (key, data, left)
      case i:Node[K,V] =>
        val (k, d, r) = i.removeMax
        (k, d, new Node(key,data,left,r))
    }
  }
}

private class Empty[K <% Ordered[K]] extends ImmutableTreeMap[K,Nothing] {
  def +[B](kv:(K,B)) = {
    new Node(kv._1, kv._2, this, this)
  }

  def -(k:K) = {
    this
  }

  def get(k:K):Option[Nothing] = {
    None
  }

  def iterator = new Iterator[(K,Nothing)] {
    def hasNext = false
    def next = null
  }
}

object ImmutableTreeMap {
  def apply[K <% Ordered[K],V](data:(K,V)*)(comp:(K,K)=>Int):ImmutableTreeMap[K,V]
      = {
    val empty = new Empty[K]()
    val d = data.sortWith((a,b) => comp(a._1,b._1)<0)
    def binaryAdd(start:Int,end:Int):ImmutableTreeMap[K,V] = {
      if(start<end) {
        val mid = (start+end)/2
        new Node(d(mid)._1,d(mid)._2,binaryAdd(start,mid),binaryAdd(mid+1,end))
      } else empty
    }
    binaryAdd(0,data.length)
  }
}
```

Most of the code here is located in the Node class. This code is similar to what we had for the mutable version, but there are some differences worth noting. One big difference is that there are no loops in this code. Instead, everything is done with recursion. Also, instead of checks for null, there are matches that look for Empty. This tree does not contain null references. This use of match is the proper style in Scala for determining types. There are methods called isInstanceOf[A] and asInstanceOf[A], but their use is strongly discouraged as it commonly leads to runtime errors.

The Empty type is remarkably simple, but we make it anyway because it greatly simplifies the recursion and streamlines the code to avoid having the tree branches be null terminated. Unlike the List, where Nil could be an object, Empty has to be a class. This is because

`ImmutableTreeMap` is invariant on type parameter `K`. This is a concept that is covered in appendix B, but what matters here is that there is no type that we can pick that will work for all key types. This could lead to significant wasted memory if every leaf of every tree referred to a different instance of `Empty`. Indeed, in that situation, the empties would take nearly as much memory as the rest of the trees. However, as this code is written, every tree that is made with the `apply` method of the `ImmutableTreeMap` object makes a single `Empty` object and that is shared as the terminator for everything in that tree, even as more elements are added and the number of leaves grows. The fact that both `Node` and `Empty` are private means that the `apply` method is the only way to instantiate one of these trees.

Another difference between this tree and the last one is that this tree requires `K <: Ordered[K]` instead of passing in a comparator. This is because each node in an immutable tree is a fully functional and independent tree of its own. Using a comparator function would require that all nodes keep a reference to that function. That is an overhead that we would rather avoid if possible. This was not an issue for the mutable BST as many nodes existed in each tree and only the enclosing tree needed to keep track of the comparator.

The immutable tree does have somewhat higher overhead for modifying operations. However, the overall performance is still $O(\log(n))$. The difference is that any time nodes are added or changed, everything must be replaced on the path from that node to the root. That allows any earlier versions of the tree to remain intact. So the immutable BST not only does $O(\log(n))$ comparisons and memory reads for the `+` and `-` operations, it also creates $O(\log(n))$ new nodes for those operations. This is still far superior to the $O(n)$ nodes that have to be created for random modifications of immutable lists.

29.4 End of Chapter Material

29.4.1 Summary of Concepts

- A tree is a structure with nodes and edges that go from one node to another. All nodes have exactly one incoming edge except the root which has none.
  - The terminology for describing trees is a combination of genealogical and biological.
    - Parent node has an edge that refers to the child node.
    - Children of the same parent are called siblings.
    - Everything below a node are called descendants while those in line above are called ancestors.
    - Nodes with no children are called leaves.
    - Number of edges down from the root is called the depth.
    - Number of edges up from lowest leaf is called the height.
    - Number of descendants plus self is the size of a node.
  - There are multiple ways to traverse a tree.
    - Depth first traversals handle the children before they handle later siblings.
      - Work well with recursion.
      - Pre-order: Handle a node before the children.
      - Post-order: Handle a node after the children.

      · In-order: Handle a node between the children. This makes most sense for trees limited to two children per node.

    * Breadth-first traversal runs across an entire generation before going to the next one. This needs to be implemented with a loop and a queue.

  - A tree makes a convenient way of representing a parsed formula. The recursive parsing code written earlier can be easily modified to produce this structure. The tree structure allows much faster evaluation than reparsing the string.

- A binary search tree, BST, is an efficient structure for storing values when there is a complete ordering on some key type.

  - Each node in a BST can have a left child and/or a right child.
  - The key of the left child must be smaller than the key of the parent and the key of the right child must be greater than the key of the parent.
  - Getting, adding, and removing from a balanced tree can all be done in $O(\log(n))$ operations.
  - A standard BST can degrade to a linked list. Such highly unbalanced trees give $O(n)$ behavior.

29.4.2 Exercises

1. Test the relative speed of the recursive parsing evaluation method we wrote in the last chapter to that with the tree built in this chapter. Modify the code from the last chapter to include variables and call the function on a formula a sufficient number of times to see your machine pause. Repeat the same thing using the tree version from this chapter, but build one `Formula` object and do the evaluations with different variable values.

2. Add a method to the formula node that will return a simplified version of that node. Simplification should include the following:

   - Addition with zero simplifies to the other value.
   - Subtracting zero simplifies to the other value.
   - Multiplication by one simplifies to the other value.
   - Division by one simplifies to the other value.
   - Multiplication by zero simplifies to zero.
   - Division with a zero numerator simplifies to zero.
   - Any operation between two numbers can become a number with the result.

   Write the code recursively so that the simplification happens after recursion when the stack pops back up so that when a node is simplified, it takes into account the simplified children.

   For a bit of an extra challenge, you could try to identify other simplifications and implement them as well.

3. Add a method to the formula node that will return the derivative of that node. Given the operations we have, this should require little more than remembering the product rule and the quotient rule.

4. Combine exercises 2 and 3 so that you have a derivative method that simplifies the expression as far as possible.

5. Alter the `-=` method of the `TreeMap` so that it is not recursive and only reassigns references when a child node actually changes. Make sure to test your code to insure that you have dealt with all possible cases. Once you feel confident it works, compare the complexity of the code and the execution speed to the recursive solution.

29.4.3 Projects

1. If you are writing a MUD, you should use a BST-based map for finding rooms by unique names.

2. If you are writing the web spider, make a site tree and display it graphically. Allow the user to specify what page to use as the root. Link in other pages so that depth is the minimum click count from the root. If a page can be reached in the same click count from the root, favor the one that occurred earlier in the page of lowest depth.

3. If you are writing a graphical game, use a BST-based map as a way for you to quickly look up or jump to different elements of your game.

4. If you are doing the math worksheet, this chapter provides you with the ability to parse your formulas to an evaluation tree, which can make processing a lot faster. You can also use the tree as a means of doing symbolic manipulation. You will want to have simplification for this to work, but you can easily write code to take the derivative of a tree and produce a new tree. With a bit more effort, you can write code to do anti-derivatives/indefinite integrals for some functions. Note that not all functions have anti-derivatives with a closed form so some things simply won't work.

   If you are looking for even more of a challenge, try to do symbolic definite integrals and simplify them in a way that makes reasonable sense.

5. The expression trees can also be used to good effect in a Photoshop® program. You can give the user the ability to make custom filters by specifying a function that is used to combine the colors in pixels around the current pixel. Given the number of pixels in any real image, this would have been unacceptably slow with our earlier `String` parsing evaluation. The tree-based evaluation makes it feasible. You can decide what syntax you will have the user use to specify different adjacent pixel Red, Green, Blue (RGB) values.

6. Fast evaluation of user-entered formulas also provides a significant benefit for the simulation workbench project. You can use custom forces to specify force functions for particle interactions, or as part of random number generation for discrete event simulations.

Chapter 30

Regular Expressions and Context-Free Parsers

Parsing text is a task that comes up in a lot of different programs, whether it is pulling data out of text files, dealing with user commands, evaluating formulas, or understanding programming languages, some form of processing is typically needed when dealing with text values. Our approach to this so far has been fairly *ad hoc*. When faced with a particular format of input, we have written specialized code to handle it. The one exception to this was using the built-in parser to deal with files that were formatted in EXtensible Markup Language (XML). In this chapter we will learn about some of the formal approaches that computer scientists have developed to deal with parsing text. These will give you that ability to deal with more complex forms of text input in fairly easy, standard, and flexible ways.

30.1 Chomsky Grammars

Our discussion begins with the theoretical basis. The hierarchy of grammars that underpins what will be discussed in this chapter as well as a fair bit of the theory of languages in Computer Science appeared in an article written by the linguist Noam Chomsky and was co-developed by Marcel-Paul Schützenberger [4].

In a formal sense, a Chomsky grammar consists of four elements:

- a finite set of non-terminal symbols, typically denoted by capital letters,
- a finite set of terminal symbols, typically denoted by lowercase letters,
- a finite set of production rules, each of which has a left-hand side a right-hand side that are composed of those symbols,
- a special non-terminal denoted as the start symbol, often the character S.

Each grammar defines a language that is all the strings of terminal symbols that can be generated by that grammar. The way a string is generated is that a section of the string matching the left-hand side of a production is converted to the right-hand side of that same production. At any time, any production can be selected for any symbol or sequence of symbols in the current string. This process continues until the string we are left with contains only terminal symbols.

To illustrate this, consider the grammar that has the set S, A, B as non-terminal symbols, with S as the start symbol, a, b as terminal symbols, and the following productions.

$$\begin{aligned}S &\rightarrow AB\\ A &\rightarrow aA\\ A &\rightarrow a\\ B &\rightarrow bB\\ B &\rightarrow b\end{aligned}$$

To see what language this grammar generates, we can run through a series of productions to generate one string.

$$\begin{array}{ll} S & \text{apply } S \rightarrow AB\\ AB & \text{apply } A \rightarrow aA\\ aAB & \text{apply } A \rightarrow aA\\ aaAB & \text{apply } B \rightarrow bB\\ aaAbB & \text{apply } A \rightarrow a\\ aaabB & \text{apply } B \rightarrow b\\ aaabb & \end{array}$$

If you play with this grammar for a little while, you should be able to see that the language it generates is one or more a symbols followed by one or more b symbols.

For the sake of brevity, when the same left-hand side can produce multiple different values on the right-hand side, it is common to list them on a single line using a single pipe, $|$, between them. This can shorten the above set of productions down to the following form.

$$\begin{aligned}S &\rightarrow AB\\ A &\rightarrow aA \mid a\\ B &\rightarrow bB \mid b\end{aligned}$$

While this is a very simple example, hopefully it gives you the idea of how these systems work. Anyone who completed the L-System projects in chapters 14 and 15 will find the Chomsky grammars to be somewhat familiar. L-Systems are a different type of grammar. The rules are slightly different, but the general ideas are very similar.

Perhaps the most interesting thing about these grammars, is that by limiting the complexity of productions in certain ways, one can subdivide all possible grammars into a hierarchy where each level in the hierarchy is fundamentally more powerful than the one below it. There are four specific levels of capability in the Chomsky hierarchy. We will look briefly at each of these in the following four subsections starting with the least powerful set of grammars.

30.1.1 Regular Grammars

In a regular grammar, all productions must have the form of a single non-terminal going to a terminal ($A \rightarrow a$) or a single non-terminal going to a terminal and a non-terminal ($A \rightarrow aB$). With some slight modifications to the example grammar made above, we can produce a regular grammar that generates the same language.

$$\begin{aligned}
&S \rightarrow A \\
&A \rightarrow aA \mid B \\
&B \rightarrow bB \mid b
\end{aligned}$$

Run through some examples with this grammar to prove to yourself that it does generate the language of one or more a symbols followed by one or more b symbols.

Given these restrictions on the rules, the evolution of a string under a regular grammar is very straightforward. At any given time there is only one non-terminal, and it is the last symbol in the string. All productions either grow the length of the string by one, inserting a new terminal in front of a non-terminal, or they replace the one non-terminal by a terminal and finish the process.

Despite these limitations, regular grammars are still fairly powerful. The primary limitation on them is that they have no memory. For example, it is impossible to make a regular grammar that generates the language that has a certain number of a symbols followed by the same number of b symbols. We refer to this language as $a^n b^n$. This is due to the fact that the system has no memory of how many a symbols it has generated and can not enforce that the number of b symbols matches. This is true at least for an arbitrary number of symbols. By employing multiple non-terminals, it is possible to make a grammar that can match a and b symbols up to a certain number, but no finite regular grammar can do this for an arbitrary value of n.

30.1.2 Context-Free Grammars

The second rung up the Chomsky hierarchy is the context-free grammars. The productions in a context free grammar have a single non-terminal on the left-hand side and ant string of terminals and/or non-terminals on the right-hand side ($A \rightarrow \gamma$). The string on the right can potentially be the empty string, typically written as ϵ.

The fact that context-free grammars are fundamentally more powerful than regular grammars can be seen with the following example grammar. The non-terminal set includes only the start symbol, S. The terminal set is a, b. The productions are $S \rightarrow aSb \mid ab$. It should not take long to convince yourself that this grammar produces the language $a^n b^n$ for $n \geq 1$. Context-free grammars can do this because they effectively have a limited memory in the form of a stack. Due to the limitations of the stack, however, there is no context-free grammar that generates the language $a^n b^n c^n$. This is because the work of making sure that there are n b symbols consumes the contents of the stack and it is then impossible to produce the same number of c symbols.

The term context-free indicates the limitation of these grammars. While it is possible to have productions that produce whatever you want, the nature of the production can not depend on the context of the non-terminal that appears on the left-hand side. Despite this limitation, context-free grammars are remarkably useful in the field of Computer Science. The syntaxes of programming languages are typically specified as context-free grammars. In that situation, the symbols are the tokens of the language. The vast majority of sentences in natural language can also be generated by context-free grammars.

30.1.3 Context-Sensitive Grammars

The next step up the hierarchy removes the limitation that choice of production can not depend on what is around the non-terminal. The general form of the rule is $\alpha A\beta \rightarrow \alpha\gamma\beta$, where α, β, and γ are all arbitrary strings of terminals and non-terminals and are potentially empty. Note that the α and β are not really involved in the production. Instead, they simply limit when the production can be applied.

Context-sensitive grammars are significantly harder to work with than context-free grammars and the number of applications that require context-sensitivity is fairly limited. For that reason, we will not go into any detail on their usage.

30.1.4 Recursively Enumerable Grammars

The most powerful class of grammars in the Chomsky hierarchy is the set of recursively enumerable grammars. There are no restrictions on the productions of these grammars. You can have anything on the left-hand side produce whatever you want on the right-hand side. The truly remarkable thing about this class of grammars is that it is a full model of computation. Anything that is computable, in the theoretical sense of the Turing machine or the lambda calculus, can be computed with a properly designed recursively enumerable grammar.

This brief introduction to Chomsky grammars and the Chomsky hierarchy serves as a foundation for the rest of this chapter. However, it is a much broader topic and interested readers should do a bit more exploration of the topic. If nothing else, a proper course on the theory of Computer Science should cover these topics in significantly greater detail than has been presented here.

30.2 Regular Expressions

The theoretical underpinnings of Computer Science are extremely important for a variety of reasons, and they play a significant role in major developments in the field. However, the primary focus of this book is on the development of problem solving and programming skills. For that reason, we now turn our focus to the more practical applications of grammars and languages, beginning with regular expressions.

As the name implies, regular expressions are related to regular grammars. They provide a simple mechanism for matching and parsing text. As with regular grammars, they have their limitations. They are still quite powerful and for the problems they are good at, they can make life a whole lot easier.

Regular expressions are typically written as strings with a particular format where certain characters have special meaning. They are used by quite a few different tools, including some that you have been using for a while now. Both `grep` and the search feature of `vi` make use of regular expressions. Their tight integration into the programming language Perl was a significant part of why many people used that language for small text processing tasks. Even the `split` method we have been using on strings uses regular expressions.

While most of the aspects of regular expressions are fairly uniform, there are some features that differ from one implementation to another. Regular expressions in Scala are supported by the `scala.util.matching.Regex` class and companion object and they are built on top of the Java implementation in `java.util.regex`. The following subsections will go through a number of the major features of regular expressions. For a more definitive description of the features that are available, one can look at the Application Programming Interface (API) documentation for `java.util.regex.Pattern`. To turn a `String` into a `Regex` in Scala, simply call the `r` method.

```
scala> val str = "This is a string."
str: java.lang.String = This is a string.

scala> val regex = "This is a regular expression.".r
regex: scala.util.matching.Regex = This is a regular expression.
```

30.2.1 Characters and Character Classes

The reason that you have been able to use regular expressions in programs without even knowing you were using regular expressions is that most characters in regular expressions act just like themselves. So the regular expression "Fred" matches the string "Fred" and nothing else. However, imagine that you want to find all the method invocations in a Scala program that are called on objects ending with "ing" using `grep`. Here is some of that output from attempting this on the `Drawing.scala` file with the matching sections in bold.

```
  > grep ing. scalabook/drawing/Drawing.scala
import scala.swing._
import javax.swing.event._
* This type represents a Drawing that you can have open in the program.
class Drawing(private val root:DrawTransform,private[drawing]
val vars:mutable.Map[String,Double]) extends Serializable {
@transient private var lTree = new javax.swing.JTree(root)
root.drawing = this
if(lTree==null) lTree = new javax.swing.JTree(root)
private[drawing] def refresh {
private[drawing] def refreshTree {
case m:javax.swing.tree.DefaultTreeModel => m.reload
```

Clearly this finds everything that we wanted, but it is matching a lot of other things as well, things that do not include any period. This is a result of the fact that a period is a wild card that can match any character[1] in regular expressions. So what we really asked for is any sequence of four characters that starts with "ing". In order to force the period to actually match only a period, we need to put a backslash in front of it.[2] For all the special characters in regular expressions, you can force them to be treated as normal strings by putting a backslash in front of them. The fact that normal string literals in Scala also treat the backslash as an escape character becomes rather a pain when dealing with regular expressions. For this reason, it is standard to use the triple-quote string literals for regular expressions.

The period is actually one of many character classes in regular expressions. A character class matches one character as long as it comes from the right set. In the case of a period, the set is anything. You can also make your own character classes by putting characters you want to match in square brackets. For example, `[abc]` will match an 'a', a 'b', or a 'c'. A slightly more complex example is the regular expression `b[aei]d` which will match the

[1]In some usages the period will not match a newline.

[2]Like regular string literals in Scala, the Linux command-line treats a backslash specially. As such, you have to include two backslashes on the command-line for this to work.

strings “bad”, “bed”, or “bid”. For consecutive characters in ASCII, you can use a hyphen. So [0-9] matches any digit and [a-zA-Z] matches any letter, upper of lowercase. If you put a ^ as the first character in the square brackets, the character class matches anything except what is specified. So [^aeiou] will match anything that is not a lowercase vowel. Of course, this means that square brackets are special characters in regular expressions and much be escaped if you want to use them for purposes other than defining a character class. The same is true for the caret in certain usages and for the minus sign inside of a character class.

Some character classes are used so frequently that there are shortcut names for them. In addition to the period, the following are defined in the Java regular expression library.

- \d - a digit, same as [0-9]
- \D - not a digit, same as [^0-9]
- \s - a whitespace character
- \S - a non-whitespace character
- \w - a word character, same as [a-zA-Z0-9]
- \W - not a word character, same as [^a-zA-Z0-9]

If you spend much time working with regular expressions these character classes will likely be committed to memory.

30.2.2 Logical Operators and Capturing Groups

You have already seen that characters or character groups that are adjacent to one another match consecutive characters. So the regular expression cat can be read as 'c' and 'a' and 't'. Character classes can be used in any place that would take an individual character as well. If adjacency is like “and”, how does one say “or”? With a single pipe, |?[3] So the regular expression cat|dog will match either “cat” or “dog”. As with logical expressions, the “and” operation has higher precedence than the “or” does.

Sections of a regular expression can also be grouped using parentheses. This gives you a way to control the parts of the regular expression you are taking the “or” of, such as a(bc|de)f, it lets you bundle sections together for the quantifiers discussed in the next subsection, and it allows you to capture sections of the match. When you put parentheses in a regular expression, they define a capturing group. The code keeps track of all of the sections of the string that wind up being in parentheses so that you can pull them back out. These groups are numbered starting at one and they are ordered by the location of the opening parentheses. So it does not matter how log groups are or how they are nested, it is only the location of the opening parentheses that matters.

Consider the simple example of a phone number in the format “(xxx) yyy-zzzz” where you want to capture each of the groups denoted by different letters and capture the standard 7-digit part of the number independently of the area code. To do this, you could use the regular expression \((\d\d\d)\) ((\d\d\d)-(\d\d\d\d)). This regular expression has four capturing groups in it numbered 1 through 4. The first one is the area code. The second one is the 7-digit phone number. The third and fourth groups are the subparts of the 7-digit number. They are numbered by the opening parentheses. Here is that same regular expression, with subscripts to make that clear, \(($_1$\d\d\d)\) ($_2$($_3$\d\d\d)-($_4$\d\d\d\d)). This

[3] Hopefully the reader appreciates the fact that there is some similarity in the symbols used for expressing similar ideas across different tools.

example regular expression also illustrates that because parentheses are special characters used for grouping in regular expressions, they have to be escaped if you want to match one.

30.2.3 Greedy Quantifiers

The real power of regular expressions comes in with quantifiers. Quantifiers allow you to have some control over the number of times an expression will occur. The default quantifiers are called "greedy" quantifiers because they will match as many characters as possible while still allowing the whole pattern to be matched. There are six options for the greedy quantifiers.

- X? - Matches the pattern X or nothing. Think of this as being 0 or 1 times.
- X* - Matches nothing or the pattern X repeated any number of times. Think of this as 0 or more times.
- X+ - Matches the pattern X 1 or more times.
- $X\{n\}$ - Matches the pattern X repeated exactly n times.
- $X\{n,\}$ - Matches the pattern X repeated n or more times.
- $X\{n,m\}$ - Matches the pattern X repeated at least n times, but not more than m times.

The regular expression for the phone number could have been expressed with quantifiers as `\((\d{3})\) ((\d{3})-(\d{4}))`. These quantifiers also allow us to express the languages for the grammars described in subsection 30.1.1. For example `a+b+` matches one or more "a"s followed by one or more "b"s. Note that there is not a good way, using these quantifiers, to allow the number of characters of each type to match without specifying how many times they occur.

This should now make it clear why the argument passed to split when we wanted to break a string on spaces was `" +"`. The `+` here is a quantifier so that this matches one or more spaces. Just in case the user also used tabs, it might be safer to use `"\s+"`.

There are also options for quantifiers that are not greedy. The interested reader is directed to the Java API and other sources for information on reluctant and possessive quantifiers.

30.2.4 Boundary Requirements

The last option of regular expressions that we will consider here is boundary requirements. In the general usage, regular expressions can be used to match any part of a string and can be used to do multiple matches inside of a single string. There are situations where you want to restrict the generality of the matches. For this you can put characters at the beginning of the end of the regular expression to specify where it needs to begin or end. The following as some of the allowed options.

- `^` - As the first character in a regular expressions it requires the match to start at the beginning of a line.
- `$` - As the last character in a regular expression it requires the match to end at the end of a line.
- `\b` - Can be placed at the beginning or end of a regular expression forcing the match to start or end on a word boundary.

- `\B` - Like `\b` except it forces a non-word boundary.
- `\A` - As the first character in a regular expressions it requires the match to start at the beginning of the input.
- `\G` - As the last character in a regular expression it requires the match to end at the end of the previous match.
- `\z` - As the last character in a regular expression it requires the match to end at the end of the input.

So if you want to match words that have four characters in them you could use the regular expression `\b\w{4}\b`.

30.2.5 Using Regular Expressions in Code

Now that you know the basics of regular expressions we can see how they are used in Scala. We already saw that you can produce a `scala.util.matching.Regex` by calling the `r` method on a string and that triple-quote strings are particularly useful for defining regular expressions because they do not handle the backslash as an escape character. The next question is what can you do with one of these `Regex` objects.

The two most useful methods in the `Regex` class are `def findAllIn(source: CharSequence): MatchIterator` and `def replaceAllIn(target: CharSequence, replacer: (Match) => String): String`. The type `CharSequence` is a suptertype of `String` that is more flexible. For our examples we will just be using normal strings. The `Match` and `MatchIterator` types are declared in the `Regex` companion object. A `Match` object defines basic information from a single match of the regular expression to a string. The API can give you full details, but one of the things you will want to do a lot is find the parts of the string that were in the different capturing groups. This can be done with the `def group(i: Int): String` method. The argument is the number of the group you want the text for. If nothing matched that group then the method returns `null`.

The `MatchIterator` extends `Iterator[String]` with `MatchData`. It also has a method called `matchData` that returns an `Iterator[Match]`. Putting these things together we can take a block of text that has several embedded phone numbers along with the regular expression we made above and pull out just the parts of the phone numbers.

```
scala> val phoneNumbers = """For help you can try to following numbers:
     | Main line: (123) 555-4567
     | Secondary line: (123) 555-7022
     | Fax: (123) 555-5847"""
phoneNumbers: java.lang.String =
For help you can try to following numbers:
Main line: (123) 555-4567
Secondary line: (123) 555-7022
Fax: (123) 555-5847

scala> val phoneRegex = """\((\d{3})\) ((\d{3})-(\d{4}))""".r
PhoneRegEx: scala.util.matching.Regex = \((\d{3})\) ((\d{3})-(\d{4}))

scala> for(m <- phoneRegex.findAllIn(phoneNumbers).matchData) {
     | println(m.group(1)+" "+m.group(3)+" "+m.group(4))
     | }
123 555 4567
123 555 7022
```

```
123 555 5847
```

The `for` loop runs through the `Matches` produced by calling `MatchData` on the results of `findAllIn`. The groups with indexes 1, 3, and 4 are printed from each match.

Hopefully it is clear how the use of regular expressions made this mode much easier to write. Finding the phone numbers embedded in that text would have been rather challenging using other techniques that we have covered. Phone numbers have a nice, well-defined format to them, but it is not a specific string sequence that we can search for.

The `replaceAllIn` method has a second argument that is a function which takes a `Match` and returns a `String` that should replace the match in the first argument. We could use this method to the phone numbers, replacing their last four digits with the character `'x'` using the following line.

```
scala> phoneRegex.replaceAllIn(phoneNumbers,m => {
     | "("+m.group(1)+") "+m.group(3)+"-xxxx"
     | })
res2: String =
For help you can try to following numbers:
Main line: (123) 555-xxxx
Secondary line: (123) 555-xxxx
Fax: (123) 555-xxxx
```

Passing in the `Match` gives you the ability to use the details of the match in determining the replacement string. In this case, we wanted to preserve the first six digits so groups 1 and 3 were used in the result. Again, if you think for a while about what would have been required to do this transformation using other techniques, you should realize that power provided by regular expressions.

There is one other approach to using regular expressions in Scala: patterns. We have previously seen patterns used with tuples, arrays, lists, and XML. They work with regular expressions as well. When a regular expression is used as a match on a string, it pulls out the groups and only works if there is a match for the whole string. The following code can be put into the `Commands.scala` file in place of more traditional string parsing code that we had before that split on the first space.

```
val CommandSplit = """\s*(\w+)(\s+(.*))?\s*""".r

def apply(input:String,drawing:Drawing):Any = {
  val CommandSplit(command,_,rest) = input
  if(commands.contains(command)) commands(command)(rest,drawing)
    else "Not a valid command."
}
```

This does what we did previously with a call to `indexOf`, but this code is a bit more robust in how it handles whitespace, particularly leading whitespace. With the old code, if the user put spaces in front of the command, it would fail. This fixes that problem by leading with `\s*`.

You might notice something else odd about this code. The variable `CommandSplit` starts with a capital letter. This is contrary to standard naming practices on how everything has been named in this text. The reason for this is the use in the pattern. Recall that names beginning with lowercase letters in patterns are assumed to be variable names that should be bound to values in the match. This is the role played by `command` and `rest` in this example and that behavior is essential to this working. However, we want `CommandSplit` to refer to the variable declared above the method. We want to do this using a variable with a name that starts with a lowercase letter, we have to use backticks in the pattern like this.

```
val `commandSplit`(command,_,rest) = input
```

The choice to make variables used in patterns start with capital letters is simply one of style. You can decide which style you prefer.

The place where regular expression patterns are probably most useful is in `for` loops. Recall that what you type before `<-` in a `for` loop is matched as a `val` pattern. Anything that does not match is skipped. This is remarkably helpful for regular expressions if you are running through a collection of strings where some might not match what you are looking for.

To see this, consider having a file where some of the lines are numbered and others are not. You only care about the ones that are numbered and they all have, for the first non-space characters, digits followed by a period. After the period is text you care about. You want to built a `Map[Int,String]` that lets you quickly look up a string based on the number of its line. The following code will do this for you.

```
val NumberedLine = """\s*(\d+)\.(.+)""".r
val source = Source.fromFile(fileName)
val lines = source.getLines
val numberedLines = (for(NumberedList(num,text) <- lines) yield {
  num -> text
}).toMap
source.close
```

The `for` loop will only execute on lines matching the pattern and skip anything that does not match. This produces compact and fairly easy to maintain code.

30.2.6 Drawback of Regular Expressions

The primary drawback of regular expressions is that they can be cryptic and somewhat difficult to maintain. Indeed, languages that highlight the use of regular expressions are often derided for their lack of readability. Coming up with exactly the right regular expression for a problem can also be quite challenging. This is expressed in this oft-repeated quote.

> Some people, when confronted with a problem, think
> "I know, I'll use regular expressions." Now they have two problems.
>
> - Jamie Zawinski (1997 Usenet post)

There is inevitably some truth to this quote. However, hopefully the examples above have shown you that in situations where regular expressions work well, they can make life much easier. Just do not try to use them to solve every problem you encounter.

30.3 Context-Free Parsers

What options do you have if you need to parse text that is beyond the power of regular expressions? Scala includes the package `scala.util.parsing` which has a number of different subpackages that provide different parsing capabilities. We will focus on the `scala.util.parsing.combinator` package which gives us the ability to parse context-free grammars.[4]

The introduction to context-free, CF, grammars given in subsection 30.1.2 used a notation fitting with theoretical Computer Science. Like mathematicians, theoretical computer scientists are quite happy to name things using single letters or symbols. In practical application, it is typically helpful to use longer, more informative names. The symbology is also altered a bit when expressing CF grammars for things like programming languages. To illustrate this, we will go back to the example of formula parsing that we have utilized in the last two chapters. The CF grammar for a basic mathematical expression using only the four primary operators can be written this way.

$$form ::= term\ \{(" + " \mid " - ")\ term\}$$
$$term ::= factor\ \{(" * " \mid "/")\ factor\}$$
$$factor ::= floatingPointNumber \mid "(" \ form\ ")"$$

In this notation, the curly braces indicate elements that can be repeated zero or more times. It is also possible to use square brackets to indicate things that are optional.

This grammar says that a *formula* is a *term* followed by zero or more + or − signs followed by terms. The *term* is a *factor* followed by factors with ∗ or / between them. Lastly, the *factor* can be a number or a full formula in parentheses. Play with this to convince yourself that it can represent the full range of possibilities that are desired. It is also worth noting that this grammar does proper order of operations and `*` and `/` are bound a level below the `+` and `-`.

Using the combinator parser library, this converts fairly directly to the following Scala code.

```
object FormulaP extends JavaTokenParsers {
  def form:Parser[Any] = term ~ rep(("+" | "-") ~ term)
  def term:Parser[Any] = factor ~ rep(("*" | "/") ~ factor)
  def factor:Parser[Any] = floatingPointNumber | "(" ~ form ~ ")"
}
```

Each production in the original grammar becomes a method that returns a `Parser`, a class found insider `scala.util.parser.combinator.Parsers`. Currently this is a `Parser[Any]` to be completely general. Consecutive symbols in the original grammar are joined using the operator. The curly braces are replaced with a call to `rep`. Had there been square brackets, they would have been replaced with a call to `opt`. There is also a method called `repsep` that can be used for alternating productions. This includes things like comma-separated lists.

Using these rules, you can take any CF grammar that you might be interested in using and build a parser for it with little effort. You can use any of these parsers by calling the `parseAll` method that `FormulaP` gets via `JavaTokenParsers` in this example.[5] The following main will do just that and print the value output produced by the parser.

```
def main(args:Array[String]) {
  println(parseAll(form,"3+5*2").get)
}
```

[4]Technically there are limitations on the grammars that can be parsed with these parsers, but it is not a limitation on power. Grammars just have to be written in a way that facilitates the parsing. The details of this are beyond the scope of this book.

[5]This example uses `JavaTokenParsers` because that is where `floatingPointNumber` is defined. It also defines `ident`, which we will use later for identifier names.

30.3.1 Default Output

If you run that main method you get the following output.

```
((3~List())~List((+~(5~List((*~2))))))
```

The string parses, but the default output is a bit challenging to understand. To do so, we need to know what the default output of different parser elements are.

- String or regular expression - Gives back the String that was matched.
- P~Q - Gives back ~(p,q), where p is the match for P and q is the match for Q. Note that ~(p,q) prints out as p~q.
- rep(P) - Give back lists of the matches: List(p1,p2,p3,....
- repsep(P,Q) - Give back lists of the matches: List(p1,p2,p3,.... Note that while Q can be a full parser, the output is ignored. This is why it is best used for things like comma-separated lists where you do not care about the separators.
- opt(P) - Gives back an Option on the match. If that optional part was not used you get None. Otherwise you get Some(p)

Using this information and the original grammar, you can figure out how this parser produced the output that it did.

30.3.2 Specified Output

While the default output provides the full information from the parse, and you could run through it using match statements to extract the desired information, it is often more useful to specify your own output from the parser. This is done using the ^^ method. The signature of ^^ in Parser[+T] is def ^^[U](f: (T) => U): Parser[U]. So it takes a function that converts from the normal output of this parser to some other type, U and gives back a Parser[U].

We can utilize this in the parser we wrote above to make it so that it outputs a numeric solution instead of the default parse output we had above. The first step in doing this is to change the return types of the various productions to Parser[Double]. That will introduce errors that we can fix by introducing the appropriate transformations. The resulting code looks like this.

```
object FormulaP extends JavaTokenParsers {
  def form:Parser[Double] = term ~ rep(("+" | "-") ~ term) ^^ {
    case d ~ lst => lst.foldLeft(d)((n,t) => if(t._1=="+") n+t._2 else n-t._2)
  }
  def term:Parser[Double] = factor ~ rep(("*" | "/") ~ factor) ^^ {
    case d ~ lst => lst.foldLeft(d)((n,t) => if(t._1=="*") n*t._2 else n/t._2)
  }
  def factor:Parser[Double] = floatingPointNumber ^^ (_.toDouble) | "("~> form <~")"

  def main(args:Array[String]) {
    println(parseAll(form,"3+5*2").get)
  }
}
```

The simplest application of ^^ in this example is after floatingPointNumber. That is a Parser[String] so the normal output is just a String that we can convert to a Double with a call to `toDouble`. The other two productions follow the `^^` with partial functions that get converted to full functions for us. Each of these has a case that match a number followed by a tilde and a list. The number will be the result of the first `term` or `factor`. The list will be the result of `rep`, which is a `List[ [String,Double]]`. To get an answer, we need to run through the list accumulating a value. At each step, the proper operation should be applied to the current value and the parse of what follows the operator. This can be done in short form using `foldLeft`. Note that the ~ type works with infix pattern matching and has _1 and _2 methods like a tuple for getting to the first and second elements.

The parser for `factor` throws in something else to keep this code compact. It uses operators where a `<` or `>` is put before or after the ~. These operators give you a shorthand way of ignoring parts of the match in the output. The direction the arrow points is toward the thing you actually care about. Without these, we would have had to follow that last case with a `^^` operator and a function that pulls off the parentheses and discards them.

With these changes, you can run this program and see that now the output is `13.0`. Compare this to our original recursive parser that worked with these same operations and you will find that this is significantly shorter. However, this version is also completely opaque to anyone who is not familiar with CF grammars or combinator parsers.

There are, of course, many other options and possibilities with combinator parsers. Looking through the API at classes such as `scala.util.parser.combinator.Parsers` and `scala.util.parser.combinator.Parsers.Parser` can give you a quick view of some of the possibilities.

30.4 Project Integration

Now it is time to put everything together, including some concepts from the last chapter. The `FormulaP` class can be expanded to parse to a tree instead of a `Double`. This will allow the use of variables with good speed. To show the true power of the combinator parsers, this implementation will include not only the four basic math operations and a few extra functions, it will also include exponentiation and an `if` conditional. These two features present interesting challenges. Exponentiation is left associative while the other operators are right associative. In order to have an `if`, we have to have comparison operators that give back Booleans. It would also be nice to have the standard Boolean operators as well. All of this needs to be integrated into a system that preserves proper order of operation.

To start off, we need to develop a CF grammar that will parse this little language that we are trying to implement. Here is the one we will use in the code.[6]

$$
\begin{array}{l}
cond ::= \text{"}if(\text{"}\ bform\ \text{")"}\ cond\ \text{"}else\text{"}\ cond \mid form \\
form ::= term\ \{(\text{"} + \text{"} \mid \text{"} - \text{"})\ term\} \\
term ::= exp\ \{(\text{"} * \text{"} \mid \text{"}/\text{"})\ exp\} \\
exp ::= func\ \{\text{""}\ func\} \\
func ::= \text{"}sin\text{"} \mid \text{"}cos\text{"} \mid \text{"}tan\text{"} \mid \text{"}sqrt\text{"}\ \text{"("}cond\text{")"} \mid factor \\
factor ::= floatingPointNumber \mid \text{"("}\ form\ \text{")"} \\
bform ::= bterm\ \{\text{"}\|\text{"}\ bterm\}
\end{array}
$$

[6]This grammar includes some of those elements that go beyond the scope of this book. For example, if you change the first production to $cond ::= form \mid \text{"}if(\text{"}\ bform\ \text{")"}\ cond\ \text{"}else\text{"}\ cond$ it runs into problems because the string "if" is a valid identifier which matches $form$.

$bterm ::= bnot \{"\&\&"\ bnot\}$
$bnot ::= "!("bform")" \mid bcomp$
$bcomp ::= cond\ ("==" \mid "!=" \mid "<=" \mid ">=" \mid "<" \mid ">")\ cond$

We convert this grammar into code, add some `Node` traits and classes, then make the parsers output nodes. The result is the following file.

```
package scalabook.util

import scala.util.parsing.combinator._

class FormulaP(val formula:String) {
  private val root = FormulaP parseAll(FormulaP.cond,formula) get

  def eval(vars:collection.Map[String,Double]) = root eval vars
}

object FormulaP extends JavaTokenParsers {
  def apply(f:String) = new FormulaP(f)

  def eval(f:String, vars:collection.Map[String,Double] = null) =
    new FormulaP(f) eval vars

  private def cond:Parser[DNode] = "if(" ~> bform ~ """\)\s*""".r ~ cond ~
      """\s*else\s*""".r ~ cond ^^ {
        case b ~ _ ~ e1 ~ _ ~ e2 => new IfNode(b,e1,e2)
      } | form
  private def form:Parser[DNode] = term ~ rep(("+" | "-") ~ term) ^^ {
    case d ~ lst => new LeftAssocBinaryOpDNode(d,lst)
  }
  private def term:Parser[DNode] = exp ~ rep(("*" | "/") ~ exp) ^^ {
    case d ~ lst => new LeftAssocBinaryOpDNode(d,lst)
  }
  private def exp:Parser[DNode] = func ~ rep("^" ~> func) ^^ {
    case d ~ lst => new PowBinaryOpDNode(d,lst)
  }
  private def func:Parser[DNode] = """(sin|cos|tan|sqrt)\(""".r ~ cond <~ ")" ^^ {
    case f ~ n => new FunctionDNode(f,n)
  } | factor
  private def factor:Parser[DNode] = floatingPointNumber ^^ (s => new
      NumNode(s.toDouble)) |
    ident ^^ (s => new VarNode(s)) | "(" ~> cond <~ ")"

  private def bform:Parser[BNode] = bterm ~ rep("||" ~> bterm) ^^ {
    case b ~ lst => new LeftAssocBinaryOpBNode(b,lst,_ || _)
  }
  private def bterm:Parser[BNode] = bnot ~ rep("&&" ~> bnot) ^^ {
    case b ~ lst => new LeftAssocBinaryOpBNode(b,lst,_ && _)
  }
  private def bnot:Parser[BNode] = "!(" ~> bform <~ ")" ^^ (b => new BNotNode(b)) |
      bcomp
  private def bcomp:Parser[BNode] = cond ~ ("""[=!><]=|<|>""".r) ~ cond ^^ {
    case c1 ~ op ~ c2 => new CompNode(c1,op,c2)
  }
```

```
private trait DNode {
  def eval(vars:collection.Map[String,Double]):Double
}

private trait BNode {
  def eval(vars:collection.Map[String,Double]):Boolean
}

private class LeftAssocBinaryOpDNode(first:DNode, restStr:List[~[String,DNode]])
     extends DNode {
  val rest = for(~(op,n) <- restStr) yield (op match {
    case "+" => (_:Double)+(_:Double)
    case "-" => (_:Double)-(_:Double)
    case "*" => (_:Double)*(_:Double)
    case "/" => (_:Double)/(_:Double)
  }, n)
  def eval(vars:collection.Map[String,Double]):Double =
    rest.foldLeft(first eval vars)((d,t) => {
      t._1(d,t._2 eval vars)
    })
}

private class PowBinaryOpDNode(first:DNode, rest:List[DNode]) extends DNode {
  def eval(vars:collection.Map[String,Double]):Double =
    math.pow(first eval vars, rest.foldRight(1.0)((n,d) => math.pow(n eval
         vars,d)))
}

private class NumNode(num:Double) extends DNode {
  def eval(vars:collection.Map[String,Double]):Double = num
}

private class VarNode(name:String) extends DNode {
  def eval(vars:collection.Map[String,Double]):Double = vars(name)
}

private class FunctionDNode(name:String, arg:DNode) extends DNode {
  val f:Double=>Double = name match {
    case "sin(" => math.sin
    case "cos(" => math.cos
    case "tan(" => math.tan
    case "sqrt(" => math.sqrt
  }
  def eval(vars:collection.Map[String,Double]):Double = f(arg eval vars)
}

private class IfNode(cond:BNode, e1:DNode, e2:DNode) extends DNode {
  def eval(vars:collection.Map[String,Double]):Double =
    if(cond eval vars) e1 eval vars else e2 eval vars
}

private class LeftAssocBinaryOpBNode(first:BNode, rest:List[BNode],
    op:(Boolean,Boolean)=>Boolean) extends BNode {
  def eval(vars:collection.Map[String,Double]):Boolean =
    rest.foldLeft(first eval vars )((tf,b) => op(tf,b eval vars ))
```

```
  }

  private class BNotNode(arg:BNode) extends BNode {
    def eval(vars:collection.Map[String,Double]):Boolean = !(arg eval vars)
  }

  private class CompNode(left:DNode, compStr:String, right:DNode) extends BNode {
    val comp:(Double,Double)=>Boolean = compStr match {
      case "<" => _<_
      case ">" => _>_
      case "<=" => _<=_
      case ">=" => _>=_
      case "==" => _==_
      case "!=" => _!=_
    }
    def eval(vars:collection.Map[String,Double]):Boolean =
      comp(left eval vars,right eval vars)
  }
}
```

One of the first things to note about this code is that regular expressions are used in the parsers to keep things short and flexible. The other thing to note is that there are two top level node traits called `DNode` and `BNode`. These both have an `eval` method, but they differ in that they return a `Double` and a `Boolean`, respectively.

The binary operator nodes are a bit more complex than their counterparts from chapter 29. This is because these have the ability to take many operators that are at the same precedence level. Despite some parts of it being complex, the entire file and code for this is under 120 lines of code and it contains a significant amount of functionality. What is more, it is fairly easy to extent, particularly for adding single operator functions. This `FormulaP` class can now be substituted for the `Formula` class developed in the last chapter.

30.5 End of Chapter Material

30.5.1 Summary of Concepts

- Chomsky Grammars
  - Formal approach to generating languages and parsing languages. Form a hierarchy with four different levels of complexity. A grammar is defined as a set of terminals, a set of non-terminals, a set of productions, and a start symbol.
  - Regular grammars are the lowest level. Productions restricted to the form $A \rightarrow a \mid aB$.
  - Context-free grammars are next up the level of complexity. Their productions have to be of the form $A \rightarrow \gamma$.
  - Context-sensitive grammars can take the symbols around a non-terminal into account when determining if a production is allowed. The productions must have the form $\alpha A\beta \rightarrow \alpha\gamma\beta$.
  - Recursively enumerable grammars have no restriction on their productions. They are a complete model of computation.

- Regular expressions are a syntax for string parsing that have a power roughly equal to regular grammars.
  - Character classes can represent a set of characters that can be matched. You can build character classes with square brackets. There are also a number of built-in character classes for common used sets of characters.
  - Quantifiers specify that a character of grouping can be present certain numbers of times. Quantifiers include `?`, `*`, and `+`.
  - Strings can be turned into `Regex` using the `r` method. The `Regex` object has methods called `findAllIn` and `replaceAllIn`. They can also be used at patterns.
- Context-Free Parsers
  - The combinator parser library in Scala makes it easy to build a parser from a CF grammar.
  - Default output of parsers include a mix or strings, `~` objects, and lists. This can be challenging to work with.
  - Using the `^^` operator you can force parsers to output specific types that you want.

30.5.2 Exercises

1. Write a simple program that will take a Chomsky grammar and allow the user to generate strings from the language.
2. For each of the following grammars, try to figure out what languages they generate. For those that are regular, create an appropriate regular expression. For that that are context free, write appropriate parsing code.
   - $S \rightarrow A$
$A \rightarrow aA|B$
$B \rightarrow bB|b$
   - $S \rightarrow aA$
$A \rightarrow aA|bC$
$C \rightarrow cC|c$
   - $S \rightarrow ABC$
$A \rightarrow aA|a$
$B \rightarrow bB|b$
$c \rightarrow cC|c$
   - $S \rightarrow AB$
$A \rightarrow aA|a$
$B \rightarrow AbB|b$
   - $S \rightarrow eLAe$
$eLA \rightarrow eRA$
$ALA \rightarrow AAL$
$ARe \rightarrow ALAe$
$ARA \rightarrow RAA$
3. Try to come up with, or look up, a simple context-free grammar for English. Write a parser for it and include a simple dictionary.

4. Write a grammar for roman numerals. Have it parse to a tree such that the output has an `eval` method that will return the numeric value as an `Int`.

30.5.3 Projects

1. If you have been working on the MUD project, this chapter gives you the ability to use commands with significantly greater complexity. This can be done with either regular expressions or a parser. You can decide which you want to use, but the suggestion would be to write a simple grammar for more complex commands that you could put into your MUD and then write a parser that will take those and build usable expression trees from them.

2. Regular expressions and parsers can be extremely useful in pulling data out of text files for the web spider. You likely wrote up a fair bit of *ad-hoc* code for getting the data you want out of HTML files. Convert that code to using a parser with regular expressions.

3. Both the networked and non-networked graphical game programs avoid having significant text input. However, in both cases, there are often reasons when it is nice to be able to give commands to the program that go beyond what a normal user should be able to do in order to help with testing and debugging.

   For this project, you should add a small command language to your game. If there is a good reason to do this for normal users, feel free to do so. Networked games could integrate this with chatting features. Non-networked games could put it into the Graphical User Interface (GUI), or you could use the logging socket that was added back in chapter 23.

4. The material from this chapter can be particularly beneficial to the math worksheet project. Using combinatorial parsers, you can create a more complex, powerful formula parser, introduce additional commands with relative ease, and even consider putting in a simple programming language. You can also include conditionals expressions in your parser so that your mathematical functions can use recursive definitions.

   For all of these additions, you should have the parser build an expression tree for faster evaluation. Add whatever set of features you feel takes the project in the direction you want to go.

5. Photoshop® is another project that included mathematical formulas, and which can benefit from having commands in a simple language. Given what you learned in this chapter, you should be able to add much more powerful parsing to your formulas that are used with filters.

6. The simulation workbench project will also benefit from having real formulas. All of the applications of formulas from chapter 29 apply when using a combinatorial parser as well. In addition, the combinatorial parser makes to reasonable to extend formulas into more of a real language.

Chapter 31

Spatial Trees

A lot of the data that you deal with in certain fields of computing is spatial. There are certain data structures that you can use to make the process of finding elements in certain parts of space more efficient. These spatial data structures can increase the speed of certain operations dramatically, especially when there are large numbers of elements involved. This chapter will consider a few of the many options for doing this.

31.1 Spatial Data and Grids

Examples of spatial data can be bodies in a physics simulation or characters spread out across a large world in a game. The way these bodies or agents behave is generally influenced by the other things that are near them. This makes the problem of finding nearby neighbors an important one. To illustrate the nature of this problem, let us consider a simplified example. You are given N points in some M dimensional space. For visualization purposes, we will typically treat $M = 2$, but it could be higher. We want to find all pairs of points (p_1, p_2) such that the distance between them is less than a given threshold, $|p_2 - p_1| < T$.

The easiest-to-write solution to this problem is a brute force approach that runs through all particles in one loop, with a nested loop that runs through all the particles after that one in the sequence and calculates and compares distances. If we model the points as anything that can be converted to `Int => Double` we can write a class that serves as the interface for all the approaches to basic neighbor visits that we will write.[1]

```
package scalabook.adt

abstract class NeighborVisitor[A <% Int => Double](val dim:Int) {
  def visitAllNeighbors(tDist:Double, visit:(A,A) => Unit)
  def visitNeighbors(i:Int, tDist:Double, visit:(A,A) => Unit)
```

[1]This is written as a class instead of a trait to keep the code simpler. The `<%` symbol gets expanded out to include an implicit argument. As traits can not take arguments, we would have to get into the more complex details of this in order to use a trait. See appendix B for details.

```
  var distCalcs = 0
  def dist[A <% Int => Double](p1:A, p2:A):Double = {
    distCalcs += 1
    math.sqrt((0 until dim).foldLeft(0.0)((d,i) => {
      val di = p1(i)-p2(i)
      d+di*di
    }))
  }
}
```

This class has a dist method as all of its subtypes will need to perform this calculation. It also takes an argument for the dimensionality of the space the points should be considered for. There are two methods defined here. One of them does a complete search across all the pairs of particles. The other does a search relative to just one point. The member named distCalcs is there simply for the purposes of benchmarking our different methods to see how many times they have to calculate the distance between points.

The brute force implementation of this looks like the following.

```
package scalabook.adt

class BruteForceNeighborVisitor[A <% Int => Double](
    d:Int,
    val p:IndexedSeq[A]
    ) extends NeighborVisitor[A](d) {

  def visitAllNeighbors(tDist:Double, visit:(A,A) => Unit) {
    for{i <- 0 until p.length
        val pi = p(i)
        j <- i+1 until p.length
        val pj = p(j)
        if dist(pi,pj)<=tDist} visit(pi,pj)
  }

  def visitNeighbors(i:Int, tDist:Double, visit:(A,A) => Unit) {
    val pi = p(i)
    for{j <- 0 until p.length
        val pj = p(j)
        if dist(pi,pj)<=tDist} visit(pi,pj)
  }
}
```

The method for finding all pairs has an outer loop using the variable i that goes through all the indices. Inside that it runs through all the points after the current point in the sequence. This way it does not deal with the same pair of particles, in opposite order, twice. The creation of the variables pi and pj where they are in the code is intended to prevent unnecessary indexing into the IndexedSeq p.

While this code is easy to write it if fairly easy to see that the number of calls to dist scales as $O(N^2)$. This is acceptable for smaller values of N, but it makes the approach infeasible when N gets large.[2] For that reason, we need to use alternate approaches that allow neighbors to be found with less effort than running through all of the other particles.

[2]The author does large-scale simulations where N typically varies from 10^5 to more than 10^8. He once needed to check if a bug had caused any particles to be duplicated. As a first pass, he wrote a short piece of code like this to do the check. When it did not finish within an hour, the author took the time to do a little calculation of how long it would take. That particular simulation had a bit over 10^7 particles. The

The simplest approach to doing this is a regular spatial grid. Using this approach, you break the space that the particles are in down into a grid of regularly spaced boxes of the appropriate dimension. For each box in the grid, we keep a list of the particles that fall inside of it. Searching for neighbors then requires only searching against particles that are on lists in cells that are sufficiently nearby. Here is code for an implementation of `NeighborVisitor` which uses this method.

```
package scalabook.adt

import collection.mutable

class RegularGridNeighborVisitor[A <% Int => Double](
    val p:IndexedSeq[A]
    ) extends NeighborVisitor[A](2) {

  private val grid = mutable.ArrayBuffer.fill(1,1)(mutable.Buffer[Int]())
  private var currentSpacing = 0.0
  private var min = (0 until dim).map(i => p.foldLeft(1e100)((d,p) => d min p(i)))
  private var max = (0 until dim).map(i => p.foldLeft(-1e100)((d,p) => d max p(i)))

  def visitAllNeighbors(tDist:Double, visit:(A,A) => Unit) {
    if(tDist<0.5*currentSpacing || tDist>5*currentSpacing) rebuildGrid(tDist)
    val mult = math.ceil(tDist/currentSpacing).toInt
    val offsets = Array.tabulate(2*mult+1,2*mult+1)((i,j) => (i-mult,j-mult)).
      flatMap(i => i).filter(t => t._2>0 || t._2==0 && t._1>=0)
    for{cx <- grid.indices
        cy <- grid(cx).indices
        (dx,dy) <- offsets
        val gx = cx+dx
        val gy = cy+dy
        if gx>=0 && gx<grid.length && gy>=0 && gy<grid(gx).length
        i <- grid(cx)(cy).indices
        val pi = p(grid(cx)(cy)(i))} {
      if(dx==0 && dy==0) {
        for{j <- i+1 until grid(cx)(cy).length
            val pj = p(grid(cx)(cy)(j))
            if dist(pi,pj)<=tDist} visit(pi,pj)
      } else {
        for{j <- grid(gx)(gy)
            val pj = p(j)
            if dist(pi,pj)<=tDist} visit(pi,pj)
      }
    }
  }

  def visitNeighbors(i:Int, tDist:Double, visit:(A,A) => Unit) {
    if(tDist<0.5*currentSpacing || tDist>5*currentSpacing) rebuildGrid(tDist)
    val mult = math.ceil(tDist/currentSpacing).toInt
    val offsets = Array.tabulate(2*mult+1,2*mult+1)((i,j) => (i-mult,j-mult)).
      flatMap(i => i)
    val cx = ((p(i)(0)-min(0))/currentSpacing).toInt
    val cy = ((p(i)(1)-min(1))/currentSpacing).toInt
```

answer was measured in weeks. Needless to say, he took the time to implement one of the methods from this chapter instead.

```
    val pi = p(i)
    for{(dx,dy) <- offsets
        val gx = cx+dx
        val gy = cy+dy
        if gx>=0 && gx<grid.length && gy>=0 && gy<grid(gx).length
        j <- grid(gx)(gy)
        if i!=j
        val pj = p(j)
        if dist(pi,pj)<=tDist} visit(pi,pj)
  }

  /**
   * Rebuild the grid to a size that is appropriate for the searches. Note that
   * this method was not written for efficiency. A true implementation should
   * include a rewrite of this method.
   */
  def rebuildGrid(spacing:Double) {
    min = (0 until dim).map(i => p.foldLeft(1e100)((d,p) => d min p(i)))
    max = (0 until dim).map(i => p.foldLeft(-1e100)((d,p) => d max p(i)))
    val cells = (0 until dim).map(i => ((max(i)-min(i))/spacing).toInt+1)
    if(grid.size<cells(0)) grid.append(mutable.Buffer.fill(cells(0)-grid.size)
        (mutable.ArrayBuffer.fill(cells(1)) (mutable.Buffer[Int]())):_*)
    else if(grid.size>cells(0)) grid.trimEnd(grid.size-cells(0))
    for(col <- grid) {
      if(col.size<cells(1))
          col.append(mutable.Buffer.fill(cells(1)-col.size)(mutable.Buffer[Int]()):_*)
      else if(col.size>cells(1)) col.trimEnd(col.size-cells(1))
      col.foreach(_.clear)
    }
    for(i <- p.indices) {
      val cx = ((p(i)(0)-min(0))/spacing).toInt
      val cy = ((p(i)(1)-min(1))/spacing).toInt
      grid(cx)(cy) += i
    }
    currentSpacing = spacing
  }
}
```

Clearly this code is longer and more complex than the brute force approach. However, there are many situations for which this will produce $O(N)$ behavior for the method that visits all neighbor pairs and $O(1)$ behavior for only the neighbors of a single particle.

This grid is hard coded to support only two dimensions. While not impossible, it is significantly more challenging to write code that uses arrays of arbitrary dimensions. Beside that point, regular grids do not work well in high dimensions. Imagine you have a n-dimensional cube and your search radius is about $1/100th$ the size of the cell so you want 100 divisions in each dimension. For a 2-D space this gives a modest $10,000 = 10^4$ cells in the grid. For a 3-D space you get a cube with $1,000,000 = 10^6$ cells. As the dimensions go up, this continues to grow as 10^{2n}. This causes significant problems above three dimensions, especially as most problems will need more than 100 bins per dimension. We could write a 3-D grid, but there really are not many applications where it helps to go above that.

The situations when a grid approach is ideal are those where the particles are fairly uniformly distributed, the search radius is small compared to the whole area, and the area increases with N, not the density. In other words, if the number of neighbors for every particle is fairly similar and when N is increased, the points are added in such a way that

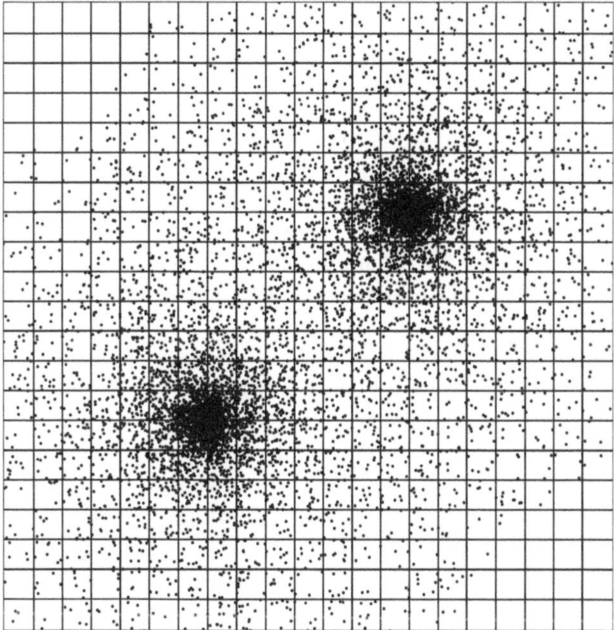

FIGURE 31.1: This figure shows an example set of points with a very non-uniform spatial distribution. A regular grid is laid over this to make clear how the point density in the grid cells will vary dramatically. The grid size shown here works well for many of the low-density regions where there are only a handful of particles per cell. A grid that only had a few particles per cell in the high-density regions would be a very fine grid with an extremely large fraction of the cells being empty in the low-density areas.

the number of neighbors stays the same, then the grid is pretty much an optimal structure. To give an idea of how good the grid is, a simple run was done with 1000 points spread randomly and uniformly across a square region, an ideal configuration for a grid. The brute force approach always does $N * (N - 1)/2 = 499500$ distance calculations. The number for the grid varies with the random distribution of the particles and the search size. However, using a search distance $1/100th$ the size of the point distribution was never seen to need more than 500 distance comparisons to find between 100 and 200 neighbor pairs.

It is also important to note that building the grid is an $O(N)$ operation. So assuming that the input for the program you are working on fits the conditions listed above, you can build a grid and find all the neighbors for all the particles in $O(N)$ operations. That is a dramatic improvement over the $O(N^2)$ behavior of the brute force approach. The brute force approach starts to become intractable around 10^6 particles and even before that it is slow on most machines for $N = 10^5$. With the grid approach, speed is likely not the limiting factor. However many points you can deal with in the memory of the machine, you can probably process them as well.

31.2 Quadtrees and Octrees

What about situations for which the grid is not well suited? The most obvious example is when the points are not well distributed. Consider the distribution of points shown in figure 31.1. Here there are two regions of very high-density surrounded by reduced density around that. The sample grid drawn on the region would work fairly well for the low-density regions as each cells has only a few points in it. However, the grid is not helping much in the high-density regions. The grid based search would still be running through a very large number of particle pairs. However, if the grid were drawn with a size appropriate for the high-density regions, the low-density regions would be mostly empty cells and lots of time could be wasted processing them. It is even possible that the memory overhead of keeping all those empty cells could be a problem.

The way around this is to use a tree. Just like the Binary Search Tree (BST) is a more dynamic and adjustable structure than a sorted array, spatial trees have the ability to adjust resolution to the nature of the data in different areas. In some ways, you could picture a BST tree as a spatial tree in a 1-D space. There is a complete ordering of value along a given line. Some thought has to be given to how this can be extended to deal with systems where there are multiple, orthogonal orderings. Many different approaches to this have been developed, we will consider two of them explicitly in this chapter.

The easiest spatial tree to understand is probably the quadtree. There are multiple types of quadtree. We will focus on a region-based quadtree. For this tree, each node represents a square region of space. We start with a bounding square that contains all of the points we are dealing with. That region is represented by the root of the tree. Each node can have four children. The children each represent the equally sized quadrants of of their parent. With this tree the points all go into the leaves. How many points are allowed in each leaf varies with the application. Empirical testing can be used to find what works best. When there are more points in a region than is allowed, that region splits into the four children and the points are moved down.[3]

The results of building a quadtree with some non-uniform data is shown in figure 31.2. This tree has roughly 1000 data points in it. You can see that in regions where there are few points, the divisions are large. However, in the higher density sections of the tree, the cells divide down to very small sizes to isolate the points.

The code for this tree is shown below. Like the grid, the quadtree is used for a 2-D spacial division and that is hard written into the code. The tree has a supertype called `Node` with two subtypes called `LNode` and `INode` for leaf node and internal node, respectively. The way this code is set up, only the leaf nodes hold particles. For this reason, it makes sense to use the approach of having two subtypes as they store very different data other than the location information. The `INode` also has additional functionality for splitting particles across the children and building the lower parts of the tree. The method `childNum` returns a number between 0 and 3 that is the index for a child based on location. The `groupBy` collection method is called as an easy way to get the different particles in each quadrant.

```
package scalabook.adt

class QuadtreeNeighborVisitor[A <% Int => Double](
    val p:IndexedSeq[A]
```

[3]It is worth noting again that this is something of a standard implementation of a region based quadtree. Different aspects can be varied depending on the application. We will see an alternative approach later in this chapter for the closely related octree.

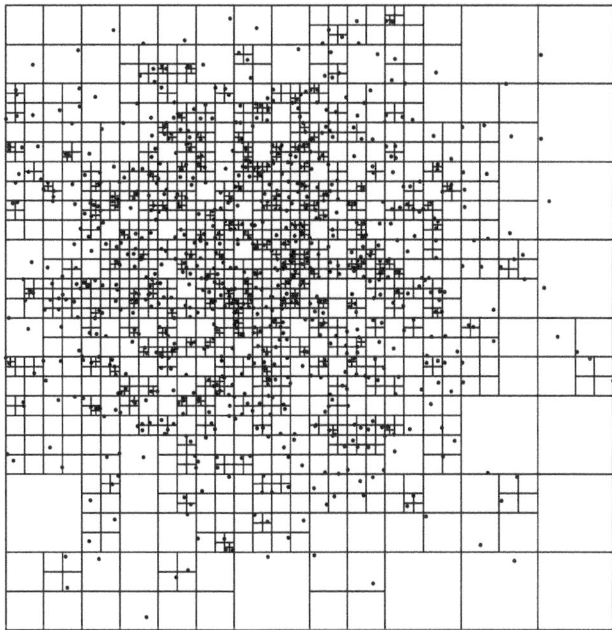

FIGURE 31.2: This is a quadtree with a number of non-uniformly distributed points to illustrate how the space is partitioned. For this example, each node is only allowed to hold a single point. The key advantage of the tree-based approach is that the resolution does not have to be uniform. Higher-density regions can divide the space much more finely.

```
  ) extends NeighborVisitor[A](2) {

  private class Node(val cx:Double, val cy:Double, val size:Double)
  private class LNode(x:Double, y:Double, s:Double, val pnts:IndexedSeq[Int])
      extends Node(x,y,s)
  private class INode(x:Double, y:Double, s:Double, pnts:IndexedSeq[Int]) extends
      Node(x,y,s) {
    val children = {
      val groups = pnts.groupBy(pn => childNum(p(pn)(0),p(pn)(1)))
      val hs = s*0.5
      val qs = s*0.25
      val ox = cx-qs
      val oy = cy-qs
      Array.tabulate(4)(i => makeChild(ox+hs*(i%2),oy+hs*(i/2),hs,
          if(groups.contains(i)) groups(i) else IndexedSeq[Int]()))
    }
    def makeChild(x:Double, y:Double, s:Double, pnts:IndexedSeq[Int]):Node =
      if(pnts.length>maxPoints) new INode(x,y,s,pnts)
      else new LNode(x,y,s,pnts)
    def childNum(x:Double, y:Double):Int = (if(x>cx) 1 else 0)+(if(y>cy) 2 else 0)
  }

  private val maxPoints = 1
  private val root = {
    val minx = p.foldLeft(1e100)((mx,pnt) => pnt(0) min mx)
    val maxx = p.foldLeft(-1e100)((mx,pnt) => pnt(0) max mx)
```

```
    val miny = p.foldLeft(1e100)((my,pnt) => pnt(1) min my)
    val maxy = p.foldLeft(-1e100)((my,pnt) => pnt(1) max my)
    val cx = 0.5*(minx+maxx)
    val cy = 0.5*(miny+maxy)
    val s = (maxx-minx) max (maxy-miny)
    if(p.length>maxPoints) new INode(cx,cy,s,p.indices)
    else new LNode(cx,cy,s,p.indices)
  }

  def visitAllNeighbors(tDist:Double, visit:(A,A) => Unit) {
    for(i <- 0 until p.length) {
      val pi = p(i)
      def recur(n:Node):Unit = n match {
        case ln:LNode =>
          ln.pnts.filter(j => j>i && dist(pi,p(j))<=tDist).foreach(j =>
              visit(pi,p(j)))
        case in:INode =>
          val x = pi(0)
          val y = pi(1)
          if(x+tDist>in.cx) {
            if(y+tDist>in.cy) recur(in.children(3))
            if(y-tDist<=in.cy) recur(in.children(1))
          }
          if(x-tDist<=in.cx) {
            if(y+tDist>in.cy) recur(in.children(2))
            if(y-tDist<=in.cy) recur(in.children(0))
          }
      }
      recur(root)
    }
  }

  def visitNeighbors(i:Int, tDist:Double, visit:(A,A) => Unit) {
    val pi = p(i)
    def recur(n:Node):Unit = n match {
      case ln:LNode =>
        ln.pnts.filter(j => j!=i && dist(pi,p(j))<=tDist).foreach(j =>
            visit(pi,p(j)))
      case in:INode =>
        val x = pi(0)
        val y = pi(1)
        if(x+tDist>in.cx) {
          if(y+tDist>in.cy) recur(in.children(3))
          if(y-tDist<=in.cy) recur(in.children(1))
        }
        if(x-tDist<=in.cx) {
          if(y+tDist>in.cy) recur(in.children(0))
          if(y-tDist<=in.cy) recur(in.children(2))
        }
    }
    recur(root)
  }
}
```

The construction of the tree itself happens in the declaration of `root`. The type of quadtree must have a bounds containing the points so that is calculated first. After that is done, the proper node to is instantiated. that node serves as the root, and if it is an `INode`, it builds the rest of the tree below it. While this tree does not take advantage of it, one of the big benefits of the region-based quadtree is that it is fairly easy to write a mutable version that allows the user to add points one at a time as long as the user specifies an original bounding region.

The `visitAllNeighbors` and `visitNeighbors` methods include nested recursive functions that run through the tree looking for neighbors. The behavior on an `INode` is such that recursive calls are only made on children that the search area actually touches. When run on the same type of uniform random configuration used for the grid above, this tree did a bit fewer than twice as many distance calculations as the grid. This was still vastly better than the brute force approach and remarkably close to the grid considering the configuration of points used was ideal for the grid.[4]

As with the grid, the quadtree is inherently 2-D. A 3-D version that divides each node into octants is called an octree. We will build one of those in a later section. Also like the grid, this type of tree does not scale well to higher dimensions. This is due to the fact that the number of children contained in each child node scales as 2^n for an n dimensional space. Given that the number of internal nodes scales as the number of points, having large internal nodes that store many children becomes inefficient. There are alternate ways of storing children, but if you really need a high-dimensional space, the next tree we will discuss is your better bet.

Before leaving the topic of the quadtree, it is also worth noting that this region-based quadtree is inherently unbalanced. The depth of the tree is significantly larger in regions of high-density. In practice, this does not introduce significant inefficiency, but it it worth noting as it is a real different from the BST tree.

31.3 kD-Trees

What should you do if you have to deal with a high-dimensional space, want a balanced spatial tree, and/or have dramatically different sizes in different dimensions that you want to resolve as needed? A possible solution to all of these is the use of a kD-tree. The "kD" in the name means k-dimensional so the name itself points to the fact that this tree is intended to scale to higher dimensions.

The way that a kD-tree works is that internal nodes have two children split across a plane that is perpendicular to one axis. Put more simply, you pick a dimension and a value, anything less than or equal to the value in that dimension goes to the left and anything greater than it goes to the right. This behavior should be fairly easy to understand as it is remarkably similar to a BST tree.

How you pick the dimension and the split value can vary greatly between applications. We will use a technique that tries to maximize the efficiency of each split and guarantees the tree is balanced. The split dimension will be the dimension with the largest spread. To pick a value, we will use the median value in that dimension. That way, the number of points on the left and the right will be as close to equal as possible, resulting in a balanced tree.

[4]The version of `visitAllNeighbors` shown here is very similar to putting `visitNeighbors` in a loop. This approach has one drawback that the recursion goes through the higher nodes many times, duplicating work. It is possible to write a function that recurses over two `Node` arguments that improves on this. It was not shown here because it is significantly more complex and difficult for the reader to understand.

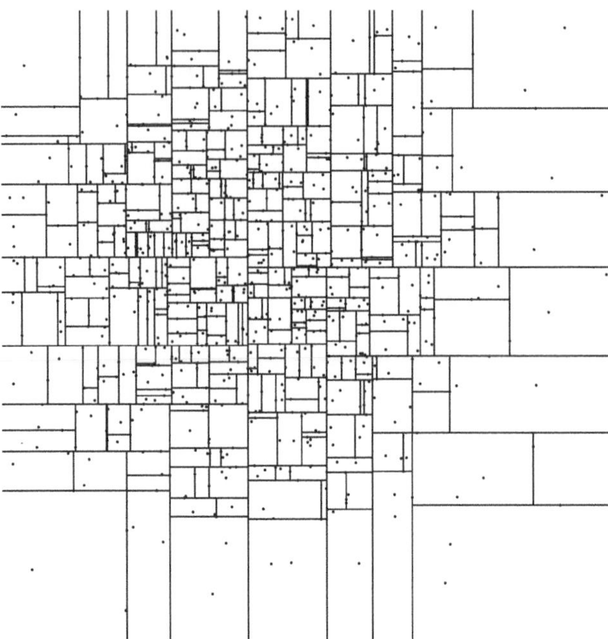

FIGURE 31.3: This is a kD-tree with a number of non-uniformly distributed points to illustrate how the space is partitioned. For this example, each node can hold up to three points. While this example only uses two dimensions, it is simple to scale the kD-tree up higher without adding overhead.

Figure 31.3 shows a 2-D tree generated using this approach allowing three points per leaf. The fact that there is no bounding box here is intentional. The kD-tree puts in divisions that span as far as needed. The ones at the ends are effectively infinite. The internal ones are cut off by other splits.

A sample code implementing `NeighborVisitor` using a kD-tree is shown here. This class is not hard coded to a specific dimensionality as there is nothing that ties it to a particular number of dimensions. There are many similarities between this code and the quadtree code so we will focus on differences. The first is that the `IndexedSeq` that is passed in is converted to a private `Array`. This forces the additional requirement of a manifest, but it greatly improves the efficiency of building the tree.

```
package scalabook.adt

class KDTreeNeighborVisitor[A <% Int => Double : Manifest](
    d:Int,
    val pIn:IndexedSeq[A]
    ) extends NeighborVisitor[A](d) {

  private val p = pIn.toArray
  private val maxPoints = 3

  class Node
  private class LNode(val pnts:Seq[Int]) extends Node
  private class INode(start:Int, end:Int) extends Node {
    val splitDim = {
      val min = (0 until dim).map(i => p.view.slice(start,end).
```

```
      foldLeft(1e100)((m,v) => m min v(i)))
    val max = (0 until dim).map(i => p.view.slice(start,end).
      foldLeft(-1e100)((m,v) => m max v(i)))
    (0 until dim).reduceLeft((i,j) => if(max(j)-min(j)>max(i)-min(i)) j else i)
  }
  val (splitVal,left,right) = {
    val mid = start+(end-start)/2
    indexPartition(mid,start,end)
    (p(mid)(splitDim),makeChild(start,mid+1),makeChild(mid+1,end))
  }
  def indexPartition(index:Int, start:Int, end:Int) {
    if(end-start>1) {
      val pivot = if(end-start<3) start else {
        val mid = start+(end-start)/2
        val ps = p(start)(splitDim)
        val pm = p(mid)(splitDim)
        val pe = p(end-1)(splitDim)
        if(ps<=pm && ps>=pe || ps>=pm && ps<=pe) start
        else if(ps<=pm && pm<=pe || ps>=pm && pm>=pe) mid else end-1
      }
      val ptmp = p(pivot)
      p(pivot) = p(start)
      p(start) = ptmp
      var (low,high) = (start+1,end-1)
      while(low<=high) {
        if(p(low)(splitDim)<=ptmp(splitDim)) {
          low += 1
        } else {
          val tmp = p(low)
          p(low) = p(high)
          p(high) = tmp
          high -= 1
        }
      }
      p(start) = p(high)
      p(high) = ptmp
      if(high<index) indexPartition(index,high+1,end)
      else if(high>index) indexPartition(index,start,high)
    }
  }
  def makeChild(s:Int, e:Int):Node = {
    if(e-s>maxPoints) new INode(s,e) else new LNode(s until e)
  }
}

private val root = new INode(0,p.length)

def visitAllNeighbors(tDist:Double, visit:(A,A) => Unit) {
  for(i <- 0 until p.length) {
    val pi = p(i)
    def recur(n:Node):Unit = n match {
      case ln:LNode =>
        ln.pnts.foreach(j => if(j>i && dist(p(j),pi)<=tDist) visit(pi,p(j)))
      case in:INode =>
        if(pi(in.splitDim)-tDist<=in.splitVal) recur(in.left)
```

```
        if(pi(in.splitDim)+tDist>in.splitVal) recur(in.right)
      }
      recur(root)
    }
  }

  def visitNeighbors(i:Int, tDist:Double, visit:(A,A) => Unit) {
    val pi = p(i)
    def recur(n:Node):Unit = n match {
      case ln:LNode =>
        ln.pnts.foreach(j => if(j!=i && dist(p(j),pi)<=tDist) visit(pi,p(j)))
      case in:INode =>
        if(pi(in.splitDim)-tDist<=in.splitVal) recur(in.left)
        if(pi(in.splitDim)+tDist>in.splitVal) recur(in.right)
    }
    recur(root)
  }
}
```

The real meat of this code is in the creation of an `INode`. All that is passed in is a start and an end index into the array to represent what points are supposed to be under this node. All manipulations are done in place reordering elements in the array `p`.

The first thing done upon construction of an `INode` is to find the dimension in which the specified points have the largest spread to make it the split dimension. After this is the declaration and assignment to `splitVal`, `left`, and `right`. Most of the work done here occurs in the method `indexPartition`. The purpose of this method is to arrange the elements of the array along the split dimension such that the point which should be at `index` is located there and all the points below or above it along the appropriate dimension are above or below it in the array. In this case we happen to always be finding the median, but the code itself is more general than that. This function can find the element at any given index and do so in $O(N)$ time.

The way this is done is that the function works very much like a quicksort. The base case is a single element. If there are more elements, a pivot it selected and that pivot is moved into place just like with would be in a quicksort. Where this differs it that it only recurses at most once. If the desired index is below the pivot, it recurses on the lower part. If it is above the pivot, the method recurses on the upper part. If the pivot happens to be at the desired index, the method returns. To see how this approach gives $O(N)$ performance, simply recall that $1+\frac{1}{2}+\frac{1}{4}+\frac{1}{8}+\ldots=2$. As long as the pivot is always close to the center, this is roughly the way the amount of work behaves with each call.

It is interesting to note that because the elements in the `p` array are actually reordered, the `LNode` sequences of particle indices are always contiguous ranges. The visit methods are fairly straightforward. They are actually a fair bit simpler than their counterparts in the quadtree because there are only two children and a single split.

Running this code through the same 2-D tests that were done for the grid results in about four times as many distance calculations as we had for the quadtree. This is in part due to the fact that the leaves have three points in them. It is also somewhat due to the fact that nodes in the kD-tree can be elongated, even when the overall distribution of points is not. This is still far fewer distance calculations than the brute force approach and you would have to do empirical tests to determine which option was truly best for your application.

Another issue that impacts both the quadtree and the kD-tree is the problem of identical points. The quadtree will run into problems if you have any cluster of points with more than `maxPoints` with a spacing that is extremely small. Such a cluster will lead to very deep leaves as the nodes have to split repeatedly to resolve those points. The kD-tree does not suffer from tight clusters in the same way, but both will crash as there are more than `maxPoints` that all have the same location. The fact that those points can not be split, regardless of how far down the tree goes means that the code will try to create an infinite number of `INodes` to break them apart. The quadtree will run into this same problem if points are placed outside bounds, particular outside the corners.

31.4 Efficient Bouncing Balls

One example of an application where spatial data structures can be extremely beneficial is the bouncing balls problem that we made a `Drawable` for previously. For the 20-ball simulation that was performed previously, any spatial data structure is overkill. However, if the number of balls were increased by several orders of magnitude, a spatial data structure would become required.

If we were using a spatial data structure, the question of search radius arises. A single search radius that is safe for all particles is twice the maximum speed multiplied by the time step plus twice the maximum particle radius. As these values can be easily calculated in $O(N)$ time, it is not a problem to add them to the code. It would be slightly more efficient to use a unique search radius for each particle that uses the speed and size of that particle plus the maximum values to account for the worst possibility for the other particle.

There are few changes that need to be made in the code for using a spatial data structure. You can find the original code in chapter 26 on page 687. Anyplace there is a call to `findEventsFor`, there is a need for the spatial data structure. The `findEventsFor` method takes a `Seq[Int]` with the indices of the particles that are to be searched against. The current implementation is brute force. It passes all indices after the current one in the timer when doing the complete search to initialize the queue and then all values other than the two involved in the collision for searching after a collision. The first one should use `visitAllNeighbors` with a visitor that calls the `collisionTime` method to see if the particles collide and adds them to the priority queue as needed. The call to `findEventsFor` is still needed to get wall collisions though it can be called with an empty list of other particles to check against as the call to `visitAllNeighbors` will have done that. The search for a single particle should use `visitNeighbors` to find the `against` list.

It is interesting to note the involvement of the timestep in the search radius. The code for the bouncing balls uses a time of 1.0 as the length of the timestep. This occurs in `findEventsFor` where there is a cutoff on the time for events to be added to the queue. If this were shortened, the particles would move less between frames being drawn. For the old method, there is not a significant motivation to do this other than to have the gravity integration be more accurate. For the spatial data structures, it can dramatically speed up the processing of finding collision events as it reduces the area that has to be searched.

31.5 End of Chapter Material

31.5.1 Summary of Concepts

- Spatial data comes in many forms and is associated with many areas of computing.

- Brute force techniques for doing things like finding nearby points in space tend to scale as $O(N^2)$.
- Spatial data structures can improve the scaling dramatically.
  - Grids are simple and for low dimension, fairly uniform data, they are ideal.
  - Region-based quadtrees and octrees divide a region that encompasses all the data into successively smaller, uniformly sized parts. This provides variable resolution for the places where it is needed.
  - The kD-tree scales to arbitrarily high dimensions. Each internal node splits the data at a particular value with lower values going to the left and higher values to the right.

31.5.2 Exercises

1. Write unit tests for the different spatial data structures that were presented in the chapter.
2. Write a quadtree that includes an `addPoint` method.
3. Parallelize any of the trees were presented in this chapter so that searches happen across multiple threads. If you want a bit more challenge, make it so that construction the the tree also occurs in parallel.
4. Pick a spatial tree of your choosing and do speed tests on it. One way you can do this is to put random points in the tree and then search for all pairs below a certain distance. For each tree that was described, you can vary the number of points that are allowed in a leaf node. Do speed testing with different numbers of points in the data set as well as different numbers of points allowed in leaves.

   It is also interesting to compare the relative speed of different trees of the same dimension.

31.5.3 Projects

1. Pretty much any data can be viewed as existing in a space of the appropriate dimensions. If you ware working on the web spider, you should do exactly that to find things that are closely related in some way. What you do this on might depend on what data you are collecting. An obvious example would be if you were skimming box scores to a sport, you could make a space that has a different dimension for each significant statistical category. You could put all the entries you have in a kD-tree of that many dimensions, then do searches for everything within a certain distance of a particular point. This could show you things like players or teams that are comparable to one another in different ways.

   If the data you are storing is not numerical, it is still possible to assign spatial values based on text or other information. For example, you could select a set of words that are relevant for what you are looking at and use a function of the number of times each word appears as a value for a different axis. If you have 100 different words of interest, this would mean you have a 100-D space that the pages would go into.[5] These can be

[5]Using word counts for the value is typically not ideal because the length of a page can dramatically impact where it appears in your space. For that reason you often want to do something like take the log of the word count or the relative fraction of the number of instances of the words.

put into the kD-tree and that can be used to find pages that are similar according to different metrics.

Even if you are cataloging images, you can associate a sequence of numeric values with images. The numeric values could relate to factors like size, color, brightness, saturation, contrast, or many other things. The image is made of a lot of RGB values so there are many functions you could devise that produce something interesting.

2. The MUD project is probably the one that has the least obvious use for spatial trees. After all, the maps in a MUD are completely text based, not spatial. Despite this, you can use an approach similar to what is described above for the web spider with the MUD as well. Characters, including player controlled and non-player controlled, likely have stats that indicate how powerful they are in different ways. These stats can be used to place them in a higher-dimensional space that you can map out with a kD-tree to enable functions like giving players the ability to find potential opponents or partners who are of similar capabilities.

   It is also possible you might want to use a word count type of approach for a MUD, but this is likely to be less useful as you need many more data points than you have dimensions, and unless you have been very busy, the room count for your MUD is not likely to be all that high.

3. Graphical games, both networked and not, have an automatic spatial nature to them. While your game might not really need a spatial data structure for locating things efficiently, you should put one in to give yourself experience with them. If you did create a game with a really big world, these spatial data structures could provide significant benefits.

4. One interesting type of analysis that you could put into the math worksheet project is the ability to measure the fractal dimension of a set of data. There are many different formal definitions of fractal dimensions. We will use one called the correlation dimension. The correlation is determined using the correlation integral is formally defined as:

$$C(\varepsilon) = \lim_{N \to \infty} \frac{g(\varepsilon)}{N^2},$$

   where $g(\varepsilon)$ is the number of pairs of points with distance less than infinity. The correlation dimension can be found from using the relationship:

$$C(\varepsilon) \sim \varepsilon^{\nu},$$

   where ν is the correlation dimension and is equal to the slope of the correlation integral on a log-log plot. Finding this value requires two steps. First, we need to calculate $C(\varepsilon)$. Once we have that, we need to take the log of both x and y values for all points on the curve and then do a linear fit to those points to find the slope.

   Actually calculating $C(\varepsilon)$ is $O(N^2)$ in the number of points, making it infeasible for a large data set. As the quality of the answer improves as $N \to \infty$, it is better to have very large data sets and so a complete, brute force calculation becomes infeasible. That is where the spatial tree comes in. The number of points in nodes of different sizes provides an approximation of the number of pairs of points with a separation that length or shorter. To make this work, you need to use a tree where all the nodes at a particular level are the same size. This applies to our quadtree and octree constructions from the chapter. It can also apply to a kD-tree. The tree is constructed in a different way where splits are done down the middle spatially instead of at the median point. Once you have built such a tree, you can use the approximation that

$$C(\varepsilon) \sim \sum_{\text{cells of size } \varepsilon} n_{cell}^2.$$

This says that the number of pairs separated by ε or less goes as the sum over all cells of size ε of the number of cells in each cell squared. This technically over counts connections in a cell by a factor of two, but it does not count any connections between cells. In addition, any multiplicative error will be systematic across all levels so the value of ν will not be altered.

Once you have a set of points representing $C(\varepsilon)$, the last thing to do is to perform a linear fit to the points $(\log(\varepsilon), \log(C(\varepsilon)))$. A linear least squares fit can be done by solving a linear equation much like what was done in project 6 (p.416). The idea is that you want to find optimal coefficients, c_i, to minimize the sum of the squares of the following differences for your data points, $(x_j, y_j) = (\varepsilon_j, C(\varepsilon_j))$.

$$y_j - (c_1 f_1(x_j) + c_2 f_2(x_j) + c_3 f_3(x_j) \cdots).$$

This can be done by solving the equation $A^T A x = A^T y$, where

$$A = \begin{bmatrix} f_1(x_1) & f_2(x_1) & f_3(x_1) & \cdots \\ f_1(x_2) & f_2(x_2) & f_3(x_2) & \cdots \\ f_1(x_3) & f_2(x_3) & f_3(x_3) & \cdots \\ \vdots & \vdots & \vdots & \ddots \end{bmatrix}$$

and y is the column vector of your y_j values.

For a simple linear equation, there are only two terms so $f_1(x) = x$ and $f_2(x) = 1$. That would reduce down to a 2x2 matrix that is fairly easy to solve. For this problem, you can typically reduce the problem even further by making the assumption that $C(0) = 0$. This is true as long as no two points every lie at exactly the same location. In that case, $A^T A = \sum_{j=1}^{N} x_j^2$ and $A^T y = \sum_{j=1}^{N} x_j y_j$. The value of the one coefficient can be found by simple division.

5. If you have been working on the Photoshop® project, you can use a 2-D kD-tree to create an interesting filter for images. The kD-tree filter breaks the pixels in an image up into small regions and colors each leaf a single color. This can give something of a stained glass effect with rectangular regions. You can choose how to do the divisions to give the effect that you want. One recommendation is to make cuts in alternating dimensions such that you balance the brightness or the total intensity of a color channel on each side. So if the first cut is vertical, you find the location where the total amount of red to the left of the cut and the total to the right of the cut are as close as possible. Then you do horizontal cuts on each side of that following the same algorithm. After a specified number of cuts, you stop and average the total color of all pixels in each region. Use that average color to fill that region.

6. Spatial trees can be used to improve collision handling in the simulation workbench. Project 6 (p.416) described how collision finding is $O(n^2)$ by default and looked at a way to improve that for most timesteps. Spatial data structures can also be used to reduce the number of pairs that have to be checked. The simplest approach just uses the tree to search for particles that are within a particular search radius of the particle in question. The search radius is given by $\Delta v \Delta t + 2R_{max}$, where Δv is the relative velocity of particles, Δt is the time step, and R_{max} is radius of the largest particle. This should take the number of pairs checked down from $O(n^2)$ to $O(n \log(n))$.

Chapter 32

Binary Heaps

Back in chapter 26 we looked at the priority queue Abstract Data Type (ADT). In that chapter, the implementation that was written used a sorted linked-list. While it was easy to write, this implementation has the downside of an $O(n)$ `enqueue` method. This gives overall performance that is $O(n^2)$ for the size of the queue. In this chapter we will look at a different implementation that provides $O(\log(n))$ performance for both `enqueue` and `dequeue`, the heap.

32.1 Binary Heaps

There are quite a few different styles of heaps. A common theme is that they have a tree structure. The simplest, and probably most broadly used, is the binary heap. As the name implies, it is based on a binary tree type of structure where each node can potentially have left and right children. For a binary tree to be a binary heap it has to have two properties:

- Complete - The tree fills in each level from left to right. There are never gaps in a proper binary heap.
- Heap Ordering - Priority queues require that elements have a complete ordering. The rule for a binary heap is that parents always have a priority that is greater than or equal to their children.

The heap ordering means that the highest priority element is always at the root of the tree. An example is shown in figure 32.1.

32.1.1 Binary Heaps as Arrays

While we conceptually think of the heap as a binary tree, we do not implement it that way. It is far more efficient to implement it as an array. This only works because the heap is complete. Figure 32.2 shows the same heap as in figure 32.1 with subscripts that number the nodes in breadth-first order. Looking closely at the subscripts you should see that there

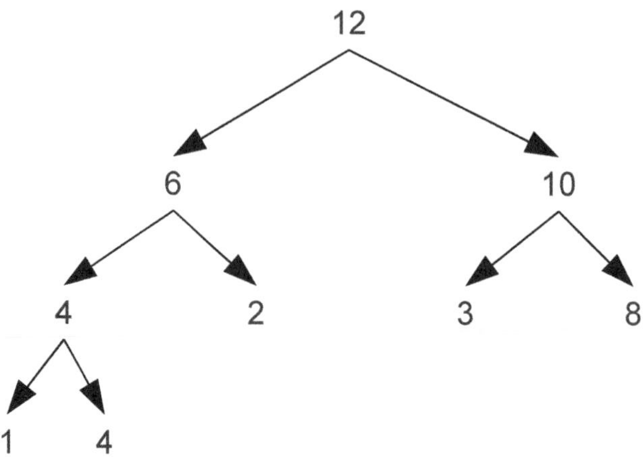

FIGURE 32.1: This figure shows an example of a binary heap. The numbers are priorities with higher values being higher priority. Note that there is no problem with duplicate priorities, unlike keys in a BST.

is a nice mathematical relationship between parents and children. Specifically, if you divide an index by two, using integer division, the result is the index of the parent. The reverse relationship is that the children of a node have indices that are twice the node's index and twice the node's index plus one.

The figure shows the array leaving the 0 index spot empty. It is possible to implement a heap with all the elements shifted down one spot so that there is not a blank. However, this requires putting `+1` and `-1` in the code at many different points. This complicates the code making it harder to understand and maintain. While it removes a little waste in memory, it introduces an overhead in every call on the priority queue. For that reason, and to keep the code simpler, the implementation shown here will leave the first element of the array as a default value.

32.2 Heaps as Priority Queues

To use the binary heap as a priority queue we have to be able to implement `enqueue`, `dequeue`, `peek`, and `isEmpty`. The last two are simple, the highest priority element is always the root so `peek` simply returns it. The `isEmpty` method just checks if there is anything present on the heap. The `enqueue` and `dequeue` are a bit more complex because they have to change the structure of the heap.

We start with `enqueue` where an element is added to the queue. Adding to a Binary Search Tree (BST) was just a task of finding the proper place to put the new element. For a heap, the job is a bit different. The heap has to remain complete so when a new element is added, there is no question about what set of nodes have to be occupied. The next open space in the row that is currently filling in or the first in the next row when the lowest row is complete has to be filled. However, the new element might not belong there. The way we deal with this is to put a "bubble" in the position that has to be filled and let it move up

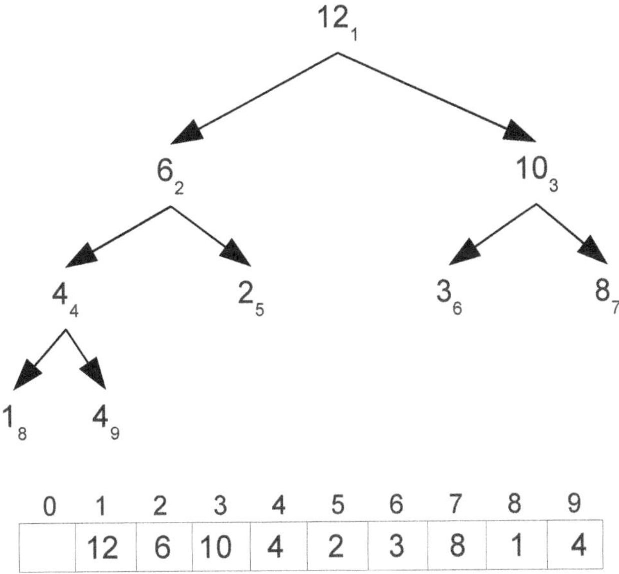

FIGURE 32.2: This figure shows the same heap as in figure 32.1 with subscripts that number the nodes in breadth-first order. Below that is an array with the elements at those locations.

until it reaches the location that the new element belongs. As long as the new element has a higher priority than the parent of the bubble, the parent is copied down and the bubble moves up. An example of adding a new element with a priority of 9 to the example heap is shown in figure 32.3.

A call to `dequeue` should remove the element from the root of the tree as that will always have the highest priority. However, to keep the tree complete, the location that needs to be vacated is at the end of the last row. To deal with this, we pull the last element out of the heap and put it in a temporary variable then put a placeholder we will call a "stone" at the root. This stone then sinks through the heap until it gets to a point where the temporary can stop without breaking heap-order. When moving down through the heap there is a choice of which child the stone should sink to. In order to preserve heap-order, it has to move to the higher-priority child as we are not allowed to move the lower-priority child into the parent position above its higher-priority sibling. This process is illustrated in figure 32.4.

Code to implement these operations using an array-based binary heap is shown below. This extends the `PriorityQueue[A]` type that was created in chapter 26. Like the sorted linked-list implementation, a comparison function is passed in to provide the complete ordering of elements for instances of this class. The only data needed by this class is an array we call `heap` and an integer called `end` that keeps track of where the next element should be added. These are created with 10 elements to begin with, one of which will never be used as index zero is left empty, and the other is set to be 1.

```
package scalabook.adt
```

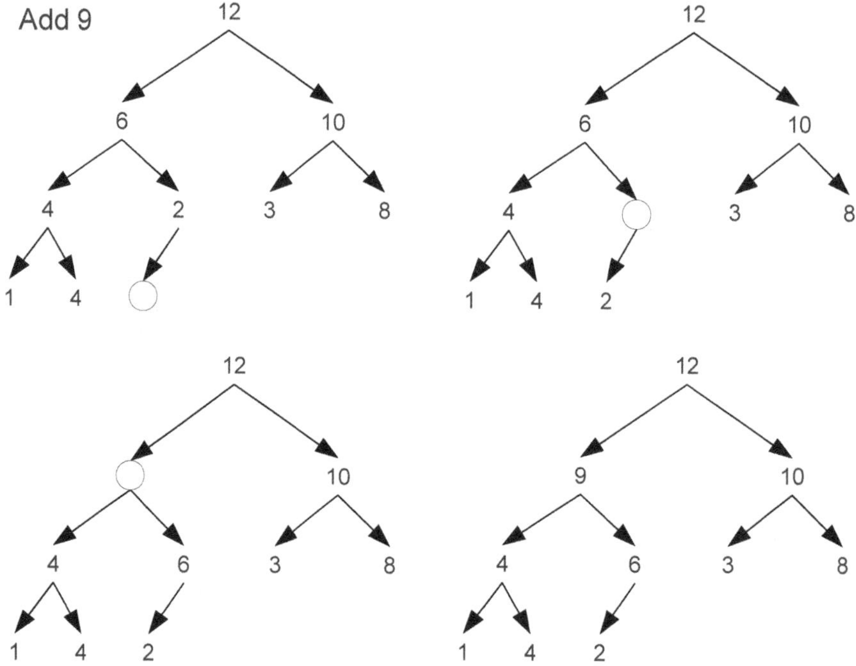

FIGURE 32.3: Adding a new element with a priority of 9 to the heap shown in figure 32.1 goes through the following steps. A bubble is added at the next location in a complete tree. It moves up until it reaches a point where the new value can be added without breaking the heap-order.

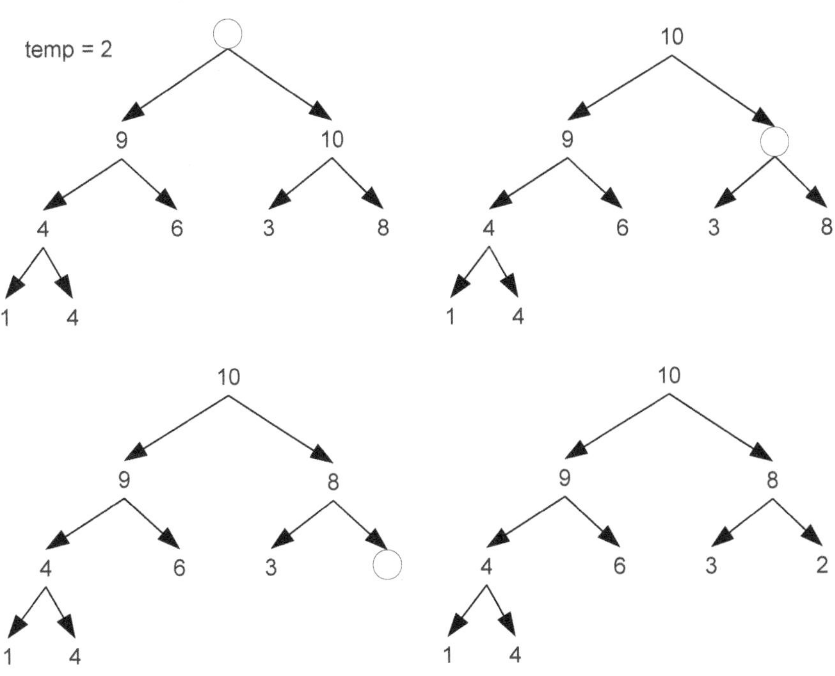

FIGURE 32.4: Calling dequeue on the heap at the end of figure 32.3 forces the tree to go through the following steps. The last element is moved to a temporary variable and a stone is placed at the top of the heap. This stone sinks down until it reaches a point where the temporary can stop without breaking heap-order. The stone always sinks in the direction of the higher-priority child.

```
class BinaryHeapPriorityQueue[A : Manifest](comp: (A,A)=>Int) extends
    PriorityQueue[A] {
  private var heap = new Array[A](10)
  private var end = 1

  def enqueue(obj:A) {
    if(end>=heap.length) {
      val tmp = new Array[A](heap.length*2)
      Array.copy(heap,0,tmp,0,heap.length)
      heap = tmp
    }
    var bubble = end
    while(bubble>1 && comp(obj,heap(bubble/2))>0) {
      heap(bubble) = heap(bubble/2)
      bubble /= 2
    }
    heap(bubble) = obj
    end += 1
  }

  def dequeue():A = {
    val ret = heap(1)
    end -= 1
    val temp = heap(end)
    heap(end) = heap(0) // Clear reference to temp
    var stone = 1
    var flag = true
    while(flag && stone*2<end) {
      var greaterChild = if(stone*2+1<end && comp(heap(stone*2+1),heap(stone*2))>0)
        stone*2+1 else stone*2
      if(comp(heap(greaterChild),temp)>0) {
        heap(stone) = heap(greaterChild)
        stone = greaterChild
      } else {
        flag = false
      }
    }
    heap(stone) = temp
    ret
  }

  def peek:A = heap(1)

  def isEmpty:Boolean = end==1
}
```

The `enqueue` method begins with a check to see if there is space in the array for a new element. If not, a bigger array is created. As with other similar pieces of code written in this book, the size of the array is doubled so that the amortized cost of copies will be $O(1)$. The rest of the `enqueue` method makes an `Int` called `bubble` that will walk up the heap to find where the new element should go. It is initialized to `end`, the first unused location in the array. The main work is done in a `while` loop that goes as long as the bubble is greater than an index of 1 or the parent of `bubble`, located at index `bubble/2`, has a higher-priority than the new element. When the loop terminates, the new element is placed at the location of the `bubble` and `end` is incremented.

The `dequeue` method starts off by storing the current root in a variable that will be returned at the end of the method. It then decrements `end`, stores the last element in the heap in a temporary, and clears out that element with the default value stored at `heap(0)`. A variable called `stone` is created that starts at the root and will fall down through the heap. There is also a `flag` variable created to indicate when the `while` loop should terminate. This style is helpful when the determinant of whether to stop the loop is complex and would be difficult to put in a single line. In this case part of the determination is easy, if the stone has gotten to a leaf, it must stop there. Otherwise, it requires checking the temporary against the child or children to see if the temporary has lower priority than the higher-priority child. That would require a complex Boolean expression that would be hard to read. The use of a `flag` generally improves the readability and maintainability of the code.

Inside the loop, the first task is determining which child has a higher priority. The check in the `while` loop condition guarantees that there is at least one. If there is a second, and it has higher priority, it is the one we want, otherwise we want the first child. Then that child is compared to the `temp` to see if the stone should keep sinking. If it should not keep sinking then the loop needs to stop so `flag` is set to `false`. When the loops terminates, the temporary is moved to the location where the `stone` stopped and the original root value is returned.

Using the image of the tree we can argue that both `enqueue` and `dequeue` are always $O(\log(N))$. This can also be argued from the code. The variable `end` is effectively n. In the `enqueue` method the `bubble` starts at `end` and is divided by 2 in every iteration through the loop. This can only happen $\log_2(n)$ times before it gets to 1. With `dequeue` the `stone` starts at 1 and each iteration it changes to either `stone*2` or `stone*2+1`. This too can only happen $\log_2(n)$ times before `stone` gets to `end`.

Testing this code is a simple matter of copying the test for the sorted linked-list priority queue and changing what type is instantiated. It can also be used to improve the performance of any other code that requires only the priority queue interface to operate. For example, the cell division example from chapter 26 could be altered and would gain a significant performance boost for large populations.

Priority Queue with Frequent Removes

Back in subsection 26.2.2, when the sorted linked-list was used for an event-based simulation of collisions, we had to add a method called `removeMatches`. This method was needed because whenever a collisions between two bodies is processed, all future events involving either of those bodies needs to be removed from the queue. The new velocity of the body will make those events invalid. This method was not hard to add to the sorted linked list implementation and the fact that it was $O(n)$ was not a problem as the `enqueue` method was also $O(n)$.

In the last chapter we saw how adding spatial data structures can improve the overall performance of a time step from $O(n^2)$ to $O(n\log(n))$ or even, in certain ideal cases, $O(n)$. However, the sorted linked-list priority queue will still limit performance to $O(c^2)$, where c is the number of collisions in the time step. That number generally does increase with the number of particles so using a sorted linked list is still a potential performance bottleneck.

We have just seen that a heap-based priority queue gives $O(\log(n))$ performance for the main operations, which would provide $O(n\log(n))$ overall performance for the number of elements added and removed. This could provide a significant performance improvement for collisions if we could implement `removeMatches` on the heap. Unfortunately, removing an element from the heap is $O(\log(n))$ and can not be improved significantly when done in bulk. Fortunately, collision handling has aspects that are predictable and can be used to make an alternate implementation.

What is important about collision handling is that from one time step to the next, the number of collisions is fairly consistent and the distribution of times for collisions is fairly uniform. This points to an implementation that combines arrays and linked lists.[1] Let us say that in the previous time step there were c collisions for n particles and each time step lasts Δt time units. We start with an array that spans time from 0 until Δt with c elements that each represent $\frac{\Delta t}{c}$ time. The array references nodes for linked lists. There is another array with n elements, one for each particle in the simulation, that also keeps references to nodes. Each node is part of two or three orthogonal doubly-linked lists and stores information about an event. One doubly-linked list going through a node is sorted and based on time. The others are not sorted, represent particles, and there is one for each particle involved. So particle-particle collisions are part of two of these and particle-wall collisions are part of two.

A graphical representation of this structure is shown in figure 32.5. The array across the top is for time. The first cell keeps the head of the list for any events between 0 and $\frac{\Delta t}{c}$. The second is for events between $\frac{\Delta t}{c}$ and $2\frac{\Delta t}{c}$. The array on the right has one cell for each particle in the simulation and the list keeps track of any collisions that particle is involved in. The lists are doubly linked so that any element can be removed without knowing about the one before it in any given list. This figure is greatly simplified in many ways to make it understandable. In a real situation the number of elements in each array should be at least in the thousands for this to be relevant. In addition, while the vertical lists are sorted and there will be a nice ordering for elements, the horizontal links will often double back around and produce a far more complex structure than what is shown here.

So what is the order of this data structure? Linked lists are inherently $O(n)$ for random access. This includes adding to sorted linked lists. On the other hand, adding to the head of a linked list is $O(1)$. In the worst case, all the events could go into a single, long linked list. That would result in $O(n^2)$ performance. Such a situations is extremely unlikely given the nature of collision events. We expect roughly c events that are roughly evenly distributed. That would result in sorted linked lists of an average length of 1 and it should be very rare to have a length of longer than a few. The `dequeue` operation has to remove all the other events for any particles involved, but that is required by any implementation and the advantage of having doubly-linked lists by particles is that this can be done as fast as possible considering only the nodes involving those particles have to be visited. So the expected time for all operations is optimal. For `dequeue` is is always $O(1)$, the implementation simply keeps track of the element in the time array that has the first element. The `enqueue` could be $O(n)$, but we expect $O(1)$ performance because the time lists are short. The `removeParticle` method will perform $O(n)$ operations in the number of events for that particle, but this too should be a small number.

You might wonder why, if this data structure is so great, it is not used more generally. The reason is that it is not generally good. It requires knowing that there should be about c events, they will all lie in a certain finite time range, and they will be evenly distributed. Change any of those conditions and this structure can be worse than a standard sorted linked list. If you ever wonder why you should bother to learn how to create data structures, like linked lists, when there are perfectly good implementations in libraries, here if a reason why. Library priority queues are typically implemented

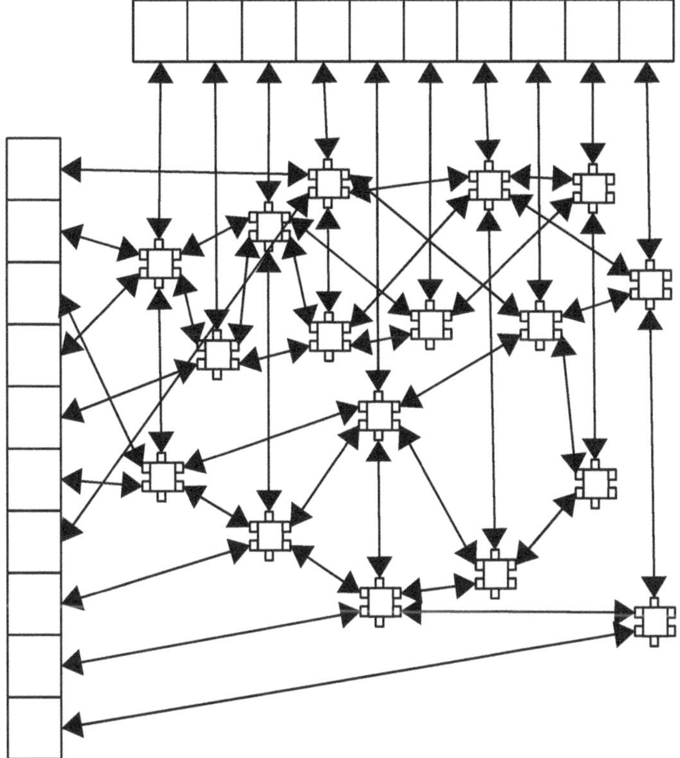

FIGURE 32.5: This shows a pictorial representation of the alternate data structure that can be used as an efficient priority queue for collision handling. The top array keeps the heads of lists for different time brackets during the time step. The left array is the heads of lists for different particles. All lists are doubly linked so that nodes can be removed without finding the one before it.

with a heap. For the general case, that is the ideal structure. Not all problems fit the general case though. In this situation, a heap is probably worse than a sorted linked list and the only way you can get significantly better performance is if you write your own data structure.

32.3 Heapsort

Any priority queue implementation can be turned into a sort. Simply add all the elements to the queue and remove them. They will come off in sorted order. Insertion sort is roughly what you get if you do this with a sorted array or linked list. The priority queue described in the above aside gives something like a bucket sort. If you build a sort from a heap-based priority queue, you get the heapsort. The fact that the ideal heap uses an array means that heapsort can be done completely in place. Simply build the heap on one side of the array and extract to the other.

Here is an implementation of a heapsort. This builds the heap using value as priority so higher values will dequeue earlier. They are then dequeued to fill in the array starting with the end.

```
def heapsort[A](a:Array[A])(comp:(A,A)=>Int) {
  for(end <- 1 until a.length) {
    val tmp = a(end)
    var bubble = end
    while(bubble>0 && comp(tmp,a((bubble+1)/2-1))>0) {
      a(bubble) = a((bubble+1)/2-1)
      bubble = (bubble+1)/2-1
    }
    a(bubble) = tmp
  }
  for(end <- a.length-1 until 0 by -1) {
    val tmp = a(end)
    a(end) = a(0)
    var stone = 0
    var flag = true
    while(flag && (stone+1)*2-1<end) {
      val greaterChild = if((stone+1)*2<end &&
          comp(a((stone+1)*2),a((stone+1)*2-1))>0)
        (stone+1)*2 else (stone+1)*2-1
      if(comp(a(greaterChild),tmp)>0) {
        a(stone) = a(greaterChild)
        stone = greaterChild
      } else {
        flag = false
      }
    }
    a(stone) = tmp
  }
}
```

You should notice similarities between parts of this code and the heap implementation. The first `for` loop contains code that looks like `enqueue`. The second loop contains code that looks like `dequeue`. Some modifications had to be made because the array that is passed in does not include a blank element at the front.

This sort is $O(n\log(n))$, the same as for merge sort and quicksort. Unlike those two, the heapsort is not recursive. It is also stable, like the mergesort, and can be done in place, like the quicksort. These factors make it a nice sort for a number of different applications.

32.4 End of Chapter Material

32.4.1 Summary of Concepts

- Binary heaps are one type of data structure that allow quick retrieval of a maximum priority element.

  - They are based on a binary tree structure.
  - They are always complete and have heap-ordering.
  - Numbering tree elements beginning with one going in breadth-first order gives a simple mathematical relationship between indices of parents and children.
  - This leads to the most efficient implementation using an array because the heap is always complete.

- Binary heaps can be used to implement efficient priority queues. Both `enqueue` and `dequeue` are always $O(\log(n))$ because the heap is a complete tree.
- The heap can be used as the basis for a sort called the heapsort.

32.4.2 Exercises

1. Perform a speed test to compare the speed of the sorted list-based priority queue from chapter 26 using a sorted list to the heap version presented in this chapter. Use exponentially spaced numbers of elements to get a broad range of n values.
2. Implement a binary heap that does not leave the 0 element empty. Test and debug it to make sure with works.
3. If you did the exercise above, you can now do a speed comparison of the heap code in the text to that which starts the heap elements at the zero index.
4. The last aside in the chapter describes an alternate implementation for a priority queue that works well for the collision simulation where scheduled events have to be removed regularly. Write code for this and compare its speed to other priority queue implementations that you have.

32.4.3 Projects

For this chapter, you should convert whatever you did in chapter 26 from the implementation you used in that chapter to one of the ones presented in this chapter. For the simulation workbench, if you have done a collisional simulation, consider doing the special priority queue that makes it easy to remove events.

Chapter 33

Direct Access Binary Files

Back in chapter 9 we learned about basic text files. These had the advantage of being simple. They were reasonably easy to program and the files themselves could be edited with a standard text editor. Unfortunately, they did not have any inherent meaning and they were very brittle to edit. Small mistakes could easily invalidate the whole file. These problems were corrected by using XML starting with chapter 14. Both flat text and XML have the downfall that they are slow and bloated in size. In these areas, XML is worse than flat text. To fix those problems we need to move in the opposite direction and go to binary files. Unfortunately, this move takes away the advantage that text and XML have that they can be read and edited easily with text editors. We saw how we could read and write in native binary in chapter 22 with the `DataInputStream` and `DataOutputStream` as well as how whole objects could be serialized to binary using `ObjectInputStream` and `ObjectOutputStream`.

In this chapter, we are going to take another step with binary files to make it so that they can be used like large arrays with the ability to access and read values at random locations. This makes many more things possible with files and is an essential part of applications that need to deal with extremely large data sets.

33.1 Random Access Files

Back in chapter 22 we focused on streams. The term stream matches their nature. Like water flowing by, data from a stream can be read as it goes by, but once it has gone by you can not bring it back. You also can not choose to just be at whatever position you want. For this capability we turn to a different class in the `java.io` package, `RandomAccessFile`. Looking at this class in the API will show you that it includes all the read methods from `DataInputStream` and the write methods from `DataOutputStream`. It is immediately clear that this one class gives you the ability to do both input and output in binary format. The real power of this class comes from the `seek(pos:Long)` method, which can jump to any location in a file. The partner of `seek` is `getFilePointer():Long`, which returns the

Type	Size in Bytes
Byte	1
Short	2
Int	4
Long	8
Float	4
Double	8
Boolean	1
Char	2

TABLE 33.1: This table shows how many bytes are written to disk for each of the basic value types using either `DataOutputStream` or `RandomAccessFile`.

current location in the file. Putting this all together means that you can jump around in a file and read or write data in any way that you want.

To create a new `RandomAccessFile`, you pass arguments of the file to open, either as a `java.io.File` or a `String`, and a `mode:String`. The `mode` specifies how that file is going to be accessed. It is only allowed to have one of the following values.

- "r" - Open the file for reading only. If this is used and any of the write methods are invoked, an `IOException` will be thrown.
- "rw" - Open the file for reading and writing. This mode will try to create a file if it does not exist.
- "rws" - Open the file for reading and writing with the additional requirement that calls that modify the file in any way synchronously go to the storage device. This will slow things down as it makes buffering of disk activity impossible, but is important for critical applications that need the file to be in a consistent configuration regardless of what happens to the application.
- "rwd" - Much like "rws", but only requires synchronization when the contents of the file are altered.[1]

If you use any other value for the `mode` an exception will be thrown. We will typically use "rw" as the mode so it will look like this.

```
val raf = new RandomAccessFile("binaryFile.bin","rw")
```

33.1.1 Fixed Record Length

The simplest usage of a binary file is with fixed length records. This is where you store data in blocks of the same size. This is fairly easy when your data is composed of a fixed number of fixed size elements like `Int` and `Double`. In that situation you simple add up the sizes of all the elements. Table 33.1 shows the sizes of the basic types in Scala in bytes written to a file. So if the record to be written was a point in 3-D space that has an integer value associated with it, you would write out three `Doubles` and one `Int` using a total of $3 * 8 + 4 = 28$ bytes.

Many types of data will make the process of determining a length a bit more complex.

[1]Files also have metadata associated with them. The "rws" mode will force synchronization when either metadata or contents are changed.

For example, table 33.1 is missing the `String` type. That is because strings will take up different amounts of space based on how long they are. If you want to use a fixed length record file with records that include strings, you have to determine a maximum length that you will allow your string to be, then write it out as a sequence of `Char` or `Byte`. Which you pick will depend on whether you expect to have characters outside of standard English in the text. Normal English characters are part of lower ASCII which fits completely in a `Byte`. If the application is going to be internationalized or include special symbols from Unicode, you should probably stick with a `Char`. You can use the `getBytes():Array[Byte]` method of `String` if you choose to use the `Byte` approach. In your code you can decide how to deal with strings that are too long, whether you want to throw an exception or silently truncate them. Either way, you have to enforce a maximum size, make sure that you are able to read back in what you write out, and never write data beyond the record boundary.

Records that contain collections have to be treated much like `Strings`. You pick a maximum number of elements that are going to be stored to file. Calculate the size of an element and multiply by the maximum number you will store to determine the size contribution of the collection to the total record size.

To apply this idea, we will write a class that uses a fixed length record random access file as the backing for a `mutable.IndexedSeq` so that data can be preserved from one run to the next. If you attach to the same file in separate runs you will automatically get access to the same data. Using a file also allows you to have an extremely large collection without consuming that much memory. The downside is that accessing this collection will be significantly slower.

```
package scalabook.adt

import collection.mutable
import java.io.{File, DataOutput, DataInput, RandomAccessFile}

class FixedRecordSeq[A](
    file:File,
    recordLength:Int,
    reader:DataInput => A,
    writer:(DataOutput,A) => Unit) extends mutable.IndexedSeq[A] {

  private val raf = new RandomAccessFile(file,"rw")

  def apply(index:Int):A = {
    raf.seek(recordLength*index)
    reader(raf)
  }

  def update(index:Int, a:A) {
    raf.seek(recordLength*index)
    writer(raf,a)
  }

  def length:Int = (raf.length()/recordLength).toInt

  def close() = raf.close()
}
```

This class inherits from `mutable.IndexedSeq`, it has fundamentally different behavior for the update method. Assigning out of bounds on this collection does not throw an exception.

Instead, it grows the collection to the size of that index. This does not technically contradict the API documentation for the supertype, but it is something that users should be aware of.

The apply method seeks to the position in the file given by the index times the record length, then does a read. This simplicity is the real advantage of fixed record length files. Given an index, a simple mathematical operation gives the location of the data in the file. The update method is equally simple with a call to seek followed by a call of the write function.

To make it clear how this class would be used we can look at some unit tests for it.

```
package test.scalabook.adt

import org.junit._
import org.junit.Assert._
import scalabook.adt._
import java.io.{File, DataInput, DataOutput}

class TestFixedRecordSeq {
  private var file:File = null
  private case class PointRec(x:Double, y:Double, z:Double, d:Int)
  private case class StudentRec(first:String, last:String, grades:List[Int])

  @Before def setFile {
    file = new File("TextFile.bin")
    file.deleteOnExit()
    if(file.exists()) file.delete()
  }

  @Test def emptyStart {
    val frs = new FixedRecordSeq[Int](file, 4, din => din.readInt(), (dout,i) =>
        dout.writeInt(i))
    assertTrue(frs.length==0)
  }

  @Test def write4 {
    val frs = new FixedRecordSeq[Int](file, 4, din => din.readInt(), (dout,i) =>
        dout.writeInt(i))
    frs(0) = 3
    frs(1) = 8
    frs(2) = 2
    frs(3) = 9
    assertEquals(4, frs.length)
    assertEquals(3, frs(0))
    assertEquals(8, frs(1))
    assertEquals(2, frs(2))
    assertEquals(9, frs(3))
  }

  @Test def write100 {
    val frs = new FixedRecordSeq[Double](file, 8, din => din.readDouble(), (dout,d)
        => dout.writeDouble(d))
    val nums = Array.fill(100)(math.random)
    for(i <- nums.indices) frs(i)=nums(i)
    assertEquals(nums.length,frs.length)
    for(i <- nums.indices) assertEquals(nums(i),frs(i),0)
```

```
}

private def readPoint(din:DataInput):PointRec = {
  new PointRec(din.readDouble(),din.readDouble(),din.readDouble(),din.readInt())
}

private def writePoint(dout:DataOutput,p:PointRec) {
  dout.writeDouble(p.x)
  dout.writeDouble(p.y)
  dout.writeDouble(p.z)
  dout.writeInt(p.d)
}

@Test def writePoints {
  val frs = new FixedRecordSeq[PointRec](file, 3*8+4, readPoint, writePoint)
  val pnts = Array.fill(10)(new PointRec(math.random, math.random, math.random,
      util.Random.nextInt))
  for(i <- pnts.indices) frs(i) = pnts(i)
  assertEquals(pnts.length,frs.length)
  for(i <- pnts.indices) assertEquals(pnts(i),frs(i))
}

@Test def rewritePoints {
  val frs = new FixedRecordSeq[PointRec](file, 3*8+4, readPoint, writePoint)
  val pnts = Array.fill(10)(new PointRec(math.random, math.random, math.random,
      util.Random.nextInt))
  for(i <- pnts.indices) frs(i) = pnts(i)
  assertEquals(pnts.length,frs.length)
  for(i <- pnts.indices) assertEquals(pnts(i),frs(i))
  for(i <- 0 until pnts.length/2) {
    val index = util.Random.nextInt(pnts.length)
    pnts(index) = new PointRec(math.random, math.random, math.random,
        util.Random.nextInt)
    frs(index) = pnts(index)
  }
  for(i <- pnts.indices) assertEquals(pnts(i),frs(i))
}

private def readStudent(din:DataInput):StudentRec = {
  val buf = new Array[Byte](20)
  din.readFully(buf)
  val first = new String(buf.takeWhile(_>0))
  din.readFully(buf)
  val last = new String(buf.takeWhile(_>0))
  val grades = List.fill(10)(din.readInt()).filter(_ > Int.MinValue)
  new StudentRec(first,last,grades)
}

private def writeStudent(dout:DataOutput,s:StudentRec) {
  dout.write(s.first.take(20).getBytes().padTo(20,0.toByte))
  dout.write(s.last.take(20).getBytes().padTo(20,0.toByte))
  s.grades.padTo(10,Int.MinValue).foreach(dout.writeInt)
}

@Test def writeStudents {
```

```
    val frs = new FixedRecordSeq[StudentRec](file, 20+20+4*10, readStudent,
        writeStudent)
    val students = Array.fill(100){
      val first = Array.fill(util.Random.nextInt(15))
          (('a'+util.Random.nextInt(26)).toChar).mkString
      val last = Array.fill(util.Random.nextInt(15))
          (('a'+util.Random.nextInt(26)).toChar).mkString
      val grades = List.fill(util.Random.nextInt(5)+6) (60+util.Random.nextInt(40))
      new StudentRec(first,last,grades)
    }
    for(i <- students.indices) frs(i) = students(i)
    assertEquals(students.length,frs.length)
    for(i <- students.indices) assertEquals(students(i),frs(i))
  }

  @Test def writeStudentsLong {
    val frs = new FixedRecordSeq[StudentRec](file, 20+20+4*10, readStudent,
        writeStudent)
    val students = Array.fill(100){
      val first = Array.fill(10+util.Random.nextInt(15))
          (('a'+util.Random.nextInt(26)).toChar).mkString
      val last = Array.fill(10+util.Random.nextInt(15))
          (('a'+util.Random.nextInt(26)).toChar).mkString
      val grades = List.fill(util.Random.nextInt(5)+6) (60+util.Random.nextInt(40))
      new StudentRec(first,last,grades)
    }
    for(i <- students.indices) frs(i) = students(i)
    assertEquals(students.length,frs.length)
    for(i <- students.indices) {
      if(students(i).first.length>20 || students(i).last.length>20) {
        assertEquals(students(i).copy(first=students(i).first.take(20),
            last=students(i).last.take(20)), frs(i))
      } else {
        assertEquals(students(i),frs(i))
      }
    }
  }
}
```

The first tests work with simple `Int`s. The others work with two different case classes. The first case class, `PointRec`, has the structure described above with three `Double`s and an `Int`. The task of reading and writing the `Int` type is so simple that function literals can be used there. For the case classes, helper functions were used. The ones for `PointRec` are both simple and straightforward.

The second case class, `StudentRec`, includes two `String`s and a `List[Int]`. Having three elements that all have variable sizes is pretty much a worst case scenario for a fixed record length file. The read and write methods determine how the records will be truncated. In this case, both strings are truncated to 20 characters and the list of grades is truncated to 10 grades. When writing, anything shorter than that, the extra spots are padded with placeholders to complete the record. The last test includes strings that are longer than 20 characters to make sure that the truncation works properly.

33.1.2 Indexed Variable Record Length

A bit more work must be done if it is not possible to settle on a single fixed record length. There are a few reasons that you might run into this. One is that you can not afford to lose any data and you can not put any safe upper boundary on the size of some part of the record. A second is that you can not lose data and the average case is much smaller than the maximum and you either can not or are unwilling to waste the space that you would have to if you used a fixed record length.

In these situations you have to use a variable record length. The problem with this is that you can not simply calculate the potion in the file to go and read the nth element. One way to deal with this is to keep a separate index that has fixed length records which gives locations for the variable length record data. We will do this, and to keep things as simple as possible, we will put the index values in a separate file. Here is a sample implementation of such a sequence. It is built very much like the fixed record length version except that instead of a record size there is a second file passed in.

```
package scalabook.adt

import collection.mutable
import java.io._

class VariableRecordSeq[A](
    index:File,
    data:File,
    reader:DataInput => A,
    writer:(DataOutput,A) => Unit) extends mutable.IndexedSeq[A] {

  private val indexFile = new RandomAccessFile(index,"rw")
  private val dataFile = new RandomAccessFile(data,"rw")

  def apply(index:Int):A = {
    indexFile.seek(12*index)
    val pos = indexFile.readLong()
    dataFile.seek(pos)
    reader(dataFile)
  }

  def update(index:Int, a:A) {
    val baos = new ByteArrayOutputStream()
    val dos = new DataOutputStream(baos)
    writer(dos,a)
    val outData = baos.toByteArray()
    val (pos,len) = if(index<length) {
      indexFile.seek(12*index)
      val p = indexFile.readLong()
      val l = indexFile.readInt()
      if(baos.size()<=l) (p,l) else (dataFile.length(),outData.length)
    } else (dataFile.length(),outData.length)
    dataFile.seek(pos)
    dataFile.write(outData)
    indexFile.seek(12*index)
    indexFile.writeLong(pos)
    indexFile.writeInt(len)
  }
```

```
  def length:Int = (indexFile.length()/12).toInt

  def close() = {
    indexFile.close()
    dataFile.close()
  }
}
```

Using an index file makes the apply method a bit more complex. It starts with a seek in the index file and a read of the position of where the data is in the main data file. That value is then passed to a seek on the data file and the value we want is read from that file. You can see from this code that the index file is basically a fixed record length file with a record length of 12. The value of 12 comes from the fact that we are storing both a Long for the position in the data file and an Int for the length of the data record in that file. Storing the length is not required, but it allows us to make the update a bit more efficient in how it stores things in the data file.

Most of the effort for using variable record length files goes into the update method. When a new value is written to file, if it fits in the space of the old value, it should be written in that same space. If it is a completely new record, or if the new value needs more space than the old one, the easiest way to deal with this is to put it at the end of the data file. These criteria are easy to describe, but they have one significant challenge, you have to know the length of what will be written before it is written to file. What is more, we really do not want to force the user of this class to have to pass in an additional function that calculates the size of a record. To get get around this, the update method makes us of a ByteArrayOutputStream. This is an OutputStream that collects everything that is written to it in an Array[Byte]. This is wrapped in a DataOutputStream and passed to the writer function before anything else is done.

Doing this at the beginning of update allows us to get an array of bytes that we need to write to the data file without having an extra function for size or writing to the real data file. After this a little logic is performed to see if the data can be written over an old value or if it needs to go at the end of the file. If this is a new record, or the length of outData is longer than the length of the space in the data file, it has to be put at the end of the data file and the length is the length of this output. Otherwise it goes at the old location and we use the old length because that is the amount of space we can safely use there.

The update method ends by jumping to the proper location in the data file and writing the array out to file, then jumping to the correct position in the index file and writing the position and length there. This code will work with the same test code as the fixed record length version with only a few changes to readStudent, writeStudent, and the last test. There is no longer a reason to cut data down or pad things in this version. Here is the modified code with a stronger version of the last test to make certain the changing records works.

```
  private def readStudent(din:DataInput):StudentRec = {
    val first = din.readUTF()
    val last = din.readUTF()
    val num = din.readInt()
    val grades = List.fill(num)(din.readInt())
    new StudentRec(first,last,grades)
  }

  private def writeStudent(dout:DataOutput,s:StudentRec) {
    dout.writeUTF(s.first)
    dout.writeUTF(s.last)
```

```
    dout.writeInt(s.grades.length)
    s.grades.foreach(dout.writeInt)
  }

  @Test def rewriteStudents {
    val frs = new VariableRecordSeq[StudentRec](iFile, dFile, readStudent,
        writeStudent)
    val students = Array.fill(100){
      val first = Array.fill(10+util.Random.nextInt(15))
          (('a'+util.Random.nextInt(26)).toChar).mkString
      val last = Array.fill(10+util.Random.nextInt(15))
          (('a'+util.Random.nextInt(26)).toChar).mkString
      val grades = List.fill(util.Random.nextInt(5)+6) (60+util.Random.nextInt(40))
      new StudentRec(first,last,grades)
    }
    for(i <- students.indices) frs(i) = students(i)
    assertEquals(students.length,frs.length)
    for(i <- students.indices) assertEquals(students(i),frs(i))
    for(i <- 0 until students.length/2) {
      val index = util.Random.nextInt(students.length)
      students(index) = {
        val first = Array.fill(10+util.Random.nextInt(15))
            (('a'+util.Random.nextInt(26)).toChar).mkString
        val last = Array.fill(10+util.Random.nextInt(15))
            (('a'+util.Random.nextInt(26)).toChar).mkString
        val grades = List.fill(util.Random.nextInt(5)+6)
            (60+util.Random.nextInt(40))
        new StudentRec(first,last,grades)
      }
      frs(index) = students(index)
    }
    for(i <- students.indices) assertEquals(students(i),frs(i))
  }
```

Notice that this code uses **readUTF** and **writeUTF**. We could not do this with the fixed record length file because the number of bytes it writes to file can vary by string contents, not just length. This freedom to use things that we do not know the exact size of can be quite useful. With a little modification, this class could be changed to work with serializable objects. We do not have strong control over the number of bytes written when an object is serialized, but using this approach that is not a problem.

This code does have a weakness though. Every time you update a record with a longer record, the old position in the data file becomes wasted space. Nothing will ever be written there again. Over time, this wasted space can build up and become problematic. To get around that weakness, you can add another method that will build a new data file without the gaps and get rid of the old one. This is left as an exercise for the reader.

33.2 Linked Structures in Files

The two file-backed sequences that we just wrote behave like arrays. They have fast direct access, but trying to delete from or insert to any location other than the middle will

Header				Data				
head	tail	length	first free	next	data	next	data	...
Long	Long	Long	Long	Long	record length	Long	record length	

FIGURE 33.1: The binary file holding the linked list has the following format. The beginning of the file is a header with four values written as `Longs`. They are the position of the head node, the position of the tail node, the number of nodes, and the position of the first free node.

be very inefficient. This discussion should remind you of a similar one in chapter 25 that motivated our initial creation of a linked list. We now want to do that same thing, but in a file. That way we can have the benefits of data retention between runs and large data sets along with the ability for inserts and deletes.

When we built the linked list in memory, the links were created by references from one node to another. The nature of those references is not explicit in Scala. However, it is likely that they are stored as addresses in memory. Languages like C and C++ explicitly expose that detail in the form of constructs called pointers. A pointer stores an address in memory, and allows mathematical operations on it that move the pointer forward or backward through the memory. The languages that expose pointers are, in many ways, ideal for binary files and random access. This is because they allow direct copies of memory to and from a disk. For most values, this is ideal and highly efficient. It does not work at all for pointers. Pointers are attached to memory and do not have meaning when they are separated from the memory of the execution of the program. The addresses of values will change from one run of a program to another and there is no simple way to go from an address in memory to another location in a file to recover the meaning of the link. For this reason, pointers have to be replaced with something that keeps a meaning in the file. The same is true of references.

Whether it is referred to as a pointer or a reference, the equivalent in a file is a position or something roughly equivalent to it. Either one would be stored as an integer value. If it were a true position in the file, it should probably be stored as a `Long`.

There is another detail when doing a linked list, or other linked structure, that we were able to ignore when working in memory, but which becomes fairly significant when we are responsible for dealing with things on our own in a file. The detail is what to do about space in the file that was part of a node that has been deleted. A very sloppy implementation could ignore this and always add new records to the end of the file. Such an implementation would require an extra function to compact the file that would need to be called fairly often. A much more efficient approach is to keep track of those places in the file using a separate linked list that we will call the "free list". The name comes from the fact that it is a linked list of nodes that are free to be used again.

Putting an array into a file is quite straightforward. That is why the code for the fixed record length sequence was so simple. Doing the same for a linked list requires a bit more planning as the linked list needs to keep track of certain information. Think back to our singly-linked list implementation in chapter 25 and the values it stored. It included a head reference, a tail reference, and a length. Our file version will need those as well as the head of the free list. These values will be written at the front of the file. After that will be the nodes, each having a next position at the beginning. This format is drawn out in figure 33.1.

To take advantage of the functionality of a linked list, the code will implement the `Buffer` trait. To make things a bit simpler, it will assume a fixed record size. The class keeps a single `RandomAccessFile` as well as variable to store the current head, tail, length, and first free position. These variables are not directly accessed. Instead, there are methods for getting and setting each. The setters include code that seeks to the proper part of the beginning of the file and writes the new value to the file. This way the variables and the contents of the file should always agree. There is also a method to get the position of a new node. If there are no free nodes, it returns the end of the file. Otherwise it will return the first free node, but before doing that it jumps to the node, reads its next reference, and stores that as the new `firstFree`.

```
package scalabook.adt

import collection.mutable
import java.io.{File, DataOutput, DataInput, RandomAccessFile}

class FixedRecordList[A](
    file:File,
    reader:DataInput => A,
    writer:(DataOutput,A) => Unit) extends mutable.Buffer[A] {

  private val raf = new RandomAccessFile(file,"rw")
  private var (localHead,localTail,localLen,localFirstFree) = {
    if(raf.length()>=3*8) {
      raf.seek(0)
      val h = raf.readLong()
      val t = raf.readLong()
      val l = raf.readLong()
      val ff = raf.readLong()
      (h,t,l,ff)
    } else {
      raf.seek(0)
      raf.writeLong(-1)
      raf.writeLong(-1)
      raf.writeLong(0)
      raf.writeLong(-1)
      (-1L,-1L,0L,-1L)
    }
  }

  private def lhead:Long = localHead

  private def lhead_=(h:Long) {
    raf.seek(0)
    raf.writeLong(h)
    localHead = h
  }

  private def length_=(len:Long) {
    raf.seek(8*2)
    raf.writeLong(len)
    localLen = len
  }

  private def ltail:Long = localTail
```

```
private def ltail_=(t:Long) {
  raf.seek(8)
  raf.writeLong(t)
  localTail = t
}

private def firstFree:Long = localFirstFree

private def firstFree_=(ff:Long) {
  raf.seek(3*8)
  raf.writeLong(ff)
  localFirstFree = ff
}

private def newNodePosition:Long = if(firstFree == -1L) raf.length else {
  val ff = firstFree
  raf.seek(ff)
  firstFree = raf.readLong()
  ff
}

def +=(elem:A) = {
  val npos = newNodePosition
  raf.seek(npos)
  raf.writeLong(-1L)
  writer(raf,elem)
  if(lhead == -1L) {
    lhead = npos
  } else {
    raf.seek(ltail)
    raf.writeLong(npos)
  }
  ltail = npos
  length += 1
  this
}

def +=:(elem:A) = {
  val npos = newNodePosition
  raf.seek(npos)
  raf.writeLong(lhead)
  writer(raf,elem)
  lhead = npos
  if(ltail == -1L) ltail = npos
  length += 1
  this
}

def apply(n:Int):A = {
  if(n>=length) throw new IllegalArgumentException("Requested index "+n+
      " of "+length)
  var i = 0
  var pos = lhead
  while(i<=n) {
```

```
    raf.seek(pos)
    pos = raf.readLong()
    i += 1
  }
  reader(raf)
}

def clear() {
  raf.seek(ltail)
  raf.writeLong(localFirstFree)
  localFirstFree = lhead
  lhead = -1
  ltail = -1
  length = 0
}

def insertAll(n:Int, elems:Traversable[A]) {
  if(n>length) throw new IllegalArgumentException("Insert at index "+n+
      " of "+length)
  var i = 0
  var (prev,next) = if(n==0) (-1L,lhead) else {
    var (pp,nn) = (lhead,-1L)
    while(i<n) {
      raf.seek(pp)
      if(i<n-1) pp = raf.readLong()
      else nn = raf.readLong()
      i += 1
    }
    (pp,nn)
  }
  if(prev != -1L) raf.seek(prev)
  for(elem <- elems) {
    val npos = newNodePosition
    if(prev == -1L) {
      lhead = npos
      prev = npos
    } else raf.writeLong(npos)
    raf.seek(npos+8)
    writer(raf,elem)
    raf.seek(npos)
  }
  if(next == -1L) ltail = raf.getFilePointer()
  raf.writeLong(next)
  length += elems.size
}

def iterator = new Iterator[A] {
  var pos = lhead
  def hasNext = lhead > -1L
  def next = {
    raf.seek(pos)
    pos = raf.readLong()
    reader(raf)
  }
}
```

```
  def length:Int = localLen.toInt

  def remove(n:Int):A = {
    if(n>=length) throw new IllegalArgumentException("Remove index "+n+
        " of "+length)
    var i = 0
    var pos = lhead
    var last,next = -1L
    while(i<=n) {
      raf.seek(pos)
      if(i==n) {
        next = raf.readLong()
      } else {
        last = pos
        pos = raf.readLong()
      }
      i += 1
    }
    val ret = reader(raf)
    if(last == -1L) {
      lhead = next
    } else {
      raf.seek(last)
      raf.writeLong(next)
    }
    if(pos == ltail) {
      ltail = last
    }
    length -= 1
    ret
  }

  def update(n:Int, elem:A) {
    if(n>=length) throw new IllegalArgumentException("Updating index "+n+
        " of "+length)
    var i = 0
    var pos = lhead
    while(i<=n) {
      raf.seek(pos)
      pos = raf.readLong()
      i += 1
    }
    writer(raf,elem)
  }

  def close() = raf.close()
}
```

Going through this code, you will notice that there are many places that have a call to `seek` followed by either a `readLong` or a `writeLong`. This is because the first eight bytes in each node are used to store the position of the next value.

As with a normal linked list, this code has quite a few `while` loops that run through the list counting. It also has the overhead of boundary cases that deal with the head and tail references. Those special cases, combined with the fact that file access is more verbose

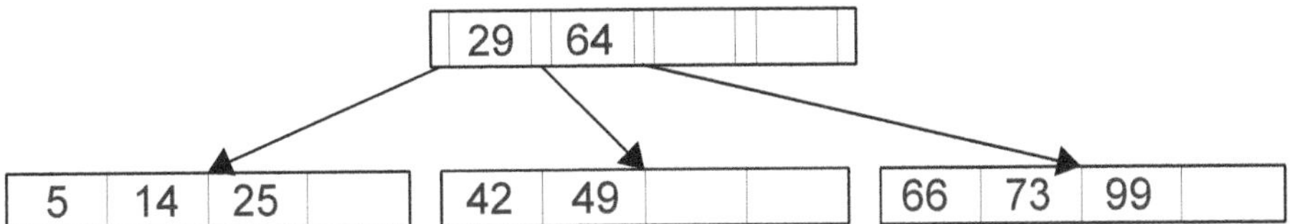

FIGURE 33.2: The nodes of a B-tree can each store a certain number of records. The number of records is selected to match an optimal size for disk reads and writes. Non-leaf nodes have references to children at the beginning, end, and between each record. The records in a node are sorted by a key. In addition, the nodes in children have to have key values that fall between the keys of the records they are between, or just below and above for the beginning and end children. B-trees are always balanced and require few disk accesses to find a record, but they do potentially waste some space.

than accessing memory makes this code fairly long though the length is largely made up of short, simple lines.

We stated above that this code uses a fixed record length, but there is no record length passed in. The way this code works, it does not have to know the length, but it will fail if a fixed length is not used for all nodes. The assumption of a fixed length is made in the fact that this code assumes that any freed node can be used for any new data. This assumption breaks down if different pieces of data have different lengths. To make this code work with different length records requires three changes. First, each node must keep a length in bytes in addition to the next reference. Second, the `writer` must be called using a `ByteArrayOutputStream` as was done earlier. Third, the `newNodePosition` must be passed the size that is required and it will have to walk the free list until it finds a node that is big enough. If none are found, it will use the end of the file, even when nodes have been made free. Making these alternations is left as an exercise for the reader.

B-Trees

The file-based linked list works, but it is, in many ways, a horrible way to do things in a file. Disk reads and writes are extremely slow, especially done in small pieces. Disks work much better if you read a large chunk at a time. All the seeking in this code, and the fact that over time the list is likely to evolve such that consecutive nodes are very distant in the file will cause very poor performance and likely make this unusable for most applications.

To deal with the special requirements of disk access, certain data structures have been developed that are specifically designed to access large chunks of data at a time and to do few such accesses. One of the easiest such structures is the B-tree, which was originally described in 1972. [3] The B-tree is an extension on the idea of a Binary Search Tree (BST). To make it more efficient for disk access, each node can hold multiple records and for non-leaves, there are children nodes not only for lower and higher keys, but also between each of the records. A sample B-tree is shown in figure 33.2.

Unlike a normal BST, a B-tree is always balanced. In addition, new nodes are added higher up the tree, instead of at the leaves. By default, a new record is added in a leaf.

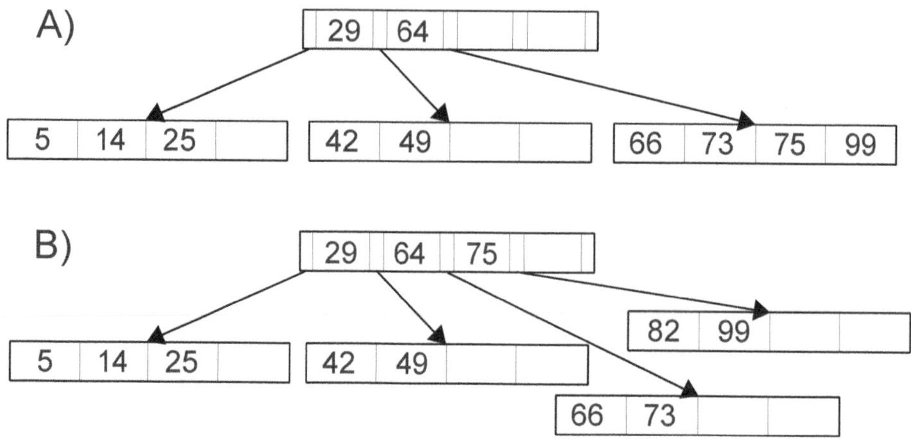

FIGURE 33.3: Part A on the top shows the tree from figure 33.2 with the addition of a record whose key value is 75. The 75 is simply placed in sorted order in a leaf because there is still space left in that node. For part B of this figure, a record with a key value of 82 has been added. It would go right after the 75 record, but there is no more space in that node. Of the five records that would be in the node, the 75 is the median so it is moved into the parent and the other records are split into two different nodes. This process can propagate to higher levels in the tree if the parent is also full. A new level is created when the root node fills and has to split.

As long as that leaf still has spots left, that new record is inserted in the proper location and the add is done. This is illustrated in figure 33.3-A. If the leaf node where the new record belongs is already full, then a new node has to be made. This is done by splitting that one node into two, pushing the median record up into the parent, and putting links to the two children nodes on either side. This is shown in figure 33.3-B.

Real B-trees will typically have many more records per node than what is drawn in these figures. The fact that their average splitting factor is very high means that they can hold an extremely large number of records in very few levels. To see this, consider a B-tree with 200 records per node that is roughly half filled. That means that each node has roughly 100 records in it and each internal node has roughly 100 children.[2] So the top level would hold 100 records. The next would hold 10,000. The next would hold 1,000,000. This is still $O(\log(n))$ behavior, but it is technically $O(\log_{100}(n))$, which grows very slowly. A tree with a billion records will only barely crack five levels. This is significant, because each level winds up being a `seek` and a `read` when trying to find records.

[2]The rough approximation of this argument allows us to ignore the fact that a node with 100 records would have 101 children.

33.3 End of Chapter Material

33.3.1 Summary of Concepts

- Direct access, or random access files allow you to jump from one position to another, reading and writing.
- This functionality allows them to be used as array-like data structures with extremely large capacities.
- It is easiest to deal with files where all records have the same length as that allows us to easily calculate where any given record is in the file.
- Variable record lengths can be used, but they require more bookkeeping. This is typically done by having a fixed record length index that tells us the position of elements in the data file.
- We can mimic references using file positions to create linked data structures in files.
  - Putting header information in a file allows us to keep track of things like the head or tail of a list, or the root of a tree.
  - Additional information can be stored with each record to include references to other nodes in the data structure.
  - While we can code a linked list in a direct access file, it is very inefficient because it requires a large number of disk accesses.
  - Other linked structures exist specifically to get around the problem of slow access times. The B-tree is an example of one.

33.3.2 Exercises

1. Write code to do speed comparisons of different file-backed sequences (array-like and linked list-like) to one another and compare them to memory-backed sequences.
2. Write a `compact` method for the `VariableRecordSeq`. This method should be called to eliminate blank space in the data file. It does this by creating a new data file and moving records from the old one to the new one, one at a time, updating the index for the new positions appropriately. Once this is done, the old data file can be deleted and the name of the new one can be changed to match the original name.
3. Make appropriate changes to the `FixedRecordList` to make a `VariableRecordList`.
4. Write a `linearize` method for the linked list that will create a new file where all the nodes have been reordered to the order they appear in the list and all free nodes have been removed.
5. Write a B-tree that only works in memory.
6. Convert your B-tree from the previous exercise so that it works using disk access.

33.3.3 Projects

All of the projects in this chapter have you writing direct access binary versions of some of your saved files for your project.

1. One of the things you have probably noticed if you are working on the MUD project is that saving individual players out to disk either requires separate files for each player, leading to a very large number of files, or it requires writing out all all the players just to save one. A direct access binary file can fix this. You might have to make some sacrifices, such as specifying a maximum number of items that can be saved. Write code to store players in a fixed length record and change the saving and loading code so that only single records are read or written as needed.

2. Depending on what data you are storing, there might be very different things that you put in a direct access binary file for the Web Spider. Things like box scores can be made much more efficient using a binary format instead of text. Images will be a bit more challenging. If you really want a challenge, try to put your data in a B-tree. The web spider is a perfect example of a project that uses enough data that having things stored on disk in a B-tree instead of in memory can be very beneficial.

3. If you have a graphical game that has different boards or screens, edit the storage so that it uses a direct access file to efficiently get to any screen in $O(1)$ time.

4. The math worksheet project can benefit from saving the worksheet in a binary file so that saving can be done on individual sections of the worksheet, instead of the whole thing.

5. If you are doing the Photoshop® project, you can put layers in variable-sized records of a binary file. Note that you can not use default serialization for images so you will likely have to write some form of custom serialization.

6. If you are doing the simulation workbench, you should make it so that large particle simulations write out using a custom binary format where the data for each particle is written out as the proper number of binary `Doubles`. This will allow you to write tools for reading things in that only read specific particles.

Chapter 34

Actors

Back in chapter 21 we explored the concepts of multithreading and looked at both the foundation concepts as well as several easier ways to get our programs to utilize multiple threads. The reality of the situation is that programs need to be multithreaded to take full advantage of modern processors, even on mobile devices like tablets and smartphones. The addition of parallel collections in Scala 2.9 makes it remarkably easy to parallelize programs as long as they fit the model of having multiple pieces of data that can be operated on independently. If you code in a completely functional way with no mutable data, it is also possible to break work across many threads with little challenge.

Unfortunately, not all problems are well suited for functional approaches, nor is everything data parallel in the manner needed for parallel collections. The creators of Scala are well aware of the challenges of parallelism and they have worked to make available as many alternative solutions as possible, working under the assumption that there is no "silver bullet" to solve the problem.

In earlier chapters, anything that did not fit into the parallel collections model was dealt with by going down to Java libraries either in `java.util.concurrent` or all the way down to `java.lang` with the creation of individual threads and explicit synchronization. These approaches are difficult and error prone. For that reason, Scala includes one other model of parallelism that can help with problems that are not data parallel, the actor model.

34.1 Actor Model

The idea of the Actor model is that many problems can be solved by individual actors who work independently of one another and communicate through messages. A classic example of this would be an assembly line. Each worker on the line is an actor. Unlike the parallel collection model where each thread is doing roughly the same thing to different data, the workers on an assembly line can be doing very different things to the pieces they are working on. Those pieces get passed from one worker to the next going down the line to produce a final product. This mental image works fairly well for describing the actor model.

Each actor can be working on its own, independent of the others, and doing a specialized task that perhaps only it does.

The actions of an actor are motivated by messages received from other actors. In the assembly-line analogy, a message could be an object that moves from one station to a bin by the next one or perhaps a shout giving people information. Either way, the message simply passes information from one actor to another. If that information comes in the form of a physical object, the actor that does the passing stops interacting with the physical object after passing it on.

The way the actor model works in threads and how it prevents standard problems like race conditions is dependent on the condition that when something is passed in a message, the sender stops changing it. You can picture each actor as having its own thread.[1] Each actor can also have mutable data that it keeps privately. This is where the actor model differs from a pure functional approach. Mutability is allowed, but it needs to be internal to one actor at a time. Actors receive messages to act on instead of having methods called on them. The difference is subtle, but significant. A method call passes control of a thread from one object to another. A message that is sent from one actor to another does not give the receiving actor control over the thread of the sender. It is just information that is being sent, not thread control. This prevents race conditions and deadlocks because mutable data is localized to a single actor/thread.

A big part of the advantage of using actors, and why they are generally easier and less error prone than using thread level controls, is that you can think about your problem at a higher level. Instead of thinking about threads and synchronization, you approach the problem as having different actors that each work mostly independently and communicate with one another. Each actor can have a state and mutate that state. They each behave differently depending on the state that they are in. Indeed, you probably should not spend much time thinking about the fact that actors can be performing actions in parallel when you are working on an actor-based program. That part will simply fall out and multiple threads will automatically be used when it fits.

The actor model of parallelism is not unique to Scala. The Erlang programming language is probably the one most commonly associated with actors. Erlang is a functional language that does pretty much everything through actors. The Scala libraries even borrow from Erlang syntax so that sending messages to actors in Scala looks the same as it would in Erlang.

34.2 `scala.actors` Package

The standard actor implementation for Scala is found in the `scala.actors` package. This package was briefly mentioned in chapter 21 as a simple way to spawn threads with or without getting a return value using `Futures.future` or `Actor.actor` respectively. While the use of `Actor.actor` technically makes an actor, it has very limited functionality because the body is not in the scope of a new class that is a subtype of the `Actor` class. To make fully functional actors we need to create a new class that extends `Actor`.

The `Actor` type has one abstract method in it that must be implemented by all concrete

[1]As we will see later in the chapter, you often do not want this extreme view to be the reality as there could be millions of actors and you do not really want to have millions of threads. Still, as the programmer will not be controlling the threads manually, the only safe assumption is that each actor could be acting on a separate thread.

subtypes, `act():Unit`. As the name implies, this method defines what the actor will do when it starts running. This is demonstrated in the following REPL sequence.

```
scala> import actors._
import actors._

scala> val actor = new Actor {
     |   def act {
     |     println("Hello world!")
     |   }
     | }
actor: java.lang.Object with scala.actors.Actor = $anon$1@368e091

scala> actor.start
res0: scala.actors.Actor = $anon$1@368e091

scala> Hello world!
```

Note that the print statement is not actually executed until the actor is told to start. When it is told to start, the line is printed using a separate thread from the REPL. In this case, that caused it to be printed after a prompt. What was done here is basically a longer version of this.

```
scala> Actor.actor{ println("Hello world!") }
Hello world!
res1: scala.actors.Actor = scala.actors.Actor$$anon$1@7c7bdb49
```

When you call `Actor.actor`, the argument becomes the body of a new subtype of `Actor` and the `start` method is called on an instance of that type.

You can only call the `start` method once. Further calls to that method will not cause anything to happen. To make the `act` method occur again, the `restart` method can be called.

34.2.1 The `receive` Method and Messages

Used in the way shown above, an actor is little more than a separate thread for some action to take place in. To make more of the actor model we need to pass messages between actors and have the actors react to those messages. Messages are sent using an exclamation point using the form *actor* ! *message*. The message has type `Any` so you can send anything that you wish as a message. There are some things, like case classes that work better for reasons that we will discuss here.

One way to have an actor get a message is by having it call `receive`. The `receive` method takes a single argument of a partial function. That partial function should have cases for the messages that the actor should handle with that `receive`. The fact that `receive` takes a partial function with cases is one of the first reasons why case classes make sense as messages. Anything that you can do pattern matching on makes sense when you are writing a partial function. This includes literals for numbers, strings, and tuples. The problem with using such built-in types is that the type itself does not carry meaning and this can lead to problems as an application grows. Using meaningful types can be just as important as using meaningful names for variables, methods, or types themselves.

Simple examples of actors are a bit lacking in their illustrative power. They are somewhat like little examples of object-orientation. They do not sufficiently motivate the topic. For that reason, we will instead look at pieces of code that are related to the larger example with

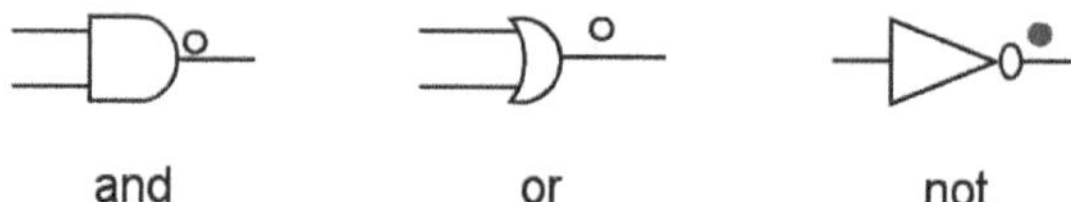

FIGURE 34.1: These are the standard graphical representations of "and", "or", and "not" gates.

project integration that will be fully fleshed out in section 34.3. The example code we will develop in that section uses actors to model elements in circuits at the level of logic gates. This will be put in the drawing program and use the `Clickable` trait like the `DrawGraph` did so that the user can build circuits graphically.

Computer circuits are built using the basic component of a transistor. Transistors can be connected in ways that produce the behavior of the logic operations that we have been using in logic operations since chapter 4: and, or, and not. These are commonly represented in graphical diagrams using the shapes shown in figure 34.1. The solid lines coming in and out of the gates represent wires that can be at a voltage that represents 0 or 1.

The simulation will use actors for the gates and wires. A basic implementation of a `Wire` class along with case classes for the types of messages that would be sent to the wires is shown below. The wire has a `Boolean` state for the voltage that is on it, values for the Cartesian position it should be drawn at, and a `Buffer` of the gates that it connects to.

```
private class Wire(private[this] var state:Boolean,
    private[this] var x:Double,
    private[this] var y:Double) extends Actor{
  private[this] val outs = mutable.Buffer[Gate]()
  def act {
    while(true) {
      receive {
        case SetWire(v) =>
          if(v!=state) {
            state = v
            outs foreach (_ ! WireChange(state,this))
            drawing.refresh
          }
        case MoveTo(nx,ny) =>
          x = nx
          y = ny
        case GetPos => sender ! (x,y)
        case GetPosAndState => sender ! PositionAndState(x,y,state)
        case ConnectWireOutput(g) =>
          outs += g
          g ! WireChange(state,this)
        case DisconnectOut(g) => outs -= g
      }
    }
  }
}

private case class SetWire(value:Boolean)
private case class MoveTo(x:Double,y:Double)
```

```
private case class ConnectWireOutput(g:Gate)
private case class DisconnectOut(g:Gate)
private case object GetPos
private case object GetPosAndState
private case class PositionAndState(x:Double,y:Double,state:Boolean)
```

The only method in the `Wire` actor is `act`. This method follows something of a standard form for many actors when using `receive`. There is a `while` loop that goes until the actor is no longer needed with a call to `receive` inside of that. In this case the loop goes forever because the circuit is only supposed to stop running if the whole application is stopped. Such a loop is needed because each call to `receive` handles only a single message.

In this case, there are six different cases for six different types of messages that can be passed to a wire. The first sets the state. If there is a change, this change is passed on to all the connected gates. Note the use of `!` to send a `WireChange` message to each gate. This is significant. Actors should only communicate using messages, not direct method calls. Direct method calls can lead to a breakdown of the safety of the actor approach with the race conditions that come with that. To make it extremely explicit that nothing other than the current actor can change the value of `x`, `y`, `state`, or what is stored in `outs`, all four member values are declared to be `private[this]`. This notation was briefly described in an aside on page 399. It means that those members can only be accessed by the object in which they exist. Not even other objects of the same type can see or change those values. All four of those values are mutable, which makes them potential causes of race conditions. When done properly, the actor model will limit modifications of such mutable values to a single thread. Declaring them `private[this]` helps to enforce the access requirements we want to maintain.

This goal of limiting access to a single thread is another reason for using case classes for messages. Data that is sent in a message is implicitly known to two different actors. If that data is mutable, it then becomes possible for two actors to make changes to the data and cause race conditions. The easiest way to prevent this is to use immutable data for messages. One of the features of case classes is that they are immutable. So if all your messages are case classes with immutable contents or basic immutable types like `Int`, `Double`, `String`, or tuples, you do not have to worry about two actors modifying the same data through a message.

While immutable messages are preferred because they keep life simpler, it is possible to use mutable messages. The assembly line example, if taken literally, is an example of mutable messages. The items that are passed from one station to another are the messages. If the workers are doing something to alter those items, then they are mutable. This situation can be replicated in code, but you have to make sure that when a mutable object is passed off, the actor sending the message stops modifying it. Ideally, it would erase all references to it just to be safe.

Two of the messages that are handled have code to send a message back to `sender`. The name `sender`, when used in this context, does exactly what you would expect. This is how we return a value from a message handler. The `!` does not set up anything to deal with a response. It gives back a value of type `Unit`. For messages that need a response we can use one of two other sending operators. The `!!` operator returns a `scala.actors.Future[Any]`. This is the same `Future` type that we saw in section 21.5 for calls to `Futures.future`. There is an `apply` method that gives back the response, blocking if none is available. If you do not want to risk blocking, the `isSet:Boolean` method will tell you whether or not the value is ready. We could use this with a `GetPos` message with code like this.

```
val fut = comp !! GetPos
val (ccx:Double,ccy:Double) = fut()
```

The call to `comp !! GetPos` gives us back a `Future[Any]` that we call `fut`. The next line gets the value of that future and matches it to a tuple of `Doubles`. This is hard coded to a pattern with a tuple of `Doubles` because we feel secure that is what we will get as the response to a `GetPos` message. If we were worried about other possibilities it would be appropriate to use a `match` and have a `case _ =>` option that gives back a default value instead of throwing a pattern match exception.

Futures can be really helpful when there are things that need to be done between the point where you start the calculation and when you want to use the result. This is not one of those situations. Here we automatically apply the future to get the value. This makes the code longer and forces the creation of a `Future` object that is not really needed. For this type of situation the `!?` operator is a preferable option. Unlike `!` and `!!`, `!?` is blocking. So when you use it, the thread you use it in will block until the actor gets to the message and responds. This is exactly what happened above with the call to `fut()` so it is not a problem. Using this in our code segment simplifies it to this

```
val (ccx:Double,ccy:Double) = comp !? GetPos
```

One word of caution here, if you use `!?` and send a message to an actor that does not respond to it, the thread will block forever. It is basically like starting a calculation that includes an infinite loop.

34.2.2 `react`, `loop`, and `loopWhile`

The `receive` method works fine for many applications and would probably be fine for what we are doing here. However, it does not scale to programs with large numbers of actors. The reason is that `receive` blocks the actors thread until a message arrives that the partial function can deal with. Blocking the thread means that every actor that uses `receive` will need its own thread. There might be some actors for which you want to reserve a thread and then this is fine, or perhaps your total number of actors is small and having a thread for each is not a problem. If your applications needs thousands of actors or more, the overhead of having that many threads will be a huge drag on the application. Full threads simply have too much overhead.

The way to get around this is to use `react` instead of `receive`. The `react` method is called much like `receive`. You pass it a partial function that includes the handlers for the different messages that you want to deal with. Looking closely at the Application Programming Interface (API) entries will show you that there is a difference between the two.

```
def receive[R](f: PartialFunction[Any, R]): R
def react(handler: PartialFunction[Any, Unit]): Nothing
```

The `receive` method is parametric on the return type. You can use it as an expression and it will give you the return of the partial function when an appropriate message is received. The return type of `react` is `Nothing`. You might recall this type from figure 19.1 on page 453. It is at the very bottom of the Scala type hierarchy, a subtype of every other type. There is no instance of type `Nothing`, which means that a method of that type can not return. Indeed, that is the behavior of `react`.

When you call react, it never comes back. Anything that you put after the call to `react` will never happen. It is completely unreachable. The advantage of this is that when you use `react` in an actor, the actor library can take that thread and use it for something else before or after an appropriate message is arrives. If you have thousands of actors that all use `react`, they can do their work sharing a number of threads that is more appropriate for the machine that you are working on instead of being forced to give a thread to each and every actor.

Unfortunately, the fact that the method never returns causes a problem for the `while` loop idiom that was shown earlier. The `while` loop repeats in a normal thread. If you call `react` in a `while` loop, the partial function to handle the messages will only be called once, even if you tried to make an infinite loop with `while(true)`. To deal with this problem, there are two other methods that are part of `Actor`, `loop` and `loopWhile`. If you are going to use `react`, you should probably use one of these. The `while(true)` idiom from above would be replaced with code like this.

```
def act {
  loop {
    react {
      case ...
    }
  }
}
```

The standard `loop` will cause the `react` to be executed until the application terminates. If you have a condition on which the actor should stop acting, you can use `loopWhile` as shown here.

```
def act {
  var flag = true
  loopWhile(flag) {
    react {
      case ...
    }
  }
}
```

Note that the syntax for `loopWhile` is identical to a standard `while` loop. This is possible thanks to three elements of the Scala syntax. First, both the condition and the body of the loop use pass-by-name semantics so that the code is executed multiple times. Second, the two arguments are passed separately using curried arguments. Third, Scala allows any argument list with only one argument to be surrounded by curly braces instead of parentheses. This is only done for the body here, but it gives the call the same structure as the loop that is part of the language syntax.

As with `react`, neither `loop` nor `loopWhile` return. If you put code after either of them it is unreachable and will never be executed.

Domain Specific Languages (DSLs)

Specialized problem areas can often benefit from having their own programming language to work with that has exactly the constructs that are needed for that area. The field of scientific computing has plenty of examples of whole languages developed primarily for solving specific types of problems. The languages built into tools like MATLAB®, Mathematica®, Maple®, and many others demonstrate this quite well. These Domain Specific Languages, DSLs, can make work in the field move much faster. However, creating a full language implementation is a difficult task that takes a lot of effort. Doing it well so that the resulting language is robust and has good performance is even harder.

Scala was designed to be highly scalable. One of the results of that effort is the ability to create libraries that look like they are language features and to create DSLs inside of Scala. The actor library is an example of this. Programming actors is programming Scala, but if your only goal was to develop actor parallel tools, you could learn a particular subset of Scala and the actor library and get up and running fairly quickly. The applications you write in this way will benefit from the effort that has gone into making Scala robust and get the performance of Scala at the same time. Writing DSL libraries in Scala takes a deep knowledge of the Scala language using advanced features that are not covered in detail in this book, but it is still a lot easier than creating your own complete language.

If you want a DSL that is dynamic, that you can nest inside of an application, and you do not care so much about performance, the combinator parsers that were discussed in chapter 30 can be helpful. As we saw in that chapter, if you can write a Context-Free (CF) grammar for a language, you can get a combinator parser up and running fairly quickly. That language can then allow your users to write simple logic inside of your running programs.

34.3 Circuit Simulation

As was mentioned earlier, we will integrate actors into the drawing framework using the example of a circuit simulation. The gates in a circuit have the ability to do logical operations on incoming voltages. Gates do not flip instantly though. When the potential on an input wire changes, there is a short delay before the output wire adjusts to the appropriate voltage. We could simulate this process using a discrete event simulation like those we discussed in chapter 26. While that is a good rigorous way of doing this, it is an inherently serial approach while circuits are really very parallel. The approach taken here uses actors to capture that parallelism.

This simulation is not going to be a highly efficient use of a multicore machine. That is largely because of the graphical updating. Instead, this is an example of using actors for event-based programming. Messages are passed between actors to represent events that occur in the circuit. For example, a wire is told when the gate that sets its voltage changes. That wire then sends messages to all the gates it is an input to about this new value. The goal here is to think about the circuit at the higher level of gates and wires and not worry about things like queues or threads.

The new `DrawCircuit` class will extend the `Clickable` trait that we first used with the `DrawGraph`. A fair bit of code is modeled after that class so that the user can add gates and switches, then connect wires by clicking on the drawing. There is a significant difference though. All values in the circuit components will be edited though messages. This includes the position of gates. In addition, the states of wires are queried using messages. This might seem like a small detail, but it makes the application more thread-safe. Recall that Graphical User Interface (GUI) events happen in a separate thread. When GUI events mutate values that are used by other threads, race conditions can occur. These race conditions are extremely hard to duplicate because they require user interaction. Actors give us a way of fixing a problem that you might not have even realized we had in code that was written earlier.

The code for `DrawCircuit` is shown below. At a bit over 400 lines it is one of the longest pieces of code we have seen in the book. Data representing the circuit structure is passed in and built into an array named `comps`. There is also a mutable `Int` that can be edited through the GUI named `speedMult`. This is a speed multiplier for the animation of the circuits. Smaller values shorten the delay between when a wire value changes and when the gates actually flip. There are also a number of different transient variables that are used for the graphical user interaction.

```
package scalabook.drawing

import java.awt.{Graphics2D,Color}
import java.awt.geom._
import scala.swing._
import scala.swing.event._
import actors.Actor
import collection.mutable

class DrawCircuit(p:DrawTransform,
    compData:Seq[DrawCircuit.CompData],
    wireData:Seq[DrawCircuit.WireData],
    private var speedMult:Int) extends DrawLeaf with Clickable {
  val parent = p
  private val comps = mutable.Buffer[CircuitComponent]()
  def setupComponents {
    import DrawCircuit._

    val gridLinks = for(cd <- compData) yield {
      val (cc,ins) = (cd match {
        case SwitchData(x,y,s) => (new Switch(s,x,y),Seq[Int]())
        case GateData(t,x,y,in) => t match {
          case "or" => (new OrGate(x,y),in)
          case "and" => (new AndGate(x,y),in)
          case "not" => (new NotGate(x,y),in)
        }
      })
      comps += cc
      cc.start
      (cc,ins)
    }
    for((cc,in) <- gridLinks)
      for(i <- in.indices; if in(i)>=0) {
        comps(in(i)) match {
          case w:Wire => cc ! ConnectGateInput(20*i-10,w)
        }
      }
  }

  setupComponents

  @transient private var propPanel:Component = null
  @transient private var speedField:TextField = null
  @transient private var clickAction:(MouseEvent,Double,Double)=>Unit = null
  @transient private var hoverComp:CircuitComponent = null
```

```
@transient private var startComp:CircuitComponent = null
@transient private var (dx,dy) = (0.0, 0.0)
@transient private var (sx,sy) = (0.0, 0.0)

def draw(g:Graphics2D) {
  val posAndState = comps.map(_ !? GetPosAndState match {
    case pas@PositionAndState(x,y,s) => pas
  })

  def drawGate(gate:Gate) = {
    val (gx,gy) = gate !? GetPos match {
      case (x:Double,y:Double) => (x,y)
    }
    g.setPaint(if(gate==hoverComp) Color.red else Color.black)
    gate match {
      case og:OrGate =>
        val orPath = new GeneralPath()
        orPath.moveTo(-15+gx,-10+gy)
        orPath.quadTo(5+gx,-10+gy,15+gx,0+gy)
        orPath.quadTo(5+gx,10+gy,-15+gx,10+gy)
        orPath.quadTo(0+gx,0+gy,-15+gx,-10+gy)
        orPath.closePath()
        g.fill(orPath)
      case ag:AndGate =>
        g.fill(new Rectangle2D.Double(gx-15,gy-10,20,20))
        g.fill(new Arc2D.Double(gx-5,gy-10,20,20,-90,180,Arc2D.PIE))
      case ng:NotGate =>
        val triangle = new GeneralPath()
        triangle.moveTo(-15+gx,-10+gy)
        triangle.lineTo(10+gx,0+gy)
        triangle.lineTo(-15+gx,10+gy)
        triangle.closePath()
        g.fill(triangle)
        g.fill(new Ellipse2D.Double(gx+10,gy-2.5,5,5))
    }
    val (ins,outs) = gate !? GetWires match {
      case (i:List[Wire],o:List[Wire]) => (i,o)
    }
    for((iw,i) <- ins.zipWithIndex; if iw!=null) {
      val index = comps.indexOf(iw)
      g.setPaint(if(posAndState(index).state) Color.green else Color.black)
      g.draw(new Line2D.Double(posAndState(index).x, posAndState(index).y,
          gx-15, gy+20*i-10))
    }
    for((ow,i) <- outs.zipWithIndex) {
      val index = comps.indexOf(ow)
      g.setPaint(if(posAndState(index).state) Color.green else Color.black)
      g.draw(new Line2D.Double(posAndState(index).x, posAndState(index).y,
          gx+15, gy+20*i))
    }
    if(startComp!=null) {
      g.setPaint(Color.black)
      g.draw(new Line2D.Double(sx,sy,dx,dy))
    }
  }
```

```
  def drawWire(i:Int) {
    val PositionAndState(x,y,s) = posAndState(i)
    g.setPaint(if(s) Color.green else Color.black)
    g.fill(new Ellipse2D.Double(x-3,y-3,6,6))
    if(comps(i)==hoverComp) {
      g.setPaint(Color.red)
      g.fill(new Ellipse2D.Double(x-2,y-2,4,4))
    }
  }

  def drawSwitch(sw:Switch,i:Int) {
    val PositionAndState(x,y,s) = posAndState(i)
    g.setPaint(if(s) Color.green else Color.black)
    g.fill(new Rectangle2D.Double(x-4,y-4,8,8))
    if(sw==hoverComp) {
      g.setPaint(Color.red)
      g.fill(new Rectangle2D.Double(x-3,y-3,6,6))
    }
    val wIndex = comps.indexOf(sw.out)
    g.setPaint(if(posAndState(wIndex).state) Color.green else Color.black)
    g.draw(new Line2D.Double(posAndState(wIndex).x, posAndState(wIndex).y, x, y))
  }

  for((comp,i) <- comps.zipWithIndex) comp match {
    case g:Gate => drawGate(g)
    case w:Wire => drawWire(i)
    case s:Switch => drawSwitch(s,i)
  }
}

def propertiesPanel() : Component = {
  if(propPanel == null) {
    import BorderPanel.Position._
    propPanel = new BorderPanel {
      val radioButtons = Seq(
        new RadioButton { action = Action("Add Switch") {
            clickAction = (e,x,y) => e match {
              case mc:MouseClicked =>
                val sw = new Switch(false,x,y)
                comps += sw
                sw.start
              case _ =>
            }
        } },
        new RadioButton { action = Action("Add And Gate") {
            clickAction = (e,x,y) => e match {
              case mc:MouseClicked =>
                val and = new AndGate(x,y)
                comps += and
                and.start
              case _ =>
            }
        } },
        new RadioButton { action = Action("Add Or Gate") {
```

```
      clickAction = (e,x,y) => e match {
        case mc:MouseClicked =>
          val or = new OrGate(x,y)
          comps += or
          or.start
        case _ =>
      }
  } },
  new RadioButton { action = Action("Add Not Gate") {
      clickAction = (e,x,y) => e match {
        case mc:MouseClicked =>
          val not = new NotGate(x,y)
          comps += not
          not.start
        case _ =>
      }
  } },
  new RadioButton { action = Action("Connect Wire") {
      clickAction = (e,x,y) => e match {
        case mp:MousePressed =>
          if(hoverComp!=null) {
            startComp = hoverComp
            val (ccx,ccy) = hoverComp !? GetPos match {
              case (x:Double,y:Double) => (x,y)
              case _ => (0.0,0.0)
            }
            sx = ccx
            sy = ccy
            dx = ccx
            dy = ccy
          }
        case mp:MouseDragged =>
          if(startComp!=null) {
            dx = x
            dy = y
          }
        case mp:MouseReleased =>
          (startComp,hoverComp) match {
            case (null,_) =>
            case (_,null) =>
            case (w:Wire,g:Gate) =>
              g ! ConnectGateInput(y,w)
            case _ =>
          }
          startComp = null
        case _ =>
      }
  } },
  new RadioButton { action = Action("Move") {
      clickAction = (e,x,y) => e match {
        case mp:MouseDragged =>
          if(hoverComp!=null) hoverComp ! MoveTo(x,y)
        case mp:MouseReleased =>
          if(hoverComp!=null) hoverComp ! MoveTo(x,y)
        case _ =>
```

```
          }
        } },
        new RadioButton { action = Action("Flip Switch") {
            clickAction = (e,x,y) => e match {
              case mp:MouseClicked =>
                if(hoverComp!=null) {
                  hoverComp match {
                    case s:Switch => hoverComp ! FlipSwitch
                    case _ =>
                  }
                }
              case _ =>
            }
        } }
      )
      val group = new ButtonGroup(radioButtons:_*)
      layout += new GridPanel(radioButtons.length+1,1) {
        radioButtons.foreach(rb => contents += rb)
        speedField = new TextField(speedMult.toString) {
          listenTo(this)
          reactions += {
            case ed:EditDone => speedMult = text.toInt
          }
        }
        contents += speedField
      } -> North
    }
  }
  propPanel
}

override def toString() = "Circuit"

def toXML : xml.Node = {
  <drawCircuit speed={speedMult.toString}>
    {comps.map(_.toXML)}
  </drawCircuit>
}

def react(evt:Event,cx:Double,cy:Double) {
  evt match {
    case me:MouseEvent =>
      hoverComp = null
      var lastDist = 1e100
      for(comp <- comps) {
        val (ccx,ccy) = comp !? GetPos match {
          case (x:Double,y:Double) => (x,y)
        }
        val (dx,dy) = (cx-ccx, cy-ccy)
        val dist = math.sqrt(dx*dx+dy*dy)
        if(dist<15 && dist<lastDist) {
          hoverComp = comp
          lastDist = dist
        }
      }
```

```
        if(clickAction!=null) clickAction(me,cx,cy)
        drawing.refresh
      case _ =>
    }
  }

  private trait CircuitComponent extends Actor {
    def toXML:xml.Node
  }

  private class Wire(private[this] var state:Boolean,
      private[this] var x:Double,
      private[this] var y:Double) extends CircuitComponent {
    private[this] val outs = mutable.Buffer[Gate]()
    def act {
      loop {
        react {
          case SetWire(v) =>
            if(v!=state) {
              state = v
              outs foreach (_ ! WireChange(state,this))
              drawing.refresh
            }
          case MoveTo(nx,ny) =>
            x = nx
            y = ny
          case GetPos => sender ! (x,y)
          case GetPosAndState => sender ! PositionAndState(x,y,state)
          case ConnectWireOutput(g) =>
            outs += g
            g ! WireChange(state,this)
          case DisconnectOut(g) => outs -= g
        }
      }
    }
    def toXML:xml.Node = <wire index={comps.indexOf(Wire.this).toString}
        x={x.toString} y={y.toString} state={state.toString}/>
  }

  private abstract class Gate(inWires:Int,outWires:Int,
      private[this] var x:Double,
      private[this] var y:Double) extends CircuitComponent {
    def act {
      loop {
        react {
          case WireChange(v,w) =>
            Actor.actor {
              Thread.sleep(delay*speedMult)
              for((iw,i) <- inputs.zipWithIndex; if iw==w) inputValues(i) = v
              setOutputs(inputValues,outputs)
            }
          case MoveTo(nx,ny) =>
            x = nx
            y = ny
          case GetPos => sender ! (x,y)
```

```
      case GetPosAndState => sender ! PositionAndState(x,y,false)
      case GetWires => sender ! (inputs.toList,outputs.toList)
      case ConnectGateInput(my,w) =>
        val index = (my-y+19).toInt/20 min inputs.length-1 max 0
        if(inputs(index)!=null) inputs(index) ! DisconnectOut(this)
        inputs(index) = w
        w ! ConnectWireOutput(this)
    }
  }
}

def toXML:xml.Node = <comp type={gateType} x={x.toString} y={y.toString}>
    { inputs.map(g => <in>{comps.indexOf(g).toString}</in>) }
  </comp>

private[this] val inputs = Array.fill(inWires)(null:Wire)
private[this] val inputValues = Array.fill(inWires)(false)
private[this] val outputs = Array.tabulate(outWires)(i => new
    Wire(false,x+25,y+20*i))
outputs.foreach(o => { o.start; comps += o })
val delay:Int
val gateType:String
def setOutputs(ins:IndexedSeq[Boolean],outs:IndexedSeq[Wire]):Unit
}

private class OrGate(x:Double,y:Double) extends Gate(2,1,x,y) {
  val delay = 20
  def setOutputs(ins:IndexedSeq[Boolean],outs:IndexedSeq[Wire]) {
    outs(0) ! SetWire(ins(0) || ins(1))
  }
  val gateType = "or"
}

private class AndGate(x:Double,y:Double) extends Gate(2,1,x,y) {
  val delay = 15
  def setOutputs(ins:IndexedSeq[Boolean],outs:IndexedSeq[Wire]) {
    outs(0) ! SetWire(ins(0) && ins(1))
  }
  val gateType = "and"
}

private class NotGate(x:Double,y:Double) extends Gate(1,1,x,y) {
  val delay = 5
  def setOutputs(ins:IndexedSeq[Boolean],outs:IndexedSeq[Wire]) {
    outs(0) ! SetWire(!ins(0))
  }
  val gateType = "not"
}

private class Switch(private[this] var state:Boolean,
    private[this] var x:Double,
    private[this] var y:Double) extends CircuitComponent {
  val out = new Wire(state,x+20,y)
  out.start
  comps += out
```

```
    def act {
      loop {
        react {
          case FlipSwitch =>
            state = !state
            out ! SetWire(state)
            drawing.refresh
          case MoveTo(nx,ny) =>
            x = nx
            y = ny
          case GetPos => sender ! (x,y)
          case GetPosAndState => sender ! PositionAndState(x,y,state)
        }
      }
    }
    def toXML:xml.Node = <comp type="switch" x={x.toString} y={y.toString}
        state={state.toString}/>
  }

  private case class WireChange(value:Boolean,wire:Wire)
  private case class SetWire(value:Boolean)
  private case class MoveTo(x:Double,y:Double)
  private case class ConnectGateInput(y:Double,w:Wire)
  private case class ConnectWireOutput(g:Gate)
  private case class DisconnectOut(g:Gate)
  private case object GetPos
  private case object GetPosAndState
  private case object GetWires
  private case object FlipSwitch
  private case class PositionAndState(x:Double,y:Double,state:Boolean)
}

object DrawCircuit {
  def apply(p:DrawTransform) = new DrawCircuit(p,Seq(),Seq(),20)

  def apply(p:DrawTransform,n:xml.Node) = {
    val compData = (n \ "comp").map(cn => {
      val x = (cn \ "@x").text.toDouble
      val y = (cn \ "@y").text.toDouble
      (cn \ "@type").text match {
        case "switch" => SwitchData(x,y,(cn \ "@state").text.toBoolean)
        case t => GateData(t, x, y, (cn \ "in").map(_.text.toInt))
      }
    })
    val wireData = (n \ "wire").map(wn => {
      val x = (wn \ "@x").text.toDouble
      val y = (wn \ "@y").text.toDouble
      val index = (wn \ "@index").text.toInt
      val state = (wn \ "@state").text.toBoolean
      WireData(index,x,y,state)
    })
    val speed = (n \ "@speed").text.toInt
    new DrawCircuit(p,compData,wireData,speed)
  }
```

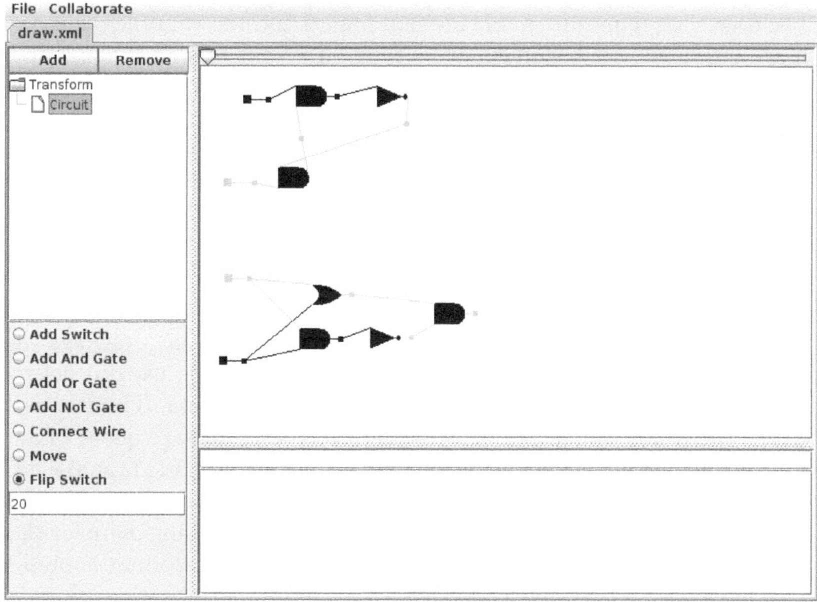

FIGURE 34.2: This is a screen shot of the drawing program with a `DrawCircuit` added and two different circuits. Each circuit has two switches on the left side represented by squares. Clicking a square can toggle them on or off. You can think of them as being inputs A and B. The bottom circuit is an "exclusive or". It implements the logic A xor B = (A or B) and not (A and B). The top circuit does not have an output. However, when both switches are turned on, a signal propagates around the loop.

```
  case class WireData(index:Int, x:Double, y:Double, state:Boolean)
  class CompData
  case class GateData(gType:String, x:Double, y:Double, ins:Seq[Int]) extends
      CompData
  case class SwitchData(x:Double, y:Double, state:Boolean) extends CompData
}
```

If you run this you can make some simple circuits like the ones shown in figure 34.2 then flip the switches and watch signals propagate through.

The elements of this code that are new center on the use of actors. Instead of calling methods on any element of `comps`, messages are passed. When values are needed immediately for things like drawing, the message is sent with `!?`.[2]

In addition to the `Wire` type there is a `Gate` with several subtypes and a `Switch` that are also subtypes of `CircuitComponent`. Each of the `Gate` subtypes has to define a `delay` value. This is proportional to how long it will take between the time an input wire changes and when the output wire changes. Note that the `WireChange` message handling in the `Gate` class includes a call to `Thread.sleep` to affect the delay. Because of this delay, and the desire to not block the actor thread, this was put in a call to `Actor.actor` so that it would occur in a separate thread and allow the actor to handle other messages during that time.

The way the actor message handling is written here should almost make it look like

[2]It should be noted that during the development of this code, that lead to the application hanging more than once when the `!?` operator was used with an actor that had not been started or did not respond to that message.

methods. The case class names resemble method names and the variables that are bound in the pattern are much like the arguments to a method. What follows that is the body. The message passing approach is more dynamic than normal method calls. This can have some benefits, but it has the downside that the compiler will not find as many errors for you. Sending a message to an actor that does not respond to that message type is not an error. If you use `!` or `!!` to send the message it will not even cause a runtime error because messages that are not dealt with simply sit in the actors message queue. This also means a lot more `match` expressions in the code in places where inclusion polymorphism and subtyping would have been employed using standard approaches.

There is one other, very significant aspect of this code that should be noted. As it is written, the `DrawCircuit` type does not serialize. This means that drawings that include `DrawCircuit` can not be saved to binary or zip, nor can they be used in network collaborations as those features rely on object streams and serialization. The reason for this is that the `Actor` type is not serializable. More specifically, the message queue of an actor can not be serialized. Even if we make `CircuitComponent` extends `Serializable`, attempts to serialize will still fail. We could get around this by making `comps` transient and overriding the default serialization. Given the current length of this code, adding more for that purpose was not seen as a good idea. The XML support for saving and loading is present so that drawings with circuits can be saved and do not have to be redrawn every time.

Akka

The actors discussed in this chapter use a library that comes with Scala in the standard libraries. This library is fairly simple to use and can meet most demands. There is also a `scala.actors.remote` package that was not discussed here that allows you to have actors distributed across multiple machines. Despite all of this, there is another actor library written for Scala that is recommended for really large actor-based projects or ones where performance is critical. This is the Akka library, `http://akka.io/`.

Akka is viewed as one of the primary extensions to the base Scala install, as signified by its addition to the Typesafe stack, `http://typesafe.com/`. Akka actors have a lower overhead than Scala actors by virtue of not use the same mailbox type of structure of messages. It is also designed to be fault tolerant so that when things go wrong, the program will continue running. Support for remote actors is present in Akka as well for applications that need to scale beyond a single computer.

34.4 End of Chapter Material

34.4.1 Summary of Concepts

- The actor model is a different approach to concurrency and parallelism in programming.
  - You can imagine each actor at having its own thread.
  - Actors communicate by sending messages, not calling methods.

  - Actors can have mutable data. As long as they do not share mutable data there is no chance for race conditions.

- Scala comes with support for actors in the `scala.actors` package.
  - To create fully functional actors, make classes that extend the `Actor` trait.
  - Those classes should implement the `act` method. This is what is called when an actor is started.
  - To handle messages, the `act` method should include a call to `receive` or `react`, typically inside of a looping structure.
    * Both `receive` and `react` take a single argument that is a partial function. It should have cases for the different messages that actor should be sent and appropriate handling code.
    * The `receive` method if a standard method that blocks until an appropriate message is sent to that actor. It holds onto the actors thread and returns a value so it can be used as an expression.
    * The `react` method does not return. Anything after it is unreachable. This allows thread sharing among actors. You need to use `loop` or `loopWhile` to make it repeat. A simple `while` loop will not work.
  - It is ideal for messages to be immutable and work with pattern matching. For this reason, case classes are often used to create different types of messages.

- An example of an actor-based event simulation was written for circuits at the logic gate level.

34.4.2 Exercises

1. Write a simple actor that uses `receive` without a loop. Start the actor and verify that it will only handle a single message without a `restart`.

2. Take your actor from the previous exercise and put the call to `receive` in a `while` loop then verify that it can now handle multiple messages.

3. Write an actor that has the ability to know about another actor of that same type that can cooperatively count down. The actor should handle two messages. One will pass it an actor that will be its friend. The other handles a numeric value and if it is greater than zero it prints the value and then sends a message tot he friend with a value one less than what it printed. Make two of these actors and have them count down. To verify that both actors are involved in the conversation, you can give them names that are printed along with the numeric values.

4. Use the actor you wrote for the last exercise with the name extension and have 3 or 4 of them arranged so that they count down in a loop.

5. Edit the code for the graphical display of circuits in this chapter to clean it up and make it look better.

6. Look up a 4-bit adder and build one in the circuit framework. Watch it work as you change input switches.

7. Make a new `Gate` type with 8 inputs and 8 outputs for a full byte adder.

8. Edit the code to make the circuit simulation from this chapter serializable using default serialization.

34.4.3 Projects

Any of the projects can be reworked so that they employ actor parallelism. The descriptions here list some different approaches you might consider using, based on which project you have been developing.

1. In the MUD project, it is possible to convert all of the major mutable entities in the game over to use actors. This includes not only characters, but rooms as well. All actions that occur in the MUD turn into messages. The advantage of doing this is that players can have their actions happen in parallel and the system will scale to take advantage of many threads, without you, as the programmer, having to worry about synchronization on the mutable data.

   There should also be an actor that acts as a timer for actions that are not motivated by the players. That actor could include a priority queue of actions, or it could simply run through and send messages to all the various non-player characters telling them that they should perform their action. You have a bit of flexibility in the details. As long as you follow the general rule that you never allow two actors to have access to the same mutable data.

2. The actor model can be used as an alternate way of organizing the processing in the web spider. You will have actor types for reading pages, parsing pages, storing pages, etc. You can also use actors for the queues for handling requests for new pages to work on.

3. Graphical games can also be parallelized using the actor model. Here again you have to make certain that all mutable data in contained inside of actors that oversee access to them. One significant limitation in doing this for a graphical game is that the graphics themselves are still a single-threaded bottleneck. As such, the benefit you see will depend a lot on the style of the game and whether there is a significant amount of processing that happens outside of the graphics.

4. Elements of the math worksheet can also be converted to actors to allow parallel processing. Some care must be taken to make sure that requests for information from different places do not lead to the revised system doing more work than is needed.

5. All the mutable elements of a Photoshop® project can be converted to actors. Any filters or other processing that requires significant resources can also be moved into actors to take full advantage of parallelism.

6. This chapter showed a simulation built on the concept of actors. If you have been writing the simulation workbench, you can put this or other simulations into place that are also based on actors.

Chapter 35

Augmenting Trees

Chapter 29 covered the basics of trees and how to use a binary search tree (BST) to implement the map Abstract Data Type (ADT). This implementation gave us the possibility of $O(\log(n))$ performance for all the operations on that ADT. Unfortunately, it also allowed the possibility of $O(n)$ performance if the tree were built in a way that made it unbalanced. In this chapter we will see how we can fix that flaw by adding a bit more data to nodes and using the data to tell us when things are out of balance so that operations can be performed to fix the situation. In addition, we will see how we can use a similar approach to let us use a tree as the storage mechanism behind a sequence ADT implementation that is $O(\log(n))$ for direct access, inserting, and removing.

35.1 Augmentation of Trees

The general idea of this chapter is that we are going to add some additional data into our tree nodes to give the tree additional capabilities. If you go back and look at the implementation of the BST, each node kept only a key, the data, the left child, and the right child. Augmenting data can deal with data values stored in subtrees or they can deal only with the structure of the tree itself.

There is one key rule for augmented data, it must be maintainable in $O(\log(n))$ operations for any of the standard methods. If this is not the case, then the augmentation will become the slowest part of the tree and it will become the limiting factor in overall performance. Fortunately, any value that you can calculate for a parent based solely on the values in the children will have this characteristic assuming it does not change except when tree structure changes. You can be certain such a value is maintainable in $O(\log(n))$ operations because operations like `add` and `remove` from a tree only touch $O(\log(n))$ nodes as they go from the root down to a leaf. Maintaining the augmented values only requires updating nodes going back from the leaf up to the root.

Example values include tree properties like height and size. The height of a node is the maximum of the height of the two children plus one. The size is the sum of the sizes of the two children plus one. Any value where the value of the parent can be found using a binary operation like `min`, `max`, `+`, or `*` or the children will generally work. Even more complex values that require more than one piece of information can be calculated. Consider the example of an average of value in a subtree. An average can not be calculated as the simple binary combination of two averages. However, the average is calculated as the quotient of two sums. The first sum is the sum of all the values you want to average. The second sum is how many there are. This is basically size, which we just described as being a sum over the children plus one. If you annotate nodes with both of those values, it is simple to produce the average for any subtree.

35.2 Balanced BSTs

The first example of any annotated tree that we will consider is a form of self-balancing BST called an AVL tree. Recall from chapter 29 that the primary shortcoming of BST was that is could become unbalanced and then exhibit $O(n)$ performance. Whether the BST was balanced or not was completely dependent on the order in which keys were added and removed. This is something that is very hard to control. A self-balancing tree does a bit of extra work to ensure that the tree always remains properly balanced, regardless of the order of the operations that are performed on it. The AVL tree was the first such tree to be described and is fairly simple to understand. [10]

The idea behind an AVL tree is that we want the height of the two children of any given node to be within one of each other. So if one child has a height of 5, the other child has to have a height of 4, 5, or 6 to be acceptable. If it does not meet this condition, the tree has to be adjusted to bring it back into compliance. This requirement of children being within one of each other in height keeps the tree as close to balanced as possible. However, it is rather restrictive and leads to a fair bit of adjustments. For that reason, it is more common for libraries to use other self-balancing trees like a red-black tree.

The AVL tree works exactly like a normal BST for searches. It is only for operations that add or remove nodes that heights have to be recalculated and adjustments have to be made. As long at the AVL condition is maintained, we can be certain that all the standard methods will complete in $O(\log(n))$ operations.

35.2.1 Rotations

The typical approach to re-balancing trees that have gotten unbalanced is through rotations. The idea of a rotation is to pull one of the children of a node up above the parent and swap on grandchild across to the other child in such a way that the proper BST ordering is preserved while pulling a grandchild that has grown too tall up to restore balance. The two basic rotations, to the right and the left are shown in figure 35.1. A single rotation to the right can be used when the outer left grandchild has gotten too tall so that the height of the left child is two greater than the right child. Conversely, a single left rotation can fix the problem when the outer right grandchild has gotten too tall. These types of imbalance can occur when an add operation adds a new child to a subtree that increases its total height.

Note that the subtrees A, B, and C in figure 35.1 are shown without any detail other than one of them is taller than the other two. This is intentional. We do not care about the internal structure of those subtrees, only their overall heights.

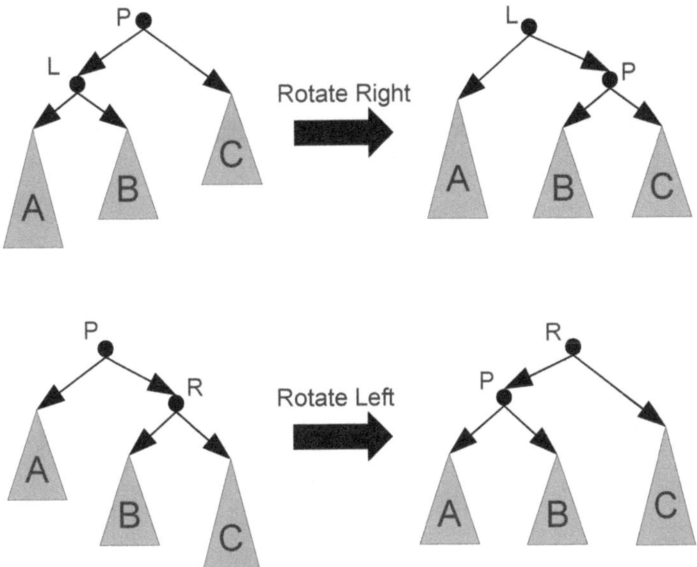

FIGURE 35.1: This figure shows single rotations to the left and the right. These rotations can be used when a node becomes unbalanced because the outer grandchild on the left or the right grows too tall.

In the situation where an internal grandchild, that is to say the right child of the left child or the left child of the right child, has grown to make the tree unbalanced two rotations are required. The first rotation is applied to the child that has the greater height, moving it outward. The second rotation then moves the parent to the opposite side. Figure 35.2 shows how this works. Note that each step in these double rotations is one of the single rotations shown in figure 35.1. This is significant in code as well because after the single rotations have been implemented, it is a simple task to use them to build a double rotation.

35.2.2 Implementation

To implement an AVL tree, we will use the immutable BST written in chapter 29 as a base and edit it appropriately. Making a mutable AVL tree is left as an exercise for the reader. Like that tree, we have a public type that represents a tree with no functionality. We also have two private classes that implement the public type. The top-level type has two additions for functionality, `rotate` and `height`, as well as one for debugging purposes, `verifyBalance`. The `rotate` method is supposed to check the balance and perform appropriate rotations if needed. The `height` member stores the height of the node. It is an abstract `val` so the subtypes have to provide a value. Code for this is shown here.

```
package scalabook.adt

abstract sealed class ImmutableAVLTreeMap[K <% Ordered[K],+V] extends Map[K,V] {
  def +[B >: V](kv:(K,B)):ImmutableAVLTreeMap[K,B]
  def -(k:K):ImmutableAVLTreeMap[K,V]
  def rotate:ImmutableAVLTreeMap[K,V]
  val height:Int
  def verifyBalance:Boolean
```

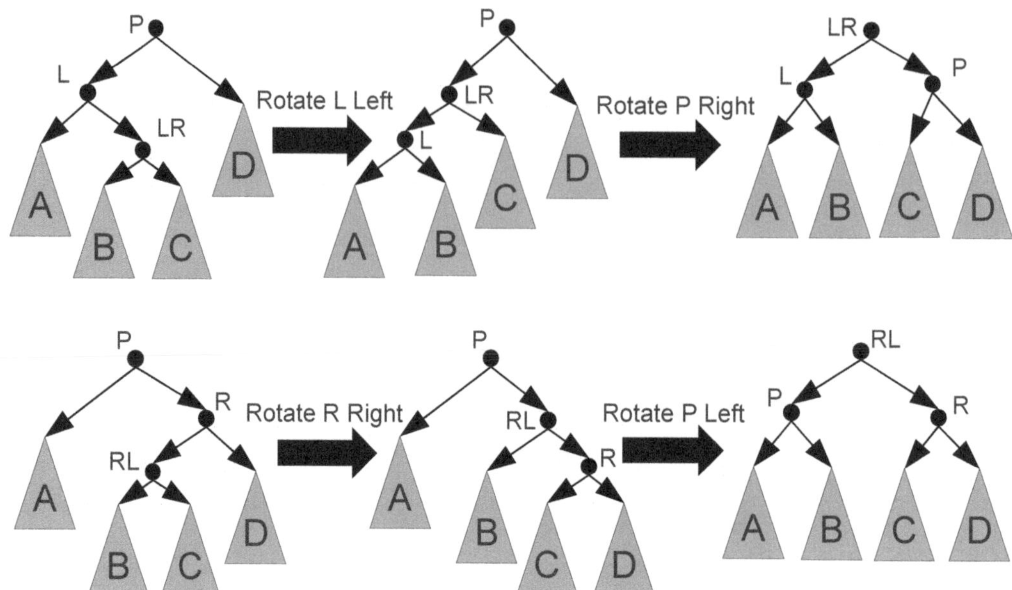

FIGURE 35.2: This figure shows double rotations to the left-right and the right-left. These rotations can be used when a node becomes unbalanced because the inner grandchild on the left or the right grows too tall. Note that each of this is simply two applications of the rotations from figure 35.1.

```
}

private final class AVLNode[K <% Ordered[K],+V](private val key:K,
    private val data:V, private val left:ImmutableAVLTreeMap[K,V],
    private val right:ImmutableAVLTreeMap[K,V]) extends ImmutableAVLTreeMap[K,V] {

  val height = (left.height max right.height)+1

  def +[B >: V](kv:(K,B)) = {
    val ret = if(kv._1 == key) new AVLNode(kv._1, kv._2, left,right)
      else if(kv._1 < key) new AVLNode(key, data, left+kv, right)
      else new AVLNode(key, data, left, right+kv)
    ret.rotate
  }

  def -(k:K) = {
    val ret = if(k == key) {
        left match {
          case _:AVLEmpty[K] => right
          case i:AVLNode[K,V] => {
            right match {
              case _:AVLEmpty[K] => left
              case _ => {
                val (k, d, newLeft) = i.removeMax
                new AVLNode(k, d, newLeft, right)
              }
            }
```

```
      }
    }
  } else if(k < key) {
    new AVLNode(key, data, left-k, right)
  } else {
    new AVLNode(key, data, left, right-k)
  }
  ret.rotate
}

def get(k:K):Option[V] = {
  if(k==key) Some(data)
  else if(k<key) left.get(k)
  else right.get(k)
}

def iterator = new Iterator[(K,V)] {
  val stack = new ListStack[AVLNode[K,V]]
  pushRunLeft(AVLNode.this)
  def hasNext:Boolean = !stack.isEmpty
  def next:(K,V) = {
    val n = stack.pop()
    pushRunLeft(n.right)
    n.key -> n.data
  }
  def pushRunLeft(n:ImmutableAVLTreeMap[K,V]) {
    n match {
      case e:AVLEmpty[K] =>
      case i:AVLNode[K,V] =>
        stack.push(i)
        pushRunLeft(i.left)
    }
  }
}

def rotate:ImmutableAVLTreeMap[K,V] = {
  if(left.height>right.height+1) {
    left match {
      case lNode:AVLNode[K,V] => // Always works because of height.
        if(lNode.left.height>lNode.right.height) {
          rotateRight
        } else {
          new AVLNode(key,data,lNode.rotateLeft,right).rotateRight
        }
      case _ => null // Can't happen.
    }
  } else if(right.height>left.height+1) {
    right match {
      case rNode:AVLNode[K,V] => // Always works because of height.
        if(rNode.right.height>rNode.left.height) {
          rotateLeft
        } else {
          new AVLNode(key,data,left,rNode.rotateRight).rotateLeft
        }
      case _ => null // Can't happen.
```

```
    }
  } else this
}

def verifyBalance:Boolean = {
  left.verifyBalance && right.verifyBalance && (left.height-right.height).abs<2
}

private def removeMax:(K,V,ImmutableAVLTreeMap[K,V]) = {
  right match {
    case e:AVLEmpty[K] => (key, data, left)
    case i:AVLNode[K,V] =>
      val (k, d, r) = i.removeMax
      (k, d, new AVLNode(key,data,left,r).rotate)
  }
}

private def rotateLeft:AVLNode[K,V] = right match {
  case rNode:AVLNode[K,V] =>
    new AVLNode(rNode.key,rNode.data,new
        AVLNode(key,data,left,rNode.left),rNode.right)
  case _ => throw new IllegalArgumentException("Rotate left called on node with
      empty right.")
}

private def rotateRight:AVLNode[K,V] = left match {
  case lNode:AVLNode[K,V] =>
    new AVLNode(lNode.key,lNode.data,lNode.left,new
        AVLNode(key,data,lNode.right,right))
  case _ => throw new IllegalArgumentException("Rotate right called on node with
      empty left.")
}
}

private final class AVLEmpty[K <% Ordered[K]] extends
    ImmutableAVLTreeMap[K,Nothing] {
  def +[B](kv:(K,B)) = {
    new AVLNode(kv._1, kv._2, this, this)
  }

  def -(k:K) = {
    this
  }

  def get(k:K):Option[Nothing] = {
    None
  }

  def iterator = new Iterator[(K,Nothing)] {
    def hasNext = false
    def next = null
  }

  def rotate:ImmutableAVLTreeMap[K,Nothing] = this
```

```
  def verifyBalance:Boolean = true

  val height:Int = 0
}

object ImmutableAVLTreeMap {
  def apply[K <%
      Ordered[K],V](data:(K,V)*)(comp:(K,K)=>Int):ImmutableAVLTreeMap[K,V] = {
    val empty = new AVLEmpty[K]()
    val d = data.sortWith((a,b) => comp(a._1,b._1)<0)
    def binaryAdd(start:Int,end:Int):ImmutableAVLTreeMap[K,V] = {
      if(start<end) {
        val mid = (start+end)/2
        new AVLNode(d(mid)._1,d(mid)._2,binaryAdd(start,mid),binaryAdd(mid+1,end))
      } else empty
    }
    binaryAdd(0,data.length)
  }
}
```

The additions to the `AVLEmpty` type are simple. The `height` is always 0, the `rotate` returns `this` because it is always balanced, and given that the `verifyBalance` method always returns true.

The internal node type is slightly more complex. The `height` is set to the obvious value of `(left.height max right.height)+1`. The `verifyBalance` method checks the heights of the children and recursively calls itself on those children. The real meat of the change is in the `rotate` method. This method checks the height of the children and if they are out of balance it performs one or two rotations to return a node that has a balanced subtree. The rotations themselves are implemented in private methods called `rotateLeft` and `rotateRight`.

There are a number of matches done in the different rotate methods. This is due to the fact that they have to get hold of `left` and `right` grandchildren. This requires a match because the top-level type does not have `left` and `right`. This makes sense as they are not defined on `AVLEmpty`. The way the code is written, it should not be possible to those methods to be called with an `AVLEmpty` node. The rotation methods are written to throw exceptions should that happen.

The only other changes are the additions of calls to rotate in the `+` and `-` methods. These calls occur popping back up the stack so the entire tree is rebalanced every time. Testing for this class begins with test code from the original BST. To ensure that it is working, a tree must be built using `+` then checked with `verifyBalance`. The shortcut approach that is implemented in the companion object was created to always produce a perfectly balanced tree so using it does not actually test the rotation code.

35.3 Order-Statistic Trees: Trees as Sequences

The AVL tree used an augmentation of height to enable self-balancing. Augmenting by size can produce a tree that works as a sequence that is fairly efficient for both random access as well as random insertion and removal. By fairly efficient, we mean $O(\log(n))$ because it will keep the height augmentation and be a balanced tree.

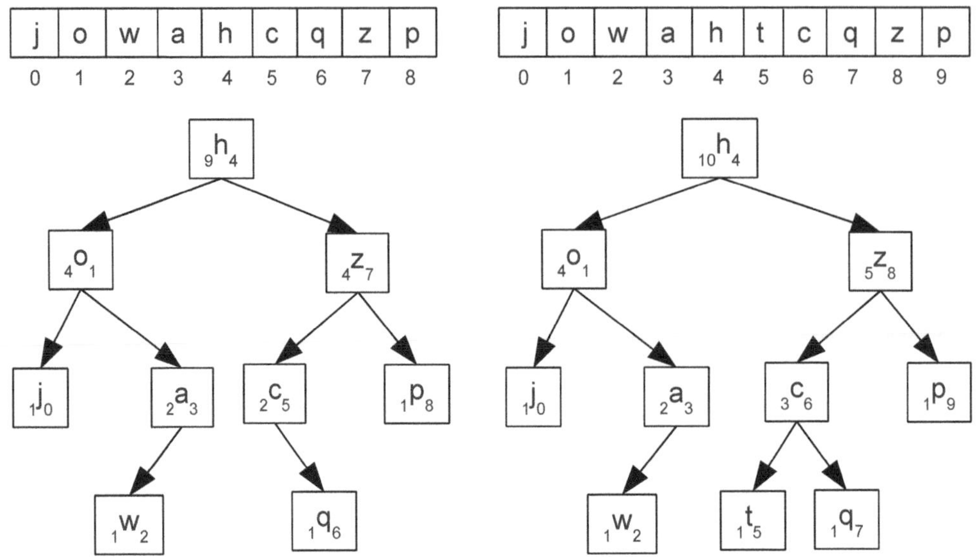

FIGURE 35.3: This figure shows two arrays and possible tree representations for them. The second was arrived at by adding the letter 't' at index 5 of the first. In the tree, each node has the value, a letter, drawn large with a preceding number that is the size of the node and a following number that is the index.

Both the BST and the AVL tree have been used to implement the map ADT. The type we are describing here is a sequence. This is significant because the map ADT needs a key value that has an ordering by which data can be accessed. In a sequence, things are accessed by an index, not a key. What is more, the index of an element can change if other items are inserted or removed before it. The key for a map has to be immutable. Using mutable keys is a great way to produce errors with a map. So how does this work, and what does the size augmentation have to do with it?

Assume a binary tree where we say that the index is the key. This is much like a sequence where you look up values by the index. Figure 35.3 shows what this might look like. Letters have been used as the data here. On the left there is a sequence with nine characters in it and a balanced tree that has the same nine characters with numeric annotations. The number before the letter is the size of the node. The number after it is the index. On the right the letter 't' has been inserted at index 5. Note how this changes the index values of all the letters which had indexes greater than or equal to 5.

To use the tree as a sequence, we have to have an efficient way of finding nodes based on index. That would allow us to to direct access, as well as finding nodes or positions for inserting and removing. The key to associating index with nodes is in the size. The first indication of this is that the size of the node for 'o' is the same as the index of the root, 'h'. This is not coincidence. The size of 'o' is the number of nodes that are "before" the root in the sequence. Since indexes begin at 0, the index of the root should be the size of the left child. This logic holds throughout the tree so having the size allows us to do a binary search through an index.

To see this, assume that we want to find the item at index 6. We start by looking at the size of the left child. It is 4, which is less than 6. This tells us that what we are looking for is on the right side of the tree. This is the signature of a binary search. We do one comparison and it allows us to throw out half the data. So we go to the right child. We are not looking for index 6 on the right child though. Instead, we are looking for index 1. The left child was the first 4 and the root itself was another. So we have thrown out the first 5 elements and we subtract that from the index we are looking for.

So after one comparison we are on the subtree with 'z' at the top looking for index 1. The size of the left child is 2. This is larger than the index we are looking for, so we go to the left. When we move to the left, the index stays the same because we are not jumping past any elements. After two comparisons we are on 'c' looking for index 1. The left child of 'c' is empty and therefore has a size of zero. That is smaller than the index we are looking for so we go to the right and subtract 1 from the index we are looking for because we skipped the node with 'c'. We are now on the node with 'q' looking for index 0. The left child of 'q' is empty, giving it size zero, the same as the index we are looking for, and therefore 'q' is the value we are looking for.

This process illustrates how storing the size only can let us find values at indexes in a number of operations proportional to the height of the tree. That is slower than for an array, but significantly faster than what a linked-list would provide.

What would happen if we looked at index 6 again after adding the 't'? Again, we start at 'h' and see the left child has size 4. Then we move to 'z' on the right looking for index 1. The left child has size 3 so we move to the left, still looking for index 1. The left child of 'c' has a size of 1 meaning that 'c' is the node we are looking for.

This same type of searching could be done with an index annotation. However, an index annotation can not be maintained efficiently. To see this, look closely at the differences between the numbers annotating the nodes in figure 35.3. The size values only changed for the nodes with 'c', 'z', and 'h'. That is the path from the new node up to the root, a path that will have length $O(\log(n))$ in a balanced tree. The index annotation has changed for 'c', 'q', 'z', and 'p'. The is everything after the inserted value. In this case the difference is 3 to 4, but you can picture putting the new value at index 1 to see that the number of indexes that need to change scales as $O(n)$. This tells us that we want to augment our nodes with size, not index values.

The code below is an implementation of an order-statistic tree that extends the `mutable.Buffer` trait. This class uses internals classes for `Node` and `INode` as well as an internal object declaration for `Empty`. While this mirrors some of the style of the immutable tree in the last section, this tree is mutable and everything in `INode` is a `var`. While this uses a similar class structure to the immutable AVL tree, it has a very different implementation with more overloaded methods and a lot fewer uses of `match`.

The code for the `Buffer` methods themselves are generally short, as are the methods in `Empty`. The majority of the code is in `INode`, specifically in the methods for removing and rotating. The use of `Empty` with methods for `+`, `+:`, and `insert` keep the insertion code short. The recurring theme we see in this code is comparing an index to the size of the left child.

```
package scalabook.adt

import collection.mutable

class OrderStatTreeBuffer[A] extends mutable.Buffer[A] {
  abstract private class Node {
    def size:Int
    def height:Int
    def +(elem:A):Node
    def +:(elem:A):Node
    def apply(n:Int):A
```

```
    def remove(n:Int):(A,Node)
    def removeMin:(A,Node)
    def update(n:Int, elem:A):Unit
    def insert(n:Int, elem:A):Node
    def rotate:Node
  }

  private class INode(var data:A, var left:Node, var right:Node) extends Node {
    var size = 1+left.size+right.size
    var height = 1+(left.height max right.height)

    private def setSizeAndHeight {
      size = 1+left.size+right.size
      height = 1+(left.height max right.height)
    }

    def +(elem:A):Node = {
      right += elem
      setSizeAndHeight
      rotate
    }

    def +:(elem:A):Node = {
      left += elem
      setSizeAndHeight
      rotate
    }

    def apply(n:Int):A = {
      if(n==left.size) data
      else if(n<left.size) left(n)
      else right(n-left.size-1)
    }

    def remove(n:Int):(A,Node) = {
      val ret = if(left.size==n) {
        val d = data
        if(right==Empty) (data,left)
        else if(left==Empty) (data,right)
        else {
          val (newData,r) = right.removeMin
          right = r
          data = newData
          setSizeAndHeight
          (d,this)
        }
      } else if(left.size>n) {
        val (d,l) = left.remove(n)
        left = l
        setSizeAndHeight
        (d,this)
      } else {
        val (d,r) = right.remove(n-left.size-1)
        right = r
        setSizeAndHeight
```

```
    (d,this)
  }
  (ret._1,ret._2.rotate)
}

def removeMin:(A,Node) = if(left==Empty) (data,right) else {
  val (d,l) = left.removeMin
  left = l
  setSizeAndHeight
  (d,rotate)
}

def update(n:Int, elem:A) {
  if(n==left.size) data = elem
  else if(n<left.size) left(n) = elem
  else right(n-left.size-1) = elem
}

def insert(n:Int, elem:A):Node = {
  if(n<=left.size) left = left.insert(n,elem)
  else right = right.insert(n-left.size-1,elem)
  setSizeAndHeight
  rotate
}

def rotate:Node = {
  if(left.height>right.height+1) {
    left match {
      case lNode:INode => // Always works because of height.
        if(lNode.left.height>lNode.right.height) {
          rotateRight
        } else {
          left = lNode.rotateLeft
          rotateRight
        }
      case _ => null // Can't happen.
    }
  } else if(right.height>left.height+1) {
    right match {
      case rNode:INode => // Always works because of height.
        if(rNode.right.height>rNode.left.height) {
          rotateLeft
        } else {
          right = rNode.rotateRight
          rotateLeft
        }
      case _ => null // Can't happen.
    }
  } else this
}

private def rotateLeft:INode = right match {
  case rNode:INode =>
    right = rNode.left
    rNode.left = this
```

```
      setSizeAndHeight
      rNode.setSizeAndHeight
      rNode
    case _ => throw new
      IllegalArgumentException("Rotate left called on node with empty right.")
  }

  private def rotateRight:INode = left match {
    case lNode:INode =>
      left = lNode.right
      lNode.right = this
      setSizeAndHeight
      lNode.setSizeAndHeight
      lNode
    case _ => throw new
      IllegalArgumentException("Rotate right called on node with empty left.")
  }
}

private object Empty extends Node {
  val size = 0
  val height = 0
  def +(elem:A):Node = new INode(elem,Empty,Empty)
  def +:(elem:A):Node = new INode(elem,Empty,Empty)
  def apply(n:Int):A = throw new
    IllegalArgumentException("Called apply on Empty.")
  def remove(n:Int):(A,Node) = throw new
    IllegalArgumentException("Called remove on Empty.")
  def removeMin:(A,Node) = throw new
    IllegalArgumentException("Called removeMin on Empty.")
  def update(n:Int, elem:A) = throw new
    IllegalArgumentException("Called update on Empty.")
  def insert(n:Int, elem:A):Node = new INode(elem,Empty,Empty)
  def rotate:Node = this
}

private var root:Node = Empty

def +=(elem:A) = {
  root = root+elem
  this
}

def +=:(elem:A) = {
  root = elem+:root
  this
}

def apply(n:Int):A = {
  if(n>=root.size) throw new
    IndexOutOfBoundsException("Requested index "+n+" of "+root.size)
  root(n)
}

def clear() {
```

```
    root = Empty
  }

  def insertAll(n:Int, elems:Traversable[A]) {
    var m = 0
    for(e <- elems) {
      root = root.insert(n+m,e)
      m += 1
    }
  }

  def iterator = new Iterator[A] {
    var stack = mutable.Stack[INode]()
    def pushRunLeft(n:Node):Unit = n match {
      case in:INode =>
        stack.push(in)
        pushRunLeft(in.left)
      case Empty =>
    }
    pushRunLeft(root)
    def hasNext:Boolean = stack.nonEmpty
    def next:A = {
      val n = stack.pop
      pushRunLeft(n.right)
      n.data
    }
  }

  def length:Int = root.size

  def remove(n:Int):A = {
    if(n>=root.size) throw new
      IndexOutOfBoundsException("Remove index "+n+" of "+root.size)
    val (ret,node) = root.remove(n)
    root = node
    ret
  }

  def update(n:Int, newelem:A) {
    if(n>=root.size) throw new
      IndexOutOfBoundsException("Update index "+n+" of "+root.size)
    root(n) = newelem
  }
}
```

Note that the rotations in this code look a bit different. This is because this code does mutating rotations where references are reset instead of building new nodes. That requires changing two references, them making sure that augmenting value are correct.

It is worth noting that the author wrote this code first without the rotations. It was a non-AVL binary tree that implemented the `Buffer` trait. That code was tested to make sure that it worked before it was refactored to do the AVL balancing. A non-AVL version of this data structure is likely to be little better than a linked list unless you happen to be doing a lot of random additions. With the rotations, the code behavior looks identical from the outside, but we can be certain of $O(\log(n))$ performance for any single data element operation.[1] The tree and rotations do add overhead though, so unless your collection needs to be quite large and you really need a lot of random access as well as random addition and deletion, this probably is not the ideal structure.

This data type is yet another example of why you need to know how to write your own data structures. The `scala.collection.mutable` package includes implementations of `Buffer` that are based on arrays and linked lists. It does not include an implementation based on an order-statistic tree. If your code needs the uniform $O(\log(n))$ operations, you really need to be capable of writing it yourself.

35.4 Augmented Spatial Trees

Binary trees are not the only trees that can be augmented. There are very good reasons why you might want to augment spatial trees as well. The data you augment a spatial tree with is typically a bit different from that in a normal tree. For example, you might store average positions of points in a subtree or the total mass of particles in a subtree. In other situations, you want to know the extreme values in a tree for different directions. In the case of the collisions that we looked at in chapter 31, the tree nodes could store information on maximum velocity and radius in a subtree so that searches could be done using a search radius that adjusts to the properties of the local population. This can make things much faster in situations where one particle happens to be moving unusually fast or is significantly larger than the others.

To illustrate an augmented spatial tree, the final example in this book is going to be a bit different. We will create an octree that is intended to provide fast calculations for things like collisions or ray intersections. This tree will be different from our previous spatial trees in a few ways. Obviously, it will be augmented with some other data. We will use the minimum and maximum values in x, y, and z for the augmentation. In addition, objects that are added to the tree will be able to go in internal nodes, not just leaves. The augmentation is intended to give us an estimate of "when" a cell becomes relevant.[2] The ability to store values high up in the tree is intended to let this structure work better for data sets where objects are of drastically different sizes. Something that is very large will have the ability to influence a lot more potential collisions or intersections than something very small. By placing objects at a height in the tree that relates to their size, we allow searches to "hit" them early and not have the bounds on lower levels of the tree altered by one large object. This is not the same as putting in a node that completely contains them. That would put any object, even small ones, that happen to fall near a boundary high up in the tree.

The code for such a tree is shown below. The class is called a `IntersectionOctree`. There is a companion object that defines two traits that need to be extended by any code wanting to use this structure. One has methods needed by object that are put in the tree and the other is a type that can be used to check for intersections based on different criteria. The type parameter on `IntersectionOctree` must inherit from the former. When an octree is built, the calling code needs to give it a sequence of objects, a center for the octree, a size for the octree, and a minimum size to split down to. This last parameter will prevent the tree from becoming extremely deep is a very small object is added.

---

[1] The `insertAll` method is $O(m\log(n))$ where m is the number of elements being added and n is the number of elements in the collection.

[2] The quotes around "when" signify that we want to know a threshold value for a parameter, and that parameter might not be time.

```
package scalabook.adt

import collection.mutable

class IntersectionOctree[A <: IntersectionOctree.IntersectionObject](
    objects:Seq[A], centerX:Double, centerY:Double, centerZ:Double,
    treeSize:Double,minSize:Double) {

  private class Node(val cx:Double,val cy:Double,val cz:Double,val size:Double) {
    val objs = mutable.Buffer[A]()
    var children:Array[Node] = null
    var min:Array[Double] = null
    var max:Array[Double] = null

    import IntersectionOctree.ParamCalc

    def add(obj:A) {
      if(obj.size>size*0.5 || size>minSize) objs += obj
      else {
        if(children==null) {
          val hsize = size*0.5
          val qsize = size*0.25
          children = Array.tabulate(8)(i =>
            new Node(cx-qsize+hsize*(i%2), cy-qsize+hsize*(i/2%2),
                cz-qsize+hsize*(i/4), hsize))
        }
        children(whichChild(obj)).add(obj)
      }
    }

    private def whichChild(obj:A):Int = {
      (if(obj(0)>cx) 1 else 0)+(if(obj(1)>cy) 2 else 0)+(if(obj(2)>cz) 4 else 0)
    }

    def finalizeNode {
      if(children!=null) children.foreach(_.finalizeNode)
      min = (0 to 2).map(i =>
        objs.view.map(_.min(i)).min min
        (if(children==null) 1e100 else children.view.map(_.min(i)).min)).toArray
      max = (0 to 2).map(i =>
        objs.view.map(_.max(i)).max max
        (if(children==null) -1e100 else children.view.map(_.max(i)).max)).toArray
    }

    def findFirst(pc:ParamCalc[A]):Option[(A,Double)] = {
      val o1 = firstObj(pc)
      val o2 = firstChildObj(pc)
      (o1,o2) match {
        case (None,_) => o2
        case (_,None) => o1
        case (Some((_,p1)),Some((_,p2))) => if(p1<p2) o1 else o2
      }
    }

    private def firstObj(pc:ParamCalc[A]):Option[(A,Double)] = {
```

```
      objs.foldLeft(None:Option[(A,Double)])((opt,obj) => {
        val param = pc(obj)
        if(opt.nonEmpty && opt.get._2<param) opt else Some(obj,param)
      })
    }

    private def firstChildObj(pc:ParamCalc[A]): Option[(A,Double)] = {
      if(children==null) None else {
        val cparams = for(c <- children; val p=pc(min,max); if
            c!=Double.PositiveInfinity) yield c -> p
        for(i <- 1 until cparams.length) {
          val tmp = cparams(i)
          var j = i-1
          while(j>0 && cparams(j)._2>tmp._2) {
            cparams(j+1) = cparams(j)
            j -= 1
          }
          cparams(j+1) = tmp
        }
        var ret:Option[(A,Double)] = None
        var i=0
        while(i<cparams.length && (ret.isEmpty || cparams(i)._2>ret.get._2)) {
          val opt = cparams(i)._1.findFirst(pc)
          if(opt.nonEmpty && (ret.isEmpty || opt.get._2<ret.get._2)) ret = opt
          i += 1
        }
        ret
      }
    }
  }

  private val root = new Node(centerX,centerY,centerZ,treeSize)

  objects.foreach(root.add)
  root.finalizeNode

  def findFirst(pc:IntersectionOctree.ParamCalc[A]):Option[(A,Double)] = {
    root.findFirst(pc)
  }
}

object IntersectionOctree {
  trait IntersectionObject extends (Int => Double) {
    def apply(dim:Int):Double
    def min(dim:Int):Double
    def max(dim:Int):Double
    def size:Double
  }

  /**
   * The A => Double should calculate the intersect parameter for an object.
   * The (Array[Double],Array[Double) => Double should calculate it for a box
   *   with min and max values
   */
  trait ParamCalc[A <: IntersectionOctree.IntersectionObject] extends
```

```
    (A => Double) with ((Array[Double],Array[Double])=>Double)
}
```

The way the tree is defined, it is mutable with an add method. However, before any tests can be done for intersections, the finalizeNode method has to be called on the tree. As it is written, all objects are added and the tree is finalized at the original construction.

The finalizeNode method has the task of setting the minimum and maximum bounds. This could be done as particles are added, but then higher-level nodes in the tree have their values checked and potentially adjusted for every particle that is added. Doing a finalize step at the end is more efficient.

The only method defined in IntersectionOctree is findFirst. This method is supposed to take a subtype of the ParamCalc type, which can calculate intersection parameters for planes and objects in the tree. The work for this method is done in a recursive call to on the root. The version in the node finds the first object hit as well as the first object in a child node hit. It gives back the first of all of these.

Where the tree comes into play is in firstChildObj. It starts by finding the intersection parameter for the bounds of each child node. These nodes can be overlapping. If any nodes are not intersected, they will not be considered at all. The nodes that are intersected are sorted by the intersection parameter using an insertion sort. This is one of those situations where an "inferior" sort can provide better performance. There are never more than eight items so normal order analysis is not meaningful. After being sorted, the code runs through them in order, calling findFirst on each. This is done in a while loop so that it can stop if ever an intersection is found that comes before the intersection with a bounding region. The idea here is that if something is hit with a parameter lower than even getting into a bounding box, nothing in that bounding box could come before what has already been found. The way this works, the recursion is pruned first by boxes that are never hit and second by boxes that are hit after something in a closer box. For large data sets this can reduce the total work load dramatically.

This tree requires the definition of two types to be useful. Those types would depend on the problem you want to solve. It is possible to use this for collision finding, but that might be better done with a method that instead of finding the first intersection, finds all intersections up to a certain time. A problem that truly needs the first intersection is ray tracing. Here are implementations of a ray and a sphere to work with the tree. This is the code the data structure was tested with.

```
case class Ray(p0:Array[Double],d:Array[Double]) extends
    IntersectionOctree.ParamCalc[Sphere] {
  def apply(s:Sphere):Double = {
    val dist = (p0 zip s.c).map(t => t._1-t._2)
    val a = d(0)*d(0)+d(1)*d(1)+d(2)*d(2)
    val b = 2*(d(0)*dist(0)+d(1)*dist(1)+d(2)*dist(2))
    val c = (dist(0)*dist(0)+dist(1)*dist(1)+dist(2)*dist(2))-s.rad*s.rad
    val root = b*b-4*a*c
    if(root<0) Double.PositiveInfinity
    else (-b-math.sqrt(root))/(2*a)
  }

  def apply(min:Array[Double], max:Array[Double]):Double = {
    if((p0,min,max).zipped.forall((p,l,h) => p>=l && p<=h)) 0.0 else {
      val minp = (0 to 2).map(i => (min(i)-p0(i))/d(i))
      val maxp = (0 to 2).map(i => (max(i)-p0(i))/d(i))
      var ret = Double.PositiveInfinity
      for(i <- 0 to 2) {
```

```
          val first = minp(i) min maxp(i)
          if(first>=0 && first<ret) {
            val (j,k) = ((i+1)%3,(i+2)%3)
            if(first>=(minp(j) min maxp(j)) && first<=(minp(j) max maxp(j)) &&
               first>=(minp(k) min maxp(k)) && first<=(minp(k) max maxp(k))) ret =
                    first
          }
        }
        ret
      }
    }
  }

  class Sphere(val c:Array[Double],val rad:Double) extends
      IntersectionOctree.IntersectionObject {
    def apply(dim:Int):Double = c(dim)
    def min(dim:Int):Double = c(dim)-rad
    def max(dim:Int):Double = c(dim)+rad
    def size:Double = rad*2
  }
```

The Ray type does the real work here. The first method takes a sphere and returns the parameter for when the ray intersects the sphere or positive infinity if it does not intersect. The second method takes arrays of min and max values for a bounding box and determines the parameter for the intersection of a ray and the box. If the ray starts in the box it returns 0.0.

The Sphere class does little more than return basic information about a sphere. However, it would be fairly simple to make a whole hierarchy of geometric types and add appropriate code into the first apply method in Ray for intersecting that geometry to turn this into a useful data structure for a fully functional ray tracer.

35.5 End of Chapter Material

35.5.1 Summary of Concepts

- Augmenting trees is the act of storing additional data in the nodes that can be used to provide extra functionality. The data can relate to the tree structure or the information stored in the tree.
- Augmenting values must be maintainable in $O(\log(n))$ operations or better for standard operations in order to be useful.
- By augmenting with information related to the height of a node we can make self-balancing binary trees.
  - The AVL tree enforces a rule the heights of siblings can not differ by more than one.
  - When we find a parent with children that break the height restriction, the problem is corrected with the application of rotations.
  - An implementation was shown based on the immutable BST. It can also be done in a mutable manner.

- Augmenting a binary tree with size produces and order-statistic tree which can be used as a sequence.
  - The index is like a key, but it is not stored so that inserts and removals can change the index.
  - Nodes can be found by index using a binary search where the desired index is compared to the size of the left child.
- Spatial trees can also be augmented.
  - This is often done using data stored in the tree.
  - An example was given where nodes store boundaries for objects inside of them to do more efficient searching for intersections.

35.5.2 Exercises

1. Write a mutable AVL tree.
2. Compare the speed of the AVL tree to the normal BST for random data as well as sequential or nearly sequential data.
3. Write a BST that is augmented with number of leafs.
4. Compare the speed of different buffer implementations. Use the `ArrayBuffer` and `ListBuffer` that are part of the Scala libraries. Use the order-statistic tree written in this chapter as well. Compare methods of adding, removing, and indexing values for collections of different sizes.
5. Write `findUntil` for the octree. This is a method that will find all collisions up to a certain parameter value instead of just the first one. This is useful for finding collisions that can occur during a timestep.

35.5.3 Projects

1. Use an order-statistic tree in place of a `Buffer` in your MUD. If you can think of another use for an augmented tree, implement that as well. You might consider the min-max tree described below in project 4.
2. If you have been working on the web spider, see if you can come up with an augmented tree structure that helps to partition and give additional meaning to your data. If you can not think of anything, simply use the order-statistic tree in place of a `Buffer`.
3. If you are writing a graphical game where characters move around and there are obstacles, you could build an augmented spatial tree to help expedite the process of finding safe paths through the world. Such a tree would keep track of significant points in a map and store path information between those points. When the code needs to get a character from one location to another, it should do a recursive descent of the tree with each node holding augmentation data specifying what children are significant for moving in certain ways.

   The advantage of this is that exhaustive searches can be done in local spaces in the leaves of the tree. Those exhaustive searches are likely costly recursive algorithms which benefit greatly from only covering a small area. When moving over a larger area, paths are put together from one node to another. the recursive descent can prune out children that are known to not be significant.

4. There is a commonly used construct in Artificial Intelligence (AI) call the min-max tree, which is an augmented tree that compares moves from different players to help a given player find the optimal move. Such a tree could be implemented in any game where you want to give computer-controlled players reasonable intelligence. The idea of a min-max tree is that you assign values to and ending where each player wins. So if player A wins, that has a value of 1. If player B wins, that has a value of -1. In games that allow a tie, 0 represents a tie. In this tree, player A wants to pick paths that maximize the value and player B wants to pick paths that minimize the value. Since each player generally alternates moves, there are alternating levels that are maximizing and minimizing values. Hence the name min-max tree.

   If you can build a tree for all possible moves in a game, you can tell exactly who should win in a well-played game.[3] For most games of any significance, you can not build this entire tree. Instead, you build it down to a certain level, and at that level you stop and apply a heuristic evaluation of the relative positions of the players. The heuristic should give a value between -1 and 1 based upon which player seems to be in the better position to win. No heuristic will be perfect, but using the min-max tree to evaluate many heuristics you can generally get a reasonable prediction of what the ideal move is at any given time.

   How this works for games like checkers or chess is fairly obvious, but you are likely doing something different. In that case there are choices of where to move, what action to take, what to build, etc. Those can also be chosen between using such a tree. Note that if the number of options is really large, the tree becomes huge. You will probably have to do a little work to keep the tree from blowing up by trimming out really bad choices early on.

5. If you have been working on the math worksheet, you can consider changing a `Buffer` implementation so that it uses an order-statistic tree. If you have added any features that would benefit from an augmented tree, you should write code for that.

6. Use the `IntersectionOctree` to add a ray-tracing element to your Photoshop® program.

7. The simulation workbench has two different augmentations that you can make to spatial trees to improve performance. For the collision tree, the search distance can vary with location. The earlier implementation, used a uniform search distance, which should be the maximum for any place in the tree. You can use a localized value by augmenting the tree.

   Recall that the search radius is given by $\Delta v \Delta t + 2R_{max}$, where Δv is the relative velocity of particles, Δt is the time step, and R_{max} is radius of the largest particle. Technically these values vary for different subtrees. A more efficient implementation can be written that uses local values for those. Augment the tree with the values of Δv and R_{max} in each subtree.

   A tree can also be used to improve the efficiency of gravity calculations. The technique of gravity trees was originally developed by Barnes and Hut [2]. The original work

[3] Doing this for tic-tac-toe is an interesting exercise for the student. The result is that a well-played game always ends in a tie.

used an octree, but many newer implementations use a kD-tree. The idea of a gravity tree code is that while the gravity from nearby masses needs to be handled exactly, the precise locations of individual bodies is not significant for things that are far away. For example, when calculating the gravity due to the Jupiter system on a mass, if the mass is near Earth, you do not need to do separate calculations for all the moons. Doing a single calculation for the total mass of the system at the center of mass of the system will be fine. However, if you are at the location of one of the outer satellites of Jupiter, the exact positions of some of the moons could be significant.

To quickly approximate the force of gravity from distant collections of mass, the tree can be augmented with information about all the particles in a subtree. The simplest augmentation is that each node stores the total mass and center of mass location of that subtree. To calculate the gravity force on a particle, the code does a recursive descent of the tree. At any given node, a check is done comparing the size of the node to its distance from the particle currently under consideration. If the condition $d < \beta s$, where d is the distance between particle and node, s is the size of the node, and β is a parameter to adjust accuracy,[4] then the recursion must go down to the children. If that condition is not met, the force can be calculated against the center of mass of that node and the children can be ignored.

[4]For a monopole tree as described here, the value of β should be between 0.1 and 0.3. Smaller values made the calculations more accurate, but increase the amount of work that is done.

Chapter 36

Wrapping Up

We have reached the end of our tour of programming. While this book has covered a good number of different topics, this is only a basic introduction to programming, which is but a stepping-stone into the field of Computer Science. The goal of this book was to lay a foundation that you can build on going forward. Now it is your job to continue the construction process.

36.1 What You Have Learned

The early parts of this book focused primarily on basic constructs used to tell a computer how to do things. You learned the use of conditional logic and looping constructs that could be put around simple sequential commands to express actions that you wanted to have the machine carry out. You learned how to break problems into pieces. At first the approach was to build separate functions that handled one part of the problem at a time, and ways of grouping data that made sense. Later on you saw that problems can be decomposed into objects that combine data with functionality. This learning to break problems apart should have had an impact on how you looked at many things through your life and in your studies outside of Computer Science.

A further step in how you approached problems came in the form of thinking about abstractions. It was possible to write a solution to one problem, but so much more could be gained if you tune that solution so that parts of it can be made to work in many different problems. The desire to abstract code appeared in many different contexts through the book with abstractions that varied over types, functionality, or both.

The modern world is teaming with data. Data is produced by all types of devices during all types of activities. You have learned how to give instructions to a computer to make it process data for you. There are limits on how much you can do by hand, but a computer can read and process millions or even billions of values quickly. You now know how to pull this data from files or over networks. You can deal with data in flat text, XML, or binary. You can use higher-order constructs to do the processing and apply multithreading to speed it up.

The also have the ability to write programs that display Graphical User Interfaces (GUIs) and draw custom graphics. You can write programs that talk to one another over networked sockets. When it really comes down to it, you have learned the basic tools to make a computer do pretty much anything. You have the foundation to create any application you can dream up.

36.2 The Possibilities/A Pet Project

So where should you go from here? Remember that programming is an art. You gain skill primarily by applying that art. If you do not apply it, you will lose your skill. As such, you need to exercise this art. There are many ways you can do this. The most important thing for you to do though is to pick a project. It does not have to be a really big project. If it is small, when you complete it you will simply pick another one. However, having a project outside of your normal course work gives you a great way to further develop your skills. It also allows you to build a portfolio of your work.

When you pick a project, you will need to look into what it takes to do it. Not all projects will favor you working in Scala. That is not a problem. Learning another language is a great exercise. It is always easier to learn the second than it was to learn the first and having knowledge of multiple languages often helps you to see problems from different perspectives. There are many types of projects that you can do in Scala though. Consider these possibilities.

36.2.1 General 2-D Application

What you know how to do best at this point is write applications. If you can think of something that interests you that works as an application, run with it. Whether it is something useful or just something for fun, there are vast possibilities in what you can get a computer to do. Take advantage of your ability to do graphics and GUIs. Make something networked or push the parallelism to fully utilize newer hardware.

36.2.2 3-D Application

If you are playing with graphics, you might want to do something in 3-D. You can render things in 3-D with ray tracing or other techniques that work in 2-D. However, if you are willing to do a bit of extra reading, which you should be, you can use Scala to write full 3-D applications through some different Java libraries.

The most standard 3-D graphics library is OpenGL®. There are several Java-based wrappers for OpenGL. The most established of these is called JOGL, the Java bindings for OpenGL. OpenGL takes advantage of 3-D hardware to make 3-D rendering fast.

The primary downfall of OpenGL is that it is a low-level API. That means you are setting up geometry and giving instructions to render it in a way that is close to what the machine wants. There are also higher level libraries like Java3D®. These use scene graph representations and can often be easier to use when you want to get things up and running.

New libraries are being written all the time. As of this writing, there are at least two different 3-D graphics libraries being developed specifically for Scala, including Sgine. More options will inevitably become available over time.

36.2.3 Mobile

Not all applications have to run on desktop machines the way we have been working either. Mobile has become a huge market and you can do mobile development in Scala as well. The Android® platform from Google® uses Java for development by default, but it can work just as well with compiled Scala code as well. There is a project called "Scala for Android" that is being developed to help make it easier to put Scala programs on Android phones. This opens up phone and tablet options. The openness of the Android environment means that you can easily develop something to share with others.

36.2.4 Web

Even on the desktop, many people do not use standard applications most of the time. Many of the things that you do on a computer have moved into the web browser. Rich web applications provide a very desktop-like feel with responsive user interactions, but they actually occur in the browser with information being stored out on the web.

You can use Scala for web development as well. Most of the large, well-known companies that use Scala at the time of this writing are in the web application field. The Lift® framework was the first web framework written specifically for Scala. By the time this book is published, version 2.0 of the Play! framework should also be available and should be part of the Typesafe stack. Version 2.0 of Play! has been written from the ground up for Scala and to take advantage of how Scala works. Typesafe support means that you can expect it to integrate nicely with other major pieces of the Scala tool set.

36.2.5 GPGPU

In chapter 21 we looked at how you can use Scala to take advantage of multicore processors. New computers have more than just multicore processors though, they also have programmable graphics cards or processing units built into their CPUs that include graphics card style circuitry. In fact, a high-end gaming machine generally have more brute force computing power on the graphics card than in the normal processor. It is possible to write programs that take advantage of this processing power and do number crunching on graphics cards.

One standard for doing this type of work is OpenCL®. It is a library/language from the same group that created OpenGL. Just as JOGL is a Java binding for OpengGL, there is a Java binding for OpenCL called JavaCL that you can use to do low-level coding for Graphics Processing Units (GPUs) in Java and, therefore, in Scala as well. In addition, there is also a Scala addition currently under development called ScalaCL. Using ScalaCL, you can make collections that are similar to the parallel collections, but instead of using multiple threads on multicore machines, they take advantage of GPUs.

36.2.6 AI, NLP, and Machine Learning

There are also libraries available that can get you up and running quickly on Artificial Intelligence (AI), Natural Language Processing (NLP), and machine learning. Remember that any library written for Java can be called easily in Scala as well. With just a little searching you can find libraries that parse English and help you derive meaning from plain written text or that can be trained to identify patterns in images.

36.2.7 Open-Source and More

If you are not certain about a program you can always look to see what other people are doing. Consider joining an open-source project. Contributions to open-source not only help out lots of people, they look great on resumes. There are open-source projects working on almost anything you can imagine and likely many things you have not imagined. You just have to go looking for them.

The bottom line is that the computer is a remarkably powerful machine and you have just taken your first steps into a world where you can control these machines to do what you want. Embrace this capability and take some time to make something wonderful.

36.3 End of Chapter Material

36.3.1 Exercises

1. Back in project 2 (p.14) you were asked to compare programming to three other activities based on an extremely limited introduction. Now that you have significantly more experience, revisit that same question.

2. Look up one of the technologies mentioned in this chapter and install whatever additional files/projects you need to use it.

3. Try writing a small example program for the technology you selected for the last exercise.

36.3.2 Projects

At this point you should have built at least one project which includes a significant amount of code and functionality. Now is the time to finish it off. Add any features that you had been putting off earlier and polish it up so that it has the desired functionality.

Appendix A

Quick Preview of Java

Scala has many of its roots in the Java programming language beyond compiling to the Java Virtual Machine (JVM) and having the ability to call on Java library functions. For this reason, it should not be too hard to pick up Java after you have learned Scala. This appendix is designed to help you in doing exactly that. To accomplish that goal, we will run through a number of different simple examples of Java code and look at the ways in which they are different from Scala code. Given that you are already familiar with Eclipse, this will be done using that platform. Simply switch to the Java perspective, which is the default for Eclipse, make a project, and try things out.[1]

A.1 Hello World

We will start with Java that same way we started with Scala, using the essential "Hello World" application. Here is what that code looks like in Java. It looks like a Scala application. In Java, there is no scripting model, nor is there an REPL. As such, this is as simple as it gets.

```
package javacode;

public class HelloWorld {
   public static void main(String[] args) {
      System.out.println("Hello World!");
   }
}
```

Just from this little example though, there are a number of things that are obviously different. First, Java does not infer semicolons. You have to type them. In this code we see one semicolon at the end of the package declaration and another at the end of the print statement.

[1]The Eclipse plug-in for Scala will allow you to put Java code into a Scala project for a mixed language project. Technically that is what was done when writing this appendix. If you want to learn Java though you should probably make a clean break.

Going into the `class` you see that it starts with the `public` keyword. Scala does not have a `public` keyword as that is the default visibility. Java also has keywords for `private` and `protected` visibility. In Java `public` visibility is not the default visibility. Instead, the default in Java is called package private. That is what we would get in Scala if we said `private[javacode]` because this `class` is in the `javacode package`. Java does not give you the fine grained control over visibility that Scala does, so the `public`, `private`, `protected`, and package private options are all you have.

In Java it is very strongly recommended that data members never be made `public`. Unlike Scala, you can not change a `public` data member into appropriate methods to include checking of values without breaking code. As such, the default for all data members should absolutely be `private` when you are coding in Java. Part of this is also the fact that method calls in Java always require parentheses and member data access never has them. So in Java, the syntax always tells you if you are using a method or a data member. Scala went the other way so that the implementation can be changed without breaking code.

The class declaration has curly braces after it. Curly braces in Java group code just like in Scala, but they are a group statement, not an expression. In Java, a code block using curly braces will not give you a value.

Looking inside the class at the one method, the declaration of `main` includes a number of differences as well. The first is the `static` keyword, which does not exist in Scala. Java does not have singleton `objects`. The closest approximation to them in Java is `static` data members and methods. The `static` keyword in Java implies that the thing being declared is associated with the `class` instead of being associated with the instances of that `class`. They are very much like the members of a companion `object` except that there is no companion `object` encapsulating them which reduces their flexibility a bit. The `main` method has to be `static` in Java for the same reason that they had to go in a singleton `object` in Scala. Without that, it could not be invoked without some `class` being instantiated first.

After the `static` keyword in another word we did not see in Scala, `void`. This is a type in Java that represents nothing. The equivalent of `void` in Scala is `Unit`. The difference is that there is no instance of `void` while there is an instance of `Unit` represented by `()`. Also not that `void` comes before the name of the method, not at the end after a colon as types did in Scala. All Java types precede the names they are associated with. One advantage of this is that Java does not need the `val`, `var`, or `def` keywords for those declarations. The type tells the language you are declaring something. The downside is that you have to specify a type. Java does not do type inference for you the way that Scala does.

After the method name, `main`, is `String[] args`. Based on your knowledge of Scala you should realize that `args` is the name of a formal parameter and you can figure out that `String[]` specified a type that is equivalent to the Scala type `Array[String]`. We will come back to the Java array syntax in a bit, but the short version here is that Java uses square brackets only for arrays, not type parameters, and you specify a type that is an array of another type by putting square brackets after the type. The formal argument shows again that the types in Java precede the names they are associated with.

Inside the `main` method is a single call to `println`. By default, there is nothing set up in Java to bring the `println` method into scope. Instead, you specify the full name, `System.out.println`. Breaking this down, `System` is a `class` in `java.lang`, which is `imported` by default. There is a `static` member in `System` named `out` that is a `java.io.PrintStream`, which has `println` methods for different types.

A.2 Arrays, Primitives, and More

The next example is a bit more significant. This creates and array of 1000 integers, fills it with the first 1000 prime numbers, then adds them up and prints the sum.

```
package javacode;

import static java.lang.System.out;

public class PrimeArray {
    private static boolean isPrime(int n) {
        for(int i = 2; i*i<=n; ++i) {
            if(n%i==0) return false;
        }
        return true;
    }

    public static void main(String[] args) {
        int[] primes = new int[1000];
        int pos = 0;
        int i = 2;
        while(pos<primes.length) {
            if(isPrime(i)) {
                primes[pos] = i;
                pos++;
            }
            i++;
        }
        int sum = 0;
        for(int j=0; j<primes.length; ++j) {
            sum += primes[j];
        }
        out.println(sum);
    }
}
```

To shorten the call to `System.out.println` a bit, the file has a `static import` at the top. Normal `imports` in Java can only import the contents of packages. The `static import` also has the ability to import `static` members. Note that the Scala version does not distinguish between types of uses. Also, Java `import` statements need to be at the top of the file and they lack much of the flexibility of those in Scala. If you want to `import` all of the contents of a package in Java use * instead of _.

The `class` in this file has two methods in it called `isPrime` and `main`. Both are `static`. The `main` method has to be static so that it functions as an entry point into an application. The `isPrime` method needs to be `static` here because only a `static` method can be called from another `static` method without instantiating a class to call it on. It also should be `static` because it does not use any information from any instance of the `class`.

Looking in the `isPrime` method shows a few more differences between Java and Scala. There are two more types that appear here, `boolean` and `int`. These should look familiar to you. They are just like what you are used to in Scala except for the fact that they start with a lowercase letter. You might have noticed this about the `void` type as well, though it is not true about the `String`. This is more than just a matter of capitalization. It has semantic implications as well. Java is not as object-oriented as Scala. The instances of the basic types in Java are not objects. They are called primitives and they do not have any methods. This is why their types start with lowercase letters. `String` is a `class` and instances of `String` are objects.

The `for` loop should stand out as another significant difference. The `for` loop in Scala is technically a for-each loop that runs through the contents of a collection. Java's basic `for` loop is more of a dressed up `while` loop. In the parentheses there are three parts, separated by semicolons. The first part is an initializer where you set things up. This happens once, right when the loop is reached. If variables are declared there, as is the case in this example, that variable has a scope only through the loop.

The second part is a condition, just like what you would put in a `while` loop. It is pre-check and the loop will iterate until the condition is false. This is why the Java `for` loop can be described as a dressed-up `while` loop. The way in which the end case is determined is just like a `while` loop. The dressing is what comes before and after the semicolons around the condition.

The last part of what appears in the parentheses of a `for` loop is an iterator. This code happens after the body of the loop is evaluated and before the next condition check. In this case, the expression is `++i`. This is a pre-increment of the variable `i`. The `++` operator in Java does an increment by 1. It can be placed before or after the variable. When it is before, the result of that expression is the incremented value. When placed after the expression the result is the value before the increment.

There is another type of `for` loop in Java which functions as a for-each loop. The syntax is `for(`*type name*`:`*iterable*`)`, where *type* is the type in the collection, *name* is the variable name you want, and *iterable* is an object of type `Iterable`.

One last thing to note about the `for` loop is that it is never an expression. It never gives back a value. There is no equivalent to `yield`. The `while` loop in Java looks just like what you are used to in Scala.

The `isPrime` method also uses the `return` keyword. This is a valid usage in Scala, but you typically do not need it because every block is an expression with a value equal to that of the last statement in the block. In Java, blocks are not expressions, so every method that returns a value needs to specifically include a `return` statement at the end. If a `return` statement is reached inside the method, it will terminate the execution of that method and immediately jump back out to where it was called from.

The `main` method demonstrates Java's array syntax. The `primes` variable has the type `int[]`. You can see the array itself is built using the `new` syntax. Inside of the `while` loop you can see that the indexing of the array is also done using square brackets, passing in the index you want to access. Arrays in Java are objects, so you can ask them how long they are using `length`. Note that `length` here is a data member. You know this in Java because there are no parentheses after it. The `length` of a `String`, however, is gotten with a method and must always have empty parentheses after it.

A.3 File Names and Interfaces

To illustrate a few more features of Java we will use another favorite example, the bank account. We begin by introducing a supertype for bank accounts.

```
package javacode;
```

```
public interface BankAccount {
   String getName();
   int getBalance();
}
```

This supertype is made using a Java construct called an `interface`. You can think of an `interface` as being like a `trait` that has no data members and where everything is abstract. Basically, an `interface` is a completely abstract supertype. An `interface` provides subtyping only, there is no code that goes into the subtypes. Java also only allows single inheritance of `classes`. This makes some things simpler as Java has no need to linearize virtual method resolution. The downside is that `interfaces` never have all that many methods as no one would want to ever implement an `interface` that forces them to write many tens of methods.

This `interface` will be inherited by a `CheckingAccount class`. Note that when inheriting from an `interface` in Java you use the `implements` keyword. When you inherit from a `class` you use the same `extends` keyword that you are used to from Scala. Multiple interfaces can follow `implements` and they are separated by commas.

```
package javacode;

public class CheckingAccount implements BankAccount {
   private final String name;
   private int balance = 0;

   public CheckingAccount(String n,int b) {
      name = n;
      balance = b;
   }

   @Override
   public String getName() {
      return name;
   }

   @Override
   public int getBalance() {
      return balance;
   }

   public boolean deposit(int amount) {
      if(amount>=0) {
         balance += amount;
         return true;
      } else {
         return false;
      }
   }

   public boolean withdraw(int amount) {
      if(amount>=0 && amount<=balance) {
         balance -= amount;
         return true;
      } else {
         return false;
      }
```

```
    }
}
```

There are many things in this class that are worth noticing. Before looking at the details though, we should note something that is not at all obvious here. The `BankAccount interface` is located in a file called "BankAccount.java". The `CheckingAccount class` is in a separate file called "CheckingAccount.java". In Scala it was recommended that separate `class`, `trait`, and `object` declarations be put in separate files with names that match. In Java, this is not just recommended, it is required. There can be only one top-level `public` declaration per file in Java, and its name must match the name of the file.

So what are some of the differences between this code and what you would have written in Scala? For one thing, the `name` member is preceded by the `final` keyword. In Scala, `final` means that something can not be overridden in a subtype. When applied to a method, it has that same meaning in Java. However, when used in front of a variable declaration, it means that the value of that variable can not be changed after it is set. A `final` variable in Java is like a `val` declaration in Scala. Without `final`, it is like you are using `var`.

There is something else worth noting about `name`, it is not given a value. In Scala, any member without a value is considered abstract. This is not the case in Java. You can declare variables either locally or as class members without providing an initial value. For local variables, Java will check that they are not used before initialization and give you an error if that happens. For member variables there is no error. Instead, they are given a default value. For primitives the default is 0, 0.0, or false, depending on the exact type. For object types, typically called reference types in Java, they are given the value of `null`. This is a very common source of errors in Java and that is one of the reasons Scala forces you to initialize variables.

A.4 Constructors and `@Override`

With this example it also starts to become obvious that there are no arguments to the class. Argument lists are not required for Scala, but in this situation they would clearly have been useful. There are no argument lists for `class`es in Java. To get information into an object when it is instantiated in Java, you write CONSTRUCTORS. A constructor is a method that gets called when an object is instantiated. By default, Java creates a constructor that takes no arguments. If you include any constructor of your own, the default one is no longer created for you. A constructor looks like a method except that it has no return type, not even `void`, and the name matches the name of the class. In `CheckingAccount` you can see that there is a constructor directly after the two lines declaring the member data. This constructor takes initial values for the member data and sets it.

There are some things worth noting here. First, writing a little `class` in Java with member data that can be set at construction is much more verbose in Java than it is in Scala. The class needs to contain declarations of the data members and a constructor that sets them instead of just listing them with `val` or `var` in an argument list. One could argue that the payoff of this is that the syntax for multiple constructors is a bit cleaner.

After the constructor are the two methods required by the supertype along with two others for doing deposits and withdraws. These should all be fairly straightforward at this point. The only thing really new is the `@Override` annotation. Scala made `override` a keyword in the language. Java did not do that originally, but added the annotation in later. This is not required in Java, but for the same reasons it is required in Scala, it is strongly recommended that you include it in Java. Eclipse will add it for you if you have it put in methods required by a supertype.

A.5 Generics and Polymorphism

To make the accounts useful, they need to be put into a bank. This is something that could involve a large amount of code, but we will stick to a very short example to demonstrate the basic concepts. Here is simple `Bank class` that keeps track of accounts using a Java collection.

```
package javacode;

import java.util.*;

public class Bank {
    private List<BankAccount> accounts = new ArrayList<BankAccount>();

    public void makeNewCheckingAccount(Scanner sc) {
        System.out.println("Who is the new checking account for?");
        String name = sc.next();
        System.out.println("What is the starting balance in cents?");
        int bal = sc.nextInt();
        accounts.add(new CheckingAccount(name,bal));
    }

    public static void main(String[] args) {
        Scanner sc = new Scanner(System.in);
        Bank bank = new Bank();
        bank.makeNewCheckingAccount(sc);
        for(BankAccount ba:bank.accounts) {
            System.out.println(ba.getName()+" "+ba.getBalance());
        }
    }

}
```

The accounts are specifically stored in a `java.util.List`. The closest equivalent to this in Scala is the `mutable.Buffer`. Like the `Buffer`, the `List` is implemented using both an array and a linked list. You can see here that we have gone with the array-based implementation.

The fact that Java used square brackets for arrays means that something else must be used for type parameters. These are called GENERICS in Java and they are put inside of angle braces, the characters we typically read as less than and greater than. As with Scala, you can make Java `classes` or methods generic. The generic type is typically inferred for methods, but not for `classes`.

If you want to make a collection or just a reference that can hold "anything" in Java, the common supertype is `Object`. Technically, `Object` is equivalent to `AnyRef`. The fact that primitives are not objects means they are not instances of any class. They are not involved in supertype/subtype relationships. To get around this, Java includes wrapper types like `java.lang.Integer` and does something called autoboxing to wrap up primitives in a way that makes life easier on the programmer. Technically Scala was doing this same thing for efficiency, but it hides the primitives from the programmer completely.

A.6 Lacking Functional Aspects

It was mentioned above that Java is not as object-oriented as Scala. As of Java 7, Java is not functional at all. There are no function literals in Java. Functions are not objects and can not be passed around freely. There is no pass-by-name semantics in Java. There are no curried methods and you do not really have higher-order methods. The closest thing to a function literal in Java is an anonymous inner class that inherits from an `interface` with a single method.

Java 8 is likely to change some of this as function literals are being added to Java in the form of lambda expressions. It isn't yet clear what impact that will have on making the language as a whole more functional. It is likely that methods like `map` and `filter` will start to appear in the standard Java collection library, but they are not present yet.

A.7 Much More

There is a lot more to Java than what has been presented here. This has been shown here. In some ways, Java is fairly simple and straightforward, but in other ways it is not. The language specification for Java is more than three times longer than that for Scala and Java has significantly more keywords than Scala. Fortunately, there are many resources you can use to help you learn Java that are available online along with many books on the subject. What is more, through exploring the Java application programming interface (API) to use with Scala programs, you already have something of a head start on the learning process.

Appendix B

Advanced Scala Concepts

Scala is an interesting language in that it has a fairly simple specification with few keywords, yet it allows for remarkable power. This book has avoided going into detail on a number of different features of the language. After all, the goal was to teach you to program, not to be an expert in Scala. The purpose of this appendix is to give you a brief introduction to many of the features that were either glossed over or left out completely. For more details you can search for the Scala language specification. Language specifications are notoriously hard to read so you might want to consider "Programming in Scala" by Odersky et al. [9].

B.1 abstract type Members

Back in chapter 5 we briefly introduced the concept of a `type` declaration that can be used to give a different name to a type. At that time, the advantage was the ability to give shorter and more meaningful names to tuples that were grouping data for you. With the introduction of `case classes` in chapter 10, the need for meaningful names on data groups was addressed in a superior way because it allows you to name the members and have stronger type checking.

The main benefit of `type` declarations in Scala come about when you put `abstract type` declarations in a supertype. You can then use that type through the code and know that it will be bound to something meaningful in the subtypes. You can also provide type bounds on the abstract type using `<:` or `>:` after the name you are giving the `type` and before the type that provides the bounds. Without bounds, you can only use the type in a way that is safe with Any.

B.2 Enumerations

Many languages include a construct called an enum, short for enumeration. This is different from the java.util.Enumeration type which we discussed in chapter 19. That type is similar to an iterator. A standard enumeration is used to define a small set of values like the colors of a stop light. Scala does not provide this as a language feature. Instead, there is a class called Enumeration that you can extend to get this type of functionality.

You should create an object that extends Enumeration like the following.

```
object LightColors extends Enumeration {
  type LightColors = Value // Not required.
  val Red, Yellow, Green = Value
}
```

The type declaration in this is not required, but it gives you a good name to use to refer to the enumeration type. Any code that is going to use this enumeration a lot should probably include the line import LightColors._.

B.3 implicits

There have been a few points in the book where implicit conversions have been mentioned. The idea of this feature is that if you call a method on an object that is not defined on that object or if you pass an object to a method that is not of the correct type, Scala will check to see if there is an implicit conversion in scope that would make it work.

These implicit conversions provide a lot of power to Scala. Both the Array and String types are basically Java types. They do not really have methods like map and filter. Despite this, you have been able to call map and filter and both Arrays and Strings. The reason this has worked is that there are implicit conversions that are imported by default that convert those types to ArrayOps and StringOps respectively and those types have the methods we associate with Scala collections.

B.3.1 Basic implict Converstions

You can make your own implicit conversions by having a "function" in scope that takes the type you want to convert from, returns the type you want to convert to, and includes the implicit keyword before def. The "function" was put in quotes because in general this is actually a method, but it needs to be brought into scope so that no object name has to be provided.

To illustrate the creation of an implicit conversion as well as motivate their existence, consider the following class and companion object.

```
class Vect3D(val x:Double,val y:Double,val z:Double) {
```

```
  def +(v:Vect3D) = Vect3D(x+v.x,y+v.y,z+v.z)
  def -(v:Vect3D) = Vect3D(x-v.x,y-v.y,z-v.z)
  def *(c:Double) = Vect3D(c*x,c*y,c*z)
  def /(c:Double) = Vect3D(c/x,c/y,c/z)
  def dot(v:Vect3D) = x*v.x+y*v.y+z*v.z
  def cross(v:Vect3D) = Vect3D(y*v.z-z*v.y,z*v.x-x*v.z,x*v.y-y*v.x)
}

object Vect3D {
  def apply(x:Double,y:Double,z:Double) = {
    new Vect3D(x,y,z)
  }
}
```

This class defines a `class` to represent vectors in 3-space with some appropriate operations. Most of the things that you would want to do are present here. There is one thing that is missing though. If v is a `Vect3D`, you can do `v*3` to scale the components of the vector up by 3. However, you can not do `3*v`. To see why, remember that `v*3` is seen by Scala as `v.*(3)`. It is calling the * method on the v instance of the `Vect3D class`. That works because such a method has been defined. On the other hand, `3*v` is trying to call a `*(Vect3D)` method on a number. That method was not part of the standard libraries so it does not work.

The solution to this is an `implicit` conversion. While `Double` does not have the needed * method, we can convert the `Double` to some other type that does. To make this work, we can add the following code to the companion `object`.

```
implicit def doubleToScaling(c:Double):VectScaler = new VectScaler(c)

class VectScaler(c:Double) {
  def *(v:Vect3D):Vect3D = Vect3D(c*v.x,c*v.y,c*v.z)
}
```

The first line of this is the implicit conversion from `Double` to the type `VectScaler`. Below that is the definition of `VectScaler`. It is a simple `class` that contains only the method that we need here for multiplication against a `Vect3D`. To use this, you need to `import` that method into the local scope. This can be done with `import Vect3D._`.

You should notice that what this did was to basically add a new method to the `Double` class. The ability to make it appear that a type has methods that was not originally part of it has been called "pimping an interface" because you are adding new features to provide functionality that is more to your liking.

B.3.2 Rules and Limits on `implicits`

While `implicit` conversions bring a lot of power, they can make code harder to read and understand if they are abused. To help prevent this, Scala has rather strict rules on when `implicit` conversions will be invoked. First, they must be in scope. If calling the conversion function would require any specifiers before the name, it will not be used as an `implicit` in that part of the code. Second, `implicit` conversions are not nested. If it would require two conversions to get from the declared type to a type that will work, Scala will not do that. Similarly, if there is ambiguity where two different implicit conversions could make the call work, neither will be used and you will get an error.

These various rules and limitations combine to make Scala `implicits` safer. They should not be invoked in ways that you are not expecting or would find overly confusing. What is more, Integrated Development Environment (IDE) plug-ins have the ability to show you what implcit conversions are being used at different points in your code so that it is even more clear.

B.3.3 implicit Parameters

It is possible to get Scala to implicitly pass extra parameters into a method as well. This can be done for multiple reasons, including passing `implicit` type conversions. Scala will only implicitly include the last argument list to a method or function. Typically, methods that have `implicit` parameters are curried. For the last parameter list to be supplied implicitly the argument list must be labeled as `implicit` and there must be a value matching each of the types in that parameter list which is in scope and labeled as `implicit`. Note that this implies that you can make `implicit val` declarations as well as `implicit def` declarations.

If you look in the Application Programming Interface (API), for example at `scala.collection.Seq`, you will see many examples of methods that include `implicit` parameters. In the collections library the `implicit` parameter list is almost always a single parameter of the type `CanBuildFrom`.

B.3.4 The Meaning of <%

When you put `<%` as a bound on a type parameter, Scala switches it to a normal type bound and adds a curried `implicit` parameter to the method. That extra parameter will be filled in with an appropriate `implicit` conversion from the current scope if one is available.

B.4 sealed classes

Many functional languages have a construct called algebraic types. This allows you to say that a type can be represented by a small number of different possible structures/values. If all you want are separate values, the `Enumeration` described above is the way to go. When the different possibilities have different structure to them, the normal object-oriented way to represent that is with inheritance. The one shortcoming of inheritance in this case is that anyone can add other subtypes at a later date, and that might be a problem. If you make a `class final`, no subtypes are allowed. In this case you want to be able to limit the subtypes to a specific number.

This mechanism is provided in Scala with the `sealed` keyword. A `sealed class` can be extended, but all the subtypes have to appear in the same file as the original declaration. That gives you much stronger control over what subtypes exist.

There is another benefit to a class being `sealed` in the form of error checking for `match` expressions. If you do a `match` on an expression whose type is a `sealed class`, Scala will produce syntax errors if you do not have cases that can match all subtypes. This completeness checking makes it so that you should not get `MatchExceptions` when a type comes in that you were not expecting.

The standard usage of a `sealed class` is to use a `sealed abstract class` that is completely empty at the top of the hierarchy, then have various `case classes` or `case` objects[1] which extend the first class. The use of case types makes it easier to take full advantage of the pattern matching.

B.5 Covariant, Contravariant, and Invariant Types

Chapter 25 brought up the concept of covariant types in relation to immutable linked lists. This deals with how Scala treats type parameters when it comes to determining subtype relationships for the main type. Imagine you have the following declarations.

```
class Stuff[A](val a:A)
class MoreStuff[A](a:A, val b:A) extends Stuff[A](a)
```

You can see from this that MoreStuff is a subtype of Stuff. However, this ignores the fact that Stuff and MoreStuff are not well-defined types on their own. In reality, you have to provide a type parameter to have a well defined type. In this case we can say that MoreStuff[String] is a subtype of Stuff[String]. The question is, is MoreStuff[String] and subtype of Stuff[AnyRef]? You might be tempted to say that it is, but as written, it is not. If you try to pass an instance of MoreStuff[String] into a function/method that accepts Stuff[AnyRef], Scala will complain. If you put the types above into the REPL, you can quickly test that with the following.

```
scala> def hasNull(s:Stuff[AnyRef]) = s.a==null
hasNull: (s: Stuff[AnyRef])Boolean

scala> val s = new Stuff("Hi")
s: Stuff[String] = Stuff@c8d310f

scala> hasNull(s)
<console>:11: error: type mismatch;
 found   : Stuff[String]
 required: Stuff[AnyRef]
Note: String <: AnyRef, but class Stuff is invariant in type A.
You may wish to define A as +A instead. (SLS 4.5)
              hasNull(s)
                      ^


scala> val ms = new MoreStuff("hi","mom")
ms: MoreStuff[String] = MoreStuff@35e20aca

scala> hasNull(ms)
<console>:12: error: type mismatch;
 found   : MoreStuff[String]
 required: Stuff[AnyRef]
Note: String <: AnyRef, but class Stuff is invariant in type A.
You may wish to define A as +A instead. (SLS 4.5)
              hasNull(ms)
                      ^
```

---

[1]You use case classes when your type has arguments. If there are now arguments, you should use a case object.

This happens because the default behavior is for type parameters to be invariant. That means that types with different type parameters are always unrelated, regardless of the inheritance relationships between the main types or the type parameters.

The type MoreStuff does not even have to be involved. You might expect that Stuff[String] should be a subtype of Stuff[AnyRef]. However, if A is invariant that is not the case either, as shown here.

```
scala> val s = new Stuff("Hi")
s: Stuff[String] = Stuff@c8d310f

scala> hasNull(s)
<console>:11: error: type mismatch;
 found   : Stuff[String]
 required: Stuff[AnyRef]
Note: String <: AnyRef, but class Stuff is invariant in type A.
You may wish to define A as +A instead. (SLS 4.5)
             hasNull(s)
                     ^
```

As the error message above indicates, this could be fixed by using the type parameter +A instead of just A. That makes the type parameter covariant. When the type parameter is covariant, the full types will be subtypes if the parameters are subtypes and the main types are the same or subtypes. In this particular example just adding a + will make this work.

```
class Stuff[+A](val a:A)
```

With this one character, everything else from above will work.

So you might wonder why invariance is the default instead of covariance. To make that clear, consider the following change to Stuff and a function that works on it.

```
class Stuff[A](var a:A)
def changeStuff(s:Stuff[AnyRef]) = s.a = List(1,2,3)
```

Now that the a member is a var, things are a bit different. Consider what happens if Scala let you pass in an instance of Stuff[String]. In that object, a has to be a String. This function would try to assign to it a List[Int], an operation that should clearly fail. The switch to a var makes covariance for this class unsafe. Scala can even tell you that.

```
class Stuff[+A](var a:A)
<console>:7: error: covariant type A occurs in contravariant position in type A of
    value a_=
       class Stuff[+A](var a:A)
             ^
```

As the error message indicates, it is the assignment method into the var that is really causing the problem. This also shows you that Scala can tell if it is safe for a type to be covariant or contravariant. It is always safe to be invariant. That is why invariant is the default. The algorithm that Scala uses to determine what is safe is beyond the scope of this appendix, but a quick rule of thumb is that when the type is passed into a method that is a contravariant position. When they are returned from methods that is a covariant position. For a type parameter to be either covariant or contravariant, it can not occur in a position of the other end. Since a public var implicitly makes one of each, any type of a public var has to be invariant.

Contravariance is the opposite of covariance. When the type parameter is `-A`, then `Stuff[Type1]` is a subtype of `Stuff[Type2]` only if `Type2` is a subtype of `Type1`. This probably seems counterintuitive, but there are times when it is very useful. The most obvious example is the `Function1` type in the Scala library. This type represents a function that takes one argument and returns a value.

```
trait Function1 [-T1, +R]
```

This is basically what we have been writing as `(T1) => R`. Here the return type is covariant and the input type is contravariant. To understand why this is, imagine a place where you say you need a function of the type `(Seq[Int]) => Seq[Int]`. What other types would work there? For the return type, anything that returns a subtype of `Seq[Int]` will work fine. For the input type, you can use any function that takes a supertype of `Seq[Int]` will work because you know it will be safe to pass it a `Seq[Int]`.

B.6 Extractors

Pattern matching has been used in many places in this book and hopefully you have become aware of the power that it can provide. One thing you might have noticed that was missing was the ability for you to make your own `class`es capable of doing pattern matching. You can make your own types that do pattern matching by writing extractors. An extractor is an `object` that includes either the `unapply` or the `unapplySeq` methods. Most of the time you will do this in a companion `object` that also has an `apply` method so that the usage resembles that of `case classes`.

B.6.1 `unapply`

You should put an `unapply` method in an `object` when you know how many fields should be pulled out of the pattern. The parameter for `unapply` is the value that would be matched on in the pattern. It should have a type that is appropriate for what you want to be working with. The result type of `unapply` depends on how many fields it should give back.

- No values results in `Boolean`. It is `true` if there was a match and `false` if there was not.
- One value of type `A` results in `Option[A]`. The result will be `None` if there was a match or `Some` with the value of the result if there was a match.
- For two or more values the result type should be an `Option` of a tuple with the proper number of types. No match will result in `None` while a match will result in `Some` of a tuple filled with the proper values.

To illustrate this, we could put an `unapply` method in the `Vect3D` companion `object` that was used earlier to show the use of `implicit` conversions.

```
def unapply(str:String):Option[(Double,Double,Double)] = {
  val s = str.trim
  if(!s.startsWith("(") || !s.endsWith(")")) None
  else {
```

```
    val parts = s.substring(1,s.length-1).split(",")
    if(parts.length!=3) None
    else
      Some(parts(0).trim.toDouble,parts(1).trim.toDouble,parts(2).trim.toDouble)
  }
}
```

This example will pattern match on a `String` that should have three numbers, separated by commas, inside of parentheses. It gives back three values that are the numbers of the three components. A simple usage of this might look like the following.

```
"(1,4.5,83)" match {
  case Vect3D(x,y,z) => println(x+" "+y+" "+z)
}
```

Note that while you can make `unapply` methods that take different inputs and return whatever, you can not make methods that differ only in their return type. So in this example, you can not also have a second version that creates a `Vect3D` instance. You have to pick between the two.

B.6.2 unapplySeq

There are also situations where the number of fields in the pattern match can vary from one match to another. Examples of this include matching on `List` or `Array`, or the matching or regular expressions where each group defines a field. To get this functionality, you need to implement the `unapplySeq` method in your `object`. The return type of this method can be an `Option[Seq[A]]`, for whatever type `A` you want. It can also have an `Option` of a tuple where the last element of the tuple is a `Seq[A]`. The first version could match any number of things, but they will all have the same type. The second version can force a certain number of fields of different types at the beginning of the pattern followed by the sequence of unknown length.

B.7 lazy Member Data

`val` and `var` declarations in `class`, `trait`, and `object` declarations can be labeled as `lazy`. This changes when the value of that member data will be set. Normally member data is initialized when the object is instantiated. `lazy` members are not initialized until they are used.

This is particularly helpful if the value requires a significant calculation or requires significant resources such as memory. In that situation, you do not want to go through the effort of making the value unless it is actually needed. If the object can go through its life without ever needing that member, setting it to be `lazy` will make it so those resources are never consumed.

B.8 scala.collection.immutable.Stream

There is another type in the Scala collections that was not used in this book, the scala.collection.immutable.Stream type. Stream is a subtype of Seq and you access members of it just like you would any other Seq using an integer index. What makes Stream interesting is that it is a lazy sequence. The last section described how the lazy keyword modifies members so that they are not calculated until they are needed. This same idea can be applied to a sequence as well. The values at different indexes are not calculated until they are needed. Once calculated they are remembered so they do not have to be calculated again, but if you never use an index, it will never be calculated.

At first glance this might seem like just an interesting way to save memory. However, it can be used for more than that. The fact that the values of a Stream are not absolutely calculated and stored means that it is possible to make Streams of infinite length. The Scala API includes an example that builds an infinite Stream of prime numbers. If you look at that you will notice that it uses the method Stream.cons instead of the :: method to add elements to the front of the stream. It also adds elements using a recursive definition of everything after the head. This is required for an infinite Stream because the tail of the Stream never really exists in full.

If you have a block of code that is going to be working with Streams s lot, you should include import Stream._. One of the useful things this brings into scope is an implicit conversion from Stream to Stream.ConsWrapper. This will allow you to use the #:: and #::: operations to cons and concatenate Streams. This allows you to write the following code to define Fibonacci numbers.[2]

```
import Stream._
def fibFrom(a:BigInt, b:BigInt):Stream[BigInt] = a #:: fibFrom(b, a+b)
```

To test this you can do the following.

```
fibFrom(1,1).take(100).toList
```

You can use this type of approach generally for any place that you need to define something that is accurately represented by an infinite sequence. It is worth noting though that you have to be careful with these. Many of the methods in the collection API keep going until they get to the end of the collection. Needless to say, that is not a good thing to do with an infinite collection.

B.9 Nested Types

We have seen quite a few places in this book where classes are written inside of other classes. What we did not generally do is refer to these types from outside of the outer class. This was in large part because none of the code we wrote really needed to do this. In fact, in most places where we did nest types in this way, we made the nested type private so that it could not be seen from the outside. For example, the Node type in a linked list should not be known to outside code as that is an implementation detail. However, there

[2]This code is presented using Int in the Scala collections API description gives at http://www.scala-lang.org/.

are instances where those nested types need to be used by outside code, and there are some significant details to it that are worth mentioning.

When you put a `class` or `trait` inside of another `class` or `trait`, every instance of that type gets its own version. If you use standard "dot" notation to refer to the type, it needs to be on an object, not the `class`, and the type you get is specific to that object. You can also refer to the general type across all instances with the `class` name followed by a `#`. The following code demonstrates this.

```
class NestedClasses(x:Double,y:Double) {
  val root = new Plus(new Num(x),new Num(y))

  trait Node {
    def eval:Double
  }

  class Plus(left:Node,right:Node) extends Node {
    def eval = left.eval+right.eval
  }

  class Num(n:Double) extends Node {
    def eval = n
  }
}

object NestedClass {
  def main(args:Array[String]) {
    val a = new NestedClasses(4,5)
    val b = new NestedClasses(8.7,9.3)

    a.root match {
      case n:a.Num => println("A number")
      case n:a.Node => println("Not a number")
    }

    def evalNode(n:NestedClasses#Node) = n.eval
    def evalNodeA(n:a.Node) = n.eval
    def evalNodeB(n:b.Node) = n.eval

    println(evalNode(a.root))
    println(evalNodeA(a.root))
    println(evalNodeB(a.root)) // This is a type mismatch.
    println(evalNode(b.root))
    println(evalNodeA(b.root)) // This is a type mismatch.
    println(evalNodeB(b.root))
  }
}
```

The `class` at the top contains types for a little expression tree. This is greatly simplified for this example. In the companion `object` is a `main` method that makes two instances of the top `class` called `a` and `b`. A `match` on `a.root` shows how you can refer to types inside of the object `a`. If you try to put something in a `case` that is not one of the subtypes of `Node` inside of `a`, Scala gives you an error saying it can not be that type.

After the `match` are three `def` declarations of local functions. The first is written to take a general `Node` from inside of any instance of `NestedClasses`. The other two are specific to `a` and `b`. There are six calls where these three methods are invoked on the `root` objects of both `a` and `b`. Two of these result in errors as `a.root` is a mismatch with `b.Node` and `b.root` is a mismatch with `a.Node`.

B.10 Self Types

There are times when you will create a `trait` which is intended to be mixed in with some other `trait` or `class`. Occasionally, inside of the mix-in `trait`, you will need to make reference to members of the type it is being mixed in to. Conceptually this makes sense as you know that this will always be safe. However, the compiler does not know that this will be safe. To get around this, you have to have a way to tell the compiler that the particular trait will only be used in situations where it is part of some other type. This can be done using a self type.

The syntax of a self type is as follows.

```
trait A {
  this: B =>
  ...
}
```

This tells Scala that any object that includes the `trait A` must include the type `B` above it in the resolution linearization. With this in place, the body of `A` can include references to members of `B`.

One common place that this is used in Scala is in something called the Cake pattern. Interested readers can use that term as a jumping-off point for searching for more information.

B.11 Making Executable JAR Files

Scala programs compile to bytecode, not an actual executable. If you want to be able to give your program to someone else for them to run, you need to package it in a proper form. For code compiled to Java bytecode, that format is an executable JAR file. You can make these using the `jar` command from the Java Development Kit (JDK). Complete instructions can be found on the book's website as the ideal set of steps is likely to change over time.

Appendix C

Glossary

Abstract Lacking a definition. A `val`, `var`, `def`, or `type` declaration in a `class` or `trait` can be left without a definition, making it abstract. If a `class` contains any members that are abstract, then it must be declared to be abstract. Abstract types, whether `class` or `trait`, can not be instantiated. Only subtypes that have defined all the abstract members can be instantiated.

Abstraction The concept of writing code so that certain parts of what happens are not concrete until the usage point. Functionality can be abstracted by passing in functions. Types can be abstracted through subtyping or type parameters.

Actor An actor is an object that receives messages and acts on them. It should be able to process those messages in a manner that is independent of what thread is doing the work.

Affine Transform This is a term from graphics that is defined as any transformation of the space where parallel lines are preserved. The basic affine transforms are translation, rotation, scale, and shear. Any combination of these is also an affine transform.

Argument A value that is passed into a function of method.

Array A basic collection of values that is a sequence represented by a single block of memory. Arrays have efficient direct access, but do not easily grow or shrink.

Class A construct that works as the blueprint for objects.

CPU The term Central Processing Unit is used to describe the primary computing element in a computer. This is where most of the work for the programs you are writing takes place.

Concrete This is a term used to describe a type in which all members are defined and that can be instantiated. Non-abstract classes are concrete.

Concurrent Concurrent execution is when two parts of a program as executing in a manner that appears to be simultaneous to the user and where the programmer has limited control over what is happening at any given time. A concurrent program does not have to be parallel. Concurrent programs include those where two threads of control swap between one another, but are never active at the same time.

Conditional An expression of the `Boolean` type that is used to determine if code is executed.

Embarrassingly Parallel This is a term used to describe a problem that breaks into problems that are completely independent. Such problems can be easily parallelized because the different pieces don't really have to communicate with one another.

Enumeration A type which can only take on one value from a small set of options. These are implemented in Scala by making an object that extends the `scala.Enumeration` type.

Event Thread This is a special thread created by a GUI to handle events, including the repainting of the screen.

Expression A sequence of tokens that has a value and a type.

File An independent grouping of information in the static storage of a computer.

Function This is a concept from mathematics of something that maps from values in a certain domain to values in a certain range. In programming, functions can serve this same purpose, but they also more generally group statements together under a particular name to allow you to break code into pieces and give it meaningful names.

Grammar A formal specification for a set of languages. Chomsky grammars are composed of sets of terminal and non-terminal tokens, a non-terminal start token, and a set of productions. There are other types of grammars, including L-systems, that follow slightly different rules.

GUI Short for Graphical User Interface. This is an interface that includes graphical elements that the user interacts with using a pointing device possibly in addition to keyboard input.

Higher-Order Function This is a term used to describe a function whose inputs or outputs include function types.

if A simple conditional construct that picks between one of two options based on whether or not a Boolean expression is true. In Scala, the `if` is a valid expression, but can also be used as statements.

Instantiation The act of creating an object that is an instance of a particular type.

Iteration The act of running through steps or elements one at a time.

Linked List A manner of storing data where the data items are in nodes and the nodes know about one another, but nothing knows where everything is. This form of storage makes random insertion and removal efficient, but requires walking the list for random access.

List An abstract data type that stores items by a numeric index and allows random access, insertion, and removal.

Loop A construct designed to execute code repeatedly.

Map This is an abstract data type that associates values with unique keys. In an ideal implementation, it should be efficient to store or retrieve values based on their keys. The critical aspect of a Map is that keys can be whatever type you want to use and they do not have to be contiguous.

Multithreading The act of having a program that allows more than one thread of control to be active at a time. This effectively allows multiple instructions to execute simultaneously. This is a form of shared-memory parallelism.

Parallel In the context of programming, this is when two or mote things are executing at the same time. This generally requires multiple computational elements such as cores, processors, or even full computers.

Parameter This is a placeholder for a value that will be passed into a function.

Polymorphism The ability of a piece of code to work with more than one type.

Priority Queue An abstract data type where items can be added or removed. The order of removal depends both on a priority and the order of arrival such that higher-priority items always come off first and items that have been on the priority queue win ties in priority.

Profiling The act of running a program to see how resources are used. Often this is done to see how many times different methods are called and what parts of the code the it spends the most time in.

Queue An abstract data type where items can be added and removed. They are removed in the same order they are added. This is often called " first-in, first-out" or FIFO.

Recursion When a function or method calls itself. In mathematical terms, this is when a function is defined in terms of itself.

Refactor The act of changing code to improve some aspect of it, such as how it does something, without changing what it does.

Regular Expression A formal specification for describing fairly simple sequences of characters. These are largely based on the regular grammars of the Chomsky hierarchy.

Search The process of looking for a particular element or its position in a collection.

Sequence (Seq) A type of collection characterized by items having a particular order and being referred to by an integer index.

Set A collection of unique items. The ideal implementation of a `Set` allows for efficient checking to see if an element belongs to it, to add elements to it, or to remove elements from it.

Signature The signature of a method/function includes the name along with the parameter types and the return type. From the signature you can see how a method/function should be called. The return type also lets you know what you can do with the result of the call.

Socket A construct in computer software that is used to support communication between machines.

Sort The act of putting the items in a sequence into the appropriate order according to some comparison function.

Stack An abstract data type with the ability to add and remove items. It is characterized by the fact that items are removed in the inverse order in which they were added. This is often called "last-in, first-out" or LIFO.

Statement A set of tokens that represents a complete command to a language. In Scala, any expression can be used as a statement.

Stream A data source or sink that does not allow random access. The name should invoke the image of water flowing past. Once a certain time has passed, it can not be returned to.

Syntax Error This is an error you are notified of during the compile stage which exists because you wrote code that violates the syntax of the language. These are typically the easiest errors to deal with because the compiler can give you an informative message with a line number. If you are working in an IDE, syntax errors will often be shown with a red underscore, much like spelling errors in a word processor.

Thread A single unit of control in a program that shares memory with other threads.

Tree A linked data structure that forms a hierarchy where nodes at higher levels know about a subset of the nodes in the level below them. Each node in a tree can only be reached from a single node in the level above it.

Token The smallest element of a programming language that has meaning on its own and which changes meaning if altered or broken apart with whitespace.

Type A construct that specifies a set of values and the operations that can be performed on them.

Variable A construct that associates a name to a reference to a value.

XML This is short for eXtensible Markup Language. XML is a standard, plain text format that can be used to represent various forms of data in a tree-like structure. It is called extensible because users are allowed to develop their own tags to represent data in the manner they want.

Bibliography

[1] Gene M. Amdahl. Validity of the single processor approach to achieving large scale computing capabilities. In *Proceedings of the April 18-20, 1967, spring joint computer conference*, AFIPS '67 (Spring), pages 483–485, New York, NY, USA, 1967. ACM.

[2] Josh Barnes and Piet Hut. A hierarchical O(N log N) force-calculation algorithm. *Nature*, 324:446–449, December 1986.

[3] Rudolf Bayer and Edward M. McCreight. Organization and maintenance of large ordered indices. *Acta Inf.*, 1:173–189, 1972.

[4] Noam Chomsky. Three models for the description of language. *IRE Transactions on Information Theory*, 2:113–124, 1956. `http://www.chomsky.info/articles/195609--.pdf` – last visited 14th January 2009.

[5] Martin Fowler. *Refactoring: Improving the Design of Existing Code.* Addison-Wesley, Boston, MA, USA, 1999.

[6] Erich Gamma, Richard Helm, Ralph E. Johnson, and John Vlissides. *Design Patterns: Elements of Reusable Object-Oriented Software.* Addison-Wesley, Reading, MA, 1995.

[7] Donald Knuth. *Structured Programming with Go To Statements*, pages 257–321. Yourdon Press, Upper Saddle River, NJ, USA, 1979.

[8] Michael P. Marder. *Research Methods for Science.* Cambridge University Press, New York, NY, USA, 1st edition, 2011.

[9] Martin Odersky, Lex Spoon, and Bill Venners. *Programming in Scala: A Comprehensive Step-by-Step Guide, 2nd Edition.* Artima Incorporation, USA, 2nd edition, 2011.

[10] Robert Sedgewick. *Algorithms.* Addison-Wesley, Reading, Mass, 1983.

[11] Herb Sutter. The free lunch is over: A fundamental turn toward concurrency in software. *Dr. Dobbs Journal*, 30(3):202–210, 2005.

Index